GEOMETRIC PHASES IN PHYSICS

ADVANCED SERIES IN MATHEMATICAL PHYSICS

Also in the Series

Volume 1: Mathematical Aspects of String Theory
edited by S-T Yau (*Harvard*)

Volume 2: Bombay Lectures on Highest Weight Representations of Infinite Dimensional Lie Algebras
by V G Kac (*MIT*) & A K Raina (*Tata*)

Volume 3: Kac-Moody and Virasoro Algebras: A Reprint Volume for Physicists
edited by P Goddard (*Cambridge*) & D Olive (*Imperial College*)

Volume 4: Harmonic Mappings, Twistors and σ-Models
edited by P Gauduchon (*CNRS, Paris*)

Volume 7: Infinite Dimensional Lie Algebras and Groups
edited by V Kac (*MIT*)

Volume 8: Introduction to String Field Theory
by W Siegel (*SUNY*)

Volume 9: Braid Group, Knot Theory and Statistical Mechanics
edited by C N Yang (*SUNY*) & M L Ge (*Nankai Inst*)

Advanced Series in Mathematical Physics
Vol. 5

GEOMETRIC PHASES IN PHYSICS

Alfred Shapere
Frank Wilczek

World Scientific
Singapore • New Jersey • London • Hong Kong

Published by

World Scientific Publishing Co. Pte. Ltd.,
P O Box 128, Farrer Road, Singapore 9128
USA office: 687 Hartwell Street, Teaneck, NJ 07666
UK office: 73 Lynton Mead, Totteridge, London N20 8DH

The editors and publisher are grateful to the authors and the following publishers for their assistance and permission to reproduce the reprinted papers found in these volumes:

American Institute of Physics (*J. Chem. Phys.*);
American Physical Society (*Phys. Rev., Phys. Rev. Lett.*);
Institute of Physics (*J. Phys. A*);
Kyoto University (*Prog. Theor. Phys.*);
Macmillan Magazines Ltd (*Nature*);
North-Holland Physics (*Nucl. Phys., Phys. Lett.*);
Raman Research Institute (*Collected Works of S. Pancharatnam*);
Redakcja Acta Physica Polonica (*Acta Phys. Polonica*);
Springer-Verlag (*Commun. Math. Phys.*);
Taylor & Francis Ltd (*Mol. Phys.*);
The Faraday Society (*Disc. Farad. Soc.*);
The Royal Society (*Proc. Roy. Soc. London*).

While every effort has been made to contact the publishers of reprinted papers prior to publication, we have not been successful in a few cases. Where we could not contact the publishers, we have acknowledged the source of the material. Proper credit will be given to these publishers in future editions of this work after permission is granted.

Library of Congress Cataloging-in-Publication Data

Geometric phases in physics / edited by F. Wilczek & A. Shapere
p. cm. -- (Advanced series in mathematical physics; vol. 5)
1. Geometry. 2. Holonomy groups. 3. Mathematical physics.
I. Wilczek, Frank. II. Shapere, A. III. Series.
QC20.7.G44G46 1989
530.1'5 -- dc20 89-14624
ISBN 9971-50-599-1
ISBN 9971-50-621-1 (pbk)

Printed in Singapore by Utopia Press.

Preface

During the last few years, considerable interest has been focused on a complex of physical ideas that share a common mathematical theme, the concept of holonomy. The recent flurry of activity began in 1984 with a paper by Michael Berry. He showed that the adiabatic evolution of energy eigenfunctions, with respect to a time-dependent quantum Hamiltonian $H(t)$, contains a phase of deeply geometrical origin (now known as "Berry's phase") in addition to the familiar dynamical phase

$$\exp -\frac{i}{\hbar}\int E(t)\,dt\,.$$

The additional phase approaches a finite, non-zero limit as the the Hamiltonian is taken more and more slowly around a closed path in its parameter space. Berry's observation, although basically elementary, seems to be quite profound. Multiplicative phases—or more generally group transformations—with similar mathematical origins have been identified and found to be important in a startling variety of physical contexts, ranging from nuclear magnetic resonace to low Reynolds number hydrodynamics to quantum field theory. It now seems clear that Berry captured a particularly fruitful concept, of wide applicability.

There are several reasons for the impact of Berry's work. Of course, the inherent universality and beauty of geometric phases has played a role, but it is also worth mentioning some of the extrinsic factors which made it "the right concept at the right time." One factor was undoubtedly surprise—the surprise of the physics community that such a simple and fundamental aspect of the adiabatic theorem had been overlooked for so many years. Another reason for its impact was the emergence of gauge theories of the interactions of elementary particles. Many gauge theoretic ideas appear in the study of geometric phases, unencumbered by the complexities usually associated with relativistic quantum field theory. Conversely, ideas associated with geometric phases clarify some subtle issues in quantum field theory—as we shall find in Chapters 5 and 7. In an era of increasing separation between everyday reality and the more theoretical branches of physics, it has been refreshing and comforting to come across a concept that both helps to explain some of the more abstruse ideas of quantum field theory, and leads to effects that can be readily measured in a laboratory.

Any judgement as to the value of an essentially mathematical concept, proposed for use as a tool in physics, should be at least partly based on its usefulness in practice. It is all too easy to believe that merely by adopting a new language one begins to make novel observations, but we trust that a perusal of the the contents of this book will suffice to show that many genuinely new insights have been gained. Although it is at present used on a relatively modest scale, we believe that the concept of a geometric phase, repeating the history of the group concept, will eventually find so many realizations and applications in physics, that it will repay study for its own sake, and become part of the *lingua franca.*

The immediate origin of this book was a workshop held on "Non-integrable Phases in Dynamical Systems" in Minneapolis on 1-3 October 1987, under the aegis of the new Theoretical Physics Institute. The enthusiastic response of the participants, and the variety and quality of work presented, led us to think that a book on the subject would make a useful addition to the physics literature. The present volume contains reprints of papers we think are particularly important or instructive, together with several contributions which have not appeared in print before. The articles are arranged by subject in nine chapters, each of which begins with an introduction where we attempt to weave the varied material into a reasonably coherent whole.

We do not purport to have made a comprehensive collection of relevant articles. Our choice of papers is more a function of our own particular interests and ignorance than anything else. Nevertheless, we hope that our book will serve as a useful introduction to an emerging field, and will stimulate its readers to seek out further material in their own areas of interest.

We are grateful to Michael Berry for his superb introductory survey and Daniel Arovas for his comprehensive article on fractional statistics and the fractional quantum Hall effect. We also wish to thank our editor, P.H. Tham, without whose hard work and assistance this book would not have been completed. The cover photo, "Calcutta Staircase 1988," was taken by Catherine Shapere.

A Reader's Guide

Probably only the most adventurous readers will be motivated to read every chapter of this book, but many readers may be interested in browsing through unfamiliar territory. As an aid to field theorists who want to understand how non-Abelian gauge potentials apply to molecular systems, and to assist physical chemists in appreciating the connection between the molecular Aharonov-Bohm effect and gauge anomalies, we have prepared introductions to each of the chapters. The introductions serve several purposes. They provide some elementary background to topics covered in the chapter, touching at least briefly on each included article. They also try to put each chapter in perspective relative to the rest of the book, and to suggest further directions for research, when possible. It is our hope that by bringing together applications of geometric phases from a variety of fields, this book will inspire continued cross-fertilization between widely separated areas of physics.

Very roughly, the chapter dependence is as follows:

```
          1
          ↓
          2
          ↓
4   ←     3    →   8
          ↓   ↘
6   ←     5        9
          ↓
          7
```

The first four chapters are of general interest. The first chapter includes two survey articles, by Berry [1.1] * and Jackiw [1.2], which we recommend to all readers. Both articles are relatively non-technical, and should help orient readers to the following three chapters. In addition, Berry's article includes original material on the natural metric on the projective Hilbert bundle and a detailed study of non-adiabatic corrections to the phase precession of the classical pendulum.

Chapter 2 contains some of the pre-1984 material which anticipated Berry's work on the quantal adiabatic phase, drawn from molecular physics and optics. This historical material is important and interesting in its own right—indeed, Pancharatnam's thirty-year-old paper on phase shifts of polarized light [2.1] has laid the groundwork for some modern optics experiments to measure geometric phases [4.2], and provides the basis for a recent extension of Berry's phase to non-closed paths [3.5]. The development over twenty years of phase concepts in molecular physics led to the use of gauge potentials in Born-Oppenheimer Hamiltonians several years before Berry's paper, and has borne a rich field of continuing activity.

* The notation [*M.N*] refers to the *N*th article in chapter *M*.

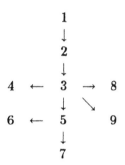

The general foundations of our subject are laid out in Chapter 3. It contains Berry's original paper [3.1] and covers many subsequent extensions and elaborations of Berry's phase, such as Wilczek and Zee's non-abelian phase for degenerate Hamiltonians [3.3] and Aharonov and Anandan's phase for general cyclic evolution [3.4]. Its introduction includes, in an appendix, a pedagogical discussion of the mathematical context of Berry's phase, which may provide useful background for some of the more mathematical articles. The final article [3.7], which has not appeared previously, is a general discussion of the Born–Oppenheimer approximation and its field theory analogues, with phase effects and non-adiabatic corrections taken into account.

Basic applications of Berry's phase are treated in the following chapter, with articles drawn from optics, magnetic resonance, and molecular and atomic physics, from both the experimental and the theoretical literature. NMR and optics have provided some of the most successful tests of Berry's phase in macroscopic systems to date. In the fully quantum mechanical context of molecular physics, phase effects can lead to energy splittings and can shift quantum numbers, that have been observed experimentally.

The remaining five chapters are concerned with more specialized applications. Chapters 5, 6, and 7, respectively on fractional statistics, the quantized Hall effect, and anomalies and Wess–Zumino terms, are about geometric phases in many–body systems and quantum field theories. All three contain extensive introductions to aid the uninitiated. We would recommend reading Chapter 5 first, since the concept of fractional statistics plays a fundamental role in the theory of the fractional Hall effect, and is closely associated with Wess–Zumino terms. In fact, the boundaries between these three chapters are not too sharply defined—for instance, the review article by Arovas contains much general material on fractional statistics, although its main focus is on the fractional Hall effect. Also, these chapters touch on several topics which are not evident from their titles. The chapter on fractional statistics includes a paper by Laughlin on high–temperature superconductivity, and Chapter 6, on the quantized Hall effect, contains articles on two other two-dimensional systems—a network of current loops enclosing magnetic flux, and a Bloch electron in a in a transverse magnetic field.

Geometric phases also appear in classical systems. Hannay's angles are classical correspondants of Berry's phase, and can appear in any classical system described by action–angle variables, in response to adiabatic variation of the Hamiltonian. Another type of classical phase occurs in describing the motion of deformable bodies, which is especially useful in studying systems that are invariant under reparameterizations of time. In particular, the motion of a self–propelled body at low Reynolds number and the rotation of a self–deforming body in space can be described in terms of a gauge field over the space of shapes. These examples are all discussed in Chapter 8.

Finally, the last chapter contains Berry's elegant paper on higher–order corrections to the adiabatic approximation. It is our belief that there is much room for further research in this area. Indeed, the subject of geometric phases in physics is far from closed, so perhaps it is fitting that we end on an unresolved note.

CONTENTS

* Original Contribution.

* Original Contribution.

* Original Contribution.

GEOMETRIC PHASES IN PHYSICS

Chapter 1

INTRODUCTION AND OVERVIEW

* Original Contribution.

1

Introduction and Overview

The two articles contained in this chapter form the proper introduction to the main contents of the book. Here, we briefly discuss the general notion of holonomy, and illustrate it with a few primeval examples.

A phase is, for our purposes, not a state of matter but a complex number of unit modulus, an element of the group $U(1)$. We shall use the term somewhat loosely to encompass elements of matrix groups as well, such as $U(N)$. The phases we shall be interested in are often associated with cyclic evolution of a physical system. More specifically, we shall find that the cyclic variation of external parameters often leads to a net evolution involving a phase depending only on the *geometry* of the path traversed in parameter space. In other words, this phase is independent of how fast the various parts of the path are traversed. For non–cyclic evolution, the extra phase will depend on the endpoints of the path. The phase is non-integrable; it can not be written as a function just of the endpoints, because it depends on the geometry of the path connecting them as well.

The natural mathematical context for geometric phases is the theory of $U(N)$ fiber bundles. There one defines a phase, known as a holonomy,* that depends on the geometry of a loop, and is independent of any coordinate choice. (For a brief introduction to the world of fiber bundles, connections, and holonomy, see the introduction to Chapter 3.)

Examples of geometric phases abound in many areas of physics. Many familiar problems that we do not ordinarily associate with geometric phases may be phrased in terms of them. Often, the result is a clearer understanding of the structure of the problem, and an elegant expression of its solution.

Consider, for example, the precession of a Foucault pendulum. Standard treatments[1] calculate the rate of precession of a pendulum in a frame rotating with the surface of the earth in terms of the Coriolis force, but a much simpler and more geometric explanation may be given as follows. Suppose

* This phase is also, and perhaps more properly, known as an *an*holonomy. We shall adhere to the established mathematical terminology, although this other usage is quite common in the physics literature.

a pendulum is transported along a closed loop C, in the gravitational field of a point mass, and that the period and amplitude of its swing are small compared to the typical time and distance scales of the transport motion. We may assume that the loop lies on the surface of a sphere concentric with the mass, although this assumption is not necessary. Now when the pendulum returns to its initial position, its invariant plane will have rotated by some angle. For example, for transport around the sphere at a constant latitude of θ (relative to the north pole), a straightforward calculation shows that the net rotation will be $2\pi \cos\theta$ radians. A remarkable feature of this result is that it is independent of the rate at which different parts the loop are traversed (provided that the traversal is slow). This is a consequence of the fact that the Coriolis force is proportional to the velocity of transport, so that its integrated effect is invariant under rescalings of time. (Velocity-dependent forces, like the magnetic force on a moving charge, tend to be associated with geometric phases.) How does the pendulum precess as it is taken around a general path C? For transport along the equator, the pendulum will not precess. This may be seen from a symmetry argument. The rate of precession does not depend on the direction of the pendulum's swing, so we may assume that the invariant plane lies in a north–south direction; then any precession would break the reflection symmetry between the northern and southern hemispheres, so the pendulum must not precess at all. (Alternatively, the Coriolis force at the equator always points vertically, and cannot torque the pendulum's invariant plane.) Now if C is made up of geodesic segments, the precession will all come from the angles where the segments meet; the total precession is equal to the net deficit angle, which in turn equals the solid angle enclosed by C modulo 2π. Finally, we can approximate any loop by a sequence of geodesic segments, so the most general result (on or off the surface of the sphere) is that the net precession is equal to the enclosed solid angle. This result may seem rather esoteric, but its generality and geometric nature suggest its depth. In fact, the mathematics describing it is essentially identical to that describing the motion of a charged particle in the field of a magnetic monopole, as well as interesting molecular, NMR, and optical systems.

A fundamental example of holonomy, involving a non-abelian symmetry, lies at the heart of general relativity. If a reference frame is parallel-transported around a closed loop in spacetime, then it is well known that the initial and final frames will not coincide. For causality's sake, we should really compare the result of parallel transport along two timelike curves with common endpoints. The final frames will be related by a Lorentz transformation, that is to say, an element of $SO(3,1)$. The failure of the frames to coincide is an $SO(3,1)$ holonomy. It may be used to measure the local curvature of spacetime. Thus, if we want to know the component $R^{\alpha}_{\beta\mu\nu}$ of the Riemann tensor, we should take a loop Γ in the $\mu\nu$-plane, enclosing an infinitesimal area $dx^{\mu}dx^{\nu}$. The resulting holonomy M^{α}_{β} (an element of the Lie algebra of $SO(3,1)$) from parallel transport around Γ will be proportional to the area enclosed, to lowest order, and the constant of proportionality will be just the curvature component that we want:

$$M^{\alpha}_{\beta} = R^{\alpha}_{\beta\mu\nu} \cdot dx^{\mu} dx^{\nu}.$$

One last example concerns the motion of charged particles in strong magnetic fields. As is well known, in a constant and uniform magnetic field, a charged particle will move in a circle, or more precisely (in three dimensions) on a circular helix whose axis is parallel to the magnetic field direction. The motion is called cyclotron motion and the orbits cyclotron orbits, in reference to the use of such motion to guide particles at high-energy accelerators. A fundamental topic in plasma physics, with many applications in astrophysics, is how this motion is perturbed by various other effects such as inhomogeneities or time dependence in the magnetic field, electric fields, or gravitational forces. Insofar as the magnetic field is strong and reasonably homogeneous (*i.e.*, if it does not vary significantly over the radius of a cyclotron orbit), to a first approximation the motion is still cyclotron motion about the field lines, of gyrofrequency $\Omega(x) = eB(x)/mc$. The angular position of a particle relative to its guiding center axis will to lowest order be equal to the time integral of $\Omega(x(t))$. However, if the field lines are curved on a large scale, there will be both corrections to the angular position and drift corrections to the location of the guiding center.

Recently, Littlejohn [2] has introduced some fresh ideas and techniques, involving geometric phases, into this old subject. This has enabled him to derive the higher-order drifts much more easily and systematically than was previously possible. In fact, even without entering into any details one can see how geometric phases enter naturally into corrections to purely cyclotronic motion. As particles rotate about their guiding centers, they are "acquiring phase"—that is, their circular motion can be parametrized by an angle that increases with time. To a first approximation, the rate of phase accumulation is the local cyclotron frequency, uniquely determined by the local value of the magnetic field. However, if our particle returns to its starting point after following a closed field line, or after drifting through some loop, the total angle through which it has turned is not just the time integral of the local frequency, but contains an additional piece depending only on the geometry of the path through the space of possible magnetic field vectors. This additional phase, which is the leading correction to the total angle, is yet another example of holonomy in a purely classical system.

We hope that the above examples, which are concrete and readily visualized, will help to put in perspective some of the more abstract applications that follow.

---

[1] Keith R. Symon, *Mechanics*, 2nd edition (Reading: Addison–Wesley, 1960).

[2] Robert G. Littlejohn, "Phase anholonomy in the classical adiabatic motion of charged particles," *Phys.Rev.* **A38** (1988) 6034; "Geometry and guiding center motion," in *Contemporary Mathematics*, edited by J.E. Marsden (Providence: American Mathematical Society, 1984), Vol. 28, p.151.

The Quantum Phase, Five Years After

M. V. Berry

H.H.Wills Physics Laboratory
Tyndall Avenue, Bristol BS8 1TL, U.K.

(Received 28 April 1988.)

ABSTRACT

Classical parallel transport of vectors is described in a manner immediately generalizable to parallel transport of quantum states in parameter space. The associated anholonomy is the geometric phase. One realization of parallel transport is by adiabatic cycling of the parameters. The phase is the flux of a 2-form. The 2-form is equivalent to the antisymmetric part of a gauge-invariant quantum geometric tensor. The symmetric part of this tensor gives a natural metric on parameter space. If the parameters are themselves regarded as dynamical variables, their adiabatic dynamics are influenced by a gauge field depending on both parts of the tensor. Corrections to the geometric phase (of higher order in an adiabatic parameter) can be obtained by successive transformations to moving frames, thereby generating a renormalization map of circuits in the space of Hamiltonians; the iterates diverge in a universal way. This quantum renormalization is illustrated by classical Newtonian and Hamiltonian renormalizations for a pendulum with changing frequency. To conclude, there are some historical remarks about geometric phases.

1. Introduction

The kind invitation to write this survey article provides two welcome opportunities. First, to present the fundamentals of the subject in a new perspective, reflecting some of the many recent developments and including some new material; and second, to make some historical remarks, drawing attention to important early works and describing the genesis of my own ideas in this field.

Two concepts are crucial to the understanding of this dusty corner of quantum theory which the brooms of our understanding are beginning to disturb. They are *anholonomy* and *adiabaticity*.

Anholonomy is a geometrical phenomenon in which nonintegrability causes some variables to fail to return to their original values when others, which drive them, are altered round a cycle. The simplest anholonomy is in the parallel transport of vectors, two examples being the change in the direction of swing of a Foucault pendulum after one rotation of the earth, and the change in the direction of linear polarization of light along a twisting ray [1][2] or coiled optical fibre [3-6] whose direction is altered in a cycle. The anholonomy to be described here is quantum-mechanical, and concerns the phase of a state which is parallel-transported round a cycle [7]. Parallel transport of a quantum state will here be introduced as a simple generalization of parallel transport of a vector.

Adiabaticity is slow change and therefore denotes phenomena at the border between dynamics and statics. Adiabatic change provides the simplest (but not the only [8]) way to make quantum parallel transport happen. The variables which are cycled are parameters in the Hamiltonian of a system. If the cycling is slow, the adiabatic theorem [9] guarantees that the system returns to its original state. But it usually acquires a nontrivial phase, a manifestation of anholonomy, and this is the phenomenon of interest here.

2. Classical Parallel Transport

It is convenient to begin by obtaining the law for the ordinary parallel transport of a vector over the surface of a sphere, expressing it in a form enabling instantaneous generalization to quantum mechanics. Let the unit vector $\mathbf{e}$ be transported by changing the unit radius vector $\mathbf{r}$ (Fig.1) and making two demands: that $\mathbf{e} \cdot \mathbf{r}$ must remain zero and that the orthogonal triad (frame) containing $\mathbf{e}$ and $\mathbf{r}$ must not twist about $\mathbf{r}$, *i.e.*, $\boldsymbol{\Omega} \cdot \mathbf{r} = 0$ where $\boldsymbol{\Omega}$ is the angular velocity of the triad. These conditions define parallel transport of $\mathbf{e}$ and lead to the law

$$\dot{\mathbf{e}} = \boldsymbol{\Omega} \wedge \mathbf{e} \qquad \text{where} \qquad \boldsymbol{\Omega} = \mathbf{r} \wedge \dot{\mathbf{r}} \tag{1}$$

This law is nonintegrable; when $\mathbf{r}$ returns to its original direction after a circuit C on the sphere, $\mathbf{e}$ does not return (in spite of never having been twisted) but has turned through an angle $\alpha(C)$ which is the anholonomy now to be determined. Define $\mathbf{e}' \equiv \mathbf{r} \wedge \mathbf{e}$ (so that $\mathbf{r}$, $\mathbf{e}$, $\mathbf{e}'$ form an orthogonal triad) and the complex unit vector

$$\boldsymbol{\psi} \equiv (\mathbf{e} + i\mathbf{e}')/\sqrt{2} \tag{2}$$

in the plane perpendicular to $\mathbf{r}$. In terms of $\boldsymbol{\psi}$, the parallel transport law (1) (which holds for $\mathbf{e}'$ as well as $\mathbf{e}$) takes the simple form

$$\mathrm{Im}\,\boldsymbol{\psi}^*\cdot\dot{\boldsymbol{\psi}} = 0 \qquad i.e., \qquad \mathrm{Im}\,\boldsymbol{\psi}^*\cdot d\boldsymbol{\psi} = 0 \tag{3}$$

where $d\boldsymbol{\psi}$ is the change in $\boldsymbol{\psi}$ resulting from a change $d\mathbf{r}$.

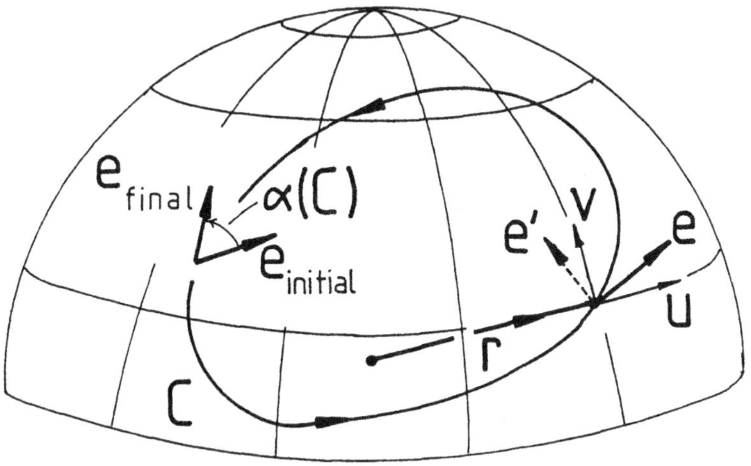

Figure 1. Rotation by $\alpha(C)$ after parallel transport of vector $\mathbf{e}$ round circuit C on a sphere.

To find $\alpha(C)$ we chart the passage of $\mathbf{e}$ and $\mathbf{e}'$ relative to a local basis of unit vectors $\mathbf{u}(\mathbf{r}), \mathbf{v}(\mathbf{r})$ (Fig.1) defined at each point on the sphere. For example, we could choose $\mathbf{u}$ and $\mathbf{v}$ to lie along the parallel of latitude θ and meridian of longitude ϕ at $\mathbf{r} = (\sin\theta\cos\phi,\ \sin\theta\sin\phi,\ \cos\theta)$, *i.e.*,

$$\mathbf{u} = (-\sin\phi,\ \cos\phi,\ 0)\ ,\quad \mathbf{v} = (-\cos\theta\cos\phi,\ -\cos\theta\sin\phi,\ \sin\theta). \tag{4}$$

Specifying a local basis is equivalent to specifying the complex unit vector

$$\mathbf{n}(\mathbf{r}) \equiv (\mathbf{u}(\mathbf{r}) + i\mathbf{v}(\mathbf{r}))/\sqrt{2} \tag{5}$$

If the angle between the transported $\mathbf{e}$ and the local $\mathbf{u}$ is $\alpha(t)$, (2) and (5) give

$$\boldsymbol{\psi} = \mathbf{n}\exp(-i\alpha) \tag{6}$$

whence (3) gives the anholonomy as

$$\begin{aligned}\alpha(C) &= \oint d\alpha = \mathrm{Im}\oint \mathbf{n}^* \wedge \cdot\, d\mathbf{n} \\ &= \mathrm{Im}\iint_{\partial S=C} d\mathbf{n}^* \cdot d\mathbf{n}\end{aligned} \tag{7}$$

where in the last equality Stokes' theorem has been used and the integral is over the area on the sphere bounded by C. It is an important result that the integrand in (7) is independent of the choice of local basis $\mathbf{u}$, $\mathbf{v}$: a change in this choice can be represented by a rotation $\mu(\mathbf{r})$ which induces the gauge transformation

$$\mathbf{n}(\mathbf{r}) \to \mathbf{n}'(\mathbf{r})\exp\{i\mu(\mathbf{r})\} \tag{8}$$

under which $d\mathbf{n}^* \wedge \cdot\, d\mathbf{n}$ is invariant.

In terms of arbitrary parameters X_1, X_2 specifying $\mathbf{r}$ (*i.e.*, position on the sphere), Eq. (7) can be written explicitly as

$$\alpha(C) = \mathrm{Im}\iint_{\partial S=C} dX_1\, dX_2\, (\partial_1\mathbf{n}^* \cdot \partial_2\mathbf{n} - \partial_2\mathbf{n}^* \cdot \partial_1\mathbf{n}) \tag{9}$$

where ∂_j denotes $\partial/\partial X_j$. The choice $X_1 = \theta$, $X_2 = \phi$, together with (4), yields the integrand $d\theta\, d\phi\, \sin\theta$, which is simply the area element on the sphere, leading to the old result that the anholonomy $\alpha(C)$ is the *solid angle* subtended by C at the centre of the sphere.

3. Quantum Parallel Transport

To make the generalization to quantum mechanics, we replace the complex unit vector $\boldsymbol{\psi}$ by a normalized quantum state $|\psi\rangle$, *i.e.*, a unit vector in a Hilbert space, and position $\mathbf{r} = (X_1, X_2)$ on the sphere by position $X = (X_1, X_2, \ldots)$ in a space of parameters governing the physical system represented by $|\psi\rangle$. At each X, $|\psi\rangle$ is defined up to a phase (just as $\mathbf{e}$ was defined up to a rotation at each $\mathbf{r}$). Then a natural transport law [10] governing the phase of $|\psi\rangle$ as X varies is provided by reinterpreting (3) as the connection

$$\mathrm{Im}\,\langle\psi|d\psi\rangle = 0. \tag{10}$$

Like (3), this law is nonintegrable: when X is taken round a circuit C, $|\psi\rangle$ returns with a changed phase. This change is the *quantum geometric phase* $\gamma(C)$; thus

$$\langle\psi_{\mathrm{initial}}|\psi_{\mathrm{final}}\rangle = \exp\{i\gamma(C)\}. \tag{11}$$

To find γ we again introduce a local basis by choosing at each X a definite (and so of course single-valued) state $|n(X)\rangle$, relative to which $|\psi\rangle$ is defined by

$$|\psi\rangle = |n(X)\rangle \exp(i\gamma) \tag{12}$$

Then (10) gives

$$\begin{aligned} \gamma(C) &= \oint d\gamma = -\text{Im} \oint \langle n|dn\rangle \\ &= -\text{Im} \iint_{\partial S=C} \langle dn| \wedge |dn\rangle \equiv - \iint_{\partial S=C} V(X). \end{aligned} \tag{13}$$

The integrand $V = \text{Im}\,\langle dn \wedge dn\rangle$ is the *phase 2-form*, whose flux through C gives the geometric phase. V is invariant under the gauge transformation

$$|n(X)\rangle \to |n'(X)\rangle \equiv |n(X)\rangle \exp\{i\mu(X)\} \tag{14}$$

For this mathematics to represent physics, it must be possible to implement the connection (10) by the Schrodinger equation

$$i\hbar|\dot{\Psi}\rangle = \hat{H}|\Psi\rangle \tag{15}$$

governing the evolution of any state $|\Psi\rangle$. A simple way [7] is to incorporate the parameters X into the Hamiltonian and change them slowly. Then the adiabatic theorem guarantees that in the absence of degeneracies (a restriction that can be removed [46]) $|\Psi\rangle$ will cling to one of the eigenstates of $\hat{H}(X(t))$, defined by

$$\hat{H}(X)\,|\psi\rangle = E_n(X)\,|\psi\rangle \tag{16}$$

The adiabatic ansatz

$$|\Psi\rangle \approx |\psi\rangle \exp\left\{-\frac{i}{\hbar}\int_0^t dt'\, E_n(X(t'))\right\} \tag{17}$$

then gives the connection (10) immediately upon projecting (15) onto $|\psi\rangle$. The state $|n(X)\rangle$ in the 2-form (13) is any solution of (16) with a definite phase at each X.

Because of (17), the total phase change of $|\Psi\rangle$ includes a dynamical part as well as the $\gamma(C)$ being studied here. Thus

$$\langle\Psi_{\text{final}}|\Psi_{\text{initial}}\rangle = \exp\{i(\gamma_d + \gamma C)\} \tag{18}$$

where, for a circuit that takes a time T,

$$\gamma_d = -\frac{1}{\hbar}\int_0^T dt\, E_n(X(t)) \tag{19}$$

One might say that γ_d and $\gamma(C)$ give the system's best answers to two questions about its adiabatic circuit. For γ_d the question is: how long did your journey take? For $\gamma(C)$ it is: where did you go?

Aharanov and Anandan [8] give a different interpretation of parallel transport. They regard the parameters X as labelling the state, rather than $\hat{H}$, so that $X_1, X_2, \ldots$ are coordinates in the *projective Hilbert space* that includes all quantum states, but where states differing only in phase (or normalization) are represented by the same point. Then a state $|\Psi\rangle$ evolving under Eq. (15) (not necessarily adiabatically) so as to return in T to the same X acquires a phase (18), with geometric part (13) (where the phase of $|n(X)\rangle$ is an arbitrary function of X) and dynamical part given by

$$\gamma_d = -\frac{1}{\hbar}\int_0^T dt\,\langle\Psi|\hat{H}|\Psi\rangle \tag{20}$$

instead of Eq. (19). The relation between the two approaches is that in the adiabatic case X parameterizes that part of the projective Hilbert space corresponding to the nth eigenstate of the chosen family of Hamiltonians $\hat{H}(X)$.

Several experiments have measured the geometric phase for particles, with spin 1/2 (neutrons [11]), spin 1 (photons [3]) and spin 3/2 (chlorine nuclei [12]). These depend on the result [7] that when $\hat{H}$ is a rotationally symmetric function of the spin, *i.e.*,

$$\hat{H} = F(\boldsymbol{\sigma}\cdot\mathbf{X}) \tag{21}$$

where $\mathbf{X} = (X_1, X_2, X_3)$ and $\boldsymbol{\sigma} = (\sigma_1, \sigma_2, \sigma_3)$ is the vector spin operator, the geometric phase for the state with spin component n along $\mathbf{X}$ is

$$\gamma_n(C) = -n\,\Omega(C) \tag{22}$$

where $\Omega(C)$ is the solid angle subtended by C at $\mathbf{X} = 0$.

These experiments all employ a superposition of eigenstates, rather than a single one, so that

$$\begin{aligned} |\Psi_{\text{initial}}\rangle &= \sum_n a_n|n\rangle \\ |\Psi_{\text{final}}\rangle &= \sum_n a_n|n\rangle \exp\{i(\gamma_{dn} + \gamma_n(C))\} \end{aligned} \tag{23}$$

At the end, that is after X has been cycled, an observable $\hat{A}$, which does not commute with the final $\hat{H}$, is measured (for example with a polarizer). Thus

$$\begin{aligned} \langle\hat{A}\rangle = \sum_n |a_n|^2\langle n|\hat{A}|n\rangle &+ 2\,\mathrm{Re}\sum_{m\neq n} a_n^* a_m\langle m|\hat{A}|n\rangle \\ &\times \cos\Big\{[\gamma_{dn} + \gamma_n(C)] - [\gamma_{dm} + \gamma_m(C)]\Big\}. \end{aligned} \tag{24}$$

The oscillatory terms reveal $\gamma_n(C)$. This scheme has proved more convenient than the earlier suggestion [7] of splitting an ensemble of systems (*e.g.*, a beam of particles) into two subensembles, one being driven by an $\hat{H}$ which is cycled and the other by an $\hat{H}$ which is not, and then recombining the subensembles to detect $\gamma(C)$ by interference. (That is, instead of using one state and two Hamiltonians it is preferable to use two states — at least — and one Hamiltonian.)

Hannay [13] found an analogue of the geometric phase for *classical* systems. This was based on the simple observation that a quantum system in an eigenstate is an oscillator (because of the time factor $\exp(-iE_nt/\hbar)$, so that classical oscillators should exhibit similar anholonomy when parameters that govern them are cycled. The phase is now an angle, which may be an angle in space, like that of a wheel, or — more commonly — an abstract angle variable in phase space as with a harmonic oscillator. If the classical system is multiply periodic (integrable) for all X, with N freedoms (that is, coordinates $\mathbf{q} = (q_1 \cdots q_N)$ and momenta $\mathbf{p} = (p_1 \cdots p_N)$ and Hamiltonian $H(\mathbf{q}, \mathbf{p}; X)$, its orbit for fixed X winds round an N-torus [14] in phase space, with N angle variables $\boldsymbol{\theta} = (\theta_1 \cdots \theta_N)$ increasing uniformly. Conjugate to $\boldsymbol{\theta}$ are N adiabatically conserved actions $\mathbf{I} = (I_1 \cdots I_N)$ which label the torus. After a slow cycle of X the angles have acquired shifts which contain a geometric as well as a dynamical part. For a spinning particle [15-17] this classical anholonomy is the angle shift given by ordinary parallel transport of a vector.

Underlying Hannay's angles is a *classical 2-form*. This is the classical limit of the phase 2-form in Eq. (13), and semiclassical asymptotics [18] provides the expression

$$V(X) \underset{\hbar \to 0}{\longrightarrow} -\langle d\mathbf{p} \wedge \cdot d\mathbf{q} \rangle / \hbar \tag{25}$$

whose symbols should be interpreted as follows. The wedge product $\wedge$ links the d's in parameter space. The scalar product $\cdot$ links $\mathbf{p}$ and $\mathbf{q}$. $\langle\ \rangle$ denotes an average over the angles on the torus labelled $\mathbf{I}$ which at X corresponds [19] to the quantum state $|n\rangle$, *i.e.*, $\langle\ \rangle = \int_0^{2\pi} d\theta_1 \cdots d\theta_N / (2\pi)^N$. $d\mathbf{q}$ is the coordinate displacement linking corresponding points (labelled by the same θ) on the tori $\mathbf{I}$ at X and $X + dX$, and similarly for $d\mathbf{p}$.

It is amusing to note that if the $2N$ variables $\mathbf{q}$ and $\mathbf{p}$ are replaced by the N complex variables

$$\mathbf{n} = (n_1 \cdots n_N) \equiv (\mathbf{q} + i\mathbf{p})/\sqrt{2\hbar} \tag{26}$$

then (25) takes the form

$$V(X) \underset{\hbar \to 0}{\longrightarrow} \operatorname{Im} \langle d\mathbf{n}^* \wedge \cdot d\mathbf{n} \rangle \tag{27}$$

which bears a close formal resemblance both to the quantum expression (13) and the geometrical formula (7).

If the classical motion is not multiply periodic, that is if it is wholly or partly chaotic, the question of the classical limit of V is more delicate. It is tempting to claim that the limit is (25) for nonintegrable as well as integrable motion, but it is difficult to interpret the average $\langle\ \rangle$ and the displacements $d\mathbf{q}$ and $d\mathbf{p}$. In one of several interpretations, obtained by a semiclassical argument (not yet published) in collaboration with M. Wilkinson, $\langle\ \rangle$ denotes a time average over all points on an infinite orbit, and $d\mathbf{q}$ and $d\mathbf{p}$ link simultaneous points on the orbits for X and $X + dX$. For nonintegrable systems, however, it is not easy to express this result by replacing $\langle\ \rangle$ by a phase-space integral over the manifold explored by the orbit, because it is not clear what are then the 'corresponding points', linked by $d\mathbf{q}$ and $d\mathbf{p}$, on the manifolds for X and $X + dX$ (for an ergodic system these are the two constant-energy surfaces).

4. The Quantum Geometric Tensor

The central mathematical object underlying the quantum phase is the 2-form $V = \operatorname{Im}\langle dn \wedge dn\rangle$. This is equivalent to an antisymmetric second-rank tensor field $V_{ij}(X)$ on the parameter space (or projective Hilbert space) with a quantum state $|n(X)\rangle$ defined at each point, namely

$$V_{ij}(X) = \operatorname{Im}\{\langle\partial_i n|\partial_j n\rangle - \langle\partial_j n|\partial_i n\rangle\} \tag{28}$$

This tensor is invariant under the gauge transformation (14), but it is not the only such invariant tensor. More general is the *quantum geometric tensor*

$$T_{ij}(X) \equiv \langle\partial_i n|\left(1 - |n\rangle\langle n|\right)|\partial_j n\rangle \tag{29}$$

which is Hermitian, *i.e.*, $T_{ij} = T^*_{ji}$. The projector $|n\rangle\langle n|$ is essential to the gauge invariance. The imaginary part of T_{ij} is simply $V_{ij}/2$, so we can write

$$T_{ij} = g_{ij} + iV_{ij}/2 \tag{30}$$

where g_{ij} is the real symmetric tensor field $\operatorname{Re} T_{ij}$.

We know the quantum meaning of V_{ij}: its flux gives the phase $\gamma(C)$. Therefore, it is natural to ask whether g_{ij} has significance. The answer is that g_{ij} provides a natural means of measuring distances along paths in parameter space; it is the *quantum metric tensor*. To understand why, observe that a natural measure of the squared distance between two nearby quantum states is the deviation from unity of their scalar product. If the states are $|1\rangle$ and $|2\rangle$ this gives, for the distance between the corresponding points X_1 and X_2 in parameter space,

$$\Delta s_{12}^2 = 1 - |\langle 1|2\rangle|^2 \tag{31}$$

Taking the limit $1 \rightarrow 2$, and using the fact that all states are normalized, we obtain (using the summation convention for repeated indices i and j)

$$\begin{aligned} ds^2 &= \langle dn|\,(1 - |n\rangle\langle n|)\,|dn\rangle = \langle\partial_i n|\,(1 - |n\rangle\langle n|)\,|\partial_j n\rangle\, dX_i dX_j \\ &= T_{ij}\, dX_i dX_j = g_{ij}\, dX_i dX_j \end{aligned} \tag{32}$$

as claimed. The quantum tensor was introduced in an interesting paper by Provost and Vallee [50].

From its structure, g_{ij} can never give a negative ds^2: in fact it is a positive semidefinite metric. Along a finite path (not necessarily closed) between $|1\rangle$ and $|2\rangle$, the quantum distance is

$$s_{12}(C) = \int_1^2 (g_{ij}\, dX_i dX_j)^{1/2}. \tag{33}$$

Page [33] and Bouchiat and Gibbons [41] give explicit forms for some metrics on the full Hilbert and projective Hilbert spaces.

The simplest example is a 2-state system, for which $\hat{H}$ has the form (21), with $\hat{\boldsymbol{\sigma}}$ the 3 Pauli matrices. If we take X as a unit vector, specified by parameters θ, ϕ (polar angles), the eigenstates are

$$|+\rangle = \begin{pmatrix} \cos(\theta/2)\, e^{i\phi/2} \\ \sin(\theta/2)\, e^{-i\phi/2} \end{pmatrix}, \quad |-\rangle = \begin{pmatrix} \sin(\theta/2)\, e^{i\phi/2} \\ -\cos(\theta/2)\, e^{-i\phi/2} \end{pmatrix} \tag{34}$$

For both of these, (32) gives $ds^2 = d\theta^2 + \sin\theta\, d\phi^2$, and this is the natural metric on the sphere of parameters (which in this case is also the projective Hilbert space).

Some interesting questions are suggested by this identification of g_{ij} as a metric on parameter space:

(i) Do the geodesics, and in particular the shortest paths, connecting non-neighbouring states $|1\rangle$ and $|2\rangle$ have physical significance? One possibility, suggested by the work of Pancharatnam [20][21], is that the geodesics are the special paths along which the state preserves its phase in the sense that $\langle 1|2\rangle$ is real. This is true for the 2-state system just discussed, but seems to fail otherwise (probably for reasons of codimension). It is worth remarking that as $2 \rightarrow 1$ the overlap $\langle 1|2\rangle$ is real to second as well as first order in dX, for any path whatever.

(ii) Can the geodesics be chaotic? This would require parameters X and states $|n(X)\rangle$ for which the Riemann curvature defined in terms of g_{ij} is negative (at least in some places) and the space is compact.

(iii) Do *families* of geodesics (for example those issuing in different directions from the same point) exhibit the generic caustic singularities classified by catastrophe theory [22][23]? Do any such caustics have physical meaning? In 2-state systems the geodesics from X focus nongenerically at the antipodal point on the sphere, where the state is orthogonal to $|n(X)\rangle$, but again this appears to be a special situation.

(iv) Is there any meaning or interest in *quantizing* the geodesic motion in parameter space, for example by taking as Hamiltonian the Laplace-Beltrami operator $g^{-1/2}\partial_i g^{-1/2} g_{ij}\partial_j$ (where $g \equiv \det g_{ij}$)? Such quantizations are different from that described in the next section.

5. Dynamics of the Parameters

Until now we have regarded X as classical parameters which can be altered arbitrarily and which are unaffected by the quantum system they drive. But no physical action is unilateral and in reality X are themselves dynamical variables of a 'heavy' system coupled to the 'light' system (what we have so far called 'the' system) and therefore subject to reaction from it. Indeed the earliest application of the adiabatic theorem was the Born-Oppenheimer theory of molecules, in which X are coordinates describing the positions of the (heavy) nuclei and the light system is the electrons. Recently it has been pointed out [24–27] that in lowest order the reaction of the light system on the heavy dynamics is through a gauge field consisting of a vector potential whose curl is the phase 2-form V, and a scalar potential. Here I will show that what the gauge field really depends on is the quantum geometric tensor T_{ij} of section 3.

Let the heavy momenta, conjugate to X_i, be P_i. Then a fairly general nonrelativistic quantum Hamiltonian for the coupled system is

$$\hat{H}_{\text{tot}} = \tfrac{1}{2}\sum_{ij} Q_{ij}\hat{P}_i\hat{P}_j + H(\hat{\xi};\hat{X}), \tag{35}$$

in which Q_{ij} is an inverse mass tensor, $\hat{\xi}$ are the dynamical variables of the light system (coordinates, momenta, spins, . . .) and H our previous Hamiltonian in which the X were regarded as parameters and which has eigenstates $|n(X)\rangle$ and energies $E_n(X)$. In the position representation for the heavy system, that is $\hat{P}_i = -i\hbar\partial_i$, the adiabatic ansatz is to write the full quantum state in the separated form

$$\langle X|\Psi\rangle \approx \Psi_{\text{heavy}}(X)|n(X)\rangle \tag{36}$$

and to consider the effective Hamiltonian governing Ψ_{heavy} to be

$$\hat{H}_{\text{eff}} = \langle n(X)|\hat{H}_{\text{tot}}|n(X)\rangle. \tag{37}$$

In $\hat{H}_{\text{eff}}$ the reaction of the light on the heavy system comes from the action of the gradient operators $\hat{P}_i$ on the X-dependence of $|n\rangle$. A straightforward calculation gives

$$\hat{H}_{\text{eff}} = \tfrac{1}{2}\sum_{ij} Q_{ij}\left\{\hat{P}_i - A_i(\hat{X})\right\}\left\{\hat{P}_j - A_j(\hat{X})\right\} + \Phi(\hat{X}) + E_n(\hat{X}) \tag{38}$$

where

$$A_i(X) = i\hbar \langle n|\partial_i n\rangle \tag{39}$$

and

$$\Phi(X) = \frac{\hbar^2}{2}\sum_{ij} Q_{ij}\, g_{ij}(X) \tag{40}$$

Here the emphasis is on the gauge potentials Φ and A_i — the scalar $E_n(X)$ is the 'potential surface' studied in conventional Born-Oppenheimer theory. Although (38) is a quantum Hamiltonian it can be used in suitable circumstances to calculate the *classical* motion of the heavy system, which will be affected by the fields A_i and Φ.

The physical effects of the vector potential A_i depend only on the 'magnetic' field

$$F_{ij} = \partial_i A_j - \partial_j A_i = -\hbar V_{ij} \tag{41}$$

(including its singularities and values in inaccessible regions — I am not denying the Aharonov-Bohm effect for heavy systems!). Thus the 'magnetic' field seen by the heavy system is the antisymmetric part of the quantum geometric tensor. The symmetric part of T_{ij} determines the 'electric' potential via Eq. (40). For an isotropic mass tensor, *i.e.*, $Q_{ij} = \delta_{ij}/M$, Φ depends on $Tr\ g_{ij}$. It is a curious asymmetry that the 'electric' field depends on the *gradients* of g_{ij} whereas the 'magnetic' field depends on V_{ij} itself.

The singularities of the gauge field are the *degeneracies* X^* of the spectrum, where $E_n(X^*) = E_{n\pm1}(X^*)$. It is already known [7] that the 'magnetic' field V_{ij} (2-form) has monopole singularities. From the definition (29) of T_{ij} it is clear that g_{ij} has similar singularities, so that the 'electric' field near X^* is an inverse-*cube* force.

The situation near a degeneracy can be described by a special case of a simple model, which is of independent interest (and which has been studied from a different viewpoint by Anandan and Aharonov [28]), where the spin s of one (light) particle is coupled to the spatial coordinates of a second otherwise free (heavy) particle. Thus

$$\hat{H}_{\text{tot}} = \frac{1}{2M}\hat{P}^2 + F(\hat{\mathbf{X}}\cdot\hat{\boldsymbol{\sigma}}) \tag{42}$$

Near a degeneracy the appropriate model is a 2-state light system, so that we should take $s = \frac{1}{2}$, with linear coupling $F \propto \mathbf{X}\cdot\boldsymbol{\sigma}$.

The eigenvalues of $\mathbf{X}\cdot\hat{\boldsymbol{\sigma}}$ are nX, where $X \equiv |\mathbf{X}|$ and $-s \le n \le s$. The quantum tensor for the state $|n\rangle$ can be shown to be

$$T^n_{ij}(X) = \frac{1}{2X^2}\left\{\left(s(s+1) - n^2\right)\left(\mathbf{e_i}\cdot\mathbf{e_j} - (\mathbf{e_i}\cdot\mathbf{x})(\mathbf{e_j}\cdot\mathbf{x})\right) \mp in(\mathbf{e_i}\wedge\mathbf{e_j})\cdot\mathbf{x}\right\} \tag{43}$$

where $\mathbf{x} = \mathbf{X}/|\mathbf{X}|$ and $\mathbf{e_i}$ is the unit vector along the i direction. The metric tensor g_{ij} has a zero eigenvalue, corresponding to radial parameter displacements, which simply scale H leaving the states $|n\rangle$ unaffected: radial motions cover zero distance.

From Eqs. (38)-(40), the *classical* Newtonian equation for the heavy parti cle involves the Lorentz force from the magnetic monopole and the 'electric' force

$$-\nabla_{\mathbf{X}}\Phi(\mathbf{X}) = -\frac{\hbar^2}{2M}\nabla_{\mathbf{X}} \operatorname{Tr} g_{ij} = \frac{\hbar^2(s(s+1)-n^2)}{MX^3}\mathbf{x} \qquad (44)$$

This is of centrifugal type, and repels the parameters from a degeneracy (becoming significant at a distance of order $M^{-1/3}$), thereby tending to preserve the validity of the adiabatic approximation. We obtain, when the light particle is in the nth spin state,

$$M\ddot{\mathbf{X}} = \frac{S_z}{2X^3}\dot{\mathbf{X}} \wedge \mathbf{X} + \frac{(S^2 - S_z^2)}{MX^4}\mathbf{X} - \frac{nF'(nX)}{X}\mathbf{X} \qquad (45)$$

where $S_z \equiv n\hbar$ and $S^2 \equiv \hbar^2 s(s+1)$. This describes integrable motion, with conserved energy and modified angular momentum $M\mathbf{X} \wedge \dot{\mathbf{X}} - S_z\mathbf{X}/X$.

6. Adiabatic Renormalization

Now we return to the adiabatic scenario of section 3 and realize that γ_d and $\gamma(C)$ in Eq. (18) are but the first two terms in an infinite series involving powers of an adiabatic slowness parameter ϵ, influencing the dynamics through $\hat{H}$ whose time-dependence enters in the combination ϵt. The dominant term is γ_d (Eq. 19) and is of order ϵ^{-1}. The next term is $\gamma(C)$, whose unique feature — and the reason for its being called geometric — is that it is independent of ϵ, and so depends only on the sequence of Hamiltonians along the circuit and not on its time history.

This uniqueness is not threatened by the observation that transformation to a moving frame (a common practice in problems involving spin [11]) can make $\gamma(C)$ appear 'dynamical' by making it emerge from a correction to the energy rather than as anholonomy: the geometric structure of $\gamma(C)$ is independent of how it is derived.

Transformations to moving frames have however another interest, in that they form the basis of a renormalization (iteration) technique for generating higher-order corrections to the phase. Details of the technique have been published elsewhere [29]; here I will outline the central idea, and give an example.

Let the Hamiltonian $\hat{H}_0(t)$ generating the quantum motion be cyclic, in the sense that $\hat{H}_0(+\infty) = \hat{H}_0(-\infty)$, and let it have instantaneous eigenstates $|n_0(t)\rangle$ and energies $E_0(n,t)$. The evolving state $|\Psi_0(t)\rangle$ is determined by

$$i|\dot{\Psi}_0(t)\rangle = \hat{H}_0(t)|\Psi_0(t)\rangle \qquad (46)$$

with the initial condition

$$|\Psi_0(-\infty)\rangle = |n_0(-\infty)\rangle \equiv |N\rangle \tag{47}$$

After the cycle, *i.e.*, at $t = +\infty$, $|\Psi_0\rangle$ will have returned only approximately to $|N\rangle$, so a phase can be defined precisely by

$$\gamma \equiv \text{Im} \log \langle N|\Psi_0(+\infty)\rangle - \gamma_d \tag{48}$$

The geometric phase $\gamma(C)$ (Eq. 13) is $lim_{\epsilon \to 0}\gamma$. The aim is to obtain increasingly accurate approximations to $\gamma - \gamma(C)$. It is worth emphasizing that the non-aim is the determination of the nonadiabatic transition probability $1 - |\langle N|\Psi_0(+\infty)\rangle|^2$, because this is the usual objective of adiabatic theory, and that the non-method is perturbation theory, because this is the usual technique [30][49].

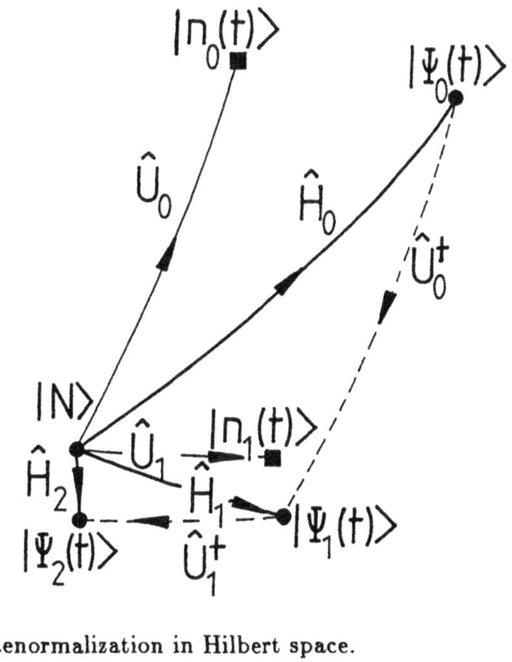

Figure 2. Renormalization in Hilbert space.

To explain the method used instead, we refer to Fig. 2. When ϵ is small we expect $|\Psi_0(t)\rangle$ to be close to $|n_0(t)\rangle$. This suggests that defining a unitary transformation $\hat{U}_0(t)$ by

$$|n_0(t)\rangle = \hat{U}_0(t)|N\rangle \tag{49}$$

will be useful. The inverse operator $U_0^\dagger$ sends $|n_0(t)\rangle$ back to $|N\rangle$, that is, it freezes the moving eigenstate. Therefore $\hat{U}_0^\dagger$ should almost freeze the evolving state $|\Psi_0(t)\rangle$, and so we define

$$|\Psi_1(t)\rangle \equiv \hat{U}_0^\dagger|\Psi_0(t)\rangle. \tag{50}$$

We are attempting to follow $|\Psi_0(t)\rangle$ by transforming to a moving frame. The Hamiltonian governing $|\Psi_1\rangle$ is

$$\hat{H}_1 = \hat{U}_0^\dagger \hat{H}_0 \hat{U}_0 - i\hat{U}_0^\dagger \dot{\hat{U}}_0, \tag{51}$$

in which the second term is the quantum analogue of the inertial forces generated classically by transforming to a moving frame.

Now that the original problem has been reduced to one of the same form but involving $|\Psi_1\rangle$ and $\hat{H}_1$ instead of $|\Psi_0\rangle$ and $\hat{H}_0$, it is natural to iterate the process by defining $|\Psi_2\rangle \equiv \hat{U}_1^\dagger|\Psi_1\rangle$, where $\hat{U}_1^\dagger$ freezes the eigenstates $|n_1\rangle$ of $\hat{H}_1$. This defines a *renormalization map* $\hat{H}_k \to \hat{H}_{k+1}$ in Hamiltonian space. The form of the map is simple when when written in a basis of initial states (which are unaffected by renormalization) and with the phases of the eigenstates chosen so that they are parallel-transported, *i.e.*, $\langle n_k|\dot{n}_k\rangle = 0$:

$$\langle M|\hat{H}_{k+1}|N\rangle = E_k(n,t)\delta_{MN} - i\frac{\langle m_k(t)|\dot{\hat{H}}_k(t)|n_k(t)\rangle}{E_k(m,t) - E_k(n,t)}(1-\delta_{MN}) \tag{52}$$

The kth approximant $\gamma^{(k)}$ to the phase is obtained by neglecting the off-diagonal terms in $\hat{H}_{k+1}$. $\gamma^{(k)}$ is the sum of the phase anholonomies of the Hamiltonians $\hat{H}_0 \ldots \hat{H}_k$ (arising from the continuation of $|n_k(t)\rangle$ from $t=-\infty$ to $t=+\infty$ and reflected as phase factors $\langle N|U_k(+\infty)|N\rangle$), together with an additional term involving E_k [29]. (A contrary choice of phases, *i.e.*, $|n_k(+\infty)\rangle = |n_k(-\infty)\rangle$, gives $\langle N|U_k(+\infty)|N\rangle = 1$, but now the diagonal terms in Eq. (52) contain extra terms $-i\langle n_k|\dot{n}_k\rangle$ and all corrections — including $\gamma^{(0)} = \gamma(C)$ as mentioned previously — appear dynamical.)

Each renormalization produces a new Hamiltonian which over $-\infty < t < +\infty$ traverses a loop in Hamiltonian space. If the renormalizations converged, successive loops would get smaller (by a factor ϵ each time). But this does not, and indeed cannot, happen. If it did, $\langle\Psi(-\infty)|\Psi(+\infty)\rangle$ would have modulus unity, contradicting the existence of transitions to other states. The accumulation of inertial forces in successive renormalizations defeats our attempts to follow the motion, which slips out of control, causing the scheme to diverge.

Nevertheless, the corrections generated by renormalization do get smaller at first, and enable γ to be determined with an error of order $\exp(-1/\epsilon)$, which occurs after $k \sim 1/\epsilon$ renormalizations. A detailed exploration [29] of 2-state systems (the simplest nontrivial case, for which the geometry of the loop map can be made explicit) reveals that the Hamiltonian loops (which lie on a 2-sphere) get smaller and then larger in a universal way (that is, almost always independent of the form of the initial loop).

This procedure is typical of asymptotic procedures and occurs also in the more usual adiabatic perturbation theory. It prompts interesting questions. What is the dynamical significance of the moving frame that produces the best approximant to γ, generated by $\hat{U}_{k\sim 1/\epsilon}\hat{U}_{k-1}\cdots\hat{U}_0$? Can the exponential residue $\gamma - \gamma^{(k)}$ be more closely approximated by generalizing the Borel (or some other) resummation method [47]?

It is instructive to illustrate adiabatic renormalization with the *classical* problem which gave birth to the entire subject, namely the Ehrenfest-Einstein pendulum [31] whose frequency is slowly changed. Newton's equation is

$$\partial_t^2 x(t) + \omega^2(t)\, x(t) = 0 \tag{53}$$

in which the frequency $\omega(t)$ is a smooth nonzero function with $\omega(+\infty) = \omega(-\infty) \equiv \omega_\infty$. The same equation describ es the (time-independent) quantum mechanics of a beam of particles with energy E encountering a potential well or hill $V(x)$ such that $E > V(x)$ for all x.

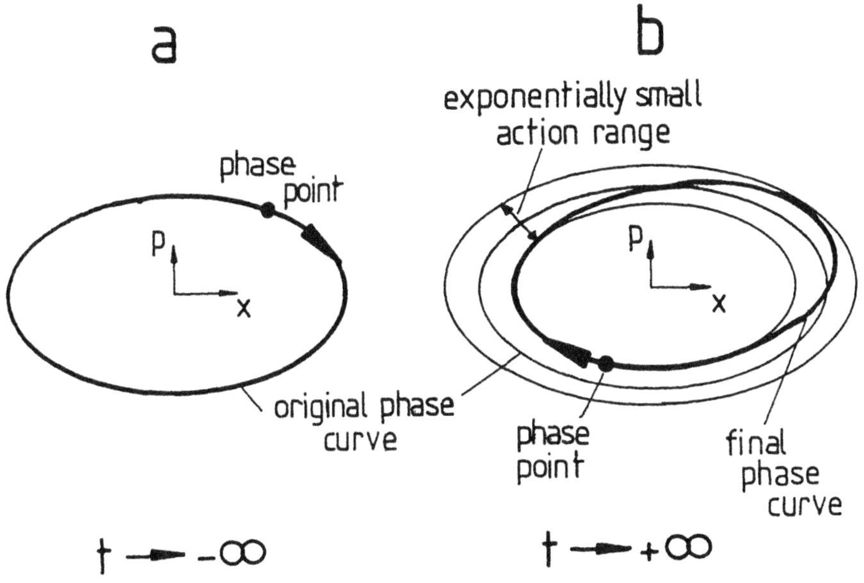

Figure 3. (a) Initial and (b) final phase portraits for slowly–altered pendulum.

Consider motion in the phase plane with variables x and $p = \dot{x}$. Initially, *i.e.*, as $t \to -\infty$, each phase point moves round an ellipse with frequency ω_∞ (Fig. 3a). The subsequent motion lies on a curve that at each instant approximates one of the elliptical contours of the Hamiltonian

$$H(x, p, t) = \tfrac{1}{2}\left[p^2 + \omega^2(t) x^2\right] \tag{54}$$

at that time. These subsequent ellipses have approximately the same area as the original one, because the adiabatically-conserved action is *area*$/2\pi$.

As $t \to +\infty$, then, the phase point is close to its original ellipse and we can ask: where is it on the ellipse, *i.e.*, what is its phase?

This would be a question about Hannay's angle were it not for the fact that this is a classical problem without anholonomy, so that the angle we seek consists entirely of nonadiabatic corrections. By identifying the solution of (53) with quantum transmitted and reflected waves, it can be shown that the oscillation which begins as

$$x + ip/\omega_\infty = A \exp\left\{-i\Big(\int_0^t dt'\omega(t') + \sigma\Big)\right\} \qquad (t \to -\infty) \qquad (55)$$

ends as

$$x + ip/\omega_\infty = A \exp\left\{-i\Big(\int_0^t dt'\omega(t') + \sigma\Big)\right\} [T^{-1} + RT^{-1}\exp(2i\sigma)] \qquad (t \to +\infty) \qquad (56)$$

where A is a real constant and R and T are the complex quantal reflection and transmission coefficients.

Therefore the phase shift depends on the initial phase σ, but this dependence is slight because R is exponentially small in the slowness parameter ϵ (if ω depends on ϵt). In any case, we can define a phase by averaging over σ, with the exact result

$$\gamma \equiv -\frac{1}{2\pi}\int_{-\pi}^{\pi} d\sigma \lim_{t\to\infty}\left[\mathrm{Im}\log\left(x + ip/\omega_\infty\right) + \int_0^t dt'\,\omega(t')\right] = \mathrm{Im}\log T. \quad (57)$$

Thus 'Hannay's angle' is here the phase of the transmission coefficient. The final action I also depends on σ, but the range is $(I_{\mathrm{max}} - I_{\mathrm{min}})/I_{\mathrm{initial}} = 4|R|/|T|^2$ which again is of order $\exp(-1/\epsilon)$; the whole initial ellipse of phase points evolves ultimately into one exponentially close to it and deforming periodically with frequency ω_∞ (Fig. 3b). Newtonian renormalization of (53) is based on the transformation

$$x(t) \equiv \frac{x_1(t_1)}{\omega^{1/2}(t)}; \qquad t_1 \equiv \int_0^t dt'\,\omega(t') \qquad (58)$$

whose new coordinate satisfies

$$\partial_{t_1}^2 x_1(t_1) + \omega_1^2(t_1)x_1(t_1) = 0 \qquad (59)$$

where

$$\omega_1^2(t_1(t)) = 1 + \omega^{-3/2}\partial_t^2\omega^{-1/2}. \qquad (60)$$

Clearly $\omega_1 \approx 1$ if ω varies slowly.

Renormalization consists of iterating this transformation, the aim being to freeze the frequency. The kth approximant for γ is obtained by approximating $\omega_{k+1} \approx 1$, so

$$\gamma^{(k)} = \int_{-\infty}^{\infty} (dt_{k+1} - \omega(t)\, dt) = \int_{-\infty}^{\infty} dt\, \omega \left(\prod_{j=1}^{k} \omega_j - 1 \right) \tag{61}$$

Thus

$$\begin{aligned} \gamma^{(0)} &= 0 \qquad\qquad \text{(no anholonomy in this problem)} \\ \gamma^{(1)} &= \int_{-\infty}^{\infty} dt\, \omega(t) \Big(\omega_1(t_1(t)) - 1 \Big) \end{aligned} \tag{62}$$

etc. $\gamma^{(1)}$ is of order ϵ.

An equivalent *Hamiltonian* renormalization is produced by iteration of the canonical transformation generated by

$$S(x, p_1, t) = x\, p_1 \omega^{1/2} - x^2 \partial_t \omega / 4\omega. \tag{63}$$

This gives

$$x_1 = x\, \omega^{1/2}\ ; \qquad p_1 = p\, \omega^{-1/2} + x\, \partial_t \omega / 2\omega^{3/2} \tag{64}$$

and hence the transformed Hamiltonian

$$\overline{H}(x_1, p_1, t) = H + \partial_t S = \tfrac{1}{2}\omega(t)\, (p_1^2 + \omega_1^2 x_1^2) \tag{65}$$

where ω_1 is given by (60). Rescaling time to t_1 as defined in (58) now gives

$$H_1(x_1, p_1, t_1) = \tfrac{1}{2}\, (p_1^2 + \omega_1^2(t_1)\, x_1^2(t_1)) \tag{66}$$

which is the first renormalization of the original Hamiltonian (54). The aim of subsequent renormalizations is to freeze the Hamiltonian into one whose contours are circles.

I have expressed these classical iteration schemes in terms of the renormalization of Newton's or Hamilton's equations in order to illustrate the idea behind the quantum renormalization described earlier. But they can be shown to be equivalent to the following fairly conventional WKB-like [32] procedure (to be contrasted with an unconventional WKB analysis by Wilkinson [45] which, unlike this one, does involve anholonomy). Write the exact solution of Eq. (53) as

$$x(t) = \Omega^{-1/2}(t) \cdot \exp\left\{ i \int_0^t dt'\, \Omega(t') \right\}. \tag{67}$$

Then the 'frequency' $\Omega(t)$ satisfies

$$\Omega^2(t) = \omega^2(t) + \Omega^{1/2}(t)\, \partial_t^2 \Omega^{-1/2}(t). \tag{68}$$

In terms of Ω, the phase shift is, exactly,

$$\gamma = \int_{-\infty}^{\infty} dt\,[\Omega(t) - \omega(t)]. \tag{69}$$

Successive approximants are obtained by the iteration

$$\Omega^{(0)} = \omega\,; \qquad \Omega^{(k+1)} = \left[\omega^2 + (\Omega^{(k)})^{1/2}\,\partial_t^2 (\Omega^{(k)})^{-1/2}\right]^{1/2}. \tag{70}$$

The inevitability and universality of the divergence of these schemes can be demonstrated by considering high-order iterations of Eq. (60), for which

$$\omega_k(t) \equiv 1 + \delta_k(t) \tag{71}$$

and $\delta_k \ll 1$. Then $t_{k+1} \approx t_k$ (cf.Eq. (58)), and Eq. (60) can be written approximately as

$$\delta_{k+1}(t) \approx -\tfrac{1}{4}\partial_t^2 \delta_k(t). \tag{72}$$

The asymptotics of this recursion as $k \to \infty$ can be estimated by Fourier analysis, on the assumption that $\delta_0(t)$ is a real function of $\tau \equiv \epsilon t$, analytic in a strip about the real τ axis with its nearest singularities at $\tau_1 \pm i\tau_2$. Then with $\xi \equiv (\epsilon t - \tau_1)/\tau_2$ it is possible to show that

$$\delta_k(t) \underset{k\to\infty}{\longrightarrow} \left[\frac{A\epsilon^{2k}(2k)!}{4^k \tau_2^{2k+1}}\right]\left[\frac{\cos\left\{(2k+1)\cos^{-1}(1+\xi^2)^{-1/2}\right\}}{(1+\xi^2)^{k+1/2}}\right] \tag{73}$$

where A is a constant.

The first factor in (73) shows the divergence: $\epsilon^{2k}(2k)!$ decreases until $k \sim \tau_2/\epsilon$, when $\delta_k \sim \exp(-2\tau_2/\epsilon)$, and then increases until $\delta_k \sim 1$, when the scheme breaks down. The second factor is the universal function describing the asymptotic 'frequency.'

7. Historical Remarks

First I consider the important special case where the transported states $|\psi\rangle$ can be represented by wavefunctions that are *real*. Then the only possible phase factors associated with a circuit C are ± 1. It follows [7] from the result (22) for spins that the factor is -1 when C encloses a degeneracy X^* of the spectrum to which $|\psi\rangle$ belongs; otherwise, it is $+1$. The peculiarity of this case is that parallel transport (10) is the only possible smooth continuation law, rather than a mathematically natural choice, concordant with quantum dynamics, from a infinity of possibilities.

Eigenfunctions can always be made real if their Hamiltonian matrix is real symmetric rather than complex Hermitian (this is the case when there is (bosonic) time-reversal symmetry [34]). Thus the phase law states that an eigenfunction of a real symmetric matrix depending on parameters *changes sign* under smooth continuation round a degeneracy. This result is so simple – it holds even for 2×2 matrices – as to deserve mention in elementary expositions of matrix theory, but I have not found it in any such text. Arnold [14] is aware of the sign change, and attributes it to Uhlenbeck [35] in 1976. It was already known to theoretical chemists: Herzberg and Longuet-Higgins [36] gave an explicit statement in 1963. But the sign change (for 2×2 matrices) was implicit in work of Darboux [37] as long ago as 1896. This concerns the differential geometry of surfaces, and is worth describing.

Darboux considered a curved surface described locally by its deviation $z(X_1, X_2)$ from the plane $X = (X_1, X_2)$. Then the 2×2 real symmetric curvature matrix at X is

$$H_{ij}(X) = \partial_i \partial_j z(X). \tag{74}$$

The two eigenvalues are the principal curvatures at X, and the corresponding eigenvectors give the (orthogonal) directions of the lines of curvature at X. Degeneracies are *umbilic points*, where the surface is locally spherical (two curvatures equal). Umbilics are singularities of the net of curvature lines. The sign-change rule states that a line of curvature turns by π in a circuit of an umbilic: the Poincaré index of the tensor field (74) is $\pm\frac{1}{2}$. Fig. 4 shows how this happens for the three generic patterns [38][39] of curvature lines near an umbilic; the star has index $-\frac{1}{2}$, and the lemon and monstar have index $+\frac{1}{2}$. Star and lemon singularities occur as disclinations in liquid crystals [48].

The full phase — rather than the impoverished special case of the sign change for real matrices — was anticipated at least twice. First, in the mid-1950's, Pancharatnam [20][21][40] studied the 2-state Hermitian case in the context of the polarization states of light travelling in a fixed direction. The parameter space is the surface of the Poincaré sphere. Pancharatnam introduced the useful idea of defining two different states $|1\rangle$ and $|2\rangle$ as 'in phase' if the intensity of their superposition is maximal, a condition equivalent to their overlap $\langle 1|2\rangle$ being real and positive. This defines a connection between the corresponding parameters X_1 and X_2 as the state $|2\rangle$ obtained from $|1\rangle$ by phase-preserving transport along the shorter geodesic arc between X_1 and X_2. He discovered that the connection is nontransitive: a circuit $X_1X_2X_3X_1$ produces a state differing from $|1\rangle$ by precisely the same phase anholonomy [21] (minus half the solid angle of the circuit) as that given by parallel transport.

Second Mead [24] and Mead and Truhlar [42], studying adiabatic theory for molecules, made two important advances. They showed how the sign-change rule for degeneracies would induce modifications in the nuclear dynamics and hence change the vibration-rotation spectrum. And they realized that in the absence of time-reversal symmetry the nuclear dynamics would be influenced by the vector potential (39) and the corresponding 'magnetic' field (41), for which they gave a general formula.

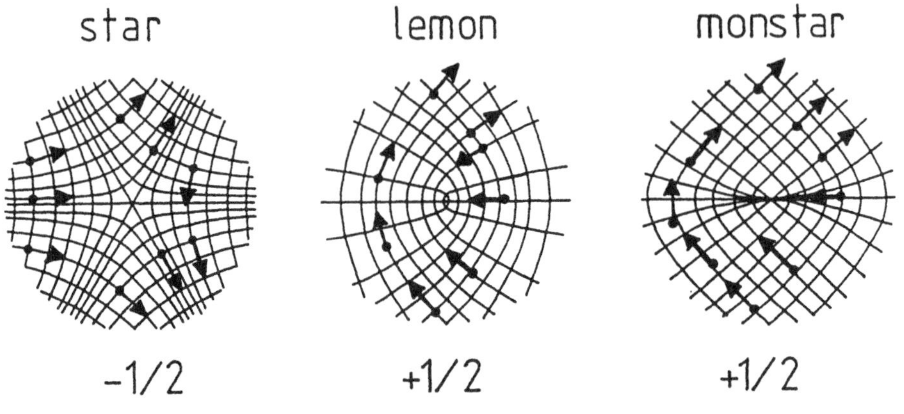

Figure 4. The generic patterns of lines of curvature near an umbilic point on a surface, illustrating the reversal ($\pm 1/2$ index) round the singularity.

My involvement with this subject began in 1979 with the appreciation [43] that degeneracies play a part in determining the fine-scale statistics of energy levels of quantum systems whose classical counterparts are nonintegrable. The systems under study possessed time-reversal symmetry and so their states should change sign round degeneracies. Seeking to display some degeneracies and their sign changes, M. Wilkinson and I [44] made a detailed investigation of the spectra of vibrating triangles as a function of angles (two parameters).

After a seminar reporting this work in the spring of 1983 at the Georgia Institute of Technology, R. Fox asked me, "what happens to the sign change if a magnetic field is switched on?", and this question led directly to the discovery of the phase and its 2-form several weeks later. Only when the work was written in first draft was I made aware (by E.Heller) of the papers by Mead and Truhlar. In August 1983, after my paper [7] had been submitted for publication, I described the phase to B.Simon, who instantly saw its relationship to Hermitian line bundles and Chern classes. His paper [10] directed many people towards this subject, thereby provoking the considerable activity of which this book is a partial record. But thanks to a referee's delay and an accident of astronomy, his paper appeared in 1983, mine in 1984.

Acknowledgments

I thank John Hannay for many inspirations in discussions of phase matters over several years. No military agency supported this research.

References

[1] Landau, L.D., Lifshitz, E.M., and Pitaevskii, L.P. 1984 *Electrodynamics of Continuous Media, 2nd ed.* (vol. 8 of Course of Theoretical Physics). Oxford: Pergamon Press.

[2] Born, M., and Wolf, E. 1956 *Principles of Optics.* London: Pergamon Press.

[3] Chiao, R.Y. and Wu, Y.S. 1986 *Phys. Rev. Lett.* **57**, 933–936. Tomita, A. and Chiao, R.Y. 1986 *Phys. Rev. Lett.* **57**, 937–940.

[4] Segert, J. 1987 *Phys.Rev.* **A36**, 10–16.

[5] Haldane, F.D.M. 1896 *Optics Letters* **11**, 730-732.

[6] Berry, M.V. 1987 *Nature* **326**, 277-278.

[7] Berry, M.V. 1984 *Proc.Roy.Soc.London* **A392**, 45-57.

[8] Aharonov, Y. and Anandan, J. 1987 *Phys. Rev. Lett.* **58**, 1593–1596.

[9] Born, M. and Fock, V. 1928 *Z.Phys.* **51**, 165–169.

[10] Simon, B. 1983 *Phys. Rev. Lett.* **51**, 2167–2170.

[11] Bitter, T. and Dubbers, D. 1987 *Phys. Rev. Lett.* **59**, 251-254.

[12] Tycko, R. 1987 *Phys. Rev. Lett.* **58**, 2281–2284.

[13] Hannay, J.H. 1985 *J.Phys.* **A18**, 221-230.

[14] Arnold, V.I. 1978 *Mathematical Methods of Classical Mechanics.* New York: Springer.

[15] Berry, M.V. 1986 "Adiabatic Phase Shifts for Neutrons and Photons," in *Fundamental Aspects of Quantum Theory,* eds. V.Gorini and A.Frigerio, NATO ASI series vol.144, 267-278. New York: Plenum.

[16] Cina, J. 1986 *Chem.Phys.Lett.* **132**, 393-95.

[17] Littlejohn, R.G. 1984 *Contemp.Math.* **28**, 151.

[18] Berry, M.V. 1985 *J.Phys.* **A18**, 15-27.

[19] Berry, M.V. 1983 *Semiclassical Mechanics of Regular and Irregular Motion*, in *Chaotic Behavior of Deterministic Systems.* Les Houches Lecture Series *XXXVI*, eds G.Iooss, R.H.G.Helleman and R.Stora. Amsterdam: North-Holland. pp.171-271.

[20] Pancharatnam, S. 1956 *Proc.Ind.Acad.Sci.* **A44**, 247-262. Pancharatnam, S. 1975 *Collected Works of S Pancharatnam*, Oxford: University Press.

[21] Berry, M.V. 1987 *J.Mod.Optics* **34**, 1401-1407.

[22] Poston, T. and Stewart, I.N. 1978 *Catastrophe Theory and its Applications*. London: Pitman.

[23] Berry, M.V. and Upstill, C. 1980 *Prog.Optics* **18**, 257–346.

[24] Mead, C.A. 1980 *Chem.Phys.(Netherlands)* **49**, 23-32, 33-38.

[25] Moody, J., Shapere, A., and Wilczek, F. 1986 *Phys.Rev.Lett.* **56**, 893-896.

[26] Jackiw, R. 1988 *Comm.At.Mol.Phys.* **20**, 71.

[27] Zygelman, B. 1987 *Phys.Lett.A.* **125**, 476–481.

[28] Anandan, J. and Aharonov, Y. 1988 *Phys. Rev. Lett.*, in press.

[29] Berry, M.V. 1987 *Proc.Roy.Soc.* **A414**, 31-46.

[30] Garrison, J.C. 1986 Preprint UCRL–94267 from Lawrence Livermor e Laboratory.

[31] Ehrenfest, P. 1916 *Ann.d.Physik* **51**, 327-352.

[32] Berry, M.V. and Mount, K.E. 1972 *Reps.Prog.Phys.* **35**, 315–397.

[33] Page, D.H. 1987 *Phys.Rev.* **A36**, 3479–3481.

[34] Porter, C.E. 1965 *Statistical Theories of Spectra: Fluctuations*. New York: Academic Press.

[35] Uhlenbeck, K. 1976 *Am.J.Math.* **98**, 1059–1078.

[36] Herzberg, G. and Longuet-Higgins, H.C. 1963 *Disc.Far.Soc.* **35**, 77-82.

[37] Darboux, G. 1986 *Leçons sur la Théorie Générale des Surfaces*, vol.4, Paris: Gauthier-Villars, note VII.

[38] Porteous, I.R. 1971 *J.Diff.Geom.* **5**, 543–564.

[39] Berry, M.V. and Hannay, J.H. 1977 *J.Phys.* **A10**, 1809–1821.

[40] Ramaseshan, S. and Nityananda, R. 1986 *Current Science (India)* **55**, 1225-26.

[41] Bouchiat, C. and Gibbons, G.W. 1988 *J.Phys. France* **49**, 187-199.

[42] Mead, C.A. and Truhlar, D.G. 1979 *J.Chem.Phys.* **70**, 2284-2296.

[43] Berry, M. 1981 *Ann.Phys.(N.Y.)* **131**, 163-216.

[44] Berry, M.V. and Wilkinson, M. 1984 *Proc.Roy.Soc.Lond.* **A392**, 15–43.

[45] Wilkinson, M. 1984 *J.Phys.* **A17**, 3459-3476.

[46] Wilczek, F. and Zee, A. 1984 *Phys. Rev. Lett.* **52**, 2111-2114.

[47] Dingle, R.B. 1973 *Asymptotic Expansions: Their Derivation and Interpretation*, New York: Academic Press.

[48] Frank, F.C. 1958 *Faraday Soc.Disc.* **25**, 19–28.

[49] Bender, C.M. and Papanicolaou, N. 1988 *J.Phys.France* **49**, 561-566.

[50] Provost, J.P. and Vallee, G. 1980 *Comm.Math.Phys.* **76**, 289–301.

International Journal of Modern Physics A, Vol. 3, No. 2 (1988) 285–297

THREE ELABORATIONS ON BERRY'S CONNECTION, CURVATURE AND PHASE*†

R. JACKIW

Center for Theoretical Physics, Laboratory for Nuclear Science and Department of Physics, Massachusetts Institute of Technology, Cambridge, Massachusetts 02139 U.S.A.

Received 23 October 1987

We discuss how symmetries and conservation laws are affected when Berry's phase occurs in a quantum system: symmetry transformations of coordinates have to be supplemented by gauge transformations of Berry's connection, and consequently constants of motion acquire terms beyond the familiar kinematical ones. We show how symmetries of a problem determine Berry's connection, curvature and, once a specific path is chosen, the phase as well. Moreover, higher order corrections are also fixed. We demonstrate that in some instances Berry's curvature and phase can be removed by a globally well-defined, time-dependent canonical transformation. Finally, we describe how field theoretic anomalies may be viewed as manifestations of Berry's phase.

We frequently analyze a physical system that is naturally and conveniently divided into two parts. We deal first with the motion of one set of variables, keeping the others fixed but arbitrary, and then complete the analysis of the whole by allowing variation of the previously fixed coordinates. The initially fixed variables we shall call *slow*, those whose motion is analyzed first are the *fast* variables—the terminology reflects the fact that molecular physicists and quantum chemists have used this decomposition for a long time in their Born–Oppenheimer studies of molecules. After resolving the dynamics of the fast variables, one is left with an effective action governing the slow variables. It has been established that this effective dynamics frequently involves an external vector potential **A** which is induced by the fast variables. Moreover, the effective equation of motion satisfied by the slow variables involves only the curl of the vector potential—a magnetic-like field **B**. Consequently, there is a gauge invariance in the description and only the gauge invariant portion of **A** leads to physical effects. This was first seen in Born–Oppenheimer studies of molecules,[1] and analogous effects were also found in quantum field theory.[2] With Berry's beautiful analysis,[3] we appreciate the full quantum mechanical generality of the phenomenon. The induced vector potential **A** is now called *Berry's connection*, the induced magnetic-like field **B** is *Berry's curvature*, while a line integral of the connection is *Berry's phase*.

* This work is supported in part by funds provided by the U.S. Department of Energy (D.O.E.) under contract number DE-AC02-76ER03069.

† Lecture delivered at "Non-Integrable Phase in Dynamical Systems," Theoretical Physics Institute, University of Minnesota, Minneapolis, Minnesota, October 1987.

Here, I shall speak on three topics: how symmetries and conservation laws are affected by Berry's connection; how higher order effects, beyond Berry's, may be determined with the help of symmetries; and finally, how Berry's connection manifests itself in quantum field theory through the anomaly phenomenon.

Let me begin by setting some notation. The complete Hamiltonian for the fast $(\mathbf{p}, \mathbf{r})$ and slow $(\mathbf{P}, \mathbf{R})$ variables is

$$H = \frac{\mathbf{P}^2}{2M} + \frac{\mathbf{p}^2}{2\mu} + V(\mathbf{R}, \mathbf{r}). \tag{1}$$

The sub-Hamiltonian governing the fast variables depends parametrically on the slow coordinates

$$h(\mathbf{R}) = \frac{\mathbf{p}^2}{2\mu} + V(\mathbf{R}, \mathbf{r}), \tag{2}$$

and possesses "instantaneous" eigenfunctions $|n; \mathbf{R}\rangle$ and eigenvalues $\varepsilon_n(\mathbf{R})$, which also depend on $\mathbf{R}$

$$h(\mathbf{R})|n; \mathbf{R}\rangle = \varepsilon_n(\mathbf{R})|n; \mathbf{R}\rangle. \tag{3}$$

These states can give rise to Berry's connection,[3] defined by

$$\mathbf{A}_n(\mathbf{R}) = \langle n; \mathbf{R}| i\nabla_{\mathbf{R}} |n; \mathbf{R}\rangle. \tag{4}$$

In the above, I have taken the nth eigenstate to be non-degenerate. Otherwise, one works within the degenerate subspace, and Berry's connection becomes a matrix in this space—an induced non-Abelian Yang-Mills-like potential.[4]

The effective Hamiltonian for slow motion, in the Born–Oppenheimer approximation, which results when off-diagonal matrix elements $\langle n; \mathbf{R}| i\nabla_{\mathbf{R}} |n'; \mathbf{R}\rangle$ are ignored, reads

$$H_{\text{eff}} = \frac{1}{2M}(\mathbf{P} - \mathbf{A}_n(\mathbf{R}))^2 + \varepsilon_n(\mathbf{R}). \tag{5}$$

Correspondingly, the effective Lagrangian is

$$L_{\text{eff}} = \tfrac{1}{2}M\dot{\mathbf{R}}^2 + \dot{\mathbf{R}} \cdot \mathbf{A}_n(\mathbf{R}) - \varepsilon_n(\mathbf{R}). \tag{6}$$

Equations (5) and (6) show that the fast system induces into the slow dynamics a potential energy $\varepsilon_n(\mathbf{R})$, and a velocity dependent interaction arising from Berry's connection $\mathbf{A}_n$, which enters dynamics only through the curvature $\mathbf{B}_n$ associated with $\mathbf{A}_n$.

$$\mathbf{B}_n = \nabla \times \mathbf{A}_n - i\mathbf{A}_n \times \mathbf{A}_n . \tag{7}$$

The last term is present in the non-Abelian matrix case.

Gauge invariance of physical processes means that physical content is not affected by gauge transformations that change **A** according to

$$\mathbf{A}(\mathbf{R}) \to g^{-1}(\mathbf{R})\mathbf{A}(\mathbf{R})g(\mathbf{R}) - ig^{-1}(\mathbf{R})\nabla g(\mathbf{R}). \tag{8}$$

[Henceforth we suppress the label "n" on the induced quantities.] Transformation (8) arises when the instantaneous eigenfunctions are redefined by R-dependent phase factors, which are not fixed by the instantaneous eigenvalue equation (3). Formula (8) is presented for the general case, where the connection **A** is matrix valued, belonging to the Lie algebra of some group, and g is the matrix representation for a group element. For example, **A** could be a Hermitian 2×2 matrix, and then g is a unitary 2×2 matrix belonging to U(2). In the simplest, Abelian case, **A** is a 1×1 matrix—a function—everything commutes and g belonging to U(1) is given by $e^{i\theta}$, so that (8) reduces to the electromagnetic gauge transformation

$$\mathbf{A} \to \mathbf{A} + \nabla\theta . \tag{9}$$

The gauge transformation (8) may also be presented in infinitesimal form. Writing

$$g = I + i\Theta + \cdots \tag{10}$$

where Θ is a Hermitian matrix in the Lie algebra, we deduce from (8) that the infinitesimal transformation is

$$\delta\mathbf{A} = \nabla\Theta - i[\mathbf{A}, \Theta] . \tag{11}$$

In the Abelian case, when **A** and Θ are functions, the commutator vanishes, and (11) reduces to (9).

We are now at our first topic: Let us suppose that the full Hamiltonian (1) possesses a symmetry and consequently there exist constants of motion commuting with H. We expect that in the effective description, the symmetry is not lost and there are effective constants of motion that commute with H_{eff}. The question is how do the symmetries manifest themselves and how are the constants of motion modified in the presence of Berry's connection.

Let us for definiteness concentrate on rotational symmetry and on the associated angular momentum constant of motion **J**. Under rotations the coordinate **R** transforms according to $R^i \to \Lambda^{ij} R^j$ with Λ^{ij} a special orthogonal matrix. We certainly know what it means for the effective potential energy to be invariant against rotations: $\varepsilon(\mathbf{R})$ should depend only on the magnitude $R = |\mathbf{R}|$. Analytically, this requirement of rotational invariance is presented as

$$\varepsilon(\Lambda\mathbf{R}) = \varepsilon(\mathbf{R}). \tag{12}$$

For a vector quantity, like Berry's connection $\mathbf{A}$, the analogous requirement recognizes the vectorial nature of $\mathbf{A}$

$$\mathbf{A}(\Lambda\mathbf{R}) = \Lambda\mathbf{A}(\mathbf{R}). \tag{13}$$

Equivalently,

$$\Lambda\mathbf{A}(\Lambda^{-1}\mathbf{R}) = \mathbf{A}(\mathbf{R}). \tag{14}$$

Vector functions satisfying (14) have the form $\mathbf{A}(\mathbf{R}) = \mathbf{R}A(R)$, where A is a scalar function of $R = |\mathbf{R}|$. However, such connections are not physically interesting because the associated curvature (7)—the magnetic-like field—vanishes.

For a nontrivial realization of rotational invariance, we need to recognize that symmetries should be looked for in the physical content of the formalism, and we must remember that the connection has unphysical components, because a gauge transformation can always be performed without affecting physical content.

The rotational invariance requirement for a vector connection can be weakened from (14), while still retaining the full force of rotational symmetry for physical processes. Rather than demanding that $\Lambda\mathbf{A}(\Lambda^{-1}\mathbf{R})$ equal $\mathbf{A}(\mathbf{R})$, as in (14), we allow a gauge transformation to intervene.

$$\Lambda\mathbf{A}(\Lambda^{-1}\mathbf{R}) = g^{-1}(\mathbf{R})\mathbf{A}(\mathbf{R})g(\mathbf{R}) - ig^{-1}(\mathbf{R})\nabla g(\mathbf{R}). \tag{15}$$

For most purposes, it is sufficient to consider the infinitesimal version of (15). When we define

$$\Lambda^{ij} = \delta^{ij} - \varepsilon^{ijk}n^k + \cdots \tag{16a}$$

so that an infinitesimal rotation of $\mathbf{R}$ is

$$\delta\mathbf{R} = \mathbf{n} \times \mathbf{R}, \tag{16b}$$

then (11) and (15) combine to

$$\mathbf{n} \times \mathbf{A} - (\mathbf{n} \times \mathbf{R}\cdot\nabla)\mathbf{A} = \nabla\Theta - i[\mathbf{A}, \Theta]. \tag{17}$$

Equation (17) may also be presented as a condition on the curvature $\mathbf{B}$, (7). An algebraic rearrangement of (17), which uses various vector identities, casts (17) in the form

$$(\mathbf{n} \times \mathbf{R}) \times \mathbf{B} = \nabla W - i[\mathbf{A}, W], \tag{18a}$$

$$W = \Theta + \mathbf{n} \times \mathbf{R}\cdot\mathbf{A}. \tag{18b}$$

To summarize: the effective Born–Oppenheimer Hamiltonian (5) and the effective Born–Oppenheimer Lagrangian (6) are rotationally invariant, provided, in addition to (12)—which expresses rotational symmetry of the instantaneous energy—the Berry connection **A** satisfies (15) or (17), equivalently the Berry curvature satisfies (18)—this assures rotational symmetry up to a gauge transformation of the connection. Moreover, one can prove with general quantum mechanical reasoning that when the fast Hamiltonian $h(\mathbf{R})$ is invariant against rotations of the fast variables, supplemented by a rotation of the slow background variable **R**, then the instantaneous eigenstates $|n; \mathbf{R}\rangle$ transform under rotations precisely in such a way that the connection defined in (4) satisfies (15).[5] Therefore, as expected, rotational symmetry is not lost in the effective Hamiltonian, but is realized in a nontrivial fashion, when a gauge connection is present.

In the known models with nonvanishing Berry's connection, one can verify (17) or (18). Consider the example,[3]

$$h(\mathbf{R}) = \mathbf{S} \cdot \mathbf{R}(t) \tag{19}$$

where **S**, playing the role of the fast variables $(\mathbf{p}, \mathbf{r})$, is an angular momentum operator that obeys the SO(3) Lie algebra

$$[S_i, S_j] = i\varepsilon_{ijk} S^k . \tag{20}$$

h is invariant against simultaneous rotations of **S** and of **R**, the slow background. The instantaneous eigenvalues are mR, with m ranging in unit steps from $-S$ to S. The model gives rise to Berry's curvature, which is proportional to a magnetic monopole field[3]

$$\mathbf{B} = -m \frac{\hat{\mathbf{R}}}{R^2} . \tag{21}$$

It is straightforward to verify that (18) is satisfied with

$$W = m \hat{\mathbf{R}} \cdot \mathbf{n} . \tag{22}$$

B in Eq. (21) is an Abelian curvature. In a non-Abelian U(2) example, relevant to diatoms with λ degeneracy, Berry's connection is a 2×2 matrix, which depends on a parameter κ[6]

$$\mathbf{A}(\mathbf{R}) = \frac{1 + \kappa}{2} \frac{\hat{\mathbf{R}} \times \sigma}{R} . \tag{23}$$

The σ matrices are Pauli matrices. Rotational invariance holds: **A** satisfies (17) and Θ is given by[7]

$$\Theta = \tfrac{1}{2} \sigma \cdot \mathbf{n} . \tag{24}$$

In fact one may reverse the reasoning and determine the connection or curvature from the requirements of rotational symmetry. One views (18) as an equation for **B** and finds the most general solution in terms of W. The point is that integrability requirements on (18) sharply limit the allowed solutions. Thus the magnetic monopole in (21) is the *unique* Abelian rotationally symmetric curvature, while the configuration (23) is one of the few non-Abelian 2×2 matrix solutions. In this way knowledge of the symmetry limits the curvature, apart from the gauge freedom.

A further remark: because invariance of H_{eff} is achieved when a rotation is supplemented by a gauge transformation, constants of motion commuting with H_{eff} are modified.[8] Recall that a constant of motion is also the generator of the infinitesimal transformation. In our case the infinitesimal rotation of coordinates must be supplemented by an infinitesimal gauge transformation of the connection. Consequently, the conserved angular momentum **J** arising from rotational symmetry in the $\hat{\mathbf{n}}$ direction is the conventional $\mathbf{n} \cdot (\mathbf{R} \times \mathbf{P})$ supplemented by Θ, the gauge transformation generator. Equivalently,

$$\mathbf{J} = \mathbf{R} \times M\dot{\mathbf{R}} + \frac{\partial W}{\partial \mathbf{n}} \tag{25}$$

where we have used (18b) and the fact that $\mathbf{P} = M\dot{\mathbf{R}} + \mathbf{A}$. [Of course, Θ and W are linear in $\mathbf{n}$.] The extra term in (25) puts into evidence the physical significance of the gauge transformation that accompanies the rotational coordinate change: $\frac{\partial W}{\partial \mathbf{n}}$ is the angular momentum stored in the gauge field, that summarizes the effect of the fast variables on the slow ones. This is the reason why in the presence of a magnetic monopole [or of the Berry curvature (21)] the angular momentum contains in addition to the kinematical term the further $m\hat{\mathbf{R}}$ [see (22)]. Also this is why the U(2) non-Abelian connection (23) requires that the kinematical angular momentum be supplemented by $\sigma/2$ [see (24)].

We have discussed in detail how rotational invariance and its associated conserved quantity/generator are affected by Berry's connection. However the same ideas can be applied to invariance under arbitrary coordinate transformations. When the complete Hamiltonian is invariant against some coordinate transformation, the effective Hamiltonian for slow motion retains this property, but the background gauge field induced by the fast variables is invariant only up to a gauge transformation,[5] whose infinitesimal generator supplements the relevant constant of motion.[8] The problem of determining all connections/curvatures that are invariant, up to a gauge transformation, against an arbitrary coordinate transformation has been solved by mathematicians and physicists.[9]

There is one more model worth mentioning: the generalized one-dimensional harmonic oscillator, with time-varying parameters

$$\begin{aligned} h &= \tfrac{1}{2}(\alpha(t)p^2 + \beta(t)(pq + qp) + \gamma(t)q^2), \\ \alpha &> 0. \end{aligned} \tag{26}$$

Berry's connection is nonvanishing.[10] The three operators p^2, $pq + qp$ and q^2 close upon commutation on the Lie algebra of SO(2, 1), which is like the angular momentum SO(3) algebra, except some crucial signs are reversed. This may be presented by redefining the fast variables as

$$T_1 = \tfrac{1}{4}(q^2 - p^2), \qquad T_2 = \tfrac{1}{4}(pq + qp), \qquad T_3 = \tfrac{1}{4}(q^2 + p^2) \tag{27}$$

and the parameters

$$R^1 = \gamma - \alpha, \qquad R^2 = 2\beta, \qquad R^3 = \gamma + \alpha. \tag{28}$$

The Hamiltonian (26) takes the form

$$h = T_i R^i(t), \tag{29}$$

with

$$[T_i, T_j] = i\varepsilon_{ijk} T^k. \tag{30}$$

Formulas (29) and (30) are analogous to (19) and (20) of the SO(3) case, except that here a metric g_{ij} intervenes in the ijk group indices, and introduces the crucial sign differences from SO(3)

$$g_{ij} = \mathrm{diag}(-1, -1, 1). \tag{31}$$

The general discussion applies, and the curvature is immediately predicted to be the unique SO(2, 1) covariant,[5]

$$B^i \propto \frac{R^i}{(R^j R_j)^{3/2}} \tag{32}$$

where $R^j R_j = (R^3)^2 - (R^1)^2 - (R^2)^2 = 4(\alpha\gamma - \beta^2)$ is assumed to be positive. Equation (32) coincides with the result of the explicit calculation based on (25).[10]

In spite of the formal similarity to the SO(3) problem, there is a crucial difference which has not been remarked upon in the literature. The curvature may be removed, not of course, by a gauge transformation, but by a globally well-defined canonical transformation, or what is equivalent by dropping a total time derivative from the Lagrangian that corresponds to the Hamiltonian (26)

$$L \equiv p\dot{q} - h = \frac{1}{2\alpha}\dot{q}^2 - \frac{1}{2}\left(\gamma - \frac{\beta^2}{\alpha} - \frac{d}{dt}\frac{\beta}{\alpha}\right)q^2 - \frac{d}{dt}\frac{1}{2}\left(\frac{\beta}{\alpha}q^2\right). \tag{33}$$

When the last term is dropped, as it can be without affecting dynamics, one is left with an equivalent theory, described by the Hamiltonian

$$\bar{h} = \frac{\alpha}{2}p^2 + \frac{1}{2}\left(\gamma - \frac{\beta^2}{\alpha} - \frac{d}{dt}\frac{\beta}{\alpha}\right)q^2, \tag{34}$$

which does not lead to a Berry connection, but is canonically equivalent to h, with the time-dependent canonical transformation $p \to p + \frac{\beta}{\alpha}q$, $q \to q$. Notice $\frac{\beta}{\alpha}$ is non-singular, because α is assumed never to vanish. Hence this canonical transformation is globally defined on the parameter space. On the other hand, an analogous transformation for the SO(3) example,[3] which would rotate $\mathbf{S}\cdot\mathbf{R}(t)$ into a fixed direction, cannot be globally defined.

The reason for this difference between the SO(3) and SO(2, 1) examples derives from the fact that the parameter space [at fixed R] of the SO(3) model is the surface of a sphere, while that of the SO(2, 1) model is one sheet of the hyperboloid R^iR_i = constant. Unlike the former, the latter is topologically trivial, and homotopic to the Euclidean plane.[11]

Symmetry considerations may also be used to calculate quickly and efficiently higher order corrections to Berry's phase, and this brings us to the second topic in my lecture. First we need to define what we wish to calculate in higher order. I have already indicated in Eq. (6) that the instantaneous energy eigenvalue and Berry's connection are two contributions to the effective Lagrangian induced by the fast variables onto the slow ones. Let us therefore consider the complete effective action I_{eff} induced by the fast variables. This quantity may be defined as follows. Consider the time-dependent Schrödinger equation with the time-dependent Hamiltonian $h(\mathbf{R}(t))$

$$i\partial_t|\psi;t\rangle = h(\mathbf{R}(t))|\psi;t\rangle. \tag{35}$$

We take the two [in, out] solutions of (35) $|\psi^{(\pm)};t\rangle$ that satisfy the initial and final conditions, respectively

$$\lim_{t\to t_i}|\psi^{(+)};t\rangle = |n;\mathbf{R}(t_i)\rangle, \tag{36a}$$

$$\lim_{t\to t_f}|\psi^{(-)};t\rangle = |n;\mathbf{R}(t_f)\rangle. \tag{36b}$$

For simplicity we shall assume $\mathbf{R}(t_i) = \mathbf{R}(t_f)$, but this can be relaxed. The effective action is defined from the in-out matrix element,

$$I_{\text{eff}} = -i\ln\langle\psi^{(-)};t|\psi^{(+)};t\rangle. \tag{37}$$

[Alternatively, I_{eff} may be given by a Feynman path integral over the fast variables.] I_{eff} is a functional of $\mathbf{R}(t)$, but it is independent of t because $|\psi^{\pm};t\rangle$ satisfy the Schrödinger equation. Hence the overlap in (37) may be evaluated variously at $t = t_i$, where $I_{\text{eff}} = -i\ln\langle\psi^{(-)};t_i|n;\mathbf{R}(t_i)\rangle$ or at $t = t_f$, where $I_{\text{eff}} = -i\ln\langle n;\mathbf{R}(t_f)|\psi^{(+)};t_f\rangle$.

One may expand I_{eff} in a series of terms with increasing time derivatives of $\mathbf{R}$. The zeroth order term has no time derivatives; it is what would survive of I_{eff} when $\mathbf{R}(t)$ is time-independent. The first order term has a single time derivative; the second order term involves two time derivatives, etc. We now see that the instantaneous energy eigenvalue is the (negative) zeroth order term; the Berry phase gives the first order term, linear in time derivatives; higher orders are to be determined—compare (6).

$$I_{eff} = -\int dt\, \varepsilon(\mathbf{R}(t)) + \int dt\, \dot{\mathbf{R}}(t)\cdot\mathbf{A}(\mathbf{R}(t))$$

$$+\frac{1}{2}\int dt\, \dot{R}^i(t) M^{ij}(\mathbf{R}(t))\dot{R}^j(t) + \cdots \tag{38}$$

The coefficient tensors A^i and M^{ij} must be transverse to R^i. This is so because a purely radial time dependence, $\mathbf{R}(t) = \hat{R}R(t)$ with time-independent $\hat{R}$, gives rise only to the first term in (38). Equation (38) is an asymptotic series, it produces a real I_{eff} but misses imaginary parts, which describe decay. [Note that a second order term of the form $\int dt\, \ddot{\mathbf{R}}(t)\cdot\mathbf{N}(\mathbf{R}(t))$ is equivalent, by an integration by parts, to the last term in (38); to justify dropping endpoint contributions, the motion must be periodic, alternatively take $t_{i,f} = \mp\infty$, where everything vanishes.]

Let me show how all higher order terms in (38) can be computed from the knowledge of the zeroth order, instantaneous eigenvalue. The discussion is confined to the SO(3) example. Observe that the generator of rotations on the fast variables S, is obtained from $h(\mathbf{R})$ by differentiating it with respect to $\mathbf{R}$,

$$\mathbf{S} = \frac{\partial h(\mathbf{R})}{\partial \mathbf{R}}. \tag{39}$$

This operator satisfies,

$$i[h(\mathbf{R}), \mathbf{S}] = \mathbf{R}\times\mathbf{S}, \tag{40a}$$

which is the Heisenberg picture equation for $\mathbf{S}(t)$,

$$\frac{d}{dt}\mathbf{S}(t) - \mathbf{R}(t)\times\mathbf{S}(t) = 0. \tag{40b}$$

The above non-conservation equation for $\mathbf{S}(t)$ reflects the fact that rotations on the fast variables S, are not symmetry operations, unless supplemented by a rotation of the external parameters.

Next I define the in-out matrix element of S,

$$\mathbf{J}(t) = \langle\psi^{(-)};t|\mathbf{S}|\psi^{(+)};t\rangle/\langle\psi^{(-)};t|\psi^{(+)};t\rangle \tag{41}$$

which by virtue of (40a) also satisfies (40b). Finally observe that $\mathbf{J}(t)$ is given by a functional derivative with respect to $\mathbf{R}(t)$ of I_{eff} [compare with (39)],

$$\mathbf{J}(t) = -\frac{\delta I_{\text{eff}}}{\delta \mathbf{R}(t)}. \tag{42}$$

Thus combining (40) with (42) we arrive at a condition on I_{eff}

$$\frac{d}{dt}\frac{\delta I_{\text{eff}}}{\delta \mathbf{R}(t)} - \mathbf{R} \times \frac{\delta I_{\text{eff}}}{\delta \mathbf{R}(t)} = 0. \tag{43}$$

Equation (43) may be applied iteratively to (38), and then equal orders of time derivatives of $\mathbf{R}(t)$ are equated. Varying I_{eff} gives

$$\mathbf{J}(t) = m\hat{\mathbf{R}}(t) - \dot{\mathbf{R}}(t) \times \mathbf{B}(\mathbf{R}(t)) - \frac{\delta}{\delta \mathbf{R}(t)}\frac{1}{2}\int d\tau\, \dot{R}^i(\tau) M^{ij}(\mathbf{R}(\tau))\dot{R}^j(\tau) + \cdots \tag{44}$$

and (43) requires

$$m\frac{d}{dt}\hat{\mathbf{R}} = -\mathbf{R} \times (\dot{\mathbf{R}} \times \mathbf{B}), \tag{45a}$$

$$\frac{d}{dt}(\dot{\mathbf{R}} \times \mathbf{B}(\mathbf{R})) = \mathbf{R} \times \frac{\delta}{\delta \mathbf{R}}\frac{1}{2}\int d\tau\, \dot{R}^i M^{ij} \dot{R}^j, \tag{45b}$$

etc. Equation (45a) determines $\mathbf{B}$: the known formula (21) is regained; Eq. (45b) gives a new result: the second order term in the derivative expansion, which is the first correction to Berry's phase. One finds

$$M^{ij} = \frac{m}{R^3}(\delta^{ij} - \hat{R}^i\hat{R}^j). \tag{46}$$

Clearly higher order terms may be similarly computed. The result (46) has been verified by a direct evaluation of the in, out matrix element.[12] Of course a specific value for I_{eff} is obtained only when a specific path $\mathbf{R}(t)$ is chosen and the time integral is performed.

My third and last topic concerns the role that Berry's phase plays in modern quantum field theory, where it gives another point of view on the anomaly phenomenon. This subject will be discussed in greater detail in another lecture.[13] Here I give an introductory description, because I believe that this peculiar feature of second quantized field theories is probably unfamiliar to many of you.

I begin by reminding that the quantum mechanical revolution has not erased our reliance on the earlier classical physics. Indeed when proposing a theory, we begin with classical concepts and construct models according to the rules of classical, pre-quantum physics. We know, however, such classical reasoning is not in complete accord with quantum reality. Therefore, the classical model is reanalyzed by the rules of quantum physics, which comprise the true laws of nature, i.e., the model is *quantized*. For a long time it was believed that symmetries of a theory are not affected by the transition from classical to quantum rules. However, more recently we have learned that this is not so. In a quantized theory, some symmetries of classical physics may disappear because symmetry violating processes, which are not seen classically, can occur when the analysis is conducted with quantum effects taken into consideration. Such tenuous symmetries are said to be *anomalously* broken. Although present classically, they are absent from the quantum version of the theory, unless the model is carefully arranged to avoid this effect.

Anomalously or quantum mechanically broken symmetries play a crucial role in our present-day theories of elementary particles. In some instances they save the models from possessing too much symmetry, which would not be in accord with experiment. In other instances, the desire to preserve a symmetry in the quantum theory places strong constraints on model building and gives experimentally verified predictions. For example, the equality in the number of quarks and leptons is understood in these terms. Also, the present-day excitement about strings derives from the fact that only very few string models can be adjusted to avoid quantum mechanical, anomalous breaking of those symmetries that make string theory free of the infinities plaguing conventional field theories. Thus the number of consistent string models appears very limited, and a limitation of theoretical possibilities is what every model builder looks for. Anomalous symmetries are also beginning to play a role in other branches of physics, like condensed matter.

For a specific example of this phenomenon, consider massless fermions moving in an electromagnetic field described by electromagnetic potentials. Since massless fermions possess a well-defined helicity, we shall consider fermions with only one helicity. Such systems are an ingredient in theories of quarks and leptons. Moreover, they also arise in condensed matter physics, not because one is dealing with massless, single-helicity particles, but because a well-formulated approximation to some many-body Hamiltonians can result in a first order matrix equation which is identical to the equation for single-helicity massless fermions, i.e. a massless Dirac equation for a spinor ψ.

As a first quantized theory, the system is gauge covariant, in that a gauge transformation on the electromagnetic potential can be compensated by a change of the wavefunction, ψ. Moreover, the norm of the wavefunction is time-independent: $N = \int \psi^\dagger \psi$, $\frac{d}{dt} N = 0$. So far there are no surprises.

To construct a quantum field theory from the above, the model is second quantized: the wavefunction ψ is promoted to an operator Ψ and the state space for the theory is a many-particle Fock space. Moreover, the ground state of the second quantized field theory has to be the filled *Dirac sea* so that the negative energy solutions of the first quantized Dirac equation are eliminated. Of course all states are functionals of the background electromagnetic potential in which the fermions move.

One expects that the second quantized theory also possesses gauge invariance, and that as a consequence of gauge invariance the total charge to which the first quantized norm is promoted, $Q = \int \Psi^\dagger \Psi$, is conserved. In fact, this is not true: the charge is not conserved; rather one finds

$$\frac{d}{dt} Q \propto \int \mathbf{E} \cdot \mathbf{H} \tag{47}$$

where **E** and **H** are the background electric and magnetic fields, and correspondingly gauge invariance is lost.[15]

There are many ways of arriving at the result where the gauge symmetry in this model is anomalously broken. In one physically transparent argument, it is established that the process of filling the negative energy sea to define the field theoretic ground state necessarily violates gauge invariance.[14]

Berry's ideas provide another viewpoint. The fermion field operators Ψ are thought of as fast variables, the analogs of **p** and **r**, or of **S** in the example (19). The background electromagnetic potential is viewed as an external parameter; it is the analog of **R**. The Fock-space state vectors are functionals of the background potential; they are analogs of $|n; \mathbf{R}\rangle$. An analysis shows that when the background electromagnetic potentials are gauge transformed—this can be an adiabatic change—the states acquire a phase variation and in this way lose electromagnetic gauge invariance. Thus the anomaly phenomenon is a manifestation in quantum field theory of Berry's phase, $\int d\mathbf{R} \cdot \mathbf{A}(\mathbf{R})$.[16]

Symmetry in quantum theory can also be seen through the realization of the symmetry algebra in the canonical commutation relations. Correspondingly, when the symmetry is anomalously or quantum mechanically broken, the algebra acquires a dynamical modification. As is seen from (6), the Berry connection induces velocity-dependent interactions, which modify the relation between canonical momentum and velocity. Consequently, the commutator of velocity components acquires a quantum mechanical correction.

$$[M\dot{R}^i, M\dot{R}^j] = i\varepsilon^{ijk} B^k \tag{48}$$

Anomalous commutators—an important chapter in the anomaly story[14]—are also connected to the Berry curvature.

References

1. H. Longuet-Higgins, U. Opik, M. Pryce and R. Sack, *Proc. Roy. Soc.* **A224** (1958) 1; H. Longuet-Higgins, *Adv. Spectrosc.* **2** (1961) 429, *Proc. R. Soc.* **A344** (1975) 147; M. Child and H. Longuet-Higgins, *Phil. Trans. R. Soc.* **254** (1961) 259; G. Herzberg and H. Longuet-Higgins, *Disc. Faraday Soc.* **35** (1963) 77; M. O'Brien, *Proc. R. Soc.* **A281** (1984) 323; A. Stone, *Proc. R. Soc.* **A351** (1976) 141; C. Mead. *J. Chem. Phys.* **70** (1979) 2276, **72** (1980)

3839, *Chem. Phys.* **49** (1980) 23, 33; C. Mead and D. Truhlar, *J. Chem. Phys.* **70** (1979) 2284, (E)**78** (1983) 6344.

2. R. Jackiw, in *E. Fradkin Festshrift*, I. Batalin, C. Isham and G. Vilkovisky, eds. (Adam Hilger, Bristol, 1987). See also M. Asorey and D. Mitter, *Phys. Lett.* **153B** (1985) 147; Y.-S. Wu and A. Zee, *Phys. Lett.* **B258** (1985) 157.
3. M. V. Berry, *Proc. R. Soc.* **A392** (1984) 45, and lecture delivered at this conference.
4. F. Wilczek and A. Zee, *Phys. Rev. Lett.* **52** (1984) 2111.
5. L. Vinet, University of Montreal preprint, August (1987).
6. J. Moody, A. Shapere and F. Wilczek, *Phys. Rev. Lett.* **56** (1986) 893. Formula (23) is presented in a gauge which differs from the one used by Moody et al.
7. R. Jackiw, *Phys. Rev. Lett.* **56** (1986) 2779.
8. R. Jackiw and N. Manton, *Ann. Phys.* (NY) **127** (1980) 257.
9. H. Wang, *Nagoya Math. J.* **13** (1958) 1; J. Harnad, S. Shnider and L. Vinet, *J. Math. Phys.* **21** (1980) 2719; R. Forgács and N. Manton, *Comm. Math. Phys.* **72** (1980) 15. For a review and more details see R. Jackiw, *Acta Phys. Austriaca*, Suppl. **XXII** (1980) 383.
10. M. V. Berry, *J. Phys. A* **18** (1985) 15; J. Hannay, *J. Phys. A* **18** (1985) 221.
11. P. Gerbert, MIT preprint CTP # 1537, October (1987). See also E. Gozzi and W. Thacker, *Phys. Rev.* **D35** (1987) 2398.
12. P. Gerbert, Ref. [11]; S. Iida, private communication.
13. G. Semenoff, lecture delivered at this conference.
14. For a review, see *Current Algebra and Anomalies*, S. Treiman, R. Jackiw, B. Zumino and E. Witten, eds. (Princeton University Press/World Scientific, Princeton NJ/Singapore, 1985).
15. D. Gross and R. Jackiw, *Phys. Rev.* **D6** (1972) 477. For details see Ref. [14].
16. P. Nelson and L. Alvarez-Gaumé, *Comm. Math. Phys.* **99** (1985) 103; H. Sonoda, *Phys. Lett.* **156B** (1985) 220, *Nucl. Phys.* **B266** (1986) 410; A. Niemi and G. Semenoff, *Phys. Rev. Lett.* **55** (1985) 927, **56** (1986) 1019, *Phys. Lett.* **B175** (1986) 439; A. Niemi, G. Semenoff and Y.-S. Wu, *Nucl. Phys.* **B276** (1986) 173. For reviews of this approach to anomalies, see G. Semenoff in *Super Field Theory*, H. Lee, V. Elias, G. Kunstatter, R. Mann and K. Viswanathan, eds. (Plenum, New York, 1987); A. Niemi in *Workshop on Skyrmions and Anomalies*, M. Jezabek and M. Praszalowicz, eds. (World Scientific, Singapore, 1987).

Chapter 2

ANTICIPATIONS

2

Anticipations

The discovery of geometric phases in physics was not a sudden event. As we have seen previously, isolated examples of holonomy have been known in a few areas of physics for many years. In the quantum mechanics of molecules, and in the analysis of polarized light, a number of examples were analyzed using concepts equivalent to those we would now recognize as geometric phases, well before the recent upsurge of interest. Clearly, Berry's phase could have been discovered long before it was; the relevant quantum mechanics had been in place for fifty years. But the impact of Berry's paper was prepared by a decade of increasing interest in geometric and topological ideas in physics. This interest, in turn, was stimulated by the rise of fundamentally geometric gauge theories of elementary particles, and by an increasing recognition of the role of broken symmetry states in physics. The latter often involve complicated order parameters capable of supporting topologically non-trivial excitations (solitons) such as magnetic monopoles.[1] Also setting the stage for Berry's work, and for its impact on the experimental communities, was the development of modern technologies of lasers and magnetic resonance, which have facilitated exquisitely controlled experiments that clearly observe the effects of geometric phases.

One great virtue of the more modern work has been its clear recognition of the universality of the ideas, and of the rich variety of contexts in which geometric phases manifest themselves. Nevertheless, it is of more than historical interest to read some of the papers that, while partially and from limited perspectives, anticipated some of the modern developments. We believe that the articles presented here remain fresh, provide interesting examples of the general ideas, and—as we shall see—contain ideas that may not have been, or are only now beginning to be, fully appreciated.

Particularly remarkable is the work of Pancharatnam on phase shifts in polarized light. This work preceded Berry's by nearly three decades. Yet already in 1956, and in relative isolation from the Western scientific community, Pancharatnam found and focused on the "right" question. He asked himself whether polarized light would, after a cyclic series of changes of polarization, have acquired a phase—over and above that associated with free propagation over the same path. Such a phase, he realized, would have a direct physical meaning. Indeed, it could be easily measured by an interference experiment.

To answer his question, Pancharatnam had to define how the phase of polarized light changes upon passage of the light through an analyzer. That is, he needed a way of defining the relative phase of two different polarization states—in modern language, he needed a connection. The choice of a relative phase between states of orthogonal polarization is arbitrary, in the very same way that the choice of phases for orthogonal quantum mechanical states is arbitrary. However, there is a natural notion of relative phase for nonorthogonal states. Precisely, if $|A\rangle$ and $|B\rangle$ are the states before and after passage through the analyzer, then they are defined to be in phase when the interference between them

$$(\langle A| + \langle B|)((|A\rangle + |B\rangle) = 2 + 2|\langle A|B\rangle| \,\mathrm{Re}\,\langle A|B\rangle \tag{2.1}$$

is of the largest possible intensity. Now, suppose that three sequential changes of polarization are made, from $|A\rangle$ to $|B\rangle$ to $|C\rangle$ and back to a state $|A'\rangle$ of the initial polarization, such that each successive state is in phase with the previous one. Pancharatnam showed that the phase difference between the initial and final states is then

$$\langle A|A'\rangle = \exp(-i\Omega_{\mathrm{ABC}}/2)\,, \tag{2.2}$$

where Ω_{ABC} is the solid angle subtended by the geodesic triangle ABC on the Poincaré sphere. Properly interpreted, this result is mathematically identical to Berry's result for the adiabatic phase of a spin-$\frac{1}{2}$ particle in a slowly changing magnetic field, which is also equal to one-half the solid angle subtended by the magnetic field in parameter space.

We have included Pancharatnam's original paper [2.1] for its historical value, although the reader will probably find Berry's account of it [2.2] far more readable. Further optical applications will be covered in papers [4.1] and [4.2].

In a quantum mechanical context, the 1963 work of Herzberg and Longuet-Higgins on polyatomic molecules [2.3] was probably the first to show that non-integrable phases can arise in the Born-Oppenheimer (B-O) approximation. Actually, the phase that these authors found is always equal to ± 1, because their wavefunctions are everywhere real, but it closely anticipates the general case. Let us describe it briefly, in context.

Much earlier, Wigner and von Neumann[2] had considered a two-level system with a real 2×2 Hamiltonian H:

$$H = \begin{pmatrix} H_{11} & H_{12} \\ H_{12} & H_{22} \end{pmatrix} \tag{2.3}$$

In order for H to have degenerate eigenvalues, it must satisfy two conditions

$$\begin{aligned} H_{11} &= H_{22} \\ H_{12} &= 0 \end{aligned} \tag{2.4}$$

So if H depends on d external parameters, such as nuclear coordinates in the Born-Oppenheimer approximation, then these conditions will be satisfied in a space of dimension $d-2$. The 2 coordinates which move us away from a degeneracy are $x = \frac{1}{2}(H_{11} - H_{22})$ and $y = H_{12}$. Teller[3] noted that near a degeneracy, the level surfaces for the eigenvalues of H, considered as a function of x and y, form a double cone with apex at the degeneracy point, namely

$$E_{\pm} = E_0 \pm \sqrt{x^2 + y^2} \tag{2.5}$$

where $E_0 \equiv \frac{1}{2}(H_{11} + H_{22})$ is the energy at the crossing.

Now to each point on this cone corresponds an energy eigenstate, which we may always chose to be real. However, as Herzberg and Longuet–Higgins pointed out, when one tries to extend this choice smoothly around a closed path which makes a single circuit around the degeneracy, the eigenfunction at the end of the circuit will be equal to minus the initial eigenfunction. Physically speaking, this means that wherever the nuclei of a molecule execute a motion which takes the [real] electronic Hamiltonian around a degeneracy n times, the electronic wavefunctions will pick up a phase $(-1)^n$ in addition to their usual dynamical phase. One important consequence of this result is that generic (as opposed to "accidental") degeneracies need not always be located at points of enhanced symmetry. Indeed, if a closed path can be found around which an eigenfunction of a real Hamiltonian changes sign, then it must enclose a degeneracy, whether or not a symmetric point is also enclosed.

What happens when the electron wavefunctions can not be taken real? This situation may arise when the Hamiltonian is intrinsically complex, for example when electronic spin interactions are included. In a beautiful 1976 paper, which the editors feel has not been sufficiently appreciated, Stone [2.4] posed this question, and found the first example of a complex, non-integrable adiabatic phase in a quantum system. He showed, quite generally, that non-integrable phases imply the existence of degeneracies, by means of the following topological argument. In the case of a complex Hamiltonian, three conditions must be satisfied at a degeneracy, in contrast to the real case above. So there is three-dimensional space of parameter values which move the Hamiltonian away from a degeneracy. In this space, let us consider the effect of transporting an eigenfunction around a closed path C, along which the eigenfunction is nondegenerate. In general, it will come back to the same eigenfunction, multiplied by a phase holonomy $\exp i\gamma(C)$. Sometimes the phase can be removed by redefining the rule of transport, but there may be global obstructions to doing this. To see how such obstructions can arise, consider transporting the eigenfunction around a sequence of paths C

which sweep out a closed surface. The phases $\exp i\gamma(C)$ will in turn sweep out a path in the unit circle of the complex plane, which will wind around the origin an integral number of times. If this winding number is nonzero, then the phases cannot be removed, and the surface must enclose a point P where the eigenfunction is discontinuous as a function of the external parameters, and hence degenerate. Because of its purely topological nature, this powerful result is immune to small perturbations, and allows one to deduce exact information from a model Hamiltonian.

There was no room in the standard Born-Oppenheimer treatment of molecules for the results of Longuet-Higgins and Herzberg, and of Stone, that nuclear wavefunctions can not be chosen to be single-valued near a degeneracy. The standard B–O effective Hamiltonian for nuclei was incomplete in its failure to predict this behavior, and without a proper effective Hamiltonian, actual computations would be difficult. This problem was solved by Mead and Truhlar [2.5], who showed that, in spite of the lack of well-defined electronic and nuclear wavefunctions, it is possible to form an effective Hamiltonian for the nuclei of a polyatomic molecule if one introduces an external "gauge potential." The sole effect of this extra term is to create a fictitious magnetic flux emanating from the crossing point, which produces the requisite phase change in the nuclear wavefunction. In brief, the origin of the gauge potential is as follows. Suppose that

$$
\begin{aligned}
H(R,r)\,\Psi &= \Big(H_N(R) + H_e(r,R)\Big)\Psi \\
&= \left\{-\frac{\hbar^2}{2M}\nabla_R^2 - \frac{\hbar^2}{2m}\nabla_r^2 + V(R,r)\right\}\Psi = E\Psi
\end{aligned}
\tag{2.6}
$$

is the molecular Schrödinger equation, where lower-case letters refer to electrons and upper-case to nuclei (and a sum over spatial indices is implicit). The electronic energy eigenfunctions $\phi_n(r,R)$ with energy $\epsilon_n(R)$ are eigenfunctions of H at fixed R, and the total wavefunction may be decomposed in terms of them as

$$
\Psi(r,R) = \sum_n \Phi_n(R)\,\phi_n(r,R)
\tag{2.7}
$$

Now in the B-O approximation, one neglects mixings between the electronic levels, so we can take for the full wavefunction just one term of the sum (2.7). Sandwiching Eq. (2.6) on the left with $\phi_n^*(r,R)$ and integrating over r then leaves an effective Schrödinger equation for the nuclear wavefunction $\Phi_n(R)$:

$$
\left\{-\tfrac{\hbar^2}{2M}(\nabla - iA)^2 + \epsilon_n(R)\right\}\Phi_n = E_n\Phi_n
\tag{2.8}
$$

where the "gauge potential" A_i is

$$
A_i \equiv i\langle\phi_n|\nabla_{R_i}|\phi_n\rangle
\tag{2.9}
$$

Aside from appearing in the Hamiltonian as a gauge potential should, A is also associated with a gauge invariance. Namely, when a redefinition of the phases of the electronic eigenfunctions is made

$$\phi_n(R) \rightarrow e^{i\lambda(R)}\phi_n(R), \tag{2.10}$$

A_i transforms as

$$A_i \rightarrow A_i - \partial_i\lambda. \tag{2.11}$$

We would like to mention that an example of a complex Hamiltonian with a topologically unremovable holonomy had been studied back in the early days of quantum mechanics by Van Vleck.[4] Without realizing its more general significance, he derived an effective nuclear Hamiltonian for the lambda-levels of a diatomic molecule, including an object that we would now recognize as a magnetic monopole gauge field. Since Van Vleck's work preceded Dirac's famous 1931 paper, we might even say that Van Vleck found the first physical application of magnetic monopoles.

Mead and Truhlar called the result of their modification of the B-O approximation the "molecular Aharonov–Bohm" effect. The name comes from a very similar but much more famous phenomenon, involving actual gauge potentials, that arises when a charged particle interacts with a solenoid. In a paper which is often referred to but rarely read, Aharonov and Bohm [2.6] found that, even if the the magnetic field vanishes outside the solenoid, the electromagnetic gauge potential does not, and its effect is to produce a phase change proportional to the enclosed flux when the wavefunction of the charged particle is transported around it. A direct physical manifestation, calculated in their paper, is the scattering for a particle of unit charge off of an infinitely thin tube of magnetic flux α. The differential cross section per unit length is

$$\frac{d\sigma}{d\theta} = \frac{\sin^2 \pi\alpha}{2\pi k \cos^2(\theta/2)} \tag{2.12}$$

and is a purely geometric quantity, depending only on the geometry of the scattering process. (The apparent angular momentum dependence merely indicates that the de Broglie wavelength sets the length scale.) In fact, the Aharonov–Bohm phase depends only on the path taken by the particle, so it is a geometric phase as well (see [3.4]). As efforts to find measurable effects associated with non-integrable phases continue, it is possible that experimental techniques used to measure the Aharonov–Bohm effect, particularly those involving scattering, will prove useful.

[1] A sampling of the main ideas behind this vast subject can be found in the following books:
P.W. Anderson, *Basic Notions of Condensed Matter Physics*, Frontiers in Physics vol. 55. London: Benjamin/Cummings, 1984.
S. Coleman, *Aspects of Symmetry*. Cambridge: University Press, 1985.
Rajaraman, *Solitons and Instantons*. Amsterdam: North-Holland, 1982.

[2] J. von Neumann and E. Wigner, *Phyzik. Z.* **30** (1927) 467.

[3] E. Teller, *J. Phys. Chem.* **41** (1937) 109.

[4] J.H. Van Vleck, *Phys. Rev.* **33** (1929) 467.

Reprinted from "The Proceedings of the Indian Academy of Sciences",
Vol. XLIV, No. 5, Sec. A, 1956

GENERALIZED THEORY OF INTERFERENCE, AND ITS APPLICATIONS

Part I. Coherent Pencils

BY S. PANCHARATNAM
(*Memoir No.* 88 *of the Raman Research Institute, Bangalore*-6)

Received October 30, 1956
(Communicated by Sir C. V. Raman)

§ 1. INTRODUCTION

THE investigations of which the results are presented in this paper arose during the study of certain specific problems in crystal optics. As investigators in this field are well aware, the simplest procedures for studying the optical properties of anisotropic media (*e.g.*, examination under the polarising microscope) generally involve the use and study of polarised light. The complexity of the peculiar interference phenomena exhibited and also of their customary theoretical analysis (by algebraic methods) become quite considerable even in the case of transparent optically active crystals like quartz—as may be seen by a reference to the treatises of Mascart (1891) and Walker (1904); this is because the two waves propagated along any direction in such a medium are no longer linearly polarised at right angles to one another, but are elliptically polarised. Nevertheless, the types of 'oppositely' polarised waves propagated in such media must be termed simple compared to the elliptically polarised waves propagated in *absorbing biaxial crystals.*

The remarkable interference phenomena exhibited by absorbing biaxial crystals may be easily studied by looking at an extended source through a plate (cut normal to an optic axis); a suitable material is the mineral iolite—which the author had the opportunity of investigating experimentally (Pancharatnam, 1955). (1) With the incident light polarised, and *even without the use of an analyser*, interference rings are seen, which are feeble but are nevertheless easily visible. (2) When, in addition, an analyser is also introduced, the biaxial interference figures seen are notably different from those seen in transparent crystals under the same conditions. (3) With the analyser in position and with no polariser—*i.e., with the incident light completely unpolarised*—feeble interference rings are again easily visible. (4) Finally, even when both analyser and polariser are absent, incipient traces of an interference pattern may be discerned.

Viewing these particular phenomena from a slightly broader perspective we see that their analysis is connected with certain general questions concerning the properties of two polarised beams travelling along the same direction. We shall now formulate these problems since they form the main content of the paper. The study of the effects with a polariser alone leads us to investigate the following questions: the interference of two coherent beams in different states of elliptic polarisation (§ 3); the resolution of any polarised beam into two beams in given states of polarisation—which occurs at the first face of the crystal plate (§ 4); and the composition of two coherent beams of different polarisation—at the second face of the plate (§§ 5, 6). The problem involved when an analyser is also introduced (keeping the incident light polarised) reduces to the following: the interference of two coherent polarised beams which are ' brought to the same state of elliptic vibration ' by the use of a suitable analyser (§ 8). In § 9 we shall consider the addition of n coherent beams in different states of polarisation.

An attempt to formulate in general terms the problems associated with (3) and (4) leads rather unexpectedly into the subject of the partial coherence of polarised beams. We shall leave the discussion of this subject for Part II.

§ 2. THE POINCARÉ SPHERE AND THE STOKES PARAMETERS

For problems—such as the one we are dealing with—where we require to handle elliptic vibrations with the same facility as linear vibrations, the indirect specification of the polarisation of an elliptic vibration by giving the equation to its rectangular components is obviously unsatisfactory: the procedure not only leads to cumbersome calculations lacking in elegance, as has been pointed out by other authors, but often ceases to give physical insight into the cause of the phenomena actually observed. Two other powerful methods for specifying the state of polarisation of a beam have been used extensively—an analytical method due to Stokes (1901), and a geometrical one due to Poincaré. The conventional theoretical presentation of the subject of the ' Stokes parameters ' may be found in Chandrasekhar (1949) and in Rayleigh (1902); that of the Poincaré sphere and some of its more well-known properties may be found in Pockels (1906), Walker (1904), Ramachandran and Ramaseshan (1952), and Jerrard (1954). In Part I of the present paper, only the Poincaré representation is used; and we may mention that this part constitutes in itself a self-contained derivation of the properties of the Poincaré sphere by a new procedure. The Stokes representation will be required only in Part II where the subject of partial polarisation naturally enters; but the entire subject of the Stokes representation is introduced there, in a new manner, through the Poincaré representation itself—by developing the ideas of Fano (1949) and Ramachandran (1952).

The state of polarisation of a completely polarised beam is directly specified by the form of the ellipse traced by the tip of the displacement vector—this being invariant for a completely polarised beam, unlike the intensity and absolute phase which may be subject to statistical fluctuations. Poincaré introduced a mapping whereby any particular form of elliptic vibration is represented by a specific point on the surface of a sphere—the points on the 'Poincaré sphere' exhausting all the conceivable forms of elliptic vibrations. The definition of the mapping allows the ellipse represented by a point P to be visualised directly in terms of the longitude 2λ and latitude 2ω of the point: for λ is the azimuth of the major axis of the elliptic vibration and $\tan \omega$ the ellipticity. Alternatively the point P may be specified in cartesian co-ordinates instead of in polar co-ordinates—with the XY plane coinciding with the equatorial plane. As Perrin (1942) had observed, the *three parameters introduced by Stokes for characterising a completely polarised beam are proportional to the cartesian co-ordinates of the point P*—the constant of proportionality being the intensity of the beam which is taken as the fourth Stokes parameter.

Two elliptical vibrations whose states of polarisation are represented by diametrically opposite points on the Poincaré sphere will be said to be oppositely polarised. The conventional procedure of decomposing any elliptic vibration into two rectangular linear vibrations (represented by two diametrically opposite points on the equator) is a particular case of decomposing it into two elliptical vibrations of opposite polarisation. The fundamental property of the Poincaré sphere relates to such a decomposition (*see* Fig. 1) and is the following:—

I. *When a vibration of intensity I in the state of polarisation C is decomposed into two vibrations in the opposite states of polarisation A and A′, the intensities of the 'A-component' and the 'A′-component' are* $I \cos^2 \frac{1}{2} CA$ *and* $I \sin^2 \frac{1}{2} CA$ *respectively.*

All the results proved in this paper are deduced from the above theorem. We shall not give the proof of the theorem since a proposition equivalent to the above has been proved by other authors (*see* § 8 below).

Since according to Theorem I, the sum of the intensities of the oppositely polarised component beams are together equal to the intensity I of the resultant beam we deduce the following well-known result.

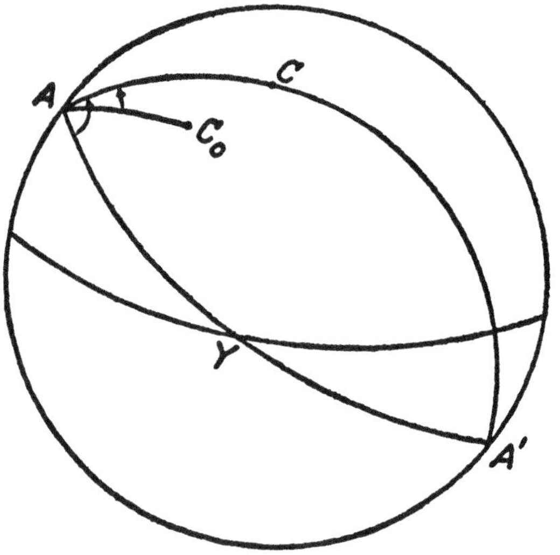

FIG. 1

II. *Two oppositely polarised beams cannot constructively or destructively interfere.*

§ 3. INTERFERENCE OF TWO NON-ORTHOGONALLY POLARISED VIBRATIONS

The interference of two *linear* vibrations not orthogonal to one another can be analysed by resolving the first vibration into two vibrations which are respectively parallel and orthogonal to the second vibration. The interference of any two elliptically polarised vibrations will now be handled in a similar manner. We shall prove the following proposition.

III. The intensity I of the beam obtained on combining two mutually coherent beams 1 and 2, of intensities I_1 and I_2 in the states of polarisation A and B respectively, will be given by the general interference formula:—

$$I = I_1 + I_2 + 2\sqrt{I_1 I_2} \cos \tfrac{1}{2}c \cos \delta \tag{1}$$

Here c is the angular separation of the states A and B on the Poincaré sphere (*see* Fig. 2). Thus $\cos^2 \frac{1}{2}c$ is a 'similarity factor' between the states of polarisation, which determines the extent of interference and which varies from unity (for identically polarised beams) to zero (for oppositely polarised beams). The significance of the 'phase difference' δ between the beams will be elucidated below.

The above relation may be obtained as follows. The vibration 2 may be replaced by two oppositely polarised vibrations—one in the state of polarisation A of the first beam, and the second in the state of polarisation A′ orthogonal to A. These vibrations (which we shall denote by 2A and 2A′ res-

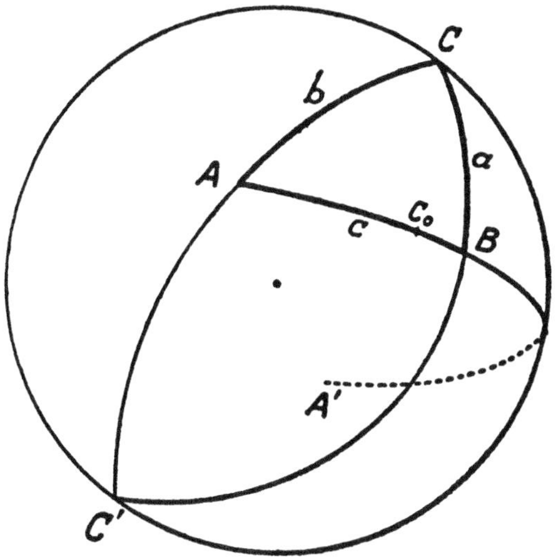

FIG. 2

pectively) will be coherent with vibration 1 and will have intensities $I_2 \cos^2 \frac{1}{2}c$ and $I_2 \sin^2 \frac{1}{2}c$ respectively (§ 2, I). The vibration 1 being in the same state of polarisation as vibration 2A (over which it has a phase advance δ, say), can be combined with it to yield a vibration (1 + 2A) of intensity

$$I_1 + I_2 \cos^2 \tfrac{1}{2}c + 2\sqrt{I_1 I_2} \cos \tfrac{1}{2}c \cos \delta$$

We are thus left with the vibration (1 + 2A) and the vibration 2A′ which are in the opposite states of polarisation A and A′. The resultant intensity I, obtained merely by adding their intensities (§ 2, II), is that given in formula (1) which we wished to deduce.

An explicit expression for the similarity factor between any two states of polarisation (in terms of the azimuths λ_1, λ_2 of the major axes and the ellipticities $\tan \omega_1$, $\tan \omega_2$ of the two vibrations) may, if necessary, be obtained from the following relation (obtained by spherical trigonometry):—

$$\cos c = \sin 2\omega_1 \sin 2\omega_2 + \cos 2\omega_1 \cos 2\omega_2 \cos 2(\lambda_2 - \lambda_1) \qquad (2)$$

It may be noted that $\cos \frac{1}{2}c$ is equal to the visibility of fringes (as defined by Michelson) obtained under the conditions $I_1 = I_2$.

In the above discussion the quantity δ has been introduced as the phase advance of the first beam over the A-component of the second beam. There are two properties of δ however which enable us to speak of it as the absolute difference of phase between the two beams themselves—though they are in different states of polarisation. In the first place we note that if the first beam is subjected to a particular path retardation relative to the second, then δ as defined above decreases by the corresponding phase angle; in the second place we note that as long as no path retardation is introduced between the two beams, any alteration of the intensities of the two beams will not change the value of δ as defined above. Hence we will be guilty of no internal inconsistency if we make the following statement by way of a definition: *the phase advance of one polarised beam over another* (not necessarily in the same state of polarisation) *is the amount by which its phase must be retarded relative to the second, in order that the intensity resulting from their mutual interference may be a maximum.*

This phase advance is identically equal to δ, and the above definition holds only for non-orthogonal vibrations.

By picturing an elliptic vibration as being made up of its rectangular components, it becomes apparent that the coherent addition contemplated in this section, of two beams in different states of polarisation will yield a resultant beam which is also elliptically polarised. The intensity of this having been determined, the next step logically would be to determine its state of polarisation. This problem we shall relegate to § 5, and take up in the next section the converse problem which is simpler to handle.

§ 4. DECOMPOSITION OF A POLARISED BEAM INTO TWO BEAMS IN GIVEN STATES OF POLARISATION

The method of approaching the general problem described in the heading is immediately made clear by regarding the following particular case, illustrated in Fig. 3. Suppose that a *linear* vibration C has to be split into two *linear* vibrations A and B (with which it makes angles $\frac{1}{2}b$ and $\frac{1}{2}a$ respectively). The intensities of the A- and B-vibrations may be obtained by the parallelogram law; more specifically, by equating the projections of the B- and C-vibrations in the direction orthogonal to the A-vibration, we obtain the intensity of the B-vibration as proportional to $(\sin \frac{1}{2}b/\sin \frac{1}{2}c)^2$, where $\frac{1}{2}c$ is the angle between the A and B vibrations.

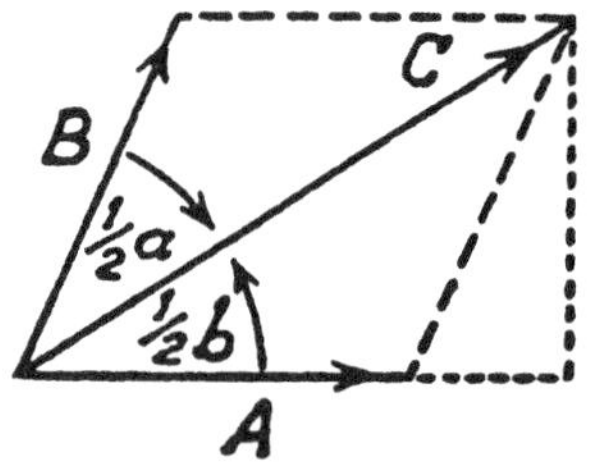

FIG. 3

By similar reasoning we shall prove the following proposition.

IV. If a beam of intensity I in the state of polarisation C is decomposed into two coherent beams 1 and 2, in the states of polarisation A and B respectively, the intensities of these beams will be given by

$$I_1 = I \cdot \frac{\sin^2 \frac{1}{2}a}{\sin^2 \frac{1}{2}c}; \qquad I_2 = I \cdot \frac{\sin^2 \frac{1}{2}b}{\sin^2 \frac{1}{2}c} \tag{3}$$

where a, b and c are the angular separations BC, CA and AB respectively (*see* Fig. 2).

To prove the proposition, we decompose each vibration into two vibrations of polarisation A and A′. The sum of the A′-components of the vibrations 1 and 2 must together be the same as the A′-component of the given vibration 3. The vibration 1, however, has no A′-component since we have chosen A′ opposite to A. Hence the A′-components of vibrations 2 and 3 are identical. Equating their intensities (as given by § 2, I) we have $I_2 \sin^2 \frac{1}{2}c = I \sin^2 \frac{1}{2}b$. Similarly, we can show that $I_1 \sin^2 \frac{1}{2}c = I \sin^2 \frac{1}{2}a$, thus establishing relations (3) above. [The expressions (3) for the intensities I_1 and I_2 could be written explicitly in terms of the constants of the elliptic vibrations by substituting the expression (2) for $\cos c$ and similar expressions, obtained by cyclic permutation, for $\cos a$ and $\cos b$. It can thus be shown that the expressions (3) are equal to the more lengthy expressions deduced directly by Stokes (*loc. cit.*).]

The beams 1 and 2 will have a definite phase relationship with one another which, as we shall see, depends in a remarkable fashion on the mutual configuration of the points A, B and C.

From (1) we have

$$\cos \delta = \frac{I - (I_1 + I_2)}{2\sqrt{I_1 I_2} \cos \frac{1}{2}c}$$

Substituting for I_1 and I_2 from (3),

$$-\cos \delta = \frac{\sin^2 \frac{1}{2}a + \sin^2 \frac{1}{2}b - \sin^2 \frac{1}{2}c}{2 \sin \frac{1}{2}a \sin \frac{1}{2}b \cos \frac{1}{2}c} \tag{4}$$

If C′ be the point diametrically opposite to C (Fig. 2), then denoting by a', b', c the sides of the triangle ABC′, we will have

$$-\cos \delta = \frac{\cos^2 \frac{1}{2}a' + \cos^2 \frac{1}{2}b' + \cos^2 \frac{1}{2}c - 1}{2 \cos \frac{1}{2}a' \cos \frac{1}{2}b' \cos \frac{1}{2}c}$$

The expression on the right-hand side is the cosine of half the solid angle subtended by the triangle C′BA at the centre of the sphere (*see* M'Clelland and

Preston, 1897, Part II, Ch. 7, p. 50, Ex. 1). Since the Poincaré sphere has unit radius, we arrive at the following unexpected geometrical result.

V. When a beam of polarisation C is decomposed into two beams in the states of polarisation A and B respectively, the phase difference δ between these beams is given by

$$|\delta| = \pi - \tfrac{1}{2}|E'| \qquad (5\,a)$$

where the angle $|E'|$ is numerically equal to the area of the triangle C′BA colunar to ABC. (E′ is also the spherical excess of the triangle C′BA, *i.e.*, the excess of the sum of its three angles over π.)

Hitherto we have fixed the state of zero phase difference between two non-orthogonally polarised beams by the criterion that the resultant intensity produced by mutual interference should be a maximum. We can equally well use the fact (shown by relation 5 *a*) that *the beams in the state of polarisation A and B will have zero phase difference when the resultant state of polarisation C* (produced by their mutual interference) *lies on the arc directly joining A and B* (for then the colunar triangle will have the area of a hemisphere). On the other hand, the beams will be opposed in phase ($\delta = \pm\pi$), when the resultant state of polarisation C lies on the *greater* segment of the great circle through A and B (colunar triangle has zero area).

The relation (5 *a*) does not give the sign of δ—the phase advance of vibration 1 over vibration 2. We shall now resolve this ambiguity. It is clear (from considerations of continuity) that the sign of δ remains the same for all positions of the resultant polarisation C lying on any one side of the great circle AB, the magnitude of δ being between 0 and π; and that δ changes sign (without discontinuity) as the resultant polarisation C crosses from one side of the great circle AB to the other. (This is forced by the physical requirement that the addition of the beams 1 and 2 with specific intensities and a specific phase relationship should lead unambiguously to a unique state of polarisation C of the resultant beam.) We can now show that the phase advance of vibration 1 over 2 will be positive if (as is drawn in Fig. 2) the point C appears to the left of AB as we proceed from A to B on the surface of the sphere. To prove this proposition it is sufficient to show that this rule of signs is true for a *particular* pair of non-orthogonal states A and B. (For it can then be shown to hold for an adjacent pair of points A and B, from continuity arguments, and hence for *any* pair of states A and B.) As a particular case which proves the general rule, we may see in Fig. 3 that if the *linear* vibration A has a positive advance of phase over the *linear* vibration B, the resultant vibration will be left-elliptic. The results of the discussion of this paragraph can also be summarised by using the sign convention that the area E′ of the triangle C′BA be counted positive only when the sequence of points C′BA are described in a counter-clockwise sense on the surface of the sphere. We can then write

$$\delta = \pi - \tfrac{1}{2}E' \tag{5 b}$$

§ 5. THE COMPOSITION OF NON-ORTHOGONALLY POLARISED PENCILS

When two coherent beams of intensities I_1 and I_2 (in the states of polarisation A and B respectively) are combined, the resultant state of polarisation C may be specified by the angular distances *b* and *a* of the point C from the points A and B (*see* Fig. 2). According to (3) these are given by

$$\sin^2 \tfrac{1}{2}a = \frac{I_1}{I} \cdot \sin^2 \tfrac{1}{2}c; \qquad \sin^2 \tfrac{1}{2}b = \frac{I_2}{I} \cdot \sin^2 \tfrac{1}{2}c \tag{6}$$

where I is the resultant intensity given by (1). On proceeding from A to B the point C appears to the left or right according as δ is positive or negative. An alternative method of finding the state C will be given in Part II.

The following geometrical facts are useful for qualitatively locating the state C. When the phase difference between the two beams is altered without altering the ratio of their intensities, the state C moves along the locus $\sin^2 \tfrac{1}{2}a/\sin^2 \tfrac{1}{2}b = \text{constant}$; this is a small circle (with its centre on the great circle through AB) which cuts the arc AB internally and externally according to this ratio (*see* M'Clelland and Preston, *loc. cit.*, Part I, Ch. III, p. 66, Ex. 1). On the other hand, when the ratio of the intensities of the beams is altered without altering their phase difference, the state C moves along the locus, $E' =$ constant; this is a small circle passing through A and B with its centre on the great circle which is the perpendicular bisector of the arc AB (M'Clelland and Preston, *loc. cit.*, Part II, § 101). The point C is thus determined by the intersection of these two families of small circles.

§ 6. THE COMPOSITION OF TWO OPPOSITELY POLARISED BEAMS

When a polarised beam is split into two orthogonally polarised components, A and A′, the fundamental property of the Poincaré sphere enunciated in § 2, I, gives us information regarding the intensities of these components but tells us nothing about their relative phase relationship. When we wish to enquire into the latter, we run into the apparent difficulty that the state of relative phase between the two beams which could be taken as the standard or ' zero ' with respect to which any additional phase differences could be measured, cannot be defined as in § 3 by their interference properties—because there is no such interference to talk of. We can however avoid the difficulty by restricting ourselves to the following query which alone is of practical importance (*see* Fig. 1).

If, after decomposing a beam of polarisation C_0 into two oppositely polarised beams of polarisation A and A′ respectively, we retard the phase of the A′-component by an amount Δ (relative to the other), and also alter their intensities in any specified manner, what will be the resultant state of polarisation C?

To find the distance of C from A (*see* Fig. 1) constitutes no problem; for, from the first theorem (§ 2, I) itself we see that *the ratio of the* (final) *intensities of the A′ and A beams must be equal to* $\tan^2 \frac{1}{2}CA$. Our question therefore really concerns the magnitude of the angle CAC_0—regarding which we shall prove the following proposition.

VI. The angle CAC_0 is equal to the phase retardation Δ introduced.

It is remarkable that this second important property of oppositely polarised vibrations follows in the ultimate analysis as a consequence of the first fundamental property itself—for we shall prove it is a limiting case of the properties of two non-orthogonally polarised vibrations as they tend towards states of opposite polarisation.

For convenience let us regard the initial and final states of polarisation C_0 and C as given (*see* Fig. 2). We first decompose the beam of polarisation C_0 into two beams of polarisation A and B respectively, where B is chosen such that the arc AB contains C_0. By introducing a phase retardation δ between these beams of polarisation A and B, altering their relative intensities suitably and then compounding them, we can produce a beam of polarisation C. According to § 4, V, we have $\delta = \pi - \frac{1}{2}E'$, where E′ is the area of the triangle C′BA. We wish to find the limit towards which δ tends, as the state of polarisation B tends towards the state A′ opposite to A. As the point B moves towards A′ and ultimately coalesces with it, the area of the triangle C′BA obviously becomes equal to the area of the *lune* enclosed between the great circular arc AC_0A' and $AC'A'$, this area being $2\angle C_0AC'$. Hence we have

$$\Delta = \pi - \angle C_0AC' = C\hat{A}C_0$$

thus proving the required proposition. It is clear from (5 *b*) that the angle CAC_0 must be measured positive in the counter-clockwise sense as indicated in Fig. 1, in order that it may have the same sign as Δ (which is the amount by which the A-component is *advanced* in phase).

The particular property of the Poincaré sphere which has led to its extensive application in tracing the passage of polarised light through transparent birefringent media follows as a corollary of Theorem VI above. On passage through any plate of such a medium, the emergent state of polarisation C can be obtained from the incident state C_0 by a *rotation* of the sphere by an anticlockwise angle Δ about the axis AA′, where A represents the state of elliptic polarisation of the faster of the two orthogonally polarised waves. (The author has not come across a general proof of this much-used property of the Poincaré sphere in any of the references quoted.)

A second corollary of the proposition VI is the following; when the ratio of the intensities of the orthogonally polarised beams is altered without altering their phase relationship, the locus of the resultant state of polarisation C is the great circular arc ACA′.

§ 7. Definition of the 'Phase Difference' Between Oppositely Polarised Vibrations

It will be a great convenience in connection with a later discussion to set up an arbitrary standard with respect to which the relative phase relationship between two orthogonal vibrations may be measured.

When two orthogonal linear vibrations of equal intensity combine to yield a linear vibration bisecting the right angle included between the directions OX and OX′ of their vibrations, we customarily say that the linear vibrations are in phase. (There will be no ambiguity regarding whether the vibrations are to be regarded in phase or opposed in phase, if we choose both the radii vectors OX and OX′ within the interval, $\pi/2 \geqslant \theta > -\pi/2$, with respect to a fixed radius vector Or on the wave-front.)

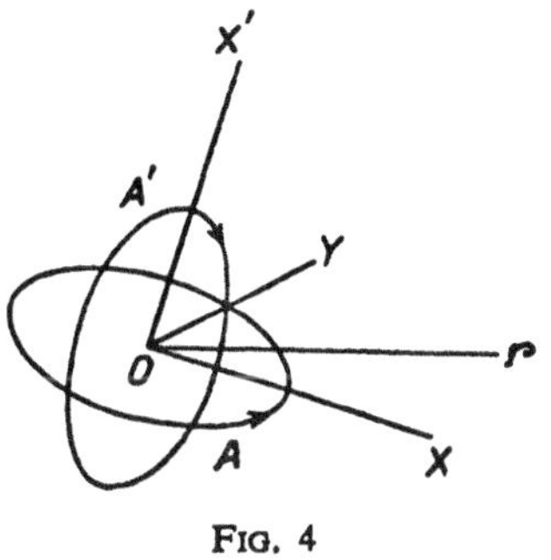

Fig. 4

In conformity with this, we may define two orthogonal elliptic vibrations of equal intensity as being in the same phase when they combine to yield a linear vibration OY bisecting the right angle included between the directions OX and OX′ of their major axes (*see* Fig. 4). The radii vectors OX and OX′ are taken parallel to the major axes of the left- and right-elliptic vibrations respectively and are chosen such that OXX′Z forms a right-handed system—OZ being the direction of propagation. In terms of the Poincaré representation (*see* Fig. 1), the above definition has the following significance. The states of polarisation of two oppositely polarised vibrations are represented by the points A and A′ on the upper and lower hemisphere respectively, while the linear vibration OY is represented by a point Y. The point Y is one of the two opposite points on the equator at an angular distance of $\pi/2$ from both A and A′; on proceeding from A to Y the upper pole appears to the left, as indicated in the figure. The A- and A′-components of the vibration Y (and hence of any vibration represented by a point on the great circular arc AYA′) are defined to be in the same phase.

The case of opposite circular vibration is not covered in the above definition. We may define two such vibrations of equal intensity to be in the same phase when they yield a linear vibration OY parallel to a fixed radius vector Or on the wave-front.

The phase advance of one polarised vibration A over an orthogonally polarised vibration A′ is then equal to the angle CAY (measured positive in the counter-clockwise sense), where C represents the resultant state of vibration obtained on compounding the two. Henceforward we may speak of any one polarised beam as having a definite phase advance δ over another coherent polarised beam without implying that the two beams are non-orthogonally polarised.

§ 8. Interference of the Components of Two Polarised Beams Transmitted by an Analyser

It is well known that light of any arbitrary elliptic polarisation C′ may be extinguished by means of an appliance consisting of a suitably oriented quarter-wave plate followed by a linear analyser at the proper setting; when light of the opposite polarisation C is incident on the same appliance the *intensity* of the transmitted light will be equal to that of the incident—as may be directly shown (Stokes, 1901), without using any property of the Poincaré sphere. Any appliance having both the above properties will be referred to as an *analyser C.*

Since light of any other polarisation P may be decomposed into two coherent beams of polarisation C and C′, the intensity transmitted by an analyser C will be equal to the C-component of the incident beam. If we assume the fundamental Theorem I, § 2, it follows that an analyser C transmits a fraction $\cos^2 \frac{1}{2}PC$ of the intensity of light of polarisation P. Conversely, one of the simplest proofs of the fundamental Theorem I, lies in the direct analytical proofs of this property of an analyser given by Fano (1949) and also by Ramachandran and Ramaseshan (1952).

As was pointed out in the Introduction, the problem to be now discussed arises when we wish to consider the phenomena exhibited by a plate of an absorbing biaxial crystal when kept between a polariser and an analyser. Emerging from the crystal plate along any particular direction will be a coherent mixture of two beams 1 and 2. Let their states of polarisation be A and B, their intensities I_1 and I_2, and let the phase advance of the first beam over the second be δ. (The state of polarisation of the resultant beam will not be required here and will *not* be denoted by C.)

In order to compute the intensity transmitted by an analyser C, we first resolve each of the two vibrations into the opposite states of polarisation C and C′ (*see* Fig. 2). The C-components of the beams of polarisation A and B will have intensities $I_1 \cos^2 \frac{1}{2}b$ and $I_2 \cos^2 \frac{1}{2}a$ respectively (§ 2, I). Since these components will have some definite phase difference δ', say, they can interfere. Hence the resultant beam obtained by combining the beams 1 and 2 will have a C-component whose intensity I_C is given by

$$I_C = I_1 \cos^2 \tfrac{1}{2}b + I_2 \cos^2 \tfrac{1}{2}a + 2\sqrt{I_1 I_2} \cos \tfrac{1}{2}a \cos \tfrac{1}{2}b \cos \delta' \quad (7)$$

Similarly the C′-components of the beams 1 and 2 will have intensities $I_1 \sin^2 \frac{1}{2}b$ and $I_2 \sin^2 \frac{1}{2}a$ respectively, and a definite phase difference δ'', say. Hence the C′-component of the resultant beam will have an intensity $I_{C'}$, given by

$$I_{C'} = I_1 \sin^2 \tfrac{1}{2}b + I_2 \sin^2 \tfrac{1}{2}a + 2\sqrt{I_1 I_2} \sin \tfrac{1}{2}a \sin \tfrac{1}{2}b \cos \delta'' \quad (8)$$

Since an analyser transmits only the intensity I_C it remains to determine δ' in terms of the phase difference between the beams 1 and 2, and the analyser position C.

Now δ' represents the phase advance of the C-component of the beam of polarisation A over the C-component of the beam of polarisation B; while δ'' is the phase advance of the C′-component of the beam of polarisation A over the C′-component of the beam of polarisation B. Hence *it follows from a consideration of VI, § 6, that* $\delta'' - \delta' = \pm \hat{C}$ (where the positive sign has to be taken if C lies on the side of AB shown in Fig. 2). Furthermore, the intensity I of the resultant beam obtained on compounding 1 and 2 is equal to the sum of the intensities of the C- and C′-components (§ 2, II).
Hence adding (7) and (8)

$$I = I_1 + I_2 + 2\sqrt{I_1 I_2}\{\cos \tfrac{1}{2}a \cos \tfrac{1}{2}b \cos \delta' + \sin \tfrac{1}{2}a \sin \tfrac{1}{2}b \cos(\delta' \pm C)\}$$

Expanding cos $(\delta' \pm C)$, it can be shown by applying the standard expressions for the spherical excess of a triangle (M'Clelland and Preston, *loc. cit.*, Part II, p. 37, Art. 104) that the above relation reduces to

$$I = I_1 + I_2 + 2\sqrt{I_1 I_2} \cos \tfrac{1}{2}c \cos(\delta' + \tfrac{1}{2}E)$$

where E represents the area of the triangle ABC itself (counted positive or negative according as the sequence of points A, B, C describe the periphery of the triangle in the counter-clockwise or clockwise sense). The intensity I is also given directly in terms of the phase difference δ between the beams 1 and 2 by the expression (1) of § 3. Comparing the two we get the value of δ' to be substituted in (7):

$$\delta' = \delta - \tfrac{1}{2}E \tag{9}$$

VII. Hence when a mixture of two coherent beams of intensities I_1 and I_2 in the states of polarisation A and B respectively, is incident on an analyser C, the transmitted intensity I_C is given by

$$I_C = I_1 \cos^2 \tfrac{1}{2}b + I_2 \cos^2 \tfrac{1}{2}a + 2\sqrt{I_1 I_2} \cos \tfrac{1}{2}a \cos \tfrac{1}{2}b \cos(\delta - \tfrac{1}{2}E) \tag{10}$$

where δ is the phase advance of the beam of polarisation A over the other, *a*, *b* and *c* being the angular sparations BC, CA and AB respectively and E the area of the triangle ABC.

It is to be noted that the above result must also hold in the limiting case when the two beams incident on the analyser are in the orthogonal states of polarisation A and A'. In this case, if AYA' be the great circular arc (Fig. 1) with respect to which the phase difference δ is measured (ref. § 7), then E (which now becomes the area of the lune AYA'CA) is equal to twice the angle CAY. The expression (9) may in this case be further simplified by the substitution $\cos \tfrac{1}{2}a = \sin \tfrac{1}{2}b$.

§ 9. The Addition of *n* Coherent Beams

Suppose a mixture of 3 coherent beams of polarisation A_1, A_2, A_3 is incident on an analyser C. Let $2\theta_i$ denote the length of the arc CA_i, δ_{ij} denote the phase *lag* of the beam *i* over the beam *j*, and E_{ij} be the area of the triangle A_iCA_j. We have obviously

$$\delta_{ij} = -\delta_{ji}; \qquad E_{ij} = -E_{ji} \tag{11}$$

The C-component of the first vibration may be written as $\sqrt{I_1}\cos\theta_1 \, \mathbf{c}\, e^{i\omega t}$ where $\mathbf{c}\, e^{i\omega t}$ is a vibration of unit intensity in the state of polarisation C†.

† The components of the vector c are the complex amplitudes of the components of the elliptic vibration.

If $A_C e^{i\omega t}$ denote the C-component of the resultant vibration obtained on compounding the three vibrations, then according to (9) we will have,

$$A = \sqrt{I_1} \cos \theta_1 + \sqrt{I_2} \cos \theta_2 \exp i (\delta_{12} - \tfrac{1}{2}E_{12}) + \sqrt{I_3} \cos \theta_3 \exp i (\delta_{13} - \tfrac{1}{2}E_{13})$$

Since the phase lag of the C-component of the second beam over the C-component of the third is according to (8), given by $(\delta_{23} - \frac{1}{2}E_{23})$ we have

$$-(\delta_{23} - \tfrac{1}{2}E_{23}) = (\delta_{12} - \tfrac{1}{2}E_{12}) - (\delta_{13} - \tfrac{1}{2}E_{13}) \qquad (12)$$

with similar expressions connecting all the δ_{ij}. Hence the intensity I_C transmitted by an analyser C being equal to AA^*, will be given by

$$I_C = \Sigma I_i \cos^2 \theta_i + \sum_{i \neq j} \sqrt{I_i I_j} \cos \theta_i \cos \theta_j \cos (\delta_{ij} - \tfrac{1}{2}E_{ij}) \qquad (13)$$

The resultant intensity I of the beam obtained on compounding the three beams will be equal to the sum of the intensities I_C and $I_{C'}$, where $I_{C'}$ is the intensity transmitted by the orthogonal analyser C′. The resultant intensity can in this manner be shown to be given by

$$I = \Sigma I_i + \sum_{i \neq j} \sqrt{I_i I_j} \cos \tfrac{1}{2}c_{ij} \cos \delta_{ij} \qquad (14)$$

It is obvious that the same argument holds when there are a mixture of *n* coherent beams; the formula (13) gives the intensity transmitted by an analyser C, while (14) gives the resultant intensity. The state of polarisation of the resultant beam will be deduced incidentally in Part II. We may here merely note that by deducing the intensity transmitted by any analyser C we have a method of deducing the resultant polarisation of the beam: for example, we could find the particular analyser for which the transmitted intensity is zero.

In conclusion the author wishes to express his deep sense of indebtedness to Professor Sir C. V. Raman, F.R.S., N.L., without whose encouragement this work could not have been written.

§ 10. Summary

The superposition of two coherent beams in different states of elliptic polarisation is discussed in a general manner. If A and B represent the states of polarisation of the given beams on the Poincaré sphere, and C that of the resultant beam, the result is simply expressed in terms of the sides, *a*, *b*, *c* of the spherical triangle ABC. The intensity I of the resultant beam is given by:

$$I = I_1 + I_2 + 2\sqrt{I_1 I_2} \cos \tfrac{1}{2}c \cos \delta;$$

the extent of mutual interference thus varies from a maximum for identically polarised beams ($c = 0$), to zero for oppositely polarised beams ($c = \pi$). The state of polarisation C of the resultant beam is located by $\sin^2 \frac{1}{2}a = (I_1/I) \sin^2 \frac{1}{2}c$ and $\sin^2 \frac{1}{2}b = (I_2/I) \sin^2 \frac{1}{2}c$. The 'phase difference' δ is equal to the supplement of half the area of the triangle C′BA (where C′ is the point diametrically opposite to C). These results also apply to the converse problem of the decomposition of a polarised beam into two others.

The interference of two coherent beams after resolution into the same state of elliptic polarisation by an elliptic analyser or compensator is discussed; as also the interference (direct, *and* after resolution by an analyser) of *n* coherent pencils in different states of polarisation.

REFERENCES

Chandrasekhar, S. .. *Radiative Transfer*, Oxford, 1949.

Fano, U. .. *J. Opt. Soc. Am.*, 1949, **39**, 859.

Jerrard, H. G. .. *Ibid.*, 1954, **44**, 634.

Mascart, E. .. *Traite d'Optique*, Paris, 1891, **2**.

M'Clelland, W. J. and Preston, T. *A Treatise on Spherical Trigonometry with Applications to Spherical Geometry*, 1897, Parts I and II, Macmillan.

Pancharatnam, S. .. *Proc. Ind. Acad. Sci.*, 1955, **42 A**, 235.

Perrin, F. .. *J. Chem. Phys.*, 1942, **10**, 415.

Pockels, F. .. *Lehrbuch der Kristalloptik*, *Teubner*, 1906.

Ramachandran, G. N. and Ramaseshan, S. .. *J. Opt. Soc. Am.*, 1952, **42**, 49.

Ramachandran, G. N. .. *J. Madras Univ.*, 1952, **22 B**, 277.

Rayleigh .. *Scientific Papers*, Cambridge, 1902, **3**, 140.

Stokes, G. G. .. *Mathematical and Physical Papers*, Cambridge, 1901, **3**, 233.

Walker, J. .. *Analytical Theory of Light*, Cambridge, 1904.

JOURNAL OF MODERN OPTICS, 1987, VOL. 34, NO. 11, 1401–1407

The adiabatic phase and Pancharatnam's phase for polarized light

M. V. BERRY
H. H. Wills Physics Laboratory,
Tyndall Avenue, Bristol BS8 1TL, England

(*Received 17 August 1987*)

Abstract. In 1955 Pancharatnam showed that a cyclic change in the state of polarization of light is accompanied by a phase shift determined by the geometry of the cycle as represented on the Poincaré sphere. The phase owes its existence to the non-transitivity of Pancharatnam's connection between different states of polarization. Using the algebra of spinors and 2×2 Hermitian matrices, the precise relation is established between Pancharatnam's phase and the recently discovered phase change for slowly cycled quantum systems. The polarization phase is an optical analogue of the Aharonov–Bohm effect. For slow changes of polarization, the connection leading to the phase is derived from Maxwell's equations for a twisted dielectric. Pancharatnam's phase is contrasted with the phase change of circularly polarized light whose direction is cycled (e.g. when guided in a coiled optical fibre).

1. Introduction

In a remarkable paper published thirty years ago, Pancharatnam ([1], reprinted in [2]) considered the phase of a beam of light whose state of polarization is made to change. His central result, when expressed symmetrically, concerned a beam that is returned to its original state of polarization via two intermediate polarizations. He showed that the phase does not return to its original value but increases by $-\Omega/2$ where Ω is the area (solid angle) spanned on the Poincaré sphere [3, 4] by the geodesic triangle whose vertices are the three polarizations. Ramaseshan and Nityananda [5] point out that Pancharatnam's study of classical light anticipated a recent general extension [6] of the adiabatic theorem of quantum mechanics, giving the phase change in the state of a system whose environmental parameters are taken slowly round a cycle. My intention here is to bring out the full originality of Pancharatnam's contribution by expressing his optics in quantum-mechanical language and clarifying the relation between his phase and the adiabatic phase.

First, it is necessary to recall the description of polarization in terms of the Poincaré sphere. For a monochromatic wave travelling in the direction z the electric displacement vector lies in the xy plane, and its state of polarization is described by the complex unit vector

$$\mathbf{d}=(d_x, d_y), \quad (\mathbf{d}^*\cdot\mathbf{d}=1). \tag{1}$$

This is equivalent to a two-component spinor

$$|\psi\rangle=\begin{bmatrix}\psi_+\\ \psi_-\end{bmatrix}, \quad \psi_\pm\equiv(d_x\pm \mathrm{i}d_y)/\sqrt{2}. \tag{2}$$

Thus $\langle\psi|\psi\rangle = 1$. Each such $|\psi\rangle$ is the eigenvector with eigenvalue $+1/2$ of some 2×2 Hermitian matrix (the polarization matrix) of the form

$$\mathbf{H}(\mathbf{r}) = \mathbf{r}\cdot\boldsymbol{\sigma} = \frac{1}{2}\begin{bmatrix} z & x-\mathrm{i}y \\ x+\mathrm{i}y & -z \end{bmatrix} = \frac{1}{2}\begin{bmatrix} \cos\theta & \sin\theta\exp(-\mathrm{i}\phi) \\ \sin\theta\exp(\mathrm{i}\phi) & -\cos\theta \end{bmatrix} \tag{3}$$

where $\boldsymbol{\sigma}$ is the vector of Pauli spin matrices and $\mathbf{r} = (x, y, z)$ is a unit vector with polar angles θ and ϕ. ($\mathbf{H}$ is the matrix representing the operator $|\psi\rangle\langle\psi| - 1/2$.)

The Poincaré sphere is the sphere with coordinates θ, ϕ. Each position $\mathbf{r}_A$ defines a matrix $\mathbf{H}$, whose eigenvector is the polarization $|\psi(\mathbf{r}_A)\rangle$ which will henceforth be written simply as $|A\rangle$. Thus

$$\mathbf{H}(\mathbf{r}_A)|A\rangle = +\tfrac{1}{2}|A\rangle. \tag{4}$$

The poles $\theta = 0$ and $\theta = \pi$ have $\psi_- = 0$ and $\psi_+ = 0$ respectively, and represent the two senses of circular polarization. Points on the equator $\theta = \pi/2$ have $|\psi_+| = |\psi_-|$ and represent the different direction of linear polarization.

2. Pancharatnam's connection

The phase difference between two distinct polarizations $|A\rangle$ and $|B\rangle$ is not determined by the eigenequation (4). The first of Pancharatnam's two main contributions was to point out that a natural convention is to measure the phase difference from a reference condition in which interference of the superposed beams $|A\rangle$ and $|B\rangle$ gives maximum intensity. This intensity is

$$(\langle A| + \langle B|)(|A\rangle + |B\rangle) = 2 + 2|\langle A|B\rangle| \cos(\mathrm{ph}\,\langle A|B\rangle) \tag{5}$$

Thus the phase difference of $|A\rangle$ and $|B\rangle$ is simply the phase, $\mathrm{ph}\,\langle A|B\rangle$, of their scalar product. The reference condition, in which $|A\rangle$ and $|B\rangle$ are defined as being in phase, corresponds to $\langle A|B\rangle$ real and positive, and will be called Pancharatnam's connection. It defines a relation between any two (non-orthogonal) states, not just neighbouring ones, and so resembles the geometric notion of distant parallelism.

Pancharatnam's second contribution was to show that his connection is non-transitive: if $|B\rangle$ is in phase with $|A\rangle$, and $|C\rangle$ with $|B\rangle$, then $|C\rangle$ need not be in phase with $|A\rangle$. Indeed if $|C\rangle$ is in phase with a state $|A'\rangle$ corresponding to the point $\mathbf{r}_A$ on the sphere, then

$$\langle A|A'\rangle = \exp(-\mathrm{i}\Omega_{ABC}/2), \tag{6}$$

where Ω_{ABC} is the solid angle of the geodesic triangle ABC on the sphere.

To relate Pancharatnam's connection and its consequence (6) to the quantum adiabatic phase, consider a two-state quantum system (e.g. a spin-1/2 particle) whose evolution is governed by the Hamiltonian operator (3), in which $\mathbf{r}$ is driven slowly round a circuit C on the sphere. Adiabatic theory based on the Schrödinger equation then shows [6] that if the system is initially in the eigenstate $|A\rangle$ with (say) the greater energy, it will remain in that eigenstate $|\psi[\mathbf{r}(t)]\rangle$ with a phase given (after subtraction of a trivial dynamical contribution) by the connection

$$\langle\psi|\partial|\psi\rangle/\partial t = 0. \tag{7}$$

This connection is mathematically natural [7–9]. For the two-state system it follows from (7) that at the end of the circuit the state $|A'\rangle$ is phase-shifted relative to the initial one, the relation being

$$\langle A|A'\rangle = \exp[-\mathrm{i}\Omega(C)/2] \tag{8}$$

where $\Omega(C)$ is the solid angle of the circuit C.

The close similarity between (6) and (8) suggests that Pancharatnam's connection is equivalent to the adiabatic connection (7). Thus if $|\psi\rangle$ is continued with the aid of (7) along the shorter geodesic arc from A to B, that is

$$|\psi(0)\rangle = |\mathrm{A}\rangle, \quad |\psi(T)\rangle = |\mathrm{B}\rangle \tag{9}$$

then $\langle \mathrm{A}|\mathrm{B}\rangle$ must be real and positive.

To show this equivalence, we introduce the unitary operator taking $|\mathrm{A}\rangle$ and $|\mathrm{B}\rangle$:

$$|\mathrm{B}\rangle = U|\mathrm{A}\rangle \tag{10}$$

For the connection (7) the operator is [10]

$$U = \mathrm{T}\exp\left[-\mathrm{i}\int_0^T \mathrm{d}t'\boldsymbol{\omega}(t')\cdot\boldsymbol{\sigma}\right] \tag{11}$$

where T denotes time-ordering and $\boldsymbol{\omega}(t)$ is the instantaneous angular velocity of a frame being parallel-transported on the sphere along the path joining A to B. For a geodesic arc,

$$\boldsymbol{\omega}(t) = \omega(t)\mathbf{n}, \tag{12}$$

where $\mathbf{n}$ is the constant unit vector perpendicular to the plane of the arc (figure 1). If the angle between A and B is $\int_0^T \mathrm{d}t\,\omega(t) = \phi_{\mathrm{AB}}$, (11) becomes

$$\begin{aligned} U &= \exp(-\mathrm{i}\phi_{\mathrm{AB}}\mathbf{n}\cdot\boldsymbol{\sigma}) \\ &= \cos(\phi_{\mathrm{AB}}/2) - \mathrm{i}\sin(\phi_{\mathrm{AB}}/2)\mathbf{n}\cdot\boldsymbol{\sigma}. \end{aligned} \tag{13}$$

Thus (10) gives

$$\langle \mathrm{A}|\mathrm{B}\rangle = \cos(\phi_{\mathrm{AB}}/2) - \mathrm{i}\sin(\phi_{\mathrm{AB}}/2)\langle \mathrm{A}|\mathbf{n}\cdot\boldsymbol{\sigma}|\mathrm{A}\rangle. \tag{14}$$

The second term vanishes because it involves the expectation value of the spin component along a direction ($\mathbf{n}$) perpendicular to that for which $|\mathrm{A}\rangle$ is an eigenstate (i.e. $\mathbf{n}\cdot\mathbf{r}_{\mathrm{A}} = 0$). Thus $\langle \mathrm{A}|\mathrm{B}\rangle = \cos(\phi_{\mathrm{AB}}/2)$, which is indeed real and positive if $|\phi_{\mathrm{AB}}| < \pi$, that is for the shorter of the geodesic arcs connecting A and B.

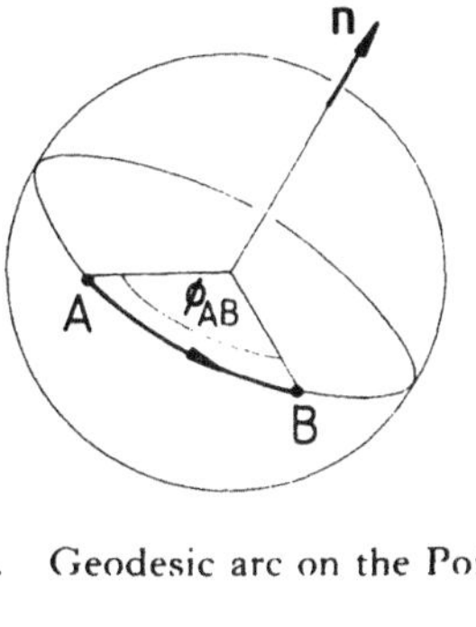

Figure 1. Geodesic arc on the Poincaré sphere.

3. Aharonov–Bohm effect on the Poincaré sphere

The solid-angle formulae (6) and (8) depend only on the connection (7), independently of its origin in adiabatic theory. Therefore the polarization changes need not be slow; they can be accomplished suddenly, for example with birefringent crystal analysers. This is how Pancharatnam [2] confirmed his theory experimentally; the phase shifts were detected by interference.

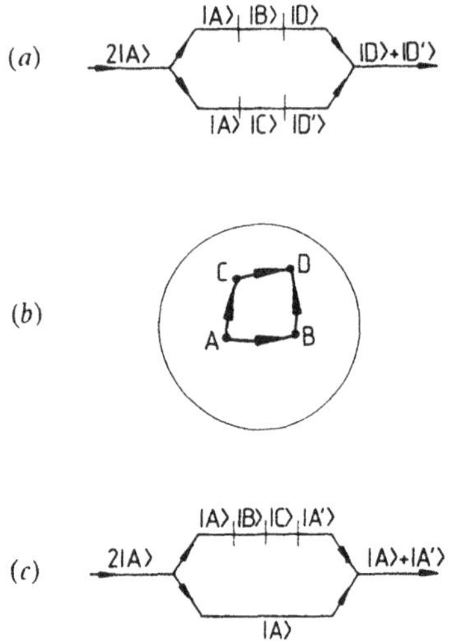

Figure 2. Aharonov–Bohm effect with polarized light: (*a*) different histories of two beams; (*b*) equivalent circuit on Poincaré sphere; and (*c*) alternative scheme, directly implementing equation (6).

It follows from (6) that if a beam $|A\rangle$ is divided into two beams (figure 2 (*a*)) which pass through different intermediate states $|B\rangle$ and $|C\rangle$ and are then analysed into a polarization D, their states $|D\rangle$ and $|D'\rangle$ will differ in phase by half the solid angle of the geodesic polygon ABCDA on the sphere (figure 2 (*b*)). (The polarizers are here assumed to produce the changes of state according to Pancharatnam's connection. In practice there will be extra phase shifts of refractive origin, but these are easy to compensate.)

This result is precisely analogous to the phase shift later predicted by Aharonov and Bohm [11, 12], according to which two electron beams develop a phase shift proportional to the magnetic flux they enclose. For polarized light the analogue of magnetic flux is the solid angle of the polygon on the Poincaré sphere. This is also a flux, namely that of an abstract monopole [6] of strength $-1/2$, situated at the centre of the sphere.

There is a simpler geometry, more directly implementing equation (6), in which one beam continues in the state $|A\rangle$ while the other returns to the same polarization A in state $|A'\rangle$ via intermediate states $|B\rangle$ and $|C\rangle$ (figure 2 (*c*)); interference between $|A\rangle$ and $|A'\rangle$ reveals the phase difference $-\Omega_{ABC}/2$. A special case, at the same time both trivial and instructive, is when A, B and C lie equally spaced on the equator of the sphere and so correspond to linear polarizations successively rotated by $\pi/3$. The geodesic triangle ABC is a hemisphere, with solid angle 2π, so that the phase shift between $|A\rangle$ and $|A'\rangle$ is π, as it must be because the effect of the three rotations is to reverse the field (1). This example resolves the paradox of polarization being governed by the same algebra as quantum states representing spin-1/2 particles, which change sign in a 2π rotation, even though photons are quantum particles with spin 1, which do not. The explanation is that a complete rotation of the spin-1/2 matrix (3) (here accomplished by increasing ϕ by 2π, with $\theta=\pi/2$) corresponds to only a half turn of the photon, that is, to a reversal of the field.

A related result of Pancharatnam [1] which can alternatively be derived and interpreted with spinors concerns the decomposition of a beam with polarization C into two beams with polarizations A and B: the relative phase Δ_{AB} of the two beams is $\pi-\Omega_{C'BA}/2$, where C′ is the antipode of C. To derive this we write the spinor decomposition as

$$|C\rangle = a|A\rangle + b|B\rangle, \tag{15}$$

and choose $|A\rangle$ and $|B\rangle$ to be in phase so that Δ_{AB} is the phase of b/a. Now the antipodal state $|C'\rangle$ is orthogonal to $|C\rangle$, so that

$$b/a = -\langle C'|A\rangle / \langle C'|B\rangle. \tag{16}$$

We can choose $|C'\rangle$ to be in phase with $|A\rangle$, so that $\langle C'|A\rangle$ is real and positive. Then

$$\Delta_{AB} = \pi - \lambda \tag{17}$$

where λ is the phase of $\langle C'|B\rangle$. Now consider the triangle BAC′B, starting in state $|B\rangle$ and ending in state $|B'\rangle$ which is in phase with $|C'\rangle$. From (6), $\langle B|B'\rangle = \exp(-i\Omega_{C'BA}/2)$, so that

$$\lambda = +\Omega_{C'BA}/2 \tag{18}$$

which with (17) is the required result.

4. Adiabatic changes of polarization

I have suggested elsewhere [13] that changes in the state of polarization can be accomplished continuously, rather than discretely, with a transparent birefringent gyrotropic medium twisted along the beam so that its dielectric tensor varies through a cycle. If the dielectric variation is slow enough, the beam remains in a polarization state determined by the local uniform medium: the displacement unit vector $\mathbf{d}(z)$ is one of the two eigenvectors of the 2×2 Hermitian submatrix $\mathbf{M}(z)$ consisting of the components transverse to z of the inverse dielectric tensor [14].

As any 2×2 Hermitian matrix can be expressed in the form (3) plus a multiple of the unit matrix, a cyclic change in $\mathbf{M}(z)$ corresponds to a circuit on the Poincaré sphere. The circuit will be accompanied by a phase factor of the form (8), provided the local eigenstates are continued by means of the analogue of (7), namely

$$\mathbf{d}^*(z)\cdot\mathbf{d}'(z) = 0, \tag{19}$$

where the prime denotes differentiation with respect to z. This connection cannot be justified by invoking the quantum adiabatic result, because this originates in the time-dependent Schrödinger equation, whereas these fields are governed by Maxwell's equations for time-independent (i.e. harmonic) waves. In this section I will outline the derivation of (19).

For waves with frequency w, Maxwell's equations reduce to the following equation for the electric displacement $\mathbf{D}(z)\exp(-iwt)$:

$$(\mathbf{M}\cdot\mathbf{D})'' + k^2\mathbf{D} = 0, \tag{20}$$

where $k = w/c$. For slowly varying $\mathbf{M}(z)$ each of the two forward-propagating adiabatic solutions has the form

$$\mathbf{D}(z) \approx a(z)\mathbf{d}(z)\exp\left[ik\int_0^z dz'\, n(z')\right] \tag{21}$$

where a is a real amplitude and n is the refractive index corresponding to one of the normalized eigenpolarizations $\mathbf{d}$, satisfying

$$\mathbf{M}\cdot\mathbf{d}=n^{-2}\mathbf{d}. \tag{22}$$

The connection satisfied by $\mathbf{d}$, and the form of $a(z)$, are determined by (20) in the adiabatic limit of large k (relative to the rate at which $\mathbf{M}$ varies). The results can however be obtained more simply from the vanishing divergence of the Poynting vector for monochromatic waves. In the present case this implies the constancy of

$$P=\mathrm{Im}\,\mathbf{D}^*\cdot\mathbf{M}\cdot(\mathbf{M}\cdot\mathbf{D})', \tag{23}$$

which can also be derived directly from (20) using the Hermiticity of $\mathbf{M}$. Substitution of the *ansatz* (21) and use of (22) gives

$$P=ka^2/n^3+(a^2/n^4)\,\mathrm{Im}\,\mathbf{d}^*\cdot\mathbf{d}'. \tag{24}$$

Constancy of the term involving k gives

$$a(z)=\text{constant}\times[n(z)]^{3/2}. \tag{25}$$

Constancy of the term independent of k now implies that $\mathrm{Im}\,\mathbf{d}^*\cdot\mathbf{d}'$ is proportional to $n(z)$, and the constant of proportionality can be identified as zero by considering stretches of the beam along which $\mathbf{M}$ is constant, for which (21) is exact with $\mathbf{d}$ (and a) constant. Thus $\mathrm{Im}\,\mathbf{d}^*\cdot\mathbf{d}'=0$, and we deduce (19) from $\mathrm{Re}\,\mathbf{d}^*\cdot\mathbf{d}'=0$ which follows from normalization.

The exact solution of (20) can be written as the sum of four terms of the form (21) with complex amplitudes $a(z)$, representing forward and backward waves with each of the two local eigenpolarizations. It is not hard to show that transitions between these waves (that is, violations of adiabaticity) are exponentially weak if $\mathbf{M}(z)$ varies analytically, the strongest being between the two waves travelling in the same direction and having amplitudes of order $\exp(-kL\Delta n)$, where Δn is the refractive-index difference between the two polarizations and L the length scale over which $\mathbf{M}$ varies.

5. Concluding remarks

Pancharatnam's phase (6) should not be confused with another optical phase given by a solid angle, namely that predicted [15] and found ([16], see also [17, 18]) in circularly polarized light propagating along a twisting path (for example by being guided in a coiled optical fibre).

In (6) the solid angle refers to a path on the Poincaré sphere representing a two-state system. The photons can be regarded as having only two states, in spite of being particles with spin 1, because the light is always travelling in the same direction and for plane waves transversality permits only helicity states, in which the spin is parallel or antiparallel to the propagation direction $\mathbf{k}$.

In the coiled light experiments, on the other hand, the direction of the light is made to change, and the solid angle refers to a path on the sphere of directions $\mathbf{k}$. Such a path represents the rotation of a full three-component spinor reflecting the unit (rather than 1/2) spin of the photon, because only in a moving frame whose z axis coincides with $\mathbf{k}$ does the zero-spin (longitudinal) spinor component vanish. In a fixed frame the spinor representing transverse light in direction **k** with polar angles θ and ϕ has the general form

$$|\psi\rangle=\frac{1}{2}\begin{bmatrix}[\alpha+\beta+(\alpha-\beta)\cos\theta]\exp(-i\phi)\\ \sqrt{2}(\alpha-\beta)\sin\theta\\ [\alpha+\beta-(\alpha-\beta)\cos\theta]\exp(i\phi)\end{bmatrix}, \qquad (26)$$

where $|\alpha|^2+|\beta|^2=1$; this involves all three components. Moreover, it is only in the adiabatic limit of slow coiling that the light is exactly transverse: for any finite rate of rotation of **k** there are non-zero longitudinal components, associated with weak non-adiabatic transitions between the two polarizations [17].

The difference between the Poincaré sphere (spin 1/2) and the sphere of directions (spin 1) is reflected in the different dependence of the phase shift on the solid angle Ω. For a polarization cycle the phase change is $-\Omega/2$, whereas for a direction cycle the phases for the two circular polarizations are $\pm\Omega$.

Acknowledgment

I thank Professor S. Ramaseshan and Professor R. Nityananda for telling me about Pancharatnam's work.

References

[1] Pancharatnam, S., 1956, *Proc. Ind. Acad. Sci.* A, **44,** 247–262.
[2] Pancharatnam, S., 1975, *Collected Works of S. Pancharatnam* (Oxford: University Press).
[3] Born, M., and Wolf, E., 1959, *Principles of Optics* (London: Pergamon).
[4] Clarke, D., and Grainger, J. F., 1971, *Polarized Light and Optical Measurement* (Oxford: Pergamon).
[5] Ramaseshan, S., and Nityananda, R., 1986, *Current Science, India*, **55,** 1225–1226.
[6] Berry, M. V., 1984, *Proc. R. Soc., Lond.* A, **392,** 45–57.
[7] Simon, B., 1983, *Phys. Rev. Lett.*, **51,** 2167-2170.
[8] Aharonov, Y., and Anandan, J., 1987, *Phys. Rev. Lett.*, **58,** 1593–1596.
[9] Segert, J., 1987, *Phys. Rev.* A, **36,** 10–16.
[10] Berry, M. V., 1987, *Proc. R. Soc., Lond,* (to be published).
[11] Aharonov, Y., and Bohm, D., 1959, *Phys. Rev.*, **115,** 485–491.
[12] Olariu, S., and Popescu, I., 1985, *Rev. mod. Phys.*, **57,** 339–436.
[13] Berry, M. V., 1986, *Fundamental Aspects of Quantum Theory, NATO ASI series Vol. 144,* edited by V. Gorini and A. Frigerio (New York: Plenum), pp. 267–278.
[14] Landau, L. D., Lifshitz, E. M., and Pitaevskii, L. P., 1984, *Electrodynamics of Continuous Media* (second edition) (Pergamon: Oxford).
[15] Chiao, R. Y., and Wu, Y. S., 1986, *Phys. Rev. Lett.*, **57,** 933–936.
[16] Tomita, A., and Chiao, R. Y., 1986, *Phys. Rev. Lett.*, **57,** 937–940.
[17] Berry, M. V., 1987, *Nature, Lond.*, **326,** 277–278.
[18] Kitano, M., Yabuzaki, T., and Ogawa, T., 1987, *Phys. Rev. Lett.*, **58,** 523.

Intersection of Potential Energy Surfaces in Polyatomic Molecules

By G. Herzberg * and H. C. Longuet-Higgins †

Received 28th January, 1963

It is shown that in polyatomic systems the conical intersections described by Teller [4] will occur not only where demanded by symmetry but also in certain non-symmetrical systems. Other kinds of intersection are also described, and it is suggested that " near-intersections " are likely to be as important in polyatomic as in diatomic systems.

1. Introduction

In diatomic molecules the potential energy curves of two states will only intersect if the states differ in symmetry or in some other essential characteristic.[1-3] However, an analogous statement is *not* true of polyatomic systems: [4] two potential energy surfaces of a polyatomic molecule can in principle intersect even if they belong to states of the same symmetry and spin multiplicity. This sentence leaves open the question whether such intersections ever occur in polyatomic systems. We have therefore tried to find some examples of intersections between states of the same species, and this paper presents some miscellaneous results.

2. General considerations

We first outline Teller's analysis of the case in which one may neglect the spin terms in the electronic Hamiltonian, so that the electronic wave function may always be taken in real form. We imagine, following von Neumann and Wigner,[3] that all but two of the solutions of the electronic wave equation have been found, and that φ_1 and φ_2 are any two functions which, together with the found solutions, constitute a complete orthonormal set. Then it must be possible to express each of the two remaining electronic wave functions in the form

$$\psi = c_1\varphi_1 + c_2\varphi_2, \tag{2.1}$$

where, in an obvious notation,

$$\begin{bmatrix} H_{11}-E, & H_{12} \\ H_{21}, & H_{22}-E \end{bmatrix}\begin{bmatrix} c_1 \\ c_2 \end{bmatrix} = 0, \tag{2.2}$$

all quantities in this equation being real.

In order that (2.2) shall have degenerate solutions it is necessary to satisfy two independent conditions, namely,

$$H_{11} = H_{22},\ H_{12}(= H_{21}) = 0, \tag{2.3}$$

and this requires the existence of at least two independently variable nuclear co-ordinates. In a diatomic molecule there is only one variable co-ordinate—the interatomic distance—so the non-crossing rule follows; but in a system of three or more atoms there are enough degrees of freedom for the rule to break down.

* National Research Council, Ottawa, Canada.
† University Chemical Laboratory, Cambridge.

Following Teller we denote the two independent co-ordinates by x and y, and take the origin at the point where $H_{11} = H_{22}$ and $H_{12} = 0$. The secular equations may then be cast in the form:

$$\begin{bmatrix} W+h_1x-E, & ly \\ ly, & W+h_2x-E \end{bmatrix}\begin{bmatrix} c_1 \\ c_2 \end{bmatrix} = 0 \tag{2.4}$$

or

$$\begin{bmatrix} W+(m+k)x-E, & ly \\ ly, & W+(m-k)x-E \end{bmatrix}\begin{bmatrix} c_1 \\ c_2 \end{bmatrix} = 0, \tag{2.5}$$

where $m = \frac{1}{2}(h_1+h_2)$, $k = \frac{1}{2}(h_1-h_2)$. The eigenvalues are

$$E = W+mx \pm \sqrt{(k^2x^2+l^2y^2)}, \tag{2.6}$$

and this is the equation of a double cone with vertex at the origin.

This result was obtained by Teller,[4] but he did not draw attention to the following property of the wave function near the origin. Define an angle θ by the equations

$$kx = R\cos\theta,\ ly = R\sin\theta, \tag{2.7}$$

where

$$R = \sqrt{(k^2x^2+l^2y^2)} > 0. \tag{2.8}$$

Taking the lower root of (2.5), namely,

$$E = W+mx-R, \tag{2.9}$$

we deduce that on the lower sheet of the energy surface the coefficients c_1 and c_2 satisfy

$$\begin{bmatrix} R+R\cos\theta, & R\sin\theta \\ R\sin\theta, & R+R\cos\theta \end{bmatrix}\begin{bmatrix} c_1 \\ c_2 \end{bmatrix} = 0. \tag{2.10}$$

It follows that

$$\frac{c_1}{c_2} = \frac{-\sin\theta}{1+\cos\theta} = \frac{\cos\theta-1}{\sin\theta} = -\tan\tfrac{1}{2}\theta. \tag{2.11}$$

Hence, if ψ is to be real, like φ_1 and φ_2, we must have

$$c_1 = \sin\tfrac{1}{2}\theta, \qquad c_2 = -\cos\tfrac{1}{2}\theta, \tag{2.12}$$

or

$$c_1 = -\sin\tfrac{1}{2}\theta, \qquad c_2 = \cos\tfrac{1}{2}\theta. \tag{2.13}$$

In either case, as we move round the origin keeping R constant and allowing θ to increase from 0 to 2π, both c_1 and c_2 change sign, and so does ψ. This result is a generalization of one which has been proved [5] in connection with the Jahn-Teller effect, [6] where one also encounters a conically self-intersecting potential surface. It shows that a conically self-intersecting potential surface has a different topological character from a pair of distinct surfaces which happen to meet at a point. Indeed, if an electronic wave function changes sign when we move round a closed loop in configuration space, we can conclude that somewhere inside the loop there must be a singular point at which the wave function is degenerate; in other words, there must be a genuine conical intersection, leading to an upper or lower sheet of the surface, as the case may be.

3. THREE HYDROGEN-LIKE ATOMS

A useful illustration of the above generalizations is a system of three hydrogen atoms near the vertices of an equilateral triangle. If the internuclear distances are a, b and c, a convenient set of internal co-ordinates are

$$x = (b+c-2a)/\sqrt{6},$$
$$y = (b-c)/\sqrt{2}$$
and
$$z = (a+b+c)/\sqrt{3}. \tag{3.1}$$

According to both the valence-bond theory and the molecular orbital theory, the ground state is of species $^2E'$ in the D_{3h} configuration. It therefore exhibits the Jahn-Teller effect,[6] and for given z the surface $E(x, y)$ is a double cone with vertex at the origin. It may be argued that this example does not really contravene the non-crossing rule, because the degeneracy arises from symmetry; but we can use it for constructing a non-trivial example, in the following way.

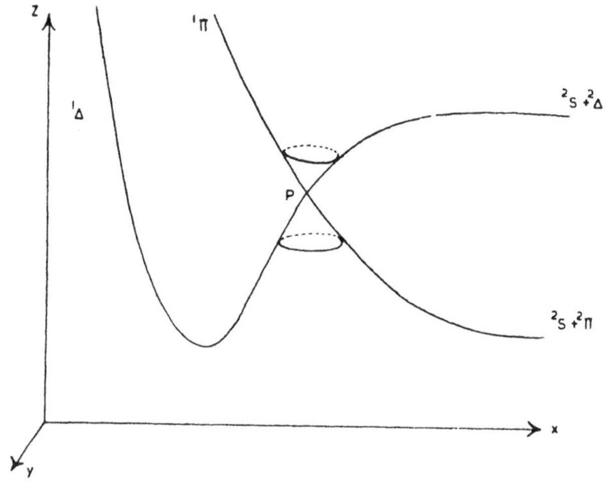

FIG. 1.

Let φ_A be the valence-bond wave function for a situation in which the electron on A has spin " up " and the electrons on B and C are spin-paired; let φ_B and φ_C be similarly defined, so that

$$\varphi_A + \varphi_B + \varphi_C = 0. \tag{3.2}$$

(This is a standard result of the simple valence-bond theory.) We now take the system round a continuous loop in (x, y) space, starting with B close to C and A far away (see fig. 1). From the arguments of § 2 the wave function must have changed sign on completion of the loop; in fact it evolves as shown in fig. 1:

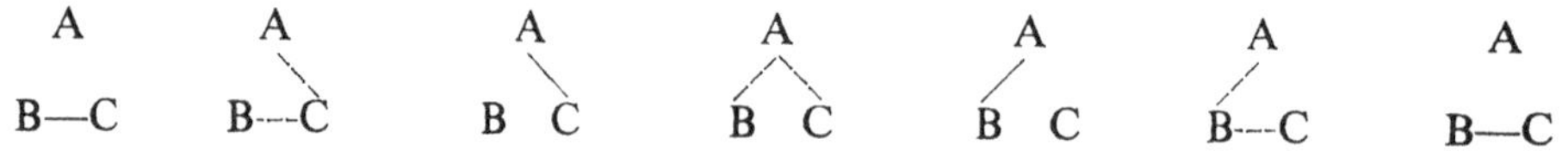

$$\varphi_A \qquad \frac{\varphi_A - \varphi_B}{\sqrt{3}} \qquad -\varphi_B \qquad \frac{-\varphi_B + \varphi_C}{\sqrt{3}} \qquad \varphi_C \qquad \frac{\varphi_C - \varphi_A}{\sqrt{3}} \qquad -\varphi_A$$

We have now confirmed the change of sign required by the general arguments of § 2. However, the valence-bond theory requires the wave function to evolve in this way even when the atoms A, B and C are *not* chemically identical; they might, for example, be atoms of Li, Na and K respectively. The conical intersection implied by the London-Eyring formula [7, 8]

$$E = Q \pm \sqrt{\left[\tfrac{1}{2}(J_{AB}-J_{BC})^2+\tfrac{1}{2}(J_{BC}-J_{CA})^2+\tfrac{1}{2}(J_{CA}-J_{AB})^2\right]} \tag{3.3}$$

must therefore be a real one unless, which is most unlikely, the theory is qualitatively in error as to the way in which the sign of the wave function is affected by taking the system round the cycle depicted in fig. 1.

We conclude that in a triangular system of three hydrogen-like atoms the lowest doublet state is linked with an excited doublet by a conical intersection even when all three atoms are dissimilar. Both states are symmetric with respect to the molecular plane (species $^2A'$), so in this case at least symmetry cannot be held responsible for the existence of the intersection!

4. OTHER TRIATOMIC SYSTEMS

In linear molecules all the electronic states are orbitally degenerate except the Σ states, but this degeneracy is removed when the molecule is bent.[9] For example, a Π state of a linear molecule splits into one state which is symmetric and another which is antisymmetric about the plane of the bent molecule. It is therefore not unusual to find that the ground state of a bent molecule is adiabatically correlated with an excited state of different symmetry, obtained by straightening the molecule out and bending it again in a different plane. We might speak of a " glancing intersection " between the potential functions of the two states.

It is also possible, however, for an A' (or A'') state of the bent molecule to be adiabatically linked to another A' (or A'') state, through a conical intersection in the linear configuration. An example is provided by HNO, which we now consider briefly.

Bancroft, Hollas and Ramsay [10] have shown that the ground state of HNO is of species $^1A'$, symmetric with respect to the molecular plane. When the molecule is straightened out, this state will pass adiabatically into the lowest singlet state of the linear molecule; analogy with the isoelectronic molecule O_2 strongly suggests that the lowest singlet state of linear HNO is a $^1\Delta$ state. If the H atom is now pulled away along the NO axis, the adiabatic correlation rules imply that the products are a ^2S hydrogen atom and a $^2\Delta$ NO molecule. But the ground state of NO is $^2\Pi$ not $^2\Delta$, so the linear dissociation curve must somewhere cross the $^1\Pi$ curve for $H(^2S)+NO(^2\Pi)$, at a point P, say.

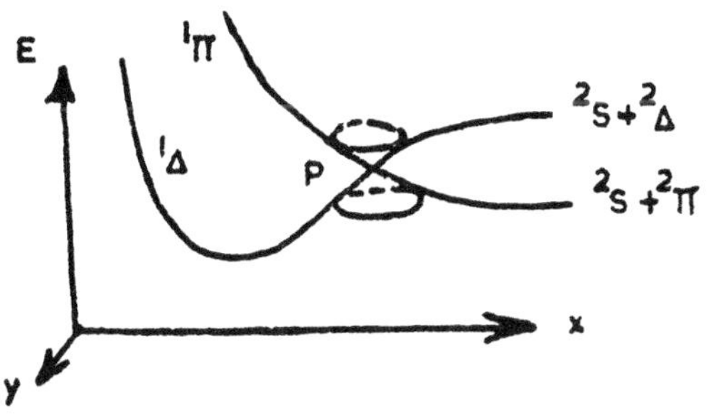

FIG. 2.

We now consider what happens when we take the HNO molecule at the point P and bend it out of a straight line. The resulting perturbation is necessarily symmetric with respect to the plane of bending; it can mix together the A' components of the $^1\Pi$ and $^1\Delta$ states, or their A'' components, but cannot mix the A' component of one with the A'' component of the other. Restricting attention to the A' components, and denoting the bending co-ordinate by y, we arrive at a situation of exactly the type described in § 2; the states φ_1 and φ_2 are the A' components of the $^1\Pi$ and $^1\Delta$ states, and x is an in-line co-ordinate measured from the point P. If we extend the potential energy functions for the linear situation by introducing y as a co-ordinate perpendicular to the plane of the paper, we can depict the resulting conical intersection in the manner shown in fig. 2 (which is not intended to be drawn to scale). This diagram shows that there is no difficulty in forming HNO adiabatically from H and NO in their ground states, provided that the H atom does not approach the NO molecule along its axis.

5. NEAR-INTERSECTIONS

So far we have considered only genuine intersections, such as the conical intersections described by Teller or the glancing intersections associated with the Renner effect. But almost equally important are " near-intersections " where two potential surfaces nearly meet and only just avoid crossing because of a weak interaction at the point of closest approach. Near-intersections of this kind are well known in diatomic molecules; a classic example is the near-intersection between the lowest " ionic " and " covalent " states of an alkali halide molecule. In NaCl, for example, this occurs at a rather large interatomic distance, at which the resonance integral for electron transfer between the two atoms is very small. In essence, the smallness of H_{12} is due to the considerable difference in electron distribution between φ_1 and φ_2. We shall now examine briefly a somewhat analogous case in a polyatomic system.

Douglas [11] has recently studied in detail the strong first ultra-violet absorption system of NH_3, whose upper state is [12] a $^1A_2''$ state of the planar molecule (symmetry group D_{3h}). In molecular orbital terms this state results from exciting one of the unshared $a_2''2p$ electrons of planar NH_3 to the Rydberg orbital $3sa_1'$. The absorption system shows strong signs of predissociation, but in the first few members of the main vibrational progression of ND_3 rotational structure is clearly discernible.

Now it seems likely that the predissociation of excited NH_3 is produced by interaction with a state arising from a hydrogen atom and an NH_2 radical in their ground states. We therefore consider the approach of an H atom to an NH_2 radical to form a planar NH_3 molecule. The ground state of NH_2 is $^2B_1$, antisymmetric with respect to the plane of the bent radical. There are two electrons in the non-bonding hybrid orbital on the N atom, and these will impede the approach of the third H atom. The singlet state formed by the combination of $H(^2S)$ and $NH_2(^2B_1)$ should therefore be repulsive, at least in the early stages of formation. But this state has the same multiplicity and symmetry (with respect to the group C_{2v}) as the observed $^1A_2''$ state of NH_3. Why, then, is the interaction between these two states so weak that the latter is only predissociated and not completely disrupted by the former?

The answer, possibly, is that although the two states have the same symmetry and multiplicity, a substantial electronic rearrangement is needed to convert one into the other: the electron on the approaching H atom must be transferred to a Rydberg 3s orbital on the NH_2 radical before the lone pair on the N atom can form a satisfactory bond to the H nucleus. It seems likely, therefore, that the matrix element H_{12} between the observed $^1A_2''$ state of NH_3 and the singlet state arising from $H(^2S)+NH_2(^2B_1)$ is very small, so that the potential surfaces for these two states, while not actually intersecting, approach so closely that all the usual conditions are satisfied for a predissociated absorption spectrum.

6. GENERAL CONCLUSIONS

Our conclusions are somewhat diverse, like the facts of molecular spectroscopy. First, the conical intersections described by Teller [4] will occur not only in situations where symmetry demands them, but in asymmetrical systems such as a set of three dissimilar hydrogen-like atoms. This follows from the fact that conical intersections differ topologically from accidental meetings of potential surfaces, and cannot be " abolished " by making infinitesimal changes in the electronic Hamiltonian.

Secondly, in non-linear triatomic molecules two states of the same species are sometimes connected through a conical intersection occurring in the linear configuration; this kind of situation is likely to occur in many cases of interest.

Lastly, narrowly avoided crossings may well be as important in polyatomic as in diatomic systems; their occurrence has nothing to do with symmetry, but is due to a substantial difference in electronic character between the two states involved.

We are indebted to Dr. A. E. Douglas for kindly allowing us to quote his unpublished results.

1 Hund, *Z. Physik.*, 1927, **40,** 742.
2 Franck and Haber, *Berliner Akademieberichte* XIII, 1931.
3 von Neumann and Wigner, *Physik. Z.*, 1927, **30,** 467.
4 Teller, *J. Physic. Chem.*, 1937, **41,** 109.
5 Longuet-Higgins, Öpik, Pryce and Sack, *Proc. Roy. Soc. A*, 1958, **244,** 1.
6 Jahn and Teller, *Proc. Roy. Soc. A*, 1937, **161,** 220.
7 London, *Z. Elektrochem.*, 1929, **35,** 522.
8 Glasstone, Laidler and Eyring, *The Theory of Rate Processes* (McGraw-Hill, New York, 1941).
9 Renner, *Z. Physik*, 1934, **92,** 172.
10 Bancroft, Hollas and Ramsay, *Can. J. Physics*, 1962, **40,** 377.
11 Douglas, this discussion.
12 Walsh and Worsop, *Trans. Faraday Soc.*, 1961, **57,** 345.

Proc. R. Soc. Lond. A. **351**, 141–150 (1976)
Printed in Great Britain

Spin-orbit coupling and the intersection of potential energy surfaces in polyatomic molecules

By A. J. Stone
University Chemical Laboratory, Lensfield Road, Cambridge

(*Communicated by H. C. Longuet-Higgins, F.R.S. – Received* 23 *April* 1976)

Longuet-Higgins' theorem, which shows that the existence of intersections between potential energy surfaces may be deduced from the behaviour of the wavefunction at points remote from the intersection, is generalized to cover cases where the Hamiltonian is complex. It is concluded that an intersection due to symmetry in one region of nuclear configuration space may imply that the same surfaces intersect over a region of higher dimension and lower symmetry where their wavefunctions belong to the same symmetry species. It is shown that this behaviour occurs in d^1 octahedral complexes in the presence of spin-orbit coupling.

1. Introduction

Longuet-Higgins (1975) showed that if the wavefunction of a given electronic state in the Born–Oppenheimer approximation changes sign when transported adiabatically round a loop in nuclear configuration space, then the state must become degenerate with another at some point within the loop. His proof assumed implicitly, however, that the wavefunction was essentially real everywhere, so that the result does not apply when spin-orbit coupling is taken into account. I seek here to remedy this deficiency.

We begin by reviewing Longuet-Higgins' proof. The electronic wavefunction $\psi(q; Q)$ depends parametrically on the nuclear configuration Q. As Q is varied, the form of $\psi(q; Q)$ will change, and if the state is non-degenerate it will change continuously with Q. Only if $\psi(q; Q)$ is degenerate with another state can it change discontinuously with Q. We assume initially, as a hypothesis to be refuted, that no degeneracies occur, so that the wavefunction is everywhere a continuous function of the Q.

Nevertheless, if the nuclei are taken round a loop in nuclear configuration space, the wavefunction may, when the nuclei are returned to their original configuration Q_0, have suffered a change of phase $Z = e^{i\Omega}$, because if $\psi(q; Q_0)$ is a solution of the electronic wave equation then so is $e^{i\Omega}\,\psi(q; Q_0)$. If the function is real everywhere then Ω must be zero modulo π, and one may divide the closed paths in Q space into those for which $Z = +1$ and those for which $Z = -1$.

Let S be any simply connected surface in nuclear configuration space, bounded by a closed loop L. Then Longuet-Higgins' theorem states that if $\psi(q; Q)$ changes sign when transported adiabatically round L, then there must be at least one point on S at which $\psi(q; Q)$ is discontinuous, implying that its potential energy surface intersects that of another electronic state.

To prove it, let l_1 be any line in S which bisects the area enclosed by L, and let L_1 and M_1 be the two loops so created (figure 1). Then if L is sign-reversing and $\psi(q; Q)$ continuous, it follows that *either* L_1 *or* M_1 must be sign-reversing. Supposing without loss of generality, that L_1 is sign-reversing, then it in turn can be bisected into a sign-reversing loop L_2 and a sign-preserving one M_2; and we can construct an infinite sequence of successively smaller sign-reversing loops, which will converge on some point P. In the immediate neighbourhood of P it is then possible to induce a finite change in $\psi(q; Q)$ by taking it round an arbitrarily small loop in Q space, so that contrary to hypothesis, $\psi(q; Q)$ must be discontinuous at P and hence degenerate with another state.

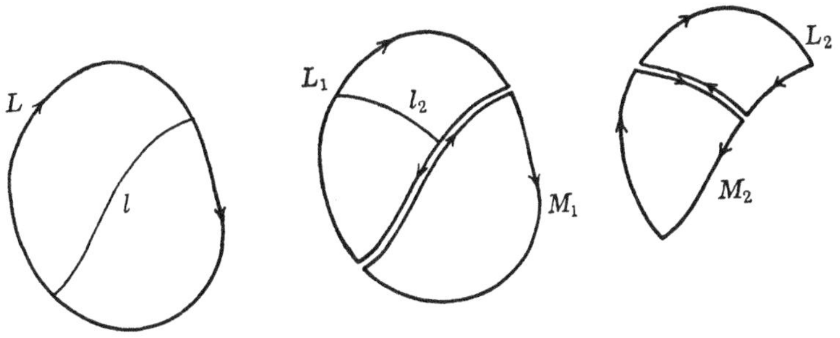

FIGURE 1

Note before we proceed that S may be *any* simply connected surface spanning the loop L. Consequently, for a triatomic, with three internal degrees of freedom, there must be a line of degeneracy, while a polyatomic with s internal degrees of freedom, if it has degeneracy of this type at all, will exhibit degeneracy over a manifold of dimension $s-2$ in nuclear configuration space.

THE EFFECT OF SPIN-ORBIT COUPLING

It is easy to see that if the wavefunction $\psi(q; Q)$ is not essentially real (that is, if it cannot be made real by multiplying by a complex number independent of the q) then the above proof fails. If the wavefunction changes by a phase factor $e^{i\pi}$ when transported round the loop L, then the phase changes round the loops L_1 and M_1 can be $e^{i\Omega_1}$ and $e^{i\Omega_1'}$, respectively, where Ω_1 and Ω_1' can be any numbers satisfying $\Omega_1 + \Omega_1' = \pi$. A sequence of loops $L_1, L_2, \ldots,$ will cause phase changes $e^{i\Omega_1}, e^{i\Omega_2}, \ldots,$ and there is no reason in general why Ω_n should not tend to zero as the size of the loop tends to zero. Consequently the *reductio ad absurdum* fails.

Indeed, the very notion of the phase change round a loop needs more careful definition. If $\psi(q; Q)$ satisfies the electronic wave equation, then so does

$$\exp(i\omega(Q))\,\psi(q; Q)$$

where $\omega(Q)$ is any function of nuclear configuration. Before proceeding further, let us therefore require that the overlap integral between the wavefunctions for two neighbouring configurations along a path shall be real to first order:

$$\mathrm{Im}\,\langle\psi(q;\,Q)|\psi(q;\,Q+\mathrm{d}Q)\rangle = O(\mathrm{d}Q^2). \tag{1}$$

Then the phase change Z_L or the phase angle Ω_L associated with adiabatic transport of the wavefunction round a loop L, during which the wavefunction changes, subject to condition (1), from $\psi(q;\,Q_0)$ to $\psi_L(q;\,Q_0)$, is given by

$$Z_L = \exp(\mathrm{i}\Omega_L) = \langle\psi(q;\,Q_0)|\psi_L(q;\,Q_0)\rangle. \tag{2}$$

That the topological proof as it stands must fail can also be seen if we recall the conventional approach (von Neuman & Wigner 1929; Teller 1937) to the non-crossing rule, where we examine a two-level system using an arbitrary basis in which the Hamiltonian matrix is

$$H = \begin{pmatrix} H_{11} & H_{12} \\ H_{21} & H_{22} \end{pmatrix}. \tag{3}$$

The splitting in energy between the two eigenstates is then

$$\tfrac{1}{2}[(H_{11}-H_{22})^2+4\,|H_{12}|^2]^{\frac{1}{2}}, \tag{4}$$

and if the matrix H is real, there are two conditions for an intersection:

$$\left.\begin{aligned} H_{11}-H_{22} &= 0, \\ H_{12} &= 0. \end{aligned}\right\} \tag{5}$$

However if H may be complex, then there are three:

$$\left.\begin{aligned} H_{11}-H_{22} &= 0, \\ \mathrm{Re}\,(H_{12}) &= 0, \\ \mathrm{Im}\,(H_{12}) &= 0. \end{aligned}\right\} \tag{6}$$

Longuet-Higgins (1975) has shown that notwithstanding recent criticism of this analysis (Naqvi & Byers Brown 1972) the conclusions are correct, and his discussion applies, *mutatis mutandis*, to the complex case.

Thus we expect degeneracies only over a $(s-3)$-dimensional manifold, compared with the $s-2$ of the real case (Nikitin 1974). For a diatomic, with its single degree of freedom, the non-crossing rule applies *a fortiori*; for a triatomic we expect intersections of potential energy surfaces only at isolated points, and the same applies to any three dimensional subspace of the nuclear configuration space of any larger molecule.

With this established let us now examine in detail a model example. We shall make fairly drastic approximations, as this example serves merely to set the scene for the theorem of the next section.

We consider a single electron moving in the field of three identical nuclei arranged roughly in an equilateral triangle in the xy plane. If we take a 2p orbital on each atom, directed towards the centre of the triangle, we can construct two antibonding orbitals, which are degenerate if the triangle is equilateral. If the three atomic orbitals are ϕ_1, ϕ_2, and ϕ_3, the molecular orbitals can then be expressed as $\mathrm{Re}(\exp(\mathrm{i}\alpha)\psi_+)$ and $\mathrm{Im}(\exp(\mathrm{i}\alpha)\psi_+)$, where

$$\psi_+ = \sqrt{\tfrac{1}{3}}\,(\phi_1+\omega\phi_2+\omega^*\phi_3), \tag{7}$$

$\omega = \exp(2\pi\mathrm{i}/3)$ and α is arbitrary.

To deal with the distorted molecule we take an origin such that the angles between the position vectors of the nuclei are always $2\pi/3$, and the lengths of the vectors are

$$\left.\begin{aligned} r_1 &= r_0+\rho\cos\theta,\\ r_2 &= r_0+\rho\cos(\theta-2\pi/3),\\ r_3 &= r_0+\rho\cos(\theta+2\pi/3),\end{aligned}\right\} \tag{8}$$

so that the components of the E' distortion are

$$\left.\begin{aligned} Q_x &= \sqrt{\tfrac{1}{6}}(2r_1-r_2-r_3) = \sqrt{\tfrac{3}{2}}\,\rho\cos\theta,\\ Q_y &= \sqrt{\tfrac{1}{2}}\,(r_2-r_3) = \sqrt{\tfrac{3}{2}}\,\rho\sin\theta.\end{aligned}\right\} \tag{9}$$

Then if the resonance integral is assumed to depend linearly on the distance between the nuclei, the perturbed eigenfunctions, to zeroth order, are easily shown to be

$$\left.\begin{aligned} \psi_1 &= \mathrm{Re}\,[\exp(\tfrac{1}{2}\mathrm{i}\theta)\,\psi_+],\\ \psi_2 &= \mathrm{Im}\,[\exp(\tfrac{1}{2}\mathrm{i}\theta)\,\psi_+],\end{aligned}\right\} \tag{10}$$

where θ is no longer arbitrary but is the distortion coordinate. If now the nuclei are taken round a loop on which r_0 and ρ remain constant, while θ increases from 0 to 2π, it is evident from equation (10) that the wavefunctions change sign. The behaviour of ψ_1, which has the lower energy, is illustrated in figure 2.

Now we introduce the spin-orbit coupling, described by the perturbation operator

$$V = [\xi(r_1)\,\boldsymbol{l}_1+\xi(r_2)\,\boldsymbol{l}_2+\xi(r_3)\,\boldsymbol{l}_3]\cdot\boldsymbol{s}, \tag{11}$$

where r_1 is the distance of the electron from nucleus 1 and $\boldsymbol{l}_1 = \boldsymbol{r}_1\wedge\boldsymbol{p}$. This introduces an off-diagonal matrix element of the form

$$\langle\psi_1|V|\psi_2\rangle = \mathrm{i}\gamma,$$

where γ is approximately proportional to the integral

$$-\mathrm{i}\,\langle\phi_2|\,\xi(r_1)\,l_{1z}\,|\phi_1\rangle = \hbar\,\langle\phi_2|\,\xi(r_1)\,|\phi_1'\rangle,$$

where ϕ_1' is the p orbital obtained by rotating ϕ_1 through $\frac{1}{2}\pi$ about the z-axis. Note that the sign of this integral, as well as its magnitude, will depend on the distance between the atoms.

If the wavefunctions ψ_1 and ψ_2 have energies $E_0-\delta$ and $E_0+\delta$ respectively in the absence of the spin-orbit coupling, a simple calculation now shows that the perturbed wavefunctions are

$$\left.\begin{aligned} \psi_1' &= \psi_1\cos\alpha+\psi_2\,\mathrm{i}\sin\alpha,\\ \psi_2' &= \psi_1\,\mathrm{i}\sin\alpha+\psi_2\cos\alpha,\end{aligned}\right\} \tag{12}$$

where $\tan 2\alpha = \gamma/\delta$. Alternative forms are

$$\left.\begin{aligned}\psi_1' &= \sqrt{\tfrac{1}{2}}\,(\cos\alpha+\sin\alpha)\,\mathrm{e}^{\mathrm{i}\frac{1}{2}\theta}\psi_+ + \sqrt{\tfrac{1}{2}}\,(\cos\alpha-\sin\alpha)\,\mathrm{e}^{-\mathrm{i}\frac{1}{2}\theta}\psi_-,\\ \psi_2' &= \sqrt{\tfrac{1}{2}}\,(\cos\alpha-\sin\alpha)\,\mathrm{e}^{\mathrm{i}\frac{1}{2}\theta}\psi_+ - \sqrt{\tfrac{1}{2}}\,(\cos\alpha+\sin\alpha)\,\mathrm{e}^{-\mathrm{i}\frac{1}{2}\theta}\psi_-,\end{aligned}\right\} \tag{13}$$

where $\psi_- = \psi_+^*$. These new wavefunctions are essentially complex, and do not satisfy equation (1) with respect to variation of θ. We need to use the functions

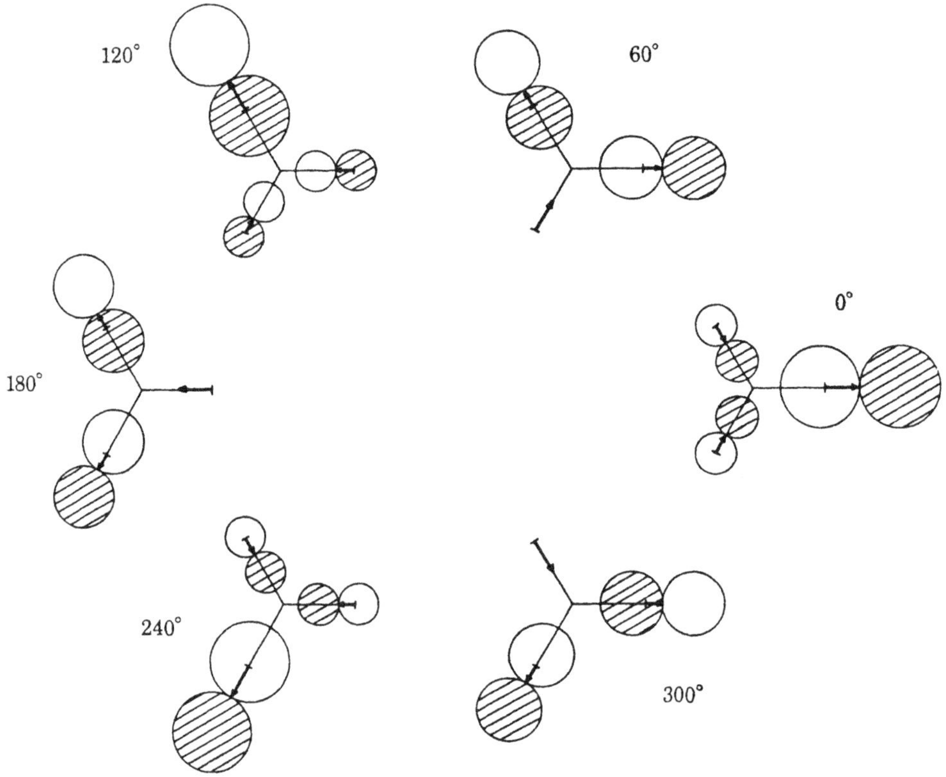

FIGURE 2. Behaviour of a molecular orbital of a nearly-trigonal triatomic molecule when the nuclei are taken round a loop in nuclear configuration space which encloses an intersection.

$\Psi_1 = \exp(\mathrm{i}\omega(\theta))\,\psi_1'$, choosing the function $\omega(\theta)$ so as to satisfy equation (1), or equivalently,

$$\left\langle \Psi_1(\theta)\middle|\frac{\partial}{\partial\theta}\Psi_1(\theta)\right\rangle = 0. \tag{14}$$

This yields a trivial differential equation for $\omega(\theta)$ whose solution is $\omega = -\frac{1}{2}\theta\sin 2\alpha$; and by equation (2) the phase change round a loop in which θ changes by 2π is

$$Z = \exp(\mathrm{i}\Omega) = \langle \Psi_1(\theta)|\Psi_1(\theta+2\pi)\rangle = \exp[\mathrm{i}\pi(1-\sin 2\alpha)]. \tag{15}$$

We can now see what happens to Longuet-Higgins' sign reversal, and also see how to generalize his theorem. If α is small ($\gamma \ll \delta$), then the phase angle Ω is close to π, and a sign reversal occurs. As the loop is contracted towards a point where there was originally an intersection ($\delta = 0$), α approaches $\pm \frac{1}{4}\pi$ and the phase angle approaches 0 or 2π – the sign reversal disappears. Only if $\gamma = 0$ exactly does the phase angle remain at π as the loop contracts.

Note now what happens if the sign of γ can be reversed by changing another nuclear coordinate – in the present example the coordinate r_0. If γ is initially positive then for a small loop round the original intersection we find $\alpha \approx +\frac{1}{4}\pi$ and $\Omega \approx 0$. If we increase the size of the loop until $\delta \gg \gamma$ then Ω increases towards π. If now we change r_0 so that γ becomes negative, then Ω increases beyond π ($\sin 2\alpha < 0$) and now decreasing the size of the loop again α approaches $-\frac{1}{4}\pi$ and $\Omega \approx 2\pi$. Thus the phase point $Z = \exp(\mathrm{i}\Omega)$ traverses the unit circle in the complex plane, completing one circuit round the origin. The sequence of loops traces out a closed surface in nuclear configuration space, and somewhere inside this surface is a point where $\gamma = 0$ and the original intersection persists. It should be remembered, incidentally, that this intersection cannot be attributed to symmetry; it is an intersection of two Kramers' doublet surfaces, making a fourfold degeneracy in all, and trigonal symmetry does not cause any additional degeneracies in odd-electron systems beyond the Kramers' degeneracy.

A topological test for intersections

We are now in a position to generalize Longuet-Higgins' theorem.

Consider a simply connected surface S, enclosing a volume V in nuclear configuration space, and suppose that $\psi(q; Q)$ is an eigenfunction of the electronic Hamiltonian which varies continuously with Q over this region. Consider a sequence of loops $L_0, L_1, \ldots, L_N$ which trace out the surface S (figure 3), and with each loop L_r associate a phase point $Z_r = \exp(\mathrm{i}\Omega_r)$ on the unit circle in the complex plane according to equations (1) and (2). N can be arbitrarily large. The initial and final points Z_0 and Z_N must be arbitrarily close to $+1$ if we let L_0 and L_N become arbitrarily small. As we follow the sequence of loops the phase point Z may encircle the origin one or more times (in which case we call S a *phase-rotating* surface) or it may return to its initial value without completing a circuit of the origin (in which case we call S *phase-preserving*). The phase-rotating or phase-preserving property is a property of the surface rather than of the particular sequence of loops; for we can transform one such sequence into another by distorting the loops continuously, so that the points Z_r move continuously. A change from phase-rotating to phase-preserving however calls for a discontinuous change in the path traced out by the phase point, and hence requires a discontinuous change in form of the wavefunction, which by hypothesis does not occur.

We now show by *reductio ad absurdum* that the volume enclosed by a phase-rotating surface S must contain at least one point where the wavefunction $\psi(q; Q)$ is discontinuous, implying that its potential energy surface intersects that of another state.

Let s_1 be any surface which bisects the volume V, and let S_1 and T_1 be the two closed surfaces so formed. If S is phase-rotating then either S_1 or T_1 (or possibly both) must be phase-rotating. This is most easily seen if the sequence of loops is chosen so that one of them, L_K say, is the intersection of S and s_1 (see figure 3). Suppose, without loss of generality, that S_1 is phase-rotating. Then by bisecting it in a similar manner we can create two further surfaces S_2 and T_2, one of which is phase-rotating. In this way we can construct a sequence of successively smaller phase-rotating

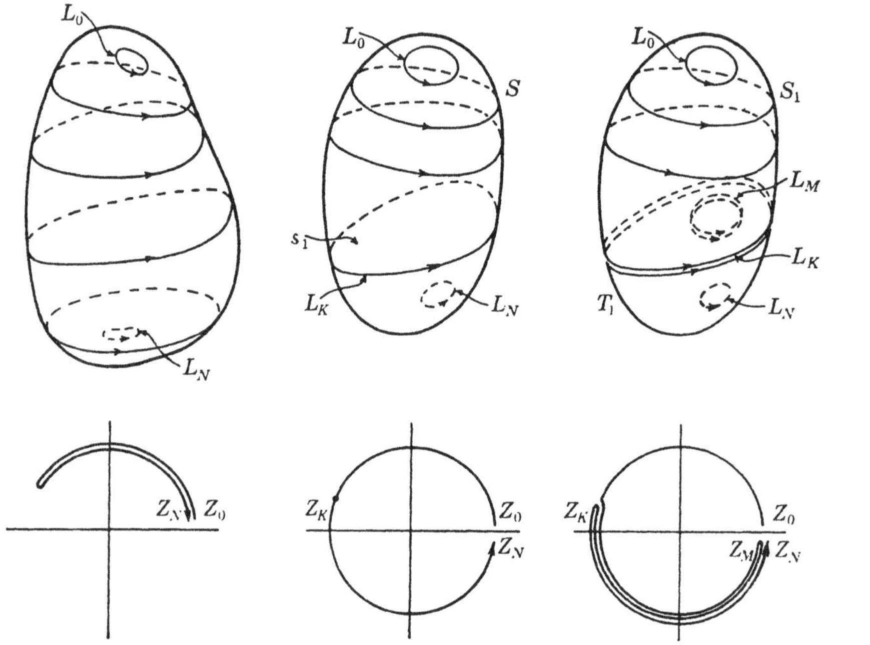

FIGURE 3. (*a*) A phase-preserving surface. (*b*) A phase-rotating surface. (*c*) Bisection of a phase-rotating surface.

surfaces, converging on some point P within V. In the immediate neighbourhood of this point we can therefore induce a finite change in $\psi(q; Q)$ by taking it round an arbitrarily small loop in nuclear configuration space. It follows that $\psi(q; Q)$ must, contrary to hypothesis, be discontinuous in Q at the point P, and that its potential surface must become degenerate there with that of some other state.

A number of important results follow from this theorem:

(1) Where there are more than three internal nuclear coordinates, the theorem applies to any volume V spanning the given surface S; so there is at least one point of intersection in any such volume, and the points of intersection form an $(s-3)$-dimensional manifold when there are s degrees of freedom, as expected.

(2) If the Hamiltonian is changed slightly (but not enough to change the wave-functions much at points remote from an intersection, so that the phase-rotating

property is unaffected) the theorem shows that there is still an intersection within the original volume. In this rather vague sense degeneracies are shown to be *stable* with respect to small additional perturbations; they are not split but merely shifted. This applies equally well to perturbations which cause a lowering of symmetry, so that symmetry is not a prerequisite for degeneracies.

(3) Actual intersections occur over $(s-3)$-dimensional manifolds and are unlikely to be encountered when the system is following a trajectory in s dimensions. However, their presence implies near-degeneracy over a significant part of nuclear configuration space, with the consequent likelihood of crossing from one surface to another.

(4) The role of symmetry in the matter of degeneracies in polyatomic systems is now seen to be rather modest. A system with s degrees of freedom will exhibit degeneracies over a manifold of dimension $s-3$, irrespective of symmetry; the imposition of symmetry on the nuclear configuration, while it may lead to degeneracies, will usually reduce the number of independent degrees of freedom by far more than three, except in quite small molecules. Consequently, degeneracies due to symmetry will usually occur over manifolds of dimension less than $s-3$. On the other hand, an intersection which is attributable to the symmetry in one particular configuration may, if a phase-rotating surface can be constructed around it, imply degeneracy in configurations with lower symmetry.

An illustration of the theorem

Some of the points made in the previous section may now be illustrated by another example – this time a less bizarre one. Let us consider a d^1 octahedral complex with six identical closed-shell ligands, such as ReF_6. The ground state in $\mathcal{O}_h$ symmetry is $^2T_{2g}$, but the Jahn–Teller theorem applies and the degeneracy is split by E_g distortions. Such distortions preserve $\mathcal{D}_{2h}$ symmetry, and the three components of T_{2g} belong to three different representations of $\mathcal{D}_{2h}$. Consequently the potential energy surface consists approximately of three intersecting paraboloids (each of which is double because of the spin degeneracy) as illustrated in figure 4.

If spin-orbit coupling is introduced, the $^2T_{2g}$ representation splits into E_g'' and U_g' of $\mathcal{O}_h^*$ (Griffith's 1961 notation) with degeneracies 2 and 4 respectively. When the symmetry is lowered to $\mathcal{D}_{2h}^*$ the six components span $E_g'+E_g'+E_g'$, $\mathcal{D}_{2h}^*$ having no representation with degeneracy higher than 2. The intersections between the paraboloids now disappear, and the two lower sheets of the surface are linked by a conical intersection at the symmetrical configuration, while the upper sheet is separated completely. The octahedral symmetry is preserved only along the single A_{1g} normal coordinate; all other distortions lower the symmetry sufficiently to remove all degeneracies higher than the twofold Kramers' degeneracy, or so conventional group theory would lead us to expect.

However explicit calculation contradicts this expectation. Let us take the basis functions $|xy^{\alpha}\rangle$, $|yz^{\beta}\rangle$ and $\mathrm{i}\,|xz^{\beta}\rangle$; we need not consider the other component $i\,|xy^{\beta}\rangle$,

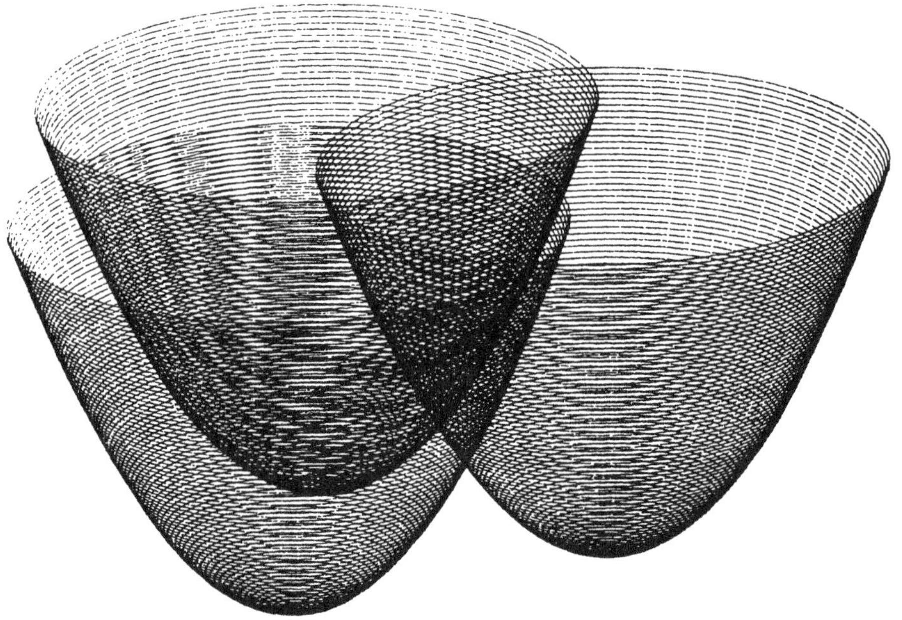

FIGURE 4. Potential energy surfaces of an octahedral complex in a T_{2g} state under E_g distortions.

$-i\,|yz^{\alpha}\rangle$ and $|xz^{\alpha}\rangle$ of each Kramers' doublet. The matrix of the perturbation arising from the E_g distortion and the spin-orbit coupling is

$$\begin{array}{ccc} |xy^{\alpha}\rangle & |yz^{\beta}\rangle & i\,|xz^{\beta}\rangle \end{array}$$

$$\begin{pmatrix} \frac{1}{2}k[(\xi-X_1)^2+(\eta-Y_1)^2] & \lambda_3 & \lambda_2^* \\ \lambda_3^* & \frac{1}{2}k[(\xi-X_2)^2+(\eta-Y_2)^2] & \lambda_1 \\ \lambda_2 & \lambda_1^* & \frac{1}{2}k[(\xi-X_3)^2+(\eta-Y_3)^2] \end{pmatrix}$$

Here ξ and η are the two E_g normal coordinates:

$$\xi = 12^{-\frac{1}{2}}(r_1+r_2+r_3+r_4-2r_5-2r_6),$$
$$\eta = \tfrac{1}{2}(r_3+r_4-r_1-r_2),$$

where r_1 to r_6 are the distances of the ligands from the metal along the $\pm x$, $\pm y$ and $\pm z$ axes respectively; (X_1, Y_1), (X_2, Y_2) and (X_3, Y_3) are the positions of the minima of the three paraboloids, which are at

$$(X_1, Y_1) = (\rho_0, 0),$$
$$(X_2, Y_2) = (-\tfrac{1}{2}\rho_0, +\sqrt{\tfrac{3}{4}}\rho_0),$$
$$(X_3, Y_3) = (-\tfrac{1}{2}\rho_0, -\sqrt{\tfrac{3}{4}}\rho_0),$$

for some value of ρ_0, and λ_1, λ_2 and λ_3 are the spin-orbit coupling matrix elements, which are all real, and all equal when $\xi = \eta = 0$.

Explicit calculation using numerical values for the parameters (e.g. $k = 2, \rho_0 = 1$, $\lambda_1 = \lambda_2 = \lambda_3 = 0.2$) shows that loops of radius ρ round the origin in the (ξ, η) plane are sign-reversing on the two lower sheets of the surface, for any ρ. Now we may introduce a different distortion, along the xy component of the T_{2g} normal coordinate. This introduces a nonzero matrix element, V say, between $|xz\rangle$ and $|yz\rangle$, so that λ_1 must be replaced by $\lambda_1 + iV$. Now the following surface is found to be phase-rotating:

(i) $V = V_1 > 0$; ρ increases from 0 to ρ_1.
(ii) V decreases from V_1 to $V_2 < 0$; $\rho = \rho_1$.
(iii) $V = V_2$; ρ decreases to zero.

Thus the intersection satisfies the requirements of the theorem, and hence it persists over a manifold of dimension $s-3 = 12$, although the symmetry only calls for an intersection along the A_{1g} normal coordinate – i.e. over a manifold of dimension 1.

Finally we may lower the symmetry to $\mathscr{D}_{2h}$ by supposing that the ligands on the three axes are different, and hence that λ_1, λ_2 and λ_3 are all different. The above surface is still found to be phase-rotating, if the value of ρ_1 is large enough; and for values of $\lambda_1 = 0.25$, $\lambda_2 = 0.15$, $\lambda_3 = 0.2$ construction of successively smaller phase-rotating surfaces exhibits an intersection near $\xi = 0.04686$, $\eta = -0.04235$, which cannot be attributed to symmetry. Furthermore, as we have seen, other distortions do not necessarily remove this degeneracy, which must persist over a domain of dimension $s-3 = 12$ for a molecule with seven atoms.

It should be emphasized that a numerical calculation cannot, by itself, conclusively prove that an intersection occurs even in the model example discussed, let alone in the real molecule. However, in conjunction with the theorem proved earlier, it can do so. Furthermore, the existence of an intersection does not depend on particular details of the model or on precise numerical values, and there is every reason to expect similar intersections to occur in real molecules.

References

Griffith, J. S. 1961 *Theory of transition metal ions*. Cambridge University Press.
Longuet-Higgins, H. C. 1975 *Proc. R. Soc. Lond.* A **344**, 147.
Naqvi, K. R. & Byers Brown, W. 1972 *Int. J. Quantum Chem.* **6**, 271.
Nikitin, E. E. 1974 (trans. M. J. Kearsley) *Theory of elementary atomic and molecular processes in gases*. Oxford Clarendon Press.
Teller, E. 1937 *J. phys. Chem.* **41**, 109.
von Neumann, J. & Wigner, E. P. 1929 *Z. Phys.* **30**, 467.

On the determination of Born–Oppenheimer nuclear motion wave functions including complications due to conical intersections and identical nuclei

C. Alden Mead and Donald G. Truhlar

Department of Chemistry, University of Minnesota, Minneapolis, Minnesota 55455
(Received 20 September 1978)

We show how the presence of a conical intersection in the adiabatic potential energy hypersurface can be handled by including a new vector potential in the nuclear-motion Schrödinger equation. We show how permutational symmetry of the total wave function with respect to interchange of nuclei can be enforced in the Born–Oppenheimer approximation both in the absence and the presence of conical intersections. The treatment of nuclear-motion wave functions in the presence of conical intersections and the treatment of nuclear-interchange symmetry in general both require careful consideration of the phases of the electronic and nuclear-motion wave functions, and this is discussed in detail.

I. INTRODUCTION

The Born–Oppenheimer adiabatic separation of electronic and nuclear motion provides a widely used framework for interpreting molecular energy levels and collision processes.[1-7] In this article we address two complications which arise in using the Born–Oppenheimer adiabatic separation, viz., how to enforce the correct permutational symmetry on the total wave function in the presence of identical nuclei and how to obtain a continuous single-valued total wave function in the presence of conical intersections of the adiabatic potential energy hypersurfaces. These complications may arise in many contexts in spectroscopy and collision theory. As an example where both effects must be considered we discuss the case of the $H + H_2$ reaction.

In recent years, accurate three-dimensional quantum mechanical calculations have become feasible for atom–diatom scattering systems, and calculations have actually been performed for the H_3 system (collision of H with H_2).[8,9] These calculations utilize accurate calculated potential energy functions and the Born–Oppenheimer approximation. One purpose of the present paper is to call attention to an effect which appears to have been overlooked in these calculations and in all formalisms[10,11] which have been proposed for treating atom–molecule collisions. This effect may have a nonnegligible effect on results for $H + H_2$ collisions and other cases where the relevant potential energy hypersurface involves a conical intersection, that is,[12-14] an intersection in which the splitting is a linear (not quadratic as in glancing intersections) function of coordinates describing the deviation from the point of intersection.

Both for making the total wave function single valued and continuous in the presence of conical intersections and for making the total wave function have the correct symmetry with respect to nuclear interchanges in the presence or absence of conical intersections, one must carefully specify the phase conventions for electronic and nuclear wavefunctions. In Sec. II we show that in many cases, e.g., for an even number of electrons or for neglect of electronic spin for any number of electrons, in the absence of electronic degeneracy for any configuration of the nuclei, the phase of the electronic wave function may be chosen in such a way that the electronic wave function is a continuous, single-valued function of nuclear coordinates and that the nuclear wave function is determined by the usual nuclear Schrödinger equation. In Sec. III we consider the case of degeneracy in more detail and show that a problem exists only for the case of conical (as opposed to glancing) intersections. We show how the effect of conical intersections can be included in the nuclear-motion Schrödinger equation by including a vector potential. We discuss the form of this potential and the effect of gauge transformations upon it in Sec. III. This potential would have an effect on $H + H_2$ collisions because the ground state of the H_3 system becomes degenerate at the equilateral triangle configuration and this intersection of the two lowest potential energy surfaces is conical.[15-20] It has also been shown[21,22] that the ground state for a system of any three hydrogenlike atoms (e.g., $H + Na_2$) exhibits a conical intersection, though not necessarily at the equilateral configuration. Other cases where conical intersections must be considered for atom–diatom collisions are also known.[23]

The nature of the complication in the Born–Oppenheimer separation when conical intersections occur is as follows: Herzberg and Longuet-Higgins[21] have shown, tacitly making the assumption that the phase of the electronic and nuclear wave functions are chosen in the usual way (i.e., to make the first derivative term in the nuclear Schrödinger equation vanish), that the electronic wave function (considered as a function of the nuclear coordinates) undergoes a change in sign when one traverses a closed path in nuclear configuration space around the curve along which the two potential surfaces intersect conically. It obviously follows (though to our knowledge no one has drawn this conclusion before) that there must be a compensating sign change on the part of the nuclear wave function if the full wave function is to be continuous and single valued. In Refs. 8–11, however, the first derivative term is neglected, but the nuclear function is chosen to be single valued and continuous.

The electronic eigenvalue equation and normalization determine the electronic wave function only up to a phase factor which may be an arbitrary function of the nuclear coordinates, and it should be possible to choose this phase factor in such a way as to cancel the sign change in traversing a closed path. However, as discussed in Sec. II, the phase factor is conventionally chosen, for convenience, in such a way as to make the first derivative term in the Born–Oppenheimer nuclear Hamiltonian vanish. Part of the price of this choice is, as we shall see, that in some cases there can be a net change in going around a closed loop. In Sec. III A, we consider the case of three nuclei (not necessarily assumed identical) and recapitulate the argument of Herzberg and Longuet-Higgins, showing that the conventional choice of phase factor leads to their change of sign. One can indeed make the electronic wave function continuous and single valued by another choice of phase factor, but at the cost of introducing the vector potential mentioned above into the effective nuclear Hamiltonian. It is a matter of convenience which choice one makes, but either one leads to a correction to the calculations of Refs. 8 and 9. In Sec. III B, we consider in more detail the case where all three nuclei are identical. The simplifications achievable in this case allow for a more complete treatment. Section III also includes a discussion of the nature of the effect and the circumstances under which it is likely to be important.

Section IV is concerned with enforcing the correct permutational symmetry with respect to nuclear interchanges. As a first step in scattering problems involving identical nuclei, one may solve the nuclear-motion Schrödinger equation under the assumption of distinguishable nuclei, but the total wave function for a collision involving n identical nuclei must be either even or odd under transposition of the space coordinates and spin of any two nuclei according as the nuclei are bosons or fermions, respectively.[24] Enforcing this requirement on a molecular scattering wave function involves the recognition that both the electronic and nuclear parts of the total wave function depend on the space coordinates of the nuclei. Thus even if one uses the Born–Oppenheimer adiabatic representation of the wave function (in which the total wave function is a product of an electronic part, which depends parametrically on nuclear coordinates, and a nuclear part) one cannot determine the correct permutational symmetry of the total wave function with respect to interchange of nuclei by considering the nuclear wave functions alone.

The fact that one must consider both the electronic and nuclear wave functions in order to impose the correct permutational symmetry with respect to interchange of identical nuclei is well known in the spectroscopy of diatomic molecules.[24a,25] Effects of the indistinguishability of identical nuclei have also been observed in cross sections for atom–atom collisions[26] and they can be understood in terms of the same symmetry considerations as are applied to bound states of diatomics. The effects of indistinguishability on collisions involving three or more identical nuclei are considered in Sec. IV. Some examples of systems which have been studied experimentally where these effects are important are ortho-para hydrogen conversion,[8,9,11,27] quantum-symmetry interference effects in nonreactive scattering of H by H_2,[8,28] reactions of oxygen atoms with oxygen molecules,[29] and H_2–H_2 scattering.[30] Similar considerations apply to certain bound-state problems, e.g., the determination of Born–Oppenheimer adiabatic wave functions for nuclear motion in bound states of $^3He_3$, $^4He_3$, $^7Li_3$, $^{16}O_3$, etc.

Micha[31] has considered the role of permutational symmetry for collisions involving three identical nuclei in general, but the only special case for which antisymmetrization procedures have been explicitly worked out is $H + H_2$.[11a,27,32] Even in this case, all previous treatments have simply enforced the correct permutational symmetry for three identical nuclei on the nuclear wave function without considering the electronic wave function. In Sec. IV A we show rigorously how to treat this problem in the Born–Oppenheimer approximation when the potential energy hypersurface has no conical intersections and the number of electrons is even or electronic spin is neglected. Scoles[28] has argued that the correct treatment involves consideration of the details of electronic exchanges during the collision, but our rigorous treatment involves only the symmetries of the initial and final states.

Section IV A treats the permutation symmetry with respect to nuclear interchanges in the case of no conical intersections and Sec. IV B treats it for the case of conical intersections.

II. PHASE FACTOR

In the Born–Oppenheimer approximation, the wave function corresponding to the ith electronic state is written (in a mixed notation which is convenient for our purposes) as

$$\Psi = \psi(\mathbf{R}) \,|\, \alpha_i(\mathbf{R})\rangle \,, \tag{1}$$

where $\mathbf{R}$ denotes the nuclear coordinates arranged into a $3n$-dimensional vector, n is the number of nuclei, and $|\alpha_i(\mathbf{R})\rangle$ is the ith member of the set of orthonormal eigenkets of the electronic Hamiltonian $\hat{H}(\mathbf{R})$, i.e.,

$$\hat{H}(\mathbf{R}) \,|\, \alpha_i(\mathbf{R})\rangle = U_i(\mathbf{R}) \,|\, \alpha_i(\mathbf{R})\rangle \tag{2}$$

and

$$\langle \alpha_i(\mathbf{R}) \,|\, \alpha_j(\mathbf{R})\rangle = \delta_{ij} \tag{3}$$

with the inner product, of course, referring to the Hilbert space of the electronic degrees of freedom only.

To obtain the equation for the nuclear wave function we retain all derivatives of the electronic wave function with respect to nuclear coordinates. This is sometimes called the Born–Huang approximation,[7c] but we call it the Born–Oppenheimer approximation. By this we mean simply that all coupling to other electronic states is neglected. In this approximation and in a coordinate system scaled so that the effective mass of every nucleus is M, the nuclear wave function $\psi(\mathbf{R})$ satisfies (in units with $\hbar = 1$)[1-6]

$$\left\{ -\frac{1}{2M}\nabla_R^2 + U_i(\mathbf{R}) - \frac{1}{M}\mathbf{F}(\mathbf{R}) \cdot \nabla_R - \frac{1}{2M}G(\mathbf{R}) \right\} \psi(\mathbf{R}) = E\psi(\mathbf{R}) \,, \tag{4}$$

where

$$\mathbf{F}(\mathbf{R}) = \langle \alpha_i(\mathbf{R}) \,|\, \nabla_R \alpha_i(\mathbf{R})\rangle \tag{5}$$

and

$$G(\mathbf{R}) = \langle \alpha_i(\mathbf{R}) | \nabla_R^2 \alpha_i(\mathbf{R}) \rangle \tag{6}$$

with the ∇_R operator referring to the nuclear coordinates. The term in G is often omitted from (4) as being small, and the term in $\mathbf{F}(\mathbf{R})$ is generally not included when coupling to excited electronic states is neglected. Setting $\mathbf{F}(\mathbf{R})$ equal to 0 for all $\mathbf{R}$ is justified by the fact that (2) and (3) determine $|\alpha_i(\mathbf{R})\rangle$ only up to a phase factor $e^{if(\mathbf{R})}$. It follows immediately from (3) that $\mathbf{F}(\mathbf{R})$ must be pure imaginary so it must vanish if $|\alpha_i(\mathbf{R})\rangle$ is chosen as real[1-6]; if it does not vanish, then multiply by a phase factor for which $i\nabla_R f(\mathbf{R})$ equals $-\mathbf{F}(\mathbf{R})$. If $\mathbf{F}(\mathbf{R})$ is now recalculated with the newly defined $|\alpha_i(\mathbf{R})\rangle$, it vanishes. In calculations, it is generally assumed that the phase factor has been chosen in this way, so the term involving $\mathbf{F}(\mathbf{R})$ is left out.[1-7] However, if $\mathbf{F}$ has nonzero curl, it cannot be made to vanish everywhere by a phase factor with a single-valued $f(\mathbf{R})$. The phase factor $f(\mathbf{R})$ can still be determined along any path so as to make $\mathbf{F}(\mathbf{R})$ vanish, but there may be a net change in phase of $\psi(\mathbf{R})$ on traversing a closed path.

To investigate this further, consider an infinitesimal change of the nuclear coordinates in (say) the x direction:

$$\hat{H}(\mathbf{R}) - \hat{H}(\mathbf{R} + \delta x \mathbf{e}_x) \cong \hat{H}(\mathbf{R}) + \hat{v}_x , \tag{7}$$

where δx is a small number, $\mathbf{e}_x$ is a unit vector, and

$$\hat{v}_x = \frac{\partial \hat{H}}{\partial x} \delta x . \tag{8}$$

Then

$$|\alpha_i(\mathbf{R})\rangle - |\alpha_i(\mathbf{R} + \delta x \mathbf{e}_x)\rangle . \tag{9}$$

For convenience of notation we no longer indicate the dependence on $\mathbf{R}$ explicitly. Then (9) may be written

$$|\alpha_i\rangle - |\alpha_{ix}\rangle \equiv |\alpha_i\rangle + |\delta_x \alpha_i\rangle . \tag{10}$$

We choose $|\delta_x \alpha_i\rangle$ in such a way that (2) and (3) are still satisfied, and also so that $\mathbf{F} = 0$, i.e., $\langle \alpha_i | \delta_x \alpha_i \rangle = 0$. These requirements determine $|\delta_x \alpha_i\rangle$ uniquely:

$$|\delta_x \alpha_i\rangle = \sum_{j \neq i} |\alpha_j\rangle \frac{\langle \alpha_j | \hat{v}_x | \alpha_i \rangle}{U_i - U_j} . \tag{11}$$

We now make an infinitesimal displacement in the y direction, adding a term

$$\hat{v}_y = \frac{\partial \hat{H}}{\partial y} \delta y \tag{12}$$

to the Hamiltonian. The eigenket now undergoes the infinitesimal change

$$|\alpha_{ix}\rangle - |\alpha_{ixy}\rangle = |\alpha_{ix}\rangle + |\delta_y \alpha_i\rangle \tag{13}$$

with

$$|\delta_y \alpha_i\rangle = \sum_{k \neq i} |\alpha_{kx}\rangle \frac{\langle \alpha_{kx} | \hat{v}_y | \alpha_{ix} \rangle}{U_i - U_k} \tag{14}$$

to lowest order in infinitesimal quantities. $|\delta_y \alpha_i\rangle$ has been chosen to be orthogonal to $|\alpha_{ix}\rangle$, but it is not necessarily orthogonal to $|\alpha_i\rangle$. By (14) we have

$$\langle \alpha_i | \delta_y \alpha_i \rangle = \sum_{k \neq i} \langle \alpha_i | \alpha_{kx} \rangle \frac{\langle \alpha_{kx} | \hat{v}_y | \alpha_{ix} \rangle}{U_i - U_k} . \tag{15}$$

Using (10) and (11) with $i = k$ to obtain $|\alpha_{kx}\rangle$ then yields

$$\langle \alpha_i | \delta_y \alpha_i \rangle = -\sum_{k \neq i} \frac{\langle \alpha_i | \hat{v}_x | \alpha_k \rangle \langle \alpha_k | \hat{v}_y | \alpha_i \rangle}{(U_i - U_k)^2} \tag{16}$$

to lowest order in infinitesimals. If the displacements had been carried out in reverse order, the order of $\hat{v}_x$ and $\hat{v}_y$ in (16) would be reversed, which, because the operators involved are hermitian, converts (16) into its complex conjugate. Thus, going around a closed path yields four contributions, two of which are orthogonal to $|\alpha_i\rangle$ and the other two of which add up to the difference of two contributions of the form (16). For an infinitesimal closed path involving these displacements, therefore, the nuclear wave function undergoes the change

$$|\alpha_i\rangle - |\alpha_i\rangle + |\delta\alpha_i\rangle , \tag{17}$$

where

$$\langle \alpha_i | \delta\alpha_i \rangle = 2i \,\mathrm{Im} \sum_{k \neq i} \frac{\langle \alpha_i | \hat{v}_y | \alpha_k \rangle \langle \alpha_k | \hat{v}_x | \alpha_i \rangle}{(U_i - U_k)^2} . \tag{18}$$

For a closed path, however, the only change can be in the phase factor, since (2) and (3), which determine everything else about $|\alpha_i\rangle$, have returned to their original form. We therefore conclude that, for an infinitesimal closed path,

$$|\alpha_i\rangle - e^{i\delta f} |\alpha_i\rangle \tag{19}$$

where

$$\delta f = 2 \,\mathrm{Im} \sum_{k \neq i} \frac{\langle \alpha_i | \partial\hat{H}/\partial y | \alpha_k \rangle \langle \alpha_k | \partial\hat{H}/\partial x | \alpha_i \rangle}{(U_i - U_k)^2} \delta x \delta y . \tag{20}$$

For closed paths that are not infinitesimal, (20) has to be integrated over some surface enclosed by the path.

The phase change given by (20) will vanish if the matrix elements can be considered real, and this will be the case if electronic spin is neglected, or if the total spin is an integer.[33] The case where half-odd integer electronic spin is included will be taken up in another paper. There is, however, another case where (20) breaks down, namely that of degeneracy. One normally handles this by choosing linear combinations of the degenerate eigenkets so that off-diagonal elements of the perturbation vanish between them, but this cannot in general be done simultaneously for both $\hat{v}_x$ and $\hat{v}_y$, so for this case the derivation of (20) is invalid and one must start over in order to determine the phase change for a closed path enclosing an intersection of two potential surfaces, with the phase being determined at each point along the path so that $\mathbf{F} = 0$.

III. CONICAL INTERSECTION

A. General

In this section, we consider a system of three nuclei, A, B, C, not necessarily identical. If the total electronic spin is not integer, it is ignored, with the understanding that it may be brought in, if desired, at a later stage of the calculation.

As internal coordinates, we choose the squares of the three internuclear distances R_{AB}^2, R_{BC}^2, R_{CA}^2. In a Cartesian coordinate system in these variables, the

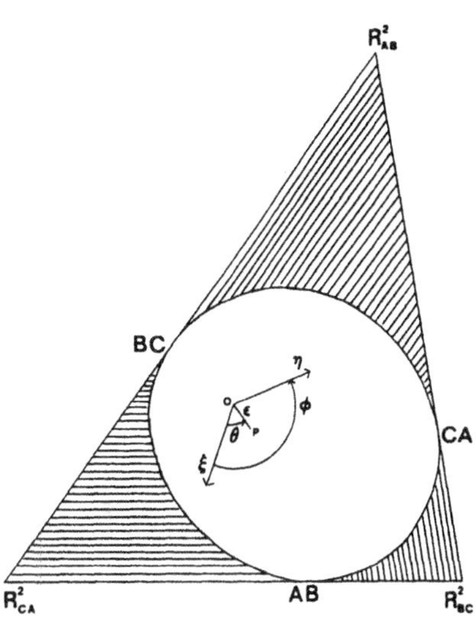

FIG. 1. Plane cross section through first octant of Cartesian coordinate system with coordinates R^2_{AB}, R^2_{BC}, R^2_{CA}. The apex marked R^2_{AB} is the point where the cross section plane intersects the R^2_{AB} axis, and similarly for the other two apices. R^2_{AB} is zero at the base of the triangle, and elsewhere is proportional to the perpendicular distance from the base, with the other two squared distances being analogously defined. Because no one of the internuclear distances can be larger than the sum of the other two, only a portion of the first octant is physically meaningful. As shown in Appendix A, this portion is a cone, so its intersection with an arbitrary plane in an ellipse, which is shown as the unshaded region in the figure. Near the tangent point labeled AB, R^2_{AB} is small and the other two internuclear distances relatively large, so one has the molecule AB with atom C at a distance. Similar remarks apply to the other two tangent points. The cross section plane is oriented so that the conical intersection curve intersects it perpendicularly at 0. With respect to an arbitrarily chosen pair of degenerate kets $|\alpha_1(\mathbf{R}_0)\rangle$ and $|\alpha_2(\mathbf{R}_0)\rangle$ at 0, the ξ and η directions are those of the gradients of $\frac{1}{2}(\langle\alpha_2(\mathbf{R}_0)|H(\mathbf{R})|\alpha_2(\mathbf{R}_0)\rangle - \langle\alpha_1(\mathbf{R}_0)|\hat{H}(\mathbf{R})|\alpha_1(\mathbf{R}_0)\rangle)$ and $\langle\alpha_1(\mathbf{R}_0)|\hat{H}(\mathbf{R})|\alpha_2(\mathbf{R}_0)\rangle$ at 0, respectively. The angle between these directions is ϕ. P is an arbitrary point, a small distance ϵ from 0, making an angle θ with the ξ axis.

physically accessible region is a portion of the first octant, with all three coordinates positive. Figure 1 shows a cross section of this octant whose orientation is specified in the next paragraph. As shown in Appendix A, because of the triangle inequality the physical region is a cone in the first octant; a general cross section through this cone is an ellipse. This ellipse is the unshaded region of Fig. 1.

The intersection of two potential surfaces requires that the internuclear distances satisfy two conditions,[21-23] so in general the surfaces may intersect along a curve. We assume that such a curve intersects our cross section at the point 0, and that the orientation of the cross section is such that it is perpendicular to the curve at the intersection. Let $\mathbf{R}$ now denote a vector from the origin of the cartesian R^2_{AB}, R^2_{BC}, R^2_{CA} coordinate system to an arbitrary point in it and let $\mathbf{R}_0$ denote such a vector to the point 0. Denote the two degenerate states at 0 by $|\alpha_1(\mathbf{R}_0)\rangle$ and $|\alpha_2(\mathbf{R}_0)\rangle$. Choose a unit vector ξ in the direction of the gradient with respect to $\mathbf{R}$ of $\frac{1}{2}(\langle\alpha_2(\mathbf{R}_0)|\hat{H}(\mathbf{R})|\alpha_2(\mathbf{R}_0)\rangle - \langle\alpha_1(\mathbf{R}_0)|\hat{H}|\alpha_1(\mathbf{R}_0)\rangle)$ at 0, and let the magnitude of this gradient be a. Similarly, let η be a unit vector in the direction of the gradient with respect to $\mathbf{R}$ of $\langle\alpha_1(\mathbf{R}_0)|\hat{H}(\mathbf{R})|\alpha_2(\mathbf{R}_0)\rangle$, with the gradient having magnitude b. Both unit vectors will lie in the cross section, because of the way it is oriented. Let ϕ be the angle between the two unit vectors ξ and η. Now consider a point P on the cross section an infinitesimal distance ϵ from 0, and in a direction making an angle θ with ξ. We have at P:

$$\begin{aligned}
&\langle\alpha_1(\mathbf{R}_0)|\hat{H}(\mathbf{R}_P)|\alpha_1(\mathbf{R}_0)\rangle = c - a\epsilon\cos\theta \,, \\
&\langle\alpha_2(\mathbf{R}_0)|\hat{H}(\mathbf{R}_P)|\alpha_2(\mathbf{R}_0)\rangle = c + a\epsilon\cos\theta \,, \\
&\langle\alpha_1(\mathbf{R}_0)|\hat{H}(\mathbf{R}_P)|\alpha_2(\mathbf{R}_0)\rangle \\
&\quad = \langle\alpha_2(\mathbf{R}_0)|\hat{H}(\mathbf{R}_P)|\alpha_1(\mathbf{R}_0)\rangle = b\epsilon\sin\left(\frac{\pi}{2} - \phi + \theta\right) \,,
\end{aligned} \tag{21}$$

where

$$c = \tfrac{1}{2}(\langle\alpha_1(\mathbf{R}_0)|\hat{H}(\mathbf{R}_P)|\alpha_1(\mathbf{R}_0)\rangle + \langle\alpha_2(\mathbf{R}_0)|\hat{H}(\mathbf{R}_P)|\alpha_2(\mathbf{R}_0)\rangle) \,.$$

We now define new variables q, λ by means of

$$q = \sqrt{a^2\cos^2\theta + b^2\sin^2\left(\frac{\pi}{2} - \phi + \theta\right)} \;; \tag{22}$$

$$\cos\lambda = \frac{a\cos\theta}{q} \;; \tag{23}$$

$$\sin\lambda = \frac{b\sin\left(\frac{\pi}{2} - \phi + \theta\right)}{q} \,. \tag{24}$$

In terms of the new variables, the Hamiltonian matrix takes the form

$$\mathbf{H}(\mathbf{R}_P) = q\epsilon\begin{pmatrix} -\cos\lambda & \sin\lambda \\ \sin\lambda & \cos\lambda \end{pmatrix} + c\mathbf{I} \,, \tag{25}$$

where $\mathbf{I}$ is the unit matrix. The eigenvalues are $c \pm q\epsilon$, with eigenfunctions

$$\begin{aligned}
|+\rangle &= \left(\sin\frac{\lambda}{2}\right)|\alpha_1(\mathbf{R}_0)\rangle + \left(\cos\frac{\lambda}{2}\right)|\alpha_2(\mathbf{R}_0)\rangle \;; \\
|-\rangle &= \left(\cos\frac{\lambda}{2}\right)|\alpha_1(\mathbf{R}_0)\rangle - \left(\sin\frac{\lambda}{2}\right)|\alpha_2(\mathbf{R}_0)\rangle \,.
\end{aligned} \tag{26}$$

The choice of phase factor in (26) is precisely the conventional one, since one easily verifies that

$$\frac{\partial}{\partial\lambda}|+\rangle = \tfrac{1}{2}|-\rangle \,, \quad \frac{\partial}{\partial\lambda}|-\rangle = -\tfrac{1}{2}|+\rangle$$

so that

$$\langle+|\frac{\partial}{\partial\lambda}|+\rangle = \langle-|\frac{\partial}{\partial\lambda}|-\rangle = 0 \,. \tag{27}$$

When one traverses a closed path around 0, the angle λ, like θ, increases by 2π, so both the kets $|-\rangle$ and $|+\rangle$ undergo a change of sign (phase change of π). This is just the result of Herzberg and Longuet-Higgins.[21]

Note that the argument would fail for a glancing inter-

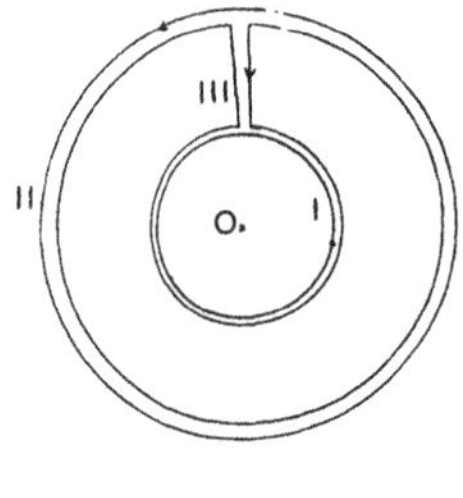

FIG. 2. Enlargement of region near 0 in Fig. 1. As shown in the text, the conventional phase convention leads to a change of sign in the electronic wave function when the small circle I is traversed. If there were not also a change in going around the larger circle II, it would follow that there is a sign change in traversing path III, which would imply that there is another conical intersection within III.

section, in which the matrix elements are quadratic, rather than linear, functions of the displacements $\epsilon \sin\theta$ and $\epsilon \cos\theta$. In this case, a rotation of π about the intersection restores all matrix elements to their original values, and thus corresponds to the full circle in the derivation we have just given. When a closed path is traversed, the matrix elements have gone through two full circles, and the wave function once again has its original value.

As argued in Refs. 21 and 12, the sign change on going around the conical intersection must hold for any closed path enclosing 0, and is not limited by the approximate nature of the eigenkets $|-\rangle$ and $|+\rangle$. In the limit as ϵ becomes arbitrarily small, the above derivation becomes rigorous, so there is certainly a change in sign for a sufficiently small circle about 0, such as circle I in Fig. 2. If there were no sign change for some other closed path around the conical intersection, such as path II in Fig. 2, then it would follow that there is a change of sign when one traverses path III, which does not enclose the intersection. But this contradicts Eq. (20), according to which there is no phase change around a closed path which does not enclose an intersection. As we have just shown, moreover, the intersection must be conical.

In order to make the electronic wave function single valued, one must draw a cut to 0, and have the function change sign on going through the cut. Since the full wave function must be continuous, this would require a compensating discontinuity in the nuclear function.

Alternatively, one could make the electronic wave function continuous and single valued by multiplying it by a phase factor such as $e^{i\theta/2}$, $e^{3i\theta/2}$, $e^{i\lambda/2}$, etc., but this leads to a nonvanishing $\mathbf{F}(\mathbf{R})$ term in the nuclear Hamiltonian. Different choices of this phase factor differ from one another by single-valued phase factors, and do not affect the total wave function. To see this, suppose that we have found a correct solution with one choice of phase factor, and we now wish to consider the change introduced when the electronic wave function is multiplied by an additional single-valued phase factor $e^{if(\mathbf{R})}$. The function $f(\mathbf{R})$ need not be single valued; it may increase by a multiple of 2π on traversing a closed path. The only effects of this change on the nuclear Schrödinger equation are that

$$\mathbf{F}(\mathbf{R}) \rightarrow \mathbf{F}(\mathbf{R}) + i\nabla f$$

and

$$G(\mathbf{R}) \rightarrow G(\mathbf{R}) + 2i\mathbf{F}(\mathbf{R}) \cdot \nabla f + i\nabla^2 f - (\nabla f)^2 \; .$$

But inspection of Eq. (4) shows that this is equivalent to

$$\nabla \rightarrow \nabla + i\nabla f \; .$$

The solution to the new nuclear Schrödinger equation can be obtained by multiplying the old nuclear wave function by e^{-if}, since

$$(\nabla + i\nabla f)(\psi e^{-if}) = e^{-if}\nabla\psi \; .$$

This leaves the full wave function unaltered.

The foregoing should not be misconstrued to mean that the solution obtained by ignoring the $\mathbf{F}(\mathbf{R})$ term and making the nuclear wave function single valued and continuous is the same as, or differs only by a phase factor from, the correct solution. We assumed in the above that we started with a correct solution, and showed that the full wave function was left unchanged by the introduction of a single-valued phase factor in the electronic part. But the solution obtained by ignoring $\mathbf{F}(\mathbf{R})$ and making the nuclear function single valued and continuous is *not* a correct solution, since the full wave function it represents is not single valued and continuous.

B. Three identical nuclei

If the three nuclei are identical, there are certain simplifications due to symmetry, making possible a more detailed treatment. There is also the added complication of having to give the full wave function the correct permutation symmetry, but we postpone consideration of this to Sec. IV, treating the nuclei here formally as distinguishable, though identical in all properties. For three identical nuclei, the curve of conical intersections is the set of equilateral triangle geometries. This curve is a straight line making the same angle with each of the R^2_{AB}, R^2_{BC}, and R^2_{CA} axes. The cross section perpendicular to the curve of intersection is one of constant $Q = R^2_{AB} + R^2_{BC} + R^2_{CA}$. As shown in Appendix A and illustrated in Fig. 3, the physical region of the cross section is a circle. Appendix A also contains a discussion of various properties of our internal coordinate system. We consider a particular value of Q and choose the degenerate kets at the intersection at 0 according to their behavior under interchange of A and B (rotation of π about the axis which passes through C and bisects a line from A to B). Let $|\alpha_1(\mathbf{R}_0)\rangle \equiv |X_{AB}\rangle$ be the degenerate ket which is left invariant by this rotation, and $|\alpha_2(\mathbf{R}_0)\rangle \equiv |Y_{AB}\rangle$ be the ket that changes sign. One could, of course, define analogous kets with respect to BC or CA, related to the ones we are using by

$$\begin{aligned} |X_{AB}\rangle &= -\tfrac{1}{2}|X_{CA}\rangle - \frac{\sqrt{3}}{2}|Y_{CA}\rangle \\ &= -\tfrac{1}{2}|X_{BC}\rangle + \frac{\sqrt{3}}{2}|Y_{BC}\rangle \; ; \end{aligned} \qquad (28)$$

$$\begin{aligned} |Y_{AB}\rangle &= -\tfrac{1}{2}|Y_{CA}\rangle + \frac{\sqrt{3}}{2}|X_{CA}\rangle \\ &= -\tfrac{1}{2}|Y_{BC}\rangle - \frac{\sqrt{3}}{2}|X_{BC}\rangle \; . \end{aligned} \qquad (29)$$

It follows from the symmetry and is proved in Appendix

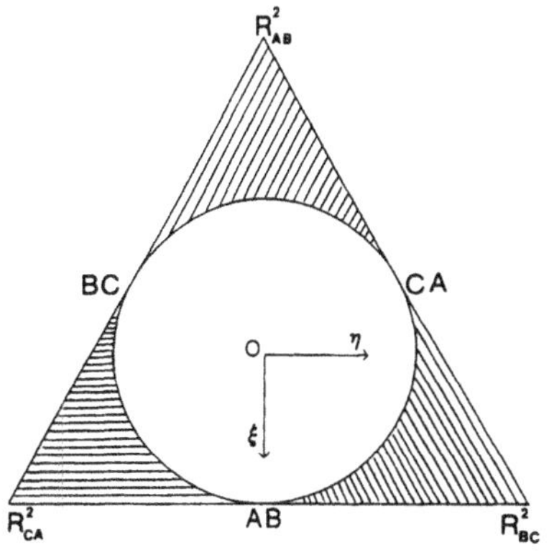

FIG. 3. Cross section of first octant in R^2_{AB}, R^2_{BC}, R^2_{CA} for the case of identical nuclei. All labels have the same significance as in Fig. 1. Because of the symmetry, the conical intersection curve is now the straight line making equal angles with the three axes, and the plane perpendicular to it intersects each axis at the same distance from the origin. The intersection with the physical region cone is now a circle, with the conical intersection at the center. As shown in Sec. III B, the ξ and η directions are now perpendicular to one another, with the ξ direction being from 0 toward the tangent point AB.

B that the directions ξ and η defined in the previous section are toward the tangent point (AB) and perpendicular to this direction, respectively, and that $b=a$. The angle λ is thus identical to θ. As one proceeds along the ξ direction, the state of lower energy must correlate with the ground state of the molecule AB, with the atom C at a distance. According to the standard analysis,[24a,25] this ground state is invariant under interchange of A and B if it is Σ_g^+ or Σ_u^-, and changes sign if it is Σ_u^+ or Σ_g^-. In the former case, which applies to H_2, therefore, the ground state is the one which approaches $|X_{AB}\rangle$ as one approaches 0 along the ξ axis, and we will assume this case for definiteness. Proceeding as in the previous section, we find for the ground state near 0:

$$|-\rangle = \cos\frac{\theta}{2}\,|X_{AB}\rangle - \sin\frac{\theta}{2}\,|Y_{AB}\rangle , \tag{30}$$

while the excited state is given by

$$|+\rangle = \sin\frac{\theta}{2}\,|X_{AB}\rangle + \cos\frac{\theta}{2}\,|Y_{AB}\rangle . \tag{31}$$

If the ground state of the diatomic molecule is Σ_u^+ or Σ_g^-, then $|+\rangle$ is the ground state and $|-\rangle$ the excited state, but everything else is unchanged. Both these kets have the "conventional" choice of phase factor, which makes $\mathbf{F}(\mathbf{R})=0$, and again we have a sign change when θ changes by 2π. This can be removed as before by a phase factor such as $e^{i\theta/2}$, $e^{3i\theta/2}$, etc. As we shall see in Sec. IV, the choice which is most convenient for the treatment of permutation symmetry is $e^{3i\theta/2}$, so this is the choice we shall make. In the next subsections we consider the nature of the effect of this on the nuclear motion problem.

C. Nature of the effect

For the sake of definiteness, and because it permits one to write down explicit expressions, we will assume in this subsection that we are dealing with identical nuclei. It will be evident, however, that the qualitative conclusions remain the same if the nuclei are not identical.

The presence of the phase factor $e^{3i\theta/2}$ means that each momentum operator $\hat{\mathbf{P}}_\gamma$ in the effective nuclear Hamiltonian, where, e.g., $\gamma = A, B, C$, is replaced by $(\hat{\mathbf{P}}_\gamma - \mathbf{a}_\gamma)$, where $\mathbf{a}_\gamma = -(\frac{3}{2})\nabla_\gamma\theta$. In the nuclear kinetic energy operator, therefore, $\hat{p}_\gamma^2$ is replaced by

$$(\hat{\mathbf{P}}_\gamma - \mathbf{a}_\gamma)^2 = \hat{p}_\gamma^2 - 2\hat{\mathbf{p}}_\gamma\cdot\mathbf{a}_\gamma + a_\gamma^2 . \tag{32}$$

In general, there is another term proportional to $(\nabla_A^2 + \nabla_B^2 + \nabla_C^2)\theta$, but it is shown in Appendix A that this is zero.

If $\mathbf{R}_A$, $\mathbf{R}_B$, and $\mathbf{R}_C$ denote the positions of the three nuclei, we see from Eq. (A15) of Appendix A that

$$\mathbf{a}_A = -\frac{6\sqrt{3}}{S^2}\{\mathbf{R}_A(R^2_{AB} - R^2_{CA}) + \mathbf{R}_B(R^2_{CA} - R^2_{BC}) + \mathbf{R}_C(R^2_{BC} - R^2_{AB})\} , \tag{33}$$

where

$$S^2 = 2\{(R^2_{AB} - R^2_{BC})^2 + (R^2_{BC} - R^2_{CA})^2 + (R^2_{CA} - R^2_{AB})^2\} \tag{34}$$

is a measure of the deviation from the equilateral triangle configuration. The corresponding expressions for $\mathbf{a}_B$ and $\mathbf{a}_C$ are obtained by cyclic permutation from (33).

In terms of the Jacobi coordinates defined in Eq. (A1) of Appendix A, the kinetic energy is

$$T = \frac{1}{2M}(\tfrac{1}{3}p_R^2 + 2p_r^2 + \tfrac{3}{2}p_\rho^2) ,$$

and

$$\mathbf{a}_{R_{c.m.}} = 0 , \tag{35}$$

$$\mathbf{a}_\rho = \frac{3\sqrt{3}}{S^2}\{(\tfrac{3}{2}r^2 - 2\rho^2)\mathbf{r} + 4(\boldsymbol{\rho}\cdot\mathbf{r})\boldsymbol{\rho}\} ; \tag{36}$$

$$\mathbf{a}_r = \frac{3\sqrt{3}}{S^2}\{(\tfrac{3}{2}r^2 - 2\rho^2)\boldsymbol{\rho} - 3(\boldsymbol{\rho}\cdot\mathbf{r})\mathbf{r}\} . \tag{37}$$

The last two equalities follow from (A18) and (A19).

We see that the extra terms behave mathematically exactly as if there were a magnetic interaction between the nuclei, with vector potential proportional to $\mathbf{a}$. Moreover, since $\mathbf{a} = -\frac{3}{2}\nabla\theta$, the pseudomagnetic field $\mathbf{h} = \nabla\times\mathbf{a} = 0$ everywhere except along the curve of intersection of the two potential surfaces. For a surface through which this curve passes, we have

$$\int \mathbf{h}\cdot d\mathcal{A} = \oint \mathbf{a}\cdot d\mathbf{l} = -\frac{3}{2}\oint d\theta = -3\pi , \tag{39}$$

where $d\mathcal{A}$ and $d\mathbf{l}$ are infinitesimal elements of surface area and linear displacement, respectively. The pseudomagnetic field $\mathbf{h}$, therefore, points in the direction of decreasing Q in our case, i.e., along the intersection curve, and is zero except along that curve, where it has a delta-function singularity. In terms of

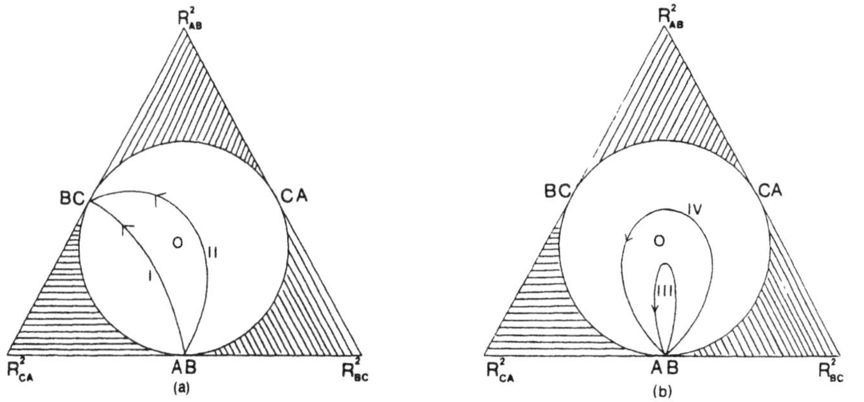

FIG. 4. (a) Cross section similar to that of Fig. 3, showing two possible trajectories, labeled I and II, for the reaction AB + C → BC + A. (b) Same cross section as in (a), now showing two possible trajectories, labeled III and IV, for nonreactive scattering of molecule AB with atom C.

the coordinates u and v of Appendix A, defined by Eqs. (A3) and (A4), which denote displacement in the ξ and η directions, respectively,

$$\int \mathbf{h}\cdot d\mathcal{A} = \int h_Q\, du\, dv ,$$

where

$$h_Q = \frac{\partial}{\partial u} a_v - \frac{\partial}{\partial v} a_u = -3\pi\delta(u)\delta(v) .$$

An analogous case in ordinary magnetism would be if one had $A_x = -y/r$, $A_y = x/r$, and $A_z = 0$, with $r = (x^2 + y^2)^{1/2}$. Here one easily verifies that $\mathbf{H} = \nabla \times \mathbf{A} = 0$ except for $x = y = 0$, but that $\oint \mathbf{A}\cdot d\mathbf{l} = 2\pi$ around any curve including the z axis so that one has $\mathbf{H} = 2\pi\delta(x)\delta(y)\mathbf{e}_z$.

Since the analogy of this effect with magnetic interactions is exact, its study can be facilitated by using our knowledge of the properties of the magnetic interaction, in particular gauge invariance. The only requirement on the vector potential **a** is the integral relation (39). If we wish, we can choose the gauge in such a way that $\mathbf{a} = 0$ everywhere except for some narrow range of θ. If, for example, our wave function is appreciable only for some range of θ centered about (say) θ_0, we can choose the gauge so that $\mathbf{a} = 0$ except near $\theta = \theta_0 \pm \pi$. The effect will thus be proportional (at most) to the amplitude of the wave function at $\theta \cong \theta_0 \pm \pi$, and will be quite negligible. We conclude that there will be only a negligible effect on the interaction between an atom and a diatom at large distances or on the vibrational state of a triatomic molecule undergoing small vibrations about an equilibrium configuration far from a conical intersection. In addition there is no effect on classical trajectories which do not pass through the curve of intersection.

If the wave function is appreciable over the entire range of θ, however, the situation is quite different, and remains so even if there is little or no probability of the configuration being close to the conical intersection. The analogous situation for ordinary magnetism has been treated by Aharonov and Bohm.[34] They considered a magnetic field confined to a narrow solenoid and showed that it would produce interference effects between trajectories passing on opposite sides of the solenoid, even if the particles were excluded from the region occupied by the solenoid itself. In our case, the role of the solenoid is played by the curve of intersection. Although the pseudomagnetic field h is zero except at the point of intersection, it may have an effect on the phase of the wave function even if the corresponding semiclassical trajectories do not pass through the intersection. For example, Fig. 4(a) shows two trajectories for the reactive process AB + C → BC + A. In path I, atom C approaches the B end of the AB molecule, then goes off with B; throughout the path, R_{CA} is the longest of the three internuclear distances. In path II, C first approaches the A end of the molecule, so that temporarily R_{CA} is the *shortest* of the three distances, before the products BC and A are finally formed. By appropriate choice of gauge, as discussed above, one can transform away the vector potential along either one of the paths, but not along both simultaneously. Thus, if either path dominates, there will negligible effect; if, however, interference between the two plays a role, the effect will be appreciable. Figure 4(b) shows the analogous situation for two nonreactive paths. In path III, R_{AB} is the shortest of the three distances throughout. In path IV, R_{AB} is temporarily the *longest* distance. Such a path is quite likely for high-energy noncollinear collisions when the atoms B and C approach closely. Again, if both paths contribute appreciably, the nature of their interference will be substantially affected by the presence of the conical intersection. In summary, we expect nonnegligible effects whenever the entire range of the angle θ plays a role. In particular, this is a strong possibility whenever noncollinear reactive events or noncollinear close encounters occur with nonnegligible probability. These considerations are, of course, somewhat qualitative. A detailed understanding of the nature and magnitude of this effect requires further study.

IV. PERMUTATIONAL SYMMETRY WITH RESPECT TO NUCLEAR INTERCHANGES

Consider scattering in a system of N electrons and n identical nuclei. The permutation symmetry group for this problem is the direct product $S_N \times S_n$, where S_N is the group of all ($N!$) permutations of the N electrons and S_n is the corresponding permutation group for the nuclei.[35] In principle, the problem can be treated in three steps, as follows:

(i) Solve for the scattering matrix (S matrix[24b]) treating *all* particles as distinguishable.

(ii) Change to a representation in terms of state vectors belonging to irreducible representations of S_N. Since the elements of S_N commute with the Hamiltonian, and hence also with the S matrix, the S matrix will now be in block diagonal form. The only block of the S matrix which is physically relevant, of course, is the block S^{Na} belonging to the totally antisymmetric representation of S_N. Moreover, it is clear that, had we limited ourselves from the beginning to state vectors totally antisymmetric in the electrons, but with the nuclei still treated as distinguishable, the resulting S matrix would be just the block S^{Na}.

(iii) We now require our basis vectors to belong also to irreducible representations of S_n. Since the permutations of S_n commute with those of S_N, this does not interfere with their belonging to irreducible representations of S_N. Each block of the previous S matrix will now be broken down into subblocks. In particular, the lone physically significant block, S^{Na}, will be further block diagonalized. The physically significant part is now the subblock of S^{Na} which also belongs to the totally symmetric or antisymmetric representation of S_n, depending on whether the nuclei are bosons or fermions, respectively. This result is clearly independent of whether S^{Na} was obtained by projecting the full S matrix onto the antisymmetric representation of S_N or by limiting oneself from the beginning to basis vectors belonging to the totally antisymmetric representation of S_N.

If Born–Oppenheimer adiabatic product wave functions are used as basis vectors, it is not sufficient that the nuclear wave function transform as a representation of S_N since the electronic part also contains the nuclear coordinates and is therefore also affected by permutations of the nuclei. What is required is that the total wave function has the correct permutational symmetry with respect to transposing the nuclei. Thus one must take account of the behavior of the electronic part of the wave function under permutations of the nuclei. We now take up the problem of how to do this, i.e., of determining the behavior of electronic wave functions under nuclear permutations.

Consider three identical nuclei, A, B, and C, with space coordinates $\mathbf{R}_A$, $\mathbf{R}_B$, and $\mathbf{R}_C$, respectively. Let q be a collective symbol for all electronic coordinates, space and spin. The electronic wave function $\langle q;R_A, R_B, R_C | \alpha_i(\mathbf{R})\rangle$ will be called $\alpha(q;R_A, R_B, R_C)$ or simply $\alpha(q;\mathbf{R})$. It is assumed to be properly antisymmetrized with respect to permutations of the electrons for each set of values of the parameters $\mathbf{R}_A$, $\mathbf{R}_B$, and $\mathbf{R}_C$. Note that the dependence of $\alpha(q;\mathbf{R})$ on electronic coordinates is defined relative to nuclear ones; if the nuclei undergo a rigid translation or rotation, then $\alpha(q;\mathbf{R})$ undergoes the same operation. Depending on the accuracy to which $\alpha(q;\mathbf{R})$ has been calculated, it may or may not contain effects of electronic spin–orbit interactions. In this paper, such interactions are always neglected if the number of electrons is odd. We also omit any physical effects of nuclear spin; these can be brought in at a later stage if desired.

A. No conical intersections present

We first consider the case of no conical intersections and either an even number of electrons or neglect of electronic spin (or both). The phase factor in $\alpha(q;R_A, R_B, R_C)$ is determined so that $\mathbf{F}(\mathbf{R})=0$. Conical intersections will be treated in Sec. IV B. The case of an odd number of electrons where electronic spin is not neglected requires explicit treatment of the Kramers degeneracy[24c] and will be discussed in a separate paper. Now consider the effect on $\alpha(q;R_A, R_B, R_C)$ of permuting nuclei A and B. The new electronic function $\alpha(q;R_B, R_A, R_C)$ is a solution of the same electronic problem as the old, and is therefore the same function (of q) as the old (because conical intersections are excluded), except perhaps for an overall multiplicative phase factor. Repeating the operation must restore the original function, since the net effect is that of traversing a closed path, and for the case we are considering the right-hand side of (20) vanishes. This leads to the conclusion that the phase factor must be ± 1, as discussed above. Applying the same reasoning to other nuclear permutations leads to the conclusion that $\alpha(q;R_A, R_B, R_C)$ must belong to a one-dimensional representation of the nuclear permutation group S_3, i.e., either to the totally symmetric or the totally antisymmetric one. As the nuclear coordinates are varied continuously, it is clear (from the fact that the electronic wave function is a continuous single-valued function of nuclear space coordinates) that the representation to which $\alpha(q;R_A, R_B, R_C)$ belongs cannot abruptly change. Therefore, it is either totally symmetric or totally antisymmetric over the entire range of R_A, R_B, and R_C.

It is clear that a similar argument applies if there are four or more nuclei present.

To determine whether a particular electronic state is totally symmetric or antisymmetric, it suffices to consider a particular nuclear configuration, most conveniently one in which two nuclei are close together forming a molecule, while the other or others are far away, permute two nuclei, and determine the result.

For example, consider the scattering of an X atom with an X_2 molecule in a Σ state. We choose the asymptotic region where the atom is far from the molecule, and permute the two nuclei making up the molecule. The effect of this on the electronic wave function is known[24a] to be a factor of $+1$ if the state is Σ_g^+ or Σ_u^- and -1 if it is Σ_u^+ or Σ_g^-. In the former two cases, the electronic term is said to be (totally) symmetric; if the latter two, (totally) antisymmetric.[24a]

The *full* scattering wave function must be chosen totally symmetric or antisymmetric under nuclear permutations, depending on whether the nuclei are bosons or fermions. The nuclear part must thus be symmetrized or antisymmetrized, as the case may be, to bring this about. From the above, one sees that, if the Born–Oppenheimer adiabatic approximation is made and if the electronic state is Σ_g^+ or Σ_u^-, the problem reduces to that of the nuclei interacting via the potential energy obtained from the electronic function, and with their normal statistics. If the state is Σ_g^- or Σ_u^+, on the other hand, the nuclear wave function must be antisymmetrized for transpositions of two bosons and symmetrized for transpositions of two fermions (sic). In this case, the problem is again that of the nuclei interacting via the effective potential, but with the statistics reversed. For non-Σ states of the diatomic, there will be one electron state of the triatomic of each type, and these become nearly degenerate as the atom–molecule distance becomes larger.

Experimentally observable cross sections for scattering processes involving identical particles can be calculated from distinguishable-particle wave functions by using these to determine distinguishable-particle scattering amplitudes and linearly combining these with appropriate coefficients.[36] In atom–molecule or molecule–molecule scattering calculations one typically treats the electronic problem first to obtain molecular electronic wave functions and potential hypersurfaces for nuclear motion. Since the electronic wave function is antisymmetrized with respect to permuting two electrons, the system at this stage of the calculation is being treated in terms of indistinguishable electrons and distinguishable nuclei. Then, if some of the nuclei are identical, one must calculate observable cross sections from the indistinguishable-electrons, distinguishable-nuclei scattering amplitudes by linearly combining them with appropriate coefficients so as to obtain indistinguishable-electrons, indistinguishable-nuclei scattering amplitudes.[32] Thus it is not necessary to explicitly calculate over all space a wave function which has correct symmetry with respect to nuclear permutations. We did consider the permutational symmetry of the wave function over all space above and it would be necessary to consider this to calculate bound-state properties from wave functions defined over all space. But for collision problems we need only the scattering amplitudes and these refer only to asymptotic states.

Consider atom–diatom collisions involving three identical nuclei. For collisions where the diatom term is symmetric,[24a] both before and after the collision, the result is simple: The electronic wave function is even under permutation of two nuclei in both the initial and the final state so the coefficients may be determined by considering only the nuclear wave function and treating fermions as fermions and bosons as bosons. For collisions in which the diatom term is antisymmetric[24a] both before and after the collision, the result is also simple: The electronic wave function is odd under permutation of two nuclei in both the initial and the final state so the coefficients may be determined by considering only the nuclear wave function and treating fermions as bosons and bosons as fermions.

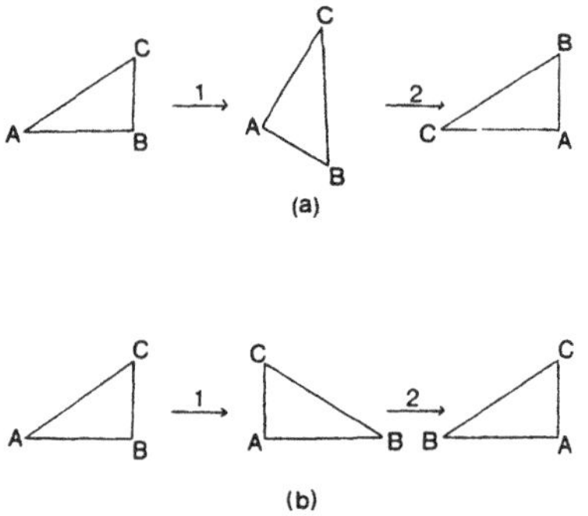

FIG. 5. (a) Two-step process for the cyclic permutation (ABC). In step 1, the values of the internuclear distances are permuted, but there is no net translation or rotation of the molecule (i.e., total linear and angular momentum are zero throughout the process). Step 2 is a rigid rotation about the center of mass into the final configuration. (b) Two-step process for interchange (AB). As in (a), step 1 is a change of internal coordinates (internuclear distances) without net translation or rotation, and step 2 is a rigid rotation about the center of mass.

B. Conical intersections present

In this section we consider how the above discussion must be changed when conical intersections are present. For concreteness we consider the most important cases

$$H(^2S) + H_2(^1\Sigma_g^+) \to H(^2S) + H_2(^1\Sigma_g^+) \quad (40)$$

and

$$H(^2S) + H_2(^1\Sigma_u^+) \to H(^2S) + H_2(^1\Sigma_u^+) \; . \quad (41)$$

But the treatment is generalizable to other cases.

Referring to the results of Sec. III B, our first task is to ascertain the behavior of the eigenkets $|-\rangle$ and $|+\rangle$ under permutations of the nuclei.

To study the behavior of $|+\rangle$ and $|-\rangle$ under the cyclic permutation (ABC), we carry out the permutation in two steps, as shown in Fig. 5(a). In the first step, the internal coordinates are changed so that R_{AB} takes the value previously held by R_{BC}, etc. This corresponds to an increase in the angle θ by $2\pi/3$. The second step is a rigid rotation without change in internal coordinates, in particular without change in θ. When the process is completed, the electronic Hamiltonian is the same as it was before, and one has not passed through any point at which electronic energies intersect, so the net effect must be to multiply each ket by a constant, which can only be ± 1, since both kets are *real* linear combinations of $|X_{AB}\rangle$ and $|Y_{AB}\rangle$. If one repeats this process twice more, one is back where one started, but θ has gone around a full circle, leading to a change of sign. It is clear, then, that the effect of a cyclic permutation is a change of sign for both eigenkets. This is explicitly verified in Appendix C.

The interchange (AB) can also be carried out in two steps, as shown in Fig. 5(b). In step 1, the values of R_{CA} and R_{BC} are interchanged, holding R_{AB} constant. This corresponds to a change in sign of θ. The second step is a rigid rotation interchanging A and B, which changes the sign of $|Y_{AB}\rangle$. The net effect, therefore, is to leave $|-\rangle$ invariant and change the sign of $|+\rangle$.

Other interchanges may be defined by combining (AB) with cyclic permutations, but there are ambiguities brought about by the fact that a rotation of 2π leads to a factor of -1. This is obviously inconvenient permutation behavior. As mentioned in Sec. III B, the simplest way to handle this problem is to introduce a phase factor $e^{3i\theta/2}$ or $e^{-3i\theta/2}$. Both $|+\rangle$ and $|-\rangle$ are now left invariant under cyclic permutations, and there is no sign change on traversing a closed path, but an interchange of two nuclei takes the phase factor into its complex conjugate and brings about a change of sign in $|+\rangle$. Accordingly, for states which asymptotically contain a Σ_g^+ or Σ_u^- diatomic state, we seek a solution of the form

$$\Psi = [\psi_+(\mathbf{R})\, e^{3i\theta/2} + \psi_-(\mathbf{R})\, e^{-3i\theta/2}]\,|-\rangle \;. \qquad (42)$$

Inserting (42) into the full eigenvalue equation, and neglecting coupling to other electronic states, we find

$$\hat{H}_+(\mathbf{R})\psi_+(\mathbf{R})\, e^{3i\theta/2} + \hat{H}_-(\mathbf{R})\psi_-(\mathbf{R})\, e^{-3i\theta/2}$$
$$= E[\psi_+(\mathbf{R})\, e^{3i\theta/2} + \psi_-(\mathbf{R})\, e^{-3i\theta/2}] \,, \qquad (43)$$

where $\hat{H}_\pm(\mathbf{R})$ is just the usual effective nuclear Hamiltonian except that $\hat{p}_\theta$, the momentum conjugate to the angle θ, is replaced by $\hat{p}_\theta \pm \frac{3}{2}$, corresponding to the $\mathbf{F}(\mathbf{R})$ term now present. To solve equation (43) we require

$$\hat{H}_+(\mathbf{R})\psi_+(\mathbf{R}) = E\psi_+(\mathbf{R}) \qquad (44a)$$

and

$$\hat{H}_-(\mathbf{R})\psi_-(\mathbf{R}) = E\psi_-(\mathbf{R}) \;. \qquad (44b)$$

An interchange of two nuclei, which reverses the sense of θ, will convert a solution of (44a) into one of (44b) and vice versa, so there is no possibility of their being in contradiction. The solution (42) must also have correct permutation symmetry. A cyclic permutation must leave Ψ invariant, and we know that it leaves $e^{\pm 3i\theta/2}|-\rangle$ invariant. It follows that both $\psi_+(\mathbf{R})$ and $\psi_-(\mathbf{R})$, both of which may have spin parts, must be invariant under cyclic permutations. An interchange must multiply Ψ by ± 1, depending on the statistics. Its effect on the electronic part is just to interchange the two phase factors. It follows that an interchange applied to $\psi_+(\mathbf{R})$ must give just $\pm\psi_-(\mathbf{R})$, depending on the statistics. Recall that an interchange applied to a solution of (44a) automatically gives a solution of (44b), so this requirement causes no difficulty. Starting with an eigenfunction $\psi(\mathbf{R})$ of $\hat{H}_+$, therefore, one obtains a Born–Oppenheimer solution to the full problem (apart from normalization) as

$$\Psi = \{[(1 + \hat{P}_{ABC} + \hat{P}_{ACB})\psi(\mathbf{R})\,|\, e^{3i\theta/2}$$
$$\pm [(\hat{P}_{AB} + \hat{P}_{BC} + \hat{P}_{CA})\psi(\mathbf{R})\,|\, e^{-3i\theta/2}\}\,|-\rangle \;, \qquad (45)$$

where the $\hat{P}$'s are permutation operators and the upper sign applies to bosons, the lower to fermions.

For the electronic state $|+\rangle$, which would be appropriate for a state which asymptotically contains a Σ_u^+ or Σ_g^- diatomic, we follow the same procedure and obtain Eq. (45) again where now the upper sign applies to fermions and the lower to bosons.

For asymptotic states in which two of the nuclei are together in a molecule while the third is at a distance, we have $\theta = 0$ or $\pm 2\pi/3$, so the phase factors are identical in this region and the amplitudes are combined in the usual[11a, 27, 32] way. For scattering states with three separated atoms, however, the phase factors must be taken into account in combining amplitudes.

V. DISCUSSION

We have shown that, in the presence of a conical intersection, if one wishes to treat the nuclear wave function as continuous and single valued, there is a term in the effective nuclear Hamiltonian, previously overlooked, which acts as if there were a delta-function magnetic field along the curve of intersection. This effect is expected to be most important when, from the path integral or classical S matrix[37] point of view, there is interference between trajectories passing on opposite sides of the conical intersection. The effect is negligible for long-range atom–diatom interactions and for small vibrations of a stable triatomic molecule about an equilibrium configuration far from the conical intersection.

We have also shown how to enforce the correct permutation symmetry on the full wave function in the case of identical nuclei, both in the presence and absence of conical intersections. In both cases, the behavior of the electronic part of the wave function under permutations of the nuclei must be carefully taken into account.

Our treatment has been restricted to the case of three nuclei, but our results have implications for systems of four or more nuclei as well. Our results are valid if electronic spin is neglected, or if the total electronic spin is integer. If it is half-odd integer, and we wish to take it into account in the calculation, the situation becomes more complicated. In the first place, each potential energy surface now corresponds to a Kramers doublet, and the intersection of two of these requires that five conditions be satisfied by the nuclear coordinates instead of two.[33] For three nuclei, therefore, there will no longer be a conical intersection. This does not mean, however, that the effect we have been discussing disappears, because with spin–orbit interaction the right-hand side of (20) no longer vanishes in the absence of degeneracy. The effect, therefore, is that the pseudomagnetic field discussed in Sec. III C is no longer confined to a curve, but is spread out somewhat. This will be discussed further in a future paper.

APPENDIX A: PROPERTIES OF INTERNAL COORDINATES AND OF THE ANGLE θ

In this appendix, for the sake of simplicity and definiteness, the treatment will be restricted to the case of three identical nuclei. The results obtained in parts 3–5 and some of the statements made in part 1 will be valid only for this case. The characterization of the physical region obtained in part 2 depends only on the existence of internal coordinates R^2_{AB}, R^2_{BC}, R^2_{CA}, and is therefore valid whether or not the nuclei are identical. In particular, the proof that the cross section of the physical region is in general an ellipse, as asserted in Sec. III A, is valid for the general case.

1. Definitions

In the following, we will be dealing with certain auxiliary coordinate systems, but with the main emphasis is on internal coordinates based on the squares of the internuclear distances. The most fundamental coordinates, of course, are the Cartesian coordinates $\mathbf{R}_A$, $\mathbf{R}_B$, $\mathbf{R}_C$ for the three nuclei. In terms of these, one defines the Jacobi coordinates (for nuclei of equal mass)

$$\mathbf{R}_{c.m.} = \tfrac{1}{3}(\mathbf{R}_A + \mathbf{R}_B + \mathbf{R}_C) ;$$
$$\mathbf{r} = \mathbf{R}_A - \mathbf{R}_B ; \qquad \text{(A1)}$$
$$\boldsymbol{\rho} = \mathbf{R}_C - \tfrac{1}{2}(\mathbf{R}_A + \mathbf{R}_B) .$$

$\mathbf{R}_{c.m.}$ represents the center-of-mass motion, and we shall always assume that it has been separated out. $\mathbf{r}$ and $\boldsymbol{\rho}$ give six coordinates representing the relative motion. These can be further rearranged so that three of the relative coordinates are Euler angles representing rigid rotations, defined so that the net angular momentum about the center of mass is expressible in terms of their time derivatives.[38] The remaining three coordinates are the true "internal" ones, on which the potential energy depends. It is convenient for our purposes to define them in terms of the squares of the internuclear distances. Accordingly, we define

$$Q = R^2_{AB} + R^2_{BC} + R^2_{CA}$$
$$= 2\rho^2 + \tfrac{3}{2}r^2 . \qquad \text{(A2)}$$
$$u = R^2_{BC} + R^2_{CA} - 2R^2_{AB}$$
$$= 2\rho^2 - \tfrac{3}{2}r^2 . \qquad \text{(A3)}$$
$$v = \sqrt{3}(R^2_{BC} - R^2_{CA})$$
$$= 2\sqrt{3}(\boldsymbol{\rho} \cdot \mathbf{r})$$
$$= 2\sqrt{3}\,\rho r \sigma , \qquad \text{(A4)}$$

where σ is the cosine of the angle between $\boldsymbol{\rho}$ and $\mathbf{r}$. In terms of the cross section shown in Fig. 3, with constant Q in the first octant of a Cartesian coordinate system with axes R^2_{AB}, R^2_{BC}, R^2_{CA}, u and v are measures of the distance from 0 in the directions ξ and η, respectively. We shall also need

$$S^2 = u^2 + v^2$$
$$= 2[(R^2_{AB} - R^2_{BC})^2 + (R^2_{BC} - R^2_{CA})^2 + (R^2_{CA} - R^2_{AB})^2] . \qquad \text{(A5)}$$

S measures the distance from 0 in the cross section of Fig. 3 and is thus a measure of the deviation from an equilateral triangle configuration. Finally, the angle θ is defined by

$$\theta = \tan^{-1}\left(\frac{v}{u}\right) . \qquad \text{(A6)}$$

2. Physical region

The entire octant of positive R^2_{AB}, R^2_{BC}, R^2_{CA} is not physically available, since no one of the internuclear distances can be greater than the sum of the other two. The boundary of the physical region is thus the linear configuration, where, e.g.,

$$R_{CA} = aR_{AB},\ R_{BC} = (1-a)R_{AB} . \qquad \text{(A7)}$$

For our coordinate system in terms of the *squares* of the internuclear distances, one verifies easily with the help of (A7) and (A2)–(A4) that on this boundary

$$u^2 + v^2 = Q^2 . \qquad \text{(A8)}$$

On a cross section of constant Q as in Fig. 3, therefore, the physical region is a circle. In three dimensions it is a cone, and the intersection of this cone with a tilted cross section, as in Fig. 1, is an ellipse, as stated in Sec. III A.

3. Properties of θ

It is a simple matter, starting with the definitions (A3) and (A4), to take the gradients of u and v with respect to the positions $\mathbf{R}_A$, $\mathbf{R}_B$, and $\mathbf{R}_C$ of the nuclei A, B, C. One easily obtains:

$$\tfrac{1}{2}\nabla_A u = 2\mathbf{R}_B - \mathbf{R}_A - \mathbf{R}_C ;$$
$$\tfrac{1}{2}\nabla_B u = 2\mathbf{R}_A - \mathbf{R}_B - \mathbf{R}_C ; \qquad \text{(A9)}$$
$$\tfrac{1}{2}\nabla_C u = 2\mathbf{R}_C - \mathbf{R}_A - \mathbf{R}_B .$$
$$\tfrac{1}{2}\nabla_A v = \sqrt{3}(\mathbf{R}_C - \mathbf{R}_A) ;$$
$$\tfrac{1}{2}\nabla_B v = \sqrt{3}(\mathbf{R}_B - \mathbf{R}_C) ; \qquad \text{(A10)}$$
$$\tfrac{1}{2}\nabla_C v = \sqrt{3}(\mathbf{R}_A - \mathbf{R}_B) .$$

From (A9) and (A10) one easily verifies that

$$\nabla^2 u = \nabla^2 v = (\nabla u) \cdot (\nabla v) = 0 , \qquad \text{(A11)}$$

and

$$(\nabla u)^2 = (\nabla v)^2 . \qquad \text{(A12)}$$

In these equations ∇^2 means $\nabla^2_A + \nabla^2_B + \nabla^2_C$ and ∇ is the nine-dimensional gradient.

Differentiation of (A6) in any coordinate system gives

$$\nabla\theta = \frac{u\nabla v - v\nabla u}{S^2} . \qquad \text{(A13)}$$

We now differentiate (A13) again, omitting terms that are zero because of (A11), and use (A5) to get the gradient of S, obtaining

$$\nabla^2\theta = -\frac{2}{S^3}(u\nabla v - v\nabla u) \cdot (2u\nabla u + 2v\nabla v)$$
$$= -\frac{4}{S^3}uv[(\nabla v)^2 - (\nabla u)^2] = 0 , \qquad \text{(A14)}$$

where the last equality follows from (A12).

To obtain an explicit expression for $\nabla\theta$ in Cartesian coordinates, we use (A9), (A10), and (A13) and find, after a little algebra,

$$\nabla_A\theta = \frac{4\sqrt{3}}{S^2}[\mathbf{R}_A(R^2_{AB} - R^2_{CA})$$
$$+ \mathbf{R}_B(R^2_{CA} - R^2_{BC}) + \mathbf{R}_C(R^2_{BC} - R^2_{AB})] , \qquad \text{(A15)}$$

with, of course, analogous expressions for B and C. In Jacobi coordinates, u, v, and hence θ are independent of $\mathbf{R}_{c.m.}$ and we find from (A3) and (A4)

$$\nabla_\rho u = 4\boldsymbol{\rho}\ ; \qquad \nabla_r u = -3\mathbf{r}\ . \tag{A16}$$

$$\nabla_\rho v = 2\sqrt{3}\,\mathbf{r}\ ; \qquad \nabla_r v = 2\sqrt{3}\,\boldsymbol{\rho}\ . \tag{A17}$$

From (A13), (A16), and (A17), we obtain

$$\nabla_\rho \theta = \frac{2\sqrt{3}}{S^2}\left[\left(2\rho^2 - \tfrac{3}{2}r^2\right)\mathbf{r} - 4(\boldsymbol{\rho}\cdot\mathbf{r})\boldsymbol{\rho}\right]\ ; \tag{A18}$$

$$\nabla_r \theta = \frac{2\sqrt{3}}{S^2}\left[\left(2\rho^2 - \tfrac{3}{2}r^2\right)\boldsymbol{\rho} + 3(\boldsymbol{\rho}\cdot\mathbf{r})\mathbf{r}\right]\ . \tag{A19}$$

4. Laplacian

In Jacobi coordinates, the contribution to the Laplacian from the relative motion is easily found to be

$$\nabla^2_{rel} = 2\nabla^2_r + \tfrac{3}{2}\nabla^2_\rho\ . \tag{A20}$$

Applying this to a function only of the internal coordinates u, v, Q, we find, using (A2)–(A4):

$$\nabla_r = 3\mathbf{r}\frac{\partial}{\partial Q} - 3\mathbf{r}\frac{\partial}{\partial u} + 2\sqrt{3}\,\boldsymbol{\rho}\frac{\partial}{\partial v}\ ; \tag{A21}$$

$$\nabla^2_r = 3\frac{\partial}{\partial Q} - 3\frac{\partial}{\partial u} + 9r^2\frac{\partial^2}{\partial Q^2} + 9r^2\frac{\partial^2}{\partial u^2} + 12\rho^2\frac{\partial^2}{\partial v^2} - 18r^2\frac{\partial^2}{\partial Q\partial u} + 12\sqrt{3}(\boldsymbol{\rho}\cdot\mathbf{r})\frac{\partial^2}{\partial Q\partial v} - 12\sqrt{3}(\boldsymbol{\rho}\cdot\mathbf{r})\frac{\partial^2}{\partial u\partial v}\ ; \tag{A22}$$

$$\nabla_\rho = 4\boldsymbol{\rho}\frac{\partial}{\partial Q} + 4\boldsymbol{\rho}\frac{\partial}{\partial u} + 2\sqrt{3}\,\mathbf{r}\frac{\partial}{\partial v}\ ; \tag{A23}$$

$$\nabla^2_\rho = 4\frac{\partial}{\partial Q} + 4\frac{\partial}{\partial u} + 16\rho^2\frac{\partial^2}{\partial Q^2} + 16\rho^2\frac{\partial^2}{\partial u^2} + 12r^2\frac{\partial^2}{\partial v^2} + 32\rho^2\frac{\partial^2}{\partial Q\partial u} + 16\sqrt{3}(\boldsymbol{\rho}\cdot\mathbf{r})\frac{\partial^2}{\partial Q\partial v} + 16\sqrt{3}(\boldsymbol{\rho}\cdot\mathbf{r})\frac{\partial^2}{\partial u\partial v}\ . \tag{A24}$$

Taking $2\nabla^2_r + \tfrac{3}{2}\nabla^2_\rho$, we find for the internal contribution to the Laplacian:

$$\nabla^2_{int} = 12\left[\frac{\partial}{\partial Q}\left(Q\frac{\partial}{\partial Q}\right) + Q\left(\frac{\partial^2}{\partial u^2} + \frac{\partial^2}{\partial v^2}\right) + \frac{\partial}{\partial Q}\left(u\frac{\partial}{\partial u} + v\frac{\partial}{\partial v}\right)\right]\ . \tag{A25}$$

If we use S and θ instead of u and v, (A25) becomes:

$$\nabla^2_{int} = 12\left[\frac{\partial}{\partial Q}\left(Q\frac{\partial}{\partial Q}\right) + \frac{Q}{S}\frac{\partial}{\partial S}\left(S\frac{\partial}{\partial S}\right) + \frac{Q}{S^2}\frac{\partial^2}{\partial\theta^2} + 2\frac{\partial}{\partial Q}\left(S\frac{\partial}{\partial S}\right)\right]\ . \tag{A26}$$

The Laplacian simplifies further if instead of S we use $s = S/Q$, and note that

$$\left(\frac{\partial}{\partial Q}\right)_S = \left(\frac{\partial}{\partial Q}\right)_s - \frac{s}{Q}\frac{\partial}{\partial s}\ .$$

In terms of Q, s, θ, the Laplacian takes on the form

$$\nabla^2_{int} = 12\left\{\frac{\partial}{\partial Q}\left(Q\frac{\partial}{\partial Q}\right) + \frac{1-s^2}{Qs}\frac{\partial}{\partial s}\left(s\frac{\partial}{\partial s}\right) + \frac{1}{2s^2}\frac{\partial^2}{\partial\theta^2}\right\} \tag{A27}$$

in which there are no cross terms.

5. Volume element

In terms of ρ, r, and σ, the internal contribution to the volume element is

$$d\tau_{int} = r^2\rho^2\,dr\,d\rho\,d\sigma\ . \tag{A28}$$

To convert this to the volume element in terms of Q, u, and v, we use (A4) to evaluate the Jacobian

$$J = \frac{(Q,u,v)}{(r,\rho,\sigma)} = \begin{vmatrix} 3r & -3r & 2\sqrt{3}\rho\sigma \\ 4\rho & 4\rho & 2\sqrt{3}r\sigma \\ 0 & 0 & 2\sqrt{3}r\rho \end{vmatrix} = 48\sqrt{3}\,r^2\rho^2\ .$$

Dividing this into (A28), we find for the internal volume element in our coordinate systems:

$$d\tau_{int} = \frac{1}{48\sqrt{3}}dQ\,du\,dv = \frac{S}{48\sqrt{3}}dQ\,dS\,d\theta = \frac{sQ^2}{48\sqrt{3}}dQ\,ds\,d\theta\ . \tag{A29}$$

APPENDIX B: CONSEQUENCES OF SYMMETRY FOR THE CASE OF THREE IDENTICAL NUCLEI

The quantity $\frac{1}{2}(\langle X_{AB}|\hat{H}|X_{AB}\rangle - \langle Y_{AB}|\hat{H}|Y_{AB}\rangle)$ is invariant under the (AB) interchange, and thus cannot have a nonvanishing gradient component in the η direction of Fig. 3, since if it did it would change sign for the (AB) interchange in which one goes from a small positive to a small negative displacement in the η direction. Its gradient is thus in the ξ direction, as stated in Sec. III B.

Similarly, the off-diagonal element $\langle X_{AB}|\hat{H}|Y_{AB}\rangle$ remains equal to zero as one proceeds along the ξ axis, since along this axis one has the isosceles triangle configuration and parity under interchange of A and B is a good quantum number. Its gradient therefore lies in the η direction, as stated.

Near the intersection, the traceless part of the Hamiltonian matrix [the part remaining when $c\mathbf{I}$ in Eq. (25) is subtracted off] for the case of identical nuclei has the form

$$\mathbf{H}_t = \begin{vmatrix} -a\cos\theta & b\sin\theta \\ b\sin\theta & a\cos\theta \end{vmatrix} \tag{B1}$$

if $|X_{AB}\rangle$ and $|Y_{AB}\rangle$ are used as basis kets. In particular, if $\theta = 0$ (isosceles triangle configuration with $R_{BC} = R_{CA}$), $\mathbf{H}_t$ has as eigenkets $|X_{AB}\rangle$ and $|Y_{AB}\rangle$, with eigenvalues $-\epsilon a$, and ϵa, respectively. If $\theta = 2\pi/3$ ($\cos\theta = -\frac{1}{2}$, $\sin\theta = \sqrt{3}/2$), we have the isosceles triangle configuration with $R_{BC} = R_{AB}$, so by symmetry the eigenkets of $\mathbf{H}_t$ must now be $|X_{CA}\rangle$ and $|Y_{CA}\rangle$, again with eigenvalues $-\epsilon a$ and ϵa. From (28) and (29) one finds

$$|X_{CA}\rangle = -|X_{AB}\rangle + \frac{\sqrt{3}}{2}|Y_{AB}\rangle\ . \tag{B2}$$

Inserting the trigonometric functions for $2\pi/3$ into (B1) and requiring that the resulting matrix have $|X_{CA}\rangle$ as given by (B2) as eigenket, with eigenvalue $-\epsilon a$, yields immediately the result $b=a$.

APPENDIX C: VERIFICATION OF EFFECT OF CYCLIC PERMUTATION ON EIGENKETS

We consider the cyclic permutation (ABC) and refer to Fig. 5(a). In step 1, the only effect is an increase of θ by $2\pi/3$, so that $\theta/2$ increases by $\pi/3$. This transforms $|-\rangle$ into

$$|-^{(1)}\rangle = \left(\tfrac{1}{2}\cos\frac{\theta}{2} - \frac{\sqrt{3}}{2}\sin\frac{\theta}{2}\right)|X_{AB}\rangle - \left(\tfrac{1}{2}\sin\frac{\theta}{2} + \frac{\sqrt{3}}{2}\cos\frac{\theta}{2}\right)|Y_{AB}\rangle = -\cos\frac{\theta}{2}|X_{CA}\rangle + \sin\frac{\theta}{2}|Y_{CA}\rangle , \quad (C1)$$

where in the second step we have used (28) and (29). Step 2 brings the CA line to where AB formerly was, thus transforming $|X_{CA}\rangle$ and $|Y_{CA}\rangle$ into $|X_{AB}\rangle$ and $|Y_{AB}\rangle$, respectively. Performing this transformation on (C1) and comparing the result to (30), one immediately verifies the change in sign. Similar considerations apply to $|+\rangle$.

[1]M. Born and K. Huang, *Dynamical Theory of Crystal Lattices* (Oxford University, New York, 1956), Appendix.
[2]A. Messiah, *Mecanique Quantique* (Dunod, Paris, 1959), Vol. 2, Chap. 18 [English transl.: A. Messiah, *Quantum Mechanics* (Wiley, New York, 1962), Vol. 2, Chap. 18].
[3]J. O. Hirschfelder and W. J. Meath, Adv. Chem. Phys. **12**, 3 (1967).
[4]S. Geltman, *Topics in Atomic Collision Theory* (Academic, New York, 1969), p. 175.
[5]W. Kolos, Adv. Quantum Chem. **5**, 99 (1970).
[6]J. C. Tully, in *Dynamics of Molecular Collisions, Part B*, edited by W. H. Miller (Plenum, New York, 1976), p. 217.
[7]See also, e.g., (a) J. H. Van Vleck, J. Chem. Phys. **4**, 327 (1936); (b) A. Dalgarno and R. McCarroll, Proc. R. Soc. London Ser. A **237**, 383 (1956); (c) C. J. Ballhausen and A. E. Hansen, Ann. Rev. Phys. Chem. **23**, 15 (1972).
[8]G. C. Schatz and A. Kuppermann, J. Chem. Phys. **62**, 2502 (1975); **65**, 4668 (1976).
[9]A. B. Elkowitz and R. E. Wyatt, J. Chem. Phys. **62**, 2504 (1975).
[10]Methods for nonreactive atom–diatom scattering are reviewed in D. Secrest, Ann. Rev. Phys. Chem. **24**, 379 (1973) and W. A. Lester, Jr., Adv. Quantum Chem. **9**, 199 (1975). For reactive scattering, see Ref. 11.
[11](a) W. H. Miller, J. Chem. Phys. **50**, 407 (1969); (b) A. B. Elkowitz and R. E. Wyatt, *ibid.* **63**, 702 (1975); (c) G. C. Schatz and A. Kuppermann, *ibid.* **65**, 4642 (1976); (d) R. B. Walker, J. C. Light, and A. Altenberger-Siczek, *ibid.* **64**, 1166 (1976).
[12]E. Teller, J. Phys. Chem. **41**, 109 (1937).
[13]G. Herzberg, *Molecular Spectra and Molecular Structure. III. Electronic Spectra and Electronic Structure of Polyatomic Molecules* (Van Nostrand Reinhold, New York, 1966), pp. 442–444.
[14]T. Carrington, Discuss. Faraday Soc. **53**, 27 (1972); Acc. Chem. Res. **7**, 20 (1974).
[15]J. O. Hirschfelder, J. Chem. Phys. **6**, 795 (1938).
[16]R. N. Porter, R. M. Stevens, and M. Karplus, J. Chem. Phys. **49**, 5163 (1968).
[17]J. L. Jackson and R. E. Wyatt, Chem. Phys. Lett. **18**, 161 (1973).
[18]B. M. Smirnov, Zh. Eksp. Teor. Fiz. **46**, 578 (1964) [English Transl.: Sov. Phys. JETP **19**, 394 (1964)].
[19]F. A. Matsen, J. Phys. Chem. **68**, 3283 (1964).
[20]E. Frenkel, Z. Naturforsch. Teil A **25**, 1265 (1970).
[21]G. Herzberg and H. C. Longuet-Higgins, Discuss. Faraday Soc. **35**, 77 (1963).
[22]H. C. Longuet-Higgins, Proc. R. Soc. London Ser. A **344**, 147 (1975).
[23]Y. T. Lee, R. J. Gordon, and D. R. Herschbach, J. Chem. Phys. **54**, 2410 (1971); J. D. McDonald, P. R. LeBreton, Y. T. Lee, and D. R. Herschbach, *ibid.* **56**, 769 (1972).
[24]L. D. Landau and E. M. Lifshitz, *Quantum Mechanics: Nonrelativistic Theory*, 2nd ed. (Pergamon, New York, 1965), pp. 209–211, (a) pp. 309–312, (b) pp. 476, 550–554, (c) pp. 206–208.
[25]G. Herzberg, *Molecular Spectra and Molecular Structure. I. Spectra of Diatomic Molecules*, 2nd ed. (Van Nostrand, Princeton, 1950), pp. 128–141, 237–240.
[26]See, e.g., the following and references therein: (a) W. Aberth, D. C. Lorents, R. P. Marchi, and F. T. Smith, Phys. Rev. Lett. **14**, 776 (1965), (b) P. E. Siska, J. M. Parson, T. P. Schafer, and Y. T. Lee, J. Chem. Phys. **55**, 5762 (1971), (c) P. Cantini, M. G. Dondi, G. Scoles, and F. Torello, *ibid.* **56**, 1946 (1972); (d) D. Dhuicq, J. Baudon, and M. Barat, J. Phys. B **6**, L1 (1973).
[27]See, e.g., the following, and references therein: (a) D. G. Truhlar and R. E. Wyatt, Ann. Rev. Phys. Chem. **27**, 1 (1976); (b) D. G. Truhlar, J. Chem. Phys. **65**, 1008 (1976).
[28]G. Scoles, Invited Lecture, Prog. Rep. Int. Conf. Phys. Electron. At. Collisions, **8**, 583 (1973).
[29]See, e.g., D. W. McCullough and W. D. McGrath, Chem. Phys. Lett. **8**, 353 (1971) and references therein. The reactions that have been studied involved $^{16}O_2$ and $^{18}O_2$ and a change of electronic state, but the present considerations would be necessary to treat theoretically the case when three identical nuclei are involved and the reaction involves only one potential energy surface.
[30]M. G. Dondi, U. Valbusa, and G. Scoles, Chem. Phys. Lett. **17**, 137 (1972); J. M. Farrar and Y. T. Lee, J. Chem. Phys. **57**, 5492 (1972).
[31]D. A. Micha, J. Chem. Phys. **60**, 2480 (1974).
[32](a) R. P. Saxon and J. C. Light, J. Chem. Phys. **56**, 3884 (1972); (b) J. D. Doll, T. F. George, and W. H. Miller, *ibid.* **58**, 1343 (1973); (c) A. Kuppermann, G. C. Schatz, and M. Baer, *ibid.* **65**, 4596 (1976).
[33]C. A. Mead, J. Chem. Phys. **70**, 2276 (1979).
[34]Y. Aharonov and D. Bohm, Phys. Rev. **115**, 485 (1959).
[35]M. Hamermesh, *Group Theory and Its Application to Physical Problems* (Addison-Wesley, Reading, MA, 1962).
[36]J. R. Taylor, *Scattering Theory* (Wiley, New York, 1972), Chap. 22, especially pp. 441–454. See also M. L. Goldberger and K. M. Watson, *Collision Theory* (Wiley, New York, 1964), especially pp. 123–151.
[37]W. H. Miller, J. Chem. Phys. **53**, 1949 (1970).
[38]I. N. Levine, *Molecular Spectroscopy* (Wiley-Interscience, New York, 1975), Chap. 5.

ERRATA

Erratum: On the determination of Born–Oppenheimer nuclear motion wave functions including complications due to conical intersections and identical nuclei [J. Chem. Phys. 70, 2284 (1979)].

C. Alden Mead and Donald G. Truhlar

Department of Chemistry, University of Minnesota, Minneapolis, Minnesota 55455

Section IV of Appendix A is incorrect and is corrected elsewhere.[1]

The parenthetical remark in the fourth paragraph about previous work on the sign change in the nuclear-motion wave function should be restricted to single-adiabatic-surface scattering problems, since the sign change is well known in two-electronic-state[2–8] and single-surface[9,10] treatments of the spectroscopy of bound Jahn–Teller systems and resonances. The conventional treatment has also been extended to scattering systems.[11,12]

The authors are grateful to Todd C. Thompson for suggesting that additional references should be given to avoid confusion.

[1]C. A. Mead, Chem. Phys. 49, 23 (1980).
[2]H. C. Longuet-Higgins, U. Öpik, M. H. L. Pryce, and R. A. Sack, Proc. R. Soc. London Ser. A 244, 1 (1958).
[3]H. C. Longuet-Higgins, Adv. Spectrosc. 2, 429 (1961).
[4]J. C. Slonczewski and V. L. Moruzzi, Physica 3, 237 (1967).
[5]R. Englman, *The Jahn–Teller Effect in Molecules and Crystals* (Wiley-Interscience, New York, 1972).
[6]P. Habitz and W. H. E. Schwarz, Theor. Chim. Acta 28, 267 (1973).
[7]A. I. Voronin, S. P. Karkach, V. I. Osherov, and V. G. Ushakov, Zh. Eksp. Teor. Fiz. 71, 884 (1976) [English translation: Sov. Phys. JETP 44, 465 (1976)].
[8]M. C. M. O'Brien and D. R. Pooler, J. Phys. C 12, 311 (1979).
[9]M. C. M. O'Brien, Proc. R. Soc. London Ser. A 281, 323 (1964).
[10]F. I. B. Williams, D. C. Krupka, and D. P. Breen, Phys. Rev. 179, 255 (1969).
[11]A. I. Voronin and V. I. Osherov, Zh. Eksp. Teor. Fiz. 66, 135 (1974) [English translation: Sov. Phys. JETP 39, 62 (1974)].
[12]S. P. Karkach, V. I. Osherov, and V. G. Ushakov, Zh. Eksp. Teor. Fiz. 68, 493 (1975) [English translation: Sov. Phys. JETP 41, 241 (1975)].

THE

PHYSICAL REVIEW

A journal of experimental and theoretical physics established by E. L. Nichols in 1893

SECOND SERIES, VOL. 115, No. 3 AUGUST 1, 1959

Significance of Electromagnetic Potentials in the Quantum Theory

Y. AHARONOV AND D. BOHM
H. H. Wills Physics Laboratory, University of Bristol, Bristol, England
(Received May 28, 1959; revised manuscript received June 16, 1959)

In this paper, we discuss some interesting properties of the electromagnetic potentials in the quantum domain. We shall show that, contrary to the conclusions of classical mechanics, there exist effects of potentials on charged particles, even in the region where all the fields (and therefore the forces on the particles) vanish. We shall then discuss possible experiments to test these conclusions; and, finally, we shall suggest further possible developments in the interpretation of the potentials.

1. INTRODUCTION

IN classical electrodynamics, the vector and scalar potentials were first introduced as a convenient mathematical aid for calculating the fields. It is true that in order to obtain a classical canonical formalism, the potentials are needed. Nevertheless, the fundamental equations of motion can always be expressed directly in terms of the fields alone.

In the quantum mechanics, however, the canonical formalism is necessary, and as a result, the potentials cannot be eliminated from the basic equations. Nevertheless, these equations, as well as the physical quantities, are all gauge invariant; so that it may seem that even in quantum mechanics, the potentials themselves have no independent significance.

In this paper, we shall show that the above conclusions are not correct and that a further interpretation of the potentials is needed in the quantum mechanics.

2. POSSIBLE EXPERIMENTS DEMONSTRATING THE ROLE OF POTENTIALS IN THE QUANTUM THEORY

In this section, we shall discuss several possible experiments which demonstrate the significance of potentials in the quantum theory. We shall begin with a simple example.

Suppose we have a charged particle inside a "Faraday cage" connected to an external generator which causes the potential on the cage to alternate in time. This will add to the Hamiltonian of the particle a term $V(x,t)$ which is, for the region inside the cage, a function of time only. In the nonrelativistic limit (and we shall assume this almost everywhere in the following discussions) we have, for the region inside the cage, $H=H_0+V(t)$ where H_0 is the Hamiltonian when the generator is not functioning, and $V(t)=e\phi(t)$. If $\psi_0(x,t)$ is a solution of the Hamiltonian H_0, then the solution for H will be

$$\psi=\psi_0 e^{-iS/\hbar}, \quad S=\int V(t)dt,$$

which follows from

$$i\hbar\frac{\partial\psi}{\partial t}=\left(i\hbar\frac{\partial\psi_0}{\partial t}+\psi_0\frac{\partial S}{\partial t}\right)e^{-iS/\hbar}=[H_0+V(t)]\psi=H\psi.$$

The new solution differs from the old one just by a phase factor and this corresponds, of course, to no change in any physical result.

Now consider a more complex experiment in which a single coherent electron beam is split into two parts and each part is then allowed to enter a long cylindrical metal tube, as shown in Fig. 1.

After the beams pass through the tubes, they are combined to interfere coherently at F. By means of time-determining electrical "shutters" the beam is chopped into wave packets that are long compared with the wavelength λ, but short compared with the length of the tubes. The potential in each tube is determined by a time delay mechanism in such a way that the potential is zero in region I (until each packet is well inside its tube). The potential then grows as a function of time, but differently in each tube. Finally, it falls back to zero, before the electron comes near the

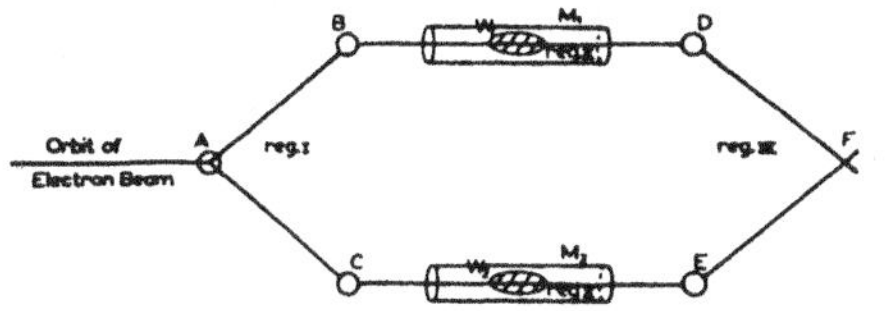

FIG. 1. Schematic experiment to demonstrate interference with time-dependent scalar potential. A, B, C, D, E: suitable devices to separate and divert beams. W_1, W_2: wave packets. M_1, M_2: cylindrical metal tubes. F: interference region.

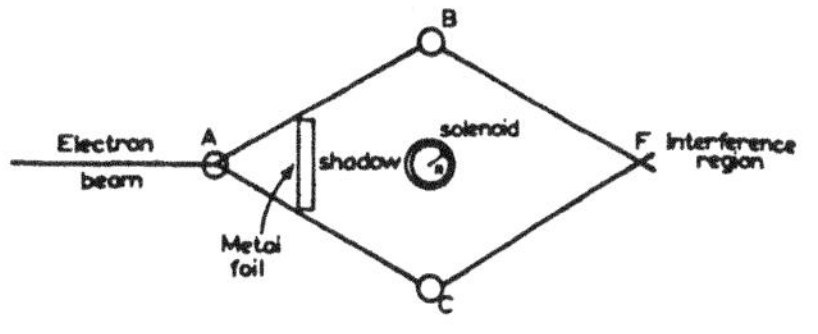

FIG. 2. Schematic experiment to demonstrate interference with time-independent vector potential.

other edge of the tube. Thus the potential is nonzero only while the electrons are well inside the tube (region II). When the electron is in region III, there is again no potential. The purpose of this arrangement is to ensure that the electron is in a time-varying potential without ever being in a field (because the field does not penetrate far from the edges of the tubes, and is nonzero only at times when the electron is far from these edges).

Now let $\psi(x,t)=\psi_1^0(x,t)+\psi_2^0(x,t)$ be the wave function when the potential is absent (ψ_1^0 and ψ_2^0 representing the parts that pass through tubes 1 and 2, respectively). But since V is a function only of t wherever ψ is appreciable, the problem for each tube is essentially the same as that of the Faraday cage. The solution is then

$$\psi=\psi_1^0e^{-iS_1/\hbar}+\psi_2^0e^{-iS_2/\hbar},$$

where

$$S_1=e\int\varphi_1dt,\quad S_2=e\int\varphi_2dt.$$

It is evident that the interference of the two parts at F will depend on the phase difference $(S_1-S_2)/\hbar$. Thus, there is a physical effect of the potentials even though no force is ever actually exerted on the electron. The effect is evidently essentially quantum-mechanical in nature because it comes in the phenomenon of interference. We are therefore not surprised that it does not appear in classical mechanics.

From relativistic considerations, it is easily seen that the covariance of the above conclusion demands that there should be similar results involving the vector potential, **A**.

The phase difference, $(S_1-S_2)/\hbar$, can also be expressed as the integral $(e/\hbar)\oint\varphi dt$ around a closed circuit in space-time, where φ is evaluated at the place of the center of the wave packet. The relativistic generalization of the above integral is

$$\frac{e}{\hbar}\oint\left(\varphi dt-\frac{\mathbf{A}}{c}\cdot dx\right),$$

where the path of integration now goes over any closed circuit in space-time.

As another special case, let us now consider a path in space only ($t=$ constant). The above argument suggests that the associated phase shift of the electron wave function ought to be

$$\Delta S/\hbar=-\frac{e}{c\hbar}\oint\mathbf{A}\cdot dx,$$

where $\oint\mathbf{A}\cdot dx=\int\mathbf{H}\cdot ds=\phi$ (the total magnetic flux inside the circuit).

This corresponds to another experimental situation. By means of a current flowing through a very closely wound cylindrical solenoid of radius R, center at the origin and axis in the z direction, we create a magnetic field, **H**, which is essentially confined within the solenoid. However, the vector potential, **A**, evidently, cannot be zero everywhere outside the solenoid, because the total flux through every circuit containing the origin is equal to a constant

$$\phi_0=\int\mathbf{H}\cdot ds=\int\mathbf{A}\cdot dx.$$

To demonstrate the effects of the total flux, we begin, as before, with a coherent beam of electrons. (But now there is no need to make wave packets.) The beam is split into two parts, each going on opposite sides of the solenoid, but avoiding it. (The solenoid can be shielded from the electron beam by a thin plate which casts a shadow.) As in the former example, the beams are brought together at F (Fig. 2).

The Hamiltonian for this case is

$$H=\frac{[\mathbf{P}-(e/c)\mathbf{A}]^2}{2m}.$$

In singly connected regions, where $\mathbf{H}=\nabla\times\mathbf{A}=0$, we can always obtain a solution for the above Hamiltonian by taking $\psi=\psi_0e^{-iS/\hbar}$, where ψ_0 is the solution when $\mathbf{A}=0$ and where $\nabla S/\hbar=(e/c)\mathbf{A}$. But, in the experiment discussed above, in which we have a multiply connected region (the region outside the solenoid), $\psi_0e^{-iS/\hbar}$ is a non-single-valued function[1] and therefore, in general, not a permissible solution of Schrödinger's equation. Nevertheless, in our problem it is still possible to use such solutions because the wave function splits into two parts $\psi=\psi_1+\psi_2$, where ψ_1 represents the beam on

[1] Unless $\phi_0=nhc/e$, where n is an integer.

one side of the solenoid[a] and ψ_2 the beam on the opposite side. Each of these beams stays in a simply connected region. We therefore can write

$$\psi_1=\psi_1^0e^{-iS_1/\hbar},\quad \psi_2=\psi_2^0e^{-iS_2/\hbar},$$

where S_1 and S_2 are equal to $(e/c)\int\mathbf{A}\cdot d\mathbf{x}$ along the paths of the first and second beams, respectively. (In Sec. 4, an exact solution for this Hamiltonian will be given, and it will confirm the above results.)

The interference between the two beams will evidently depend on the phase difference,

$$(S_1-S_2)/\hbar=(e/\hbar c)\oint\mathbf{A}\cdot d\mathbf{x}=(e/\hbar c)\phi_0.$$

This effect will exist, even though there are no magnetic forces acting in the places where the electron beam passes.

In order to avoid fully any possible question of contact of the electron with the magnetic field we note that our result would not be changed if we surrounded the solenoid by a potential barrier that reflects the electrons perfectly. (This, too, is confirmed in Sec. 4.)

It is easy to devise hypothetical experiments in which the vector potential may influence not only the interference pattern but also the momentum. To see this, consider a periodic array of solenoids, each of which is shielded from direct contact with the beam by a small plate. This will be essentially a grating. Consider first the diffraction pattern without the magnetic field, which will have a discrete set of directions of strong constructive interference. The effect of the vector potential will be to produce a shift of the relative phase of the wave function in different elements of the gratings. A corresponding shift will take place in the directions, and therefore the momentum of the diffracted beam.

3. A PRACTICABLE EXPERIMENT TO TEST FOR THE EFFECTS OF A POTENTIAL WHERE THERE ARE NO FIELDS

As yet no direct experiments have been carried out which confirm the effect of potentials where there is no field. It would be interesting therefore to test whether such effects actually exist. Such a test is, in fact, within the range of present possibilities.[2] Recent experiments[3,4] have succeeded in obtaining interference from electron beams that have been separated in one case by as much as 0.8 mm.[3] It is quite possible to wind solenoids which are smaller than this, and therefore to place them between the separate beams. Alternatively, we may obtain localized lines of flux of the right magnitude (the magnitude has to be of the order of $\phi_0=2\pi c\hbar/e\sim4\times10^{-7}$ gauss cm^2) by means of fine permanently magnetized "whiskers".[5] The solenoid can be used in Marton's device,[3] while the whisker is suitable for another experimental setup[4] where the separation is of the order of microns and the whiskers are even smaller than this.

In principle, we could do the experiment by observing the interference pattern with and without the magnetic flux. But since the main effect of the flux is only to displace the line pattern without changing the interval structure, this would not be a convenient experiment to do. Instead, it would be easier to vary the magnetic flux within the same exposure for the detection of the interference patterns. Such a variation would, according to our previous discussion, alter the sharpness and the general form of the interference bands. This alteration would then constitute a verification of the predicted phenomena.

When the magnetic flux is altered, there will, of course, be an induced electric field outside the solenoid, but the effects of this field can be made negligible. For example, suppose the magnetic flux were suddenly altered in the middle of an exposure. The electric field would then exist only for a very short time, so that only a small part of the beam would be affected by it.

4. EXACT SOLUTION FOR SCATTERING PROBLEMS

We shall now obtain an exact solution for the problem of the scattering of an electron beam by a magnetic field in the limit where the magnetic field region tends to a zero radius, while the total flux remains fixed. This corresponds to the setup described in Sec. 2 and shown in Fig. 2. Only this time we do not split the plane wave into two parts. The wave equation outside the magnetic field region is, in cylindrical coordinates,

$$\left[\frac{\partial^2}{\partial r^2}+\frac{1}{r}\frac{\partial}{\partial r}+\frac{1}{r^2}\left(\frac{\partial}{\partial\theta}+i\alpha\right)^2+k^2\right]\psi=0, \tag{1}$$

where $\mathbf{k}$ is the wave vector of the incident particle and $\alpha=-e\phi/ch$. We have again chosen the gauge in which $A_r=0$ and $A_\theta=\phi/2\pi r$.

The general solution of the above equation is

$$\psi=\sum_{m=-\infty}^{\infty}e^{im\theta}[a_mJ_{m+\alpha}(kr)+b_mJ_{-(m+\alpha)}(kr)], \tag{2}$$

where a_m and b_m are arbitrary constants and $J_{m+\alpha}(kr)$ is a Bessel function, in general of fractional order (dependent on ϕ). The above solution holds only for $r>R$. For $r<R$ (inside the magnetic field) the solution has been worked out.[6] By matching the solutions at $r=R$ it is easily shown that only Bessel functions of positive order will remain, when R approaches zero:

---

[a] Dr. Chambers is now making a preliminary experimental study of this question at Bristol.

[3] L. Marton, Phys. Rev. 85, 1057 (1952); 90, 490 (1953). Marton, Simpson, and Suddeth, Rev. Sci. Instr. 25, 1099 (1954).

[4] G. Mollenstedt, Naturwissenschaften 42, 41 (1955); G. Mollenstedt and H. Düker, Z. Physik 145, 377 (1956).

[5] See, for example, Sidney S. Brenner, Acta Met. 4, 62 (1956).

[6] L. Page, Phys. Rev. 36, 444 (1930).

This means that the probability of finding the particle inside the magnetic field region approaches zero with R. It follows that the wave function would not be changed if the electron were kept away from the field by a barrier whose radius also went to zero with R.

The general solution in the limit of R tending to zero is therefore

$$\psi=\sum_{m=-\infty}^{\infty} a_m J_{|m+\alpha|} e^{im\theta}. \tag{3}$$

We must then choose a_m so that ψ represents a beam of electrons that is incident from the right ($\theta=0$). It is important, however, to satisfy the initial condition that the current density,

$$\mathbf{j}=\frac{h(\psi^*\nabla\psi-\psi\nabla\psi^*)}{2im}-\frac{e}{mc}\mathbf{A}\psi^*\psi, \tag{4}$$

shall be constant and in the x direction. In the gauge that we are using, we easily see that the correct incident wave is $\psi_{\text{inc}}=e^{-ikx}e^{-i\alpha\theta}$. Of course, this wave function holds only to the right of the origin, so that no problem of multiple-valuedness arises.

We shall show in the course of this calculation that the above conditions will be satisfied by choosing $a_m=(-i)^{|m+\alpha|}$, in which case, we shall have

$$\psi=\sum_{m=-\infty}^{\infty}(-i)^{|m+\alpha|}J_{|m+\alpha|}e^{im\theta}.$$

It is convenient to split ψ into the following three parts: $\psi=\psi_1+\psi_2+\psi_3$, where

$$\psi_1=\sum_{m=1}^{\infty}(-i)^{m+\alpha}J_{m+\alpha}e^{im\theta},$$

$$\psi_2=\sum_{m=-\infty}^{-1}(-i)^{m+\alpha}J_{m+\alpha}e^{im\theta},$$

$$=\sum_{m=1}^{\infty}(-i)^{m-\alpha}J_{m-\alpha}e^{-im\theta}, \tag{5}$$

$$\psi_3=(-i)^{|\alpha|}J_{|\alpha|}.$$

Now ψ_1 satisfies the simple differential equation

$$\frac{\partial\psi_1}{\partial r'}=\sum_{m=1}^{\infty}(-i)^{m+\alpha}J_{m+\alpha}'e^{im\theta}$$

$$=\sum_{m=1}^{\infty}(-i)^{m+\alpha}\frac{J_{m+\alpha-1}-J_{m+\alpha+1}}{2}e^{im\theta},\quad r'=kr \tag{6}$$

where we have used the well-known formula for Bessel functions:

$$dJ_\gamma(r)/dr=\tfrac{1}{2}(J_{\gamma-1}-J_{\gamma+1}).$$

As a result, we obtain

$$\begin{aligned}\frac{\partial\psi_1}{\partial r'}&=\frac{1}{2}\sum_{m'=0}^{\infty}(-i)^{m'+\alpha+1}J_{m'+\alpha}e^{i(m'+1)\theta}\\&\quad-\frac{1}{2}\sum_{m'=2}^{\infty}(-i)^{m'+\alpha-1}J_{m'+\alpha}e^{i(m'-1)\theta}\\&=\frac{1}{2}\sum_{m'=1}^{\infty}(-i)^{m'+\alpha}J_{m'+\alpha}e^{im'\theta}(-ie^{i\theta}+i^{-1}e^{-i\theta})\\&\quad+\tfrac{1}{2}(-i)^{\alpha}[J_{\alpha+1}-ie^{i\theta}J_\alpha].\end{aligned} \tag{7}$$

So

$$\partial\psi_1/\partial r'=-i\cos\theta\psi_1+\tfrac{1}{2}(-i)^\alpha(J_{\alpha+1}-iJ_\alpha e^{i\theta}).$$

This differential equation can be easily integrated to give

$$\psi_1=A\int_0^{r'}e^{ir'\cos\theta}[J_{\alpha+1}-iJ_\alpha e^{i\theta}]dr', \tag{8}$$

where

$$A=\tfrac{1}{2}(-i)^\alpha e^{-ir'\cos\theta}.$$

The lower limit of the integration is determined by the requirement that when r' goes to zero, ψ_1 also goes to to zero because, as we have seen, ψ_1 includes Bessel functions of positive order only.

In order to discuss the asymptotic behavior of ψ_1, let us write it as $\psi_1=A[I_1-I_2]$, where

$$\begin{aligned}I_1&=\int_0^{\infty}e^{ir'\cos\theta}[J_{\alpha+1}-ie^{i\theta}J_\alpha]dr',\\I_2&=\int_r^{\infty}e^{ir'\cos\theta}[J_{\alpha+1}-ie^{i\theta}J_\alpha]dr'.\end{aligned} \tag{9}$$

The first of these integrals is known[7]:

$$\int_0^\infty e^{i\beta r}J_\alpha(kr)=\frac{e^{i[\alpha\arcsin(\beta/k)]}}{(k^2-\beta^2)^{\frac{1}{2}}},\quad 0<\beta<k,\quad -2<\alpha.$$

In our cases, $\beta=\cos\theta$, $k=1$, so that

$$I_1=\left[\frac{e^{i\alpha(\frac{1}{2}\pi-|\theta|)}}{|\sin\theta|}-ie^{i\theta}\frac{e^{i(\alpha+1)(\frac{1}{2}\pi-|\theta|)}}{|\sin\theta|}\right]. \tag{10}$$

Because the integrand is even in θ, we have written the final expression for the above integral as a function of $|\theta|$ and of $|\sin\theta|$. Hence

$$\begin{aligned}I_1&=e^{i\alpha(\frac{1}{2}\pi-|\theta|)}\left[\frac{ie^{-i|\theta|}-ie^{i\theta}}{|\sin\theta|}\right]\\&=0\quad\text{for }\theta<0,\\&=e^{-i\alpha\theta}2i^\alpha\quad\text{for }\theta>0,\end{aligned} \tag{11}$$

where we have taken θ as going from $-\pi$ to π.

[7] See, for example, W. Gröbner and N. Hofreiter, *Integraltafel* (Springer-Verlag, Berlin, 1949).

We shall see presently that I_1 represents the largest term in the asymptotic expansion of ψ_1. The fact that it is zero for $\theta<0$ shows that this part of ψ_1 passes (asymptotically) only on the upper side of the singularity. To explain this, we note that ψ_1 contains only positive values of m, and therefore of the angular momentum. It is quite natural then that this part of ψ_1 goes on the upper side of the singularity. Similarly, since according to (5)

$$\psi_2(r',\theta,\alpha)=\psi_1(r',-\theta,-\alpha),$$

it follows that ψ_2 will behave oppositely to ψ_1 in this regard, so that together they will make up the correct incident wave.

Now, in the limit of $r'\to\infty$ we are allowed to take in the integrand of I_2 the first asymptotic term of J_α,[8] namely $J_\alpha\to(2/\pi r)^{1/2}\cos(r'-\frac{1}{2}\alpha-\frac{1}{4}\pi)$. We obtain

$$I_2=\int_r^\infty e^{ir'\cos\theta}(J_{\alpha+1}-ie^{i\theta}J_\alpha)dr'\to C+D, \tag{12}$$

where

$$C=\int_r^\infty e^{ir'\cos\theta}[\cos(r'-\tfrac{1}{2}(\alpha+1)\pi-\tfrac{1}{4}\pi)]\frac{dr'}{(r')^{1/2}}\left(\frac{2}{\pi}\right)^{1/2},$$
$$D=\int_r^\infty e^{ir'\cos\theta}[\cos(r'-\tfrac{1}{2}\alpha-\tfrac{1}{4}\pi)]\frac{dr'}{(r')^{1/2}}\left(\frac{2}{\pi}\right)^{1/2}(-i)e^{i\theta}. \tag{13}$$

Then

$$C=\int_r^\infty e^{ir'\cos\theta}\left[e^{i[r'-\frac{1}{2}(\alpha+1)\pi-\frac{1}{4}\pi]}+e^{-i[r'-\frac{1}{2}(\alpha+1)\pi-\frac{1}{4}\pi]}\right]\frac{dr'}{(2\pi r')^{1/2}}$$
$$=\left(\frac{2}{\pi}\right)^{1/2}\frac{(-i)^{\alpha+1/2}}{(1+\cos\theta)^{1/2}}\int_{[r'(1+\cos\theta)]^{1/2}}^\infty\exp(+iz^2)dz$$
$$+\left(\frac{2}{\pi}\right)^{1/2}\frac{i^{\alpha+1/2}}{(1-\cos\theta)^{1/2}}\int_{[r'(1-\cos\theta)]^{1/2}}^\infty\exp(-iz^2)dz, \tag{14}$$

where we have put

$$z=[r'(1+\cos\theta)]^{1/2}\quad\text{and}\quad z=[r'(1-\cos\theta)]^{1/2},$$

respectively.

Using now the well-known asymptotic behavior of the error function,[9]

$$\int_a^\infty\exp(iz^2)dz\to\frac{i}{2}\frac{\exp(ia^2)}{a},$$
$$\int_a^\infty\exp(-iz^2)dz\to\frac{-i}{2}\frac{\exp(-ia^2)}{a}, \tag{15}$$

we finally obtain

$$C=\left[\frac{(-i)^{\alpha+1/2}}{(2\pi)^{1/2}}\frac{e^{ir'}}{[r'(1+\cos\theta)^2]^{1/2}}+\frac{i^{\alpha+1/2}}{(2\pi)^{1/2}}\frac{e^{-ir'}}{[r'(1-\cos\theta)^2]^{1/2}}\right]e^{ir'\cos\theta}, \tag{16}$$

$$D=\left[\frac{(-i)^{\alpha-1/2}}{(2\pi)^{1/2}}\frac{e^{ir'}}{[r'(1+\cos\theta)^2]^{1/2}}+\frac{i^{\alpha-1/2}}{(2\pi)^{1/2}}\frac{e^{-ir'}}{[r'(1-\cos\theta)^2]^{1/2}}\right]e^{ir'\cos\theta}(-i)e^{i\theta}. \tag{17}$$

Now adding (16) and (17) together and using (13) and (9), we find that the term of $1/(r')^{1/2}$ in the asymptotic expansion of ψ_1 is

$$\frac{(-i)^{1/2}}{2(2\pi)^{1/2}}\left[(-1)^\alpha\frac{e^{ir'}}{(r')^{1/2}}\frac{1+e^{i\theta}}{1+\cos\theta}+i\frac{e^{-ir'}}{(r')^{1/2}}\frac{1-e^{i\theta}}{1-\cos\theta}\right]. \tag{18}$$

Using again the relation between ψ_1 and ψ_2 we obtain for the corresponding term in ψ_2

$$\frac{(-i)^{1/2}}{2(2\pi)^{1/2}}\left[(-1)^{-\alpha}\frac{e^{ir'}}{(r')^{1/2}}\frac{1+e^{-i\theta}}{1+\cos\theta}+i\frac{e^{-ir'}}{(r')^{1/2}}\frac{1-e^{-i\theta}}{1-\cos\theta}\right]. \tag{19}$$

Adding (18) and (19) and using (11), we finally get

$$\psi_1+\psi_2\to\frac{(-i)^{1/2}}{(2\pi)^{1/2}}\left[\frac{ie^{-ir'}}{(r')^{1/2}}+\frac{e^{ir'}}{(r')^{1/2}}\frac{\cos(\pi\alpha-\frac{1}{2}\theta)}{\cos(\frac{1}{2}\theta)}\right]+e^{-i(r'\cos\theta+\alpha\theta)}. \tag{20}$$

There remains the contribution of ψ_3, whose asymptotic behavior is [see Eq. (12)]

$$(-i)^{|\alpha|}J_{|\alpha|}(r')\to(-i)^{|\alpha|}\left(\frac{2}{\pi r'}\right)^{1/2}\cos(r'-\tfrac{1}{4}\pi-\tfrac{1}{2}|\alpha|\pi).$$

Collecting all terms, we find

$$\psi=\psi_1+\psi_2+\psi_3\to e^{-i(\alpha\theta+r'\cos\theta)}\pm\frac{e^{ir'}}{(2\pi ir')^{1/2}}\sin\pi\alpha\frac{e^{-i\theta/2}}{\cos(\theta/2)}, \tag{21}$$

where the $\pm$ sign is chosen according to the sign of α.

The first term in equation (21) represents the incident wave, and the second the scattered wave.[10] The scattering cross section is therefore

$$\sigma=\frac{\sin^2\pi\alpha}{2\pi}\frac{1}{\cos^2(\theta/2)}. \tag{22}$$

[8] E. Jahnke and F. Emde, *Tables of Functions* (Dover Publications, Inc., New York, 1943), fourth edition, p. 138.

[9] Reference 8, p. 24.

[10] In this way, we verify, of course, that our choice of the a_m for Eq. (3) satisfies the correct boundary conditions.

When $\alpha=n$, where n is an integer, then σ vanishes. This is analogous to the Ramsauer effect.[11] σ has a maximum when $\alpha=n+\frac{1}{2}$.

The asymptotic formula (21) holds only when we are not on the line $\theta=\pi$. The exact solution, which is needed on this line, would show that the second term will combine with the first to make a single-valued wave function, despite the non-single-valued character of the two parts, in the neighborhood of $\theta=\pi$. We shall see this in more detail presently for the special case $\alpha=n+\frac{1}{2}$.

In the interference experiment discussed in Sec. 2, diffraction effects, represented in Eq. (21) by the scattered wave, have been neglected. Therefore, in this problem, it is adequate to use the first term of Eq. (21). Here, we see that the phase of the wave function has a different value depending on whether we approach the line $\theta=\pm\pi$ from positive or negative angles, i.e., from the upper or lower side. This confirms the conclusions obtained in the approximate treatment of Sec. 2.

We shall discuss now the two special cases that can be solved exactly. The first is the case where $\alpha=n$. Here, the wave function is $\psi=e^{-ikx}e^{-i\alpha\theta}$, which is evidently single-valued when α is an integer. (It can be seen by direct differentiation that this is a solution.)

The second case is that of $\alpha=n+\frac{1}{2}$. Because $J_{(n+\frac{1}{2})}(r)$ is a closed trigonometric function, the integrals for ψ can be carried out exactly.

The result is

$$\psi=\frac{i^{\frac{1}{2}}}{\sqrt{2}}e^{-i(\frac{1}{2}\theta+r'\cos\theta)}\int_0^{[r'(1+\cos\theta)]^{\frac{1}{2}}}\exp(iz^2)dz. \qquad (23)$$

This function vanishes on the line $\theta=\pi$. It can be seen that its asymptotic behavior is the same as that of Eq. (2) with α set equal to $n+\frac{1}{2}$. In this case, the single-valuedness of ψ is evident. In general, however, the behavior of ψ is not so simple, since ψ does not become zero on the line $\theta=\pi$.

5. DISCUSSION OF SIGNIFICANCE OF RESULTS

The essential result of the previous discussion is that in quantum theory, an electron (for example) can be influenced by the potentials even if all the field regions are excluded from it. In other words, in a field-free multiply-connected region of space, the physical properties of the system still depend on the potentials.

It is true that all these effects of the potentials depend only on the gauge-invariant quantity $\oint \mathbf{A}\cdot d\mathbf{x}=\int \mathbf{H}\cdot d\mathbf{s}$, so that in reality they can be expressed in terms of the fields inside the circuit. However, according to current relativistic notions, all fields must interact only locally. And since the electrons cannot reach the regions where the fields are, we cannot interpret such effects as due to the fields themselves.

[11] See, for example, D. Bohm, *Quantum Theory* (Prentice-Hall, Inc., Englewood Cliffs, New Jersey, 1951).

In classical mechanics, we recall that potentials cannot have such significance because the equation of motion involves only the field quantities themselves. For this reason, the potentials have been regarded as purely mathematical auxiliaries, while only the field quantities were thought to have a direct physical meaning.

In quantum mechanics, the essential difference is that the equations of motion of a particle are replaced by the Schrödinger equation for a wave. This Schrödinger equation is obtained from a canonical formalism, which cannot be expressed in terms of the fields alone, but which also requires the potentials. Indeed, the potentials play a role, in Schrödinger's equation, which is analogous to that of the index of refration in optics, The Lorentz force $[e\mathbf{E}+(e/c)\mathbf{v}\times\mathbf{H}]$ does not appear anywhere in the fundamental theory, but appears only as an approximation holding in the classical limit. It would therefore seem natural at this point to propose that, in quantum mechanics, the fundamental physical entities are the potentials, while the fields are derived from them by differentiations.

The main objection that could be raised against the above suggestion is grounded in the gauge invariance of the theory. In other words, if the potentials are subject to the transformation $A_\mu \rightarrow A_\mu'=A_\mu+\partial\psi/\partial x_\mu$, where ψ is a continuous scalar function, then all the known physical quantities are left unchanged. As a result, the same physical behavior is obtained from any two potentials, $A_\mu(x)$ and $A_\mu'(x)$, related by the above transformation. This means that insofar as the potentials are richer in properties than the fields, there is no way to reveal this additional richness. It was therefore concluded that the potentials cannot have any meaning, except insofar as they are used mathematically, to calculate the fields.

We have seen from the examples described in this paper that the above point of view cannot be maintained for the general case. Of course, our discussion does not bring into question the gauge invariance of the theory. But it does show that in a theory involving only local interactions (e.g., Schrödinger's or Dirac's equation, and current quantum-mechanical field theories), the potentials must, in certain cases, be considered as physically effective, even when there are no fields acting on the charged particles.

The above discussion suggests that some further development of the theory is needed. Two possible directions are clear. First, we may try to formulate a nonlocal theory in which, for example, the electron could interact with a field that was a finite distance away. Then there would be no trouble in interpreting these results, but, as is well known, there are severe difficulties in the way of doing this. Secondly, we may retain the present local theory and, instead, we may try to give a further new interpretation to the potentials. In other words, we are led to regard $A_\mu(x)$ as a physical variable. This means that we must be able to define the physical difference between two quantum states which differ only by gauge transformation. It will be shown in a future paper that in a system containing an undefined number of charged particles (i.e., a superposition of states of different total charge), a new Hermitian operator, essentially an angle variable, can be introduced, which is conjugate to the charge density and which may give a meaning to the gauge. Such states have actually been used in connection with recent theories of superconductivity and superfluidity[12] and we shall show their relation to this problem in more detail.

ACKNOWLEDGMENTS

We are indebted to Professor M. H. L. Pryce for many helpful discussions. We wish to thank Dr. Chambers for many discussions connected with the experimental side of the problem.

[12] See, for example, C. G. Kuper, *Advances in Physics*, edited by N. F. Mott (Taylor and Francis, Ltd., London, 1959), Vol. 8, p. 25, Sec. 3, Par. 3.

Chapter 3

FOUNDATIONS

* Original Contribution.

3

Foundations

We saw in the previous chapter how molecular and optical physicists originally learned to take account of geometric phases. In optics, Pancharatnam's phase led to measurable interference effects. In molecular physics, the molecular Aharonov–Bohm effect was found to have a significant effect on molecular dynamics near to a degeneracy of two electronic energy levels. It was left for Berry to fully appreciate the universal significance of these phases, to show that whenever an adiabatic approximation applies, we may expect to find a geometric phase. The crucial role of adiabaticity is to make sure that the cyclic variation of parameters leads to cyclic evolution. (If one is otherwise able to ensure a cyclic evolution, then as Aharonov and Anandan have pointed out, the adiabatic hypothesis may be dispensed with.) It is the ubiquity of the adiabatic approximation, in both theoretical and experimental physics, that has led to the wide application of Berry's basic observation.

We now come to Berry's original paper on the quantal adiabatic phase [3.1], and its subsequent elaborations and generalizations. Among the latter are Wilczek and Zee's work on non-Abelian phases associated with adiabatic evolution of degenerate hamiltonians [3.3], and Aharanov and Anandan's geometric phase for cyclic evolution, not necessarily adiabatic, in projective Hilbert space [3.4]. The final entry of this chapter is a detailed discussion of adiabatic phases and related issues in the context of the Born-Oppenheimer approximation [3.7].

The paper "Quantal phase factors accompanying adiabatic changes" [3.1], elegantly presents the key concepts surrounding what we have come to know as Berry's phase—the gauge invariance of the adiabatic phase, the expression of the phase as the integral of a two-form over an enclosed surface, the theorem on degeneracies of a complex hamiltonian. The example of a spin in a slowly changing magnetic field, which has become a paradigm in studies (and measurements) of geometric phases, is shown to lead to magnetic–monopole–like effects. Berry proposes an experiment to measure the phase, by splitting a beam of coherently polarized electrons in a magnetic field, rotating the field around one of the beams, and measuring the resulting phase difference by interference.

In the next paper [3.2], Simon gives a mathematical interpretation of Berry's phase as the holonomy of a complex line bundle. To make his paper more accessible to non-mathematicians, we include an appendix on holonomy and fiber bundles at the end of this introduction. The appendix may also help to clarify mathematical aspects of several other papers in this and later chapters.

Berry's paper assumes that the energy eigenstate undergoing adiabatic evolution is non-degenerate. If this is so, then a cyclic variation C of the external parameters will return the system to its original state, multiplied by a complex number of unit modulus, the product of a dynamical phase and a geometric phase $\exp i\gamma(C)$:

$$|\Psi(T)\rangle = \exp\left\{-\frac{i}{\hbar}\int^{T} E(t)\,dt\right\}\cdot \exp i\gamma(C)\,|\Psi(0)\rangle \tag{3.1}$$

However, when the state is degenerate over the full course of its evolution, the system need not return to the original eigenstate, but only to one of the degenerate states. As observed in the paper by Wilczek and Zee [3.3], the accumulated "phase" for an N-fold degenerate level will actually be a $U(N)$ matrix in general:

$$|\Psi_\alpha(T)\rangle = \exp\left\{-\frac{i}{\hbar}\int^{T} E(t)\,dt\right\} U_{\alpha\beta}|\Psi_\beta(0)\rangle \tag{3.2}$$

where α runs from 1 to N. The unitary matrix $U_{\alpha\beta}$ may be expressed as a time-ordered exponential integral:

$$U = \mathrm{T}\exp\int_0^{T} A(t)\,dt\,; \qquad A_{\alpha\beta} \equiv \langle\Psi_\alpha|\dot{\Psi}_\beta\rangle \tag{3.3}$$

Wilczek and Zee present several examples of systems exhibiting these non-Abelian phase factors; many more may be found in Chapter 4. Systems with global degeneracies are quite interesting experimentally, because the $U(N)$ adiabatic phase represents the leading contribution to mixings between degenerate levels.

Another important generalization of Berry's phase, one which has recently provoked a great deal of interest, is the phase of Aharonov and Anandan [3.4]. Recall that Berry studied cyclic evolution in parameter space, which in the adiabatic limit always leads to to cyclic evolution in the projective Hilbert space $\mathcal{R}$ of states $|\Psi\rangle$ modulo phases, as shown by Eq. (3.1). Thus the essential ingredient needed in order to define a geometric phase is the closed path in $\mathcal{R}$; adiabaticity is the sufficient (but not necessary) condition which guarantees the existence of such a path. Aharonov and Anandan take $\mathcal{R}$ as their starting point, and show how any closed path in

$\mathcal{R}$ has a geometric phase associated to it in a natural way. The underlying parameter space plays no fundamental role in their considerations, although of course the Aharonov–Anandan (AA) phase reduces to Berry's phase when the closed path arises from adiabatic evolution of external parameters. In their paper, they discuss three examples of cyclic evolution which are not necessarily adiabatic. One example, an electron in an external electromagnetic field, is used to demonstrate that Aharonov-Bohm effect is a special case of the AA phase. The other applications concern the much-studied system of a spin precessing in a magnetic field—in fact, the AA phase has been directly measured in magnetic resonance experiments [4.6]. In general, however, it seems to be rather difficult to ensure cyclic evolution in projective Hilbert space in the absence of an adiabatic hypothesis.

The paper [3.5] by Samuel and Bhandari shows that even the hypothesis of cyclic evolution may be dispensed with. Therein, these authors propose a further generalization of both the Berry and AA phases. Harking back to work of Pancharatnam in optics [2.2] they point out that there is a natural way to close an open path, if one is given a metric on the underlying space. Namely, one simply joins the two ends by a geodesic. So if the state of a system undergoes a non-cyclic evolution as its parameters are varied over a time interval from 0 to T, we can "close" its evolution with a geodesic (in projective Hilbert space), and obtain the corresponding geometric phase. This shows that, provided we are given a metric on the projective Hilbert space $\mathcal{R}$, and *modulo* questions of uniqueness, there is a geometric phase associated to every path, closed or not! Now over Hilbert space there is indeed a natural *gauge invariant* metric, which Berry has discussed in the introduction to this volume [1.1]. If $|dn\rangle$ is a tangent vector to $\mathcal{R}$ at $|n\rangle$, then the metric tensor is

$$ds^2 = \langle \partial_i n| \left(1 - |n\rangle\langle n|\right) |\partial_j n\rangle \, dX_i dX_j = g_{ij} \, dX_i dX_j \tag{3.4}$$

This metric is not positive definite, and does not in general determine a unique geodesic between any two points. However, as Samuel and Bhandari show, the geometric phase obtained is *independent* of the choice of geodesic. It has a nice physical interpretation, too: when the initial and final states are "in phase" (so that the geometric phase is 1), then the norm of the sum of the two states reaches a maximum. The only condition for the relative phase to be well-defined is that the initial and final states have a nonzero overlap. All this is strongly reminiscent of Pancharatnam's connection, which allows one to compare relative phases of beams of light in very different polarization states, not just nearby ones. There, the relative phase of two states is defined by parallel transporting one of the states along a geodesic on the Poincaré sphere connecting the two. Just as in the case at hand, Pancharatnam's condition for two beams of light to be in phase implies that the intensity of their sum is maximal. It is hard to imagine anything more general than the geometric phase of Samuel and Bhandari, which applies to essentially any type of quantum evolution imaginable.

The remaining two entries in this chapter treat adiabatic phases in the context of the Born-Oppenheimer approximation. Kuratsuji and Iida were, as far as we know, the first to discuss Berry's phase in a path-integral framework, as opposed to a Schrödinger description [3.6]. Such a framework lends itself well to the study of Born-Oppenheimer systems, where one is interested in separating the integrations over "fast" electronic variables and "slow" nuclear variables. Typically, one performs the electronic path integration first, in a fixed nuclear background, and making the adiabatic approximation that electronic transitions do not occur. The result is an effective action involving only the nuclear coordinates. (This procedure is discussed in detail in the final paper, by Moody, Shapere, and Wilczek [3.7].) Now when, for example, the nuclei traverse a closed path, at a typically slow nuclear velocity, the electronic wavefunction will acquire an adiabatic phase. Berry's phase must be taken into account in the nuclear path integral. As Kuratsuji and Iida point out, the appropriate modification consists of a simple addition to the nuclear effective action.

The path-integral description also makes it easy to address questions about the semiclassical limit of nuclear motion. Here Berry's phase has a strikingly direct effect on the energy levels. If C is a closed classical path and $(P(t), Q(t))$ the corresponding trajectory in phase space, then Kuratsuji and Iida, and also Wilkinson [6.6], find the following quantization rule:

$$\oint P \cdot dX = \left(n + \frac{\alpha}{4} - \frac{\Gamma(C)}{2\pi}\right) 2\pi\hbar. \tag{3.5}$$

Here $\Gamma(C)$ is the Berry's phase associated with the path and α is an integer known as the Keller-Maslov index, and determined by continuing the WKB wavefunction around the turning points.

"Adiabatic Effective Lagrangians" [3.7] is the paper referred to in [4.3] as Ref. 7, which has not appeared in print until now. Therein, the general procedure of "integrating out" fast degrees of freedom is explained, emphasizing its application to molecular systems. The Born-Oppenheimer method in both its Hamiltonian and Lagrangian forms is laid out in detail, taking into account the possibility of an adiabatic phase. As we have said, the phase enters into the nuclear effective theories as a canonically coupled background gauge potential. Perhaps surprisingly, the Born–Oppenheimer Hamiltonian and Lagrangian are not related by a Legendre transformation. An explicit expansion giving corrections to adiabatic evolution is derived, and used to give a proof of the adiabatic theorem. For smooth classical motions of the nuclei, tunneling corrections are exponentially suppressed, away from level crossings. On the other hand, corrections to phase evolution are only suppressed by powers of the expansion parameter. These may be accounted for directly by adding higher–derivative terms to the nuclear effective Lagrangian.

Appendix: Monopoles, Holonomy and Fiber Bundles

A "geometric phase" is what mathematicians would call a $U(1)$ holonomy, and the natural mathematical context for holonomy is the theory of fiber bundles. Because they will continue to arise implicitly and explicitly throughout this book, we would now like to give a brief introduction to fiber bundles and some related mathematical concepts, such as connections and parallel transport. Wherever possible, we shall try to translate mathematics into physics, and to illustrate key concepts by way of a particular physical example, the magnetic monopole.

Before saying what a fiber bundle is, let us jump ahead to explain what a fiber bundle is good for. Suppose the state of a system is described by two sets of parameters, which we shall refer to as internal and external. Suppose further that the internal parameters change in a well-defined way when we vary the external parameters. Thus, in quantum mechanics with a Hamiltonian depending on slowly-varying external parameters $R(t)$, we may think of the phase of the wavefunction as an internal parameter that depends on $R(t)$ through the energies and through the adiabatic phase

$$\exp \int A_n(R) \cdot dR \qquad (3.A.1)$$

where $A_n(R) \equiv \langle n(R)|\nabla|n(R)\rangle$. (To avoid irrelevant complications, we suppose the energy $E_n(R) = 0$.) Where does the evolving wavefunction live? It is not enough to say it is a function of R, since its phase is a function of the whole previous history. Rather, we may think of it as living on a space that looks at least locally like $\{R\} \times U(1)$, and whose global topological properties are encoded in $A_n(R)$. In fact, globally this product may be "twisted"—this will be the case if $A_n(R)$ has unremovable singularities, like the Dirac string of a magnetic monopole gauge potential. To describe such twisted products, in a way that avoids talking about singularities, we introduce the following construction.

A bundle E with fiber F is a generalization of the direct product of two spaces $M \times F$. Locally, in a small neighborhood U of any point of M, the bundle looks like $U \times F$, but globally, it may be topologically twisted. M is called the base space, and the fibers of E (which are all homeomorphic to F) may conveniently be thought of as residing vertically over the points of M, with one fiber above each point (see Fig.A.1).

The Möbius strip is a simple example of a fiber bundle. As depicted in Fig. A.2, its base space is a circle and its fiber is a line segment, which we may take to be the interval $[-1, 1]$. Every point on the circle has a neighborhood U over which the bundle is homeomorphic to $U \times [-1, 1]$, but globally there is a twist. One way to see the twist is to choose a homeomorphism i between the fiber over x_0 and $[-1, 1]$, and to try to extend it continuously around the circle. When we come back to x_0, we find that its "direction" has been

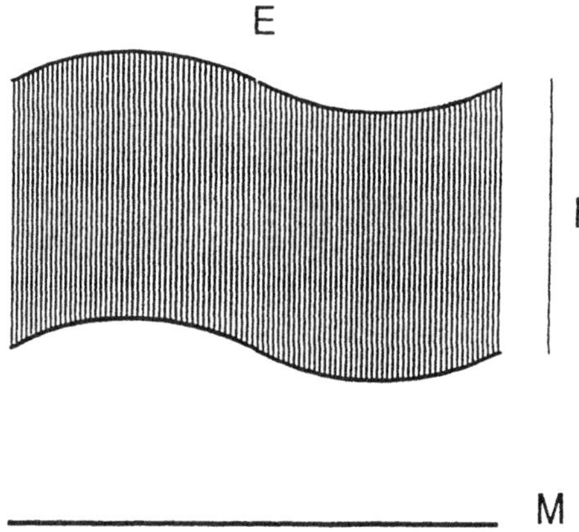

Figure A.1 A fiber bundle with total space E, fiber F, and base space M.

reversed: topologically, transporting i around the circle gives back $-i$. This minus sign is the simplest example of a non-trivial holonomy. It is essentially the same minus sign found by Herzberg and Longuet-Higgins in [2.3], that arose from transporting an eigenstate of a real Hamiltonian around a conical degeneracy.

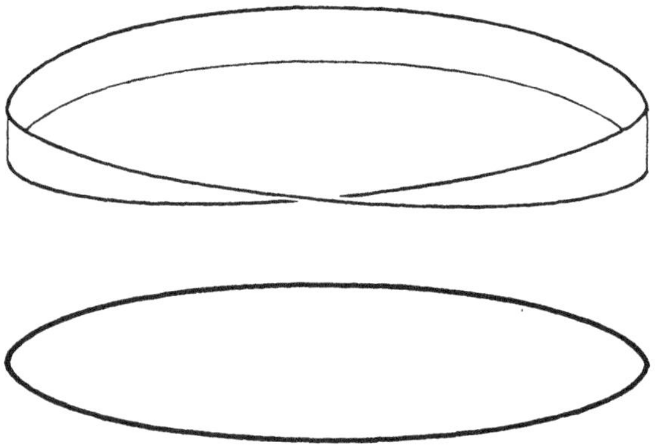

Figure A.2 The Möbius strip is one of the simplest nontrivial examples of a fiber bundle.

We shall be particularly interested in bundles that have a symmetry group G acting upon the fiber. These are properly known as G-bundles, or bundles with structure group G. When the fiber F is the group G itself, we say that we have a principal G-bundle. The action of G on the fibers is to be thought of as a change of coordinates for each fiber. For example, the fiber may be an n-dimensional complex vector space. To label a vector, we must choose a basis; but there remains an action of $U(n)$ representing our freedom to change bases. Suppose now that we cover M with a collection of small neighborhoods $\{U_i\}$, that have intersections $U_i \cap U_j \equiv U_{ij}$, and over each of which the bundle is isomorphic to $U_i \times F$. Then to sew all the mini-bundles $U_i \times F$ together to make the full bundle E, we need rules for how to identify elements of overlapping mini-bundles (*i.e.*, how to relate two different coordinate systems on the fibers residing over U_{ij}). These rules are known as ***transition functions***, and are responsible for all the global topological twisting. They are maps g_{ij} from U_{ij} into the structure group G, and may be regarded as taking elements (p, x) of $U_i \times F$ to (p, gx) in $U_j \times F$, for $p \in U_{ij}$.

Let us illustrate the above abstractions with the "original" example of a fiber bundle, known to physicists as the magnetic monopole[1] and to mathematicians as the Hopf bundle.[2] This is a bundle over the two-dimensional sphere, with fiber $U(1)$. Physically, it is related to the interaction of an electric charge moving on the surface of the sphere with a magnetic monopole charge at the center of the sphere. Now if the electric charge moves very slowly, then the classical force between the two charges will be vanishingly small. But, of course, classical electromagnetism is not the whole story. Quantum mechanically, there is a very significant interaction, found by Aharonov and Bohm [2.6], involving a geometric phase that depends only on the shape of the path taken by the charge, and not on how fast that path is traversed. The interaction may be described as follows. The magnetic charge has a field $\mathbf{B} = g\,\hat{\mathbf{r}}/r^2$, and transporting the charge around a closed loop C produces a phase change of the charged particle's wavefunction of $e^{i\Phi(C)}$, where $\Phi(C)$ is the enclosed magnetic flux (if the electric charge is equal to 1). In the absence of magnetic monopoles, $\nabla \cdot \mathbf{B} = 0$, so we can write $\mathbf{B}$ as the curl of a gauge potential $\mathbf{A}$ (also known as a vector potential or a connection) and express the enclosed flux as

$$\Phi = \int_S \mathbf{B} \cdot \mathbf{dS} = \oint_C \mathbf{A} \cdot \mathbf{dl} \tag{3.A.2}$$

The right-hand integral shows clearly that the accumulated phase depends only on the geometry of the path C. (Another advantage of using a gauge potential is that it is usually easier to do a line integral than a surface integral.)

When magnetic monopoles are present,

$$\nabla \cdot \mathbf{B} = \sum_{\text{monopoles}} g_i\, \delta^{(3)}(\mathbf{x} - \mathbf{x}_i) \tag{3.A.3}$$

where g_i is the magnetic charge at $\mathbf{x}_i$, and the gauge potential A will in general be singular. For example, for a single monopole at the origin, a common choice of gauge potential is

$$A_\phi = g\,(1 - \cos\theta) \qquad\qquad A_r = A_\theta = 0 \tag{3.A.4}$$

in spherical coordinates. Note that A_ϕ has a "Dirac string" singularity along the line $\theta = \pi$: there the magnitude of A_ϕ,

$$g^{\phi\phi} A_\phi A_\phi = \frac{g^2(1-\cos\theta)^2}{\sin^2\theta},$$

blows up. The string can be moved around by means of a gauge transformation $\mathbf{A} \rightarrow \mathbf{A} + \nabla\Lambda$, but it cannot be removed.

In order to avoid singular gauge potentials, Wu and Yang[3] introduced the following construction. They covered S^2 with two patches S^+ and S^-,

$$\begin{aligned} S^+ &\equiv \{(\theta,\phi)\ :\ 0 \le \theta < \frac{\pi}{2} + \epsilon\} \\ S^- &\equiv \{(\theta,\phi)\ :\ \frac{\pi}{2} - \epsilon < \theta \le \pi\}, \end{aligned} \tag{3.A.5}$$

two open sets that respectively contain the northern and southern hemispheres, and whose intersection is an open set containing the equator. Over each patch, the restricted bundle is isomorphic to the trivial bundle $S^\pm \times U(1)$, with elements $(\theta, \phi, e^{i\alpha_\pm})$. The transition function g_{+-} that connects the upper and lower halves will in general be of the form

$$g_{+-} : (\theta, \phi, e^{i\alpha_-}) \longrightarrow (\theta, \phi, e^{if_{+-}(\theta,\phi)} \cdot e^{i\alpha_+}) \tag{3.A.6}$$

where θ and ϕ are in $S^+ \cap S^-$. Obviously, the winding number of f_{+-} must be an integer. This integer does not depend on our patching scheme, and in fact it is a topological invariant that gives a complete topological characterization of all principal $U(1)$ bundles over the two-sphere. We shall see that this integer is proportional to the magnetic charge of the monopole.

Using the above construction, we can specify a non-singular gauge potential for the magnetic monopole field. One choice is

$$\begin{aligned} A_\phi^+ &= g(\cos\theta + 1) \qquad \text{over } S^+ \\ A_\phi^- &= g(\cos\theta - 1) \qquad \text{over } S^- \end{aligned} \tag{3.A.7}$$

These are each nonsingular in their respective domains, give the right curl, and are related on the equator $S^+ \cap S^-$ by a gauge transformation

$$A_\phi^+ = A_\phi^- + \partial_\phi(2g\phi). \tag{3.A.8}$$

This gauge transformation corresponds to having for a transition function $f_{+-}(\theta,\phi) = 2g\phi$. One may show that the flux Φ of Eq.(3.A.2), a physical quantity, is invariant under this transformation, and that it is equal to the monopole charge times the solid angle subtended by the closed path C. So as claimed, $2g$ is equal to the winding number of f_{+-}, and g must be a multiple of $1/2$. This is Dirac's famous quantization condition.

Let us now look at the charge–monopole interaction in a more general setting. To repeat, the effect of this interaction on a unit charge that is slowly transported through the field of a magnetic monopole along any path C, open or closed, is to multiply the wavefunction by a geometric phase

$$\exp i \int_C A \cdot dl \tag{3.A.9}$$

The gauge potential thus gives us a rule for lifting a curve in the charged particle's position space to a curve in the $U(1)$ bundle of position wavefunctions. This is an example of the notion of parallel transport, to which we now turn.

In general, a rule for parallel transport is a rule for changing ones fiber coordinate as one moves horizontally over the base space M. It describes a way to lift a curve in the base manifold to a curve in the bundle. This is often necessary for performing explicit computations, and it gives us valuable geometric information about the bundle, as well. It is usually most convenient to define such a rule infinitesimally, that is, as a rule for lifting vectors from M to E. If the vectors lifted are the tangent vectors of a curve C in M, their liftings may then be integrated to give a lifting of C. So let v be a vector tangent to M, and let A be a linear map taking v to a vector tangent to the fiber G, which we denote by A_v. A now specifies a rule for lifting vectors,

$$v \longrightarrow v + A_v \tag{3.A.10}$$

Locally, over an open set U where the bundle looks like $U \times F$, (v, A_v) is a vector in the tangent space of $U \times F \subseteq E$. Thus, A_v is the "vertical" component of the lifted vector. A is known to mathematicians as a connection, and is mathematically identical to a gauge potential.

There is a certain amount of freedom in choosing a connection, corresponding to our freedom to choose coordinates for the fibers. A different choice of a homeomorphism between the fiber above x and the fibering space F yields a different connection. Given one rule for lifting a curve, we obtain another lifting from M to E by multiplying each point of the lifted curve by a different group element $g(x)$. This amounts to rotating the coordinates of the fiber by $g(x)$. Geometrically, the new lifting is equivalent to the old one; we have just chosen to describe it using different coordinates. Although the choice we make is rather arbitrary, it is necessary to make one, in order to specify a lifting. Physicists call this a choice of gauge, and different choices are related by gauge transformations. Under such a transformation, A transforms just as a gauge potential should:

$$A_v(x) \longrightarrow g(x)A_v(x)g(x)^{-1} + g(x)\nabla_v g(x)^{-1} \tag{3.A.11}$$

We found an example of this type of transformation in Eq.(3.A.8), relating A^+ and A^- at the equator.

Parallel transport does not necessarily lift a closed curve C in M to a closed curve in the total space of the bundle. Generally, the initial and final points of such a curve will lie in the same fiber, and be related by an element of G (the symmetry group acting on F). This element is the *holonomy* of C. Because it actually lies in the group (as opposed to the fiber), it is independent of the choice of gauge. We can obtain an explicit expression for the holonomy in terms of the connection A: an infinitesimal motion v in M gets lifted to $v + A_v$ in E, so the initial fiber coordinate gets multiplied by the group element $1 + A_v$. The total shift in the fiber when we go around a closed curve is

$$W = \mathrm{P} \exp \oint_C A \cdot dx, \tag{3.A.12}$$

an object known in gauge theory as the Wilson line integral. This expression should be compared to Eq. (3.A.9); there are two differences that we should explain. The first is that Eq. (3.A.12) has no i in the exponent; this is because we have taken A to be Lie algebra valued, and thus anti-Hermitian, so the i has been absorbed into A. The second difference is the P in Eq. (3.A.12), which tells us how to order the non-abelian matrices $A(x)$ in expanding the exponential. The path ordering expressed by P means that all matrices A in the expansion of the exponential are to be ordered as the points in the path are, with later times on the right. If $x(t)$ is a parameterization of the curve C, the explicit expansion is

$$\begin{aligned} W &= \mathrm{P} \exp \int A \cdot \dot{x}\, dt \\ &= 1 + \int_0^1 dt\, A \cdot \dot{x} + \int_0^1 dt \int_0^t dt'\, A(t) \cdot \dot{x}(t)\ A(t') \cdot \dot{x}(t') + \cdots \end{aligned} \tag{3.A.13}$$

When the gauge potential is Abelian, the holonomy of Eq. (3.A.12) is the same as Eq. (3.A.9); everything commutes, so no path ordering is necessary. Furthermore, we can use Stokes' theorem (which does not apply in the non-abelian case) to relate the holonomy of a curve C to the magnetic flux enclosed by the curve. For our old friend the magnetic monopole, the holonomy associated to a closed curve C on the sphere is equal to $\exp ig\Omega$, where Ω is the solid angle subtended by C.

To summarize, there are gauge structures that give perfectly sensible rules for parallel transport which cannot be specified by a gauge potential A that is everywhere regular. We may specify A's on different patches, and rules (gauge transformations) to identify them on the overlaps. Fiber bundles provide the most natural context in which to view such gauge potentials, whose singularities are closely related to the topology of the associated bundle. As we have seen, magnetic monopoles of different charge g generate different $U(1)$ bundles over a sphere. On the other hand, one can show that any two gauge potentials having the same total magnetic flux out of the sphere, lead to topologically equivalent fiber bundles. Thus, the number of Dirac quanta of magnetic charge fully characterizes the topological class of the associated $U(1)$ bundle. The topological classification of fiber bundles in general is the subject of the theory of characteristic classes[4].

This completes our short introduction to fiber bundles. A complementary approach to some of the topics we have covered is contained in the article by Aitchison [7.1]. Readers desiring a more formal and detailed treatment may wish to consult Ref. 4.

---

[1] S. Coleman, "The Magnetic Monopole Fifty Years Later," in A. Zichichi, ed., *The Unity of the Fundamental Interactions* (New York: Plenum Press, 1983) pp.21–117.

[2] R. Bott and L. Tu, *Differential Forms in Algebraic Topology* (New York: Springer-Verlag, 1982).

[3] T.T. Wu and C.N. Yang, *Nucl. Phys.* **B107** (1976) 365.

[4] N. Steenrod, *The Topology of Fibre Bundles* (Princeton: University Press, 1951);
D. Husemoller, *Fibre Bundles* (New York: Springer-Verlag, 1975).

Proc. R. Soc. Lond. A **392**, 45–57 (1984)
Printed in Great Britain

Quantal phase factors accompanying adiabatic changes

BY M. V. BERRY, F.R.S.
H. H. Wills Physics Laboratory, University of Bristol, Tyndall Avenue, Bristol BS8 1TL, U.K.

(*Received* 13 *June* 1983)

A quantal system in an eigenstate, slowly transported round a circuit C by varying parameters $\boldsymbol{R}$ in its Hamiltonian $\hat{H}(\boldsymbol{R})$, will acquire a geometrical phase factor $\exp\{i\gamma(C)\}$ in addition to the familiar dynamical phase factor. An explicit general formula for $\gamma(C)$ is derived in terms of the spectrum and eigenstates of $\hat{H}(\boldsymbol{R})$ over a surface spanning C. If C lies near a degeneracy of $\hat{H}$, $\gamma(C)$ takes a simple form which includes as a special case the sign change of eigenfunctions of real symmetric matrices round a degeneracy. As an illustration $\gamma(C)$ is calculated for spinning particles in slowly-changing magnetic fields; although the sign reversal of spinors on rotation is a special case, the effect is predicted to occur for bosons as well as fermions, and a method for observing it is proposed. It is shown that the Aharonov–Bohm effect can be interpreted as a geometrical phase factor.

1. INTRODUCTION

Imagine a quantal system whose Hamiltonian $\hat{H}$ describes the effects of an unchanging environment, and let the system be in a stationary state. If the environment, and hence $\hat{H}$, is slowly altered, it follows from the adiabatic theorem (Messiah 1962) that at any instant the system will be in an eigenstate of the instantaneous $\hat{H}$. In particular, if the Hamiltonian is returned to its original form the system will return to its original state, apart from a phase factor. This phase factor is observable by interference if the cycled system is recombined with another that was separated from it at an earlier time and whose Hamiltonian was kept constant.

My purpose here is to explain how the phase factor contains a circuit-dependent component $\exp(i\gamma)$ in addition to the familiar dynamical component $\exp(-iEt/\hbar)$ which accompanies the evolution of any stationary state. A general formula for γ in terms of the eigenstates of $\hat{H}$ will be obtained in § 2. If the circuit is close to a degeneracy in the spectrum of $\hat{H}$, γ takes a particularly simple form which will be derived in § 3; this contains, as a special case, the sign change around a degeneracy of the eigenstates of a system whose Hamiltonian is real as well as Hermitian (Herzberg & Longuet-Higgins 1963; Longuet-Higgins 1975; Mead 1979; Mead & Truhlar 1979; Mead 1980*a*, *b*; Berry & Wilkinson 1984).

A particle of any spin in an eigenstate of a slowly-rotated magnetic field is another case where γ can be calculated explicitly (§ 4), and gives predictions that could be tested experimentally. This phase factor exists for bosons as well as fermions. A special case is the sign change of spinors slowly rotated by 2π, predicted by Aharonov & Susskind (1967); this will be shown to be different from the dynamical phase factors measured in experiments on precessing neutrons (reviewed by Silverman 1980).

Finally, it is shown in §5 that physical effects of magnetic vector potentials in the absence of fields, predicted by Aharonov & Bohm (1959) and observed by Chambers (1960), can be understood as special cases of the geometrical phase factor.

2. General formula for phase factor

Let the Hamiltonian $\hat{H}$ be changed by varying parameters $\boldsymbol{R} = (X, Y, \ldots)$ on which it depends. Then the excursion of the system between times $t = 0$ and $t = T$ can be pictured as transport round a closed path $\boldsymbol{R}(t)$ in parameter space, with Hamiltonian $\hat{H}(\boldsymbol{R}(t))$ and such that $\boldsymbol{R}(T) = \boldsymbol{R}(0)$. The path will henceforth be called a circuit and denoted by C. For the adiabatic approximation to apply, T must be large.

The state $|\psi(t)\rangle$ of the system evolves according to Schrödinger's equation

$$\hat{H}(\boldsymbol{R}(t))\,|\psi(t)\rangle = \mathrm{i}\hbar\,|\dot{\psi}(t)\rangle. \tag{1}$$

At any instant, the natural basis consists of the eigenstates $|n(\boldsymbol{R})\rangle$ (assumed discrete) of $\hat{H}(\boldsymbol{R})$ for $\boldsymbol{R} = \boldsymbol{R}(t)$, that satisfy

$$\hat{H}(\boldsymbol{R})\,|n(\boldsymbol{R})\rangle = E_n(\boldsymbol{R})\,|n(\boldsymbol{R})\rangle, \tag{2}$$

with energies $E_n(\boldsymbol{R})$. This eigenvalue equation implies no relation between the phases of the eigenstates $|n(\boldsymbol{R})\rangle$ at different $\boldsymbol{R}$. For present purposes any (differentiable) choice of phases can be made, provided $|n(\boldsymbol{R})\rangle$ is single-valued in a parameter domain that includes the circuit C.

Adiabatically, a system prepared in one of these states $|n(\boldsymbol{R}(0))\rangle$ will evolve with $\hat{H}$ and so be in the state $|n(\boldsymbol{R}(t))\rangle$ at t.

Thus $|\psi\rangle$ can be written as

$$|\psi(t)\rangle = \exp\left\{\frac{-\mathrm{i}}{\hbar}\int_0^t \mathrm{d}t'\,E_n(\boldsymbol{R}(t'))\right\}\exp\left(\mathrm{i}\gamma_n(t)\right)|n(\boldsymbol{R}(t))\rangle. \tag{3}$$

The first exponential is the familiar dynamical phase factor. In this paper the object of attention is the second exponential. The crucial point will be that its phase $\gamma_n(t)$ is *non-integrable*; γ_n cannot be written as a function of $\boldsymbol{R}$ and in particular is not single-valued under continuation around a circuit, i.e. $\gamma_n(T) \neq \gamma_n(0)$.

The function $\gamma_n(t)$ is determined by the requirement that $|\psi(t)\rangle$ satisfy Schrödinger's equation, and direct substitution of (3) into (1) leads to

$$\dot{\gamma}_n(t) = \mathrm{i}\,\langle n(\boldsymbol{R}(t))\,|\,\nabla_{\boldsymbol{R}}\, n(\boldsymbol{R}(t))\rangle\cdot\dot{\boldsymbol{R}}(t). \tag{4}$$

The total phase change of $|\psi\rangle$ round C is given by

$$|\psi(T)\rangle = \exp(\mathrm{i}\gamma_n(\mathrm{C}))\exp\left\{\frac{-\mathrm{i}}{\hbar}\int_0^T \mathrm{d}t\, E_n(\boldsymbol{R}(t))\right\}|\psi(0)\rangle, \tag{5}$$

where the *geometrical phase change* is

$$\gamma_n(\mathrm{C}) = \mathrm{i}\oint_{\mathrm{C}} \langle n(\boldsymbol{R})|\nabla_{\boldsymbol{R}} n(\boldsymbol{R})\rangle\cdot \mathrm{d}\boldsymbol{R}. \tag{6}$$

Thus $\gamma_n(\mathrm{C})$ is given by a circuit integral in parameter space and is independent of how the circuit is traversed (provided of course that this is slow enough for the adiabatic approximation to hold). The normalization of $|n\rangle$ implies that $\langle n|\nabla_{\boldsymbol{R}} n\rangle$ is imaginary, which guarantees that γ_n is real.

Direct evaluation of $|\nabla_{\boldsymbol{R}} n\rangle$ requires a locally single-valued basis for $|n\rangle$ and can be awkward. Such difficulties are avoided by transforming the circuit integral (6) into a surface integral over any surface in parameter space whose boundary is C. In order to employ familiar vector calculus, parameter space will be considered as three-dimensional, and this will turn out to be the important case in applications; the generalization to higher dimensions will be outlined at the end of this section.

Stokes's theorem applied to (6) gives, in an obvious abbreviated notation.

$$\gamma_n(\mathrm{C}) = -\mathrm{Im}\iint_{\mathrm{C}} \mathrm{d}\boldsymbol{S}\cdot\nabla\times\langle n|\nabla n\rangle, \tag{7a}$$

$$= -\mathrm{Im}\iint_{\mathrm{C}} \mathrm{d}\boldsymbol{S}\cdot\langle\nabla n|\times|\nabla n\rangle, \tag{7b}$$

$$= -\mathrm{Im}\iint_{\mathrm{C}} \mathrm{d}\boldsymbol{S}\cdot\sum_{m\neq n}\langle\nabla n|m\rangle\times\langle m|\nabla n\rangle, \tag{7c}$$

where $\mathrm{d}\boldsymbol{S}$ denotes area element in $\boldsymbol{R}$ space and the exclusion in the summation is justified by $\langle n|\nabla n\rangle$ being imaginary. The off-diagonal elements are obtained from (2) as

$$\langle m|\nabla n\rangle = \langle m|\nabla\hat{H}|n\rangle/(E_n - E_m), \quad m\neq n. \tag{8}$$

Thus γ_n can be expressed as

$$\gamma_n(\mathrm{C}) = -\iint_{\mathrm{C}} \mathrm{d}\boldsymbol{S}\cdot\boldsymbol{V}_n(\boldsymbol{R}), \tag{9}$$

where

$$\boldsymbol{V}_n(\boldsymbol{R}) \equiv \mathrm{Im}\sum_{m\neq n}\frac{\langle n(\boldsymbol{R})|\nabla_{\boldsymbol{R}}\hat{H}(\boldsymbol{R})|m(\boldsymbol{R})\rangle\times\langle m(\boldsymbol{R})|\nabla_{\boldsymbol{R}}\hat{H}(\boldsymbol{R})|n(\boldsymbol{R})\rangle}{(E_m(\boldsymbol{R})-E_n(\boldsymbol{R}))^2}. \tag{10}$$

Obviously $\gamma_n(\mathrm{C})$ is zero for a circuit which retraces itself and so encloses no area.

Equations (9) and (10) embody the central results of this paper. Because the dependence on $|\nabla n\rangle$ has been eliminated, phase relations between eigenstates with different parameters are now immaterial, and (as is evident from the form of (10)), it is no longer necessary to choose $|m\rangle$ and $|n\rangle$ to be single-valued in $\boldsymbol{R}$: any solutions of (2) may be employed without affecting the value of $\boldsymbol{V}_n$. This is a surprising conclusion, as can be seen by comparing (9) with (7a) which show that $\boldsymbol{V}_n$ is the curl of a vector, $\langle n|\nabla n\rangle$, and $\langle n|\nabla n\rangle$ certainly does depend on the choice of phase of the (single-valued) eigenstate $|n(\boldsymbol{R})\rangle$. The dependence on phase is of the following kind: if $|n\rangle \rightarrow \exp\{\mathrm{i}\mu(\boldsymbol{R})\}|n\rangle$ then $\langle n|\nabla n\rangle \rightarrow \langle n|\nabla n\rangle + \mathrm{i}\nabla\mu$ (in another context the importance of such gauge transformations has been emphasized by Wu & Yang (1975)). Thus the vector is not unique but its curl is. The quantity $\boldsymbol{V}_n$ is analogous to a 'magnetic field' (in parameter space) whose 'vector potential' is $\mathrm{Im}\,\langle n|\nabla n\rangle$. In Appendix A it is shown directly from (10) that $\nabla\cdot\boldsymbol{V}_n$ vanishes, thus confirming that (9) gives a unique value for $\gamma_n(\mathrm{C})$.

Using perturbation theory, Mead & Truhlar (1979) obtained essentially the formulae (9) and (10) for an infinitesimal circuit, in a study of molecular electronic states which (in the Born–Oppenheimer approximation) depend parametrically on nuclear coordinates. Their phase factor was not intended to apply to a $|\psi\rangle$ that evolves slowly under the time-dependent Schrödinger equation, but to the variation of eigenstates $|n\rangle$ under a particular phase-continuation rule in $\boldsymbol{R}$-space which can be shown to give the same result.

In parameter spaces of higher dimension, Stokes's theorem cannot be employed to transform the circuit integral (6). The appropriate generalization, provided by the theory of differential forms (see, for example, Arnold 1978, chap. 7), transforms (6) into the integral of a 2-form over a surface bounded by C. The surprising result (10) can now be expressed as follows: independently of the choice of phases of the eigenstates, there exists in parameter space a *phase 2-form*, which gives $\gamma(\mathrm{C})$ when integrated over any surface spanning C. This 2-form is obtained from (10) by replacing ∇ by the exterior derivative d and $\times$ by the wedge product $\wedge$. The validity of this generalization is consistent with the observation that in the three-dimensional version there are infinitely many choices of interpolating Hamiltonian (and hence of parameter spaces) on the surfaces bounded by C, and the geometrical phase factor is independent of the choice.

Professor Barry Simon (1983), commenting on the original version of this paper, points out that the geometrical phase factor has a mathematical interpretation in terms of holonomy, with the phase two-form emerging naturally (in the form (7*b*)) as the curvature (first Chern class) of a Hermitian line bundle.

3. Degeneracies

The energy denominators in (10) show that if the circuit C lies close to a point $\boldsymbol{R}^*$ in parameter space at which the state n is involved in a degeneracy, then $\boldsymbol{V}_n(\boldsymbol{R})$, and hence $\gamma_n(\mathrm{C})$, is dominated by the terms m corresponding to the other states involved. We shall consider the commonest situation, where the degeneracy involves only two states, to be denoted + and −, where $E_+(\boldsymbol{R}) \geqslant E_-(\boldsymbol{R})$. For $\boldsymbol{R}$ near $\boldsymbol{R}^*$, $\hat{H}(\boldsymbol{R})$ can be expanded to first order in $\boldsymbol{R}-\boldsymbol{R}^*$, and

$$\boldsymbol{V}_+(\boldsymbol{R}) = \mathrm{Im}\,\frac{\langle +(\boldsymbol{R})|\,\nabla\hat{H}(\boldsymbol{R}^*)|-(\boldsymbol{R})\rangle \times \langle -(\boldsymbol{R})|\,\nabla\hat{H}(\boldsymbol{R}^*)|+(\boldsymbol{R})\rangle}{(E_+(\boldsymbol{R})-E_-(\boldsymbol{R}))^2}. \tag{11}$$

Obviously $\boldsymbol{V}_-(\boldsymbol{R}) = -\boldsymbol{V}_+(\boldsymbol{R})$, so that $\gamma_-(\mathrm{C}) = -\gamma_+(\mathrm{C})$.

Without essential loss of generality we can take $E_{\pm}(\boldsymbol{R}^*) = 0$ and $\boldsymbol{R}^* = 0$. $H(\boldsymbol{R})$ can be represented by a 2×2 Hermitian matrix coupling the two states. The most general such matrix satisfying the given conditions depends on three parameters X, Y, Z which will be taken as components of $\boldsymbol{R}$, and by linear transformation in $\boldsymbol{R}$-space can be brought into the following standard form

$$\hat{H}(\boldsymbol{R}) = \frac{1}{2}\begin{bmatrix} Z & X - \mathrm{i}Y \\ X + \mathrm{i}Y & -Z \end{bmatrix}. \tag{12}$$

The eigenvalues are

$$E_+(\boldsymbol{R}) = -E_-(\boldsymbol{R}) = \tfrac{1}{2}(X^2 + Y^2 + Z^2)^{\frac{1}{2}} = \tfrac{1}{2}R. \tag{13}$$

Thus the degeneracy is an isolated point at which all three parameters vanish. This illustrates an old result of Von Neumann & Wigner (1929): for generic Hamiltonians (Hermitian matrices), it is necessary to vary three parameters in order to make a degeneracy occur accidentally, that is, not on account of symmetry. Alternatively stated, degeneracies have co-dimension three.

The form (12) was chosen to exploit the fact that

$$\nabla \hat{H} = \tfrac{1}{2}\hat{\boldsymbol{\sigma}}, \tag{14}$$

where the components $\hat{\sigma}_X$, $\hat{\sigma}_Y$, $\hat{\sigma}_Z$ of the vector operator $\hat{\boldsymbol{\sigma}}$ are the Pauli spin matrices. When evaluating the matrix elements in (11) it greatly simplifies the calculations to take advantage of the isotropy of spin and temporarily rotate axes so that the Z-axis points along $\boldsymbol{R}$, and to employ the following relations, which come from the commutation laws between the components of $\hat{\boldsymbol{\sigma}}$:

$$\hat{\sigma}_X|\pm\rangle = |\mp\rangle, \quad \hat{\sigma}_Y|\pm\rangle = \pm\mathrm{i}|\mp\rangle, \quad \hat{\sigma}_Z|\pm\rangle = \pm|\pm\rangle. \tag{15}$$

With these rotated axes, (11) gives

$$\left.\begin{aligned} V_{X+} &= (\mathrm{Im}\langle +|\hat{\sigma}_Y|-\rangle\langle -|\hat{\sigma}_Z|+\rangle)/2R^2 = 0, \\ V_{Y+} &= (\mathrm{Im}\langle +|\hat{\sigma}_Z|-\rangle\langle -|\hat{\sigma}_X|+\rangle)/2R^2 = 0, \\ V_{Z+} &= \mathrm{Im}\langle +|\hat{\sigma}_X|-\rangle\langle -|\hat{\sigma}_Y|+\rangle = 1/2R^2. \end{aligned}\right\} \tag{16}$$

Reverting to unrotated axes, we obtain

$$V_+(\boldsymbol{R}) = \boldsymbol{R}/2R^3. \tag{17}$$

Now use of (9) shows that the phase change $\gamma_+(\mathrm{C})$ is the flux through C of the magnetic field of a monopole with strength $-\frac{1}{2}$ located at the degeneracy. Thus we obtain the pleasant result, valid for the natural choice (12) of standard form for $\hat{H}$, that the geometrical phase factor associated with C is

$$\exp\{\mathrm{i}\gamma_{\pm}(\mathrm{C})\} = \exp\{\mp\tfrac{1}{2}\mathrm{i}\Omega(\mathrm{C})\}, \tag{18}$$

where $\Omega(\mathrm{C})$ is the *solid angle* that C subtends at the degeneracy; Ω is, in a sense, a measure of the *view* of the circuit as seen from the degeneracy. The phase factor is independent of the choice of surface spanning C, because Ω can change only in multiples of 4π (when the surface is deformed to pass through the degeneracy).

An important special case of (18) occurs when C consists entirely of *real* Hamiltonians and so is confined to the plane $Y = 0$ (cf. (12)). The energy levels $E_{\pm}$ intersect conically in the space E, X, Z, whose origin, where the degeneracy occurs, is a 'diabolical point' of the type recently studied by Berry & Wilkinson (1984) in the spectra of triangles. This illustrates the result that for real symmetric matrices, degeneracies have co-dimension two: see Appendix 10 of Arnold 1978. If C encloses the degeneracy, $\Omega = \pm 2\pi$; if not, $\Omega = 0$. Thus the phase factor (18) is

$$\exp\{i\gamma_{\pm}(C)\} = -1, \quad \text{if C encircles the degeneracy,}$$
$$= +1, \quad \text{otherwise,} \tag{19}$$

which expresses the sign changes of real wavefunctions as a degeneracy involving them is encircled, a phenomenon first described by Herzberg & Longuet-Higgins (1963). (Sign changes are not restricted to circuits involving real Hamiltonians: (18) shows that the phase factor is -1 if C lies in *any* plane through the degeneracy and encircles it.)

Confirmation of the correctness of (17) can be obtained without the rotation-of-axes trick, by a lengthy calculation of (11) involving explicit formulae for the eigenvectors $|\pm(\boldsymbol{R})\rangle$ of the matrix (12). Alternatively, direct continuation of the eigenvectors may be attempted. This cannot be accomplished for all circuits by means of (6) because it is not possible to construct eigenvectors that are globally single-valued continuous functions of $\boldsymbol{R}$; multivaluedness can be reduced to singular lines connecting the degeneracy with infinity, and in the analogue $\boldsymbol{V}(\boldsymbol{R})$ these appear as Dirac strings attached to the monopole. Such approaches obscure the simplicity and essential isotropy of the solid-angle result (17).

Using topological arguments not involving explicit formulae for $\gamma_n(C)$, Stone (1976) proved that if C is expanded from one point $\boldsymbol{R}$ and contracted on to another so as to sweep out a surface enclosing a degeneracy, then the geometrical phase factor traverses a circle in its Argand plane. This property (which follows easily from (18)), is the Hermitian generalization of the sign-reversal test for degeneracy.

4. Spins in magnetic fields

A particle with spin s (integer or half-integer) interacts with a magnetic field $\boldsymbol{B}$ via the Hamiltonian

$$\hat{H}(\boldsymbol{B}) = \kappa\hbar\boldsymbol{B}\cdot\hat{\boldsymbol{s}}, \tag{20}$$

where κ is a constant involving the gyromagnetic ratio and $\hat{\boldsymbol{s}}$ is the vector spin operator with $2s+1$ eigenvalues n with integer spacing and that lie between $-s$ and $+s$. The eigenvalues are

$$E_n(\boldsymbol{B}) = \kappa\hbar Bn, \tag{21}$$

and so there is a $(2s+1)$-fold degeneracy when $\boldsymbol{B} = 0$. (The special case $s = \frac{1}{2}$ reproduces the two-fold degeneracy considered in the last section.) We consider the components of $\boldsymbol{B}$ as the parameters $\boldsymbol{R}$ in our previous analysis, and calculate the phase change $\gamma_n(\mathrm{C})$ of an eigenstate $|n, s(\boldsymbol{B})\rangle$ of $\hat{\boldsymbol{s}}$ in the direction along $\boldsymbol{B}$, as $\boldsymbol{B}$ is slowly varied (and hence the spin rotated) round a circuit C.

The vector $\boldsymbol{V}_n(\boldsymbol{B})$ as given by (10) can be expressed by using (20) and (21) as

$$V_n(B) = \frac{\mathrm{Im}}{B^2} \sum_{m \neq n} \frac{\langle n, s(\boldsymbol{B})\,|\hat{\boldsymbol{s}}|\, m, s(\boldsymbol{B})\rangle \times \langle m, s(\boldsymbol{B})\,|\hat{\boldsymbol{s}}|\, n, s(\boldsymbol{B})\rangle}{(m-n)^2}. \tag{22}$$

To evaluate the matrix elements we again temporarily rotate axes so that the Z-axis points along $\boldsymbol{B}$, and employ the following generalization of (15):

$$\left.\begin{aligned} (\hat{s}_X + \mathrm{i}\hat{s}_Y)\,|n, s\rangle &= [s(s+1) - n(n+1)]^{\frac{1}{2}}\,|n+1, s\rangle, \\ (\hat{s}_X - \mathrm{i}\hat{s}_Y)\,|n, s\rangle &= [s(s+1) - n(n-1)]^{\frac{1}{2}}\,|n-1, s\rangle, \\ s_Z\,|n, s\rangle &= n\,|n, s\rangle. \end{aligned}\right\} \tag{23}$$

It is clear that only states with $m = n \pm 1$ are coupled with $|n\rangle$ in (22), and that V_x and V_y are zero because they involve off-diagonal elements of $\hat{s}_Z$. To find V_Z, we make use of (23) to obtain

$$\left.\begin{aligned} \langle n \pm 1, s|\, s_X\, |n, s\rangle &= \tfrac{1}{2}[s(s+1) - n(n \pm 1)]^{\frac{1}{2}}, \\ \langle n \pm 1, s|\, s_Y\, |n, s\rangle &= \mp \tfrac{1}{2}\mathrm{i}[s(s+1) - n(n \pm 1)]^{\frac{1}{2}}, \end{aligned}\right\} \tag{24}$$

then (22) gives

$$\begin{aligned} V_{Zn} &= \frac{\mathrm{Im}}{B^2}\{\langle n|\, s_X\, |n+1\rangle\langle n+1\,|s_Y\,|n\rangle - \langle n|\, s_Y\,|n+1\rangle\langle n+1|\, s_X\,|n\rangle \\ &\quad + \langle n|\, s_X\,|n-1\rangle\langle n-1|\, s_Y\,|n\rangle - \langle n|\, s_Y\,|n-1\,|\rangle\langle n-1|\, s_X\,|n\rangle\} \\ &= \frac{n}{B^2}. \end{aligned} \tag{25}$$

Reverting to unrotated axes, we obtain

$$\boldsymbol{V}_n(\boldsymbol{B}) = n\boldsymbol{B}/B^3. \tag{26}$$

Now, use of (9) shows that $\gamma_n(\mathrm{C})$ is the flux through C of the 'magnetic field' of a monopole $-n$ located at the origin of magnetic field space. Thus the geometrical phase factor is

$$\exp\{\mathrm{i}\gamma_n(\mathrm{C})\} = \exp\{-\mathrm{i}n\Omega(\mathrm{C})\}, \tag{27}$$

where $\Omega(\mathrm{C})$ is the solid angle that C subtends at $\boldsymbol{B} = 0$. Note that γ_n depends only on the eigenvalue n of the spin component along $\boldsymbol{B}$ and not on the spin s of the particle, so that γ_n is insensitive to the strength $2s+1$ of the degeneracy at $\boldsymbol{B} = 0$.

It follows from (27) that any phase change can be produced by varying $\boldsymbol{B}$ round a suitable circuit. For fermions (half-integer n), a whole turn of $\boldsymbol{B}$ (rotation through 2π in a plane, giving $\Omega = 2\pi$) produces a phase factor -1. In the special case $n = \frac{1}{2}$ this shows that the sign change of spinors on rotation and the sign change of wavefunctions round a degeneracy have the same mathematical origin. For bosons (integer n), a whole turn of $\boldsymbol{B}$ produces a phase factor $+1$. To produce a sign change, different circuits are required; if $n = 1$, for example, varying $\boldsymbol{B}$ round a cone of semiangle 60° will give $\Omega = \gamma = \pi$ and hence a phase factor -1.

The following experiment could be carried out to test the predictions embodied in (27). A polarized monoenergetic beam of particles in spin state n along a magnetic field $\boldsymbol{B}$ is split into two. Along the path of one beam $\boldsymbol{B}$ is kept constant. Along the path of the other beam, $\boldsymbol{B}$ is kept constant in magnitude but its direction is varied slowly (in comparison with the dynamical precession frequency) round a circuit C subtending a solid angle Ω, the field being generated by an arrangement enabling Ω to be changed. The beams are then combined and the count rate at a detector is measured as a function of Ω. The dynamical phase factor (the second exponential in (5) is the same for both beams because the energy $E_n(\boldsymbol{B})$ (21) is insensitive to the direction of $\boldsymbol{B}$. There will in addition be a propagation phase factor which can be made unity by adjusting the path-length of one of the beams when $\Omega = 0$. The resulting fringes occur as a consequence of the geometrical phase factor. If C is a circuit round a cone of semiangle θ, the predicted intensity contrast is

$$I(\theta) = \cos^2(n\pi(1-\cos\theta)). \tag{28}$$

I wish to emphasize that this proposed experiment is different from those carried out by Rauch *et al.* (1975, 1978) and Werner *et al.* (1975) (see Silverman 1980) with *unpolarized* neutrons in a *constant* magnetic field. Those neutrons were not in an eigenstate, and their phase changed dynamically, rather than geometrically, under the Hamiltonian (20) (with $\boldsymbol{B}$ along Z and $\hat{\boldsymbol{\sigma}}$ replacing $\hat{\boldsymbol{s}}$) according to the evolution operator

$$\exp(-\mathrm{i}\hat{H}t/\hbar) = \exp(-B\kappa t\hat{\sigma}_Z) = \cos\tfrac{1}{2}\kappa Bt\begin{bmatrix}1 & 0\\ 0 & 1\end{bmatrix} + \mathrm{i}\sin\tfrac{1}{2}\kappa Bt\begin{bmatrix}1 & 0\\ 0 & -1\end{bmatrix}. \tag{29}$$

The sign changed whenever $\frac{1}{2}\kappa Bt$ was an odd multiple of π, and this was interpreted on the basis of precession theory as corresponding to odd numbers of complete rotations about $\boldsymbol{B}$.

5. Aharonov–Bohm effect

Consider a magnetic field consisting of a single line with flux Φ. For positions $\boldsymbol{R}$ not on the flux line, the field is zero but there must be a vector potential $\boldsymbol{A}(\boldsymbol{R})$ satisfying

$$\oint_{\mathrm{C}} \boldsymbol{A}(\boldsymbol{R})\cdot \mathrm{d}\boldsymbol{R} = \Phi, \tag{30}$$

for circuits C threaded by the flux line. Aharonov & Bohm (1959) showed that in quantum mechanics such vector potentials have physical significance even though they correspond to zero field. I shall now show how their effect can be interpreted as a geometrical phase change of the type described in § 2.

Let the quantal system consist of particles with charge q confined to a box situated at $\boldsymbol{R}$ and not penetrated by the flux line (figure 1). In the absence of flux

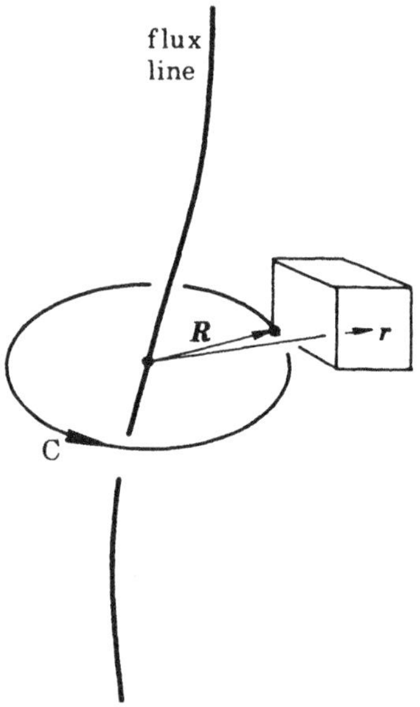

FIGURE 1. Aharonov–Bohm effect in a box transported round a flux line.

($\boldsymbol{A} = 0$), the particle Hamiltonian depends on position $\hat{\boldsymbol{r}}$ and conjugate momentum $\hat{\boldsymbol{p}}$ as follows:

$$\hat{H} = H(\hat{\boldsymbol{p}}, \hat{\boldsymbol{r}} - \boldsymbol{R}), \tag{31}$$

and the wavefunctions have the form $\psi_n(\boldsymbol{r} - \boldsymbol{R})$ with energies E_n independent of $\boldsymbol{R}$. With non-zero flux, the states $|n(\boldsymbol{R})\rangle$ satisfy

$$H(\hat{\boldsymbol{p}} - q\boldsymbol{A}(\hat{\boldsymbol{r}}), \hat{\boldsymbol{r}} - \boldsymbol{R})|\, n(\boldsymbol{R})\rangle = E_n\, |n(\boldsymbol{R})\rangle, \tag{32}$$

an equation whose exact solutions are obtained by multiplying ψ_n by an appropriate Dirac phase factor, giving

$$\langle \boldsymbol{r}|\, n(\boldsymbol{R})\rangle = \exp\left\{\frac{\mathrm{i}q}{\hbar}\int_{\boldsymbol{R}}^{\boldsymbol{r}} \mathrm{d}\boldsymbol{r}' \cdot \boldsymbol{A}(\boldsymbol{r}')\right\} \psi_n(\boldsymbol{r} - \boldsymbol{R}). \tag{33}$$

These solutions are single-valued in $\boldsymbol{r}$ and (locally) in $\boldsymbol{R}$. The energies are unaffected the vector potential.

Now let the box be transported round a circuit C threaded by the flux line; in this particular case it is not necessary that the transport be adiabatic. After completion of the circuit there will be a geometrical phase change that can be calculated from (6) and (33) by using

$$\begin{aligned}\langle n(\boldsymbol{R})|\, \nabla_{\boldsymbol{R}} n(\boldsymbol{R})\rangle &= \iiint \mathrm{d}^3\boldsymbol{r}\, \psi_n^*(\boldsymbol{r} - \boldsymbol{R}) \left\{\frac{-\mathrm{i}q}{\hbar} \boldsymbol{A}(\boldsymbol{R})\, \psi_n(\boldsymbol{r} - \boldsymbol{R}) + \nabla_{\boldsymbol{R}} \psi_n(\boldsymbol{r} - \boldsymbol{R})\right\} \\ &= -\mathrm{i}q\boldsymbol{A}(\boldsymbol{R})/\hbar. \end{aligned} \tag{34}$$

(The vanishing of the second term in braces follows from the normalization of ψ_n.) Evidently in this example the analogy between $\mathrm{Im}\langle n|\nabla n\rangle$ and a magnetic vector potential becomes a reality. Thus

$$\gamma_n(\mathrm{C}) = \frac{q}{\hbar}\oint_{\mathrm{C}} \boldsymbol{A}(\boldsymbol{R})\cdot \mathrm{d}\boldsymbol{R} = q\Phi/\hbar, \tag{35}$$

which shows that the phase factor is independent of n, and also of C if this winds once round the flux line. The phase factor could be observed by interference between the particles in the transported box and those in a box which was not transported round the circuit.

In elementary presentations of the Aharonov–Bohm effect (including its anticipation by Ehrenburg & Siday 1949), the Dirac phase factor is often invoked in comparing systems passing opposite sides of a flux line. Such invocations are subject to the objection that the wavefunction thus obtained is not single-valued. One way to avoid this objection is by summation over all contributions (whirling waves) representing different windings round the flux line (Schulman 1981; Berry 1980; Morandi & Menossi 1984). Another way, adopted in the original paper by Aharonov & Bohm, is to solve Schrödinger's equation exactly for scattering in the flux line's vector potential. The argument of the preceding paragraphs, which employs the geometrical phase factor, is a third way of obtaining the Aharonov–Bohm effect by using only single-valued wavefunctions.

Mead (1980*a*, *b*), employs the term 'molecular Aharonov–Bohm effect' in a different context, to describe how degeneracies in electron energy levels affect the spectrum of nuclear vibrations. He explains two options, both leading to the same vibration spectrum. The first option is to continue the electronic states round degeneracies (in the space of nuclear coordinates) in the manner described in this paper, thus causing the electronic wavefunctions to be multi-valued, with a compensating multi-valuedness in the nuclear states, which must be incorporated into their boundary conditions. The alternative is to enforce single-valuedness on the electronic (and hence also the nuclear) states, and this introduces a vector potential into the Schrödinger equation for nuclear motion. In general one may expect such effects whenever an isolated system is considered as being divided into two interacting parts, each slaved to a different aspect of the other (in the molecular case, electron states are slaved to nuclear coordinates, and nuclear states are slaved to the electronic states and wavefunctions). The systems considered in this paper might be regarded as a special case, in which the coupling is with 'the rest of the Universe' (including us as observers). The only role of the rest of the Universe is to provide a Hamiltonian with slowly-varying parameters, thus forcing the system to evolve adiabatically with phase continuation governed by the time-dependent Schrödinger equation.

6. Discussion

A system slowly transported round a circuit will return in its original state; this is the content of the adiabatic theorem. Moreover its internal clocks will register the passage of time; this can be regarded as the meaning of the dynamical phase factor. The remarkable and rather mysterious result of this paper is that in addition the system records its history in a deeply geometrical way, whose natural formulation (9) and (10) involves phase functions hidden in parameter-space regions which the system has not visited.

The total phase of the transported state (5) is dominated by the dynamical part, because $T \rightarrow \infty$ in the adiabatic limit, and it might be thought that this must overwhelm the geometrical phase γ_n and make its physical effects difficult to detect. This objection can be met by observing that the strengths of non-adiabatic transitions are exponentially small in T if $\hat{H}$ changes smoothly (Hwang & Pechukas 1977), so that essentially adiabatic evolution can occur even when the dynamical phase is only a few times greater than 2π.

As we saw in § 3, degeneracies in the spectrum of $\hat{H}(\boldsymbol{R})$ are the singularities of the vector $\boldsymbol{V}(\boldsymbol{R})$ (equation (10)) in parameter space, and so have an important effect on the geometrical phase factor. This is reminiscent of the part played by singularities of an analytic function, but the analogy is imperfect: if $\gamma(\mathrm{C})$ were completely singularity-determined, $\boldsymbol{V}(\boldsymbol{R})$ would be the sum of the 'magnetic fields' of 'monopoles' situated at the degeneracies (cf. (17)) and so would have zero curl, which is not the case (zero curl, unlike zero divergence, is not a property which is invariant under deformations of $\boldsymbol{R}$ space, and in the general case the sources of $\boldsymbol{V}$ are not just monopoles but also 'currents' distributed continuously in parameter space). A closer analogy is with wavefront dislocation lines, which are phase singularities of complex wavefunctions in three-dimensional position space (Nye & Berry 1974; Nye 1981; Berry 1981), that dominate the geometry of wavefronts without completely determining them.

In view of the emphasis on degeneracies as organizing centres for phase changes, it is worth remarking that close approach of energy levels is not a necessary condition for the existence of nontrivial geometrical phase factors. Indeed, our examples have shown that $\gamma(\mathrm{C})$ can be non-zero even if C involves isospectral deformations of $\hat{H}(\boldsymbol{R})$ (in the Aharonov–Bohm illustration, the levels E_n do not depend on $\boldsymbol{R}$ at all).

The results obtained here are not restricted to quantum mechanics, but apply more generally, to the phase of eigenvectors of any Hermitian matrices under a natural continuation in parameter space. Therefore they have implications throughout wave physics. For example, the electromagnetic field of a single mode travelling along an optical fibre will change sign if the cross section of the fibre is slowly altered so that its path (in the space of shapes) surrounds a shape for which the spectrum of the Helmholtz equation is degenerate (such as one of the diabolical triangles discovered by Berry & Wilkinson 1984).

I thank Dr J. H. Hannay, Dr E. J. Heller and Dr B. R. Pollard for several suggestions, and Professor Barry Simon for showing me, before publication, his paper which comments on this one. This work was not supported by any military agency.

References

Aharonov, Y. & Bohm, D. 1959 *Phys. Rev.* **115**, 485–491.
Aharonov, Y. & Susskind, L. 1967 *Phys. Rev.* **158**, 1237–1238.
Arnold, V. I. 1978 *Mathematical methods of classical dynamics.* New York: Springer.
Berry, M. V. 1980 *Eur. J. Phys.* **1**, 240–244.
Berry, M. V. 1981 Singularities in waves and rays. In *Les Houches Lecture Notes for session* XXXV (ed. R. Balian, M. Kléman & J-P. Poirier), pp. 453–543. Amsterdam: North-Holland.
Berry, M. V. & Wilkinson, M. 1984 *Proc. R. Soc. Lond.* A **392**, 15–43.
Chambers, R. G. 1960 *Phys. Rev. Lett* **5**, 3–5.
Ehrenburg, W. & Siday, R. E. 1949 *Proc. phys. Soc.* B **62**, 8–21.
Herzberg, G. & Longuet-Higgins, H. C. 1963 *Discuss. Faraday Soc.* **35**, 77–82.
Hwang, J.-T. & Pechukas, P. 1977 *J. chem. Phys.* **67**, 4640–4653.
Longuet-Higgins, H. C. 1975 *Proc. R. Soc. Lond.* A **344**, 147–156.
Mead, C. A. 1979 *J. chem. Phys.* **70**, 2276–2283.
Mead, C. A. 1980*a* *Chem. Phys.* **49**, 23–32.
Mead, C. A. 1980*b* *Chem. Phys.* **49**, 33–38.
Mead, C. A. & Truhlar, D. G. 1979 *J. chem. Phys.* **70**, 2284–2296.
Messiah, A. 1962 *Quantum mechanics*, vol. 2. Amsterdam: North-Holland.
Morandi, G. & Menossi, E. 1984 *Nuovo Cim.* B (Submitted.)
Nye, J. F. 1981 *Proc. R. Soc. Lond.* A **378**, 219–239.
Nye, J. F. & Berry, M. V. 1974 *Proc. R. Soc. Lond.* A **336**, 165–190.
Rauch, H., Wilfing, A., Bauspiess, W. & Bonse, U. 1978 *Z. Phys.* B **29**, 281–284.
Rauch, H., Zeilinger, A., Badurek, G., Wilfing, A., Bauspiess, W. & Bonse, U. 1975 *Physics Lett.* A **54**, 425–427.
Schulman, L. S. 1981 Techniques and Applications of Path Integration. New York: John Wiley.
Silverman, M. P. 1980 *Eur. J. Phys.* **1**, 116–123.
Simon, B. 1983 *Phys. Rev. Lett.* (In the press.)
Stone, A. J. 1976 *Proc. R. Soc. Lond.* A **351**, 141–150.
Von Neumann, J. & Wigner, E. P. 1929 *Phys. Z.* **30**, 467–470.
Werner, S. A., Colella, R., Overhauser, A. W. & Eagen, C. F. 1975 *Phys. Rev. Lett.* **35**, 1053–1055.
Wu, T. T. & Yang, C. N. 1975 *Phys. Rev.* D **12**, 3845–3857.

Appendix A

To show that $\gamma(\mathrm{C})$ is independent of the surface spanning C, it is necessary to prove that $\boldsymbol{V}(\boldsymbol{R})$ (equation (10)) has zero divergence. This can be accomplished by expressing $\boldsymbol{V}$ in terms of the vector Hermitian operator $\hat{\boldsymbol{B}}$ defined by

$$\hat{\boldsymbol{B}} \equiv -\mathrm{i} \sum_n |\nabla n\rangle\langle n|. \tag{A 1}$$

From (8), the off-diagonal elements of $\hat{\boldsymbol{B}}$ are

$$\langle n|\,\hat{\boldsymbol{B}}\,|m\rangle = -\mathrm{i}\,\langle m|\,\nabla H\,|n\rangle/(E_n - E_m), \quad m \neq n. \tag{A 2}$$

Thus (10) becomes

$$V = \mathrm{Im}\,\langle n|\hat{\boldsymbol{B}}\times\hat{\boldsymbol{B}}|\,n\rangle. \tag{A 3}$$

Now we can calculate the divergence:

$$\nabla\cdot V = \mathrm{Im}\,\{\langle\nabla n|\cdot\hat{\boldsymbol{B}}\times\hat{\boldsymbol{B}}\,|n\rangle+\langle n|\,\boldsymbol{B}\times\boldsymbol{B}\cdot|\nabla n\rangle+\langle n|\,\nabla\cdot(\hat{\boldsymbol{B}}\times\hat{\boldsymbol{B}})\,|n\rangle\}, \tag{A 4}$$

Use of a consequence of (A 1), namely

$$|\nabla n\rangle = \mathrm{i}\hat{\boldsymbol{B}}\,|n\rangle \tag{A 5}$$

gives

$$\nabla\cdot V = n(-\hat{\boldsymbol{B}}\cdot\hat{\boldsymbol{B}}\times\hat{\boldsymbol{B}}+\hat{\boldsymbol{B}}\times\hat{\boldsymbol{B}}\cdot\hat{\boldsymbol{B}})\,|n\rangle+\mathrm{Im}\,\langle n|\,(\nabla\times\hat{\boldsymbol{B}}\cdot\hat{\boldsymbol{B}}-\hat{\boldsymbol{B}}\cdot\nabla\times\boldsymbol{B}\,|n\rangle. \tag{A 6}$$

For the curl of $\boldsymbol{B}$, (A 1) and (A 5) give

$$\nabla\times\hat{\boldsymbol{B}} = +\mathrm{i}\sum_n\,|\nabla n\rangle\times\langle\nabla n|\; = \mathrm{i}\sum_n\hat{\boldsymbol{B}}\,|n\rangle\times\langle n|\hat{\boldsymbol{B}} = \mathrm{i}\hat{\boldsymbol{B}}\times\hat{\boldsymbol{B}}\rangle, \tag{A 7}$$

whence $\nabla\cdot V$ vanishes by the dot-cross rule for triple products.

This result is valid everywhere except at the 'monopole' singularities arising from degeneracies.

VOLUME 51, NUMBER 24 PHYSICAL REVIEW LETTERS 12 DECEMBER 1983

Holonomy, the Quantum Adiabatic Theorem, and Berry's Phase

Barry Simon
Departments of Mathematics and Physics, California Institute of Technology, Pasadena, California 91125
(Received 18 October 1983)

It is shown that the "geometrical phase factor" recently found by Berry in his study of the quantum adiabatic theorem is precisely the holonomy in a Hermitian line bundle since the adiabatic theorem naturally defines a connection in such a bundle. This not only takes the mystery out of Berry's phase factor and provides calculational simple formulas, but makes a connection between Berry's work and that of Thouless *et al.* This connection allows the author to use Berry's ideas to interpret the integers of Thouless *et al.* in terms of eigenvalue degeneracies.

PACS numbers: 03.65.Db, 02.40.+m

Vector bundles and their integral invariants (Chern numbers) are already familiar to theoretical physicists because of their occurrence in classical Yang-Mills theories. Here I want to explain how they also enter naturally into nonrelativistic quantum mechanics, especially in problems connected with condensed matter physics. If one has a Hermitian operator $\tilde{H}(x)$ depending smoothly on a parameter x, with an isolated nondegenerate eigenvalue $E(x)$ depending continuously on x, then $\{(x,\varphi)|\,\tilde{H}(x)\varphi = E(x)\varphi\}$ defines a line bundle over the parameter space. I will show that the twisting of this line bundle affects the phase of quantum mechanical wave functions.

Berry, in a beautiful recent paper,[1] discovered a striking phenomenon in the quantum adiabatic theorem.[2] That theorem says[3] that if $H(t)$, $0 \le t \le 1$, is a family of Hermitian Hamiltonians, depending smoothly on t, and if $E(t)$ is a smooth function of t which is a simple eigenvalue of $H(t)$ isolated from the rest of the spectrum of $H(t)$ for each t, then the solution $\psi_T(s)$ of the time-dependent Schrödinger equation

$$i\,d\psi_T(s)/ds = H(s/T)\,\psi_T(s) \tag{1}$$

with $\psi_T(0) = \varphi_0$ where $H(0)\varphi_0 = E(0)\varphi_0$ has the property that as $T \to \infty$, $\psi_T(T)$ approaches the eigenvector φ_1 of $H(1)$ with $H(1)\varphi_1 = E(1)\varphi_1$. Berry asked the following question: Suppose that $\tilde{H}(x)$ is a multiple-parameter family and that $C(t)$ is a closed curve in parameter space, so that $\tilde{H}(C(t)) \equiv H(t)$ obeys the hypotheses of the adiabatic theorem. Then that theorem says that φ_1 is just a phase factor times φ_0 and Berry asks, "What phase factor?" Surprisingly, the "obvious" guess

$$\varphi_1 = \exp[-i\int_0^T E(s/T)\,ds]\,\varphi_0$$

is wrong; rather, Berry finds

$$\varphi_1 = \exp[-i\int_0^T E(s/T)\,ds]\,\exp[i\gamma(C)]\,\varphi_0, \tag{2}$$

where $\gamma(C)$ is an extra phase which Berry extensively studies, and which he suggests could be experimentally measured.[4]

The purpose here is first to advertise what Berry calls a "remarkable and rather mysterious result," but more basically to try to take the mystery out of it by realizing that γ is an integral of a curvature so that Berry's phenomenon is essentially that of holonomy which is becoming quite familiar to theoretical physicists.[5] This realization will allow us a more compact formula than that used principally by Berry, and one that is easier to compute with. Most importantly, it will give a close mathematical relationship between his work and that of Thouless *et al.*,[6] so that Berry's interesting analysis of the relation of degeneracy to $\gamma(E)$ will allow us to interpret the TKN[2] integers in a new and interesting way.

To explain that γ is a holonomy, I begin by replacing $H(s)$ by $H(s)-E(s)$ which produces a trivial, computable phase change in the solution $\psi_T(s)$ of (1). Thus, without loss, we can take $E(s) \equiv 0$. Define $\eta_T(s) = \psi_T(sT)$ so that η solves

$$i\,d\eta_T(s)/ds = TH(s)\eta_T(s), \tag{3}$$

and the adiabatic theorem says that $\eta_T(s)$ has a limit $\eta(s)$ with

$$H(s)\,\eta(s) = 0 \tag{4}$$

[since we have taken $E(s) = 0$]. If now $\tilde{H}(x)$ is a multiparameter family of Hermitian Hamiltonians, so that in some region M of parameter space, 0 is an isolated nondegenerate eigenvalue, then given any curve $C(t)$ and any choice η_0 of normalized zero-energy eigenvector of $H(C(0))$ (i.e., a choice of phase), the adiabatic limit yields a way of transporting η_0 along the curve $C(t)$, i.e., a connection. In this way, (2) is just an expression of the holonomy associated to this connection. So far this is just fancy words to describe Berry's discovery. However, there is a mathematically natural connection already long known in the situation of distinguished lines in Hilbert space. For given x, let X_x denote the zero-energy eigenspace for $\tilde{H}(x)$. This yields a line bundle over the parameter space which, since it is embedded in $M \times \mathcal{H}$, has a natural Hermitian connection, studied, e.g., by Bott and Chern.[7] In this connection, one transports a vector β_0 along a curve $C(t)$ so that $\beta(t)$ obeys $\langle\beta(t+\delta t), \beta(t)\rangle = 1 + O((\delta t)^2)$. I claim that $\eta(s)$ precisely obeys this condition; formally, one can argue that

$$\left(\eta(s), \frac{d\eta(s)}{ds}\right) = \lim_{T\to\infty}\left(\eta(s), \frac{d\eta_T(s)}{ds}\right)$$
$$= \lim_{T\to\infty}(\eta(s), -iTH(s)\eta_T(s)) = 0$$

by (4). One can give a rigorous proof just using the convergence of η without worrying about the question of convergence of derivatives.[8] Thus the connection given by the adiabatic theorem [*when* $E(s)\equiv 0$] is precisely the conventional one for embedded Hermitian bundles and γ is the conventional integral of the curvature which is just the Chern class of the connection. In particular, γ only depends on the X_x's, not other aspects of $\tilde{H}(x)$. This means that one has a simple compact formula[9] for γ:

$$\gamma(C) = \int_S V,$$

where S is any oriented surface in M with $\partial S = C$ and V can be defined in terms of an arbitrary smooth choice,[10] $\varphi(x)$, of unit vectors in X_x by

$$V = i(d\varphi, d\varphi), \tag{5}$$

which is shorthand for the two-form[11]

$$V = \sum_{i<j} \mathrm{Im}(\partial\varphi/\partial x_i, \partial\varphi/\partial x_j)\, dx_i \wedge dx_j$$

written in terms of local coordinates. The formula that Berry used has the advantage over (5) of being manifestly invariant under phase changes of $\varphi(x)$,[12] but it appears to depend on details of $\tilde{H}(x)$ and not just on the spaces X_x. Moreover, since it has a sum over intermediate states, it could be difficult to compute in general, although in his examples it is easy to compute, since the sum over intermediate state in these examples is finite. Even in these examples, (5) can be very easy to use in computations.[13]

Equation (5) shows that $V = 0$ if one can choose the $\varphi(x)$ to be all simultaneously real. Thus, the phenomena we discuss here are only present in magnetic fields or some other condition producing a nonreal Hamiltonian.

As an example of significance below which is also considered by Berry,[1] let $M = R^3\backslash\{0\}$ and given $x \in M$, let $H(x) = \vec{x}\cdot\vec{L}$ where L is a spin-S spin on C^{2S+1}. Then all eigenvalues are nondegenerate and we can, for each $m = -S, -S+1, \ldots, S$, compute a $V_m(\vec{x})$ associated to the eigenvalue $|\vec{x}|m$. By rotation covariance, $V_m(\vec{x})$ must be a function of $|\vec{x}|$ times the area form $A(\vec{x})$ on the sphere of radius $\vec{x}$. Thus, we need only compute V_m at $\vec{x} = (0,0,z)$. If $|m\rangle$ is the vector with $L_z|m\rangle = m|m\rangle$, then for $\vec{x}$ near $(0,0,z)$, we can take

$$\varphi(\vec{x}) = \exp\left[i\left(\frac{x}{z}L_y - \frac{y}{z}L_x\right) + O(x^2+y^2)\right]|m\rangle$$

so that

$$d\varphi = iz^{-1}[dx\,L_y - dy\,L_x]\,|m\rangle,$$
$$(d\varphi, d\varphi) = z^{-2}dx\wedge dy\,\langle m|[L_y, L_x]|m\rangle$$
$$= iz^{-2}m\,dx\wedge dy,$$

and thus, returning to general x,

$$V_m(\vec{x}) = m|\vec{x}|^{-2}A(\vec{x}). \tag{6}$$

In particular, if S is any sphere[14] about the origin,

$$(2\pi)^{-1}\int_S V_m(\vec{x}) = 2m \tag{7}$$

is an integer. This is no coincidence: If C is a clockwise circuit around the equator of the sphere S, which breaks up S into two hemispheres S_+, S_- with $\partial S_\pm = \pm C$, then

$$\exp[i\gamma(C)] = \exp\left(i\int_{S_+}V\right) = \exp\left(-i\int_{S_-}V\right)$$

so that $\int_S V$ must be 2π times an integer. We therefore see the familiar quantization of the integral of the Chern class, V, of the bundle, as a consistency condition on the holonomy, a standard fact.

Thouless *et al.*, in their deep analysis of the quantized Hall effect,[6] considered a band of a two-dimensional solid in magnetic field, so that for each k in T^2, the Brillouin zone, the corresponding band energy is nondegenerate. If $\varphi(k)$ is the corresponding eigenvector, then $(i/2\pi)\times\int_{T^2}(d\varphi, d\varphi)$ is an integer, the TKN[2] integer of the band. Using the source[15] analogy of Berry,[1] we can "interpret" these integers. Suppose the band under consideration is the nth; and suppose an arbitrary smooth interpolation, $\tilde{H}(k)$,[16] of the band Hamiltonian $H(k)$ is given from the surface of the torus into the solid torus $\tilde{T}$, i.e., $\tilde{H}(k)$ is defined for $\tilde{k}$ in $\tilde{T}$ and equals $H(k)$ on the surface.

The Wigner–von Neumann theorem[17] says that generically,[16] the nth band is only degenerate for isolated points $\{p_i\}_{i=1}^{l}$ in $\tilde{T}$. One can define V on $\tilde{T}$ with these points removed, and since $dV=0$, Gauss' theorem assures us that the integral of V over the torus is just the same as its integral over little spheres about the degeneracies. Each sphere has a "charge" associated with it which is $\frac{1}{2}q_i$ with q_i an integer, and $\sum q_i$ is the TKN^2 integer.

It is worthwhile to expand slightly on this picture.[18] Consider a matrix family $M(\vec{x})$, depending smoothly on these parameters. If all eigenvalues of $M(\vec{x}_0)$ are nondegenerate, we say that $\vec{x}_0$ is a *regular* point. $\vec{x}_0$ is a *normal singular point* if and only if (i) only one eigenvalue is degenerate and its multiplicity is 2; (ii) the degeneracy is removed to first order for any line through $\vec{x}_0$. If 0 is a normal singular point and P is the projection onto the degenerate eigenvalues of $M(0)$, then for $\vec{x}$ near zero,

$$PM(\vec{x})P - PM(0)P = \vec{a}\cdot\vec{x}P + \vec{B}\cdot\vec{x} + O(x^2)$$

where $\vec{B}$ is a vector of traceless operators on P. Picking a basis for the range of P,[19] we can write $\vec{B}\cdot\vec{x} = \vec{\sigma}\cdot C\vec{x}$ where C is a 3×3 matrix and $\vec{\sigma}$ are the usual Pauli matrices in the basis. The condition on removal of degeneracy says that $\det(C) \neq 0$. The Hermiticity of M implies $\det(C)$ is real. I call the sign of $\det(C)$ the *signature* $\sigma(\vec{x}_0)$ of the normal singular point. With use of a deformation argument and the example discussed above (with spin $S=\frac{1}{2}$), it is not hard to show that if the nth and $(n+1)$st levels are degenerate at x_0, and if V_j is the Chern class associated to the jth level, then for a small sphere S about x_0, we have that

$$(2\pi)^{-1}\int_S V_{n+1} = \sigma(x_0); \quad (2\pi)^{-1}\int_S V_n = -\sigma(x_0).$$

If M is a smooth, regular matrix family on T^2 and $\tilde{M}$ is a smooth interpolation to $\tilde{T}$ with only normal singular points,[16] then the nth TKN^2 integer is exactly equal to a weighted sum of singular points: Points where the nth level is nondegenerate get weight zero, those where it is degenerate with the next lower level get weight $\sigma(p)$ where σ is the signature of p, and those where it is degenerate with the next higher level get weight $-\sigma(p)$.[20]

I conclude with a mathematical remark: I have shown how the Chern integers associated to certain line bundles can be understood in terms of singularities of interpolations of the bundle. It would be interesting to extend this picture to general vector bundles.

It is a pleasure to thank D. Robinson and N. Trudinger for the hospitality of the Australian National University, where this work was done, M. Berry for telling me of his work, and B. Souillard for the remark (via M. Berry) that there must be a connection between Berry's work and that of TKN^2. This research was partially supported by the National Science Foundation through Grant No. MCS-81-20833.

---

[1]M. V. Berry, to be published; see also Proceedings of the Como Conference on Chaos, 1983 (to be published).

[2]See T. Kato, J. Phys. Soc. Jpn. 5, 435–439 (1950); A. Messiah, *Quantum Mechanics* (North-Holland, Amsterdam, 1962), Vol. 2.

[3]There are, in the infinite-dimensional case, also domain conditions if $H(t)$ is unbounded. We ignore these in our discussion. For purposes of this note, one can think of finite-dimensional cases.

[4]Since $\int_0^T E(s/T)ds = T\int_0^1 E(s)ds$ will be very large in the limit $T\to\infty$ unless $\int_0^1 E(s)ds = 0$, it may be difficult to set up the experiment in such a way that the "dynamic phase" $\int_0^T E(s/T)ds$ does not wash out $\gamma(C)$.

[5]See T. Eguchi *et al.*, Phys. Rep. 66, 213 (1980); Y. Choquet-Bruhat *et al.*, *Analysis Manifolds and Physics* (North-Holland, Amsterdam, 1983).

[6]D. Thouless, M. Kohmoto, M. Nightingale, and M. den Nijs, Phys. Rev. Lett. 49, 405 (1982), designated as TKN^2.

[7]R. Bott and S. Chern, Acta Math. 114, 71 (1965).

[8]Pick f a smooth function of compact support and compute

$$\int f(s)\left(\eta(s), \frac{d\eta}{ds}\right)ds = \lim_{T\to\infty}\int f(s)\left(\eta_T(s), \frac{d\eta}{ds}\right).$$

Integrate by parts and get one term $(d\eta_T/ds, \eta)$ which is zero by (3) and (4) and one term

$$\lim_{T\to\infty}\int f'(s)(\eta_T, \eta) = \int f'(s)(\eta, \eta)ds = \int f'(s)ds = 0.$$

[9]While one gets (5) by appealing to the abstract theory, it is quite easy to compute. If $\eta(t) = e^{i\theta(t)}\varphi(t)$, then $(\eta(s), d\eta/ds) = 0$ becomes $(\varphi(t), d\varphi/dt) + i\dot{\theta} = 0$, so that $\gamma(C) = \oint_C i(\varphi, d\varphi) = \int_S i(d\varphi, d\varphi)$ by Stokes' theorem.

[10]If M has "holes," it may not be possible to choose φ globally, but since (5) is phase invariant, one need only make a choice in a neighborhood of any given point. If S is the image of a disk (as it typically is), one can always make a global choice.

[11]This formula appears as Eq. (7b) in Ref. 1 but only in passing; surprisingly, it is not used again. For example, with it, the one-page calculation in Ref. 1 that $dV=0$ is trivial from $d^2=0$.

[12]It is an easy calculation that (5) is invariant under $\varphi(x)\to e^{i\theta(x)}\varphi(x)$ since the normalization condition

$(\varphi, \varphi) = 1$ implies that $(\varphi, d\varphi) + (d\varphi, \varphi) = 0$. Avron, Seiler, and Simon [J. Avron, R. Seiler, and B. Simon, to be published; see also J. Avron *et al.*, Phys. Rev. Lett. 51, 51–53 (1983)] give a simple manifestly phase-invariant form for V; viz. $V = \frac{1}{2} i \operatorname{Tr}(dP P dP)$ where $P(x)$ is the orthogonal projection onto X_x.

[13]I emphasize that (5) holds for the factor $\gamma(C)$ in (2) even if $E(s) \neq 0$.

[14]Since $dV = 0$ (i.e., div $\tilde{V} = 0$ if $V = \tilde{V}_z dx \wedge dy + \tilde{V}_y dz \wedge dx + \tilde{V}_x dy \wedge dz$), away from $\vec{x} = 0$, (7) holds for any surface S surrounding 0.

[15]Since Berry is talking about integrating (e, de) along curves which he makes analogous to a vector potential, he talks about magnetic monopoles. Since we only care about (de, de) whose dual is divergenceless away from degeneracies, we do not use the magnetic monopole language. Since the dual may not have zero curl, electrostatic language is not appropriate. Since the sources have a sign, we still use the phrase "charge" for the coefficient of the delta function in $d[(de, de)]$ at singularities.

[16]One can show (see Avron, Seiler, and Simon, and Avron *et al.*, Ref. 12) that such interpolations exist; indeed, that in the finite-dimensional case, the set of such interpolations is a dense open set in the set of all interpolations.

[17]J. von Neumann and E. Wigner, Phys. Z. 30, 467 (1929); see also J. Avron and B. Simon, Ann. Phys. (N.Y.) 110, 85–110 (1978).

[18]These things will be discussed further in Avron, Seiler, and Simon, Ref. 12.

[19]Changing basis multiplies C by a unitary and so $\sigma(x_0)$ is independent of basis. It does depend on an orientation of R^3 (order of x, y, z) but so do the TKN^2 integers. Everything (σ, the TKN^2 integers, the sign of spanning surfaces for C) changes sign under change of orientation.

[20]The fact that the sum of the TKN^2 integers is zero in the finite-dimensional case is made particularly transparent by these relations. The number of points of degeneracies of level n and level $n+1$ is at least the absolute value of the sum of the first n TKN^2 integers.

VOLUME 52, NUMBER 24 PHYSICAL REVIEW LETTERS 11 JUNE 1984

Appearance of Gauge Structure in Simple Dynamical Systems

Frank Wilczek and A. Zee[a]
Institute for Theoretical Physics, University of California, Santa Barbara, California 93106
(Received 9 April 1984)

Generalizing a construction of Berry and Simon, we show that non-Abelian gauge fields arise in the adiabatic development of simple quantum mechanical systems. Characteristics of the gauge fields are related to energy splittings, which may be observable in real systems. Similar phenomena are found for suitable classical systems.

PACS numbers: 03.65.Bz, 11.15.-q

Gauge fields, both Abelian and non-Abelian, figure prominently in modern theories of fundamental interactions. They also arise naturally in many geometrical contexts, and are central to much of modern mathematics. In this note we point out that gauge fields appear in a very natural way in ordinary quantum mechanical problems, whose initial formulation has no apparent relationship to gauge fields. We discuss some simple model problems in detail, and sketch in a general way how observable consequences of the gauge structures might be extracted for real physical systems. Finally, analogous behavior for classical oscillators is described.

It is, of course, potentially significant for models of elementary particles that gauge fields can arise "from nowhere," but we shall not attempt specific speculations along that line here.

Adiabatic problem.—Consider problems of the following general type: We are given a family of Hamiltonians $H(\vec{\lambda})$ depending continuously on parameters $\vec{\lambda}$, all of which have a set of n degenerate levels. By a simple renormalization of the energies, we can suppose that these levels are at $E=0$. Such degeneracies typically will occur when for each fixed value of the $\vec{\lambda}$ there is a symmetry; however, there need not be a single symmetry which is valid for all $\vec{\lambda}$. For example, the symmetry might be rotation around an axis whose direction is specified by $\vec{\lambda}$. More generally, the symmetry group H responsible for the degeneracy is embedded in a larger group G in a $\vec{\lambda}$-dependent way.

By the reasoning leading to the usual adiabatic theorem,[1] if the parameters are slowly varied from an initial value $\vec{\lambda}_i$ to some final value $\vec{\lambda}_f$ over a long time interval T, and the given space of degenerate levels does not cross other levels, then solutions of

$$H(\vec{\lambda}_i)\psi=0 \tag{1}$$

are mapped onto solutions of

$$H(\vec{\lambda}_f)\psi=0 \tag{2}$$

by solving the time-dependent Schrödinger equation

$$i\,\partial\psi/\partial t=H(\vec{\lambda}(t))\psi \tag{3}$$

with the boundary conditions $\vec{\lambda}(0)=\vec{\lambda}_i$, $\vec{\lambda}(T)=\vec{\lambda}_f$.

If $\vec{\lambda}_i=\vec{\lambda}_f$, so that the initial and final Hamiltonians are identical, then it becomes possible to formulate a more refined question: Given that the n degenerate levels are mapped back onto themselves by adiabatic development, is this mapping a nontrivial transformation? We find that it is, and that to describe such transformations gauge fields are the appropriate tool.

For $n=1$, a single level, the mapping is a simple phase multiplication, or for real wave functions, a sign. These situations, corresponding to U(1) or Z_2 gauge fields, were discussed by Berry[2] and by Simon.[3]

In the problem above, choose an arbitrary smooth set of bases $\psi_a(t)$ for the various spaces of degenerate levels, so that

$$H(\vec{\lambda}(t))\psi_a(t)=0. \tag{4}$$

Such a smooth choice can always be made locally, which is sufficient for our purposes. Let us write for the solutions of the Schrödinger problem (3), with the initial condition $\eta_a(0)=\psi_a(0)$,

$$\eta_a(t)=U_{ab}(t)\psi_b(t). \tag{5}$$

In writing (5) we have assumed the adiabatic limit, which can be justified to a sufficient degree of accuracy. Our task is to determine $U(t)$. We demand that the $\eta_a(t)$ remain normalized, so that

$$0=(\eta_b,\dot\eta_a)=(\eta_b,\dot U_{ac}\psi_c)+(\eta_b,U_{ac}\dot\psi_c) \tag{6}$$

which leads, in an evident notation, to the equation

$$(U^{-1}\dot U)_{ba}=(\psi_b,\dot\psi_a)\equiv A_{ab}. \tag{7}$$

We will show that A, an anti-Hermitian matrix, plays the role of a gauge potential. Equation (7) is solved in terms of path-ordered integrals by

$$U(t)=P\exp\int_0^t A(\tau)\,d\tau. \tag{8}$$

It is remarkable that A depends only on the geometry of the space of degenerate levels. The specific form of A_{ab} computed from (7) depends, of course, upon the choice of bases $\psi_a(t)$. If one makes a different choice

$$\psi'(t) = \Omega(t)\psi(t), \tag{9}$$

then the A fields transform as

$$A'(t) = \dot{\Omega}\Omega^{-1} + \Omega A \Omega^{-1}, \tag{10}$$

i.e., as proper gauge potentials. As for ordinary gauge potentials, the path-ordered integral (8) around a closed loop transforms in a simple way under the gauge transformation (9), like (10) but without the inhomogeneous term. In particular, its eigenvalues are gauge invariant.

More generally, we can define the gauge potential A_μ everywhere on M, the space coordinatized by the parameters $\vec{\lambda} = \{\lambda^1, \ldots, \lambda^\mu, \ldots\}$. Explicitly,

$$A_\mu^T = (\psi, \partial\psi/\partial\lambda^\mu). \tag{11}$$

The ordered integral

$$U(t) = P\exp\int_0^t A_\mu(\vec{\lambda}(t))\,d\lambda^\mu \tag{12a}$$

depends only on the path and not on its parametrization. In particular, for a closed path on M one obtains the Wilson loop

$$U = P\exp\oint A_\mu\,d\lambda^\mu. \tag{12b}$$

As a simple illustration of the preceding framework, consider the generic example of a system with $(n+1)$ levels, of which n levels are degenerate (at zero energy by normalization). Let the Hamiltonian be $H = R(t)H_0R^{-1}(t)$. Here H_0 denotes an $(n+1)$-dimensional matrix with the entries $(H_0)_{ij} = 0$ unless $i = j = n+1$ and $R(t) = R(\theta(t))$ is the rotation

$$R = \exp(i\theta_n T_{n,n+1})\cdots\exp(i\theta_2 T_{2,n+1})\exp(i\theta_1 T_{1,n+1}).$$

The embedding of the relevant symmetry group SO(n) in SO($n+1$) varies with time. The parameter space M is, of course, the coset space SO($n+1$)/SO(n) = S^n. A simple evaluation of Eq. (11) gives the non-Abelian gauge potential and field strength

$$A_\mu = \pi R^{-1}(\partial R/\partial\theta^\mu)\pi, \tag{13a}$$

$$-F_{\mu\nu} = \pi R^{-1}\frac{\partial R}{\partial\theta^\mu}(1-\pi)R^{-1}\frac{\partial R}{\partial\theta^\mu}\pi - (\mu \leftrightarrow \nu), \tag{13b}$$

where π represents projection onto the first n components. Note that left and right projection π of the pure gauge $R^{-1}\partial_\mu R$ gives us a nontrivial SO(n) gauge field. For $n = 3$ we find explicitly

$$A_1 = 0, \quad A_2 = \sin\theta_1 T_{12},$$
$$A_3 = \sin\theta_1\cos\theta_2 T_{13} + \sin\theta_2 T_{23} \tag{14}$$

leading to the field strengths

$$F_{12} = \cos\theta_1 T_{12}; \quad F_{12}^0 = T_{12},$$
$$F_{23} = \cos^2\theta_1\cos\theta_2 T_{23}; \quad F_{23}^0 = T_{23}, \tag{15}$$
$$F_{13} = \cos\theta_1\cos\theta_2 T_{13}; \quad F_{13}^0 = T_{13}.$$

Since the metric structure

$$ds^2 = d\theta_1^2 + \cos^2\theta_1\,d\theta_2^2 + \cos^2\theta_1\cos^2\theta_2\,d\theta_3^2$$

is diagonal we can define

$$F_{ij}^0 \equiv -(g^{ii}g^{jj})^{1/2}F_{ij}, \tag{16}$$

the Cartesian tensor. As might have been anticipated from the simplicity of the starting Hamiltonians (13), the gauge structure is quite simple. In fact, the rotations induced by the ordered integrals (8) amount to parallel transport of tangent vectors to the sphere, with the obvious identifications. Nevertheless, this very fact shows that the example involves truly non-Abelian gauge structure.

The example can obviously be generalized. With the Hamiltonian suitably parametrized on the homogeneous space G/H, we can evaluate the gauge field at the "north pole," thus obtaining from Eq. (13) a simple expression in terms of the structure constant f of G:

$$F_{\mu\nu} = f_{\mu\nu a}\pi\lambda_a\pi. \tag{17}$$

(Here λ_a denote the generators of G; those generators not in H are labeled by a Greek index.) In particular, for the potentially physical example of a Hamiltonian with three levels, two of which are degenerate (at zero energy), and parametrized on SU(3)/SU(2) ⊗ U(1) = CP_2 we have, in the standard SU(3) notation, $F_{45} = (1+\tau_3)$, $F_{67} = (1-\tau_3)$ $F_{47} = -F_{56} = \tau_1$, and $F_{46} = F_{57} = \tau_2$.

Stationary states.—In many real systems there are fast and slow degrees of freedom, and then one may estimate the effect of the slow variables on the fast ones in the adiabatic approximation. An important familiar example is the Born-Oppenheimer treatment of molecules, and we shall use the terminology of this example for definiteness, although we have not investigated any possible applications in realistic detail. In this context, an important problem is to find the stationary states, which in general requires, of course, that we treat the slow variables quantum mechanically. In order that our previous discussion, where these variables were, of course, treated classically, apply fairly directly, let us first discuss this in the correspondence or quasiclassical limit.

Suppose that the nuclei can be described quasiclassically as undergoing a motion with period $2\pi/\omega$; i.e., let them be in a quantum state of the type

$$|s\rangle = \int_0^{2\pi/\omega} e^{-ip\omega\tau}|\vec{\lambda}(\tau)\rangle\, d\tau, \qquad (18)$$

where the label $\vec{\lambda}$ is periodic with period $2\pi/\omega$ and p is an integer. States labeled by different $\vec{\lambda}$ have negligible overlap, and $e^{-iHt}|\vec{\lambda}(t_0)\rangle = |\vec{\lambda}(t_0+t)\rangle$ to an adequate approximation, where H is the Hamiltonian for the nuclear motion calculated as if the electrons followed instantaneously. Within the stated approximations $|s\rangle$ is a stationary state, $e^{-iHt}|s\rangle = e^{-ip\omega t}|s\rangle$.

Let us suppose that for any fixed $\vec{\lambda}$ there is a symmetry guaranteeing degeneracy of two electronic levels, but that the symmetry cannot be defined independent of $\vec{\lambda}$, as in the previous discussion. As we have seen, the development of these levels in response to the motion of $\vec{\lambda}$ can involve nontrivial phases and mixings after a complete period of the motion of $\vec{\lambda}$. We can diagonalize the mixing matrix and thus find states which are multiplied by phases $\exp(i\gamma_1)$, $\exp(i\gamma_2)$, after a period. For these states, we then find the quantization condition altered to read

$$\omega' T + \gamma_i = 2\pi p$$

or

$$\omega' = p\omega - \gamma_i\omega/2\pi. \qquad (19)$$

In accordance with the correspondence principle, Eq. (19) represents the energy splittings for small p. In a more general framework one would construct an effective Lagrangian for the slow variables and treat this fully quantum mechanically. The phase we have found adiabatically represents a term in the Lagrangian *linear in* $\dot{\theta}$, where θ is the coordinate of nuclear rotation. Such a term contributes nothing to the classical equations of motion (in line with its origin as a pure phase) but does change the quantization condition. The rotational energy is altered from $n^2/2I$, $n=$ integer, to $E_n = (n-\gamma/2\pi)^2/2I$. This agrees with Eq. (19) for large quantum numbers, viz.,

$$E_{n+p_1}(\gamma_1) - E_{n+p_2}(\gamma_2) \simeq \frac{n}{I}\left(p_1 - p_2 - \frac{\gamma_1}{2\pi} + \frac{\gamma_2}{2\pi}\right) = \omega(p_1-p_2) - \omega\left(\frac{\gamma_1}{2\pi} - \frac{\gamma_2}{2\pi}\right). \qquad (19a)$$

An example of this general framework is the phenomenon of Λ doubling.[4]

Mechanical analogs.—Simple mechanical analogs exist for many of the systems discussed above. The point is that the mechanical equation $\dot{x} = Ax$, for A anti-Hermitian, becomes the Schrödinger equation for $\psi = x$.

In our study of mechanical analogs, we have uncovered other phenomena which may be interpreted as noncompact gauge fields. Consider a planar harmonic oscillator in a magnetic field perpendicular to the plane:

$$\ddot{\vec{x}} + B(t)\dot{\vec{x}} + \mu(t)\vec{x} = 0, \qquad (20)$$

$$B(t) = \gamma(t)\begin{pmatrix} 0 & 1 \\ -1 & 0 \end{pmatrix}. \qquad (21)$$

With γ and μ slowly varying, we have the approximate solution

$$\vec{x}(t) = \begin{pmatrix} 1 \\ i \end{pmatrix} a(t)\exp i\left\{\int_{t_0}^{t}\omega(\tau)\,d\tau\right\}, \qquad (22)$$

with

$$\omega^2 + \gamma\omega - \mu = 0, \qquad (23)$$

$$\dot{a}/a = -\dot{\omega}/(2\omega+\gamma) \qquad (24)$$

(the induced electric field has been ignored).

The second equation indicates that in response to an infinitely slow cyclic variation of the parameters in the (γ,μ) plane the amplitude a gets multiplied by a nontrivial path-dependent factor. Interestingly, amplification occurs despite the arbitrary slow variation and so the relevant gauge group is $GL(1,R)$, a noncompact group. More explicitly, the factor is given by

$$\exp(\oint da/a) = \exp(\oint A) = \exp(\int F),$$

with

$$A = \left[\pm \frac{1}{2(\gamma^2+4\mu)^{1/2}} - \frac{\gamma}{2(\gamma^2+4\mu)} \right] d\gamma - \frac{1}{\gamma^2+4\mu} d\mu \tag{25}$$

and the field strength

$$F = \mp (\gamma^2+4\mu)^{-3/2}. \tag{26}$$

If either γ or μ is constant the area enclosed by the closed path in the (γ,μ) plane collapses and there is no amplification. Also, note that if $\mu = \text{const}$ our system conserves $\dot{\bar{x}}^2 + \mu \bar{x}^2 = \dot{a}^2 + a^2(\omega^2+\mu)$. In the adiabatic limit, $a^2(\omega^2+\mu) = \text{const}$, in contrast to the standard adiabatic theorem $a^2(\omega+\frac{1}{2}\gamma) = \text{const}$ for the case $\gamma = \text{const}$.[5]

This material is based upon research supported in part by the National Science Foundation under Grant No. PHY77-27084, supplemented by funds from the National Aeronautics and Space Administration. We thank W. Kohn and R. Schrieffer for helpful comments.

(a)On leave from University of Washington, Seattle, Wash. 98195. Present address: Institute for Advanced Study, Princeton, N. J. 08540.

[1]M. Born and V. Fock, Z. Phys. **51**, 165 (1928); L. I. Schiff, *Quantum Mechanics* (McGraw-Hill, New York, 1955), p. 290.

[2]M. V. Berry, to be published.

[3]B. Simon, Phys. Rev. Lett. **51**, 2167 (1983).

[4]A doubling is discussed somewhat in the spirit of this note by G. Wick, Phys. Rev. **73**, 51 (1948).

[5]Taking the induced electric field into account abolishes the amplification in this particular example. Nevertheless, the phenomenon discussed does arise for differential equations describing physical systems, in particular, feedback circuits with effectively imaginary resistance or variable-length pendula acted upon by Coriolis forces.

VOLUME 58, NUMBER 16 PHYSICAL REVIEW LETTERS 20 APRIL 1987

Phase Change during a Cyclic Quantum Evolution

Y. Aharonov and J. Anandan

Department of Physics and Astronomy, University of South Carolina, Columbia, South Carolina 29208

(Received 29 December 1986)

A new geometric phase factor is defined for any cyclic evolution of a quantum system. This is independent of the phase factor relating the initial- and final-state vectors and the Hamiltonian, for a given projection of the evolution on the projective space of rays of the Hilbert space. Some applications, including the Aharonov-Bohm effect, are considered. For the special case of adiabatic evolution, this phase factor is a gauge-invariant generalization of the one found by Berry.

PACS numbers: 03.65.-w

A type of evolution of a physical system which is often of interest in physics is one in which the state of the system returns to its original state after an evolution. We shall call this a cyclic evolution. An example is periodic motion, such as the precession of a particle with intrinsic spin and magnetic moment in a constant magnetic field. Another example is the adiabatic evolution of a quantum system whose Hamiltonian H returns to its original value and the state evolves as an eigenstate of the Hamiltonian and returns to its original state. A third example is the splitting and recombination of a beam so that the system may be regarded as going backwards in time along one beam and returning along the other beam to its original state at the same time.

Now, in quantum mechanics, the initial- and final-state vectors of a cyclic evolution are related by a phase factor $e^{i\phi}$, which can have observable consequences. An example, which belongs to the second category mentioned above, is the rotation of a fermion wave function by 2π rad by adiabatic rotation of a magnetic field[1] through 2π rad so that $\phi = \pm\pi$. Recently, Berry[2] has shown that when H, which is a function of a set of parameters R^i, undergoes adiabatic evolution along a closed curve Γ in the parameter space, then a state that remains an eigenstate of $H(\mathbf{R})$ corresponding to a simple eigenvalue $E_n(\mathbf{R})$ develops a geometrical phase γ_n which depends only on Γ. Simon[3] has given an interpretation of this phase as due to holonomy in a line bundle over the parameter space. Anandan and Stodolsky[4] have shown how the Berry phases for the various eigenspaces can be obtained from the holonomy in a vector bundle. For the adiabatic motion of spin, this is determined by a rotation angle α, due to the parallel transport of a Cartesian frame with one axis along the spin direction, which contains the above-mentioned rotation by 2π radians as a special case. The result of a recent experiment[5] to observe Berry's phase for light can also be understood as a rotation of the plane of polarization by this angle α.

In this Letter, we consider the phase change for *all* cyclic evolutions which contain the three examples above as special cases. We show the existence of a phase associated with cyclic evolution, which is universal in the sense that it is the same for the infinite number of possible motions along the curves in the Hilbert space $\mathcal{H}$ which project to a given closed curve $\hat{C}$ in the projective Hilbert space $\mathcal{P}$ of rays of $\mathcal{H}$ and the possible Hamiltonians $H(t)$ which propagate the state along these curves. This phase tends to the Berry phase in the adiabatic limit if $H(t) \equiv H[\mathbf{R}(t)]$ is chosen accordingly. For an electrically charged system, we formulate this phase gauge invariantly and show that the Aharonov-Bohm (AB) phase[6] due to the electromagnetic field may be regarded as a special case. This generalizes the gauge-noninvariant result of Berry that the AB phase due to a static magnetic field is a special case of his phase. This also removes the mystery of why the AB phase, even in this special case, should emerge from Berry's expression even though the former is independent of this adiabatic approximation.

Suppose that the normalized state $|\psi(t)\rangle \in \mathcal{H}$ evolves according to the Schrödinger equation

$$H(t)|\psi(t)\rangle = i\hbar(d/dt)|\psi(t)\rangle, \tag{1}$$

such that $|\psi(\tau)\rangle = e^{i\phi}|\psi(0)\rangle$, ϕ real. Let $\Pi:\mathcal{H}\to\mathcal{P}$ be the projection map defined by $\Pi(|\psi\rangle) = \{|\psi'\rangle : |\psi'\rangle = c|\psi\rangle$, c is a complex number$\}$. Then $|\psi(t)\rangle$ defines a curve C: $[0,\tau]\to\mathcal{H}$ with $\hat{C}\equiv\Pi(C)$ being a closed curve in $\mathcal{P}$. Conversely given any such curve C, we can define a Hamiltonian function $H(t)$ so that (1) is satisfied for the corresponding normalized $|\psi(t)\rangle$. Now define $|\tilde{\psi}(t)\rangle = e^{-if(t)}|\psi(t)\rangle$ such that $f(\tau)-f(0)=\phi$. Then $|\tilde{\psi}(\tau)\rangle = |\tilde{\psi}(0)\rangle$ and from (1),

$$-\frac{df}{dt} = \frac{1}{\hbar}\langle\psi(t)|H|\psi(t)\rangle - \langle\tilde{\psi}(t)|i\frac{d}{dt}|\tilde{\psi}(t)\rangle. \tag{2}$$

Hence, if we remove the dynamical part from the phase ϕ by defining

$$\beta \equiv \phi + \hbar^{-1}\int_0^\tau \langle\psi(t)|H|\psi(t)\rangle dt, \tag{3}$$

it follows from (2) that

$$\beta = \int_0^\tau \langle\tilde{\psi}|i(d|\tilde{\psi}\rangle/dt)dt. \tag{4}$$

Now, clearly, the same $|\tilde{\psi}(t)\rangle$ can be chosen for every curve C for which $\Pi(C)=\hat{C}$, by appropriate choice of

$f(t)$. Hence β, defined by (3), is independent of ϕ and H for a given closed curve $\hat{C}$. Indeed, for a given $\hat{C}$, $H(t)$ can be chosen so that the second term in (3) is zero, which may be regarded as an alternative definition of β. Also, from (4), β is independent of the parameter t of $\hat{C}$, and is uniquely defined up to $2\pi n$ (n = integer). Hence $e^{i\beta}$ is a geometric property of the unparametrized image of $\hat{C}$ in $\mathcal{P}$ only.

Consider now a slowly varying $H(t)$, with $H(t)|n(t)\rangle = E_n(t)|n(t)\rangle$, for a complete set $\{|n(t)\rangle\}$. If we write

$$|\psi(t)\rangle = \sum_n a_n(t)\exp\left[-\frac{i}{\hbar}\int E_n dt\right]|n(t)\rangle,$$

and use (1) and the time derivative of the eigenvector equation,[7] we have

$$\dot{a}_m = -a_m\langle m|\dot{m}\rangle - \sum_{n\neq m} a_n \frac{\langle m|\dot{H}|n\rangle}{E_n - E_m}\exp\left[\frac{i}{\hbar}\int (E_m - E_n)dt\right], \tag{5}$$

where the dot denotes time derivative. Suppose that

$$\sum_{n\neq m}\left|\frac{\hbar\langle m|\dot{H}|n\rangle}{(E_n - E_m)^2}\right| \ll 1. \tag{6}$$

Then if $a_n(0) = \delta_{nm}$, the last term in (5) is negligible and the system would therefore continue as an eigenstate of $H(t)$, to a good approximation.

In this adiabatic approximation, (5) yields

$$a_m(t) \simeq \exp\left[-\int \langle m|\dot{m}\rangle dt\right] a_m(0).$$

For a cyclic adiabatic evolution, the phase $i\int_0^\tau \langle m|\dot{m}\rangle dt$ is independent of the chosen $|m(t)\rangle$ and Berry[2] regarded this as a geometrical property of the parameter space of which H is a function. But this phase is the same as (4) on our choosing $|\tilde{\psi}(t)\rangle \simeq |m(t)\rangle$ in the present approximation. But β, defined by (3), does not depend on any approximation; so (4) is exactly valid. Moreover, $|\psi(t)\rangle$ need not be an eigenstate of $H(t)$, unlike in the limiting case studied by Berry. Also, the two examples below will show respectively that it is neither necessary nor sufficient to go around a (nontrivial) closed curve in parameter space in order to have a cyclic evolution, with our associated geometric phase β. For these reasons, we regard β as a geometric phase associated with a closed curve in the projective Hilbert space and not the parameter space, even in the special case considered by Berry. But given a cyclic evolution, an $H(t)$ which generated this evolution can be found so that the adiabatic approximation is valid. Then β can be computed with the use of the expression given by Berry in terms of the eigenstates of this Hamiltonian.

We now consider two examples in which the phase β emerges naturally and is observable, in principle, even though the adiabatic approximation is not valid. Suppose that a spin-$\frac{1}{2}$ particle with a magnetic moment is in a homogeneous magnetic field $\mathbf{B}$ along the z axis. Then the Hamiltonian in the rest frame is $H_1 = -\mu B\sigma_z$, where

$$\sigma_z = \begin{bmatrix} 1 & 0 \\ 0 & -1 \end{bmatrix}.$$

Also,

$$|\psi(0)\rangle = \begin{bmatrix}\cos(\theta/2) \\ \sin(\theta/2)\end{bmatrix}$$

so that

$$|\psi(t)\rangle = \exp(i\mu Bt\sigma_z/\hbar)|\psi(0)\rangle = \begin{bmatrix}\exp(i\mu Bt/\hbar)\cos(\theta/2) \\ \exp(-i\mu Bt/\hbar)\sin(\theta/2)\end{bmatrix},$$

which corresponds to the spin direction being always at an angle θ to the z axis. This evolution is periodic with period $\tau = \pi\hbar/\mu B$. Then from (3), for each cycle, $\beta = \pi(1-\cos\theta)$, up to the ambiguity of adding $2\pi n$. Hence, β is $\frac{1}{2}$ of the solid angle subtended by a curve traced on a sphere, by the direction of the spin state, at the center. This is like the Berry phase except that in the latter case (1) the solid angle is subtended by a curve traced by the magnetic field $\mathbf{B}'(t)$ which is large [i.e., $\mu B'/\hbar \gg \omega$, the frequency of the orbit of $\mathbf{B}'(t)$] so that the adiabatic approximation is valid, and (2) $|\psi(t)\rangle$ is assumed to be an eigenstate of this Hamiltonian. Indeed, we may substitute such a Hamiltonian for the above H_1 or add it to H_1 with $\omega = 2\mu B/\hbar$, without changing β, in this approximation. The spin state will also move through the same closed curve in the projective Hilbert space as above if the magnetic field $\mathbf{B} = (B_0\cos\omega t, -B_0\sin\omega t, B_3)$ with $\cot\theta = (B_3 - \hbar\omega/2\mu)/B_0$, where $B_0 \neq 0$.[8] And β is the same for all such Hamiltonians. This illustrates the statement earlier that β is the same for all curves C in $\mathcal{H}$ with the same $\hat{C} \equiv \Pi(C)$. Also, β may be interpreted as arising from the holonomy transformation, around the closed curve on the above sphere traced by the direction of the spin state, due to the curvature on this sphere,[4] which is a rotation. By varying appropriately a magnetic field applied to the two arms of a neutron interferometer with polarized neutrons, it is possible to make the dynamical part of β [the last term in (3)] the same for the two beams.[2,4] Then the phase difference between the two beams is just the geometrical phase, which is observable in principle, from the interference pattern, even when the magnetic field is varied nonadiabatically. In particular, a phase difference of $\pm\pi$ rad would correspond to a 2π-rad rotation of the fermion wave function, which is thus observable.

As our second example, suppose that the magnetic field is $\mathbf{B}(t) = \mathbf{B}_0 + \mathbf{B}_1(t)$, where $\mathbf{B}_0$ is constant and $\mathbf{B}_1(t)$ rotates slowly in a plane containing $\mathbf{B}_0$ with $|\mathbf{B}_1(t)|$

$=|\mathbf{B}_0|$. Suppose that at time t the angle between $\mathbf{B}_1$ and $\mathbf{B}_0$ is $\pi-\theta(t)$ and the spin state $|\psi(t)\rangle$ is in an approximate eigenstate of $H(t)=\mu\mathbf{B}\cdot\sigma$, where σ^i are the Pauli spin matrices. For $0\le\theta\ll1$, the adiabatic condition (6) gives $0\le-\hbar\dot\theta/\mu B_0\theta\ll1$, assuming $\dot\theta\le0$. Hence $\theta\gg\theta_0\exp(-\mu B_0t/\hbar)>0$. So θ can never become zero. That is, if $\mathbf{B}(T)=\mathbf{0}$ for some T then the adiabatic approximation, as defined above, cannot be satisfied, regardless of how slowly $\mathbf{B}_1(t)$ rotates. However, because of conservation of angular momentum, $|\psi(t)\rangle$ remains an eigenstate of $H(t)$ even at $t=T$. But if θ changes monotonically then a level crossing occurs at the point of degeneracy $(\mathbf{B}=\mathbf{0})$ so that the energy eigenvalue corresponding to $|\psi(t)\rangle$ changes sign at $t=T$. For each rotation of $\mathbf{B}_1$ by 2π rad, $|\psi\rangle$ rotates by π rad, so that the system returns to its original state after two rotations of $\mathbf{B}(t)$. For this cyclic evolution, our $\beta=\pi$ which can be seen from the fact that a spin-$\frac{1}{2}$ particle acquires a phase π during a rotation, or that the curve $\hat{C}$ on the projective Hilbert space, which is a sphere, is a great circle, subtending a solid angle 2π at the center. This example is similar to Berry's phase in that $|\psi(t)\rangle$ is always an eigenstate of $H(t)$, even though Berry's prescription cannot be applied here because of the crossing of the point of degeneracy at which the adiabatic approximation breaks down.

Consider now a system with electric charge q for which $H=H_k(\mathbf{p}-(q/c)\hat{\mathbf{A}}(t),R_i)+q\hat{A}_0(t)$ in (1). Here, $\langle\mathbf{x}|\hat{A}_\mu(t)|\psi(t')\rangle=A_\mu(\mathbf{x},t)\psi(\mathbf{x},t')$, where $A_\mu(\mathbf{x},t)$ is the usual electromagnetic four-potential, and R_i are some parameters. Under a gauge transformation,

$$|\psi(t)\rangle\to\exp[i(q/c)\hat\Lambda(t)]\,|\psi(t)\rangle,$$

$$\hat{A}_0(t)\to\hat{A}_0(t)-c^{-1}\partial\hat\Lambda(t)/\partial t,$$

and

$$H_k(t)\to\exp[i(q/c)\hat\Lambda(t)]H_k(t)\exp[-i(q/c)\hat\Lambda(t)].$$

As before, define $|\tilde\psi(t)\rangle=e^{-if(t)}|\psi(t)\rangle$. If we require that $|\tilde\psi\rangle$ undergo the same gauge transformation as $|\psi(t)\rangle$, $f(t)$ is gauge invariant. Then, from (1),

$$\frac{df}{dt}(t)=\langle\tilde\psi(t)|\frac{d}{dt}-\frac{q}{\hbar}\hat{A}_0(t)|\tilde\psi(t)\rangle-\frac{1}{\hbar}\langle\psi(t)|H_k(t)|\psi(t)\rangle. \tag{7}$$

We consider now a cyclic evolution so that

$$|\psi(\tau)\rangle=e^{i\phi}\exp\left[-\frac{iq}{\hbar}\int_0^\tau\hat{A}_0dt\right]|\psi(0)\rangle.$$

Choose $f(t)$ so that $\phi=f(\tau)-f(0)$. Then

$$|\tilde\psi(\tau)\rangle=\exp\left[-i\frac{q}{\hbar}\int_0^\tau\hat{A}_0dt\right]|\tilde\psi(0)\rangle.$$

So we now define the gauge-invariant generalization of (3) as

$$\beta\equiv\phi+\frac{1}{\hbar}\int_0^\tau\langle\psi(t)|H_k(t)|\psi(t)\rangle dt, \tag{8}$$

which on use of (7) gives

$$\beta=\int_0^\tau\langle\tilde\psi(t)|i\frac{d}{dt}-\frac{q}{\hbar}\hat{A}_0(t)|\tilde\psi(t)\rangle dt. \tag{9}$$

Here, $|\tilde\psi(\tau)\rangle$ is obtained by parallel transport of $|\tilde\psi(0)\rangle$, with respect to the electromagnetic connection, along the congruence of lines parallel to the time axis. We could have chosen, instead, any other congruence of paths from $t=0$ to $t=\tau$ in our definition of ϕ and therefore $|\tilde\psi(\tau)\rangle$. This would correspondingly change β, which therefore depends on the chosen congruence. But, again, β is independent of ϕ and $H(t)$ for all the motions in $\mathcal{H}$ that project to the same closed curve $\hat{C}$ in $\mathcal{P}$, for a given chosen congruence. Both β and ϕ, which satisfies

$$e^{-i\phi}=\langle\psi(\tau)|\exp\left[-\frac{iq}{c}\int_0^\tau\hat{A}_0dt\right]|\psi(0)\rangle,$$

are gauge invariant. In the adiabatic limit, $|\tilde\psi(t)\rangle$ can be chosen to be an eigenstate of $H_k(t)$ and (9) is then a gauge-invariant generalization of the Berry phase.

We illustrate this by means of the AB effect.[6] Berry has obtained the AB phase from the gauge-noninvariant expression (4) with $|\tilde\psi(t)\rangle$ an eigenstate of $H(t)$, for a stationary magnetic field, in a special gauge.[9] But a gauge can be chosen so that the AB phase is included in the dynamical phase instead of the geometrical phase (4). Also, in general, there is no cyclic evolution in an AB experiment. But our β defined by Eq. (8) or (9) is gauge invariant and includes the AB phase in the special case to be described now.

Suppose that a charged-particle beam is split into two beams at $t=0$ which, after traveling in field-free regions, are recombined so that they have the same state at $t=\tau$. It is assumed here that the splitting and the subsequent evolution of the two beams occur under the action of two separate Hamiltonians. This is possible if we restrict ourselves to the Hilbert space of a subset of the degrees of freedom of a given system, as in the example considered by Aharonov and Vardi.[10] This belongs to the third example of a cyclic evolution mentioned at the beginning of this Letter. The wave function of each beam at $t=\tau$, assuming that it has a fairly well defined momentum, is

$$\psi_i(\mathbf{x},\tau)=\exp\left[-\frac{i}{\hbar}\int_0^\tau E_i\,dt\right]\exp\left[-\frac{iq}{c}\int_{\gamma_i}A_\mu dx^\mu\right]\exp\left[\frac{i}{\hbar}\int_{\gamma_i}\mathbf{p}\cdot d\mathbf{x}\right]\psi(\mathbf{x},0),\quad i=1 \text{ or } 2,$$

where γ_i is a space-time curve through the beam and $\mathbf{p}$ represents the approximate kinetic momentum of the beam. Hence on using (8), we have

$$\beta=-\frac{q}{c}\oint_\gamma A_\mu dx^\mu+\frac{1}{\hbar}\oint_\gamma \mathbf{p}\cdot d\mathbf{x}, \tag{10}$$

where γ is the closed curve formed from γ_1 and γ_2. But this is only an approximate treatment and a more careful investigation of this problem is needed.

In conclusion, we note that $\mathcal{H}^*=\mathcal{H}-\{0\}$ is a principal fiber bundle over $\mathcal{P}$ with structure group C^* (the group of nonzero complex numbers), and the disjoint union of the rays in $\mathcal{H}$ is the natural line bundle over $\mathcal{P}$ whose fiber above any $p\in\mathcal{P}$ is p itself. Then, clearly, β, given by (4), arises from the holonomy due to a connection in either bundle such that $|\psi(t)\rangle$ is parallel transported if

$$\langle\psi(t)|(d/dt)|\psi(t)\rangle=0, \tag{11}$$

i.e., the horizontal spaces are perpendicular to the fibers with respect to the Hilbert space inner product. Condition (11) was used by Simon[3] to define a connection on a line bundle over parameter space, which is different from the above bundles. The real part of (11) says that $\langle\psi(t)|\psi(t)\rangle$ is constant during parallel transport. Since this is true also during any time evolution determined by (1), we may restrict consideration to the subbundle $\mathcal{S}=\{|\psi\rangle\in\mathcal{H}:\langle\psi|\psi\rangle=1\}$ of $\mathcal{H}^*$. This $\mathcal{S}$ is the Hopf bundle[11] over $\mathcal{P}$. Then the imaginary part of (11) defines the horizontal spaces in $\mathcal{S}$ which determine a connection. This is the usual connection in $\mathcal{S}$ and $e^{i\beta}$ is the holonomy transformation associated with it. If $\mathcal{H}$ has finite dimension N then $\mathcal{P}$ has dimension $N-1$. For $N=2$, $\mathcal{P}$ is the complex projective space $P_1(C)$ which is a sphere with the Fubini-study metric[11] on $\mathcal{P}$ being the usual metric on the sphere. Opposite points on this sphere represent rays containing orthogonal states. Our geometric phase can then be obtained from the holonomy angle α associated with parallel transport around a closed curve on this sphere like in Ref. 4.

It is a pleasure to thank Don Page for suggesting the relevance of the Hopf bundle and the Fubini-Study metric to this work.

[1]Y. Aharonov and L. Susskind, Phys. Rev. **158**, 1237 (1967).

[2]M. V. Berry, Proc. Roy. Soc. London, Ser. A **392**, 45 (1984).

[3]B. Simon, Phys. Rev. Lett. **51**, 2167 (1983).

[4]J. Anandan and L. Stodolsky, Phys. Rev. D **35** 2597 (1987).

[5]R. Y. Chiao and Y.-S. Wu, Phys. Rev. Lett. **57**, 933 (1986); A. Tomita and R. Y. Chiao, Phys. Rev. Lett. **57**, 937 (1986).

[6]Y. Aharonov and D. Bohm, Phys. Rev. **115**, 485 (1959).

[7]See, for example, L. I. Schiff, *Quantum Mechanics* (McGraw-Hill, New York, 1968), pp. 289–291.

[8]An experiment of this type has been done to measure Berry's phase ($\omega\to 0$) using nuclear magnetic resonance by D. Suter, G. Chingas, R. A. Harris, and A. Pines, to be published. One of us (J.A.) wishes to thank A. Pines for a discussion during which it was realized that the same type of experiment can be used to measure the geometric phase β introduced in the present Letter for nonadiabatic cyclic evolutions as well.

[9]In this proof, in Ref. 2, the eigenfunctions, in the absence of the electromagnetic field, are in effect assumed to be real, in order that Eq. (34) is valid. Since the coefficients of the stationary Schrödinger equation are then real, it is always possible to find real solutions. Then, for any eigenfunction belonging to a given eigenvalue to be necessarily a real function multiplied by $e^{i\lambda}$ ($\lambda=$const), it is necessary and sufficient that the eigenvalue is simple. But in our treatment of the AB effect, it is not necessary to make this assumption.

[10]Y. Aharonov and M. Vardi, Phys. Rev. D **20**, 3213 (1979).

[11]See, S. Kobayashi and K. Nomizu, *Foundations of Differential Geometry* (Interscience, New York, 1969), Vol. 2.

PHYSICAL REVIEW LETTERS

VOLUME 60 | 6 JUNE 1988 | NUMBER 23

General Setting for Berry's Phase

Joseph Samuel and Rajendra Bhandari
Raman Research Institute, Bangalore 560080, India
(Received 30 November 1987)

It is shown that Berry's phase appears in a more general context than realized so far. The evolution of the quantum system need be neither unitary nor cyclic and may be interrupted by quantum measurements. A key ingredient in this generalization is the use of some ideas introduced by Pancharatnam in his study of the interference of polarized light, which, when carried over to quantum mechanics, allow a meaningful comparison of the phase between any two nonorthogonal vectors in Hilbert space.

PACS numbers: 03.65.Bz, 02.40.+m

Three years ago, Berry[1] made a rather perceptive and interesting observation regarding the behavior of quantum-mechanical systems in a slowly changing environment. If the system is initially in an eigenstate of the instantaneous Hamiltonian, the adiabatic theorem guarantees that it remains so. This, however, determines the state of the system only up to a phase. Berry asked the question "What is the phase of the system?" and got a somewhat unexpected answer. If the environment (more precisely, the Hamiltonian) returns to its initial state, the system also does, but it acquires an extra phase over and above the dynamical phase, which can be calculated and allowed for. This effect has been studied and measured in various contexts.

Simon[2] gave a simple geometrical interpretation of Berry's phase. If one regards the space of normalized states as a fiber bundle over the space of rays[3] (a ray is defined as an equivalence class of states differing only in phase), then this bundle has a natural connection. This connection permits a comparison of the phases of states on two neighboring rays. Simon observed that when the dynamical phase factor is removed, the evolution of the system as determined by the Schrödinger equation is a parallel transport of the phase of the system according to this natural connection. Berry's phase is then a consequence of the curvature of this connection.

Recently, Aharanov and Anandan[4] generalized Berry's results by giving up the assumption of adiabaticity. The key step in this work is their identification of the integral of the expectation value of the Hamiltonian as the dynamical phase. Once this dynamical phase is removed, the evolution of the phase of the system is again determined by the natural connection and one recovers Berry's phase for any cyclic evolution of the quantum system.

The purpose of this Letter is to point out that Berry's phase appears in a still more general context. The evolution of the system need be neither unitary (norm preserving) nor cyclic (returning to the original ray). This generalization is based on the work of Pancharatnam[5] on the interference of polarized light. Carrying Pancharatnam's ideas over to quantum mechanics yields a fairly general setting for a discussion of Berry's phase. We briefly describe Pancharatnam's work before developing the subject of the present paper.

Pancharatnam posed the following question: Given two beams of polarized light, is there a natural way to compare the phases of these beams? His physically motivated answer was to cause interference of these two polarized beams and regard them as "in phase" when the resultant intensity is maximum. This provides a "connection" (a rule for the comparison of phases) between any two states of polarization which are not orthogonal. (This rule breaks down for orhtogonal states. These do not interfere and the resultant intensity is insensitive to the relative phase of the two beams.) Consider three (nonorthogonal) states of polarization represented by three points 1, 2, and 3 on the Poincaré sphere. Suppose now that 1 and 2 are "in phase" and 2 and 3 are "in phase"; then 1 and 3 are not necessarily in phase. Pancharatnam showed that the excess phase of 3 over 1 is given by half the solid angle subtended by the spherical triangle 123 at the center of the Poincaré sphere. Thus the Pancharatnam connection has curvature. While Pancharatnam's studies, both theoretical and experimental, were carried out in the 1950's, the relation between his work and Berry's was pointed out only recently by Ramaseshan and Nityananda.[6] They, and subsequently Berry,[7] observed that Pancharatnam's excess phase is in fact an early example of Berry's phase. A laser interferometer experiment demonstrating Pancharatnam's excess phase has earlier been reported by us.[8] In the rest of this paper we show that carrying over Pancharatnam's ideas to quantum mechanics leads to a fruitful generalization of Berry's phase.

Consider a quantum system whose state vector $|\psi\rangle$ (an element of a Hilbert space $\mathcal{H}$) evolves according to the Schrödinger equation $i(d/dt)|\psi(t)\rangle = \hat{H}(t)|\psi(t)\rangle$. ($\hat{H}$ is a linear operator, possibly non-Hermitean.) Let us define a new state vector $|\phi(t)\rangle$, which differs from $|\psi(t)\rangle$ only in that it has had a dynamical phase factor removed:

$$|\phi(t)\rangle = \exp\left[i\int_0^t h(t')dt'\right]|\psi(t)\rangle,$$

where

$$h(t') = \langle\psi|\psi\rangle^{-1}\mathrm{Re}\langle\psi(t')|\hat{H}(t')|\psi(t')\rangle.$$

Clearly, $|\phi(t)\rangle$ satisfies the equation

$$i(d/dt)|\phi(t)\rangle = [\hat{H}(t) - h(t)]|\phi(t)\rangle.$$

Contracting this with $\langle\phi(t)|$ yields the parallel-transport law

$$\mathrm{Im}\langle\phi(t)|(d/dt)|\phi(t)\rangle = 0. \qquad (1)$$

While this law has its origin in the Schrödinger equation, it is purely geometric, as are the considerations in the rest of this paper.

Let $\mathcal{N}$ denote the set of normalizable states in $\mathcal{H}$:

$$\mathcal{N} = \{|\psi\rangle \in \mathcal{H} \mid \langle\psi|\psi\rangle \neq 0\}.$$

Let $\mathcal{R}$ be the space of rays: $\mathcal{R} = \mathcal{N}/\sim$, where $\sim$ denotes that elements of $\mathcal{N}$ which differ only by a phase are regarded as equivalent. There is a natural projection map $\pi:\mathcal{N} \to \mathcal{R}$, which maps each vector to the ray on which it lies. The triplet $(\mathcal{N},\mathcal{R},\pi)$ forms a principal fiber bundle over the base space $\mathcal{R}$ [with structure group U(1)] and the parallel-transport law (1) defines a natural connection on this fiber bundle. A connection[9] is an assignment of a "horizontal subspace" in the tangent space of each point in $\mathcal{N}$. Horizontal vectors are those that satisfy (1). Given a curve in $\mathcal{R}$, one can lift this curve up to $\mathcal{N}$ so that its tangent vector is horizontal. However, the horizontal lift of a closed curve in $\mathcal{R}$ may be open in $\mathcal{N}$. This is referred to as holonomy of the connection and provides a geometric picture of Berry's phase.

Let $|\phi(s)\rangle$ be a curve in $\mathcal{N}$. Let $|u\rangle = (d/ds)|\phi(s)\rangle$ denote the tangent vector to this curve. Let us define[10]

$$A_s = \mathrm{Im}\langle\phi|u\rangle/\langle\phi|\phi\rangle. \qquad (2)$$

Under transformations of the kind $|\phi(s)\rangle \to \exp[i \times \alpha(s)]|\phi(s)\rangle$ (referred to as gauge transformations), A_s transforms inhomogeneously,

$$A_s \to A_s + d\alpha/ds, \qquad (3)$$

like the vector potential in electrodynamics. The parallel-transport law (1) states that A_s vanishes along the actual curve $|\phi(s)\rangle$ followed by the quantum system.

Let us first consider $|\psi(t)\rangle$, a solution of the Schrödinger equation which is cyclic, i.e., returns to the initial ray at some time τ. This defines a curve in $\mathcal{N}$. The "shadow" of this curve under projection map π is a closed curve in $\mathcal{R}$. Given the closed curve $r(s)$ in $\mathcal{R}$, let us ask for the curve $|\phi(s)\rangle$ in $\mathcal{N}$ traced out by the state vector (with the dynamical phase removed). Using (1), we find that the curve is determined by the condition $A_s = 0$ along the curve. Consider the integral

$$\gamma = \oint A_s\, ds \qquad (4)$$

along the curve $|\phi(s)\rangle$ in $\mathcal{N}$ closed by the vertical curve joining $|\phi(\tau)\rangle$ to $|\phi(0)\rangle$. The segment $|\phi(s)\rangle$ represents the actual evolution of the system and along this, $A_s = 0$. The vertical contributes the phase difference between $|\phi(0)\rangle$ and $|\phi(\tau)\rangle$ and represents Berry's phase. However, the integral (4) is gauge invariant because of (3) and can be regarded as an integral on $\mathcal{R}$. With use of Stokes's theorem, γ can be expressed as

$$\gamma = \int_S F, \qquad (5)$$

where S is any surface in $\mathcal{R}$ bounded by the closed curve $r(s)$ in $\mathcal{R}$ and F is the gauge-invariant ("field strength") two-form representing the curl of A. γ depends only on the geometric curve $r(s)$ and not on the rate at which it is transversed in time. This gives the formula for Berry's phase in a (possibly nonunitary) cyclic evolution of a quantum system.

In a general evolution, the state vector may not return to the initial ray. In order to handle this situation, we need a method of comparing states on different rays for phase. This is provided by the *Pancharatnam connection*: Let $|\phi_1\rangle$ and $|\phi_2\rangle$ be any two elements of $\mathcal{N}$ which are not orthogonal. Interference of these two states by superposition yields

$$\| |\phi_1\rangle + |\phi_2\rangle \|^2 = \langle\phi_1|\phi_1\rangle + \langle\phi_2|\phi_2\rangle + 2\,\mathrm{Re}\langle\phi_1|\phi_2\rangle.$$

The modulus of the resultant vector is clearly a maximum when $\langle\phi_1|\phi_2\rangle$ is real and positive. Under this condition, $|\phi_1\rangle$ and $|\phi_2\rangle$ are said to be "in phase." More generally, if one writes the complex number $\langle\phi_1|\phi_2\rangle$ in polar form, $\rho\exp i\beta$, $\rho>0$, then the phase difference between $|\phi_1\rangle$ and $|\phi_2\rangle$ is β. The Pancharatnam connection has a clear physical basis and is more general than the natural connection since it permits a comparison of *any* two (nonorthogonal) states for phase and not just neighboring ones. In the particular case where $|\phi_1\rangle$ and $|\phi_2\rangle$ are on neighboring rays, the Pancharatnam connection reduces to the natural connection.

We now go on to express the Pancharatnam phase difference in terms of the natural connection. In order to do this, we need to explore some geometrical properties of the ray space $\mathcal{R}$. We observe that $\mathcal{R}$ has a natural metric on it, which comes from the (positive definite) inner product $\langle\,|\,\rangle$ on $\mathcal{H}$. Since each point of $\mathcal{R}$ is an entire equivalence class, it is convenient to take representative elements from $\mathcal{N}$ and make sure our considerations are gauge invariant. Let $|\phi(s)\rangle$ be a curve in $\mathcal{N}$ and $|u\rangle$ its tangent vector. Under gauge transformations, $|u\rangle$ does not transform covariantly. But its projection orthogonal to the fiber,

$$|u'\rangle=|u\rangle-|\phi\rangle[\langle\phi|u\rangle-\langle u|\phi\rangle](2\langle\phi|\phi\rangle)^{-1}$$

is gauge covariant. $|u'\rangle$ is in fact the covariant derivative

$$|u'\rangle=(d/ds)|\phi(s)\rangle-iA_s|\phi(s)\rangle.$$

$\langle u'|u'\rangle$ is gauge invariant and can be used to define a metric on $\mathcal{R}$: $dl^2=\langle u'|u'\rangle ds^2$. dl^2 is the square of the distance between points $\pi(|\phi(s)\rangle)$ and $\pi(|\phi(s+ds)\rangle)$. This metric can also be expressed with use of the density matrix $\rho=|\psi\rangle\langle\psi|$, which contains information only about the ray and not the phase. Its form is

$$dl^2=(\mathrm{Tr}\rho)^{-1}[\mathrm{Tr}(d\rho d\rho)-\tfrac{1}{2}(\mathrm{Tr}d\rho)^2].$$

This metric then determines geodesics in $\mathcal{R}$. These can be found by variation of $\int\langle u'|u'\rangle dl$, where l is an affine parameter. This yields the geodesic equation

$$\frac{D^2}{dl^2}|\phi(l)\rangle=\frac{d}{dl}|u'\rangle-iA_s|u'\rangle=0. \qquad (6)$$

Curves in $\mathcal{N}$ which satisfy this equation project down to geodesics in $\mathcal{R}$. Notice that (6) is gauge covariant and so the geodesic nature is a property of the "shadow" of the curve and not the curve itself.

The importance of geodesic curves in $\mathcal{R}$ stems from the fact that one can express the Pancharatnam phase difference as a line integral of A_s with use of the *geodesic rule*: Let $|\phi_1\rangle$ and $|\phi_2\rangle$ be any two (nonorthogonal) states in $\mathcal{N}$, with phase difference β according to the Pancharatnam connection. Let $|\phi(s)\rangle$ be any *geodesic* curve connecting $|\phi_1\rangle$ to $|\phi_2\rangle$: $|\phi(0)\rangle=|\phi_1\rangle$, $|\phi(1)\rangle=|\phi_2\rangle$. Then β is given by

$$\beta=\int A_s ds, \qquad (7)$$

where A_s is given by (2).

Proof: Let $r(s)$ be a geodesic curve in $\mathcal{R}$ joining $\pi(|\phi_1\rangle)$ to $\pi(|\phi_2\rangle)$. Consider the horizontal lift $|\bar\phi(s)\rangle$ of this curve, which starts from $|\phi_1\rangle$ $[|\bar\phi(0)\rangle=|\phi_1\rangle$, $\bar A_s=0]$. The geodesic equation (6) reduces to $(d^2/ds^2)|\bar\phi(s)\rangle=0$, whose solution is a straight line in $\mathcal{N}$. Further, $|\bar\phi(s)\rangle$ is "in phase" with $|\bar\phi(0)\rangle$. To see this, define $g(s)=\mathrm{Im}\langle\bar\phi(0)|\bar\phi(s)\rangle$. Clearly, $g(0)=0$ and $\dot g(0)=0$ since $|\bar\phi(s)\rangle$ is a horizontal curve. Now $\ddot g(s)$ can be worked out from the geodesic equation $\ddot g(s)=\mathrm{Im}\langle\bar\phi(0)|(d^2/ds^2)|\bar\phi(s)\rangle=0$, so that $g(s)$ is identically zero along the horizontal curve $|\bar\phi(s)\rangle$; hence $\langle\phi_1|\bar\phi(s)\rangle$ is real,[11] and so $|\bar\phi(s)\rangle$ and $|\phi_1\rangle$ are "in phase." To prove (7), we simply perform a gauge transformation $|\phi(s)\rangle=\exp[i\alpha(s)]|\bar\phi(s)\rangle$, where $\alpha(s)$ is chosen so that $\alpha(0)=0$, $\alpha(1)=\beta$. Then $|\phi(s)\rangle$ is still a geodesic curve (since the geodesic equation is gauge covariant) and connects $|\phi_1\rangle$ to $|\phi_2\rangle$. Using the fact that $\bar A_s$ was zero along the horizontal curve $|\bar\phi(s)\rangle$ and the behavior (3) of A_s under gauge transformations, we find that the right-hand side of (7) becomes $\int_0^1(d\alpha/ds)ds=\beta$ and (7) is verified. The geodesic rule[12] (7) is the main result of this paper.

We are now ready to show how Berry's phase also appears in a noncyclic evolution. Let the state vector $|\phi(t)\rangle$ (with the dynamical phase removed as always) evolve from $|\phi(0)\rangle$, initially, to $|\phi(\tau)\rangle$. If $|\phi(\tau)\rangle$ is not orthogonal to $|\phi(0)\rangle$, it is meaningful to ask, "What is the phase difference between them?" If we use the geodesic rule, this can be expressed as in (7). Now add to this integral the quantity $\int A_s ds$ integrated along the actual curve determined by the Schrödinger equation. Because of (1), this vanishes and the phase difference γ between $|\phi(\tau)\rangle$ and $|\phi(0)\rangle$ is expressed as an integral (4) where the contour C is given by the actual evolution $|\phi(t)\rangle$ from $|\phi(0)\rangle$ to $|\phi(\tau)\rangle$ and *back along any geodesic* curve joining $|\phi(\tau)\rangle$ to $|\phi(0)\rangle$. This expression for γ is gauge invariant and so can be regarded as defined on the base $\mathcal{R}$. So γ can be expressed as the integral (5) of the two-form F over a surface bounded by the closed curve $\pi(C)$. γ is clearly a gauge-invariant quantity and measureable. It has a purely geometric origin and depends only on the geometric path that the system traces in $\mathcal{R}$ and not on its rate of traversal.

Let us next consider a quantum system undergoing a nonunitary evolution, as happens, for example, when the system is subjected to measurements. According to the collapse postulate, the effect of the measurement on the system is described by the projection operator $P=|\psi\rangle\langle\psi|$ onto the eigenstate corresponding to the eigenvalue (the outcome of the measurement) of the operator measured. Consider a system initially in the state $|\psi_1\rangle$ on which three successive measurements are made. If the effects of these measurements are to project the state of the system onto $|\psi_2\rangle$, then onto $|\psi_3\rangle$ and back onto $|\psi_1\rangle$, the final state of the system is given by $|\psi_1\rangle\langle\psi_1|\psi_3\rangle\langle\psi_3|\psi_2\rangle\langle\psi_2|\psi_1\rangle$. (We ignore the time evolution, i.e., set $H=0$, so that we can concentrate on the effects due to projection that we are interested in.) The final and initial states have a well-defined phase difference, given by the phase of the complex number $\langle\psi_1|\psi_3\rangle\langle\psi_3|\psi_2\rangle\langle\psi_2|\psi_1\rangle$. Using the geodesic rule, we see that the phase is given by (5), where now the surface is bounded by the geodesic triangle connecting rays 1, 2, and 3. Thus Berry's phase also appears in systems subjected to quantum measurements. For a spin-$\frac{1}{2}$ (two-state) system, this formula for γ reduces to half the solid angle subtended at the center of the Poincaré sphere by the rays 1, 2, and 3. γ can be experimentally measured.

In summary, Berry's phase appears to be more general than the context in which it was discovered by Berry, i.e., for an adiabatic, cyclic, and unitary evolution. Our discussion uses a new ingredient—the Pancharatnam connection—and contains previous work as special cases. Since Berry's phase is being studied and applied in many different contexts, this generalization may be of interest.

It is a pleasure to thank S. Ramaseshan and Rajaram Nityananda for exciting our interest in Pancharatnam's work and S. Sridhar for useful discussions.

---

[1]M. V. Berry, Proc. Roy. Soc. London **392**, 45 (1984).

[2]B. Simon, Phys. Rev. Lett. **51**, 2167 (1983); M. S. Narasimhan and S. Ramanan, Am. J. Math. **83**, 563 (1961).

[3]Actually, Simon used the induced bundle on a finite-dimensional parameter space mapped injectively into the ray space. Y. Aharonov and J. Anandan, Phys. Rev. Lett. **58**, 1593 (1987), work directly with the ray space. See also Don Page, Phys. Rev. A **36**, 3479 (1987).

[4]Aharonov and Anandan, Ref. 3.

[5]S. Pancharatnam, Proc. Indian Acad. Sci. **A44**, 247 (1956), reprinted in *Collected Works of S. Pancharatnam* (Oxford Univ. Press, London, 1975).

[6]S. Ramaseshan and R. Nityananda, Curr. Sci. **55**, 1225 (1986).

[7]M. V. Berry, J. Mod. Opt. **34**, 1401 (1987). A proof of the geodesic rule is given here for two-state quantum systems.

[8]R. Bhandari and J. Samuel, Phys. Rev. Lett. **60**, 1211 (1988).

[9]S. Kobayashi and K. Nomizu, *Foundations of Differential Geometry* (Wiley, New York, 1963).

[10]The connection can be given by our specifying a connection one-form A on $\mathcal{N}$. The contraction of this one-form with a tangent vector $|u\rangle$ is denoted by A_s in the text. Note that A is a linear functional on tangent vectors over the reals (not over **C**).

[11]In fact, $\langle\tilde\phi(0)|\tilde\phi(s)\rangle$ is also positive if $|\tilde\phi(s)\rangle$ is the shortest geodesic connecting $|\tilde\phi(0)\rangle$ with $|\tilde\phi(1)\rangle$. Along this curve, $\langle\tilde\phi(0)|\tilde\phi(s)\rangle$ never vanishes and, since it was positive to start with, remains so.

[12]This rule breaks down for orthogonal states, since these are connected by a continuous infinity of geodesics, and this prescription does not give a definite answer.

Progress of Theoretical Physics, Vol. 74, No. 3, September 1985

Effective Action for Adiabatic Process

—— *Dynamical Meaning of Berry and Simon's Phase* ——

Hiroshi KURATSUJI*) and Shinji IIDA

Department of Physics, Kyoto University, Kyoto 606

(Received March 16, 1985)

By applying the path integral method to two interacting systems, it is shown that the specific phase Γ appearing in the quantum adiabatic process recently found by Berry and Simon is obtained as an additive action to the conventional dynamical action function. This scheme naturally gives a dynamical meaning to Berry and Simon's topological phase, which leads to a novel form of semiclassical quantization rule including the phase Γ.

Following an early implication in molecular physics,[1] Berry[2] recently discovered a rather unexpected phenomenon on quantum adiabatic theorem: During an excursion along a closed loop C in the external parameter space the adiabatic change of a wave function gains an extra phase in addition to the conventional dynamical phase; $\psi_n(T) = \exp[i\Gamma_n(C)]\exp[-i/\hbar\int_0^T E_n(R(t))dt]|n(R(T))\rangle$. Here Berry demonstrated that the phase $\Gamma(C)$ reflects a peculiar structure associated with the level-crossing which is inherent to the global geometric nature of the external parameter space. Subsequently Simon[3] gave this specific phase a topological meaning that it is nothing but the "holonomy" constructed from the vector bundle of parametrized wave functions and discussed a connection with the quantized Hall effect.[4] The topological concept such as vector bundle or connection has now become familiar in gauge field theory.[5] However it is rather surprising that the topological concept appears even in the usual non-relativistic quantum mechanics.

Although Berry and Simon's phase (B-S phase) has such an appealing feature, it is still concerned with the static aspect only, i.e., time development of external parameter space is given from the outset. Actually, the parameter space itself can be regarded as a dynamical object. For example, in the Born-Oppenheimer theory, the internuclear distance, which is frozen in the adiabatic process, should be regarded as a dynamical variable. Thus we are forced to inquire a dynamical meaning to Berry and Simon's topological phase. The purpose of this paper is to put forward an answer to this question. A similar dynamical argument was suggested by Mead and Truhlar[1] before Berry and Simon; they showed that this specific phase acquires a meaning of the effective vector potential in the Schrödinger equation for the nuclear motion in molecular collisions. However, the argument based on the Schrödinger equation is of essentially local nature and does not seem to be appropriate for describing the global character of this non-integrable phase. In order to push the global aspect forward, we adopt the path integral formulation for the bound state problem of interacting two systems. In this formulation, the B-S phase naturally arises as an additional action to the conventional action function

*) Present address: Department of Mathematics and Physics, Ritsumeikan University, Kyoto 603.

induced by adiabatic process. Simultaneously this topological action is shown to modify the semiclassical quantization rule for the motion of the external system.

Effective action by path integral: Consider two interacting systems, which are described by the variables conventionally called "internal" and "collective" coordinates; q and X respectively. We adopt a Hamiltonian $\hat{H}=\hat{h}(q, X)+\hat{H}_0(P, X)$, where the internal Hamiltonian $\hat{h}$ is assumed to depend on X and not on its conjugate momentum P. Let us consider the trace of the evolution operator $K(T)=\mathrm{Tr}[\exp(-i\hat{H}T/\hbar)]$, which is written as

$$K(T)=\sum_n\int\langle n(X_0), X_0|\exp[-i\hat{H}T/\hbar]|n(X_0), X_0\rangle d\mu(X_0). \tag{1}$$

In Eq. (1) one naturally picks up the transition amplitude for the quantum process starting from the initial state of product form $|n(X_0), X_0\rangle(\equiv|n(X_0)\rangle\otimes|X_0\rangle)$ and returning via closed loops C to the same state, where $|X_0\rangle$ denotes the eigenstate of $\hat{X}$ and $|n(X_0)\rangle$ is the eigenstate of $\hat{h}(q, X_0)$ at $X=X_0$ with eigenvalue $E_n(X_0)$. Then, with the aid of the time-discretization together with the completeness relation holding for X, we get

$$\langle n(X_0), X_0|\exp[-i\hat{H}T/\hbar]|n(X_0), X_0\rangle$$
$$=\int\prod_{k=1}^{N-1} d\mu(X_k)\langle n(X_0), X_0|\exp[-i\hat{H}\varepsilon/\hbar]|X_{N-1}\rangle\cdots\langle X_1|\exp[-i\hat{H}\varepsilon/\hbar]|n(X_0), X_0\rangle \tag{2}$$

with $\varepsilon=T/N$. Further noting the relation for $\varepsilon\approx 0$,

$$\langle X_k|\exp[-i\hat{H}\varepsilon/\hbar]|X_{k-1}\rangle\simeq\langle X_k|\exp[-i\hat{H}_0\varepsilon/\hbar]|X_{k-1}\rangle\exp[-i\hat{h}(X_k)\varepsilon/\hbar]$$
$$=\int dP_k\exp[(iP_k(X_k-X_{k-1})-iH_0(X_k, P_k)\varepsilon)/\hbar]\exp[-i\hat{h}(X_k)\varepsilon/\hbar], \tag{3}$$

Eq. (1) can be expressed as

$$K(T)=\sum_n\int T_{nn}(C)\exp\left[\frac{i}{\hbar}S_0(C)\right]\prod_t d\mu(X_t, P_t) \tag{4}$$

with $S_0(C)(\equiv\int(P\dot{X}-H_0)dt)$ the action for the collective motion along closed loops C. $T_{nn}(C)$ is just the internal transition amplitude and given by

$$T_{nn}(C)=\langle n(X_0)|\exp[-i\hat{h}(N)\varepsilon/\hbar]\cdots\exp[-i\hat{h}(1)\varepsilon/\hbar]|n(X_0)\rangle, \tag{5}$$

i.e., the time ordered product, where $\hat{h}(k)$ denotes the internal Hamiltonian at the point $X=X_k$ on the loop C.[6] Namely, if we denote $|\phi_n(T)\rangle$ as a solution of the time-dependent Schrödinger equation; $(i\hbar\partial/\partial t-\hat{h}(q, X_t))|\phi_n(t)\rangle=0$ with the boundary condition $|\phi_n(0)\rangle=|n(X_0)\rangle$, $T_{nn}(C)$ is written as $T_{nn}(C)=\langle n(X_0)|\phi_n(T)\rangle$.

Under the above prescription we turn to the case of the adiabatic motion where the period T is large. By inserting the completeness relation holding for the internal state on each point of external variables X_k; $\sum_{m_k}|m_k\rangle\langle m_k|=1$, Eq. (5) is written as

$$T_{nn}(C)=\sum_{m_1}\cdots\sum_{m_{N-1}}\langle n(X_0)|\exp[-i\hat{h}(N)\varepsilon/\hbar]|m_{N-1}\rangle$$

$$\cdots\langle m_k|\exp[-i\hat{h}(k)\varepsilon/\hbar]|m_{k-1}\rangle\cdots\langle m_1|\exp[-i\hat{h}(1)\varepsilon/\hbar]|n(X_0)\rangle. \tag{6}$$

In the adiabatic approximation, we pick up the quantum transition only between the states with the same quantum number n; $\langle n_k|\exp(-i\hat{h}(k)\varepsilon/\hbar)|n_{k-1}\rangle$. Then using the relation $\hat{h}(k)|n_k\rangle=E_n(k)|n_k\rangle$ ($E_n(k)$ is an energy of an adiabatic level n at $X=X_k$), we obtain $T_{nn}(C)=\exp[-i/\hbar\int_0^T E_n(X_t)\,dt]\langle n(X_0)|n(X_T)\rangle_C$. Here the overlap function $\langle n(X_0)|n(X_T)\rangle_C$ is given as an infinite product:

$$\langle n(X_0)|n(X_T)\rangle_C=\lim_{N\to\infty}\prod_{k=1}^{N}\langle n(X_k)|n(X_{k-1})\rangle, \tag{7}$$

where we adopt a phase convention $|n(X_0)\rangle=|n(X_T)\rangle$. This overlap function naturally involves the history of the excursion in the X-space which is indicated by the suffix C. Each factor $\langle n(X_k)|n(X_{k-1})\rangle$ in (7) defines a "connection" between two infinitesimally separated points X_{k-1} and X_k, hence Eq. (7) gives a finite connection along circuit C given by a set of division points $\{X_k\}$.[7] Thus, by using the approximate relation

$$\langle n(X_k)|n(X_{k-1})\rangle\simeq 1-\langle n|\partial/\partial X_i|n\rangle\Delta X_i\approx\exp[i\omega],$$

Eq.(7) is written as

$$\langle n(X_0)|n(X_T)\rangle_C=\exp[i\Gamma_n(C)] \tag{8}$$

with

$$\Gamma_n(C)=\oint_C\omega=\oint\langle n|i\partial/\partial X_k|n\rangle dX_k\,. \tag{9}$$

Equation (9) is essentially the same as the phase obtained by Berry.*) However the present derivation is quite different from Berry's and somewhat similar to Simon's[3] which is based upon the holonomy of vector bundle over X-space. Thus we arrive at the effective path integral associated with the adiabatic change of the external dynamical variable X,

$$K^{\text{eff}}(T)=\sum_n\int\exp\left[\frac{i}{\hbar}(S_n^{ad}+\hbar\Gamma_n(C))\right]\prod_t d\mu(X_t,P_t), \tag{10}$$

where $S_n^{ad}(\equiv S_0-\int_0^T E_n(X_t)\,dt)$ is the adiabatic action function. From (10) we get a natural explanation that the phase $\Gamma_n(C)$ *appears as a topological action function which is to be added to the usual dynamical action.* This is a first consequence of the paper. If we note $\omega=A_i\dot{X}_i$ dt with $A_i=\langle n|i\partial/\partial X_i|n\rangle$, the effective action in (10) can be regarded as the action function for a system in the effective "gauge field" described by the "vector potential" A_i. This result was already obtained by Mead and Truhlar[1] by using the Schrödinger equation which was described by only usual canonical variables (X, P). However, the present path integral formulation is applicable to more general external systems which are described by non-canonical variables, e.g., spin variables.[8),9]

Level-crossing structure revisited: Here we examine a specific model Hamiltonian revealing the topological meaning of the phase Γ; consider the following internal Hamiltonian:

---

*) Here, it is noted that in the present procedure the closed loop C is naturally introduced as a consequence of the trace formula, whereas in Ref. 2) it is presupposed from the outset.

$$\hat{h}(X)=\begin{pmatrix} z & x+iy \\ x-iy & -z \end{pmatrix}, \qquad X=(x, y, z)\,. \tag{11}$$

Although this model already has been studied by Berry,[2] we treat it in a different way by adopting the "coherent-state" representation,[9] which may reserve an applicability to more complicated Hamiltonians. Using two-component Pauli spinor we write the eigenvector of (11) as $|\xi\rangle=\cos(\theta/2)|-1/2\rangle+\sin(\theta/2)e^{i\phi}|1/2\rangle={}^t(\sin(\theta/2)e^{i\phi}, \cos(\theta/2))\equiv{}^t(a, b)$, which is given as the $SU(2)$ (spin) coherent-state

$$|\xi\rangle=(1+|\xi|^2)^{-1/2}\exp[\xi\hat{S}_+]|-1/2\rangle \tag{12}$$

with $\xi=\tan(\theta/2)e^{i\phi}$. The eigenvalues of (11) are calculated as

$$\lambda_\pm=\pm\sqrt{x^2+y^2+z^2}\,, \tag{13}$$

which show a remarkable feature that two levels λ_+ and λ_- cross at $X=0$, namely, the origin $X=0$ becomes a singular-point of cone type. In the following we take the lower level λ_-, for which the corresponding eigenvector is given by

$$a/b=\tan(\theta/2)e^{i\phi}=-(x+iy)/(\sqrt{x^2+y^2+z^2}+z)\,, \tag{14}$$

which yields $\phi=\beta$ and $\theta=-\alpha$, where α and β are the polar angle defined by $x=r\sin\alpha\cdot\cos\beta$, $y=r\sin\alpha\sin\beta$, $z=r\cos\alpha$. $\Gamma(C)$ is thus evaluated as

$$\Gamma(C)=\oint\frac{1}{2i}\frac{(\xi^*\nabla\xi-\text{c.c.})}{1+|\xi|^2}dX=\frac{1}{2}\oint(1-\cos\alpha)\nabla\beta dX\,, \tag{15}$$

which is written as $\oint AdX$, where the vector potential becomes

$$A_x=(\cos\alpha-1)\frac{\sin\beta}{r\sin\alpha}, \quad A_y=(1-\cos\alpha)\frac{\cos\beta}{r\sin\alpha} \tag{16}$$

and $A_z=0$. The striking point is that the negative z-axis (i.e., $\alpha=\pi$) forms a *singular line* on which A diverges.[10] This suggests that the "Dirac pole" is located at the origin as was pointed out by Berry, but the present result gives an explicit form of the specific singular nature. The phase Γ is converted into the surface integral by Stokes' theorem; $\Gamma(C)=\int_S d\omega=\frac{1}{2}\int_S\sin\alpha d\alpha\wedge d\beta$, which is just the solid angle suspended by the closed loop C. Here we note that the present path integral formalism naturally allows the "topological quantization" analogous to the Dirac quantization which is familiar in gauge theory. Consider a sphere S^2 and divide it by C into two hemispheres S and $\tilde{S}$. We can choose two different gauges such that ω is singular-free on hemispheres S or $\tilde{S}$, respectively. Then, the topological part of the propagator $K^{\text{eff}}(T)$ can be expressed in two ways according to these two choices, and the consistency condition asserts these two expressions should coincide. Namely, the relation

$$\exp\left[i\oint_C\omega\right]=\exp\left[i\int_S d\omega\right]=\exp\left[-i\int_{\tilde{S}}d\omega\right] \tag{17}$$

should hold, which leads to the quantization condition $\int_S d\omega+\int_{\tilde{S}}d\omega=\int_{S^2}d\omega=2\pi\times(\text{integer})$.

Finally we give a remark on a generalization to $n\times n$ matrix Hamiltonian; it may be simply achieved by replacing the eigenstate of form (12) by the $SU(n)$ coherent-state[9] where the parameter space becomes the complex projective space $U(n)/U(n-1)\times U(1)$ the point of which is coordinated by n-dimensional complex vector $\boldsymbol{\xi}=(\xi_1, \cdots, \xi_n)$ and the resultant phase yields

$$\Gamma(C)=\frac{i}{2}\oint_C\sum_{\mu,i}\left(\frac{\partial \log F}{\partial \xi_\mu}\frac{\partial \xi_\mu}{\partial X_i}-\text{c.c.}\right)dX_i \tag{18}$$

with $F(\equiv 1+\xi^\dagger\xi)$ being the overlap function of the coherent state. There may also occur the singularity due to level crossing leading to the general form of the topological quantization.

Semiclassical quantization rule: Now we address a question how one can look at the effect of the topological phase. The most direct way for this is to examine the energy spectra. The energy spectra is rapidly estimated by the semiclassical quantization rule[11] which is derived from the effective propagator (10). Consider the Fourier transform of $K^{\text{eff}}(T)$; $K(E)=i\int_0^\infty K^{\text{eff}}(T)\exp[iET/\hbar]dT$, where we restrict ourselves to a specific adiabatic level n. Firstly the semiclassical limit of $K^{\text{eff}}(T)$ is approximated by the method of stationary phase,

$$K^{\text{sc}}(T)\sim\sum_{P\cdot O}\exp\left[\frac{i}{\hbar}S^{ad}(C)+i\Gamma(C)-i\frac{\pi}{2}\alpha(C)\right], \tag{19}$$

where $\alpha(C)$ denotes the so-called Keller-Maslov index and $\sum_{P\cdot O}$ indicates the sum over periodic orbits. Next, taking the Fourier transform of (19) and evaluating the integral over T by the method of stationary phase, then we get

$$K^{\text{SC}}(E)\sim\sum_{P\cdot O}\exp\left[\frac{i}{\hbar}W^{ad}(E)+i\Gamma(C)-i\frac{\pi}{2}\alpha(C)\right], \tag{20}$$

where $W^{ad}(E)=S^{ad}+ET$ (action integral) and $T(E)$ is determined by the stationary phase condition $\partial/\partial T(S^{ad}+ET)=0$. Here, we restrict ourselves to the case that there appear a finite number of isolated closed orbits for each value of the energy. For this case, a semi-classical quantization condition can be written down explicitly. Namely, taking account of the contribution from the multiple traversals of basic orbits, i.e., putting $W^{ad}\to m\cdot W^{ad}$, $\alpha\to m\cdot\alpha$ and $\Gamma\to m\cdot\Gamma$ for m-times traversals and summing over m, $K^{\text{SC}}(E)$ turns out to be

$$K^{\text{SC}}(E)\sim\sum_{P\cdot O}\exp[i\tilde{W}/\hbar]\cdot\{1-\exp[i\tilde{W}/\hbar]\}^{-1} \tag{21}$$

with $\tilde{W}(E)=W^{ad}(E)+\hbar\Gamma(C)-\hbar\alpha/2\cdot\pi$. From the pole of (21) we get the formula*)

$$W^{ad}(E)=\oint PdX=\left(n+\frac{\alpha}{4}-\frac{\Gamma}{2\pi}\right)2\pi\hbar\,. \tag{22}$$

This gives the energy spectrum for the collective motion including the effect of the topological phase Γ.**) Equation (22) is the second main consequence of this paper.

We examine the above formula for a simple case. Consider the internal Hamiltonian

*) The more precise form of formula (22) includes stability exponents (see Ref. 11)).

**) A similar formula has been recently presented by Wilkinson[12] in the course of investigating the band structure of Bloch electrons in the magnetic field, which is, however, concerned with a rather specific problem. On the other hand, our formula is derived on the general framework of the adiabatic theorem.

of type (11). The level-crossing (singular point) is just located at $x=y=z=0$. Further, we restrict collective periodic orbits to circular motions around the z-axis with a radius ρ, namely collective motions occur in the plane $z=\eta=$const. Assuming that the collective Hamiltonian has a rotational symmetry, $\hat{H}_0$ expressed in the polar coordinate is essentially a function of $\hat{p}_\phi=i/\hbar\partial/\partial\phi$ alone. Then, the total Hamiltonian considered is given by

$$\hat{H}=\hat{H}_0(\hat{P}_\phi)+\begin{pmatrix} \eta & \rho e^{i\phi} \\ \rho e^{-i\phi} & -\eta \end{pmatrix}. \tag{23}$$

The B-S phase is $\Gamma_\pm(C)=\mp\pi(1-\eta/e_\pm)$ for the internal upper (lower) state where $e_\pm=\pm\sqrt{\rho^2+\eta^2}$ denotes an internal energy eigenvalue. From Eq. (22), the semiclassical quantization condition becomes

$$\frac{1}{2\pi}\oint_C P_\phi d\phi=P_\phi(E_\pm)=\left(m-\frac{\Gamma_\pm}{2\pi}\right)\hbar, \tag{24}$$

where $p_\phi(E)$ is defined as an inverse relation of $H_{\text{eff}}(P_\phi)=E$. Semiclassical energy eigenvalues are calculated as

$$\begin{aligned} E_\pm^{sc}&=H_0\left(P_\phi=m\hbar-\frac{\hbar}{2\pi}\Gamma_\pm\right)+e_\pm \\ &=H_0(P_\phi=m\hbar)+e_\pm\pm\frac{\hbar}{2}\frac{dH_0}{dP_\phi}\bigg|_{P_\phi=m\hbar}-\frac{\hbar}{2}\frac{\eta}{e_\pm}\frac{dH_0}{dP_\phi}\bigg|_{P_\phi=m\hbar}+O(\hbar^2). \end{aligned} \tag{25}$$

In the present case, $\hat{H}$ can easily be diagonalized and we get exact energy eigenvalues as

$$\begin{aligned} E_\pm&=\frac{1}{2}[H_0(P_\phi=m\hbar)+H_0(P_\phi=(m\pm1)\hbar) \\ &\quad\pm\sqrt{\{H_0(P_\phi=m\hbar)-H_0(P_\phi=(m\pm1)\hbar)\pm2\eta\}^2+4\rho^2} \\ &=H_0(P_\phi=m\hbar)\pm\sqrt{\rho^2+\eta^2}\pm\frac{\hbar}{2}\frac{dH_0}{dP_\phi}\bigg|_{P_\phi=m\hbar}\mp\frac{\hbar}{2}\frac{\eta}{\sqrt{\rho^2+\eta^2}}\frac{dH_0}{dP_\phi}\bigg|_{P_\phi=m\hbar}+O(\hbar^2). \end{aligned} \tag{26}$$

In Eq. (25), the first and the second term represent the unperturved eigenvalue of collective Hamiltonian $\hat{H}_0$ and a conventional adiabatic potential which is constant $e_\pm$ in the present case, respectively. The third and fourth terms come from the B-S phase $\Gamma_\pm(C)$. In comparison with Eq. (26), we can see that this modification of the quantization rule reproduces the exact result up to $\hbar$-order. Since higher-order terms than $\hbar$ are beyond a semiclassical approximation, eigenvalues (25) can be regarded to be exact within a semiclassical treatment in spite of adiabatic approximation. In addition to the above, Eq. (24) shows that the angular momentum quantization is modified by the existence of the level-crossing. This phenomenon is very analogous to angular momentum quantization in the presence of magnetic flux generated by a solenoid.[13] Following an analogy with this, Eq. (24) suggests that due to the elimination of internal degrees of freedom the level-crossing produces an "effective spin" for an external dynamical system, which is a sort not to be locally described but only describable by a global aspect of the internal level structure.

Final remarks: The appearance of the topological additive term in the action function may be regarded as a rather universal phenomenon whenever one deals with interacting dynamical systems in the adiabatic approximation. This viewpoint may shed a light on a wide class of theoretical problems. For example, the "anomaly" in gauge field theory may be naturally understood within the present scheme, which would indeed await a further investigation.

Acknowledgements

The authors would like to thank Dr. T. Hatsuda for a constructive and fruitful discussion. The authors also thank other members of nuclear theory group for their interest. Furthermore, they thank Professor T. Suzuki for his useful comments. One of the authors (S. I.) is indebted to Japan Society for the Promotion of Science for financial support.

References

1) C. A. Mead and D. G. Truhlar, J. Chem. Phys. **70** (1979), 2284.
2) M. V. Berry, Proc. Roy. Soc. **A392** (1984), 45.
3) B. Simon, Phys. Rev. Lett. **51** (1983), 2167.
4) D. J. Thouless, M. Kohmoto, M. P. Nightingale and M. den Nijs, Phys. Rev. Lett. **49** (1982), 405.
5) See, e.g., R. Jackiw, Lecture at Les Houches, July, 1983.
6) P. Pechukas, Phys. Rev. **181** (1969), 174.
7) H. Kuratsuji and T. Hatsuda, in *Proceedings of 13-th International Colloquium in Group Theoretical Method in Physics*, Univ. of Maryland, May 1984, ed. W. W. Zachary (World Scientific), p. 238.
8) H. Kuratsuji and T. Suzuki, J. Math. Phys. **21** (1980), 472.
9) R. Gilmore, Ann. of Phys. **74** (1972), 391.
J. R. Klauder, J. Math. Phys. **4** (1963), 1055, 1058.
10) P. A. M. Dirac, Proc. Roy. Soc. **A133** (1931), 60.
11) See, e.g., M. C. Gutzwiller, J. Math. Phys. **12** (1971), 343.
W. H. Miller, J. Chem. Phys. **63** (1975), 996.
12) M. Wilkinson, J. of Phys. **A17** (1984), 3459.
13) F. Wilczek, Phys. Rev. Lett. **48** (1982), 1144.

Note added in proof: The non-abelian extension of Berry and Simon's Phase has been recently studied by F. Wilczek and A. Zee (Phys. Rev. Lett. **52** (1984), 2111). The authors would like to thank Professor A. Zee for informing of their work.

Adiabatic Effective Lagrangians

John Moody

Department of Computer Science
Yale University
New Haven, CT 06520

Alfred Shapere and Frank Wilczek

School of Natural Sciences
Institute for Advanced Study
Princeton, NJ 08540

ABSTRACT

We discuss the general theory of effective Lagrangians and Hamiltonians in molecular physics. The Born–Oppenheimer effective Lagrangian for the nuclei involves a gauge potential, which may be nonabelian if the electronic energy levels are degenerate. We develop a systematic procedure for finding corrections to the adiabatic approximation, both perturbative and non-perturbative. The former may be incorporated directly into the effective nuclear Lagrangian.

1. The Adiabatic Approximation: General Considerations

Berry's original paper on geometric phases emphasized quantum systems influenced by external parameters [1]. He showed that when these parameters are slowly taken around a closed circuit, the wavefunction of the system may acquire a geometric phase. Although the external parameters were implicitly taken to be classical variables, many interesting applications of the same basic ideas occur in a fully quantum mechanical setting. One can form an effective Born–Oppenheimer Hamiltonian or Lagrangian for the external parameters, that incorporates the effect Berry's phase through a gauge-potential-like term [2] [3] [4]. Upon quantization, the presence of this extra term may lead to significant observable effects, such as shifted quantum numbers and level splittings.

Berry's phase is only the leading correction to the traditional Born–Oppenheimer approximation. Higher-order corrections may also be directly incorporated by adding further terms to the effective Lagrangian [5] [6].

In this paper, we hope to give a relatively unified account of the "modern" Born–Oppenheimer method. We shall discuss both the Hamiltonian and Lagrangian approaches, their relationship, and their apparent inequivalence. Our discussion must necessarily include a description of the procedure for incorporating corrections to the adiabatic approximation, and at the moment, this subject is far from closed. Accordingly, a significant portion of the paper is devoted to a discussion of the adiabatic approximation itself. It covers the Dykhne and Landau–Zener formulas, corrections to them, and geometric phases in the complex plane.

The Born–Oppenheimer approximation first arose in the context of molecular physics [7], but more generally applies whenever a system exhibits two widely separated energy scales. This approximation is often described as a separation of "slow" and "fast" variables; these are just the variables associated with the different energy scales. Quantum mechanically, the separation is made possible by the existence of a large energy gap.

In the original application to molecular physics, the gap involved is the spacing between the electronic energy levels. This gap is typically much larger than the separation between levels associated with vibrations and rotations of the nuclear degrees of freedom that do not involve re-arrangement of electronic orbitals.* Now if we want to describe the spectrum of low-energy excitations of the molecule, *i.e.*, the excitations with energy much less than the electronic energy gap, then we should be able to form a description that involves only the nuclear degrees of freedom. Indeed, at such low energies the electrons have no independent dynamics—they are "enslaved" to

* Often a small finite number of electronic states are actually or approximately degenerate. The formalism appropriate to this case is discussed further below. For now, we assume no degeneracy.

the nuclear degrees of freedom—because only one state is available to them. Therefore, it is possible to describe the low-energy excitations by an effective Lagrangian involving the nuclear degrees of freedom alone, with no explicit reference to the electrons. Of course the value of the numerical parameters appearing in this Lagrangian will depend implicitly upon the electrons.

We find this way of formulating the Born-Oppenheimer idea much more appropriate, and easier to generalize, than the usual formulation in terms of "fast" and "slow" variables. The connection between the two is as follows. Transitions to states separated by a large energy gap require large changes in frequency, and are therefore associated with "fast" variables. Rapid oscillations in time accompany such transitions, and lead to cancellations in processes whose characteristic time scale is much longer—that is, in processes associated with motion of the "slow" variables. Towards the end of this article we shall discuss the relationship between these two approaches more precisely. It is appropriate to mention one conclusion from that discussion now, however: we shall find that quantum variables can only be slow in a very weak sense. For example, in a path integral description the important space-time paths are not differentiable, and the typical velocity is strictly speaking infinite even for so-called "slow" variables. Nevertheless, not being fussy, we shall freely refer to "fast" and "slow" variables throughout this paper.

In quantum field theories containing heavy particles, there is a large gap between the ground state of these heavy particles—*e.g.*, the filled Dirac sea for heavy fermions—and the energy of any excited state. Indeed, to reach an excited state with the same quantum numbers we must in general supply at least the energy to produce a particle-antiparticle pair. Suppose now that the theory contains in addition other fields, describing lighter particles. Then we should be able to describe slow space-time variations of these other fields by an effective Lagrangian that makes no explicit reference to the heavy particles. The usual jargon is to say that we can "integrate out" the heavy particle degrees of freedom. Clearly, the formation of effective Lagrangians in quantum field theory is fully analogous to the corresponding procedure in molecular physics [8]. (But notice that in field theory the heavy particles are the "fast" degrees of freedom!)

In the derivation of effective Lagrangians, we should expect—and will find—that geometric phases occur. This is particularly clear if we think in terms of path integrals. Then along any particular path the slow degrees of freedom can be considered as external parameters governing the state of the fast ones. Therefore, the amplitude for such a path can contain a geometric phase factor of the classic type. Geometric phases of this sort are connected with some of the most subtle and interesting phenomena in quantum field theory, including the occurrence of fractional quantum numbers and anomalies [9] [10].

2. The Born-Oppenheimer Hamiltonian

We now return to the historical context of the Born–Oppenheimer approximation, to discuss the derivation of effective Hamiltonians and Lagrangians. In molecular physics, it is useful to treat the electronic and nuclear degrees of freedom as fast and slow variables, respectively. This is because the gap between electronic energy levels is typically much larger than the gap between nuclear levels, by a factor of order $(M/m)^{\frac{1}{4}}$ [7]. In the Born-Oppenheimer approximation, one solves for the electronic states in a fixed nuclear background. By the adiabatic theorem, one expects these electronic states to be approximately stationary with respect to the relatively slow motions of the nuclei. We can thus obtain an effective description for the nuclear motion, relative to a fixed electronic orbital, by integrating over electronic coordinates. We shall find that the effective nuclear Lagrangian obtained in this way involves both an ordinary potential term due to the electronic energy levels and a background gauge potential which couples to the nuclear current [2]. This gauge potential takes into account the Berry phase accumulated by the electronic wavefunctions when the nuclear coordinates change adiabatically [3].

The Born-Oppenheimer approximation begins with the full Schrödinger equation

$$(T_{\text{nuc}} + T_{\text{el}} + V)\Psi = E\Psi \tag{2.1}$$

where T_{el} and T_{nuc} are the electronic and nuclear kinetic energy terms, $V(r, R)$ contains the potential and interaction energies of the electrons and nucleons, and r and R are the electronic and nuclear coordinates. The wave function Ψ is separated into nuclear and electronic components Φ_n and ϕ_n as

$$\Psi(r, R) = \sum_n \Phi_n(R)\phi_n(r, R) \tag{2.2}$$

where the subscript n labels the electronic energy eigenstates in a fixed nuclear background. That is, $\phi_n(r, R)$ satisfies the electronic Schrödinger equation at a fixed value of R

$$[T_{\text{el}} + V(r, R)]\phi_n(r, R) = \epsilon_n(R)\phi_n(r, R) \tag{2.3}$$

In terms of the electronic eigenfunctions, the full Schrodinger equation may now be rewritten as

$$\sum_n [T_{\text{nuc}} + \epsilon_n(R)]\Phi_n(R)\phi_n(r, R) = E\sum_n \Phi_n(R)\phi_n(r, R) \tag{2.4}$$

We may now integrate out the electronic degrees of freedom to leave a system of equations for the nuclear wavefunction Ψ alone. Using bracket notation for the normalized electronic eigenstates, we get

$$\sum_n \langle \phi_m | T_{\rm nuc} \Phi_n | \phi_n \rangle + \epsilon_m(R) \Phi_m = E \Phi_m \tag{2.5}$$

The nuclear kinetic energy operator $T_{\rm nuc} = -\frac{1}{2M}\nabla^2$ (with $\hbar = 1$) operates on both the nuclear and electronic wavefunctions, $\Phi_n(R)$ and $|\phi_n(r,R)\rangle$. Thus the kinetic energy terms in (2.5) are proportional to

$$\langle \phi_m | \nabla_R^2 \Phi_n | \phi_n \rangle = \sum_k \left(\delta_{mk} \nabla_R + \langle \phi_m | \nabla_R \phi_k \rangle\right) \left(\delta_{kn} \nabla_R + \langle \phi_k | \nabla_R \phi_n \rangle\right) \Phi_n \tag{2.6}$$

The Born-Oppenheimer approximation applies when the mixing between different electronic levels is small, so that the off-diagonal matrix elements in Eq.(2.6) can be neglected. If, furthermore, the electronic states can be chosen to be real for each R, then $\langle \phi_n | \nabla_R \phi_n \rangle = 0$ and Eq.(2.6) reduces to

$$\left(-\frac{1}{2M} \nabla^2 + \sum_{k \neq n} \frac{1}{2M} \langle \phi_n | \nabla \phi_k \rangle \langle \phi_k | \nabla \phi_n \rangle + \epsilon_n(R) \right) \Phi_n = E \Phi_n \tag{2.7}$$

In this approximation, the nuclei propagate in a background potential

$$\tilde{\epsilon}_n(R) = \epsilon_n(R) + \sum_{k \neq n} \frac{1}{2M} \langle \phi_n | \nabla \phi_k \rangle \langle \phi_k | \nabla \phi_n \rangle$$

The peculiar extra term may be rewritten as follows:

$$\frac{1}{2M} \sum_{k \neq n} \left| \frac{\langle \phi_n | (\nabla H) | \phi_k \rangle}{\epsilon_n - \epsilon_k} \right|^2 \tag{2.8}$$

Hence, when the energy splittings between level n and the other levels are large, this term may be neglected. Berry has pointed out that it is proportional to the trace of the "natural metric" on projective Hilbert space [11].

However, it is not always possible to form a basis of electronic wavefunctions that are everywhere real. Furthermore, corrections to adiabatic evolution will involve mixings of electronic levels. We introduce the "gauge potential" notation

$$A_{mn} \equiv i \langle \phi_m | \nabla_R \phi_n \rangle \tag{2.9}$$

to account for both of these possibilities. Putting together Eqs.(2.4), (2.5), and (2.9), we can write a complete matrix-valued Schrödinger operator for the nuclear wave functions

$$H^{\text{eff}}_{mn} = -\frac{1}{2M}\sum_k (\delta_{mk}\nabla_R - iA_{mk}(R))\cdot(\delta_{kn}\nabla_R - iA_{kn}(R)) + \delta_{mn}\epsilon_n(R) \tag{2.10}$$

which acts on the vector Φ_n

$$H^{\text{eff}}_{mn}\Phi_n = E\Phi_m \tag{2.11}$$

(The Schrödinger operator is, of course, associated with an effective Hamiltonian after the replacement $-i\nabla_R = p_R$.)

In the Born–Oppenheimer approximation, the effect of the off–diagonal matrix elements A_{mn} which mix different energy levels is ignored. Sections 4 and 5 will be devoted to a justification of this procedure; for now we simply state the result that corrections are indeed suppressed, by a factor depending on the ratio of the typical nuclear and electronic energy splittings. Then for a nondegenerate electronic level, the effective nuclear Schrödinger operator in the Born-Oppenheimer approximation is then simply

$$H^{\text{BO}}_n = -\frac{1}{2M}(\nabla_R - iA_n(R))^2 + \tilde{\epsilon}_n(R) \tag{2.12}$$

where $A_n \equiv A_{nn}$.

Eq. (2.12) looks like the Schrödinger operator of a charged particle in the presence of a background magnetic potential. To further strengthen this analogy, the vector field A_n even transforms like a $U(1)$ gauge potential, as we shall now explain. The phase each of the wavefunctions $|\phi_n(R)\rangle$ is arbitrary, and our description of the dynamics of the nuclei must always respect this arbitrariness. The effect of a redefinition of phases of the electronic wavefunctions $|\phi_n(R)\rangle \to e^{i\lambda_n(R)}|\phi_n(R)\rangle$, is to rotate the nuclear wavefunctions oppositely

$$\Phi_n(R) \to e^{-i\lambda_n(R)}\Phi_n(R), \tag{2.13}$$

so that the full wavefunction $\Psi(r,R)$ is preserved. From Eq. (2.9), we see that the gauge potential transforms just as it should:

$$A_n(R) \to A_n(R) + \nabla_R\lambda_n(R) \tag{2.14}$$

and it is easy to see that the overall effect of the phase redefinition is to leave the nuclear Schrödinger equation invariant (including the term (2.8)).

We conclude that the nuclei behave like charged particles in a magnetic field $B = \nabla \times A_n$. Semiclassically speaking, when the nuclei go around a closed path, the wavefunction will accumulate a geometrical phase proportional to the enclosed magnetic flux. (We will be able to see this more clearly from the Lagrangian point of view discussed in the following section.) This phase is nothing but Berry's phase in quantum mechanical clothing—the phase that the evolving electron wavefunctions accumulate when their external parameters R are slowly varied has just been passed down to the nuclear wavefunctions.

The degenerate case is slightly more complicated; the evolution will generally involve $U(N)$ rotations among the N degenerate states [12] (provided there are no selection rules forbidding such rotations). In the adiabatic approximation, again restricting attention to a single energy level, we obtain an effective matrix Hamiltonian as in (2.10) with a $U(N)$ gauge potential A_{mn}. The N electronic eigenfunctions may now be regarded as an N-component vector; its "phase" is a $U(N)$ matrix.

For example, the effective nuclear Hamiltonian operator for a molecule with doubly degenerate electronic energy levels (labeled by $\uparrow$ and $\downarrow$) contains a $U(2)$ gauge potential:

$$H^{\mathrm{BO}} = -\frac{1}{2M}\left\{\nabla_R - i\begin{pmatrix} A_{\uparrow\uparrow} & A_{\uparrow\downarrow} \\ A_{\downarrow\uparrow} & A_{\downarrow\downarrow}\end{pmatrix}\right\}^2 + \epsilon(R) \qquad (2.15)$$

Such a Hamiltonian arises in considering the degenerate Λ-levels of a diatomic molecule [3]. For $\Lambda = \frac{1}{2}$, there is no choice of electronic basis states for which the $U(2)$ gauge potential becomes everywhere diagonal.

To close this section, we remark that there is a much bigger symmetry group that is always present, which mixes states of different energies. The group in question is the unitary symmetry $U(\infty)$ of the electronic Hilbert space $\mathcal{H}_R$. It is difficult to see where all this symmetry has gone in the nuclear Schrödinger equation Eq. (2.7), because in choosing a decomposition of the total wavefunction in terms of electronic energy eigenfunctions, we have "fixed the gauge" down to a product of $U(N)$ factors (one factor for each N-fold degenerate level). However, there is an alternative formulation in which the full symmetry is manifest, involving a different effective Hamiltonian. We sandwich the time-dependent Schrödinger equation

$$(T_{\mathrm{nuc}} + T_{\mathrm{el}} + V)\Psi = E\Psi \qquad (2.16)$$

between a complete set of electronic states (not necessarily energy eigenstates) to obtain a matrix Schrödinger equation analogous to Eq. (2.5)*

$$i(\partial_t - iA_0)\Phi = -\frac{1}{2M}(\nabla_{R_i} - iA_i)^2\Phi + \delta_{mn}\epsilon_n(R) \qquad (2.17)$$

where $A_0 \equiv i\langle\phi_m|\dot{\phi}_n\rangle$, with the time derivative referring to the implicit time dependence of $|\phi_n(R(t))\rangle$ (but not on the "dynamical" phase factor

* A similar but not identical nuclear Hamiltonian has been obtained by Zygelman [13].

$\exp i \int \epsilon_n$). Under arbitrary R-dependent (and possibly time-dependent) unitary rotations that preserve (2.2),

$$\begin{aligned} \phi_n(r,R) &\to \phi_m(r,R)U^\dagger_{mn}(R) \\ \Phi_n(R) &\to U_{nm}(R)\Phi_m(R) \end{aligned} \tag{2.18}$$

(where $U^\dagger = U^{-1}$), A_0 behaves like the time-component of a gauge field and Eq. (2.17) is fully $U(\infty)$-covariant. $A_\mu \equiv (A_0, A_i)$ is thus a 4-component $U(\infty)$ gauge potential.

3. Lagrangian Formulation

Often it is more convenient to work with a path-integral description. Phenomenological models are typically easier to formulate in terms of a Lagrangian, where symmetries are manifest. Non-equilibrium and non-perturbative problems, such as calculating tunneling amplitudes, may be easier to solve in the language of path integrals. In addition, as we shall see, it is much easier to incorporate corrections to the adiabatic approximation (which are higher-order in time derivatives) in an effective Lagrangian.

In models of the type we have been considering, effective Lagrangians (typically matrix-valued) for slow degrees of freedom arise naturally when one functional integrates over the fast variables. Generally, functional integration of a matrix-valued integrand requires extra care, to order the operators correctly [14]. But in the adiabatic approximation, if the fast variables are locked into a non-degenerate state, the effective Lagrangian is a scalar, and there is no time ordering to worry about.

As in the previous section, it is convenient to split the Lagrangian into slow and fast parts as follows

$$L = L_{\text{nuc}} + L_{\text{el}} \tag{3.1}$$

with

$$L_{\text{nuc}}(R) = \tfrac{1}{2}M\dot{R}^2 \tag{3.2}$$

$$L_{\text{el}}(r,R) = \tfrac{1}{2}m\dot{r}^2 - V(r,R). \tag{3.3}$$

The full time-evolution kernel which connects states at time t_0 to states at time t can be written as a Feynman sum over all paths from configuration (r_0, R_0) to configuration (r, R):

$$K(r_1,R_1,t_1;\ r_0,R_0,t_0) = \int_{R_0}^{R_1}\int_{r_0}^{r_1} \mathcal{D}[R]\mathcal{D}[r]\ \exp i\int_{t_0}^{t_1} dt\,(L_{\text{nuc}} + L_{\text{el}}(R,r)) \tag{3.4}$$

To form an effective Lagrangian for the nuclei, we now want to integrate (3.4) over the electron coordinates. The first step is to perform a Born-Oppenheimer separation on the molecular wavefunction, as we did in the last section:

$$\Psi(r,R) \equiv \sum_n \Phi_n(R)\phi_n(r,R) \tag{3.5}$$

As before, the electronic eigenstates $\phi_n(r,R)$ are solutions of the electronic Hamiltonian at nuclear configuration R, and the vector-valued nuclear states Φ_n are solutions of the matrix-valued nuclear Schrodinger equation (2.10).

In order to isolate the evolution of the nuclear wavefunctions, we now reorder the general path integral, separating the nuclear and electronic integrations:

$$K = \int_{R_0}^{R_1} \mathcal{D}[R]\, e^{i\int_{t_0}^{t_1} L_{\rm nuc}(R)dt} \int_{r_0}^{r_1} \mathcal{D}[r]\, e^{i\int_{t_0}^{t_1} L_{\rm el}(r,R)dt} \tag{3.6}$$

With respect to the decompostion of Eq.(3.5), we can express the result of the electronic path integral for a given nuclear path $R(t)$ in terms of an electronic time evolution kernel:

$$\int_{r_0}^{r_1} \mathcal{D}[r]\, e^{i\int_{t_0}^{t_1} L_{\rm el}(r,R)dt} \equiv \sum_{mn} \phi_m(r_1,R_1,t_1)\, K^{\rm el}_{mn}\, \phi_n(r_0,R_0,t_0) \tag{3.7}$$

where

$$\begin{aligned} K^{\rm el}_{mn} &= {\rm T}\ \exp\left(-i\int_{t_0}^{t_1} dt\, \left[\epsilon_m\delta_{mn} + i\langle\phi_m|\dot{\phi}_n\rangle\right]\right) \\ &= {\rm T}\ \exp\left(-i\int_{t_0}^{t_1} dt\, \left[\epsilon_m\delta_{mn} + A_{mn}(R(t))\cdot\dot{R}(t)\right]\right) \end{aligned} \tag{3.8}$$

is the evolution kernel for the electronic eigenstates. (The notation conveys that we are to take the time ordered exponential of the operator whose mn matrix element is displayed in brackets.) This expression for the kernel comes from integrating the electronic Schrodinger equation for $|\phi_n(R(t))\rangle$

$$\begin{aligned} i\frac{d}{dt}|\phi_n\rangle &= i\frac{\partial}{\partial t}|\phi_n\rangle + i|\dot{\phi}_n\rangle = \epsilon_n|\phi_n\rangle + i\dot{R}(t)\cdot\nabla_R|\phi_n\rangle \\ &= \sum_m \left[\epsilon_n\delta_{mn} + i\langle\phi_m|\nabla_R\phi_n\rangle\cdot\dot{R}\right]|\phi_m\rangle \end{aligned} \tag{3.9}$$

with respect to t. $K^{\rm el}$ just gives the usual dynamical evolution of the electronic energy eigenfunctions with an additional piece coming from the time-dependence of the eigenfunctions through $R(t)$. In the adiabatic limit, the kernel effectively diagonalizes, and the nth electronic eigenstate obeys the evolution equation

$$|\phi_n(t_1)\rangle = K^{\rm el}_{nn}(t_1,t_0)\,|\phi_n(t_0)\rangle = \exp\left(-i\int_{t_0}^{t_1} dt\,\Big[(\epsilon_n + i\langle\phi_n|\dot\phi_n\rangle)\Big]\right)|\phi_n(t_0)\rangle \tag{3.10}$$

The second term in the exponent is immediately recognized as Berry's phase.

With the electronic motion solved for, the nuclear kernel can now be extracted from the path integral:

$$\Phi_m(R_1) = \left\{\int_{R_0}^{R_1}\mathcal{D}[R]\ \mathrm{T}\exp i\,S[R]\right\}_{mn}\Phi_n(R_0) \tag{3.11}$$

where the exact effective nuclear action is

$$\begin{aligned} S^{\rm eff}_{mn}[R] &= \int_{t_0}^{t_1}\left\{\tfrac12 M\dot R^2\delta_{mn} - iA_{mn}(R(t))\cdot\dot R(t) - \epsilon_m(R)\delta_{mn}\right\}dt \\ &\equiv \int_{t_0}^{t_1} L^{\rm eff}_{mn}\,dt \end{aligned} \tag{3.12}$$

As with the exact effective nuclear Hamiltonian (2.10), the electronic energies $\epsilon_m(R)$ contribute an effective potential for the nuclei, and the velocity-dependent potential term containing A_{mn} modifies the nuclear kinetic energy. In fact, $L^{\rm eff}_{mn}$ can be obtained directly from $H^{\rm eff}_{mn}$ by a Legendre transformation, provided one orders the matrix-valued canonical momenta correctly.

When the electronic levels are nondegenerate, the effective action (3.12) diagonalizes in the Born–Oppenheimer approximation and the time ordering in Eq. (3.11) is unnecessary. For electrons in the nth energy level, the approximate effective action is

$$S^{\rm BO}_n[R] = \int_{t_0}^{t_1}\left\{\tfrac12 M\dot R^2 - iA_n(R(t'))\cdot\dot R(t') - \epsilon_n(R)\right\}dt' \tag{3.13}$$

(see also [4]). Curiously, this is *not* the effective action we would have obtained after a Legendre transformation of (2.12), because it does not include the term (2.8). Order by order, the Hamiltonian and Lagrangian formulations of the Born–Oppenheimer approximations are not equivalent.

There are at least two ways one might directly try to incorporate superadiabatic corrections in an effective Lagrangian framework. (The prefix "super-" indicates that we are looking for corrections to the adiabatic approximation; we are still concerned with the adiabatic regime.) The first is to take, in place of the scalar effective action (3.13), a matrix effective action including a small number of electronic levels, presumably those which are closest in energy to the particular level of interest. The problem with this scheme is that one still must deal with path-ordered exponentials of matrices, and operator ordering makes the quantization of the nuclear degrees of freedom quite tricky. Another approach is to expand the full time-ordered exponential (3.11) out to some finite degree, and to incorporate this expansion directly into an effective Lagrangian for the nth level, by adding extra terms to (3.13). We shall see how to do this to the lowest super-adiabatic order in the next section.

4. Classical Corrections to Adiabatic Evolution

We now embark on a detailed study of corrections to adiabatic the adiabatic approximation. In this section, our focus will be on corrections to the evolution of electronic wavefunctions with respect to smooth, classical nuclear motions. In Section 5, from the vantage point of the electronic path integral, we briefly discuss why the quantization of the nuclei fundamentally alters the nature and size of these corrections.

The total evolution of a wavefunction which begins in an energy eigenstate is best split into two parts—the amplitude to remain in that eigenstate, and the amplitude to have a transition to another eigenstate. The first part of the problem has been beautifully treated by Berry [15] by means of an iterative procedure. The essential idea is, for a given Hamiltonian $H(t)$, to perform an iterative sequence of time-dependent unitary transformations. At each step, the transformations

$$\begin{aligned} \phi^{(i)}(t) &= U_i(t)\,\phi^{(i-1)}(t) \\ H_i &= U_i H_{i-1} U_i^\dagger - i U_i \dot{U}_i^\dagger \end{aligned} \tag{4.1}$$

(with $H = H_0$) are supposed to be chosen in such a way that the evolution of the transformed wavefunction with respect to the new Hamiltonian is more adiabatic than the last. The phase evolution of the original energy eigenstate ϕ_n is obtained by evolving

$$\phi^{(i)} = U_i U_{i-1} \cdots U_0 \phi_n \tag{4.2}$$

with respect to H_i in the adiabatic approximation, and then transforming back to the original basis. This scheme is expected to converge rapidly, at least until the ith successive correction becomes comparable in magnitude to the typical amplitude for a transition to another level. At this point, the sequence of iterations begins to diverge—the expansion is asymptotic. (A similar scenario occurs in quantum field theory, where the perturbation expansion is an asymptotic series, which begins to diverge when tunneling processes become important.) Berry's procedure only gives the evolution of the *phase* of ϕ_n; changes in the magnitude of ϕ_n come from transitions to other levels, which are willfully ignored in this approximation.

In what follows, we shall consider two very different methods for calculating super-adiabatic transition amplitudes. The more straightforward approach is a version of time–dependent perturbation theory (TDPT), that explicitly separates adiabatic from super-adiabatic evolution. The other is essentially non-perturbative and involves analytic continuation into the complex time plane. The perturbative approach will enable us to give a proof of the adiabatic theorem and to find corrections. However, it is not very useful as a calculational tool, and indeed, not always very reliable. On the other hand, the non-perturbative method, embodied in Dykhne's formula and its generalizations, turns out to be quite powerful and accurate. (If some of the following material seems too abstract or technical, the reader may wish to refer to the example beginning with Eq.(4.21) for orientation.)

Adiabatic Perturbation Theory. Our discussion of time–dependent perturbation theory begins with the exact equation for evolution of the electronic wavefunction according to a time–dependent Hamiltonian. From Eqs.(3.8) and (3.9), we have

$$| \psi(t)\rangle = \sum_{mn} |\phi_m(R(t))\rangle \, U_{mn}(t,-\infty) \, \langle \phi_n(R(-\infty)) \mid \psi(R(-\infty))\rangle$$

$$U(t,-\infty) \equiv \mathrm{T}\exp \; - \; i\int_{-\infty}^{t} dt' \left[\epsilon_a \delta_{ab} + i\langle \phi_a \mid \dot{\phi}_b\rangle\right] \tag{4.3}$$

where for convenience we have taken $t_1 = -\infty$. In the nondegenerate case, the adiabatic approximation to it is

$$| \psi(t)\rangle_{\mathrm{ad}} = \sum_{n} \mid \phi_n(R(t))\rangle \left\{ \exp \; - i\int_{-\infty}^{t} dt' \left[\epsilon_n + i\langle \phi_n \mid \dot{\phi}_n\rangle\right] \right\}$$
$$\cdot \langle \phi_n(R(-\infty)) \mid \psi(R(-\infty))\rangle \tag{4.4}$$

We expect this adiabatic wavefunction to be a better and better approximation to the exact wavefunction, as the time dependence of the parameters $R(t)$ becomes slower and slower. To quantify this, we write

$$R(t,\tau) \;\equiv\; R(t/\tau)$$

and let $\psi_\tau(R(t))$ be the solution of the Schrodinger equation when the internal parameters vary as $R(t,\tau)$. Thus, the larger τ is, the more slowly the parameters R are changing. Note that replacing $R(t) \to R(t/\tau)$ in Eq.(4.3) is equivalent to rescaling time $t \to \tau t$ everywhere except inside of $R(t)$.

To compare the exact evolution to the adiabatic approximation, it is convenient to use the following general formula for untangling the time–ordered exponential integral of the sum of two operators:

$$T\exp \int (M+N) \;=\; T\exp \int M \;\cdot T\exp \int N'$$
$$N' \;=\; (T\exp \int M)^{-1} \cdot N \cdot T\exp \int M \tag{4.5}$$

which is easily proved by differentiating both sides. In our example, we wish to take

$$\begin{aligned} M_{ab} &= -i\epsilon_a\delta_{ab} + \langle\phi_a \mid \dot{\phi}_a\rangle\delta_{ab} \\ N_{ab} &= \langle\phi_a \mid \dot{\phi}_b\rangle(1-\delta_{ab}) \end{aligned} \tag{4.6}$$

so that the true evolution is governed by $M+N$, and the adiabatic evolution by M. By means of these manipulations, the corrections to the adiabatic evolution operator in Eq.(4.3) are isolated into a compact formal expression, that is:

$$\begin{aligned} \mathrm{T}\exp\int_{-\infty}^{t} dt'\; N'(t') &= \mathrm{T}\exp\int_{-\infty}^{t} dt' \Bigg\{ \langle\phi_p|\dot{\phi}_q\rangle(1-\delta_{pq}) \\ &\quad \cdot \exp\int_{-\infty}^{t'} dt'' \Big[i(\epsilon_p-\epsilon_q) - \langle\phi_p|\dot{\phi}_p\rangle + \langle\phi_q|\dot{\phi}_q\rangle \Big] \Bigg\} \end{aligned} \tag{4.7}$$

where we have used the fact that M is diagonal to get rid of some of the time ordering.

Our task is now to see how this expression approaches the identity operator as $\tau \to \infty$. Note first that rescaling the time variable does nothing to the integrals over $\langle\phi_p|\dot{\phi}_p\rangle$ and $\langle\phi_q|\dot{\phi}_q\rangle$, since the scale factors coming from the dt'' in the measure and from the time derivative in the integrand cancel. As we have seen in other contexts, these integrals have a purely geometric character. So, if we rescale the time, the only modification to Eq. (4.7) is to replace $(\epsilon_p - \epsilon_q)$ by $\tau(\epsilon_p - \epsilon_q)$. In other words, the slowness parameter appears only in an oscillatory exponential factor, which we expect will make the total integral very small as $\tau \to \infty$, *à la* the Riemann–Lebesgue lemma [16]. To get an idea of just how small, we expand the time-ordered exponential to first order:

$$\begin{aligned} T_{pq} &= \mathrm{T}\exp\int_{-\infty}^{\infty} dt'\; N'(t') \\ &= \delta_{pq} + \int_{-\infty}^{\infty} dt' \Big\{ \langle\phi_p|\dot{\phi}_q\rangle(1-\delta_{pq}) \\ &\quad \cdot \exp\int_{-\infty}^{t'} dt'' \Big[i(\epsilon_p-\epsilon_q) - \langle\phi_p|\dot{\phi}_p\rangle + \langle\phi_q|\dot{\phi}_q\rangle \Big] \Big\} + \cdots \\ &\equiv \delta_{pq} + \int_{-\infty}^{\infty} dt' \gamma_{pq}(t')\, \exp i\Delta_{pq}(t') + \cdots \end{aligned} \tag{4.8}$$

To simplify this, let us redefine the phases of the wavefunctions $|\phi_n(R(t))\rangle$ for each t so that the wavefunctions are always real along the contour of integration; then $\langle\phi_n|\dot{\phi}_n\rangle = 0$. We now make three crucial assumptions, that $\epsilon_p \neq \epsilon_q$ for all times, that γ_{pq} and $\epsilon_p - \epsilon_q$ are infinitely differentiable, and that as $t \to \pm\infty$, all derivatives of γ_{pq} and $\epsilon_p - \epsilon_q$ approach zero with sufficient rapidity. (If $R(t)$ is cyclic, the last assumption is not necessary.) Then integrating by parts, we may ignore all surface terms, and we find for the first-order term in (4.8)

$$T_{pq}^{(1)} = i \int_{-\infty}^{\infty} dt' \, \frac{\partial}{\partial t} \left(\frac{\gamma_{pq}}{\epsilon_p - \epsilon_q} \right) \, \exp i \int^{t'} dt'' \, (\epsilon_p - \epsilon_q) \tag{4.9}$$

from which it follows that

$$|T_{pq}^{(1)}| \leq \left| \int_{-\infty}^{\infty} dt' \, \frac{\partial}{\partial t} \left(\frac{\gamma_{pq}}{\epsilon_p - \epsilon_q} \right) \right| \tag{4.10}$$

The right-hand side scales like τ^{-1} when we scale $t \to \tau t$, so $T_{pq}^{(1)}$ goes to zero at least as fast as τ^{-1}. After n repeated integrations by parts, we obtain an expression that vanishes like τ^{-n}. In other words, $T_{pq}^{(1)}$ goes to zero faster than any power of τ. By a similar procedure, one may also show that the same is true for all *off-diagonal* higher-order terms in the expansion (4.8). This completes the proof of the adiabatic theorem.

The diagonal terms in (4.8) need not vanish so fast; indeed, they only vanish like powers of τ. For example, the pp component of the second-order term contains the non-oscillating piece

$$i \sum_{q \neq p} \int_{-\infty}^{\infty} dt' \, \frac{\gamma_{pq}(t') \, \gamma_{qp}(t')}{\epsilon_p - \epsilon_q} \tag{4.11}$$

which vanishes only like τ^{-1}. This first-order correction to purely adiabatic evolution may be incorporated directly into the effective Lagrangian (3.13) as a counterterm

$$S_n^{(1)}[R] = \int_{t_0}^{t_1} dt' \left(\sum_{q \neq n} \frac{A_{nq}^i A_{qn}^j}{\epsilon_n - \epsilon_q} \right) \dot{R}_i \dot{R}_j \, \delta_{mn}. \tag{4.12}$$

It modifies the metric on parameter space—which we have taken to be δ_{ij}—to

$$g_{ij} = \delta_{ij} + \frac{2}{M} \sum_{q \neq n} \frac{A_{nq}^i A_{qn}^j}{\epsilon_n - \epsilon_q}. \tag{4.13}$$

Not surprisingly, Eq.(4.11) is the same expression we would have obtained by expanding the first-order phase approximant from Berry's iteration scheme in powers of τ. When Berry's nth-order phase approximant is rearranged as an expansion in τ, it must agree with the TDPT expansion through the nth order in τ^{-1}. Both schemes diverge asymptotically. The difference with TDPT is that in principle, we might hope, it may give us information about super-adiabatic transitions.

Unfortunately, the expansion we have presented above is not very useful for actually computing off-diagonal corrections to adiabatic evolution, due to the presence of rapidly oscillating phases and the multiple integrations required to go beyond first order. And even when, say, the first order term in Eq.(4.8) can be evaluated, there may be no guarantee that the second-order term will be smaller—indeed, we shall discuss a specific example below where the higher-order corrections are larger. The moral of the story is that, in the adiabatic regime, transitions between levels are by nature non-perturbative, and attempting to treat them perturbatively is misguided.

Super-adiabatic transitions: Dykhne's formula. Much of what is known about transitions in the adiabatic limit is summarized by an elegant non-perturbative result known as Dykhne's formula [17], relating the amplitude for a transtion between two nondegenerate energy levels to the location of their common crossing point in the complex time plane. Suppose that $H(t)$ is a nondegenerate 2×2 Hamiltonian matrix, $E_1(t)$ and $E_2(t)$ are its two instantaneous energy levels, and $E_2 > E_1$ for all real times. If $E_1(t)$ and $E_2(t)$ are extendable into the complex time plane, there will typically be a point t_c where they cross. Dykhne's formula states that the transition probability to go from E_1 to E_2 as t runs from $-\infty$ to $+\infty$ is approximately

$$P_{12} \sim \exp - 2\,\mathrm{Im} \int_0^{t_c} (E_2 - E_1)\, dt \tag{4.14}$$

In general, Dykhne's formula is a good approximation when the crossing points are located far away from the real time axis. This will be true if the energy splittings and/or the typical time scale τ over which $H(t)$ changes are large. The relevant dimensionless expansion parameter is

$$\epsilon \sim \frac{\hbar}{\tau \Delta E} \tag{4.15}$$

and Dykhne's formula states that super-adiabatic transition amplitudes are of order $\mathcal{O}(\exp - \lambda/\epsilon)$ for some positive constant λ. This is the canonical form for non-perturbative corrections to an asymptotic expansion.

It is somewhat ironic that Dykhne's formula for super-adiabatic transitions may be proved by using a version of the adiabatic theorem in the complex time plane [18]. Like the real-time adiabatic theorem, this theorem describes the approximate evolution of the projection of a wavefunction onto an energy eigenstate, along a contour in the complex plane. The idea behind the proof of Dykhne's formula (which we shall only sketch here) is to find an appropriate contour in complex time such that continuation along the contour connects the two energy levels. For a non-degenerate two-level system, the crossing point will typically be a branch point of square root type for the function $(E_2 - E_1)(t)$. (If $E_2 > E_1$ on the real axis, we will take the branch point in the upper-half plane.) So if a wavefunction is initially in the eigenstate $\phi_1(-\infty)$ with energy $E_1(\infty)$, and if it is evolved along a contour C_{21} which goes over the branch point and across the cut, then one may compute its component in the direction $\phi_2(+\infty)$ (see Fig. 1). This will be related directly to the transition amplitude.

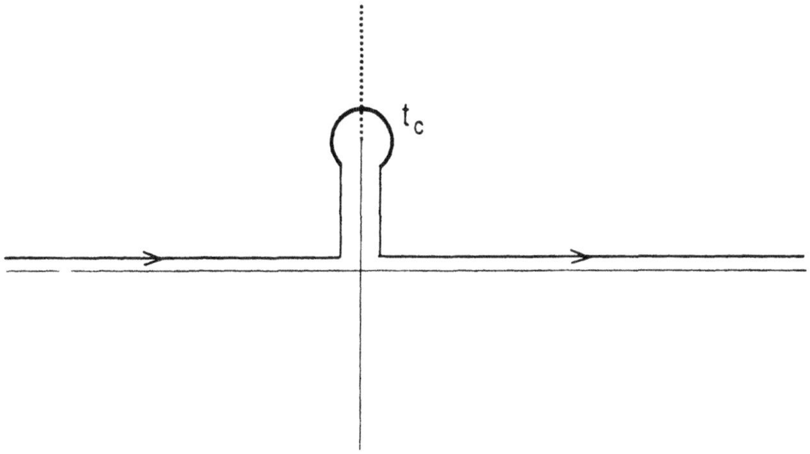

Figure 1. A contour C_{21} that connects two electronic eigenstates with energies E_1 at $t = -\infty$ and E_2 at $t = +\infty$. The function $E_1(t) - E_2(t)$ typically has a square-root branch point in the complex time plane. When the contour passes over the cut, the electron wavefunction crosses over from the E_1 to the E_2 level surface.

To be more explicit, let us choose a smooth basis of eigenstates $\phi_{1,2}(t)$ along the real time axis, and analytically continue our choice into the complex plane. Along the contour C_{21}, we also must have a smooth basis of eigenstates $|\tilde{\phi}_{1,2}\rangle$. Since ϕ_1 and ϕ_2 get interchanged in crossing the cut, it will generally not be possible to take $|\tilde{\phi}_{1,2}\rangle = |\phi_{1,2}\rangle$ everywhere. Instead, our choice along C_{21} will be as follows: to the left of the cut, we take $|\tilde{\phi}_{1,2}(t)\rangle = |\phi_{1,2}(t)\rangle$, but to the right, we take $|\tilde{\phi}_{1,2}(t)\rangle = e^{i\alpha_{2,1}}|\phi_{2,1}(t)\rangle$, where $e^{i\alpha}$ is a phase needed to make the choice of basis continuous across the cut. We shall likewise denote the energy levels along the real time axis by $E_{1,2}(t)$, and along C_{21} by $\tilde{E}_{1,2}(t)$; again because of the cut, $\tilde{E}_{1,2}(+\infty) = E_{2,1}(+\infty)$.

Finally, let $\psi(t)$ be a wavefunction evolving according to $H(t)$, initially in an eigenstate $\psi(-\infty) = \phi_1(-\infty)$. If $H(t)$ is analytic in a strip S, then

$\psi(t)$ will also be analytic and single-valued in S [19]. We may now state the result of the adiabatic theorem for evolution along C_{21}:

$$\langle\tilde{\phi}_1(+\infty)|\psi(+\infty)\rangle \simeq \exp -i\int_{C_{21}}\left[\tilde{E}_1(t)+i\langle\tilde{\phi}_1|\dot{\tilde{\phi}}_1\rangle\right]dt \tag{4.16}$$

This is precisely of the same form as Eq.(4.4), sandwiched on the left by $\langle\phi_n|$. Just as Eq.(4.4) said nothing about the "transition" components of ψ, so Eq.(4.16) is silent about the $\tilde{\phi}_2(+\infty) = e^{i\alpha}\phi_1(+\infty)$ component (which in fact is quite large). The big difference here is that the "energies" $\tilde{E}$ need no longer be real. Hence, the norm of (4.16) need not be equal to 1; in fact, in the adiabatic limit, it will be exponentially small.

We now wish to evaluate the transition probability

$$P_{21} \equiv |\langle\phi_2(+\infty)|\psi(+\infty)\rangle|^2 \tag{4.17}$$

The only part of (4.16) contributing to P_{21} comes from the imaginary part of the energy integral. We can put this into a convenient form by deforming the contour so that it follows along the real axis up to $t = 0$, then heads upward to the branch point, then returns to zero and continues along the real axis to infinity. The result is

$$P_{21} = \exp\ -2\,\mathrm{Im}\int_0^{t_c}(E_2-E_1)\,dt \tag{4.18}$$

The transition probability is only part of the story; the phase of the transition amplitude is also of interest. From Eq.(4.16), the phase of the total amplitude is

$$\begin{aligned}
&\text{phase}\left(\langle\phi_2(+\infty)|\psi(+\infty)\rangle\right)\\
&\quad= e^{i\alpha}\exp\ -i\int_{-\infty}^{0}\left[E_1+i\langle\phi_1|\dot{\phi}_1\rangle\right]dt\\
&\qquad\cdot\exp -i\int_0^{t_c}\left[\mathrm{Re}(E_1-E_2)+i\langle\phi_1|\dot{\phi}_1\rangle-i\langle\phi_2|\dot{\phi}_2\rangle\right]dt\\
&\qquad\cdot\exp\ -i\int_0^{+\infty}\left[E_2+i\langle\phi_2|\dot{\phi}_2\rangle\right]dt
\end{aligned} \tag{4.19}$$

The total phase contains, as usual, both a dynamical and a geometric component. The geometric phase itself splits into two pieces, an adiabatic phase and a *super-adiabatic* phase associated specifically with the tunneling process

$$\exp i\gamma_{1\to 2} = \exp\int_0^{t_c}\left(\langle\phi_1|\dot{\phi}_1\rangle - \langle\phi_2|\dot{\phi}_2\rangle\right)dt \tag{4.20}$$

We shall compute this phase below in the particular case of a Hamiltonian linear in t—the answer will turn out to be independent of any of the parameters appearing in $H(t)$.

The Landau-Zener formula. As an example, we now consider a Hamiltonian $H(t)$ in the vicinity of an avoided crossing, as originally studied by Landau and Zener [20] [21]. We focus upon the two levels whose energies cross, and study the equation governing their mixing in time:

$$i\frac{d\psi}{dt} = H\psi \tag{4.21}$$

$$H(t) = \begin{pmatrix} at & b \\ b & -at \end{pmatrix} = at\sigma_3 + b\sigma_1 \tag{4.22}$$

Here H is the Schrödinger operator in the two-level subspace. We have located the crossing at $t = 0$ and linearized around it, thrown away a possible constant term in the energy, and assumed $b \equiv \langle\psi_p|\dot{\psi}_q\rangle$ is real; none of these simplifications entails a loss of generality.

The eigenvalues of $H(t)$ are

$$E_{1,2}(t) = \pm\sqrt{b^2 + a^2t^2} \tag{4.23}$$

and the crossing (a square–root branch point) is located in the upper-half t-plane at $t_c = ib/a$. Hence, according to Eq.(4.18),

$$\begin{aligned} P_{21} &\simeq \exp - 2\,\mathrm{Im}\int_0^{t_c} 2\sqrt{b^2 + a^2t^2}\,dt \\ &= \exp - \pi\frac{b^2}{a} \end{aligned} \tag{4.24}$$

Dykhne's formula works amazingly well here; in fact, this is the *exact* result obtained by Zener after a much more involved analysis. Incidentally, TDPT is worse than useless here: the first–order term in Eq.(4.8) differs from the correct amplitude by a factor of π.

To find the phase of the transition amplitude requires a little more work. First we need an explicit basis of eigenfunctions: with eigenvalue $-\sqrt{b^2 + a^2t^2}$ we have

$$\phi_1 = \mathcal{N}_1 \begin{pmatrix} 1 \\ -\frac{at}{b} - \sqrt{1 + \left(\frac{at}{b}\right)^2} \end{pmatrix} \tag{4.25}$$

and with eigenvalue $+\sqrt{b^2 + a^2t^2}$,

$$\phi_2 = \mathcal{N}_2 \begin{pmatrix} \frac{at}{b} + \sqrt{1 + \left(\frac{at}{b}\right)^2} \\ 1 \end{pmatrix} \tag{4.26}$$

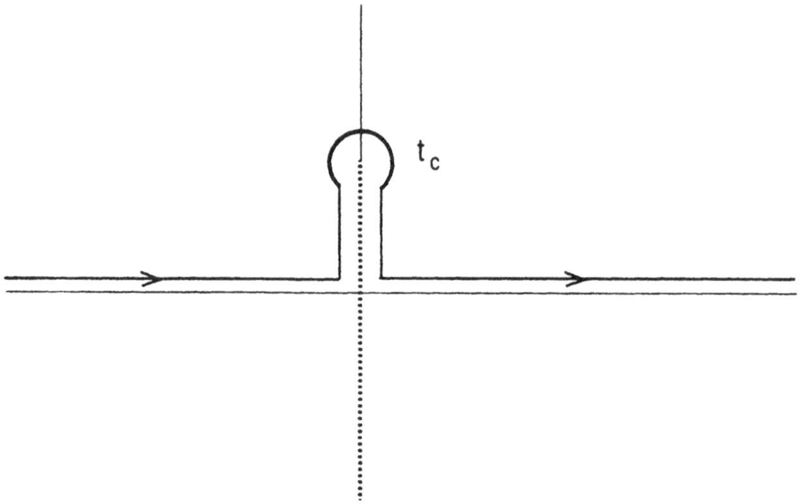

Figure 2. The same contour, from a different vantage point. This figure emphasizes that there is no discontinuity in the evolution of the electron wavefunction along the contour.

For real t, the eigenfunctions may be taken to be real, and the geometric phase receives no contribution from the integration along the real axis. For complex t, we choose the normalization $\mathcal{N}_1 = 1/\sqrt{2}$. We want to evaluate the integral

$$\int_0^{t_c} \Big(\langle \phi_1 | \dot{\phi}_1 \rangle - \langle \phi_2 | \dot{\phi}_2 \rangle \Big) \, dt = \int_C \langle \tilde{\phi}_1 | \dot{\tilde{\phi}}_1 \rangle \, dt \tag{4.27}$$

along the contour shown in Fig. 2. It is important to be sure that the eigenfunctions ϕ_1 and ϕ_2 match up precisely at t_c; for this to occur, it is necessary to take $\mathcal{N}_2$ to be $i/\sqrt{2}$. Extending this choice of phase downwards, we find that ϕ_2 is imaginary along the real t–axis. The actual computation of the geometric phase (4.20) is straightforward; the result is

$$\gamma_{1\to 2} = -i\frac{\pi}{2} \tag{4.28}$$

for a total phase of $-i$. This phase is precisely what is needed to cancel the i picked up in matching the wavefunctions at the branch point; it makes the non-dynamical part of the wavefunction real for all real times.

It is curious that the geometric phase we just computed is completely independent of a and b. In fact, we can argue that this sort of phase will arise quite generally, for Hamiltonians that are real on the real time axis. Whenever the energies cross at t_c, they will also cross at the conjugate point t_c^*. Let us join these two branch points by a cut draw contours above and below the cut, in such away that the images of the two contours are complex conjugate (see Fig. 3). It is easy to see that the total phases obtained by integrating along either contour must also be conjugate, and since the wavefunction must be single-valued, the two phases must be equal to ± 1. Furthermore, if the contour is chosen so that the total dynamical phase vanishes, then the geometric part of the wavefunction must be real, as we found in our example above. (This argument of course does not apply to complex Hamiltonians, and in general we can obtain complex geometric phases for such processes.)

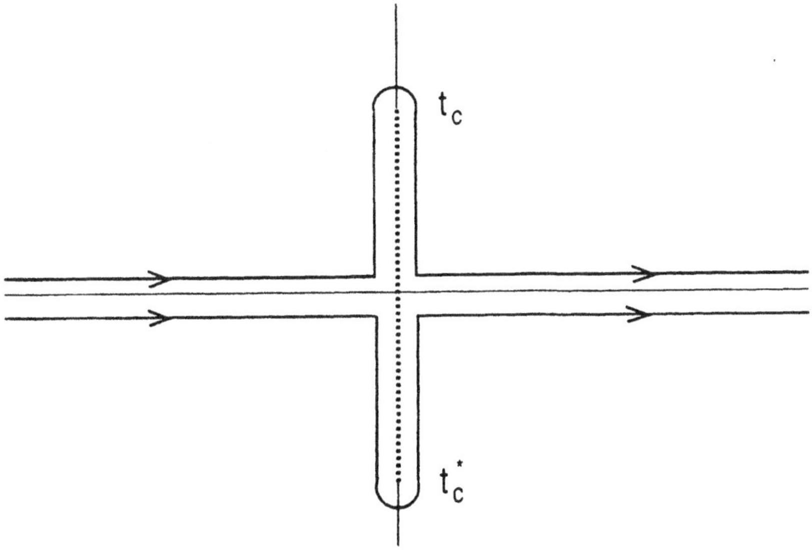

Figure 3. Real Hamiltonians will have crossings at conjugate points. The results of evolution along the lower and upper contours are conjugate, and equal.

Corrections to tunneling. We would briefly like to discuss corrections to Eqs.(4.18) and (4.19). The simplest types of corrections will come from crossing points farther from the real time axis. These will be of the same form as before, with a sum over all of the other crossing points.

Less straightforward are corrections to adiabatic evolution along a contour, modifying a super-adiabatic tunneling amplitude. To handle these, one may follow a similar prescription to the above perturbative expansion for diagonal corrections to adiabatic evolution. As in Eq.(4.7), one may separate off the adiabatic part and expand the residual path-ordered exponential; the result will be a series expansion in powers of τ. The zeroth-order term, as we have seen, is a pure geometric phase, but higher-order terms will correct both the phase and norm of the transition amplitude.

5. Quantum Corrections

In the preceding section, we discussed perturbative and non-perturbative corrections to classical adiabatic evolution. We found that corrections to the phase evolution of an energy eigenstate could be put into the form of an asymptotic expansion in powers of τ^{-1}, and that transitions could be calculated by analytically continuing a smooth nuclear path into the complex time plane.

Now we would like to understand how these conclusions are affected when the nuclear degrees of freedom are quantized. In the previous section, we assumed that the nuclear motions were infinitely differentiable, in order to derive our expansion (which involved performing successive integrations by parts) and to prove that tunneling corrections to adiabatic evolution vanish faster than any power of τ^{-1}. The delicate argument breaks down when we try to integrate over nuclear paths in quantum mechanics, because a typical path in the measure is generically *nowhere* differentiable [22]. Needless to say, for such a path, our analytic continuation method for calculating tunneling amplitudes does not apply. Furthermore, in a quantum mechanical context, there is no reason to expect the individual terms in our perturbation series to converge. Indeed, each successive term added to the effective Lagrangian, being higher-order in time derivatives than any of the preceding terms, represents a singular perturbation and diverges for a typical path. How is all this consistent with the successful use of the Born–Oppenheimer approximation in quantum mechanics? Two questions need to be asked: Do we still have a useful perturbative expansion in powers of $\dot{R}$ (and higher time derivatives) for the evolution of an eigenstate? Are tunneling processes still exponentially suppressed?

In general, this will only be true if we take matrix elements between particularly nice states. The point is the following. In passing from our effective Lagrangian to a Hamiltonian, the powers of $\dot{R}$ are converted into powers of the momentum p. (The procedure here is to treat the higher-derivative terms as perturbations, ignoring their effect on the canonical momenta, and to re-express them in terms of $p = M\dot{R}$.) Now p is an *unbounded* operator, and so our expansion "typically" diverges. However, for suitable initial and final states, p may have small matrix elements, and in that case our adiabatic expansion is useful. Similar remarks apply to semiclassical expansions around *smooth* tunneling paths.

To conclude, the validity of the adiabatic approximation in situations where the external parameters are themselves quantized is far from obvious, and should be studied on a case–by–case basis. Nevertheless, in many useful cases, the corrections are expected to be small.

We would like to thank William Bialek, Joanne Cohn, David Eliezer, Pawel Mazur, Hirosi Ooguri, and Bernard Zygelmann for useful discussions.

A significant part of this research was done at the Institute for Theoretical Physics of the University of California, Santa Barbara. This research was supported in part by the National Science Foundation under Grant No. PHY82-17853, supplemented by funds from the National Aeronautics and Space Administration. A.S. was also supported in part by the Department of Energy under Grant. No. DE-AC02-76ERO-2220

References

[1] M.V. Berry, "Quantal phase factors accompanying adiabatic changes," *Proc. R. Soc. Lond. A.* **392** (1984) 45–57.

[2] C.A. Mead and D.G. Truhlar, "On the determination of Born–Oppenheimer nuclear motion wave functions including complications due to conical intersections and identical nuclei," *J.Chem.Phys.* **70**, (1979) 2284–96.

[3] J. Moody, A. Shapere, and F. Wilczek, "Realizations of magnetic–monopole gauge fields: Diatoms and spin precession," *Phys. Rev. Lett.* **56**, (1986) 893.

[4] H. Kuratsuji and S. Iida, "Effective action for adiabatic processes," *Prog.Th.Phys.* **74**(1985) 439–445

A. Bulgac, "Effective action for nonadiabatic processes," *Phys.Rev.A* **37**, (1988) 4084.

[5] R. Jackiw, "Three Elaborations on Berry's Connection, Curvature and Phase." *Int. J. Mod. Phys.* **A3** (1988) 285–297.

[6] Ph. de Sousa Gerbert, "A systematic expansion of the adiabatic phase," MIT preprint no. CTP-1537, submitted to Nuclear Physics B.

[7] A. Messiah, *Quantum Mechanics*, vol.2 (Amsterdam: North–Holland).

[8] F. Wilczek, lectures delivered at the Theoretical Advanced Study Institute in High-Energy Physics, Univ. of Michigan, 1984.

[9] Chapters 5 and 7, this volume.

[10] M. Stone, "Born–Oppenheimer approximation and the origin of Wess–Zumino terms: Some quantum mechanical examples," *Phys.Rev.D* **33** (1986) 1191;

P. Nelson and L. Alvarez-Gaume, "Hamiltonian interpretation of anomalies," *Commun.Math.Phys.* **99**, (1985) 103-114.

[11] M.V. Berry, "The quantum phase, five years later," Chapter 1, this volume.

[12] F. Wilczek and A. Zee, "Appearance of gauge structure in simple dynamical systems," *Phys. Rev. Lett.* **52** (1984) 2111.

[13] B. Zygelman, "Appearance of gauge potentials in atomic collision physics," *Phys. Lett.* **125A**, (1987) 476.

[14] L. Schulman, *Techniques and Applications of Path Integration* (New York: Wiley, 1981).

[15] M.V. Berry, "Quantum phase corrections from adiabatic iteration," *Proc. R. Soc. Lond.* **A 414** (1987) 31–46.

[16] M. Reed and B. Simon, *Functional Analysis* (New York: Academic Press, 1972).

[17] A.M. Dykhne, "Adiabatic perturbation of discrete spectrum states," *JETP* **14** (1962) 941.

[18] J-T. Hwang and P. Pechukas, "The adiabatic theorem in the complex plane and the semiclassical calculation af nonadiabatic transition amplitudes," *J.Chem.Phys.* **67** (1977) 4640-4653.

[19] E.A. Coddington and N. Levinson, *Theory of Ordinary Differential Equations* (New York: McGraw–Hill, 1955).

[20] A. Zener, "Non-adiabatic crossing of energy levels," *Proc.Roy.Soc.A* **137**, (1932) 696.

[21] L.D. Landau and E.M. Lifshitz, *Quantum Mechanics* (Oxford: Pergamon, 1977) section 52.

[22] L.F. Abbott and M.B. Wise, "Dimension of a quantum–mechanical path," *Am.J.Phys.* **49(1)** (1981) 37–39.

Chapter 4

SOME APPLICATIONS AND TESTS

4

Some Applications and Tests

We now begin our survey of modern (1984 and later) applications of geometric phases, with examples drawn from optics, magnetic resonance, and molecular and atomic physics. These examples fall into two general classes, which could be described as quantum and classical. However, the nomenclature is misleading, since "classical" geometric phases often have quantum mechanical origins, and quantum phases may have (semi-)classical interpretations. It is perhaps more correct to make a division into macroscopic and microscopic systems.

In quantum systems with macroscopic coherence, geometric phases are observed (usually through some sort of interference) when external parameters are adiabatically varied, and may also be explained in classical terms. For example, the precession of spins in a rotating magnetic field may be described by a classical Hamiltonian for the macroscopic magnetization parameter, or by a quantum Hamiltonian operator for the microscopic spin degrees of freedom. Berry's phase in the classical description would correspond to a small additional precession equal in magnitude to the quantum geometric phase, in addition to the net "dynamical" precession obtained by integrating the instantaneous contributions of the time-varying magnetic field.

The other applications we shall encounter here occur in systems where a Born–Oppenheimer approximation applies. In molecular Born–Oppenheimer (B-O) systems, the "external parameters" are the nuclear coordinates, and Berry's adiabatic phase makes its appearance through a gauge potential in the effective nuclear Hamiltonian. We cannot control the nuclear coordinates as we can a macroscopic magnetic field, because they are not really external parameters—they are microscopic quantum spin degrees of freedom. But this direct modification of the nuclear Hamiltonian can and does lead to observable effects, such as the splitting of energy levels that would be degenerate in the absence of the gauge potential term, and the shifting of nuclear quantum numbers.

As we saw in Chapter 2, it appears that the earliest prediction and observation of a geometric phase was made in the field of optics [2.1]. This is not surprising, because optical phase effects are easy to measure, by interference or filtering. Recall that Pancharatnam was interested in the phase shift in a beam of coherent linearly polarized light resulting from a sequence of changes of polarization. This he related to the solid angle swept out by the polarization vector on the Poincaré sphere. Another geometric phase associated with polarized light has been studied more recently by Chiao and Wu[1] and by Tomita and Chiao [4.1]. These authors consider the propagation of linearly polarized light down a helically coiled optical fiber; the external parameter being varied is the direction of propagation of the light, which is always tangent to the fiber. Thus, the parameter space is a two-dimensional sphere, the space of possible directions. By adiabatically varying the direction of propagation around a closed circuit on the sphere, one induces a change in polarization that depends only on the fiber's geometry. (In this context, the condition of adiabaticity is just that the radius of curvature of the optical fiber be large compared to the wavelength of the light.) If $\mathbf{t}$ denotes the tangent vector to the fiber, then the direction of linear polarization will be rotated by an amount equal to the solid angle subtended in $\mathbf{t}$-space. This geometric phase is the same that would be accrued by a unit electric charge moving in the field of a magnetic monopole of charge of twice the minimal Dirac charge, and it is a direct consequence of the fact that the photon has spin 1. (Pancharatnam's phase, on the other hand, corresponds to a magnetic monopole of minimal charge at the center of the Poincaré sphere.) The results of the optical fiber experiment are reported in [4.1]. The subsequent paper by Berry [4.2] offers an explanation of this phenomenon in terms of classical electromagnetism, as well as estimates of violations of adiabaticity. Berry also demonstrates explicitly that the amplitude for non-adiabatic transitions is proportional to the ratio of the wavelength to the radius of curvature.

Nuclear magnetic resonance (NMR) provides another context for Berry's phase which may be described in either classical or quantum terms. We consider a coherent ensemble of spins $\mathbf{S}$ in a large static magnetic field B_z plus a perpendicular field $\mathbf{B}_\perp \equiv (B_x, B_y, 0)$ rotating with (radio) frequency ω. If γ is the gyromagnetic ratio, then classically we expect the spins to precess with frequency $\omega_0 = \gamma B_z$. Normally, one takes $\omega \approx \omega_0$, so that the oscillating field can "pump" the rotating spins. This makes it possible to study the resonant absorption spectrum of the spin system. The system is described by a time-dependent Hamiltonian with four external parameters, B_z, $\mathbf{B}_\perp$, and ω, all of which may be varied quite easily in a standard NMR setup:[2]

$$H_0 = -\gamma\, \mathbf{S}_z \cdot [B_z \hat{\mathbf{z}} + \mathbf{B}_\perp \sin \omega t] \tag{4.1}$$

Now this Hamiltonian may not appear to be slowly varying in time; its instantaneous eigenstates certainly do not evolve according to the adiabatic approximation. However, a time-independent Hamiltonian H_{rot} may be obtained by transforming to a frame rotating with frequency ω

$$H_{\rm rot} \approx -\gamma \mathbf{S} \cdot \left[B_z \left(1 - \frac{\omega}{\omega_0} \right) \hat{\mathbf{z}} + \mathbf{B}_\perp \right] \tag{4.2}$$

where $\omega_0 \equiv \gamma B_z$ and $H_{\rm rot}$ is approximate because we have thrown out a term of frequency 2ω. (This term comes from the piece of H_0 that rotates out of phase with the spins, and does not contribute significantly to the resonant absorption near ω_0.)

Now, suppose that the system is prepared in an eigenstate of $H_{\rm rot}$. When the independent parameters B_z and $\mathbf{B}_\perp$ are taken adiabatically around a closed circuit in parameter space, one finds that the evolution of the eigenstate involves a geometric phase. That is, the wavefunction accumulates a phase with a geometric component equal to the spin S times the solid angle subtended relative to the point $(0,0,\omega/\gamma)$ in the parameter space spanned by B_x, B_y, and B_z. Classically speaking, the accumulated phase corresponds to the net precession of the spins. Its geometric component may be observed by beating the precessed signal against a reference signal, or by spin–echo techniques. This experiment was proposed by Moody, Shapere, and Wilczek [4.3] and Cina,[3] and carried out by Suter, Chingas, Harris, and Pines [4.4]. The latter group has found strikingly good agreement with theoretical predictions. It would be interesting to compare their observations away from the adiabatic regime with theoretical corrections to the adiabatic approximation (see, for example, [3.7] and [9.1]).

With minor modifications, the same basic setup has also been used to measure the phase of Aharonov and Anandan [4.5]. Tycko [4.6] performed a related experiment with nuclear quadrupole resonance (NQR) which involved rotating the sample in a fixed magnetic field, instead of rotating the magnetic fields around the sample. Finally, Zee [4.7] has proposed modifications of Tycko's experiment which would enable the observation of a non-Abelian phase. Because of the high degree of control over the external parameters in NMR and NQR, and the refinement of experimental techniques, magnetic resonance has proved to be one of the most successful tests of Berry's phase.

The remaining papers in this chapter concern applications of Berry's phase in molecular and atomic physics. As we noted Chapter 2, it was Mead and Truhlar who first realized that the holonomy of electron wavefunctions in the Born-Oppenheimer approximation can be taken into account by including a background gauge potential in the the effective nuclear Hamiltonian. Since their work in 1979, the catalog of Born-Oppenheimer systems where topological considerations require the inclusion of a gauge potential has grown to include a variety of important examples.

In the paper by Moody, Shapere, and Wilczek [4.3], it is shown that the effective nuclear Hamiltonian for diatoms includes a gauge potential, which due to its topological nature can not be gauge transformed away by any choice of phases for the electronic eigenstates. For a fixed internuclear separation, the nuclear parameter space is just the two-dimensional sphere of possible orientations, and the gauge potential is a magnetic monopole potential over this sphere. The monopole possesses a quantized magnetic charge proportional to the electronic angular momentum along the internuclear axis, Λ. Its effect, as we have explained, is quite simple; it generates the geometric phase accumulated by the electronic wavefunctions in response to the "slow" nuclear motion, which for a closed nuclear path will be proportional to the solid angle swept out. Yet its presence in the nuclear Hamiltonian has a profound effect on the quantization of the nuclear degrees of freedom. When Λ is half-integral, the electronic wavefunction goes into minus itself when rotated by 2π about the internuclear axis. Of course, the ability of the electrons to have half-integral orbital angular momentum is closely related to their fermi statistics, and even after they are integrated out, they manage to leave their "fermion-ness" behind. Then, the nuclear coordinate that remains must be quantized with half-integral spin, even if the nuclei are integer-spin bosons!

Normally, to find the eigenstates of a rotationally invariant system, it is convenient to diagonalize J^2 and J_z simultaneously, and to expand state vectors in terms of spherical harmonics. An interesting feature of the nuclear Hamiltonian for the diatom is that it is not manifestly rotationally invariant, although the Hamiltonian for the complete system is. This is because the gauge potential contains (and must contain, in any gauge) a Dirac string singularity, which picks out a direction, even though the actual magnetic field is spherically symmetric. Nevertheless, all physical quantities should still be rotationally invariant, and indeed it is possible to write down a set of modified angular momentum operators that commute with the Hamiltonian. The correct modification in the background of an Abelian monopole has been known for a long time[4]

$$J_i = -i\epsilon_{ijk} r_j (\nabla_k - iA_k) - \tfrac{1}{2} r_i \epsilon_{lmn} r_l F_{mn}, \tag{4.3}$$

Once the correct operators are obtained, they may be used to construct rotationally covariant monopole harmonics, out of which eigenstates of the nuclear Hamiltonian may be built.

One might wonder how physicists and chemists could have worked with diatoms in the B-O approximation for so many years prior to the work of Mead and Truhlar, without knowing anything about the adiabatic phase. In fact, the diatom seems to be one instance where they did not erroneously throw away the "gauge potential" term; as we mentioned in Chapter 2, it might even be said that Van Vleck[5] was the true discoverer of the "molecular Aharonov–Bohm effect," since his 1929 Born–Oppenheimer treatment of the diatom includes precisely these magnetic monopole gauge potentials. Of course, he did not call them by that name, since magnetic monopoles had yet to be invented, but his nuclear Hamiltonian contains them quite explicitly,

Dirac string singularities and all. The eigenfunctions he finds are precisely the monopole harmonics.

An especially interesting case occurs when the angular momentum of a diatom along its internuclear axis is $\Lambda = \pm 1/2$. The gauge potential then is non-Abelian, because the matrix elements $\langle +1/2|\nabla| - 1/2\rangle$ are nontrivial, so the nuclear Hamiltonian is non-diagonal. Diagonalization produces a splitting of the levels with $\Lambda = \pm 1/2$. These results are not new—the effective Hamiltonian and the energy splitting were first found by Van Vleck—but we believe that their physical origin is illuminated by this more modern approach.

The next paper [4.8], by Mead, discusses the Kramers degeneracy common to all molecules with an odd number of electrons. Mead demonstrates that the nuclei move in the background of a a non-Abelian gauge potential, and proposes an experiment to measure its effect. In an electric field, the Stark effect splits the levels of a molecule with total angular momentum J into $J + 1/2$ Kramers doublets. The effective nuclear Hamiltonian for the "middle" doublet with angular momentum component $|M| = 1/2$ in the direction of the electric field includes a $U(2)$ gauge potential which mixes the $M = \pm 1/2$ states, and should lead to observable changes in M when the applied electric field is slowly rotated.

The dynamics of slow atomic collisions is another domain where the Born-Oppenheimer separation into electronic and nuclear degrees of freedom applies, so we might expect to find gauge potentials here as well. But in contrast to static molecules, the study of collisions must go beyond the adiabatic approximation. If the energy transferred in a scattering process is large compared to the splitting between the nondegenerate levels involved and other nearby levels, then it is clear that level mixings will be quite important. The standard treatment, known as the method of perturbed stationary states (PSS), is similar to the Born–Oppenheimer approximation, but the effective nuclear Hamiltonian is an $N \times N$ matrix, where N is the number of "channel" states that participate in the scattering process. It is therefore not surprising that non-abelian gauge potentials also arise in this situation. In paper [4.9], Zygelman shows that the PSS Schrödinger equation involves a *4-component* $U(N)$ gauge potential, and that it is gauge covariant under rotations of the electronic basis states.

We should mention that in all of the above cases involving doubly degenerate levels, the non-abelian part of the $U(2)$ gauge potential (corresponding to the $SU(2)$ factor of $U(2) \simeq SU(2) \times U(1)$) is topologically trivial; this is a corollary to the fact that there are no non-trivial homotopy classes of maps from the equator of the two–sphere into $SU(2)$. (The non-abelian potential is, nonetheless, physically important, since its curl is non-zero.) However, in more complicated Born–Oppenheimer systems, topologically non-trivial non-abelian gauge potentials may arise. Indeed, an example of such an object, known as an instanton to particle physicists, has recently been found, for a system with two doubly degenerate levels.[6]

Aside from indirect effects on energy level splittings referred to above, the actual experimental observation of the "molecular Aharonov-Bohm effect" has proved to be considerably subtler than the detection of geometric phases in optics and magnetic resonance, where one has direct control over the slowly varying degrees of freedom. One approach is to study the resonant absorption spectrum of a gas, and to compare the results with theoretical modifications of nuclear quantum numbers resulting from the background effective gauge potential. In the case of a simple polyatomic molecule such as Na_3, the old sign-reversal result of Herzberg and Longuet-Higgins [2.3] predicts half-integral quantization of a quantum number known as the ***pseudorotation***, associated with certain internuclear deformation modes. This prediction is confirmed in the work of Delacrétaz *et.al.* [4.10]. If and when more refined experimental techniques become available, they will be sure to reveal a rich field of applications in the detection of the effects of $U(1)$ and non-Abelian phases on molecular spectra.

---

[1] R. Chiao and Y.-S. Wu, *Phys. Rev. Lett.* **57** (1986) 933-7.

[2] A. Abragam, *The Principles of Nuclear Magnetism* (Oxford: University Press, 1961).

[3] J.A. Cina, *Chem. Phys. Lett.* **132** (1986) 393.

[4] M. Saha, *Indian J. Phys.* **10** (1936) 145.

[5] J.H. Van Vleck, *Phys. Rev.* **33** (1929) 467.

[6] J.E. Avron, L. Sadun, J. Segert, and B. Simon, *Phys. Rev. Lett.* **61** (1988).

VOLUME 57, NUMBER 8 PHYSICAL REVIEW LETTERS 25 AUGUST 1986

Observation of Berry's Topological Phase by Use of an Optical Fiber

Akira Tomita[(a)]
AT&T Bell Laboratories, Murray Hill, New Jersey 07974

and

Raymond Y. Chiao
Department of Physics, University of California, Berkeley, California 94720
(Received 28 February 1986)

We report the first experimental verification of Berry's topological phase. The key element in the experiment was a single-mode, helically wound optical fiber, inside which a photon of a given helicity could be adiabatically transported around a closed path in momentum space. The experiment confirmed at the classical level that the angle of rotation of linearly polarized light in this fiber gives a direct measure of Berry's phase. The topological nature of this effect was also verified, i.e., the rotation was found to be independent of deformations of fiber path if the solid angle of the path in momentum space stayed constant.

PACS numbers: 03.65.Bz, 42.10.Nh, 42.81.Fr

Recently, Chiao and Wu[1] have pointed out some novel and observable quantum interference phenomena which arise from Berry's phase[2] for the photon. This phase, which is similar in many respects to the Aharonov-Bohm phase, has recently appeared theoretically in many fields of physics, from high-energy physics to low (e.g., from chiral anomalies in gauge field theories to a treatment of the Born-Oppenheimer approximation).[3] Hence it is important to look for Berry's phase experimentally. The optical effects predicted by Chiao and Wu allow such observations. The Bose nature of the photon permits optical manifestations of Berry's phase on a classical, macroscopic level, unlike the case of Fermi particles. Thus an intuitive understanding of this general phase factor emerges. One of their predictions is the appearance of an effective optical activity of a helically wound, single-mode optical fiber. They showed that the angle of rotation of linearly polarized light propagating down the fiber is a direct measure of Berry's phase. This optical activity does not come from a local elasto-optic effect caused by torsional stress,[4] but rather arises solely from the overall geometry of the path taken by the light, and hence is a global topological effect. Thus this effect is independent of the detailed material properties of the fiber.

In this Letter, we report an experimental study of the optical activity arising from Berry's phase in a single-mode fiber. To explore the topological nature of this effect, we compare the results from complex paths of nonuniform helices with those from simple uniform helices. We find good agreement between the measured rotation angles and those predicted by Berry's phase in all cases. These observations confirm the topological nature of this phase, which is one of its most significant properties. The rotation angle is found to be independent of the path of the fiber in configuration space as long as the solid angle subtended by the path in momentum space stays constant. In the special case of planar paths, no significant optical rotation is observed independent of the paths's complexity. Hence the light is able to distinguish between two and three spatial dimensions. This again confirms the topological nature of the effect.

Connection of earlier observations of optical activity in fibers,[5,6] with Berry's phase and its quantal, global topological properties went unnoticed. The observation of polarization rotation ascribed to geometrical effects was previously reported by Ross,[5] who used a single-mode fiber wound in a uniform helix, and by Varnham, Birch, and Payne,[6] who fabricated a fiber with a core wound into a uniform helix. Both papers studied the case where the helix was uniform, i.e., with a constant pitch. The observations were in good agreement with a classical analysis,[5,6] which treated the rotation of the plane of polarization locally at each point along the fiber for the case of a uniform helix by use of differential geometry.

The experimental setup is schematically shown in Fig. 1(a). A He-Ne laser and a pair of linear polarizers, one at the input, the other at the output end of the fiber, were used to measure the rotation of the plane of polarization in a 180-cm-long single-mode fiber. The fiber had a conventional step-index–type profile with a relative core-cladding index difference of 0.6%, and a core diameter of 2.6 μm. Its cladding index of refraction was 1.45 and its cladding diameter was 70 μm, which was coated with uv-curable acrylate of thickness ~ 100 μm. The fiber was first inserted loosely in a Teflon sleeve in the form of a 175-cm-long tube, to minimize any torsional stress on the fiber during winding. The tube was wound helically with the output end of the fiber free to rotate. Thus care was taken not to introduce any torsional stress which might

VOLUME 57, NUMBER 8 PHYSICAL REVIEW LETTERS 25 AUGUST 1986

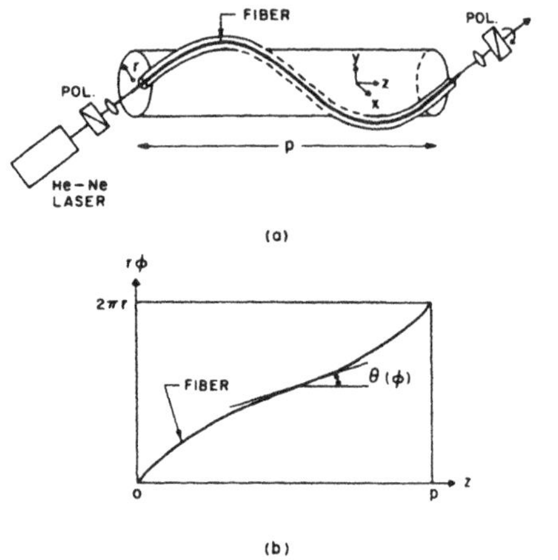

FIG. 1. (a) Experimental setup; (b) geometry used to calculate the solid angle in momentum space of a nonuniformly wound fiber on a cylinder.

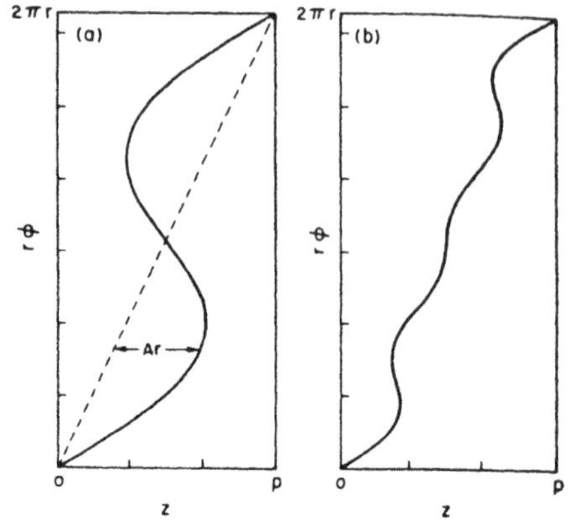

FIG. 2. (a) The solid line represents the path of the fiber on an unwrapped cylinder surface for nonuniform helices (squares in Figs. 3 and 4) with one harmonic of deformation [Eq. (5) with $A = 1.2$], and the dashed line a uniform helix ($A = 0$); (b) the path for a nonuniform helix (triangle in Figs. 3 and 4) with three harmonics of deformation [Eq. (6)].

result in a rotation of the plane of polarization due to the elasto-optic effect.[7] Also, we found that the fiber showed a negligibly small linear birefringence as long as the fiber was wound smoothly on a large enough diameter.[8] In order to form a closed path in momentum or **k** space,[1] the propagation directions of the input and output of the fiber were kept identical. In the first experiment, the fiber was wound into a uniform helix. The pitch angle of the helix θ, i.e., the angle between the local waveguide axis and the axis of the helix, was varied by attaching the Teflon sleeve along the outside perimeter of a spring, which was stretched from a tightly coiled configuration into a straight line. In this way, the pitch length p was varied, as was the radius r of the helix, but the fiber length $s = [p^2 + (2\pi r)^2]^{1/2}$, i.e., the arc length of the helix, was kept constant. The range of p was from 30 to 175 cm. Hence the diameter of the helix ranged from 55 cm down to zero. By geometry, $\cos\theta = p/s$ [see Fig. 1(b)]. The solid angle in momentum space $\Omega(C)$ spanned by the fiber's closed path C in this space, in this case a circle, is $2\pi(1-\cos\theta)$. Berry's phase, $\gamma(C) = -\sigma\Omega(C)$,[1] for a single-turn uniform helix is therefore

$$\gamma(C) = -2\pi\sigma(1 - p/s), \quad (1)$$

where $\sigma = \pm 1$ is the helicity quantum number of the photon. The quantum theory[1] predicts that $-\gamma_+(C)$, where $\gamma_+(C)$ is Berry's phase for $\sigma = +1$, is the angle Θ of rotation of linear polarization. The classical theory[5,6] predicts an angle of optical rotation in agreement with this quantum result.

In the second experiment, the fiber was wound onto a cylinder of a fixed radius to form a nonuniform helix. The procedure was first to wrap a piece of paper with a computer generated curve onto the bare cylinder. Then the Teflon sleeve with the fiber inside was laid on top of this curve. (To allow for variations in fiber path while using a fiber of fixed length, we left a straight section of fiber path at the output end, which had a variable length.) The solid angle in momentum space could then be calculated from the curve by unwrapping the paper onto a plane [see Figs. 1(b) and 2]. Let the horizontal axis of the paper, which was aligned with respect to the axis of the cylinder, be the z axis. Then the vertical axis represents $r\phi$, where r is the radius of the cylinder and $\phi = \tan^{-1}(y/x)$ is the azimuthal angle of a point on the curve with coordinates $(r\phi, z)$. The local pitch angle from Fig. 1(b),

$$\theta(\phi) = \tan^{-1}(r\, d\phi/dz), \quad (2)$$

characterizes the tangent to the curve followed by the fiber, and represents the angle between the local waveguide and the helix axes. In momentum space, $\theta(\phi + \pi/2)$ traces out a closed curve C corresponding to the fiber path on the surface of a sphere. The solid angle subtended by C with respect to the center of the sphere is given by

$$\Omega(C) = \int_0^{2\pi} [1 - \cos\theta(\phi)]\, d\phi. \quad (3)$$

VOLUME 57, NUMBER 8 PHYSICAL REVIEW LETTERS 25 AUGUST 1986

Berry's phase is then given by[1]

$$\gamma(C) = -\sigma\Omega(C). \tag{4}$$

One sees that Eq. (1) is a special case of Eqs. (3) and (4), when θ is a constant.

Figure 3 shows the measured rotation angle Θ versus the calculated solid angle $\Omega(C)$. The open circles represent the case of uniform helices, and the squares and the triangle represent nonuniform helices. The solid circles represent arbitrary planar curves formed by laying the fiber on a flat surface. The solid circle at $\Omega=0$ corresponds to a snake-like path, and the one at $\Omega \approx 2\pi$ to a loop with a crossing. The squares represent helices with a single harmonic of deformation,

$$z/r = (p/2\pi r)\phi + A\sin\phi, \tag{5}$$

where $p = 42.6$ cm and $r = 14.2$ cm, and A ranges from 0 to 1.5 in steps of 0.3 [see Fig. 2(a)]. The triangle represents a helix with three harmonics of deformation,

$$z/r = (p/2\pi r)\phi + A_1\sin\phi + A_2\sin2\phi + A_3\sin3\phi, \tag{6}$$

where $A_1 = A_2 = A_3 = 0.2$ [see Fig. 2(b)].

By inspection of Fig. 3, one sees that in all cases the measured rotations agree with the calculated magnitude of Berry's phase $|\gamma_+(C)|$ [see Eq. (4)] indicated by the solid line. The sense of the rotation, when one looks into the output end of the fiber, was found to be clockwise (i.e., dextrorotatory) for a left-handed helix, in agreement with theoretical prediction.[1]

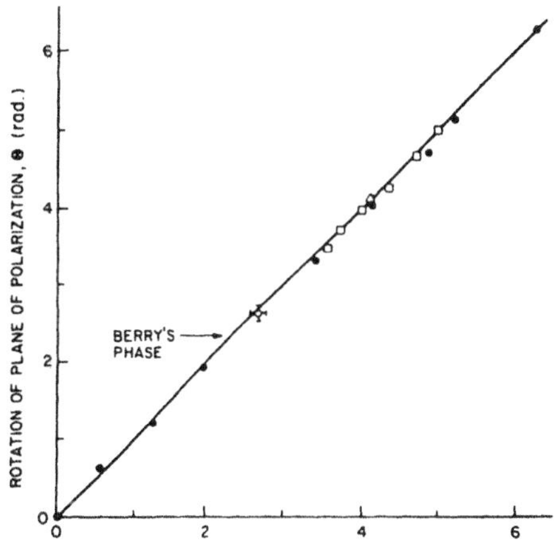

FIG. 3. Measured angle of rotation of linearly polarized light vs calculated solid angle in momentum space, Eq. (3). Open circles represent the data for uniform helices; squares and triangle represent nonuniform helices (see Fig. 2); solid circles represent arbitrary planar paths. The solid line is the theoretical prediction based on Berry's phase, Eq. (4).

The typical vertical error bar in Fig. 3 represents the dominant systematic error in this experiment, namely residual optical rotation due to torsional stress in the fiber. In separate auxiliary experiments, the optical rotation in a deliberately torsionally stressed fiber was measured, and also the residual strain, i.e., the twist of the fiber due to its rubbing against the walls of the Teflon sleeve, was measured microscopically near its free end. From these measurements, an estimate of size of the vertical error bar was determined. The typical horizontal error bar represents the uncertainty in the determination of the solid angle $\Omega(C)$ due to the fact that the fiber was free to roam within the 5-mm inner diameter of the Teflon tube. Random errors due to photon statistics were negligible compared with these systematic errors.

To check quantitatively the topological nature of the optical rotation, we replot the data in Fig. 3, as the slope $\Delta\Theta/\Delta\Omega$ of a line joining a datum point with the origin versus a deformation parameter D, onto Fig. 4. We define D as follows:

$$D = \left\{\int_0^{2\pi} [1-\cos\theta(\phi) - \Omega(C)/2\pi]^2 d\phi\right\}^{1/2} \Big/ \Omega(C). \tag{7}$$

Here D is a measure of the root mean square deviation of the fiber path from a uniform helix. By inspection of Fig. 4, one arrives at the conclusion that the specific optical rotation $\Delta\Theta/\Delta\Omega$ is in all cases independent of the deformation as quantified by the parameter D, and is therefore independent of geometry. This confirms the topological nature of Berry's phase. Since $\Delta\Theta/\Delta\Omega$ is a direct measure of σ,[1] one can view Fig. 4 as experimental evidence for the quantization of the "topological charge" of the system, which in this case is the he-

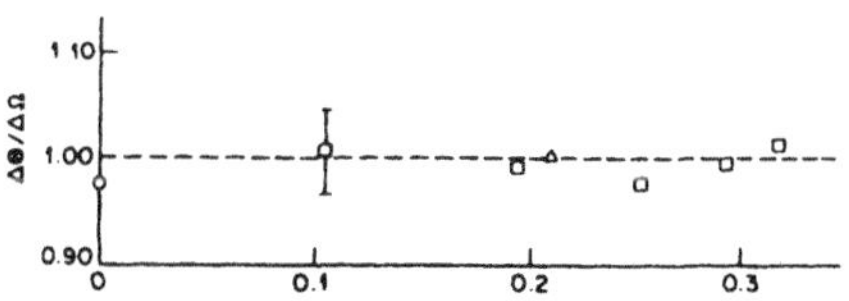

FIG. 4. The slopes $\Delta\Theta/\Delta\Omega$ of the points in Fig. 3 vs the deformation parameter D, Eq. (7), for nonuniform helices (squares and triangle). The open circle represents the average for all uniform helices. The dashed line represents the theoretical prediction.

licity of the photon, a relativistic quantum number.

The experiments reported here are essentially at the classical level, since we used an enormous number of photons in a single coherent state. Therefore at this point we can only say that we have verified the existence of Berry's phase and its topological properties at the classical level. These observations do support, however, the statement that these effects are "topological features of classical Maxwell theory which originate at the quantum level, but survive the correspondence-principle limit ($\hbar \to 0$) into the classical level."[1,9] It would be interesting to verify Berry's phase experimentally also at the quantum level, where fluctuations due to individual photons propagating inside the fiber appear. Then the truly quantum mechanical nature of this phase will become evident.

The authors thank Y. S. Wu for valuable discussions.

Note added.—After we submitted this paper, another paper by G. Delacrétaz *et al.* [Phys. Rev. Lett. 56, 2598 (1986)] was submitted and published, which independently verified the existence of Berry's phase experimentally in another context (i.e., the molecular system Na_3).

---

[a]Present address: Raychem Corporation, Menlo Park, CA 94025.

[1]R. Y. Chiao and Y. S. Wu, preceding Letter [Phys. Rev. Lett. 57, 933 (1986)]. Both in that paper and here, we adopt the sign conventions that $\Omega(C) > 0$ for a counterclockwise orientation of C with respect to the outward normal areal vector enclosed by C, by the right-hand rule in **k**-space. Thus the helix winding number $N > 0$ for a right-handed helix. Also, $\Theta > 0$ for dextrorotatory optical rotations [see F. A. Jenkins and H. E. White, *Fundamentals of Optics* (McGraw-Hill, New York, 1957), p. 572].

[2]M. V. Berry, Proc. Roy. Soc. London Ser. A **392**, 45 (1984).

[3]See Refs. 8–15 in Ref. 1.

[4]R. Ulrich and A. Simon, Appl. Opt. **18**, 2241 (1979).

[5]J. N. Ross, Opt. Quantum Electron. **16**, 455 (1984).

[6]M. P. Varnham, R. D. Birch, and D. N. Payne, in *Proceedings of the Fifth International Conference on Integrated Optics and Optical Fiber Communication and the Eleventh European Conference on Optical Communications* (Istituto Internazionale delle Communicazioni, Genova, Italy, 1985), p. 135.

[7]The straight, unstressed fiber possessed, however, an intrinsic circular birefringence due to the fiber drawing process, which produced an optical rotation of 0.436 rad/m. This number was checked by cutting the fiber into 30-cm sections. The rotation angles reported in the rest of this paper were measured with respect to the output polarization of the straight fiber as the zero reference. Also, $\kappa = 0.301$ rad/m in Eq. (10) of Ref. 1.

[8]The magnitude of the linear birefringence $n_{\parallel} - n_{\perp}$ was measured to be 1×10^{-9} and 5×10^{-9}, respectively, for the cases where the 180-cm fiber was laid straight, and where it was curved with a radius of 30 cm. The elliptical polarization of the output light of the fiber, under conditions where there were nonzero optical rotations arising from Berry's phase, was measured to be less than 2%, when the input light was essentially completely linearly polarized.

[9]J. H. Hannay, J. Phys. A. **18**, 221 (1985), has also discussed the classical limit of Berry's phase, e.g., in the case of a symmetric top. The angle Θ here is analogous to the angle that he found.

NATURE VOL. 326 19 MARCH 1987

Interpreting the anholonomy of coiled light

M. V. Berry

H. H. Wills Physics Laboratory, Tyndall Avenue, Bristol BS8 1TL, UK

Circular birefringence of purely geometric origin was recently predicted[1] and observed[2] in helically coiled monomode optical fibres, and widely reported[3-5] as a successful application to photons of a general theory[6,7] for phase shifts in adiabatically transported quantum states. However, earlier similar observations[8-10] had been interpreted not by quantum mechanics but simply as a classical anholonomy, namely parallel transport of the polarization[11]. Indeed, because the magnitude of the effect is independent of the wavelength of the light as well as Planck's constant, it might seem that 'classical' here means that not only quantum but also wave effects can be neglected. Here, I argue that these experiments, and their discrete analogues, are most appropriately described at the level of classical electromagnetism; the parallel transport law can then be derived (rather than assumed[8-11]) and nonadiabatic polarization changes calculated.

In the quantum description[1], photons in right- or left-circularly polarized light are assumed to be in the eigenstate of positive or negative helicity defined by the local tangent vector $\mathbf{t}(s)$ of the fibre (Fig. 1). Because the input and output ends of the fibre are parallel, the eigenstate is transported round a closed loop in $\mathbf{t}$ space and thereby acquires the geometrical phase shift appropriate to spin one, namely[6] minus or plus the solid angle Ω subtended by the loop at the origin of $\mathbf{t}$ space. These opposite phase shifts $\mp\Omega$ imply[1] that the direction of linear polarization will be rotated by Ω, and it is this gyrotropy that was observed[2,8-10].

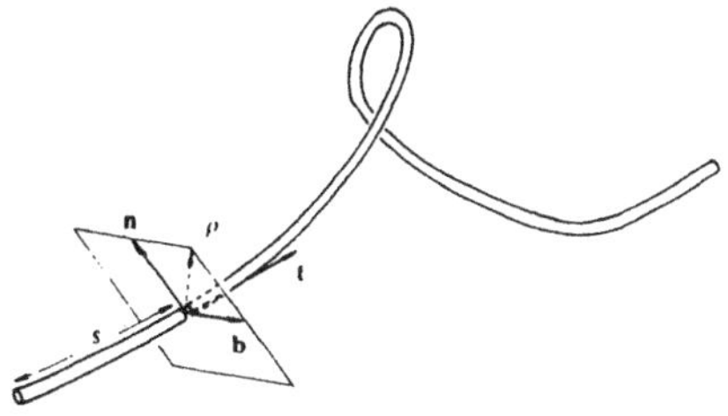

Fig. 1 Geometry of helically coiled optical fibre.

Underlying this successful prediction there are, however, several uncertainties. In the absence of any obvious governing Hamiltonian it is not clear why the phase continuation rule of the general theory[6] is applicable. Even if it is, the fact that the two helicity states are degenerate suggests that the application could equally be made to any superposition of them, yielding different results. (Relativistic arguments[12] do yield the phase continuation rule, and also uniquely select the helicity states, but are inapplicable within fibres.) Moreover, it is hard to see how this type of quantum description can provide estimates of the probability that the coiling will produce nonadiabatic transitions to the opposite helicity. In any case, the high photon flux in all the experiments so far carried out makes a quantum description unnecessary.

At the level of geometrical optics (shortwave limit) it is known[13,14] that in a medium with smoothly-varying refractive index μ the electromagnetic field vectors are parallel-transported along a light ray. This result should, however, not be invoked to explain experiments with monomode fibres, because their fields cannot legitimately be described by ray optics.

In classical electromagnetism the field is governed by Maxwell's equations, with μ depending only on perpendicular distance ρ from the fibre axis. Using the ideas of coupled local modes and the weak-guidance approximation[15], we can write the transverse electric field $\mathbf{E}$ at position s along the fibre as a superposition of fields linearly polarized along the fibre normal $\mathbf{n}(s)$ and binormal $\mathbf{b}(s)$[16]; this is an adiabatic approximation (with s playing the role that time does in quantum mechanics), valid for gently coiled fibres. Thus,

$$\mathbf{E}(\rho, s) = \exp\{i\beta s\} f(\rho)[c_1(s)\mathbf{n}(s) + c_2(s)\mathbf{b}(s)] \qquad (1)$$

where β and $f(\rho)$ are, respectively, the propagation constant and modal amplitude appropriate to the straight fibre. Substitution into Maxwell's equation gives, after some analysis (to be published elsewhere), the following effective Hamiltonian evolution equation for the polarization coefficients c_1 and c_2 in terms of the fibre curvature κ and torsion τ:

$$i\frac{\partial}{\partial s}\begin{pmatrix} c_1(s) \\ c_2(s) \end{pmatrix} = \begin{pmatrix} \kappa^2(s)/2\beta & i\tau(s) \\ -i\tau(s) & 0 \end{pmatrix}\begin{pmatrix} c_1(s) \\ c_2(s) \end{pmatrix} \qquad (2)$$

This simplest possible theory neglects radiative leaking, coupling to reflected modes and all elasto-optic effects.

Now $\beta \approx 2\pi/\lambda$ where λ is the light wavelength in the fibre cladding, and κ^{-1} and τ^{-1} are both comparable with the fibre bend distances, so in lowest approximation the term $\kappa^2/2\beta$ is negligible and the torsion terms dominate. Then, for initial linear polarization in direction α (relative to $\mathbf{n}$), (2) gives

$$c_2(s)/c_1(s) = \tan\left(\alpha - \int_{-\infty}^{s} ds'(\tau(s')\right) \qquad (3)$$

This is precisely the parallel transport law for $\mathbf{E}$, giving, after one complete helical turn, a polarization rotation equal to the

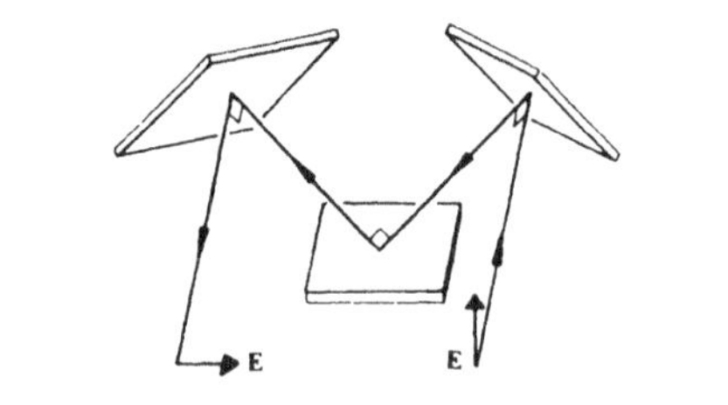

Fig. 2 Antiadiabatic rotation of polarization E by three reflections in metal mirrors.

solid angle[10]

$$\Omega = 2\pi - \int_{-\infty}^{\infty} ds'\tau(s')$$

Moreover, (2) resolves the degeneracy in favour of the helicity states: local eigenmodes (again neglecting $\kappa^2/2\beta$) are circularly polarized, that is, $c_2/c_1 = +i$, and acquire opposite phase shifts $\mp\Omega$ round the turn.

Nonadiabatic transitions are induced by the curvature term $\kappa^2/2\beta$. In lowest order this gives the probability for a change from (say) + to −, as

$$P_{\mp} = \left| \int_{-\infty}^{\infty} ds\kappa^2(s) \exp\left\{2i \int_{-\infty}^{s} ds'\tau(s')\right\}\right|^2 \Big/ 16\beta^2 \qquad (4)$$

For a helix uniformly wound on a cylinder of radius R so as to produce phase shifts $\mp\Omega$, this can be written as

$$P_{+-} = [\Omega/2\pi(2-\Omega/2\pi)]^3 \sin^2\Omega/(1-\Omega/2\pi)^2/16\beta^2R^2 \qquad (5)$$

Even for $R \sim 1$ mm this never exceeds 10^{-8}. For planar bends ($\tau = 0$) there is no polarization rotation, and the eigenmodes of (2) are linearly polarized with a tiny bend-induced shift $\kappa^2/2\beta$ in the local propagation constant of the mode polarized along **n**.

It has been suggested (J. N. Ross, personal communication and ref. 17) that the coiling of **t** can be accomplished by (at least three) discrete reflections, the resulting sudden changes in **t** being simulations of adiabatic change. But ideal mirrors (infinite conductivity) do not conserve helicity; they reverse it, and with this 'antiadiabaticity' the solid angle Ω must be accumulated with **t** replaced by −**t** on alternate segments of the light path[17]. If the light beam is reversed after three successive reflections, each changing its direction by 90° (Fig. 2), the predicted polarization rotation (see also ref. 18, pages 84–86) is also 90°. Real metal mirrors (finite conductivity) do not quite reverse helicity: the 'nonadiabatic' probability for preserving the original helicity is

$$P_{++} \approx 1/(2|\mu|^2) \qquad (6)$$

which for silver with $\mu = 0.2 + 3.44i$ is 0.042.

The simulation of true adiabatic change with discrete reflections can be achieved only by total internal reflection in a dielectric with index μ at the critical incidence angle $i = i_c = \sin^{-1}(\mu^{-1})$ (this follows from Fresnel's formulae[13]). For glass, $i_c \approx 45°$, so that rotation of 90° can be accomplished by successive reflections in three 45° right prisms arranged as in Fig. 3. However, i_c is not precisely 45° (because $\mu \neq \sqrt{2}$), and this will cause nonadiabatic helicity switching, with probability

$$P_{+-} = (\mu^2\sin^2 i - 1)/(\mu^2\tan^2 i - 1) \quad (\sin i > \mu^{-1}) \qquad (7)$$

For $i = 45°$ and glass with $\mu = 1.5$, $P_{+-} = 0.1$.

The arrangements in Figs 2 and 3 form the basis of robust, simple and (nonadiabatic conditions notwithstanding) convincing lecture demonstrations of the anholonomic transport of polarization.

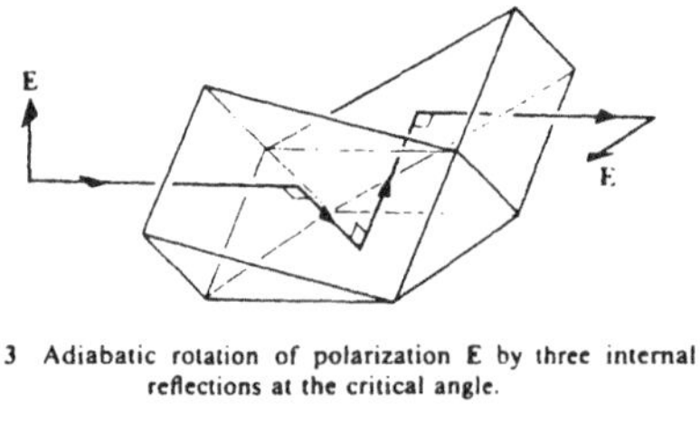

Fig. 3 Adiabatic rotation of polarization E by three internal reflections at the critical angle.

I thank Dr J. H. Hannay and Professor R. J. McCraw for helpful discussions, and to Professor M. Kitano for sending me his paper[17] before publication.

Received 2 December 1986; accepted 3 February 1987.

1. Chiao, R. Y. & Wu, Y. S. *Phys. Rev. Lett.* **57**, 933–936 (1986).
2. Tomita, A. & Chiao, R. Y. *Phys. Rev. Lett.* **57**, 937–940 (1986).
3. Maddox, J. *Nature* **323**, 199 (1986).
4. Robinson, A. L. *Science* **234**, 424–426 (1986).
5. MacCallum, M. *New Scient.* **1536**, 22 (1986).
6. Berry, M. V. *Proc. R. Soc.* **A392**, 45–57 (1984).
7. Simon, B. *Phys. Rev. Lett.* **51**, 2167–2170 (1983).
8. Ross, J. N. *Optical Quantum Electron.* **16**, 455–461 (1984).
9. Varnham, M. P., Birch, R. D. & Payne, D. N. *Tech. Dig. Int. Conf. Integrated Optics Optical Fiber Commns. Eur. Conf. Opt. Comms.* 135–138 (Instituto internazionale delle Comunicazione, Genova, 1985).
10. Varnham, M. P., Birch, R. D., Payne, D. N. & Love, J. D. *Tech. Dig. Conf. Opt. Fiber Comms. Atlanta* 68–71 (Optical Society of America, 1986).
11. Haldane, F. D. M. *Optics Lett.* **11**, 730–732 (1986).
12. Bialynicki-Birula, I. & Bialynicki-Birula, Z. *Phys. Rev. D* (in the press).
13. Born, M. & Wolf, E. *Principles of Optics* (Pergamon, London, 1959).
14. Kline, M. & Kay, I. M. *Electromagnetic Theory and Geometrical Optics* (Interscience, New York, 1965).
15. Snyder, A. W. & Love, J. D. *Optical Waveguide Theory* (Chapman and Hall, London, 1983).
16. Weatherburn, C. E. *Elementary Vector Analysis* (Bell, London, 1955).
17. Kitano, M., Yabuzuaki, T. & Ogawa, T. *Phys. Rev. Lett.* **58**, 523 (1987).
18. Chow, W. W., Gea-Banacloche, J., Pedrotti, L. M., Sanders, V. E., Schleich, W. & Scully, M. O. *Rev. mod. Phys.* **57**, 61–104 (1985).

VOLUME 56, NUMBER 9 PHYSICAL REVIEW LETTERS 3 MARCH 1986

Realizations of Magnetic-Monopole Gauge Fields: Diatoms and Spin Precession

John Moody, A. Shapere, and Frank Wilczek
Institute for Theoretical Physics, University of California, Santa Barbara, California 93106
(Received 29 July 1985)

It is found that the effective Hamiltonian for nuclear rotation in a diatom is equivalent to that of a charged particle in a background magnetic-monopole field. In certain cases, half-integer orbital angular momentum or non-Abelian fields occur. Furthermore, the effects of magnetic-monopole-like gauge fields can be experimentally observed in spin-resonance experiments with variable magnetic fields.

PACS numbers: 03.65.Ge, 11.15.Kc, 14.80.Hv, 33.10.Ev

Gauge potentials have been found to appear very naturally in the description of quantum-mechanical systems which depend upon slowly varying external parameters. Berry[1] performed a detailed study of Abelian potentials which appear when a single nondegenerate level is subjected to adiabatically varying external parameters. He showed that Abelian magnetic-monopole fields can occur near a degeneracy of the quantum levels (the monopole) in the space of external parameters. He also proposed an interference experiment using split electron beams to search for an "excess" geometrical phase induced by his gauge potentials. Simon[2] recast Berry's findings in the language of holonomy theory and made a connection to the quantum Hall effect. Finally, Wilczek and Zee[3] showed that non-Abelian gauge potentials can arise when sets of degenerate levels are subjected to adiabatically varying external parameters.

In work preceding Berry's by over twenty years, Herzberg and Longuet-Higgins[4] noticed sign ambiguities in the Born-Oppenheimer[5] wave functions of simple molecules. These sign ambiguities are associated with conical intersections in the zeroth-order electronic energy levels. To compensate for these and other, more general, phase ambiguities and thus to make the Born-Oppenheimer wave functions single valued, Mead[6] suggested adding a "vector-potential-type term" to the nuclear Hamiltonian. He refers to the effects of this term on the nuclear motion as the molecular Bohm-Aharonov effect.

The gauge potentials of Berry, Simon, and Wilczek and Zee can be generalized to systems where the slowly varying parameters are no longer external, but are themselves quantized; in the case of molecular physics, these are the nuclear coordinates. Such a description subsumes the findings of Herzberg and Longuet-Higgins and Mead. In a paper to be published elsewhere,[7] we show that the Born-Oppenheimer description of molecules can be cast in a rigorously gauge-covariant form, and discuss higher-order corrections to this description.

Diatoms as monopoles.—In the Born-Oppenheimer method,[5,7] eigenfunctions of the full Hamiltonian

$$H(R,r) = -\frac{1}{2m_N}\nabla_R^2 - \frac{1}{2m_e}\nabla_r^2 + V_N(R) + V_e(R,r) \tag{1}$$

are decomposed into electric and nuclear components:

$$\Psi^a(R,r) = \textstyle\sum_m \Phi_m^a(R)\phi_m(R,r).$$

Here R and r are nuclear and electronic coordinates, respectively. The $\phi_m(R,r)$ form a basis of electronic eigenfunctions of H for fixed R when the nuclear kinetic energy term is ignored; the energy of the state $\phi_m(R,r)$ is denoted by $\epsilon_m(R)$. The vector wave functions Φ_m^a are then acted on by the matrix-valued nuclear Hamiltonian H_{mn} which results when H is sandwiched between electronic eigenstates $|n(R)\rangle = |\phi_n\rangle$:

$$H_{mn} = -\frac{1}{2m_N}\sum_l [\nabla_R + \langle m(R)|\nabla_R|l(R)\rangle][\nabla_R + \langle l(R)|\nabla_R|n(R)\rangle] + V_N(R)\delta_{mn} + \epsilon_n(R)\delta_{mn}. \tag{2}$$

Here, H_{mn} is exact and has been written in gauge-covariant form.[8] The $\epsilon_n(R)$ act as effective potentials for the nuclear motion, while the $\langle m|\nabla_R|n\rangle$ act as effective gauge potentials and transition terms between different electronic states. In the adiabatic limit, where the nuclei move very slowly relative to the electrons, we may assume that the electrons remain in the nth level [i.e., we ignore off-diagonal transition terms in (2)], in which case the relevant Hamiltonian is approximately

$$H_n^{\text{eff}} \simeq -\frac{1}{2m_N}[\nabla_R - i\mathbf{A}(R)]^2 + V'(R), \tag{3}$$

where $\mathbf{A}(R) = \langle n|i\nabla_R|n(R)\rangle$. As noted by Mead[6] and Berry[1] and discussed in detail in Ref. 7, A transforms as a U(1) gauge potential when we locally change our choice of phases for $|n(R)\rangle$. If $|n\rangle$ belongs to an N-fold degenerate level, $\mathbf{A}$ is an $N\times N$ matrix which transforms as a U(N) gauge potential.[3,7] The approximation leading to (3) differs from the usual Born-Oppenheimer approximation where $|n(R)\rangle$ is chosen real so that $\mathbf{A}(R)=0$. As we shall see, this choice of gauge is not always possible globally.

For example, we consider the doubly degenerate Λ levels of a diatomic molecule. The effective Hamiltonian for this system is, in general, a 2×2 non-Abelian matrix determined as follows. Given the electronic eigenstates with $\Lambda=\pm l$ for the nuclear axis in the $\theta=\phi=0$ direction, we can choose a family of eigenstates adapted to nuclei pointing toward θ,ϕ by simple rotations:

$$|\pm(\theta,\phi)\rangle=e^{iJ_3\phi}e^{iJ_1\theta}e^{-iJ_3\phi}|\pm(0,0)\rangle. \tag{4}$$

According to our general prescription one then finds effective U(2) gauge potentials which are 2×2 matrices (with $\pm$ indices implicit):

$$A_\theta=\langle\theta,\phi|i\partial/\partial\theta|\theta,\phi\rangle=\langle 0,0|-e^{iJ_3\phi}J_1e^{-iJ_3\phi}|0,0\rangle=\langle 0,0|[-\cos\phi J_1+\sin\phi J_2]|0,0\rangle, \tag{5}$$

$$A_\phi=\langle\theta,\phi|i\partial/\partial\phi|\theta,\phi\rangle=\langle 0,0|[(1-\cos\theta)J_3+\sin\theta\sin\phi J_1+\sin\theta J_2]|0,0\rangle. \tag{6}$$

If $l\neq\frac{1}{2}$ we find $A_\theta=0$. A_ϕ is also readily evaluated in this case; one finds an "Abelianized" diagonal structure

$$A_\phi^{++}=l(1-\cos\theta),\quad A_\phi^{--}=-l(1-\cos\theta),\quad A_\phi^{+-}=A_\phi^{-+}=0. \tag{7}$$

More compactly, we may write $A_\phi=l(1-\cos\theta)\sigma_3$. Evaluating the field strength one finds $F_{\theta\phi}=\partial_\theta A_\phi-\partial_\phi A_\theta=l\sin\theta\sigma_3$, or for the corresponding Cartesian tensor $[F_{\theta\phi}\equiv(g^{\theta\theta})^{1/2}(g^{\phi\phi})^{1/2}]$ on the sphere simply

$$\hat{F}_{\theta\phi}=l\sigma_3. \tag{8}$$

Evidently, the nuclear coordinates act as if they parametrize the motion of a charged particle in a magnetic-monopole field.

To analyze this further, we need the angular momentum operators appropriate to the problem.[9] They are

$$\begin{aligned}\tilde{J}_x&=i\left\{\sin\phi\frac{\partial}{\partial\theta}+\cot\theta\cos\phi\frac{\partial}{\partial\phi}\right\}-l\frac{1-\cos\theta}{\sin\theta}\cos\phi\,\sigma_3,\\ \tilde{J}_y&=i\left\{-\cos\phi\frac{\partial}{\partial\theta}+\cot\theta\sin\phi\frac{\partial}{\partial\phi}\right\}-l\frac{1-\cos\theta}{\sin\theta}\sin\phi\,\sigma_3,\quad \tilde{J}_z=i\left\{-\frac{\partial}{\partial\phi}\right\}-l\sigma_3,\end{aligned} \tag{9}$$

where in each case the bracketed term is the standard or naive rotation operator. $\tilde{J}$ may be rewritten in the more compact form

$$\tilde{\mathbf{J}}=-i\mathbf{r}\times\mathbf{D}-\tfrac{1}{2}\mathbf{r}\epsilon_{ijk}r_iF_{jk}, \tag{10}$$

where $\mathbf{D}=\nabla-i\mathbf{A}$. Note that half-integer angular momentum is associated with the *orbital* nuclear motion, for l half integer.[10]

We may now write the Hamiltonian (1) as

$$H=\tfrac{1}{2}(\tilde{J}^2-l^2),$$

and construct its eigenstates by the usual procedure.[10]

The truly non-Abelian case $l=\frac{1}{2}$.—Although the algebra starting from Eq. (7) does not especially distinguish the case $l=\frac{1}{2}$, the original physical problem *is* different for this case. Specifically, in evaluating the potentials (5) and (6) we find off-diagonal terms:

$$A_\theta=\frac{1}{2}\begin{pmatrix}0 & -\kappa e^{i\phi}\\ -\kappa e^{-i\phi} & 0\end{pmatrix}, \tag{11}$$

$$A_\phi=\frac{1}{2}\begin{pmatrix}1-\cos\theta & -i\kappa e^{i\phi}\sin\theta\\ i\kappa e^{-i\phi}\sin\theta & -1+\cos\theta\end{pmatrix}, \tag{12}$$

where $\kappa=2\langle+|J_1|-\rangle$. Since the states in question are not eigenfunctions of angular momentum (but only of J_3), κ can in principle take any value; we have taken it real without loss of generality. From (11) and (12) we compute the field strength

$$\begin{aligned}F_{\theta\phi}&=\partial_\theta A_\phi-\partial_\phi A_\theta-i[A_\theta,A_\phi]\\ &=\tfrac{1}{2}(1-\kappa^2)\sin\theta\,\sigma_3.\end{aligned} \tag{13}$$

The field strength (13) vanishes for $\kappa=1$. This might have been anticipated, for the following reason: If the degenerate electron states with $\Lambda=\pm\frac{1}{2}$ actually formed a representation of the rotation group we would have $\kappa=1$. Furthermore, we could choose a fixed basis for the electron states, independent of the nuclear-axis orientation, since the two-dimensional *space* of states is rotationally invariant.

The field strength (13) superficially resembles the monopole field (8) that we encountered before, but the interpretation here is quite different. For one thing, κ is not quantized. At a deeper level, in the present case the gauge fields are *truly non-Abelian* (for

$\kappa \neq \pm 1$). While previously the potentials (7) all pointed in the same direction in internal space, here they do not. It is important to check that they cannot be made to do so by a different choice of basis, i.e., in a different gauge. Actually this already follows from the fact that the field strength is not quantized; it can also be seen by noting that the covariant derivative of the field strength, $D_\phi F_{\theta\phi}$, has a contribution from the commutator, which implies that it does *not* point in the same direction as $F_{\theta\phi}$ in internal space.

Because of the nonvanishing of $D_\phi F_{\theta\phi}$, the operators of Eq. (10) do not satisfy the angular momentum commutation relations. However, the "Abelianized" operators appropriate to the $\kappa = 0$ case, i.e., Eq. (9) with $l = \frac{1}{2}$, commute with the Hamiltonian not only for $\kappa = 0$ but generally. We do not know of a canonical procedure which yields these angular momentum operators, but we are convinced of their uniqueness (for $\kappa \neq \pm 1$). There is no nontrivial κ dependence which may be added to the operators of Eq. (10) that preserves their algebra.[11]

The implications of the equation implying conservation of $\tilde{J}$,

$$[\tilde{J}_i, H_\kappa - H_0] = 0, \tag{14}$$

can be made more transparent by defining the block-diagonal operators

$$\tilde{J}_i = \begin{pmatrix} M_i & 0 \\ 0 & N_i \end{pmatrix}, \quad H_\kappa - H_0 = \kappa \begin{pmatrix} 0 & C \\ C^* & 0 \end{pmatrix} + \frac{\kappa^2}{2}. \tag{15}$$

Here M_i and N_i each satisfy the angular momentum algebra, and (14) gives the intertwining relations

$$M_i C = C N_i, \quad C^* M_i = N_i C^*. \tag{16}$$

A direct computation gives

$$\begin{pmatrix} 0 & C \\ C^* & 0 \end{pmatrix}^2 = \tilde{J}^2 + \tfrac{1}{4}. \tag{17}$$

Thus C is a sort of a Dirac operator, roughly a local square root of the covariant spherical Laplacian. Its effect is to split the eigenvalues of H by $2\kappa(j+\frac{1}{2})$, in agreement with the classic result.[12]

Spin precession in an oscillating magnetic field.—In the preceding section, we saw how magnetic-monopole fields can arise dynamically in the context of the Born-Oppenheimer method. Berry studied the case of spins in an applied, adiabatically varying magnetic field and found another monopole, leading to a nondynamical precession effect. He showed that if the external magnetic field goes around a closed path $B(t)$ $(0 \leq t \leq T)$ enclosing a solid angle Ω, then spin-$\frac{1}{2}$ wave functions with $s_z = \pm\frac{1}{2}$ acquire geometrical phases $\exp(\pm i\Omega/2)$ in addition to the dynamical phase $\exp[-i\int_0^T E(t)dt]$. The relative phase between the two spin components implies an extra rotation of the spin by an angle Ω in the x-y plane. In B space, this mimics the effect of a magnetic monopole at $B = 0$ with unit Dirac charge on electrically charged particles following the trajectory $B(t)$. Berry proposed an experiment to observe this geometric phase by splitting an electron beam in a magnetic field, rotating the field applied to one of the components of the beam, and looking for interference when the split beam is recombined. Unfortunately, such an experiment is impractical, because of the difficulty in guaranteeing identical dynamical evolution of the two beam components.

We wish to point out that Berry's result may be extended to the case of spins in a rapidly oscillating magnetic field, thus providing a practical way to measure geometric phases within the familiar context of magnetic resonance. Specifically, let B_z be a large constant orienting field and $B_\perp(t)\sin(\omega t)$ a small, slowly modulated oscillating field with mean rf frequency ω in the x-y plane. Then regarding $B_\perp$ and ω as free parameters and varying $B_\perp$ around a closed loop will result in a geometric phase which can be observed by beating the resulting rf signal against a reference signal. Thus, an interference experiment is done upon a single sample, avoiding the problem of guaranteeing precisely equivalent dynamical evolution for two separate systems.

Consider the Hamiltonian

$$H_{\text{lab}} = -\gamma\sigma \cdot [B_z\hat{\mathbf{z}} + \mathbf{B}_\perp(t)\sin\omega t]. \tag{18}$$

We suppose operation near a resonant frequency $\omega \approx \omega_0 = \gamma B_z$, and go to the rotating frame. The appropriate effective Hamiltonian is

$$H_{\text{rot}} = -\gamma\sigma \cdot \left[B_z \left(1 - \frac{\omega}{\omega_0}\right)\hat{\mathbf{z}} + \mathbf{B}_\perp(t) \right]. \tag{19}$$

This is precisely the sort of Hamiltonian considered by Berry, with the origin (and the monopole) now located at the resonance point $\omega = \omega_0$, $B_\perp = 0$. Thus a variation of $B_\perp$ around a loop $C(B_\perp(t))$ leads to a geometric phase $\gamma(C, \omega_0)$.

The resulting geometric phase occurs together with the much larger rotation, proportional to the time, due to ordinary precession. It can be isolated by beating the precession signal against a fixed-frequency signal. Notice that this phase depends on ω_0, so that the apparent splitting of nearby resonance peaks ω_0 and ω_0' will be altered by an amount

$$\omega_0 - \omega_0' = \frac{\gamma(C, \omega_0) - \gamma(C, \omega_0')}{T}.$$

We wish to thank William Bialek and Pawel Mazur for useful discussions. This research was supported in part by the National Science Foundation under Grant No. PHY82-17853, supplemented by funds from the

VOLUME 56, NUMBER 9 PHYSICAL REVIEW LETTERS 3 MARCH 1986

National Aeronautics and Space Administration.

[1]M. Berry, Proc. Roy. Soc. London, Ser. A **392**, 45 (1984).

[2]B. Simon, Phys. Rev. Lett. **51**, 2167 (1983).

[3]F. Wilczek and A. Zee, Phys. Rev. Lett. **52**, 2111 (1984).

[4]G. Herzberg and H. C. Longuet-Higgins, Discuss. Faraday Soc. **35**, 77 (1963).

[5]M. Born and J. Oppenheimer, Ann. Phys. (Leipzig) **84**, 457 (1927). For a good recent account see H. Bethe and R. Jackiw, *Intermediate Quantum Mechanics* (Benjamin, New York, 1968), p. 177.

[6]C. Mead, Chem. Phys. **49**, 23, 33 (1980). See also references in R. L. Whetten, G. S. Ezra, and E. R. Grant, Annu. Rev. Phys. Chem. **36**, 277 (1985).

[7]J. Moody, A. Shapere, and F. Wilzcek, "Adiabatic Methods and Induced Gauge Structures in Quantum Mechanics" (to be published).

[8]See Ref. 7 for a detailed derivation and discussion of this Hamiltonian.

[9]A derivation of the modified rotation operators is given in Ref. 7. Alternative derivations can be found in M. Saha, Indian J. Phys. **10**, 145 (1936); P. Goddard and D. Olive, Rep. Prog. Phys. **41**, 1357 (1978); S. Coleman, in *The Unity of the Fundamental Interactions*, edited by A. Zichichi (Plenum, New York, 1983); R. Jackiw and N. S. Manton, Ann. Phys. (N.Y.) **127**, 257–273 (1980).

[10]Coleman, Ref. 9, and references therein.

[11]These angular momentum operators should be contrasted with the symmetry operators described in Jackiw and Manton, Ref. 9.

[12]Essentially this Hamiltonian appears already in J. H. Van Vleck, Phys. Rev. **33**, 467 (1929), published two years before Dirac's monopole paper [P.A.M. Dirac, Proc. Roy. Soc. London, Ser. A **133**, 60 (1931)].

MOLECULAR PHYSICS, 1987, VOL. 61, NO. 6, 1327–1340

Berry's phase in magnetic resonance

by DIETER SUTER, GERARD C. CHINGAS, ROBERT A. HARRIS and ALEXANDER PINES†
University of California, Berkeley, California 94720, U.S.A.

(*Received 18 March 1987; accepted 31 March 1987*)

According to Berry, quantum states of a hamiltonian which varies adiabatically through a circuit C in parameter space may acquire geometrical phase factors $\exp(i\gamma(C))$ in addition to the normal dynamical phase factors $\exp((-i/\hbar)\int E(t)\,dt)$. We present N.M.R. experiments in the rotating frame which bear out these predictions for simple conical circuits, and point out that they are related to familiar behaviour based on the classical Bloch equations and on Haeberlen–Waugh coherent averaging theory. Extensions to coupled spins and electric quadrupolar effects are discussed.

1. Introduction

A system prepared in an eigenstate of a slowly varying hamiltonian remains in an eigenstate of the instantaneous hamiltonian [1]. In 1984, Berry pointed out [2] that in a cyclic adiabatic process, that is one in which the slowly time varying hamiltonian returns to its original form via a circuit C, a quantum state may acquire a 'geometrical' phase factor $\exp(i\gamma(C))$ in addition to the 'normal' dynamical phase factor $\exp((-i/\hbar)\int E_m(t)\,dt)$. In an elegant calculation, Berry showed that if the circuit occurs in the vicity of a degeneracy of the hamiltonian in parameter space, then the geometrical phase is proportional to the solid angle Ω subtended by the circuit at the degeneracy.

As an illustrative example, Berry considered spins in a magnetic field characterized by slowly varying parameters R as depicted in figure 1. The hamiltonian for this system has a degeneracy at $R = 0$ where $B = 0$. For the simplest case of a cone, θ constant, the solid angle is $\Omega = 2\pi(1 - \cos\theta)$. Imagine that such a conical circuit is traversed adiabatically, that is with small δ, where $\delta = 2\pi/T$ and T is the period of the circuit. A spin eigenstate with magnetic quantum number m should accumulate a geometrical phase $\gamma(C) = 2\pi m(1 - \cos\theta)$ in addition to the dynamical phase $m\gamma_I \int B(t)\,dt$, where γ_I is the magnetogyric ratio.

Wilczek and co-workers [3] and Cina [4] have suggested that a manifestation of the geometrical phase should be observed in interference between eigenstates, for example in the evolution of a coherent superposition of states m and m'. Such a superposition corresponds to magnetization or to higher rank tensor coherences [5] and the phase changes of such coherences have been observed for states in N.M.R. undergoing non-adiabatic circuits [6]. Upon completion of an adiabatic circuit, a coherence should acquire a geometrical phase change or extra rotation, $\phi_g = \gamma_m(C) - \gamma_{m'}(C)$, in addition to the dynamical precession angle ϕ_d. For the case of a

† On sabbatical leave during 1987 at the E.S.P.C.I., Laboratoire Physique Quantique, 10 Rue Vauquelin, Paris Cedex 05, France. I am grateful to Professors A. P. Le Grand and P. G. de Gennes for their invitation and hospitality.

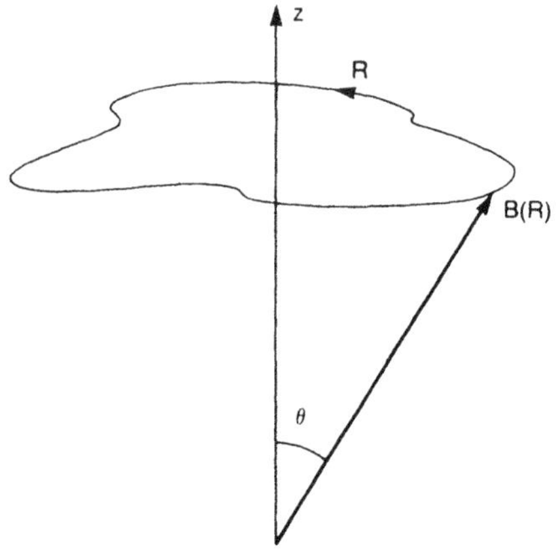

Figure 1. Berry's experiment for spins in a magnetic field. A field of magnitude B moves adiabatically through a circuit of parameters R in the laboratory. The experiments in this paper deal with simple conical circuits in the rotating frame for which $|B|$ and θ are constant. The considerations are similar if $\boldsymbol{B}$ is the axis in zero field of a dipolar coupled spin-1/2 pair or spin-1 with electric quadrupole coupling.

magnetic field of constant magnitude $|B| = \omega_d/\gamma_I$ and constant angle θ in figure 1

$$\phi_d = (m - m')\gamma_I BT = \Delta m \omega_d T, \quad (1)$$

$$\phi_g = (m - m')\Omega = 2\pi\Delta m(1 - \cos\theta). \quad (2)$$

Chiao *et al.* [7] have reported a classical optical version of such an experiment in which the plane of linearly polarized light (which corresponds to a superposition of the $m = \pm 1$ photon states) was rotated by a geometrical phase imposed by helically wound optical fibers. Tycko [8] has recently performed a nuclear quadrupole resonance experiment in which the geometrical phase of a spin-3/2 was observed during rotation of a crystal, thereby moving the quantization axis of the electric field gradient in a cone. The geometrical phase is also related to early work on fractional quantum numbers [9] in molecules and the classical work on conical intersections by Herzberg and Longuet-Higgins [10]. Indeed, Mead and Truhlar had earlier used the concept of a geometrical phase in their discussion of conical intersections [11].

In the present paper we outline N.M.R. experiments that measure the geometrical phase acquired by a spin-1/2 in a magnetic field of constant magnitude and varying direction in the rotating frame [3]. The experiments and corresponding treatment cover the range from the adiabatic limit (δ small), which yields Berry's geometrical phase, to the non-adiabatic regime characteristic of resonant processes. Such circuits are well known in N.M.R. experiments which involve precisely such combinations of static and rotating fields. We also relate the adiabatic behaviour to well-known coherent averaging phenomena in pulsed and iterative N.M.R. schemes which exploit geometrical scaling of resonance frequency differences and spin–spin couplings [12, 13].

2. N.M.R. experiment

The experiment might be conducted in the laboratory frame using the methods of pure N.Q.R. [14] or zero field N.M.R. [15], but we prefer the inherently greater sensitivity of high frequency N.M.R. in a high field magnet as suggested by Wilczek and co-workers [3]. To recall how the adiabatic circuits of figure 1 can be implemented in such circumstances, we refer to figure 2. An ensemble of spin-1/2 nuclei I is immersed in a large static magnetic field B_0 along the z axis so that their Larmor frequency is given by $\omega_0 = \gamma_I B_0$ where γ_I is the magnetogyric ratio. The spins develop an equilibrium magnetic polarization described by the reduced high temperature density operator [16]

$$\rho(0) = I_z, \tag{3}$$

where we have omitted, as usual, the unity operator and all proportionality constants.

The spins are irradiated at a frequency ω_{rf} near ω_0 with a circularly polarized radio-frequency (rf) field of magnitude B_1 such that $\gamma_I B_1 = \omega_1$. The evolving

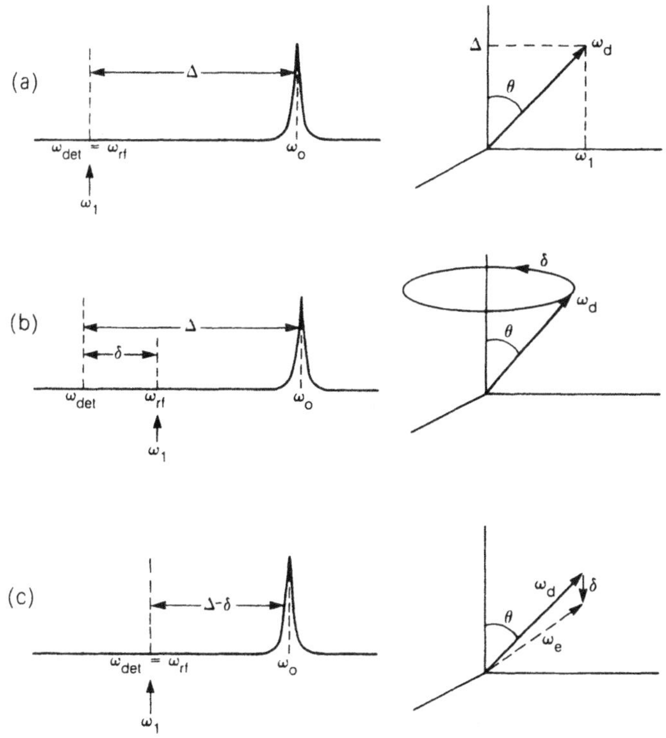

Figure 2. Rotating (detector) frame pictures of spins irradiated at frequency ω_{rf} near resonance (ω_0) with a circularly polarized radio-frequency field of magnitude $B_1 = \omega_1/\gamma_I$. A phase sensitive detector at frequency ω_{det} is responsible for recording the transverse magnetization. (*b*) shows that in the detector frame we have implemented, for high field N.M.R., a situation equivalent to the laboratory picture ($\omega_{det} = 0$) of figure 1 for simple conical circuits.

magnetization is detected with a phase sensitive detector operating at frequency ω_{det} such that

$$\omega_0 - \omega_{det} = \Delta, \tag{4}$$

$$\omega_{rf} - \omega_{det} = \delta. \tag{5}$$

We consider the case of simple conical circuits for which ω_0 and ω_1 are constant in time. In figure 2(*a*) ω_{rf} has been set equal to ω_{det} so that $\delta = 0$, whereas figure 2(*b*) reflects the general situation in which $\delta \neq 0$. The frequency ω_{det} can be thought of as the reference or zero of the frequency scale. A laboratory frame experiment of the type in figure 1 with a constant field $|B| = \omega_d/\gamma_I$ (here ω_d denotes the dynamical frequency) moving with constant θ at frequency δ around z would correspond to $\omega_{det} = 0$ in Figure 2(*b*).

In a frame of reference rotating with the detector [17], we term this the 'detector frame', the effective static field along the z axis is $B_0 - \omega_{det}/\gamma_I$ so the effective Larmor frequency is Δ, given by equation (4), and the effective rotation frequency of the rf field B_1 is δ, given by equation (5). The situation in figure 2(*b*) is thus equivalent to a magnetic field of magnitude $B = \omega_d/\gamma_I$ moving in a cone of angle 2θ around the z axis at frequency δ. The effective hamiltonian in the detector frame is given by

$$\tilde{\mathscr{H}}(t) = -\hbar\omega_d[\cos\theta I_z + \sin\theta(I_x \cos\delta t + I_y \sin\delta t)], \tag{6}$$

where

$$\omega_d = (\Delta^2 + \omega_1^2)^{1/2}, \tag{7}$$

$$\theta = \tan^{-1}(\omega_1/\Delta), \tag{8}$$

and I_α are the spin angular momentum operators. But this is precisely the arrangement corresponding to figure 1 for the case of constant $|B|$ and θ. Thus figure 2(*b*) is the high field detector frame equivalent of conical circuits in the laboratory. To implement the general circuits of figure 1 in the rotating frame ω_d and θ can be varied by modulating ω_0, ω_{rf} and ω_1.

To calculate the evolution of the magnetization in the detector frame of figure 2(*b*) it is convenient to transform temporarily to a frame rotating at frequency δ with respect to the detector frame [17] as shown in figure 2(*c*); we imagine moving the detector frequency to ω_{rf}. In the laboratory case $\omega_{det} = 0$, this corresponds to a frame rotating at frequency δ around the laboratory z axis. In this rotating frame the total effective magnetic field is static with a magnitude ω_e/γ_I where the effective frequency ω_e is given by

$$\omega_e = ((\Delta - \delta)^2 + \omega_1^2)^{1/2}. \tag{9}$$

The adiabatic limit corresponds to

$$\delta \ll \omega_d, \tag{10}$$

in which case the quantization axis remains along the direction of ω_d to first order and the magnetization precesses at frequency ω_e given by expanding equation (9) in δ/ω_d and using equations (7) and (8) to give

$$\omega_e \approx \omega_d - \delta\cos\theta. \tag{11}$$

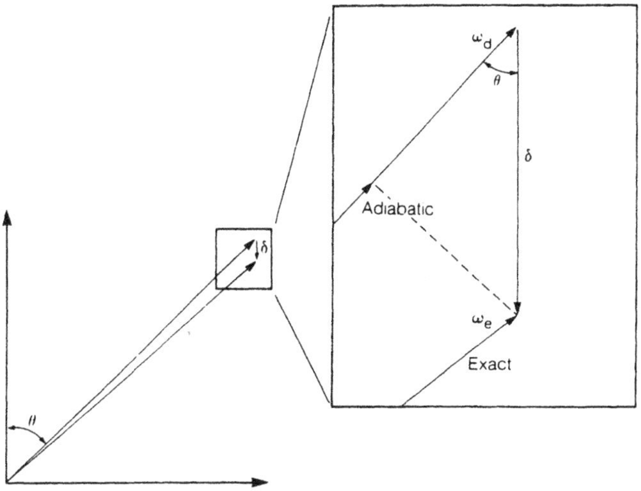

Figure 3. Expansion of figure 2(*c*) showing a view of the effective static fields in a frame rotating at frequency δ with respect to the Berry situation in figures 1 and 2(*b*). The adiabatic limit corresponds to taking the projection of δ onto ω_d.

Details of the relevant geometry and the adiabatic projection of δ onto ω_d are shown in figure 3.

The phase ϕ' accumulated after one adiabatic circuit, $\delta T = 2\pi$, in the rotating frame of figure 2(*c*) is given according to equation (11) by

$$\phi' = \omega_e T \approx \omega_d T - 2\pi \cos \theta, \quad (12)$$

which corresponds in the original detector frame of figure 2(*b*) to an accumulated phase ϕ of

$$\phi = \phi' + 2\pi = \phi_d + \phi_g = \omega_d T + 2\pi(1 - \cos \theta), \quad (13)$$

with ϕ_d and ϕ_g corresponding to the dynamical and geometrical phases of equations (1) and (2) with $\Delta m = 1$. Thus it is possible to determine Berry's geometrical phase, corresponding to 'stroboscopic' observations (once per circuit), by measuring ω_e and δ.

The exact expression for the evolving density operator in the detector frame, beginning with initial condition equation (3), is given by the well-known transient solution to the Bloch equations [17] neglecting relaxation:

$$\begin{aligned}\rho(t) = {} & \sin \theta I_x[\cos \theta(1 - \cos \omega_e t) \cos \delta t + \sin \omega_e t \sin \delta t] \\ & + \sin \theta I_y[\cos \theta(1 - \cos \omega_e t) \sin \delta t - \sin \omega_e t \cos \delta t] \\ & + I_z[\cos^2 \theta - \sin^2 \theta \cos \omega_e t], \end{aligned} \quad (14)$$

from which the adiabatic and non-adiabatic regimes can be inferred. Experimentally a linearly polarized rf field was used, invoking the rotating wave approximation. The phase sensitive detector measures $\langle I_x \rangle$, $\langle I_y \rangle$ and, upon Fourier transformation of the signal, ω_e can be determined. Experiments were performed on the proton

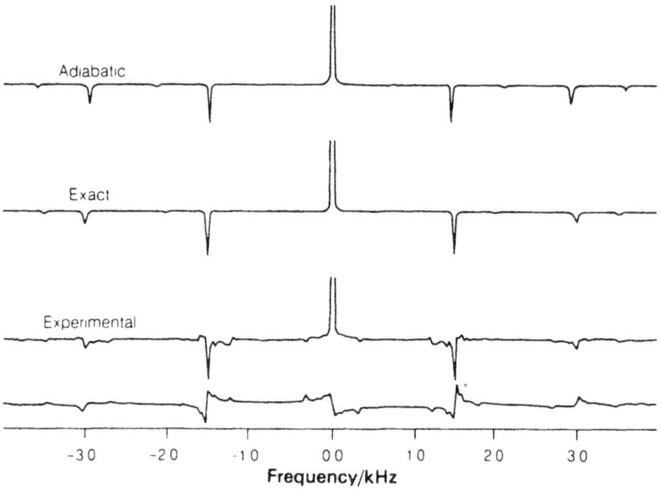

Figure 4. Fourier transform spectra of the detected transverse magnetization signal in an experiment of the type shown in figure 2(*b*) with $\theta = 62{\cdot}1°$, $\omega_d/2\pi = 1{\cdot}31$ kHz, $\delta/2\pi = -0{\cdot}33$ kHz. Adiabatic and exact simulations are shown as well as the complex experimental signal. Deviations from adiabatic behaviour are visible since $\delta \cong 0{\cdot}25\ \omega_d$.

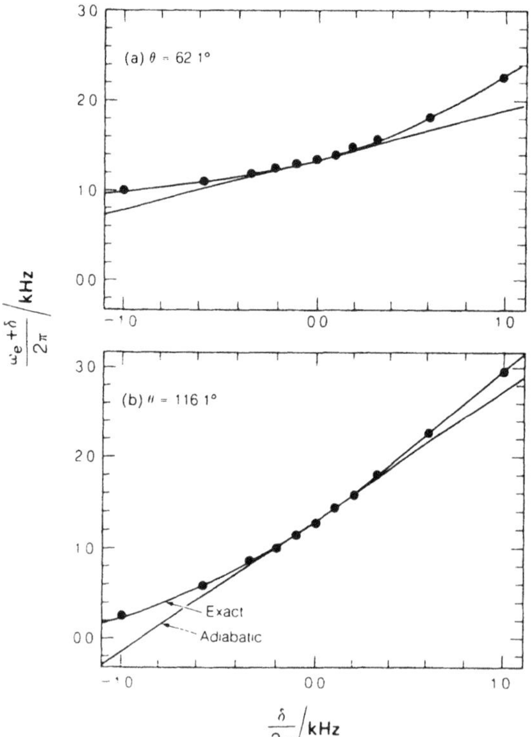

Figure 5. Experimental measurements of $\omega_e + \delta$ versus δ for two values of θ. According to equation (15), in the adiabatic regime ($\delta \ll \omega_d$) the data should conform to $\omega_e + \delta \approx \omega_d + \delta(1 - \cos\theta)$, which is shown as the straight lines. The adiabatic behaviour holds quite well for $\delta < 0{\cdot}2\ \omega_d$.

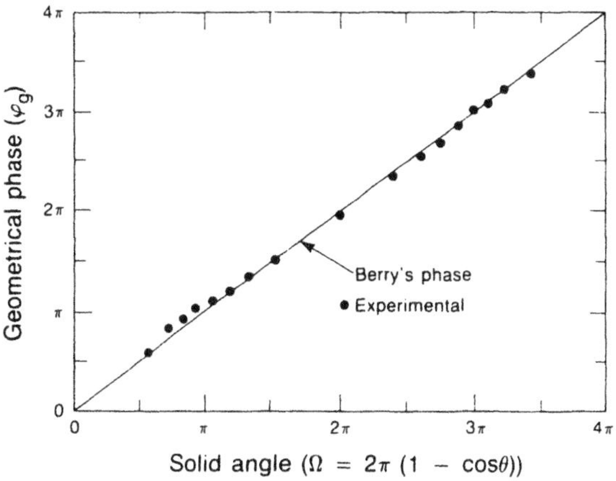

Figure 6. Plot of experimental geometrical phase (extra rotation angle of the magnetization per circuit) extracted from least squares linear fits to the adiabatic ($\delta < 0{\cdot}2\omega_d$) data of figure 5 versus solid angle. The straight line with slope 1 corresponds to Berry's geometrical phase.

spins of a water/acetone sample in a superconducting magnet with $\omega_0/2\pi = 500\,\mathrm{MHz}$, $\omega_1/2\pi = 1{\cdot}16\,\mathrm{kHz}$ and various values of θ and δ. An example of the Fourier transform of the detector signal is shown in figure 4 together with exact and adiabatic simulations for $\theta = 62{\cdot}1°$, $\omega_d/2\pi = 1{\cdot}31\,\mathrm{kHz}$ and $\delta/2\pi = -0{\cdot}33\,\mathrm{kHz}$. In figure 5 we plot $\omega_e + \delta$, as determined from spectra such as the one shown in figure 4, as a function of δ for two values of θ, using calibrated oscillators to vary δ. According to equation (11), in the adiabatic limit this should be given by

$$\omega_e + \delta \approx \omega_d + \delta(1 - \cos\theta). \tag{15}$$

Indeed, for small values of δ the data are linear, and the geometrical phase ϕ_g is determined by the slope of the adiabatic straight line in figure 5 multiplied by $T = 2\pi/\delta$ and is plotted versus the solid angle $2\pi(1 - \cos\theta)$ in figure 6. The behaviour is similar to that observed in the optical rotation [7] and N.Q.R. [8] experiments.

In these experiments the geometrical phase is present as a small factor in the presence of the large dynamical phase. The effects of the dynamical phase may be removed either by stroboscopic observation commensurate with the dynamical period, or by periodically (for example once in the middle of the circuit, or just prior to repeating the circuit) reversing the direction of B. In the latter case the dynamical phase is refocused as an echo [17(*b*)] at the end of the circuit leaving only the pure geometrical phase.

The cos θ factor which derived from the projection of the small vector δ onto the large ω_d is familiar in N.M.R. [5]. It corresponds to the secular approximation or average hamiltonian [11, 12]. What we have done in going from the laboratory or detector frame to the rotating frame is to transform a large slowly time varying hamiltonian into a large time independent hamiltonian plus a small additional term which is treated by perturbation theory. This is explained in more detail in the next section. We have used a time independent perturbation δI_z as the generator of the adiabatic circuit whereas Cina [4] used a time dependent generator orthogonal to the field ω_d at all times to express the problem as one of parallel transport [2].

The scaling of chemical shifts by cos θ, for example $1/\sqrt{3}$ at the 'magic angle' [18], is related to the frequency differences predicted by Wilczek and co-workers [3] and observed in this work. Dipolar couplings between spins and other higher rank interactions are correspondingly scaled by higher order Legendre polynomials as predicted by the coherent averaging theory of Haeberlen and Waugh [12]. The treatment and experiments in this paper are of course applicable to spins greater than 1/2 or to any multilevel quantum system which can be cast in the framework of a fictitious spin [5, 19]. Finally, we note that the effects measured in this work can be interpreted classically and are equivalent to the accumulation of phase due to Coriolis forces in accelerated frames of reference [20]. They would occur for an inclined top spinning on a slowing rotating platform. Quantum mechanical effects due to non-integral m values of the spin could be observed in adiabatic versions of the N.M.R. interference experiments [6].

3. General circuits and coherent averaging

In this section we discuss the relationship between the adiabatic theorem and well-known coherent averaging effects in N.M.R. Consider a 'large' hamiltonian $\tilde{\mathscr{H}}(t)$ with a slow time dependence characterized by a small parameter δ. We assume that $\tilde{\mathscr{H}}(t)$ is cyclic and that it goes through one cycle or circuit:

$$\tilde{\mathscr{H}}(T) = \tilde{\mathscr{H}}(0). \tag{16}$$

Such cyclic hamiltonians are familiar in pulsed N.M.R. When the time dependence is rapid then coherent averaging theory [12] is directly applicable, but here we are interested in adiabatic processes for which a change of picture may be appropriate. An eigenstate $|0\rangle$ of $\tilde{\mathscr{H}}(0)$ will evolve by the end of the circuit to

$$|T\rangle = \tilde{U}(T)|0\rangle, \tag{17}$$

where the circuit propagator $\tilde{U}(T)$ is given by

$$\tilde{U}(T) = \mathscr{T} \exp(-i/\hbar) \int_0^T \tilde{\mathscr{H}}(t)\, dt \tag{18}$$

and $\mathscr{T}$ is a time ordering operator [21]. If the change is adiabatic then

$$|T\rangle = \exp\{i(\gamma_d + \gamma(C))\}|0\rangle \tag{19}$$

where γ_d and $\gamma(C)$ are the dynamical and geometrical phases [2] respectively. Taking only the dynamical phase γ_d is tantamount to assuming that the eigenvalues of the average hamiltonian $\bar{\tilde{\mathscr{H}}}$ [12] are the same as the average eigenvalues of $\tilde{\mathscr{H}}(t)$, namely $\langle 0|\bar{\tilde{\mathscr{H}}}|0\rangle = \overline{\langle t|\tilde{\mathscr{H}}(t)|t\rangle}$. The circuit propagator $\tilde{U}(T)$ in equation (18) can be factored [22]

$$\tilde{U}(T) = R_\delta^\dagger(T)U(T), \tag{20}$$

where

$$R_\delta(t) = \mathscr{T} \exp(-i/\hbar) \int_0^t \mathscr{H}_\delta(t')\, dt', \tag{21}$$

$$U(T) = \mathscr{T} \exp(-i/\hbar) \int_0^T (\mathscr{H}(t) + \mathscr{H}_\delta(t))\, dt, \tag{22}$$

and

$$\mathscr{H}(t) = R_\delta(t)\tilde{\mathscr{H}}(t)R_\delta^\dagger(t). \tag{23}$$

The hamiltonian $\mathscr{H}_\delta(t)$ in equation (21) can be regarded as the generator of an interaction picture [1] in which the effective hamiltonian is $\mathscr{H}(t) + \mathscr{H}_\delta(t)$ and the effective circuit propagator is $U(T)$ in equation (22). In such an interaction picture

$$R_\delta(T)|T\rangle = U(T|0\rangle. \tag{24}$$

Now, the objective is to find a 'small' $\mathscr{H}_\delta(t)$ and thus an $R_\delta(t)$ such that $\mathscr{H}(t)$ in equation (23) is time independent or commutes with itself at different times. If $\mathscr{H}(t)$ is time independent (this is easily generalized to a commuting hamiltonian), namely

$$\mathscr{H}(t) = \mathscr{H}_\mathrm{d} \tag{25}$$

where $\mathscr{H}_\mathrm{d}$ is a large 'local' hamiltonian giving rise to normal dynamical evolution of the eigenstates with a characteristic frequency

$$\omega_\mathrm{d} = 2\pi/t_\mathrm{d}, \tag{26}$$

then equation (22) can be written

$$U(T) = \mathscr{T} \exp(-i/\hbar)\int_0^T (\mathscr{H}_\mathrm{d} + \mathscr{H}_\delta(t)\, dt). \tag{27}$$

Of course the choices of R_δ and $\mathscr{H}_\mathrm{d}$ are not unique and it is the different pictures and local hamiltonians which give rise to the choice of dynamical phases and fractional quantum numbers. Now, since $\mathscr{H}_\delta$ is small, $\|\mathscr{H}_\delta\| \ll \|\mathscr{H}_\mathrm{d}\|$, we use coherent averaging theory [12], retaining only the zero order average hamiltonian $\bar{\mathscr{H}}_\delta$ in the Magnus expansion [23], that is the part of $\mathscr{H}_\delta(t)$ which is secular or commutes with $\mathscr{H}_\mathrm{d}$. This is given by

$$\bar{\mathscr{H}}_\delta = \frac{1}{t_\mathrm{d}}\int_0^T\int_0^{t_\mathrm{d}} \exp((-i/\hbar)t\mathscr{H}_\mathrm{d})\mathscr{H}_\delta(t') \exp((i/\hbar)t\mathscr{H}_d)\, dt\, dt'. \tag{28}$$

$\bar{\mathscr{H}}_\delta$ is a generalization of the vector projection proportional to $-\delta \cos\theta$ of the previous section where the hamiltonians $\mathscr{H}_\mathrm{d}$ and $\mathscr{H}_\delta$ correspond to magnetic field vectors. Corrections to $\bar{\mathscr{H}}_\delta$ which correspond to non-adiabatic deviations are provided by the correction terms $\bar{\mathscr{H}}_\delta(k)$ [12, 23].

The adiabatic circuit propagator $\tilde{U}(T)$ in equation (20) is thus given by combining equations (20), (21), (27) and (28):

$$\tilde{U}(T) = R_\delta^\dagger(T)\bar{R}_\delta(T) \exp((-i/\hbar)T\mathscr{H}_\mathrm{d}), \tag{29}$$

where

$$\bar{R}_\delta(T) = \exp((-i/\hbar)T\bar{\mathscr{H}}_\delta). \tag{30}$$

The dynamical and geometrical phase factors in equation (19) can now be recognized as

$$\exp((-i/\hbar)T\mathscr{H}_\mathrm{d})|0\rangle = \exp(i\gamma_\mathrm{d})|0\rangle, \tag{31}$$

$$R_\delta^\dagger(T)\bar{R}_\delta(T)|0\rangle = \exp(i\gamma(C))|0\rangle. \tag{32}$$

Parallel transport corresponds to a choice of $\mathscr{H}_\delta(t)$ orthogonal to $\mathscr{H}_\mathrm{d}$ so that $\bar{\mathscr{H}}_\delta = 0$ and $\bar{R}_\delta(T) = 1$. The geometrical phase is then given entirely by $R_\delta^\dagger(T)$ in equation (32) acting on $|0\rangle$. This relates the adiabatic behaviour to the general case of parallel transported circuits due to Aharonov and Anandan [24].

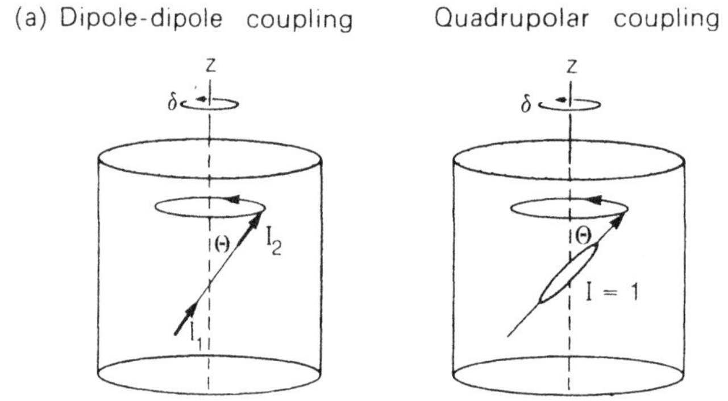

Figure 7. Adiabatic rotation of a sample can generate a geometrical phase for (*a*) dipole-coupled spins or (*b*) spins $I = 1$ subject to electric quadrupole coupling.

Equation (29) makes clear the contributions to the phase factor in a circuit. The dynamical phase of equation (31) derives from the static hamiltonian $\mathscr{H}_d$ which corresponds in the magnetic field case to a static magnetic field or ω_d in figure 2. Any mystery in the geometrical phase is a consequence of the seductive intuition that in the transformed picture, or local coordinates, the evolution, boundary conditions and quantization should be the same as if $\mathscr{H}_d$ were originally static. The terms $\bar{R}_\delta(T)$ and $R_\delta^\dagger(T)$ in equation (32) give rise respectively in the magnetic field case to the $-2\pi \cos\theta$ term of equation (12) and the 2π term of equation (13). The case $\theta = 0$ in § 2 corresponds here to $R_\delta^\dagger(T)\bar{R}_\delta(T) = 1$ which can be achieved for example (not necessarily) if

$$[\mathscr{H}_d, \mathscr{H}_\delta] = 0. \tag{33}$$

In this case equation (27) can be factored exactly and the adiabatic approximation is not necessary. The magnetic case $\theta = \pi/2$ corresponds to $\bar{R}_\delta(T) = 1$ in which case $R_\delta^\dagger(T)$ is responsible for the geometrical phase and gives rise to the familiar spinor sign changes under 2π rotations [2, 6]. Such considerations also apply to the case of molecules with coupled rotors [9].

Interesting versions of the geometrical phase occur for coupled spins or for spins greater than $\frac{1}{2}$. Suppose two spins 1 and 2 are coupled by magnetic dipolar interactions in the absence of a static external field. In the principal axis system (x', y', z') of the dipolar tensor, the coupling hamiltonian is

$$\mathscr{H}_d = \tfrac{2}{3}\hbar\omega_d[(3I_{1z'}I_{2z'} - I_1 \cdot I_2) + \eta(I_{1x'}I_{2x'} - I_{1y'}I_{2y'})]. \tag{34}$$

Similarly, for a spin $I = 1$ in the principal axis system of the electric field gradient tensor

$$\mathscr{H}_d = \tfrac{1}{3}\hbar\omega_d[(3I_{z'}^2 - I^2) + \eta(I_{x'}^2 - I_{y'}^2)]. \tag{35}$$

In the case of axial symmetry ($\eta = 0$), a component of magnetization perpendicular to the symmetry (z') axis oscillates linearly at the dynamical frequency ω_d [15] so the total magnetization evolves in a plane, the polarization plane. A superposition of the $m = \pm 1$ eigenstates, corresponding to double quantum coherence, would remain constant in time [5]. Imagine that the symmetry axis (z') is now rotated adiabatically in a cone at frequency δ, for example by physically rotating a solid

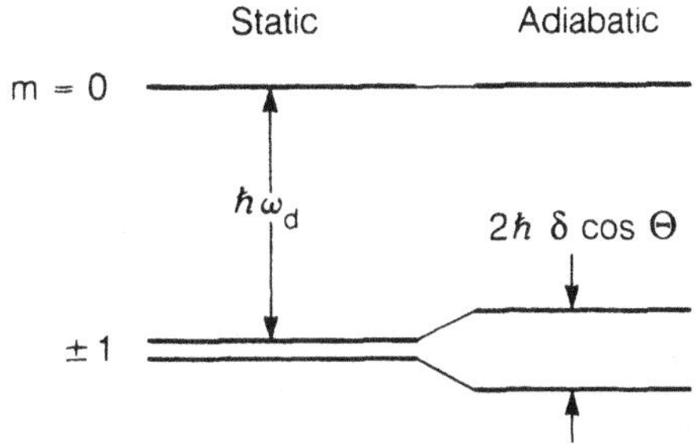

Figure 8. The degeneracy of the $m = \pm 1$ levels is lifted by sample rotation.

sample around an axis tilted at an angle θ with respect to z', as shown in figure 7. The evolution of the spin system can then be described in a frame rotating at δ with the crystal or spin pair by a time independent hamiltonian (equation (34) or (35)) with an additional term $\hbar\delta[\cos\theta(I_{1z'}+I_{2z'})+\sin\theta(I_{1x'}+I_{2x'})]$ or $\hbar\delta[\cos\theta I_{z'}+\sin\theta I_{x'}]$. The term $\bar{\mathscr{H}}_\delta$ of equation (28) is given by the projection of this term onto the symmetry axis, i.e. $\hbar\delta\cos\theta(I_{1z'}+I_{2z'})$ for the dipolar case, and $\hbar\delta\cos\theta I_{z'}$ for the quadrupolar case (see figure 8), again violating the intuitive notion that if we 'run around' with the pair of spins then the local dynamics should look the same as they would for a static pair in the laboratory. This is identical to the consideration of § 2 and so the phase at the end of the circuit is given precisely by equation (13). Thus the polarization plane of the magnetization is now rotated by the geometrical phase and the orientation of the ± 1 superposition by twice the geometrical phase when the crystal is rotated adiabatically. This is analogous to the rotation of the optical polarization plane in the experiments of Chiao and co-workers [7] and will give rise to effective Zeeman splittings in the N.M.R. experiments or powder patterns in the zero field spectra [15]. Similar considerations hold for a spin-1 with an axially symmetric electric quadrupole coupling or for spin-3/2 [8]. If the coupling for spin-1 is not axially symmetric $\eta \neq 0$, and $\delta \ll \eta\omega_d$, then we find $\bar{\mathscr{H}}_\delta = 0$ and the geometrical phase is $2\pi m$. This corresponds to the situation in which the effective rotation in parameter space occurs in a plane containing the degeneracy.

If the dipolar coupled spin pair described by equation (34) is in the presence of a large magnetic field it is known that the dynamical phase due to the secular dipolar coupling $\mathscr{H}'_d$ can be eliminated by rotating the sample adiabatically in circuits around an axis inclined at the magic angle ($\theta_m = \cos^{-1}(1/\sqrt{3})$) relative to the magnetic field [18]. Dynamical phases due to quadrupolar couplings and anisotropic chemical shifts are similarly removed. Since $\mathscr{H}'_d$ commutes with itself at different times, the geometrical phase is also zero so that only isotropic Zeeman couplings or chemical shifts remain. Instead of moving the sample relative to the field, similar effects can be achieved by rotating the field in a cone of angle $2\theta_m$ with the sample fixed [25]. In this latter case, however, there will be frequency shifts due to the geometrical phase factors given by equation (2). These will add to the isotropic shifts of the magic angle spectra.

Finally, by combining dipolar or quadrupolar couplings together with magnetic fields, for example spin $I = 1$ with

$$\mathscr{H}_d = \tfrac{1}{3}\hbar\omega_d[(3I_{z'}^2 - I^2) + 3I_{z'}], \qquad (36)$$

where, again, the z' axis is moved in a circuit, the non-abelian case of Wilczek and Zee [3] can be investigated [26]. In fact, such a possibility also exists for the $\pm\frac{1}{2}$ manifold of the spin $-\frac{3}{2}$ N.Q.R. [8], but, for the case of conical circuits, could be treated by Tycko as adiabatic Berry phases on diagonal states.

4. Conclusions

When two systems are coupled together, the eigenstates of the total hamiltonian do not in general exhibit statistical independence in the separate systems. That is, the probability amplitudes are not the product of probability amplitudes for each system. Often the states may be exactly, or approximately, separated into states which are functions of variables involving both systems. It is often the case that one or more of the degrees of freedom describing one of the systems goes through a sequence of values, for example, an angle. Although the overall wave function must be single valued when the hamiltonian repeats itself, there is no reason for the individual amplitudes which make up the products above to be single valued. This partial multivaluedness was pointed out and analysed in the early treatments of the coupling of internal rotations to overall rotations [9] showing that fractional quantum numbers were a natural consequence of multivaluedness.

An extreme example of strong coupling is the Born–Oppenheimer approximation. In the electronic wave functions, the nuclear coordinates are parameters. An overall wave function is the product of the electronic wave function times the nuclear wave function. Again there is no *a priori* reason for the electronic wave function and the nuclear wave function to be separately single valued in the nuclear coordinates. Herzberg and Longuet-Higgins [10] pointed out that near a conical intersection of a triatomic molecule the electronic wave function would be multivalued. Thus fractional quantum numbers could be expected in certain vibration states of triatomic species such as $(Na)_3$, and indeed appear to have been observed [27]. In the later 1970s and early 1980s, Mead and Truhlar [11] pointed out that the partial multivaluedness could be removed by multiplying both nuclear and electronic amplitudes by a phase which countered the multivalued real amplitudes. Thus, in the nuclear Schrödinger equation one restored single-valuedness at the cost of introducing a 'vector potential'. Mead and Truhlar dubbed this construct the molecular 'Bohm–Aharonov' [28] effect and proceeded to relate the form of the vector potential to certain circulation integrals over the nuclear potential energies. A related approach was discussed recently by Wilczek and co-workers [3].

The papers of Berry and Simon [2] are in essence a complete analysis of the time dependent adiabatic theorem. When the adiabatic eigenstates are complex and single valued, the diagonal phase factors, usually discarded [1], must be retained. Not surprisingly, the amplitudes may be multivalued. Also not surprisingly, the phase is identical to that derived by Mead and Truhlar. In fact, the additional phase required to restore the single valuedness of the eigenstates is the geometrical phase of Berry. Thus there may be observable effects in the dynamics. In particular, Berry suggested a series of experiments including one involving the polarization of light and the one investigated in the present paper involving adiabatic rotation of the spatial degrees of freedom of a magnetic system.

Although Berry used quantum mechanical arguments to obtain his phase, it is clear that the dynamical manifestation of this phase is classical in nature since the magnetization or the polarization of any fictitious spin satisfies the Bloch equations

[17], and, in fact, Cina [4] obtained the Berry phase from classical arguments. Thus for spins and other systems which undergo rotation, Berry's effect is entirely analogous to evolution under Coriolis forces in an accelerating non-planar reference frame [20]. Since the polarization of light may be described in terms of the Stokes parameters, the experiments of Chiao and co-workers [7] is another manifestation of this idea.

In this paper we have related the effects for spin-1/2 to well-known behaviour in N.M.R. involving continuous irradiation of spins near resonance. The experiments were carried out for conical circuits over a range of parameters which include both adiabatic and non-adiabatic effects. In addition we have invoked the known solutions to the Bloch equations for the evolution of magnetization with arbitrary values of the parameters, and hence were able to compare the exact results both with the experiment and with the general adiabatic theory. Finally, the relationship between geometrical phase in general adiabatic circuits and the average hamiltonian in an interaction picture was outlined. Such an approach to adiabatic N.M.R. experiments is useful because it lends to the processes a deeply geometrical view [2] in the spirit of the topological arguments of Berry and Simon.

We are grateful to Dr. J. Cina, Dr. J. Anandan and Dr. R. Tycko for communicating their work prior to publication and to Professors W. H. Miller, C. A. Mead, Y. Aharonov, C. Bouchiat and E. D. Commins for some very helpful discussions. This work was supported by the Director, Office of Energy Research, Office of Basic Energy Sciences, Materials Sciences Division of the U.S. Department of Energy under Contract No. DE-AC03-76SF00098.

References

[1] MESSIAH, A., 1962, *Quantum Mechanics* (North Holland). SCHIFF, L. I., 1968, *Quantum Mechanics* (McGraw-Hill).

[2] BERRY, M. V., 1984, *Proc. R. Soc. Lond.* A, **392**, 45. BERRY, M. V., 1985, *J. Phys.* A, **18**, 15. SIMON, B., 1983, *Phys. Rev. Lett.*, **51**, 2167.

[3] MOODY, J., SHAPERE, A., and WILCZEK, F., 1986, *Phys. Rev. Lett.*, 56, 893. WILCZEK, F., and ZEE, A., 1984, *Phys. Rev. Lett.*, **52**, 2111.

[4] CINA, J. A., 1986, *Chem. Phys. Lett.*, **132**, 393.

[5] PINES, A., 1987, *Lectures on Pulsed NMR* (Proceedings of the 100th Fermi School on Physics), edited by B. Maraviglia (North Holland), LBL Preprint No. 22316.

[6] STOLL, M. E., VEGA, A. G., and VAUGHAN, R. W., 1977, *Phys. Rev.* A, **16**, 1521. SUTER, D., PINES, A., and MEHRING, M., 1986, *Phys. Rev. Lett.*, **57**, 242.

[7] CHIAO, R. Y., and WU, Y.-S., 1986, *Phys. Rev. Lett.*, **57**, 933. TOMITA, A., and CHIAO, R. Y., 1986, *Phys. Rev. Lett.*, **57**, 937.

[8] TYCKO, R., 1987, *Phys. Rev. Lett.* (in the press).

[9] NIELSON, H. H., 1932, *Phys. Rev.*, **40**, 445. MILLER, W. H., and PINES, A. (to be published).

[10] HERZBERG, G., and LONGUET-HIGGINS, H. C., 1963, *Discuss. Faraday Soc.*, **35**, 77.

[11] MEAD, C. A., and TRUHLAR, D. G., 1979, *J. chem. Phys.*, **70**, 2284 and references therein.

[12] (a) HAEBERLEN, U., and WAUGH, J. S., 1968, *Phys. Rev.*, **175**, 453. (b) MEHRING, M., 1983, *High Resolution NMR in Solids* (Springer).

[13] CHO, H. M., TYCKO, R., PINES, A., and GUCKENHEIMER, J., 1986, *Phys. Rev. Lett.*, **56**, 1905.

[14] DAS, T. P., and HAHN, E. L. 1958, *Solid State Physics*, Suppl. I (Academic).

[15] THAYER, A. M., and PINES, A., 1987, *Accts chem. Res.*, **20**, 47. ZAX, D. B., BIELECKI, A., ZILM, K. W., PINES, A., and WEITEKAMP, D. P., 1985, *J. chem. Phys.*, **83**, 4877.

[16] GOLDMAN, M., 1970, *Spin Temperature and Nuclear Magnetic Resonance in Solids* (Oxford).
[17] (a) BLOCH, F., 1946, *Phys. Rev.*, **70**, 460. (b) HAHN, E. L., 1950, *Phys. Rev.*, **80**, 580. (c) RABI, I. I., RAMSAY, N. F., and SCHWINGER, J., 1954, *Rev. mod. Phys.*, **26**, 167.
[18] (a) LEE, M., and GOLDBERG, W. I., 1965, *Phys. Rev.* A, **140**, 1261. (b) ANDREW, E. R., BRADBURY, A., and EADES, R. G., 1958, *Nature, Lond.*, **182**, 1659. (c) LOWE, I. J., 1959, *Phys. Rev. Lett.*, **2**, 285. (d) MARICQ, M. M., and WAUGH, J. S., 1979, *J. chem. Phys.*, **70**, 3300.
[19] (a) FEYNMAN, R. P., VERNON, F. L., and HELLWARTH, R. W., 1957, *J. appl. Phys.*, **28**, 49. (b) ABRAGAM, A., 1961, *The Principles of Nuclear Magnetism* (Oxford University Press).
[20] GOLDSTEIN, H., 1980, *Classical Mechanics* (Addison-Wesley). HANNAY, J. H., 1985, *J. Phys. A*, **18**, 221.
[21] DYSON, F. J., 1949, *Phys. Rev.*, **75**, 486.
[22] WILCOX, R. M., 1967, *J. Math. Phys.*, **8**, 962. JEENER, J., and HENIN, F., 1986, *Phys. Rev.* A, **34**, 4897.
[23] MAGNUS, W., 1954, *Commun. pure appl. Math.*, **7**, 649.
[24] AHARONOV, Y., and ANANDAN, J., 1987, *Phys. Rev. Lett.* (in the press). BOUCHIAT, C. (private communication).
[25] LEE, C., SUTER, D., and PINES, A., 1987, *J. magn. Reson.*, **74** (in the press).
[26] SUTER, D., and PINES, A., 1987 (to be published). VEGA, S., and PINES, A., 1977, *J. chem. Phys.*, **66**, 5624.
[27] DELACRÉTAZ, G., GRANT, E. R., WHETTEN, R. L., WÖSTE, L., and ZWANZIGER, J. W., 1986, *Phys. Rev. Lett.*, **56**, 2598.
[28] AHARONOV, Y., and BOHM, D., 1959, *Phys. Rev.*, **115**, 485.

PHYSICAL REVIEW

LETTERS

VOLUME 58 | 1 JUNE 1987 | NUMBER 22

Adiabatic Rotational Splittings and Berry's Phase in Nuclear Quadrupole Resonance

Robert Tycko

AT&T Bell Laboratories, Murray Hill, New Jersey 07974

(Received 17 March 1987)

Sample rotation is shown to induce frequency splittings in nuclear-quadrupole-resonance spectra. The splittings are interpreted both as a manifestation of Berry's phase, associated with an adiabatically changing Hamiltonian, and as a result of a fictitious magnetic field, associated with a rotating-frame transformation. Real and fictitious fields are contrasted. Related effects are predicted in other magnetic resonance experiments that involve sample rotation.

PACS numbers: 03.65.−w, 33.30.+a

The evolution of a quantum-mechanical system under an adiabatically changing Hamiltonian is an important topic that is receiving renewed attention.[1-12] In particular, when the Hamiltonian follows a closed path in parameter space in the time interval $[0,T]$ [i.e., $\mathcal{H}(T)=\mathcal{H}(0)$] and when the system is initially in an eigenstate of $\mathcal{H}$ [i.e., $\mathcal{H}(0)|\psi(0)\rangle=E_n(0)|\psi(0)\rangle$], it has been recognized[1-3,8] that the system ends up in the state $|\psi(T)\rangle=e^{-i\beta_n}e^{-i\phi_n}|\psi(0)\rangle$. β_n, called the quantum adiabatic phase or Berry's phase, depends only on the path or certain geometric features of the path; ϕ_n, called the dynamical phase and equal to $\hbar^{-1}\int_0^T dt\,E_n(t)$, depends both on the path and on the rate at which the path is followed. Berry's phase has been invoked in discussions and experiments involving molecular spectroscopy,[3-7] atom-molecule scattering,[8] the quantum Hall effect,[2,9] and light propagation in optical fibers.[10,11] Spin problems have been used as solvable examples in theoretical treatments of Berry's phase.[1,2] Suggestions for experimental demonstrations of Berry's phase in magnetic resonance, involving *rotating external fields*, have been made.[1,7] In this Letter, an effect of Berry's phase on the magnetic resonance spectrum of a *rotating sample* is described and demonstrated with use of pure nuclear quadrupole resonance (NQR). This effect, which may be called the adiabatic rotational splitting, has important consequences in magnetic resonance in addition to serving as an interesting manifestation of Berry's phase.

Consider the experiment depicted in Fig. 1. A single-crystal sample containing nuclear spins S with an axially symmetric quadrupole coupling with coupling constant ω_Q and principal-axis direction z' is rotated about the z axis at angular velocity ω_R. The NQR spectrum is observed by applying radio-frequency (rf) pulses and detecting free-induction decay signals with a solenoid coil wound along z. Provided that $\omega_R \ll \omega_Q$, the nuclear-spin Hamiltonian is changing adiabatically. However, only the instantaneous eigenstates, and not the instantaneous energies, change with time since the NQR frequencies (in the absence of external static fields) are independent of orientation in a static sample. In other words, the dynamical phase evolution is unaffected by the rotational motion. Any observed rotational effects must arise from Berry's phase. Suppose that a superposition of two eigenstates $|\psi_1\rangle$ and $|\psi_2\rangle$ is created by a rf pulse at $t=0$. At time $mT=2\pi m/\omega_R$ later, the

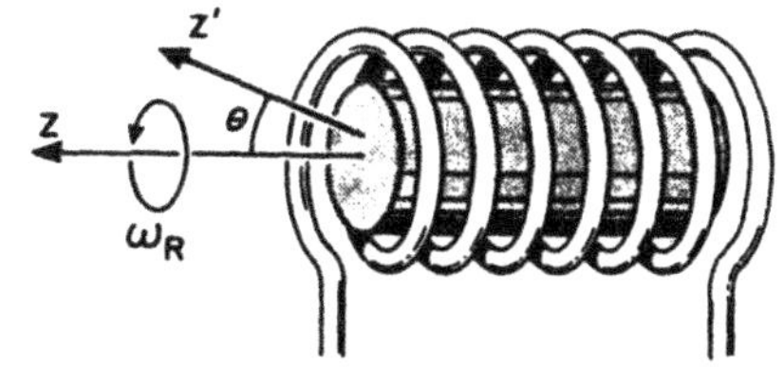

FIG. 1. Experimental arrangement for observation of adiabatic rotational splittings in NQR. z is the rotation axis and z' is the principal axis of the quadrupole coupling.

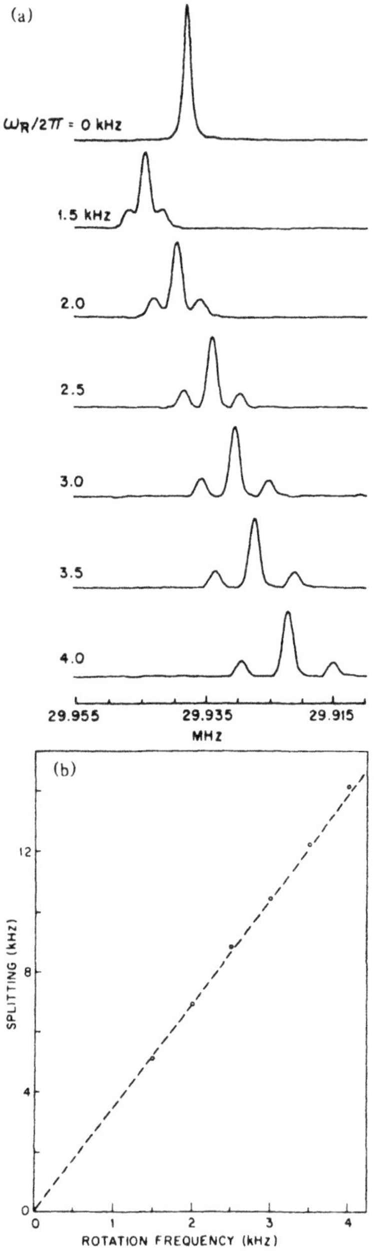

FIG. 2. (a) ^{35}Cl NQR spectra of a $NaClO_3$ crystal obtained at various sample rotation frequencies showing adiabatic rotational splittings. Ambient temperature is $23 \pm 1\,^\circ C$. Overall shifts of the resonances are due to crystal temperature variations. (b) Splitting between the countermost lines vs rotation frequencies. The dashed line has a slope of $2\sqrt{3}$.

two eigenstates acquire a relative phase difference $(\beta_1 - \beta_2)m$. The NQR signal phase is shifted by the same amount. Since a continuously increasing phase shift is equivalent to a frequency shift, Berry's phase is observable as a frequency shift of $(\beta_1 - \beta_2)/2\pi T$. In the experiments described here, $S = \frac{3}{2}$ and there is a degenerate transition that exhibits a phase shift of $-(\beta_1 - \beta_2)m$. Therefore, an adiabatic rotational splitting of the NQR line that is proportional to ω_R is expected.[13]

Figure 2(a) shows ^{35}Cl NQR spectra of $NaClO_3$ observed at various rotation frequencies. The 119-mg $NaClO_3$ crystal was oriented with a cleavage plane perpendicular to the rotation axis. At this orientation, the angle θ between z and z' satisfies $\cos^2\theta = \frac{1}{3}$ for all four Cl nuclei in the unit cell. A commercial nuclear-magnetic-resonance (NMR) probe designed for magic-angle spinning NMR experiments was used. The single NQR resonance observed when $\omega_R = 0$ splits into three lines with relative areas 0.25, 1.00, and 0.22 when $\omega_R \neq 0$. Figure 2(b) shows that the splitting between the outer lines is well described by $\sqrt{3}\omega_R/\pi$. In addition to the splittings, overall shifts of the resonances are apparent in Fig. 2(a). These are artifacts due to changes in the crystal temperature that accompany sample spinning. At $23\,^\circ C$, $\omega_Q/2\pi$ changes by 2.1 kHz/deg.[14]

The results in Fig. 2 can be predicted quantitatively from established formulas for Berry's phase.[1,2] The spin Hamiltonian is

$$\mathcal{H}(t) = \omega_Q S_{z'}^2(t), \tag{1}$$

$$S_{z'}(t) = S_z\cos\theta + S_x\sin\theta\cos\omega_R t + S_y\sin\theta\sin\omega_R t. \tag{2}$$

For reasons given below, it is convenient to use a basis of time-dependent eigenstates, written in terms of eigenstates of $S_{z'}(t)$ as follows:

$$|a\rangle = |\tfrac{3}{2}\rangle, \tag{3a}$$

$$|b\rangle = \cos(\tfrac{1}{2}\xi)\,|\tfrac{1}{2}\rangle - \sin(\tfrac{1}{2}\xi)\,|-\tfrac{1}{2}\rangle, \tag{3b}$$

$$|c\rangle = \sin(\tfrac{1}{2}\xi)\,|\tfrac{1}{2}\rangle + \cos(\tfrac{1}{2}\xi)\,|-\tfrac{1}{2}\rangle, \tag{3c}$$

$$|d\rangle = |-\tfrac{3}{2}\rangle, \tag{3d}$$

$$\tan\xi = 2\tan\theta.$$

According to Simon[2] and Berry,[1] the adiabatic phase accumulated by a spin state $|\psi\rangle$ in one rotation is

$$2\pi\int_0^\theta d\theta' \sin\theta' \langle\psi|S_{z'}|\psi\rangle.$$

Therefore,

$$\beta_a = 3\pi(\cos\theta - 1), \tag{4a}$$

$$\beta_b = -\pi[(4-3\cos^2\theta)^{1/2} - 1], \tag{4b}$$

$$\beta_c = \pi[(4-3\cos^2\theta)^{1/2} - 1], \tag{4c}$$

$$\beta_d = -3\pi(\cos\theta - 1). \tag{4d}$$

Possible phase differences, modulo 2π, are $\pm 2\sqrt{3}\pi$ and 0 at the crystal orientation used, corresponding to frequency shifts of $\pm\sqrt{3}\omega_R/2\pi$ and 0 as observed.

The observed spectra, including the relative intensities of the lines, can also be predicted from the quantum-mechanical expression for the free induction signal $F(t)$:

$$F(t) = \mathrm{Tr}\{S_z U(t)\rho(0)U(t)^{-1}\}, \tag{5}$$

$$U(t) = \mathbf{T}\exp\left[-i\int_0^t dt'\mathcal{H}(t')\right], \tag{6}$$

where $\rho(0)$ is the spin density operator after the rf pulse and $\mathbf{T}$ is the time-ordering operator. Since $\mathcal{H}(t)$ is related to $\mathcal{H}(0)$ by a rotation around z, the evolution operator $U(t)$ can be rewritten as

$$U(t) = e^{-i\omega_R S_z t} e^{-i\mathcal{H}t}, \tag{7}$$

$$\mathcal{H} = \mathcal{H}(0) - \omega_R S_z; \tag{8}$$

$\mathcal{H}$ is the Hamiltonian in a frame that rotates with the crystal. The signal becomes

$$F(t) = \mathrm{Tr}\{S_z e^{-i\mathcal{H}t}\rho(0)e^{i\mathcal{H}t}\}. \tag{9}$$

Equation (9) shows that the effect of Berry's phase on the signals can also be viewed as the effect of the "fictitious field" term $-\omega_R S_z$ that appears in the rotating frame. The adiabatic limit $\omega_R \ll \omega_Q$ corresponds to the first-order perturbation-theory limit. In that limit, and with an initial density operator $[S_{z'}(0)\cos\alpha + S_{y'}(0)\sin\alpha]^2$,

$$F(t) = -(\tfrac{2}{3})^{1/2}\sin 2\alpha(2+\cos\sqrt{3}\omega_R t)\sin 2\omega_Q t, \tag{10}$$

showing that lines are expected at $(2\omega_Q - \sqrt{3}\omega_R)/2\pi$, ω_Q/π, and $(2\omega_Q - \sqrt{3}\omega_R)/2\pi$ with relative areas 0.25, 1.00, and 0.25.

In the general case of a system with a time-dependent

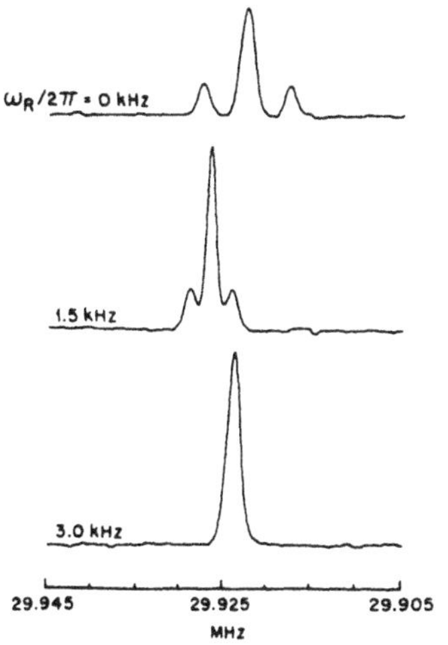

FIG. 3. As in Fig. 2(a), but with a 7.2-G static field applied along the rotation axis. The field is produced by Helmholtz coils coaxial with z and with the current direction opposite to the sample rotation direction.

Hamiltonian, the evolution operator can be written

$$U(t) = W(t)\mathbf{T}\exp\left\{-i\int_0^t dt'[A(t')+B(t')]\right\}, \tag{11}$$

$$A(t) = \sum_n \omega_n(t)\,|n(0)\rangle\langle n(0)|, \tag{12}$$

$$B(t) = i[dW^{-1}(t)/dt]W(t), \tag{13}$$

where $\{|n(t)\rangle\}$ and $\{\omega_n(t)\}$ are the instantaneous eigenstates and energies at time t and $W(t)$ is a unitary transformation between $\{|n(0)\rangle\}$ and $\{|n(t)\rangle\}$. Equation (11) is the generalization of Eq. (7), expressing $U(t)$ in terms of the Hamiltonian $A(t)+B(t)$ that appears in a frame that "follows" the time-dependent eigenstates of the original Hamiltonian. In the adiabatic limit, the generalized fictitious field $B(t)$ is a perturbation on $A(t)$ and β_n is given by

$$e^{-i\beta_n} = \langle n(0)|W(T)|n(0)\rangle \exp\left[-i\int_0^T dt\langle n(0)|B(t)|n(0)\rangle\right]. \tag{14}$$

An objection to the claim that Fig. 2 is a manifestation of Berry's phase may be raised. Since the instantaneous eigenstates of $\mathcal{H}(t)$ are two doubly degenerate pairs, it may be unclear whether a true adiabatic change is possible. However, provided that the instantaneous eigenstates are chosen so as to diagonalize S_z within each manifold of degenerate states, as has been done in Eqs. (3), transitions between degenerate eigenstates do not occur and Berry's phase remains meaningful. Wilczek and Zee[6] have treated degenerate systems in greater detail. Alternatively, a real static magnetic field H can be applied along z to lift the degeneracy of the states in Eqs. (3). As shown in Fig. 3, the linear dependence of the splitting of the NQR line on ω_R is unchanged. The real and fictitious fields can be made to cancel one another.

The equivalence in NQR between rotation and the application of a field is not complete. For example, if the rf fields are applied and detected perpendicular to the rotation axis or static field direction, different signals will be observed in the two cases unless the rf coils rotate with the sample. In addition, the spin interaction with a real field depends on the gyromagnetic ratio γ, while the effect of the fictitious field induced by rotation is independent of γ. The absolute sign of γ can then be measured by determining which sense of rotation relative to the real field direction tends to cancel the real field. Figure 3 shows that γ is positive for ^{35}Cl, as expected.

Sample spinning is commonly used in solid-state NMR to obtain narrow lines.[15] It is generally assumed that the NMR spectrum of a rapidly rotating sample is simply the orientationally averaged spectrum. The results here show that this assumption is not always correct. In particular, whenever the spin eigenstates depend on the sample orientation, adiabatic rotational splittings or shifts may appear. Such effects may be expected in overtone NMR spectra[16] and NMR spectra of dipole-coupled nuclei when ω_R is small compared to the dipole couplings. It is also anticipated that the adiabatic rotational splitting may serve as a coherence dephasing mechanism, for example in NQR of randomly tumbling particles when the rotational correlation time is large compared to ω_Q^{-1}.[17] Finally, Pines and co-workers have provided independent experimental evidence for Berry's phase effects in magnetic resonance, using rotating rf fields in high field NMR.[18,19]

I thank D. C. Douglass and T. Sleator for many informative discussions and experimental assistance, L. W. Jelinski for the generator loan of a spectrometer, and J. P. Remeika for providing the $NaClO_3$ crystal.

[1]M. V. Berry, Proc. Roy. Soc. London, Ser. A **392**, 45 (1984).

[2]B. Simon, Phys. Rev. Lett. **51**, 2167 (1983).

[3]G. Herzberg and H. C. Longuet-Higgins, Discuss. Faraday Soc. **35**, 77 (1963).

[4]C. A. Mead, Chem. Phys. **49**, 23 (1980).

[5]G. Delacrétaz, E. R. Grant, R. L. Whetten, L. Wöste, and J. W. Zwanziger, Phys. Rev. Lett. **56**, 2598 (1986).

[6]F. Wilczek and A. Zee, Phys. Rev. Lett. **52**, 2111 (1984).

[7]J. Moody, A. Shapere, and F. Wilczek, Phys. Rev. Lett. **56**, 893 (1986).

[8]C. A. Mead and D. G. Truhlar, J. Chem. Phys. **70**, 2284 (1979).

[9]D. Arovas, J. R. Schrieffer, and F. Wilczek, Phys. Rev. Lett. **53**, 722 (1984).

[10]R. Y. Chiao and Y.-S. Wu, Phys. Rev. Lett. **57**, 933 (1986).

[11]A. Tomita and R. Y. Chiao, Phys. Rev. Lett. **57**, 937 (1986).

[12]H. Kuratsuji and S. Iida, Phys. Rev. Lett. **56**, 1003 (1986).

[13]Actually, this argument only determines the frequency shifts to within multiples of $\omega_R/2\pi$. The multiples that are observed depend on the orientation of the rf coils with respect to the spinning axis.

[14]H. S. Gutowsky and G. A. Williams, Phys. Rev. **105**, 464 (1957).

[15]M. Mehring, *Principles of High Resolution NMR in Solids* (Springer, New York, 1983).

[16]R. Tycko and S. J. Opella, J. Chem. Phys., to be published.

[17]S. Alexander and A. Tzalmona, Phys. Rev. **138**, A845 (1965). This paper treats longitudinal, not transverse, relaxation caused by slow rotation in terms of an effective field.

[18]D. Suter, G. Chingas, R. A. Harns, and A. Pines, to be published.

[19]D. B. Zax, A. Bielecki, K. W. Zilm, A. Pines, and D. P. Weitekamp, J. Chem. Phys. **83**, 4877 (1985). This paper predicts effects in zero-field NMR due to rapid, nonadiabatic sample rotation.

VOLUME 60, NUMBER 13 PHYSICAL REVIEW LETTERS 28 MARCH 1988

Study of the Aharonov-Anandan Quantum Phase by NMR Interferometry

D. Suter, K. T. Mueller, and A. Pines[a]

University of California and Lawrence Berkeley Laboratory, Berkeley, California 94720

(Received 19 November 1987)

Aharonov and Anandan have recently reformulated and generalized Berry's phase by showing that a quantum system which evolves through a circuit C in projective Hilbert space acquires a geometrical phase $\beta(C)$ related to the topology of the space and the geometry of the circuit. We present NMR interferometry experiments in a three-level system which demonstrate the Aharonov-Anandan phase and its topological invariance for different circuits.

PACS numbers: 03.65.Bz, 42.10.Jd, 76.60.−k

It was shown by Berry[1] that a nondegenerate quantum state $|\psi(t)\rangle$ of a Hamiltonian $\mathcal{H}(t)$ which varies adiabatically through a circuit C in parameter space acquires, in addition to the "normal" dynamical phase

$$\gamma_d = -\frac{1}{\hbar}\int \langle\psi(t)|\mathcal{H}(t)|\psi(t)\rangle dt \qquad (1)$$

(the generalization of ωt), a phase which is related to the geometry of the circuit. Simon[2] explained that this geometrical phase could be viewed as a consequence of parallel transport in a curved space appropriate to the quantum system. Much experimental and theoretical work on "Berry's phase"[3] and its connection to the early work of Aharonov and Bohm, and Mead and Truhlar,[4] has since appeared. Recently, three important generalizations of this phase have appeared. In the first, Wilczek and Zee[5] removed the constraint of nondegenerate states and related the evolution and phases of a degenerate manifold to a non-Abelian gauge. In the second, Berry, as well as Jackiw and co-workers, removed the constraint of adiabaticity in the Hamiltonian circuit[6] and developed asymptotic expansions for the evolving (noncyclic) states and phases.

The third generalization, a fundamental one, which forms the subject of NMR experiments in this Letter and optical experiments in the accompanying Letters by Chiao and co-workers and Bhandari and Samuel,[7] is due to Aharonov and Anandan.[8] They cast the problem in terms of circuits of the quantum system itself, rather than circuits of the Hamiltonian in parameter space. It is clear from recent work that this is, in some sense, a continuous version of the phase discovered by Pancharatnam more than 30 years ago.[9] A simple formulation of the Aharonov-Anandan (AA) phase is as follows: If the density operator $|\psi\rangle\langle\psi|$ for a pure state (generalized to mixed states by superposition) undergoes a cyclic evolution through a circuit C in projective Hilbert (density operator) space $|\psi\rangle\langle\psi| \rightarrow_C |\psi\rangle\langle\psi|$, then the quantum state $|\psi\rangle$ acquires a geometrical phase $\beta(C)$ related to the (an)[1] holonomy[2] associated with parallel transport around the circuit. This phase appears in addition to the dynamical phase γ_d given by (1), where $\mathcal{H}(t)$ may be noncyclic and nonadiabatic.[8] Thus, we can write

$$|\psi\rangle \underset{C}{\rightarrow} \exp\{i[\gamma_d+\beta(C)]\}\,|\psi\rangle. \qquad (2)$$

The importance of the AA formulation is that it applies whether or not the Hamiltonian $\mathcal{H}(t)$ is cyclic or adiabatic—***the geometrical phase depends only on the cyclic evolution of the system itself.*** This establishes a simple connection of the geometrical phase to the Aharonov-Bohm effect[4] which does not invoke adiabaticity of the circuit. The Berry phase thereby emerges as a manifestation of the AA geometry in the case of adiabatic evolution. Bouchiat and Gibbons[10] have presented a thorough theoretical analysis of the AA phase for a three-level spin-1 system.

As a demonstration of the AA phase, consider the situation depicted in Fig. 1, a version of NMR interferometry related to neutron (say) interferometry.[11] The two-level system (TLS) comprising states 2 and 3 can be treated as a fictitious spin-$\frac{1}{2}$ system.[12] Time-dependent magnetic fields are applied to take the TLS through a circuit. The resultant phase factor associated with state 2 is detected by means of its effect on transition 1-2. Suppose the system begins in thermal equilibrium so that the initial density operator for the three-level system is diagonal in the eigenbase of the unperturbed Hamiltonian. The $\pi/2$ pulse applied to the 1-2 transition

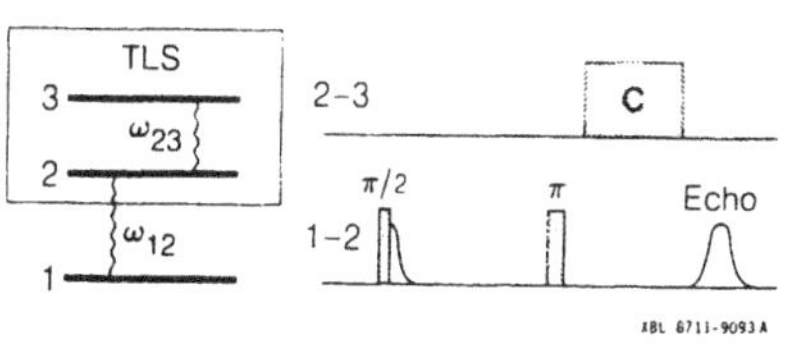

FIG. 1. Schematic representation of the NMR interferometry experiment to demonstrate the Anaronov-Anandan (AA) phase. The 2-3 two-level system (TLS) undergoes a circuit C in projective Hilbert space and the phase is determined by means of its effect on the echo produced by the coherent superposition of levels 1 and 2.

produces a coherent superposition of the two states whose phase serves as the reference for the subsequent measurement of the geometrical phase. The π pulse refocusses the 1-2 superposition as an echo[13] of transverse magnetization. This 1-2 echo corresponds to the element ρ_{12} of the density matrix,

$$\rho_{\text{echo}}=\begin{pmatrix} \rho_{11} & \rho_{12} & 0 \\ \rho_{21} & & \\ & \left(\rho^{2\text{-}3}\right) & \\ 0 & & \end{pmatrix}, \tag{3}$$

and can be observed by means of a phase-sensitive detector at frequency ω_{12}. The submatrix $\rho^{2\text{-}3}$ is proportional to $1+\mathbf{p}^{2\text{-}3}\cdot\boldsymbol{\sigma}$, the projected density matrix for the 2-3 TLS, where $\mathbf{p}^{2\text{-}3}$ is the TLS polarization vector (initially along z) and $\boldsymbol{\sigma}$ is the vector $(\sigma_x,\sigma_y,\sigma_z)$ of Pauli matrices. Imagine now that $\mathbf{p}^{2\text{-}3}$ is made to undergo a circuit, that is $\mathbf{p}^{2\text{-}3}\rightarrow c\,\mathbf{p}^{2\text{-}3}$ (and therefore $\rho^{2\text{-}3}\rightarrow c\,\rho^{2\text{-}3}$), by means of a perturbation applied selectively to the 2-3 transition. The phase factor acquired by the quantum states does not affect $\rho^{2\text{-}3}$ and is therefore unobservable in the TLS.[10] However, the phase factor acquired by state 2 does manifest itself in the phase of the 1-2 echo in the element ρ_{12} ($\rho_{12}\rightarrow c\exp\{i[\gamma_d+\beta(C)]\}\rho_{12}$) and is therefore detected by the phase-sensitive detector at frequency ω_{12}.

For the fictitious spin-$\frac{1}{2}$ TLS, projective Hilbert space corresponds to a two-sphere and the geometrical phase becomes

$$\beta(C)=m\,\Omega(C)=\pm\tfrac{1}{2}\,\Omega(C), \tag{4}$$

where m is the magnetic quantum number and $\Omega(C)$ is the solid angle subtended by the circuit C at the origin. We add that the π pulse and the echo in the 1-2 transition are not necessary in principle since the phases could be detected in the coherent signal following the $\pi/2$ pulse. In practice, however, the echo is a convenient experimental means of compensating for any extraneous dephasing due to nonuniform magnetic fields and other inhomogeneous broadening mechanisms. We also mention that $\mathbf{p}^{2\text{-}3}$ need not begin and end along z; it is necessary only that it go through a circuit.

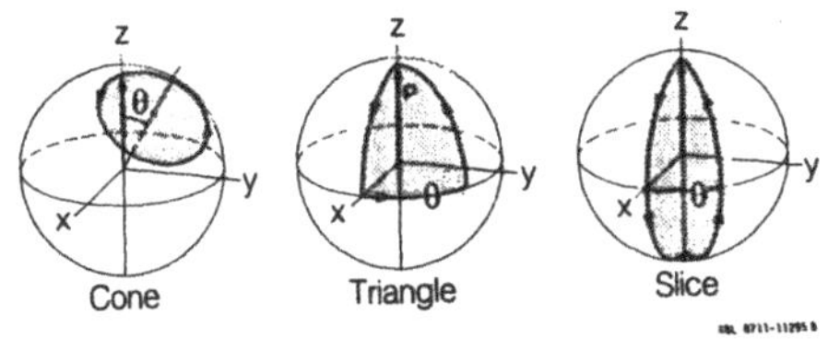

FIG. 2. Three types of circuits experienced by the polarization p ($\mathbf{p}^{2\text{-}3}$ in the text) for the TLS of Fig. 1 in the 2-3 frame of reference. The solid angles are $\Omega(C)=2\pi(1-\cos\theta)$ for the cone, $\Omega(C)=\theta$ for the triangle, and $\Omega(C)=2\theta$ for the slice.

Experiments were performed on the spin-1 manifold of two proton spins-$\frac{1}{2}$ coupled by magnetic dipolar interactions in the molecule CH_2Cl_2 oriented in a nematic liquid-crystal solvent.[14] The static magnetic field was 8.4 T and the rotating magnetic fields at frequencies $\omega_{12}=362.023524$ MHz and $\omega_{23}=362.019675$ MHz had amplitudes of 1.7 and 1.3 μT, respectively. The system was allowed to reach thermal equilibrium in the magnet and three types of circuits with a range of solid angles $\Omega(C)$ were implemented for the TLS as shown in Fig. 2, by the application of time-dependent phase-shifted magnetic fields near ω_{23}.[15] The cone circuits were induced by a magnetic field which was, in the rotating frame, tilted at an angle θ with respect to the z axis. For the spherical triangles and the slices, rectangular magnetic field pulses were applied perpendicular to the polarization vector. For the spherical triangles the pulses were a $\pi/2$ pulse along (0,1,0), a θ pulse along (0,0,1), and finally a $\pi/2$ pulse along $(\sin\theta,-\cos\theta,0)$. Similarly, the slices were generated with a π pulse along (0,1,0) followed by another π pulse along $(\sin\theta,-\cos\theta,0)$. The geometrical phase was determined by measurement of the phase relative to a reference phase (γ_d), determined by pure dynamical evolution. For the spherical triangles

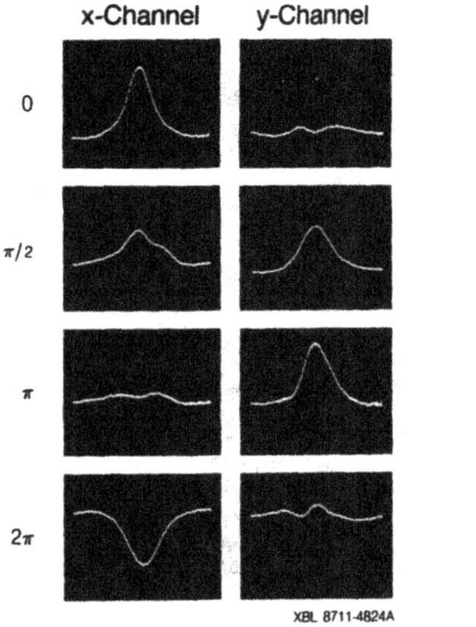

FIG. 3. Oscilloscope traces of the 1-2 echoes detected in the two orthogonal (x,y) channels of a phase-sensitive detector at frequency ω_{12}. In this case, the 2-3 TLS undergoes the slice circuits of Fig. 2, with solid angles $\Omega(C)$ equal to 0, $\pi/2$, π, and 2π.

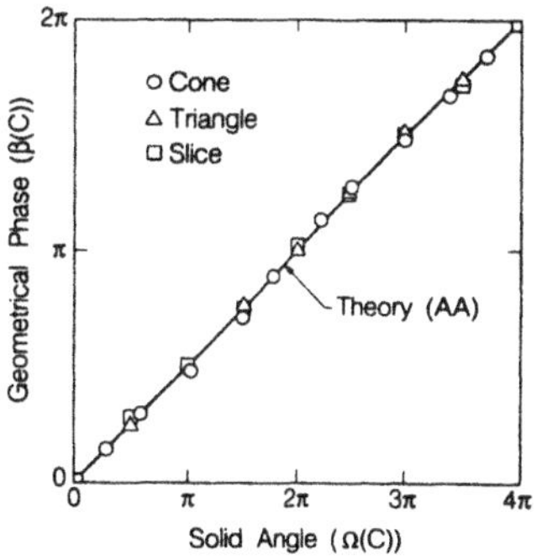

FIG. 4. Summary of experimentally determined geometrical phase $\beta(C)$ as a function of solid angle $\Omega(C)$ for the three types of circuits in Fig. 2. The solid line corresponds to the theoretical (AA) phase.

and the slice circuits the dynamical phase vanished, since in our experiment the applied field was always orthogonal to $\mathbf{p}^{2\text{-}3}$, generating parallel transport. For the cone, the dynamical phase was determined via a reference experiment that included only the component of the field parallel to $\mathbf{p}^{2\text{-}3}$.

Figure 3 shows some examples of the echoes observed, exhibiting the phase shifts induced by evolution around the slice circuits of Fig. 2, with various solid angles. The geometrical phase is given by $\beta(C) = \arctan(S_y/S_x)$ where S_x and S_y are the integrated amplitudes of the signals in the two detector channels. Figure 4 shows a plot of $\beta(C)$ versus solid angle for all three circuits of Fig. 2. It is clear that the geometrical quantum phase is proportional to the solid angle subtended by the circuit. Our results illustrate the invariance of the AA phase to details of the circuit geometry and to the (perhaps nonadiabatic and noncyclic) Hamiltonian responsible for generating the circuit.

Part of this work was done while one of us (A.P.) was on sabbatical leave at the Laboratoire de Physique Quantique, Ecole Supérieure de Physique et Chimie, Paris. Professor A. P. Le Grand and Professor P. G. de Gennes are gratefully acknowledged for their invitation and hospitality. We also thank Professor Y. Aharonov, Professor J. Anandan, Professor C. Bouchiat, Professor R. Y. Chiao, and Professor R. A. Harris for very helpful discussions. One of us (K.T.M.) is a National Science Foundation Graduate Fellow. This work is supported by the Director, Office of Energy Research, Office of Basic Energy Sciences, Materials Sciences Division of the U.S. Department of Energy under Contract No. DE-AC03-76SF00098.

[a]Address correspondence to this author, at the Department of Chemistry, University of California, Berkeley, CA 94720.

[1]M. V. Berry, Proc. Roy. Soc. London A **392**, 45 (1984), and Nature (London) **326**, 277 (1987).

[2]B. Simon, Phys. Rev. Lett. **51**, 2167 (1983).

[3]For example, R. Y. Chiao and Y.-S. Wu, Phys. Rev. Lett. **57**, 933 (1986); J. H. Hannay, J. Phys. A **18**, 221 (1985); J. Moody, A. Shapere, and F. Wilczek, Phys. Rev. Lett. **56**, 893 (1986); T. Bitter and D. Dubbers, Phys. Rev. Lett. **59**, 251 (1987); G. Delacrétaz, E. R. Grant, R. L. Whetten, L. Wöste, and J. W. Zwanziger, Phys. Rev. Lett. **56**, 2598 (1986); R. Jackiw, Comments At. Mol. Phys. (to be published); R. Tycko, Phys. Rev. Lett. **58**, 2281 (1987); C. Bouchiat, J. Phys. (Paris) **48**, 1627 (1987); D. Suter, G. Chingas, R. A. Harris, and A. Pines, Mol. Phys. **61**, 1327 (1987), and references therein.

[4]Y. Aharonov and D. Bohm, Phys. Rev. **115**, 485 (1959); C. A. Mead and D. G. Truhlar, J. Chem. Phys. **70**, 2284 (1979), and references therein.

[5]F. Wilczek and A. Zee, Phys. Rev. Lett. **52**, 2111 (1984); J. Segert, Ann. Phys. (N.Y.) (to be published); A. Zee, to be published; D. Suter and A. Pines, to be published.

[6]M. V. Berry, Proc. Roy. Soc. London A **414**, 31 (1987); experiments have been done by J. Anandan, R. Y. Chiao, G. Chingas, K. M. Ganga, R. A. Harris, H. Jiao, A. S. Landsberg, H. Nathel, A. Pines, and D. Suter, to be published; R. Jackiw, Massachusetts Institute of Technology Report No. CTP 1529, Mod. Phys. (to be published); P. de Sousa Gerbert, Massachusetts Institute of Technology Report No. CTP 1537 (to be published).

[7]R. Y. Chiao *et al.*, preceding Letter [Phys. Rev. Lett. **60**, 1214 (1988)]; presentations of M. V. Berry and R. Y. Chiao at the Theoretical Physics Workshop on Nonintegrable Phases in Dynamical Systems, Minneapolis, 30 September–2 October 1987 (unpublished); R. Bhandari and J. Samuel, second preceding Letter [Phys. Rev. Lett. **60**, 1211 (1988)].

[8]Y. Aharonov and J. Anandan, Phys. Rev. Lett. **58**, 1593 (1987).

[9]S. Pancharatnam, Proc. Indian Acad. Sci. **A44**, 247 (1956); Y. Anaronov and M. Vardi, Phys. Rev. D **21**, 2235 (1980); S. Ramaseshan and R. Nityananda, Curr. Sci. (India) **55**, 1225 (1986); M. V. Berry, J. Mod. Opt. **34**, 1401 (1987).

[10]C. Bouchiat and G. W. Gibbons, J. Phys. (Paris) (to be published).

[11]M. E. Stoll, A. J. Vega, and R. W. Vaughan, Phys. Rev. A **16**, 1521 (1977); H. Hatanaka, T. Terao, and T. Hashi, J. Phys. Soc. Jpn. **39**, 825 (1975); D. Suter, A. Pines, and M. Mehring, Phys. Rev. Lett. **57**, 242 (1986); H. Rauch, A. Zeilinger, G. Badurek, A. Wilfing, W. Bauspiess, and U. Bonse, Phys. Lett. **54A**, 425 (1975); S. A. Werner, R. Colella, A. W. Overhauser, and C. F. Eagan, Phys. Rev. Lett. **35**, 1053 (1975); U. Fano, Rev. Mod. Phys. **55**, 855 (1983).

[12]R. P. Feynman, F. L. Vernon, and R. W. Hellwarth, J. Appl. Phys. **28**, 49 (1957); M. Mehring, *Principles of High Resolution NMR in Solids* (Springer-Verlag, New York, 1983).

[13]E. L. Hahn, Phys. Rev. **80**, 580 (1950).

[14]A. Pines, in *Lectures on Pulsed NMR, International School of Physics "Enrico Fermi," Course 100*, edited by B. Maraviglia (North-Holland, Amsterdam, 1988); J. W. Emsley and J. C. Lindon, *NMR Spectroscopy Using Liquid Crystal Solvents* (Pergamon, Oxford, 1975).

[15]S. Vega and A. Pines, J. Chem. Phys. **66**, 5624 (1977).

Non-Abelian gauge structure in nuclear quadrupole resonance

A. Zee
Institute for Theoretical Physics, University of California, Santa Barbara, California 93106
(Received 26 October 1987)

We elucidate the non-Abelian gauge structure associated with nuclear quadrupole resonance. The Abelian part of this structure has been experimentally observed. The phases to be observed in various non-Abelian experiments are computed.

Berry's remarkable discovery[1,2] that gauge structures exist naturally in slowly varying quantum systems has generated a great deal of interest,[3-9] both theoretically and experimentally. In particular, it has been shown[4] that a non-Abelian gauge structure emerges if a set of quantum states remains degenerate as the Hamiltonian varies.

In an interesting recent experiment, Tycko[9] demonstrated the effect of Berry's phase on the magnetic resonance spectrum of a rotating magnetic sample by using a pure nuclear quadrupole resonance. The spin quadrupole Hamiltonian describing Tycko's experiment is effectively given by

$$\begin{aligned} H &= (\mathbf{S}\cdot\mathbf{B})^2 \\ &= (S_x \sin\theta\cos\varphi + S_y \sin\theta\sin\varphi + S_z\cos\theta)^2 B^2 \ . \end{aligned} \tag{1}$$

Besides being invariant under rotation and under time reversal ($\mathbf{S}\to-\mathbf{S}$ and $\mathbf{B}\to-\mathbf{B}$), the Hamiltonian is also invariant under the operation $\mathbf{S}\to-\mathbf{S}$. It is of course this additional symmetry (of the instantaneous Hamiltonian) that guarantees the pairwise degeneracy of states as we move about in parameter space (that is, as we vary **B** adiabatically).

In the actual experiment, a spin-$\frac{3}{2}$ Cl atom in a $NaClO_3$ crystal is used. Clearly, the four spin-$\frac{3}{2}$ states fall into two doubly degenerate sets and thus we can expect to see a non-Abelian gauge structure in accordance with the general analysis of Ref. 4. Unfortunately, the actual experimental setup was such that only Abelian phases were observed. In Tycko's experiment, the magnetic field results from an internal electric field gradient in the crystal. When the crystal is rotated, an effective magnetic field arises. In the subsequent theoretical analysis, we will just treat the equivalent problem of a spin in a rotating magnetic field and we will speak as if an external magnetic field were being rotated. In this paper, we elucidate the non-Abelian structure and comment on how this structure can be observed in experiments analogous to (but more complicated than) Tycko's.

We begin by reviewing the general framework. Let η_a be an N-fold degenerate set of orthonormal instantaneous eigenstates of the Hamiltonian $H|\eta_a\rangle = E|\eta_a\rangle$. (With no loss of generality, by replacing H by $H-E$, we can set $E=0$.) The assumption that H varies adiabatically implies that the true wave function may be expanded in terms of $|\eta_a\rangle$: $|\psi_a\rangle = \sum_b |\eta_b\rangle U_{ba}$. Plugging into $i(\partial/\partial t)|\psi_a\rangle = H|\psi_a\rangle$, we find immediately that

$$\dot U_{ba} = -\sum_c \langle \eta_b | \dot\eta_c\rangle U_{ca} \ . \tag{2}$$

Berry's essential insight was the recognition, in the nondegenerate case, that the phase $\langle\eta|\dot\eta\rangle$ cannot be trivially absorbed, as was erroneously asserted by various standard texts read by generations of physicists. Rather, a gauge structure can be defined as follows. In general, H depends on a number of parameters $x^\mu, \mu=1,\dots,p$, and H varies as the parameters $x^\mu(t)$ vary with time. We can thus define the gauge potential

$$A_{ab\mu} = \left\langle \eta_a \left| \frac{\partial}{\partial x^\mu} \right| \eta_b \right\rangle , \tag{3}$$

so that $\langle \eta_a | \dot\eta_b\rangle = \sum_\mu A_{ab\mu}(dx^\mu/dt)$ and Eq. (2) can be integrated immediately to give

$$U_{ab} = \left[\mathcal{P}\exp\left[-\int A_\mu dx^\mu \right] \right]_{ab} , \tag{4}$$

where $\mathcal{P}$ indicates to path-ordered product. In what follows, we use the notation of differential forms and write $A_{ab} = A_{ab\mu}dx^\mu$. We will also suppress the indices $a,b,\dots$, on the matrix form. The gauge group in question corresponds to the unitary freedom in choosing the basis states $|\eta_a\rangle$. With a different choice $|\eta'_a = \sum_b |\eta_b\rangle\omega_{ba}$, we have the transformed gauge potential $A' = \omega^\dagger A\omega + \omega^\dagger d\omega$. Here ω is an $N\times N$ unitary matrix. The gauge field $F = dA + A\wedge A$ transforms covariantly, of course: $F' = \omega^\dagger F\omega$. The trace of U for a closed loop is gauge invariant. Thus we have a

$U(N)=SU(N)\times U(1)$ gauge field existing on a p-dimensional space.

After a complete cycle, the physical situation returns to what it was at the start of the cycle but the states $|\eta_a\rangle$ are multiplied by the unitary matrix $U=\mathcal{P}\exp(-\oint A)$. The states evolve into linear combinations of each other. Diagonalizing U to have eigenvalues e^{iE_iT} (where T is the time taken to complete the cycle), we can also say that the corresponding eigenstate has shifted in energy by E_i.[10] The unitarity of U implies that $\sum_i E_i T$ is an integer multiple of 2π (see below).

Given this general framework, we can easily calculate the gauge potential by parametrizing the Hamiltonian in Eq. (1) as

$$H=B^2e^{-i\varphi S_z}e^{-i\theta S_y}S_z^2e^{i\theta S_y}e^{i\varphi S_z}, \tag{5}$$

keeping in mind that this set of coordinates is not everywhere defined on the 2-sphere. (This fact introduces certain subtleties to be discussed below.) Then, as noted in Ref. 4, we obtain the instantaneous eigenstates immediately as

$$|\eta_a\rangle=e^{-i\varphi S_z}e^{-i\theta S_y}|a\rangle,$$

where $|a\rangle$, $|b\rangle$, and so on, are simply (1,0,0,0) etc., in the standard spinor notation. We proceed with the calculation taking the spin S as arbitrary. [Notice as usual that for S half-integral $|\eta_a\rangle$ at $\varphi=2\pi$ is equal to (-1) times $|\eta_a\rangle$ at $\varphi=0$. The states $|\eta_a\rangle$ provide a double covering of parameter space. Thus, the phase factor U determines the phase change in addition to the phase produced by rotation through 2π.] Evaluating the general formulas (3), we find

$$A_{ab\varphi}=(-i)\langle a|(\cos\theta S_z-\sin\theta S_x)|b\rangle,$$
$$A_{ab\theta}=(-i)\langle a|S_y|b\rangle. \tag{6}$$

It is worth emphasizing that these formulas are quite general and hold whenever H has the form

$$H=e^{-i\varphi S_z}e^{-i\theta S_y}H_0e^{i\theta S_y}e^{i\varphi S_z}.$$

The "unrotated" Hamiltonian H_0 serves only to determine the degeneracy structure, namely, the gauge group. In particular, it does not matter if H_0 is S_z^2 or S_z raised to any even power. Also, H_0 can in general be a many-body Hamiltonian. The analysis also clearly does not depend on the system being nuclear. We also remark that if H_0 is anisotropic, for instance, $H_0=\alpha S_x^2+\beta S_y^2+\gamma S_z^2=\alpha' S_x^2+\gamma' S_z^2+\text{const}$, we will have to use the general rotation $e^{i\psi S_z}e^{i\theta S_y}e^{i\varphi S_z}$ and the gauge potential will exist on a three-dimensional space.

The states $|a\rangle$ can be taken to be the eigenstates $|m\rangle$ with $S_z|m\rangle=m|m\rangle$. For the problem at hand, the states $(\pm m)$ form a doubly degenerate secotr. For $|m|\neq\frac{1}{2}$, we obtain only an Abelian structure with $A_\theta=0$ and $A_\varphi=(-i)m\cos\theta$. This follows since $\mathbf{S}$ cannot cause transition with $\Delta m>1$. The corresponding field strength $F=(-i)m\,d\Omega$ (where $d\Omega=\sin\theta\,d\theta\,d\varphi$ is the invariant area element) shows that the gauge potential is that which is generated by a monopole of strength $2m$.

It may be instructive to compare the nondegenerate situation studied by Berry,[1] namely, that for the spin dipole Hamiltonian $H=\mathbf{S}\cdot\mathbf{B}$. As remarked above, the "unrotated" Hamiltonian H_0, here BS_z, serves only to determine the degeneracy structure. Thus, the formulas in (6) apply, with the indices a and b and A omitted and with the states $|a\rangle$ comprised of some eigenstates of S_z with eigenvalue m. Berry thus obtained $A_\theta=0$ and $A_\varphi=(-i)m\cos\theta$ exactly as in the preceding, except that m may be equal to $\frac{1}{2}$ in this case.

In the $|m|=\frac{1}{2}$ sector we obtain a truly non-Abelian structure with

$$A=(-i)\left[\left(\cos\theta\frac{\sigma_3}{2}-\alpha\sin\theta\frac{\sigma_1}{2}\right)d\varphi+\alpha\frac{\sigma_2}{2}d\theta\right], \tag{7}$$

with $\alpha\equiv S+\frac{1}{2}$ an integer. In this sector, while S_z is represented by $\sigma_3/2$, S_x and S_y are represented by $\alpha(\sigma_1/2)$ and $\alpha(\sigma_2/2)$, respectively. The corresponding gauge field

$$F=(-i)(\alpha^2-1)\frac{\sigma_3}{2}d\Omega$$

is formally the same as the Abelian field strength in the $|m|\neq\frac{1}{2}$ sector. Nevertheless, the gauge structure is non-Abelian. What is physically relevant is not so much F as the Wilson loop

$$\mathrm{Tr}U=\mathrm{Tr}\mathcal{P}\exp\left[-\oint A\right];$$

the path ordering is clearly essential. Note that the gauge field does not satisfy the sourceless Yang-Mills equation. In particular, $F_{\theta\varphi}$ does not depend on φ, while $[A_\varphi,F_{\theta\varphi}]$ is proportional to $\sin^2\theta\sigma_2$. Note also that F vanishes for $\alpha=1$ (that is, $S=\frac{1}{2}$) as it should since the problem collapses in that case, H_0 (and therefore H) being proportional to the unit matrix.

We now see that, since in Tycko's experiment θ is held fixed (with $\cos\theta=1/\sqrt{3}$) while φ varies cyclically, the non-Abelian character of A is lost. (The reason that the fixed θ path is followed in Tycko's experiment is simply that a rotation about a fixed axis is the only rotation of a crystal that can be performed rapidly enough to produce an observable phase shift.) A is proportional to the fixed matrix $(\cos\theta\sigma_3-\alpha\sin\theta\sigma_1)$ with eigenvalues $\pm(\cos^2\theta+\alpha^2\sin^2\theta)^{1/2}$ and thus after each rotation of φ through 2π we obtain the phase shifts $\pm\pi(\cos^2+\alpha^2\sin^2\theta)^{1/2}$ for the appropriate eigenstates, in agreement with Tycko for $\alpha=2$ (that is, $S=\frac{3}{2}$). As the magnetic field turns, the two states $|\eta_{+(1/2)}\rangle$ and $|\eta_{-(1/2)}\rangle$ rotate into a linear combination of each other. The two degenerate levels are split by $\Delta E=2\pi(\cos^2\theta+\alpha^2\sin^2\theta)^{1/2}/T$, where T is the time period over which the adiabatic variation goes through one cycle. Note that at $\theta=\pi/2$ the phase factor $e^{\pm i\pi\alpha}$ acquired in each cycle is $+1$ or -1 for $\alpha=S+\frac{1}{2}$ even or odd, respectively. Actually, there is an additional factor of (-1) to be explained later. [Also note that the + eigenstate is largely the spin up state near the north pole $(\theta \sim 0)$ and turns into the spin down state near the south pole $(\theta \sim \pi)$. The same applies for the − eigenstate.]

(There is a point of potential confusion here. It would appear naively that even in the $\alpha=1$ case the two states are split by $2\pi/T$, while, in fact, the Hamiltonian is proportional to the identity operator. The resolution is, of course, that when a quantum state has acquired a phase $e^{i\xi}$ it is the phase factor $e^{i\xi}$ and not the phase angle ξ that is defined. Thus $e^{i\pi}$ and $e^{-i\pi}$ represent the same phase. In other words, by looking at Berry's phase, the energy splitting is determined as only modulo $2\pi/T$. Note that this ambiguity is of the same order in $1/T$ as the energy splitting and so is potentially substantive. In Tycko's experiment, however, the ambiguity can be resolved, since we can follow the buildup of Berry's phase infinitesimally in time. With the fixed-θ path chosen by Tycko, the derivative of Berry's phase with respect to φ is a constant: A does not depend on φ. In other words, the physical situation in Tycko's experiment is essentially stationary. Thus we can integrate the phase change to obtain the phase shifts per cycle as cited above.)

In this gauge, the state $|m=+\frac{1}{2}\rangle$ is rotated by the matrix $\exp i\pi(\cos\theta\sigma_3-\alpha\sin\theta\sigma_1)$ into a linear combination of itself and the state $|m=-\frac{1}{2}\rangle$. It is easy enough to go to another gauge by diagonalizing the matrix

$$(\cos\theta\sigma_3-\alpha\sin\theta\sigma_1)=\omega\cos\theta\sigma_3\omega^\dagger$$

with $\omega=e^{i\chi(\sigma_2/2)}$ and $\tan\chi=\alpha\tan\theta$, that is, instead of the basic states $|\eta_a\rangle$ we choose $|\eta'_a\rangle=\sum_b|\eta_b\rangle\omega_{ba}$. Thus we compute

$$A'=\omega^\dagger A\omega+\omega^\dagger d\omega$$

$$=(-i)\left[\frac{\cos\theta}{\cos\chi}\frac{\sigma_3}{2}d\varphi+\frac{(\alpha^2-1)}{\alpha}\sin^2\chi\frac{\sigma_2}{2}d\theta\right]. \qquad (8)$$

In this basis,

$$F'=(-i)(\alpha^2-1)\left[\cos\chi\frac{\sigma_3}{2}+\sin\chi\frac{\sigma_1}{2}\right]d\Omega .$$

Note that $\cos\theta/\cos\chi=(\cos^2\theta+\alpha^2\sin^2\theta)^{1/2}$ as it should.

To observe the non-Abelian character of Berry's phase in this situation, it is necessary to vary θ as well as φ. In general, we would like to compute U for an arbitrary loop on the 2-sphere. In practice, the computation is intrinsically non-Abelian and difficult, since each segment in the time-ordered exponential loop integral does not commute with the next. The integral can, however, be evaluated by numerical methods.

At first sight, one might think that one could compute U for loops consisting of cyclic motion on the sphere, maintaining a fixed angle with some arbitrary unit vector. It would appear that the non-Abelian structure would come into play, since both θ and φ vary. However, it is clear that the original Hamiltonian is rotational invariant and the z axis is preferentially picked out only by convention. The gauge factor obtained can only differ from the gauge factor obtained by tracing the "standard" fixed-θ loop (with the appropriate θ) by an inessential similarity transformation. This rotational invariance is reflected in the invariance of gauge invariant (geometrical) quantities such as

$$\mathrm{Tr}F\wedge *F=(-)(\alpha^2-1)^2\tfrac{1}{2}\sin\theta\,d\theta\,d\varphi .$$

Under a rotation, the gauge potential A changes by a gauge transformation, of course.

We note that the gauge potential we obtain here is the same as the gauge potential obtained by Wilczek *et al.*[5] in an analysis of diatomic molecules. This is hardly surprising: although the physical situations involved are different, the mathematical steps leading to the determination of the gauge potential are the same. In more mathematical language, one would say that the rotationally invariant connection on the sphere is essentially unique. A general (but rather involved) proof has been given by Forgacs and Manton.[11] A casual proof acceptable to many physicists can be sketched as follows. Let $F=\sum_j C_j(\sigma_j/2)d\Omega$. Since

$$\mathrm{Tr}F\wedge *F=\left|\sum_j C_j\right|^2\tfrac{1}{2}d\Omega ,$$

$\sum_j C_j^2$ is invariant by assumption. By a gauge transformation, we can rotate C_j to the standard form $C_1=C_2=0$ and $C_3=\text{const}$. The loophole is, of course, the question of whether the required gauge transformation can be carried out globally.

As noted, to explore the non-Abelian structure in nuclear quadrupole resonance, we should traverse a "non-Abelian" path in which θ and φ both vary. It seems to us that such a path can be traced in an experiment with an external magnetic field by varying the field suitably. For instance, we may have crossed magnetic fields with B_z varying at a different frequency from B_x and B_y. Of course, we also do not want T to be so large that the energy splitting is infinitesimal.

To "see" the non-Abelian character of the gauge structure, we have to show how the noncommutativity produces physically observable effects. Consider three closed loops A, B, and C, that all start (and end) at the same point x_0 on the sphere. Denote the phase factor generated by traversing each of the three loops to be U_A, U_B, and U_C, respectively. Consider the composite path in which one traverses first A, then B, and finally C. After a cycle, the phase factor $U=U_CU_BU_A$ is generated. In contrast, were one to traverse the three loops in the order, first A, then C, then B, the phase factor would be $U'=U_BU_CU_A$. while $\det U=\det U'$, in general $\mathrm{Tr}U$ is not equal to $\mathrm{Tr}U'$ and thus U and U' have different eigenvalues leading to different energy shifts.[12] Of course, each of the loops A, B, and C may be traversed many times, in which case U_A, U_B, and U_C represent the phase factor for each loop raised to a suitable power.

We were told[13] that in certain types of experiments it may be easiest to trace paths made up of segments along which either θ or φ is fixed. (It is not clear to us, however, how to go around the "corners.") Theoretically, it is of course effortless to calculate U for such path segments.

We now mention some subtleties associated with the

usual difficulty with spherical coordinates, namely, that φ is not defined at the north and south poles. To see that something is remiss, consider computing

$$U=\mathcal{P}\exp\left[-\int A\right]$$

for an infinitesimal small loop circling the north pole. We find, using (7), that $U\sim e^{i(\sigma_3/2)2\pi}=(-1)$. Thus there is a singularity at the north pole. Similarly, there is a singularity at the south pole. The gauge potential in (7), call it A_E (E for equatorial), can only be defined on a coordinate patch including the equator but excluding the north and south poles.

To understand what is going on, we refer to the analogous situation of the Dirac monopole. As noted earlier, the same formulas that produce A_E also produce in the Abelian case the gauge potential $a_E=(-i)m\cos\theta\, d\varphi$ and the gauge field $f_E=da_E=(-i)m\, d\Omega$. Here too, the phase factor $\exp(-\int a_E)$ for an infinitesimal small loop circling the north pole is equal to $\sim e^{im2\pi}=(-1)$ for $m=\frac{1}{2}$. Similarly, there is a singularity at the south pole. Consider, in contrast, the gauge potential $a_S=(-i)m(1+\cos\theta)d\varphi$. Clearly, since a_S and a_E differs only by a term proportional to $d\varphi$, the corresponding gauge field f_S is identical to f_E. The same remark can be made for the gauge potential $a_N=(-i)m(-1+\cos\theta)d\varphi$. Suppose for the moment that m is not known. Since $(1+\cos\theta)$ vanishes at the north pole, a_S is defined on the sphere excluding a region around the north pole. Similarly, a_N is defined on the sphere excluding a region around the south pole. The requirement is that in the overlap region where both a_N and a_S are defined they must be related by a gauge transformation. We have $a_N-a_S=2im\,d\varphi=e^{-2im\varphi}de^{2im\varphi}$. Indeed, a_N and a_S are related by a gauge transformation effected by $e^{2im\varphi}$, provided that $e^{2im\varphi}$ is defined, that is, if $e^{2im(2\pi)}=1$. Thus m is quantized to be integer multiple of $\frac{1}{2}$. This represents one of the standard arguments[14] for the quantization of magnetic charge.

But what about a_E? Since $a_N-a_E=im\,d\varphi=e^{-im\varphi}de^{im\varphi}$, a_E is related to a_N by a gauge transformation effected by $e^{im\varphi}$. However, for m half-integral, this gauge transformation, although locally legitimate, is not globally legitimate. The phase integral $\exp(-\int a_E)$ integrated from $\varphi=0$ to 2π differs from $\exp(-\int a_N)$ by $e^{im2\pi}=(-1)$ for m half-integral.[15] One way of describing the situation is that the equatorial patch on which a_E is defined is not a legitimate patch since it is not simply connected. (Strictly speaking, the patches on which a_N and a_S are located should be subdivided into smaller patches. The overlaps between patches are also required to be patches and hence each of the overlaps has to be simply connected.)

It is now clear what the problem is with the non-Abelian gauge potential A_E. By analogy with the Abelian case, we can define $A_S=\rho^\dagger A_E\rho+\rho^\dagger d\rho$ with the globally illegitimate (but locally legitimate) gauge transformation with $\rho=e^{-i(\sigma_3/2)\varphi}$. We obtain

$$A_S=(-i)\left\{\left[(1+\cos\theta)\frac{\sigma_3}{2}-\alpha\sin\theta\left[\cos\varphi\frac{\sigma_1}{2}-\sin\varphi\frac{\sigma_2}{2}\right]\right]d\varphi+\alpha\left[\cos\varphi\frac{\sigma_2}{2}+\sin\varphi\frac{\sigma_1}{2}\right]d\theta\right\}. \tag{9}$$

The result is more complicated in the non-Abelian case but the essential feature is that the combination $(1+\cos\theta)$ ensures that A_S is not singular at the south pole. We can now also define $A_N=\rho^2A_S\rho^{2\dagger}+\rho^2d\rho^{2\dagger}$ with a globally legitimate gauge transformation with the single-valued function $\rho^2=e^{-i\sigma_3\varphi}$. For completeness, let us record that

$$A_N=(-i)\left\{\left[(-1+\cos\theta)\frac{\sigma_3}{2}-\alpha\sin\theta\left[\cos\varphi\frac{\sigma_1}{2}+\sin\varphi\frac{\sigma_2}{2}\right]\right]d\varphi+\alpha\left[\cos\varphi\frac{\sigma_2}{2}-\sin\varphi\frac{\sigma_1}{2}\right]d\theta\right\}. \tag{10}$$

The moral of the story is that we should use A_N and A_S rather than A_E. However, we note that in practice it is easier to compute with A_E. For instance, along a fixed θ path, A_N and A_S depend on φ, while A_E does not. We can always use A_E, keeping in mind that for a line segment from point 1 to point 2

$$\exp\left[-\int_1^2 A_E\right]=\rho(2)\exp\left[-\int_1^2 A_S\right]\rho^\dagger(1)\,.$$

Thus, for a closed path over whith φ does not vary by more than 2π, TrU can be safely computed with A_E. On the other hand, for a path that wraps around once from $\varphi=0$ to $\varphi=2\pi$, $\exp(-\oint A_E)$ differs from $\exp(-\oint A_S)$ [and $\exp(-\oint A_N)$] by an extra factor of $e^{-i\sigma_3\pi}=(-1)$. This factor, however, does not matter in the computation of energy splitting. Thus, in particular, contrary to what was stated earlier, the phase factor acquired in each cycle around the equator should actually be (-1) or $(+1)$ for $\alpha=$ even or odd, respectively.

The careful discussion above is necessary; otherwise, one can easily fall into confusing traps. For instance, it is contemplated experimentally[13] to have a path (the "orange slice") going from the north pole to the south pole along some longitude and then back to the north pole along some other longitude. By looking at A_E [in Eq. (7)] one would conclude erroneously that since the gauge potential is independent of φ the phase acquired on the south-bound leg cancels that acquired on the north-bound leg and the net phase factor is unity. To do the calculation correctly, we must compute with A_N (say) and take care to stay within the north patch. We will actually compute the phase factor for the "spherical triangle" path in which we start from the north pole and go south at a fixed φ_1 to latitude θ, go at fixed θ to φ_2, and then return to the north pole along longitude φ_2. The phase factor desired for the "orange slice" path is then obtained in the limit $\theta\to\pi$.

Since the final quantity $\mathrm{Tr}U_{\mathrm{ST}}=\mathrm{Tr}\exp(-\int_{ST}A_N)$ (ST stands for spherical triangle) can only depend on $\varphi_2-\varphi_1$, we take $\varphi_1=0$ and $\varphi_2=\Delta$. We read off from Eq. (10) that we have for the south-bound leg $W=e^{i\alpha(\sigma_2/2)\theta}$ and for the north-bound leg

$$(W')^{-1}=\exp\left[-i\alpha\frac{\theta}{2}(\cos\Delta\sigma_2-sim\Delta\sigma_1)\right].$$

For the east-bound leg along fixed θ, we note that A_N varies along the integration path. Fortunately, we know that

$$\begin{aligned}V&=\exp\left[-\int_1^2 A_N\right]\\&=\rho_2\exp\left[-\int_1^2 A_E\right]\rho_1^\dagger\\&=\exp-i\sigma_3\Delta/2\exp[i(\cos\theta\sigma_3-\alpha\sin\theta\sigma_1)\Delta/2]\ .\end{aligned}\tag{11}$$

We are thus to evaluate $\cos\gamma_{\mathrm{ST}}=\frac{1}{2}\mathrm{Tr}U_{\mathrm{ST}}=\frac{1}{2}\mathrm{Tr}(W')^{-1}VW$. (This implies an energy splitting of $\Delta E=2\gamma/T$, of course.) After a tedious computation, we find

$$\cos\gamma_{\mathrm{ST}}=\cos\eta\cos\frac{\Delta}{2}+\sin\eta\sin\frac{\Delta}{2}\cos(X-\alpha\theta)\ ,\tag{12}$$

where $\eta=(\cos^2\theta+\alpha^2\sin^2\theta)^{1/2}\Delta/2$ and X is as defined earlier.

Let us look at some special cases of this otherwise rather opaque expression. As $\theta\to\pi$, we have for the "orange slice" path

$$\cos\gamma_{\mathrm{OS}}=\cos^2\frac{\Delta}{2}-(-1)^\alpha\sin^2\frac{\Delta}{2}\ .\tag{13}$$

This is decidedly not unity as one would conclude erroneously from computing with A_E naively. In particular, for $\Delta=\pi$, we have the factor $(-1)^{\alpha+1}$. [This factor can of course also be read off directly with $W=(W')^{-1}=e^{i\alpha\sigma_2\pi/2}$ and $V=e^{-i\sigma_3\pi}=(-1)$.] Note that since the path is a great circle this factor must agree with the phase acquired upon going around the equator once as computed from Eq. (7) but with an extra factor of (-1) included.

For $\theta=\pi/2$, we have $\eta=\alpha\Delta/2$ and $X=\pi/2$, and thus

$$\cos\gamma_{\mathrm{ST}}^{o}=\cos\frac{\alpha\Delta}{2}\cos\frac{\Delta}{2}+\sin\frac{\alpha\Delta}{2}\sin\frac{\Delta}{2}\sin\frac{\alpha\pi}{2}\ .\tag{14}$$

For α even, this is equal to $\cos\alpha\Delta/2\cos\Delta/2$, for α odd of the form $4k+1$ with k an integer, $\cos(\alpha-1)\Delta/2$, and for α odd of the form $4k+3$, $\cos(\alpha-1)\Delta/2$. Perhaps this peculiar dependence on α can be tested by using nuclei with different spin.

In practice, it is easier to compute with A_E [or its gauge transform in Eq. (8)] than with A_N and A_S. As long as the path under consideration does not touch the poles and does not wrap around the sphere, we can safely compute with A_E. For instance, we can compute the phase factor U_{SR} obtained for the "spherical rectangle" path starting from the point (θ_1,ϕ_1), then going to (θ_2,ϕ_1) along a fixed-ϕ path, then to (θ_2,ϕ_2) along a fixed-θ path, then to (θ_1,ϕ_2) along a fixed-ϕ path, and finally back to (θ_1,ϕ_1) along a fixed-θ path. We find readily that

$$\frac{1}{2}\mathrm{Tr}U_{\mathrm{SR}}=\cos\frac{p_1\Delta}{2}\cos\frac{p_2\Delta}{2}+\sin\frac{p_1\Delta}{2}\sin\frac{p_2\Delta}{2}\cos Q\ ,\tag{15}$$

where $p_i=(\cos^2\theta_i+\alpha^2\sin^2\theta_i)^{1/2}$, $Q=\alpha(\theta_2-\theta_1)-(X_2-X_1)$, and $\Delta=\phi_2-\phi_1$. As $\theta_1\to 0$, this expression reduces to the result for "spherical triangle," Eq. (12).

Incidentally, we can now read off the rotation of states in Tycko's experiment. The appropriate phase factor as the magnetic field is moved along a fixed-θ path from $\phi=0$ to $\phi=\Delta$ is given by V in Eq. (11). In particular, starting with the initial state $|\eta_+\rangle=(1,0)$, we find the amplitude of being in the state $|\eta_-\rangle=(0,1)$ to be $(-i)\sin\eta\sin X e^{i\Delta/2}$, which we note vanishes only for some certain values of Δ, namely, for

$$\Delta=2k\pi/(\cos^2\theta+\alpha^2\sin^2\theta)^{1/2},\quad k=\text{integer}\ .\tag{16}$$

As already mentioned, V contains an extra factor of $e^{-i\sigma_3\pi}=(-1)$ for $\Delta=2\pi$.

In the Abelian situation,[1] Berry could, and wisely did, avoid these complications by immediately using Stokes's theorem to write all expressions in terms of the gauge field F. The corresponding theorem is, however, not available for non-Abelian gauge potentials.

Incidentally, our discussion also allows us to illustrate an interesting feature of non-Abelian gauge potentials. Suppose we are given the gauge field

$$F=(-i)(\alpha^2-1)\frac{\sigma_3}{2}d\Omega$$

on the sphere. We know that the field strengths F_N, F_S, and F_E calculated from A_N, A_S, and A_E, respectively are all equal to $F_N=F_S=F_E=F$, since these three gauge potentials are related by (locally legitimate) gauge transformations that commute with σ_3. (We can of course also check this equality by direct computation.) However, as already noted, $\mathrm{Tr}\mathcal{P}\exp(-\int A_E)$ differs by a factor of (-1) from $\mathrm{Tr}\mathcal{P}\exp(-\int A_{N,S})$. This represents a non-Abelian version of the Aharonov-Bohm phenomenon. Imagine removing the caps around the north and south poles. Knowing the field strength in the nonsimply corrected equatorial "patch" does not determine for us the net phase change as we go around the loop.

Consider next the gauge potential

$$A_X=(-i)(\alpha^2-1)\cos\theta\frac{\sigma_3}{2}d\phi$$

with the corresponding field strength $F_X=F_N=F_S=F_E=F$. As we go around the equator once,

$$\mathrm{Tr}\exp\left[-\int A_X\right]=\mathrm{Tr}\exp[i(\alpha^2-1)\pi\sigma_3]\ .$$

Since this differs from $\mathrm{Tr}\exp(-\int A_E)=\mathrm{Tr}\exp(i\alpha\pi\sigma_3)$, A_X is clearly not related to A_E by a globally legitimate gauge transformation. In fact, A_X is not even related to A_E by a locally legitimate gauge transformation. To see this, we consider the gauge covariant quantity $D_\phi F_{\theta\phi}=\partial_\phi F_{\theta\phi}+[A_\phi,F_{\theta\phi}]$, namely, the source current of the gauge field. Computed with A_X, this quantity is clearly zero. On the other hand, as has already been noted, it is not zero when computed with A_E. This is in contrast to the Aharonov-Bohm phenomenon, in which the two relevant gauge potentials, while not related by a globally legitimate gauge transformation, are related by a locally legitimate gauge transformation.[16]

In their original discussion on non-Abelian gauge structures, Yang and Mills spoke of the degeneracy of the proton and neutron under isospin and imagined transporting a proton from one point in the universe to another. That a proton at one point can be interpreted as a neutron at another in an isospin invariant world necessitates the introduction of a non-Abelian gauge potential. We find it amusing that this discussion can now be realized analogously in the laboratory.

We are grateful to Y. S. Wu for helpful discussions. We would also like to thank M. Goodman and A. Strominger for useful conversations and F. Wilczek for encouraging us to publish this paper. We have also benefited from conversations with A. Pines and R. Tycko. This research was supported in part by the National Science Foundation under Grant No. PHY82-17853, supplemented by funds from the National Aeronautics and Space Administration, at the University of California at Santa Barbara.

[1]M. V. Berry, Proc. Roy. Soc. London, Ser. A **392**, 45 (1984). For a review and further references, see R. Jackiw, Massachusetts Institute of Technology. Report No. MIT-CTP-1475, 1987 (unpublished).

[2]G. Herzberg and H. C. Longuet-Higgins, Discuss. Faraday Soc. **55**, 77 (1963); C. A. Mead, Chem. Phys. **49**, 23 (1980); C. A. Mead and D. G. Truhlar, J. Chem. Phys. **70**, 2284 (1979); R. L. Whetten, G. S. Ezra, and E. R. Grant, Ann. Rev. Phys. Chem. **36**, 277 (1985).

[3]B. Simon, Phys. Rev. Lett. **51**, 2167 (1983); J. E. Avron, R. Seiler, and L. G. Yaffe, Commun. Math. Phys. **110**, 33 (1987).

[4]F. Wilczek and A. Zee, Phys. Rev. Lett. **52**, 2111 (1984).

[5]J. Moody, A. Shapere, and F. Wilczek, Phys. Rev. Lett. **56**, 893 (1986).

[6]H. Kuratsuji and S. Iida, Phys. Rev. Lett. **56**, 1003 (1986).

[7]R. Y. Chiao and Y. S. Wu, Phys. Rev. Lett. **57**, 933 (1986); A. Tomita and R. Y. Chiao, *ibid.* **57**, 937 (1986); M. V. Berry, Nature **19**, 277 (1987).

[8]Y. Aharonov and J. Anandan, Phys. Rev. Lett. **58**, 1593 (1987).

[9]R. Tycko, Phys. Rev. Lett. **58**, 2281 (1987).

[10]There is some subtlety involved. The basic principle of quantum mechanics asserts that if a state is an energy eigenstate with energy E its wave function depends on time according to e^{-iEt}. Here the time dependence of the states is not of the form e^{-iEt} and in fact is subject to experimental control. With a time-dependent Hamiltonian, energy is not, strictly speaking, a useful concept because the external agent causing the time dependence is doing work on the system. Rather, one should enlarge the Hamiltonian to include the external agent which makes the Hamiltonian time dependent, as for example in the treatment of molecular dynamics in the Born-Oppenheimer approximation. We should couple in an external electromagnetic field and compute its frequency shift. Alternatively, we can quantize the effective Lagrangian describing the dynamics of θ and φ and incorporating the phase shift (as outlined in Ref. 4).

[11]M. Forgacs and N. Manton, Commun. Math. Phys. **72**, 15 (1980). We thank Y. S. Wu for calling our attention to this reference. See also E. Witten, Phys. Rev. Lett. **38**, 121 (1977); Gu Chaohao, Phys. Rep. **80**, 251 (1981).

[12]The trace of a 2×2 simple unitary matrix determines the matrix up to a unitary transformation, of course. Thus three loops are needed here. In general, only two loops are needed.

[13]A. Pines (private communication).

[14]T. T. Wu and C. N. Yang, Nucl. Phys. B **107**, 365 (1976).

[15]Another way of seeing this point is to note that (for spin $\frac{1}{2}$) the instantaneous state η defined in the text is equal to $[e^{(-i\varphi/2)}\cos(\theta/2),\ e^{(i\varphi/2)}\sin(\theta/2)]$ and thus undefined at the poles. In contrast $\eta'=e^{(i\varphi/2)}\eta$ is undefined only at the south pole.

[16]A similar phenomenon has been discussed by L. Brown and W. Weisberger Nucl. Phys. B **157**, 285 (1979); with the gauge potentials $A=(-i)(\sigma_1 dx+\sigma_2 dy)/2$ and $A'=(-i)\sigma_3(dy/2)$ in two-dimensional Euclidean space.

VOLUME 59, NUMBER 2 PHYSICAL REVIEW LETTERS 13 JULY 1987

Molecular Kramers Degeneracy and Non-Abelian Adiabatic Phase Factors

C. Alden Mead
Chemistry Department, University of Minnesota, Minneapolis, Minnesota 55455
(Received 23 March 1987)

For a molecular system with an odd number of electrons, the electronic Hamiltonian possesses a twofold Kramers degeneracy for all nuclear configurations. It is shown that this leads to a non-Abelian gauge-field term in the effective nuclear Hamiltonian, with permissible gauge transformations belonging to SU(2). This should lead to observable effects analogous to the Berry phase for the nondegenerate case. In particular, an atom in a slowly rotating electric field exhibits changes in its component of angular momentum along the field because of the non-Abelian phase factor.

PACS numbers: 03.65.Bz, 11.15.Kc, 32.60.+i, 33.10.Lb

The phase factor experienced by an eigenfunction of a parameter-dependent Hamiltonian when the parameters adiabatically traverse a closed path has recently become a topic of considerable interest. Some aspects of the phenomenon have been known for decades,[1,2] but the modern interest stems from the derivation of a general phase factor formula in the context of a molecular Born-Oppenheimer problem[3] followed by its rederivation and recasting in a more general context by Berry[4] and Simon.[5] The generalization to degenerate eigenvalues[6] leads to the replacement of the phase factor by a unitary transformation among the degenerate eigenfunctions that have the properties of a non-Abelian gauge field. This idea has been applied in the context of Born-Oppenheimer treatments of molecular systems to degenerate states of the diatomic molecule,[7] and to a model in which nuclei are constrained to move in a manifold where there is a degeneracy-producing symmetry.[8] Here I wish to point out that Kramers degeneracy in molecular systems furnishes a wide variety of systems in which non-Abelian gauge fields play a role and should lead to many applications in chemical physics and, by analogy, in other fields as well. The only previous treatment of this known to the author[9] dealt only with a special case for which the possibilities were limited. I have tried to formulate this Letter in such a way that it can serve to introduce the chemical physicist to the role of gauge fields in his area, and also be of interest to those looking for simple but nontrivial systems exhibiting gauge behavior.

We consider an electronic Hamiltonian $H(\mathbf{R})$ depending on nuclear coordinates $\mathbf{R}$. We use three-dimensional vector notation for $\mathbf{R}$, but it will be evident that the results are generalizable to any number of dimensions and to any Hamiltonian depending on continuously variable parameters. For each $\mathbf{R}$ we have the eigenvalue equation

$$H(\mathbf{R})\,|ja(\mathbf{R})\rangle = W_j(\mathbf{R})\,|ja(\mathbf{R})\rangle, \tag{1}$$

where Latin letters $j, k, \ldots$ are used to denote energy levels, and Greek letters $\alpha, \beta, \ldots$ denote states within a level if there is a degeneracy which holds for all $\mathbf{R}$. Of course, I have principally in mind the case of the twofold Kramers degeneracy.

If $\mathbf{R}$ is displaced continuously along some closed curve C starting at an origin 0, the eigenkets must change continuously in such a way as to satisfy (1) at each point. I describe this by

$$|ja(C)(\mathbf{R})\rangle = \sum_{k,\beta} |k\beta(0)\rangle\langle k\beta|S(C)(\mathbf{R})|ja\rangle. \tag{2}$$

For an infinitesimal displacement, we have

$$|ja(C)(\mathbf{R}+d\mathbf{R})\rangle = \sum_{k,\beta} |k\beta(C)(\mathbf{R})\rangle\langle k\beta|\,1+i\mathbf{G}(\mathbf{R})\cdot d\mathbf{R}\,|ja\rangle, \tag{3}$$

where $G(\mathbf{R})$ is a Hermitean vector operator defined for each $\mathbf{R}$. Combining (2) and (3), we find for the matrix representation of $S(C)(\mathbf{R})$,

$$S(C)(\mathbf{R}+d\mathbf{R}) = S(C)(\mathbf{R})[1+i\mathbf{G}(\mathbf{R})\cdot d\mathbf{R}],$$

which has the formal solution

$$S(C)(\mathbf{R}) = 1 + i\int_0^{\mathbf{R}} \mathbf{G}(\mathbf{R}')\cdot d\mathbf{R}' - \int_0^{\mathbf{R}} \left[\int_0^{\mathbf{R}'} \mathbf{G}(\mathbf{R}'')\cdot d\mathbf{R}''\right]\mathbf{G}(\mathbf{R}')\cdot d\mathbf{R}' + \ldots \tag{4}$$

Applying this to a closed curve which is an infinitesimal rectangle in the xy plane with sides Δx and Δy, and keeping only leading terms, we easily find

$$S(C) - 1 = \Delta x\,\Delta y\left[i\left(\frac{\partial G_y}{\partial x} - \frac{\partial G_x}{\partial y}\right) - [G_x, G_y]\right]. \tag{5}$$

It is convenient to divide $\mathbf{G}(\mathbf{R})$ into two parts:

$$\mathbf{G}(\mathbf{R}) = \mathbf{f}(\mathbf{R}) + \mathbf{g}(\mathbf{R}), \tag{6}$$

where $\mathbf{f}(\mathbf{R})$ is block diagonal with respect to the energy levels, and $\mathbf{g}(\mathbf{R})$ has matrix elements only between states belonging to different levels. The form of $\mathbf{g}(\mathbf{R})$ is determined by the requirement that (1) must be satisfied at all points:

$$\langle k\beta | \mathbf{g}(\mathbf{R}) | j\alpha \rangle = i[W_k(\mathbf{R}) - W_j(\mathbf{R})]^{-1} \langle k\beta(\mathbf{R}) | \nabla H(\mathbf{R}) | j\alpha(\mathbf{R}) \rangle. \tag{7}$$

There is a freedom of choice of $\mathbf{f}(\mathbf{R})$ corresponding to the freedom of choice in defining degenerate states, but the definitions clearly require that

$$\langle j\beta | \mathbf{f}(\mathbf{R}) | j\alpha \rangle = -i \langle j\beta(\mathbf{R}) | \nabla j\alpha(\mathbf{R}) \rangle. \tag{8}$$

It is easy to show that the right-hand side of (5) is identically block diagonal, as of course it should be, since the closed path must return one to the same energy level. The block-diagonal part depends on the choice of $\mathbf{f}$. If, for example, the eigenkets are required to be single-valued functions of $\mathbf{R}$, we must have $S(C) = 1$, implying

$$\frac{\partial}{\partial x} f_y - \frac{\partial}{\partial y} f_x + i[f_x, f_y] = -i[g_x, g_y]_d, \tag{9}$$

where the subscript d stands for the (block-)diagonal part. Equation (9) can be thought of as applying to the entire block-diagonal matrix or separately to each block. On the other hand, if we choose $f = 0$ corresponding to adiabatic variation of $\mathbf{R}$ with time,[4] we find

$$S - 1 = -\Delta x \, \Delta y [g_x, g_y]_d,$$

which, together with (7), gives the appropriate generalization of the Berry formula[3,4] for the nondegenerate case.

We now specialize to the case of interest for this Letter in which $H(\mathbf{R})$ commutes with the time-reversal operator T for all $\mathbf{R}$ and in which the total spin is half-odd integer (odd number of electrons) so that $T^2 = -1$. In this case, we have a twofold Kramers degeneracy for each $\mathbf{R}$.[10,11] Specializing to a particular energy level and suppressing the index j, we have two states $|a\rangle$ and $|b\rangle$ for each $\mathbf{R}$, related by

$$T|a\rangle = |b\rangle, \quad T|b\rangle = -|a\rangle. \tag{10}$$

We note from (7) that $\mathbf{g}$ anticommutes with T. With use of this along with (10) and the well-known properties of T, one easily shows that the right-hand side of (9) (restricted to our 2×2 block) is a traceless, Hermitean matrix, but is otherwise arbitrary, depending on details of the Hamiltonian. One also sees from (8) that $\mathbf{f}$ is traceless and Hermitean within the block.

We can define a gauge transformation as an $\mathbf{R}$-dependent unitary transformation $U(\mathbf{R})$ applied to the states $|a\rangle$ and $|b\rangle$:

$$|a'(\mathbf{R})\rangle = U(\mathbf{R}) |a(\mathbf{R})\rangle = \sum_\beta |\beta(\mathbf{R})\rangle \langle \beta | U(\mathbf{R}) | \alpha \rangle, \tag{11}$$

with α, β taking on the values a and b. If (10) is still to be satisfied by the primed kets, $U(\mathbf{R})$ must belong to the special unitary group SU(2) of unitary operators with unit determinant. The effect of (11) on $\mathbf{f}$ is obtained by the application of (8) to the primed kets,

$$\mathbf{f}' = U^\dagger \mathbf{f} U - i U^\dagger \nabla U, \tag{12}$$

in agreement with the usual formula for a non-Abelian gauge transfermation.

In Born-Oppenheimer approximation, we write a molecular wave function belonging to the given energy level in the form

$$\Psi = \sum_\alpha |\alpha(\mathbf{R})\rangle \psi_\alpha(\mathbf{R}).$$

Applying the nuclear gradient operator to this, with use of (8), and ignoring any coupling to other energy levels, we find

$$\nabla\Psi = \sum_\alpha |\alpha\rangle \left\{ \nabla\psi_\alpha + i \sum_\beta \langle \alpha | f | \beta \rangle \psi_\beta \right\}.$$

In the effective Hamiltonian for the nuclear wave function $\psi(\mathbf{R})$, therefore, the gradient operator ∇ must be replaced by

$$\nabla + i\mathbf{f}(\mathbf{R}),$$

in which each component of $\mathbf{f}$ is a matrix operating on the two-component column vector $\psi(\mathbf{R})$. The effective Hamiltonian for the motion of the nuclei thus contains a gauge-field term. This is the generalization of the effective vector potential (Abelian gauge field) found in the nondegenerate case.[3,12-15] The presence of such a term is unavoidable if the electronic eigenkets are to be single-valued functions of $\mathbf{R}$, in which case (9) must be satisfied.

It is evident that molecular systems with Kramers degeneracy will furnish many examples of the role of non-Abelian gauge fields and phase factors in chemical physics. The remainder of this Letter is devoted to a particular simple example which is expected to lead to easily observable effects.

Consider an atom with an odd number of electrons, in a level with total angular momentum quantum number $J = \frac{1}{2}, \frac{3}{2}, \frac{5}{2}, \ldots$. The case $J = \frac{1}{2}$ will prove to be uninteresting, but is included for completeness. The $^2P_{3/2}$ ground state of fluorine is an example. If the atom is placed in a uniform electric field $\mathbf{E}$, the degenerate level is split by the Stark Hamiltonian

$$H = A(\mathbf{E}\cdot\mathbf{J})^2, \qquad (13)$$

in which A is a constant. The components of $\mathbf{E}$ can here be considered to be continuously variable parameters on which H depends. The Hamiltonian (13) is split into $J+\frac{1}{2}$ Kramers doublets, characterized by the quantum number $|M| = \frac{1}{2}, \frac{3}{2}, \ldots, J$, denoting the absolute value of the component of angular momentum along the field, with energies given by

$$W(|M|) = AE^2|M|^2. \qquad (14)$$

Now let $\mathbf{E}$ be initially directed along the z axis, and consider an infinitesimal rotation through an angle $d\phi$, about an axis in the (y,z) plane making an angle θ with the field. The small change in H is given by

$$dH = AE^2\sin\theta(J_zJ_x + J_xJ_z)d\phi. \qquad (15)$$

If the rotation is carried out adiabatically, the eigenfunctions will follow it according to Eqs. (2)-(3), but with the block-diagonal part $f=0$, so that $\mathbf{G}=\mathbf{g}$. With use of (7), (14), (15), and some elementary angular momentum theory, one finds, for the matrix elements of $\mathbf{g}$,

$$\langle M+1|ig_\phi|M\rangle = -\langle M|ig_\phi|M+1\rangle = [(J+M+1)(J-M)]^{1/2}\sin\theta. \qquad (16)$$

Equation (16) holds for elements of g_ϕ linking states belonging to different doublets; if $|M+1| = |M|$, however, which occurs for the doublet with $|M| = \frac{1}{2}$, the matrix element of g_ϕ is zero.

The operator $ig_\phi d\phi$, which generates the adiabatic change in the eigenfunction under the rotation, is *not* the same as the operator for a rotation of the eigenfunction about the axis, the generator of which is

$$-i(J_z\cos\theta + J_y\sin\theta)d\phi. \qquad (17)$$

This is easily seen to have the same off-diagonal elements as $ig_\phi d\phi$, but also has a block-diagonal part. The effect of the infinitesimal adiabatic rotation can thus be pictured as a rotation of the eigenfunction following the field according to (17), followed by subtraction of the block-diagonal part. Relative to a coordinate system following the field, therefore, each doublet experiences the infinitesimal transformation

$$1 + dU = 1 + i(J_z\cos\theta + J_{yd}\sin\theta)d\phi, \qquad (18)$$

where J_{yd} is the block-diagonal part of J_y.

For doublets with $|M| > \frac{1}{2}$, $J_{yd}=0$, and (18) leads to a phase factor, opposite in sign for the two components of the doublet. This is of some interest, but does not exhibit non-Abelian properties.

Of greater interest to us is the doublet with $|M| = \frac{1}{2}$, for which both terms of (18) contribute. It is convenient to represent J_z and J_{yd} within this block in terms of the Pauli matrices. Elementary angular momentum theory gives the result

$$J_z = \tfrac{1}{2}\sigma_z, \quad J_{yd} = \tfrac{1}{2}(J+\tfrac{1}{2})\sigma_y. \qquad (19)$$

For $J=\frac{1}{2}$, we note that (18) simply cancels out the rotation (17), with the result that the angular momentum is simply "left behind." For $J > \frac{1}{2}$, however, this is not possible: No purely block-diagonal operator like (18) corresponds to a rotation.

There are two particularly simple cases in which (18) can be integrated to yield the result of a finite rotation.

(a) Let $\theta = \pi/2$, so that one is rotating about the y axis itself. Integrating (18) and with use of (19), we find

$$U(\phi) = \exp[\tfrac{1}{2}i(J+\tfrac{1}{2})\sigma_y\phi]$$
$$= \cos[\tfrac{1}{2}(J+\tfrac{1}{2})\phi] + i\sigma_y\sin[\tfrac{1}{2}(J+\tfrac{1}{2})\phi]. \qquad (20)$$

Equation (20) predicts a reversal of sign of angular momentum component along the field with unit probability at an angle

$$\phi_r = \pi/(J+\tfrac{1}{2}) = \begin{cases} \pi & (J=\frac{1}{2}), \\ \dfrac{\pi}{2} & (J=\frac{3}{2}), \\ \dfrac{\pi}{3} & (J=\frac{5}{2}), \\ \text{etc.} \end{cases}$$

The result for $J=\frac{1}{2}$ just corresponds to the fact that in this case the angular momentum stays put as the field is rotated, with the result that it has reversed direction relative to the field when the field has been rotated through an angle π. The results for other J, however, although easily derivable, are nontrivial and reflect the presence of the non-Abelian gauge field.

(b) Let θ be arbitrary, but rotate through an angle $\phi = 2\pi$. In this case, integration of (18) with the aid of (19) gives

$$U(\theta) = \exp\{i\pi[\sigma_z\cos\theta + (J+\tfrac{1}{2})\sigma_y\sin\theta]\} = \cos q + i(\pi/q)[\sigma_z\cos\theta + (J+\tfrac{1}{2})\sigma_y\sin\theta]\sin q, \qquad (21)$$

where

$$q = \pi[\cos^2\theta + (J+\tfrac{1}{2})^2\sin^2\theta]^{1/2}. \qquad (22)$$

VOLUME 59, NUMBER 2 PHYSICAL REVIEW LETTERS 13 JULY 1987

For $J=\frac{1}{2}$, we see from (22) that $q=\pi$, so that (21) just predicts the usual sign change after a rotation through 2π, with no reversal of direction of angular momentum. For other J, though, (21) and (22) predict a finite probability (never equal to unity) for a reversal of angular momentum direction, given by

$$P=q^{-2}[(J+\tfrac{1}{2})\pi\sin\theta\sin q]^2. \qquad (23)$$

The angular momentum reversal probability given by (23) is equal to zero for $\theta=0$ or $\pi/2$; in between, it has $J-\frac{1}{2}$ maxima and $J-\frac{3}{2}$ zeros.

The results predicted by (20) and (23) would appear to make possible a relatively simple experimental observation of the non-Abelian phase factor analogous to that already achieved for the Abelian case by observing the rotation of polarization of photons passed through a helically wound optical fiber.[16]

This research was supported by National Science Foundation Grant No. CHE-83-11450.

[1]Y. Aharonov and D. Bohm, Phys. Rev. **115**, 485 (1959).

[2]G. Herzberg and H. C. Longuet-Higgins, Discuss. Faraday Soc. **535**, 77 (1963).

[3]A. Mead and D. G. Truhlar, J. Chem. Phys. **70**, 2284 (1979).

[4]M. V. Berry, Proc. Roy. Soc. London A **292**, 45 (1984).

[5]B. Simon, Phys. Rev. Lett. **55**, 927 (1983).

[6]F. Wilczek and A. Zee, Phys. Rev. Lett. **52**, 2111 (1984).

[7]J. Moody, A. Shapere, and F. Wilczek, Phys. Rev. Lett. **56**, 893 (1986).

[8]H. Li, Phys. Rev. Lett. **58** 539 (1987).

[9]C. A. Mead, Chem. Phys. **49**, 33 (1980).

[10]E. P. Wigner, *Group Theory*, translated by J. J. Griffin (Academic, New York, 1950), Chap. 26.

[11]D. G. Truhlar, C. A. Mead, and M. A. Brandt, Adv. Chem. Phys. **33**, 295 (1975).

[12]C. A. Mead, Chem. Phys. **49**, 23 (1980).

[13]C. A. Mead, J. Chem. Phys. **72**, 3839 (1980).

[14]S. P. Keating and C. A. Mead, J. Chem. Phys. **82**, 5102 (1985).

[15]S. P. Keating and C. A. Mead, J. Chem. Phys. **86**, 2152 (1987).

[16]A. Tomita and R. Y. Chiao, Phys. Rev. Lett. **57**, 937 (1986).

Volume 125, number 9 PHYSICS LETTERS A 7 December 1987

APPEARANCE OF GAUGE POTENTIALS IN ATOMIC COLLISION PHYSICS

Bernard ZYGELMAN
Harvard-Smithsonian Center for Astrophysics, 60 Garden Street, Cambridge, MA 02138, USA

Received 21 July 1987; revised manuscript received 23 October 1987; accepted for publication 26 October 1987
Communicated by B. Fricke

The dynamical equations obtained from the method of perturbed stationary states (PSS) are shown to be formally equivalent to the dynamical equations of a particle with N internal degrees of freedom minimally coupled to $U(N)$ static gauge potentials. Several examples are given that illustrate the appearance of non-abelian and magnetic monopole gauge potentials in simple systems. Advantages of expressing the PSS equations as a gauge theory are discussed.

In a slow collision between two atoms it is convenient to think of the nuclei moving in a force field created by the charge cloud of the electrons. In the Born–Oppenheimer approximation, the charge cloud adjusts adiabatically to the position of the nuclei. If the collision velocity is large enough the relative motion of the nuclei can induce transitions from one Born–Oppenheimer (BO) state to another. The rate of the transition from state $|a\rangle$ to state $|b\rangle$ is determined by the square of the matrix element of the nuclear kinetic energy operator between these states. This picture is the physical basis for the method of perturbed stationary states developed by Bates and McCarroll [1], and has been widely used for a large variety of atomic collision problems. In our discussion we limit ourselves to diatomic collisions, we ignore the special considerations of electronic translation factors [1], and without loss of generality we neglect the spin and identity of particles. These considerations will be addressed in a future publication [2]. In the quantal version of the PSS method the total wavefunction of the system is expressed as a sum over the channel states $\phi_n(\boldsymbol{R}, \boldsymbol{r})$,

$$\Psi(\boldsymbol{R}, \boldsymbol{r}, t) = \sum_n^N \phi_n(\boldsymbol{R}, \boldsymbol{r}) F_n(\boldsymbol{R}, t) , \tag{1}$$

where $\boldsymbol{R}$, $\boldsymbol{r}$ are the coordinates of the relative nuclear motion and electrons respectively, in a fixed coordinate system whose origin is at the center of mass of the nuclei. In practice the expansion (1) is restricted to a finite set, N, of states. If we ignore mass polarization terms the hamiltonian of the system is given by

$$H = -\frac{1}{2\mu}\nabla_R^2 + H_{el}(\boldsymbol{R}, \boldsymbol{r}) \tag{2}$$

where μ is the reduced mass of the system. ∇_R is the gradient operator along the direction $\boldsymbol{R}$, and H_{el} is the fixed-nuclei electronic hamiltonian whose eigenstates $\phi_n(\boldsymbol{R}, \boldsymbol{r})$ depend parametrically on the vector $\boldsymbol{R}$ and have eigenvalues $\epsilon(R)$. Using (1) and (2) we obtain the Schrödinger equation for the amplitudes $F_i(\boldsymbol{R}, t)$,

$$\nabla_R^2 F_i - 2i \sum_j^N (\nabla_R F_j)\cdot \boldsymbol{A}_{ij} - \sum_j^N B_{ij} F_j - 2\mu \sum_j^N V_{ij} F_j + 2\mu i \frac{\partial F_i}{\partial t} = 0 , \tag{3}$$

where the matrix elements $\boldsymbol{A}_{ij}(\boldsymbol{R})$, $B_{ij}(\boldsymbol{R})$ and $V_{ij}(R)$ are given by

Volume 125, number 9 PHYSICS LETTERS A 7 December 1987

$$A_{ij}(\boldsymbol{R}) = \mathrm{i}\int \mathrm{d}^3r\, \phi_i^\dagger(\boldsymbol{R}, \boldsymbol{r})\nabla_R\phi_j(\boldsymbol{R}, \boldsymbol{r})\,, \qquad B_{ij}(\boldsymbol{R}) = -\int \mathrm{d}^3r\, \phi_i^\dagger(\boldsymbol{R}, \boldsymbol{r})\nabla_R^2\phi_j(\boldsymbol{R}, \boldsymbol{r})\,,$$

$$V_{ij}(R) = \delta_{ij}\epsilon_i(R)\,. \tag{4}$$

In most applications, the electronic eigenfunctions are constrained to be real functions for all $\boldsymbol{R}$. Under this restriction the diagonal terms of the **A** matrix vanish. Herzberg and Longuet-Higgins [3], and Mead and Truhlar [4] pointed out that phase restrictions in general cannot be imposed on the BO states of a polyatomic sytem. Moody, Shapere and Wilczek [5] showed that these restriction are also too severe for di-atom electronic eigenstates. We do not impose a definite phase on the channel states appearing in (1); the physical meaning of this will become evident below. We can consider the components F_j to belong to an N-dimensional column vector, $\mathsf{F}(\boldsymbol{R}, t)$, and the A_{ij}, B_{ij}, and V_{ij} to be the elements of $N\times N$ hermitian matrices operating on F. We can then rewrite (3) as

$$-\frac{1}{2\mu}[\mathrm{I}\nabla_R - \mathrm{i}\mathbf{A}(\boldsymbol{R})]^2\mathsf{F} + \mathsf{V}(R)\mathsf{F} - \mathrm{i}\frac{\partial \mathsf{F}}{\partial t} = 0\,, \tag{5}$$

where we have used the identity

$$B_{ij} = \sum_k A_{ik}\cdot A_{kj} + \mathrm{i}\nabla_R\cdot A_{ij}\,,$$

the sum being over a complete set of channel states, and neglected a residual term which couples the channel states in (1) to states outside this sum and is small by definition. If a complete set of states is included in (1), eq. (5) is exact.

Eq. (5) is in a form strongly suggestive of the gauge theories studied in elementary particle physcis [6]. We have a wavefunction, F, that is a vector in an N-dimensional internal space minimally coupled to a vector potential with spatial components given by the matrix **A**, and a temporal component given by the diagonal matrix V. At any point $\boldsymbol{R}$ we may define a new wavefunction F′ related to F by a unitary transformation $\mathsf{U}(\boldsymbol{R}, t)$ such that $\mathsf{F}' = \mathsf{U}(\boldsymbol{R}, t)\mathsf{F}$. This transformation is a direct product of a unimodular unitary transformation on F and a phase transformation of each component of F. Such a transformation is then an element of the group $\mathrm{U}(N)$. Under this transformation (5) will be form invariant (covariant) since the gauge potentials transform inhomogeneously according to

$$\mathbf{A}(\boldsymbol{R}) \to \mathsf{U}^{-1}\mathbf{A}(\boldsymbol{R})\mathsf{U} + \mathrm{i}\mathsf{U}^{-1}\nabla_R\mathsf{U}\,, \qquad \mathsf{V}(\boldsymbol{R}) \to \mathsf{U}^{-1}\mathsf{V}(\boldsymbol{R})\mathsf{U} - \mathrm{i}\mathsf{U}^{-1}\partial\mathsf{U}/\partial t\,. \tag{6}$$

The vector potential can be rewritten as

$$A_{ij}(\boldsymbol{R}) = \mathrm{i}\int \mathrm{d}^3r'\, \phi_i^\dagger(R, \boldsymbol{r}')\nabla_R\phi_j(R, \boldsymbol{r}')\,, \tag{7}$$

where $\phi_a(R, \boldsymbol{r}') \equiv |\phi_a\rangle$ are Born–Oppenheimer states defined in a coordinate system with the quantization axis fixed along the internuclear radius vector. In this body-fixed frame the electronic coordinates $\boldsymbol{r}'$ are implicit functions of $\boldsymbol{R}$. The BO states in this frame are characterized by their quantum number $\pm\Lambda$, the component of the electronic angular momenta along the internuclear axis, in addition to their principal quantum number assignment. For systems with no spin, Λ is an integer. Expression (7) becomes [2]

$$A_{ij}(\boldsymbol{R}) = \hat{\boldsymbol{r}}[A_R(R)]_{ij} + \frac{\hat{\boldsymbol{\theta}}}{R}[A_\theta(R)]_{ij} + \frac{\hat{\boldsymbol{\phi}}}{\sin\theta R}[A_\phi(R)]_{ij}\,,$$

$$[A_R]_{ij} = \mathrm{i}\langle\phi_i|\mathrm{d}/\mathrm{d}R|\phi_j\rangle\,, \qquad [A_\theta]_{ij} = \langle\phi_i|L_x|\phi_j\rangle\,, \qquad [A_\phi]_{ij} = \langle\phi_i|\cos\theta\, L_z + L_y \sin\theta|\phi_j\rangle\,, \tag{8}$$

where L_x, L_y, L_z are the angular momentum operators for the electrons in the body frame, and $\hat{\boldsymbol{r}}$, $\hat{\boldsymbol{\theta}}$, $\hat{\boldsymbol{\phi}}$, are the spherical basis vectors in the fixed frame. Below, we illustrate aspects of writing the PSS equation in the form (5) by considering specific cases that commonly arise in atomic collision theory.

Volume 125, number 9 PHYSICS LETTERS A 7 December 1987

The simplest manifestation of the PSS equation occurs when the sum (1) contains only one state. The matrix **A** is now an ordinary vector potential, $\boldsymbol{A}$, and V is a scalar potential $\epsilon(R)$. The Schrödinger equation (5) is covariant under a phase transformation of F. This is just the U(1) group of electromagnetism. If the channel state, ϕ_n, is a $\Sigma(\Lambda=0)$ state then the only nonvanishing term for $\boldsymbol{A}$ is the radial component

$$A(R)=\mathrm{i}\hat{r}\langle\phi_n|\mathrm{d}/\mathrm{d}R|\phi_n\rangle\,. \tag{9}$$

The curl of the vecor potential (9) vanishes at all $\boldsymbol{R}$, and the vector potential can be "gauged away" by a phase transformation [#1]. Proceeding with the transformation $F\to UF$, where $U=\exp[\mathrm{i}(\int^R \mathrm{d}R'\,A(R'))]$, and using (6) we find that the gauged vector potential vanishes. We can express the BO eigenfunctions in the body frame by $|\phi\rangle=\exp[\mathrm{i}\phi_{\mathrm{Im}}(R)]\phi_{\mathrm{Re}}(R)$ where ϕ_{Im} and ϕ_{Re} are real functions. The transformation $\mathrm{F}\to\mathrm{UF}$ is then equivalent to the requirement that ϕ_{Im} vanishes, i.e we can choose real channel states in expansion (1) for all $\boldsymbol{R}$. This is not true for the case considered below.

If the channel state in (1) has nonvanishing Λ, the vector potential (8) becomes

$$A(R)=\hat{r}A_R(R)+\hat{\phi}\frac{\cot\theta}{R}\Lambda\,. \tag{10}$$

Because of the presence of the A_ϕ component, the curl of (10) does not vanish and there exists no phase transformation that eliminates $\boldsymbol{A}(\boldsymbol{R})$ for all $\boldsymbol{R}$. Defining a magnetic field, $\boldsymbol{B}=\nabla\times\boldsymbol{A}$, we get [#2]

$$\boldsymbol{B}=\frac{\hat{r}}{R^2\sin\theta}\frac{\partial}{\partial\theta}A_\phi=-\hat{r}\frac{\Lambda}{R^2}\,. \tag{11}$$

Using (5) and (10) to obtain the equation of motion for the mean value of the relative motion vector, $\langle\boldsymbol{R}\rangle$, we get the Ehrenfest theorem,

$$\frac{\mathrm{d}\langle\boldsymbol{R}\rangle}{\mathrm{d}t}=\langle(-\mathrm{i}/\mu)(\nabla_R-\mathrm{i}\boldsymbol{A})\rangle\equiv\langle\boldsymbol{v}\rangle\,,\qquad \mu\frac{\mathrm{d}\langle\boldsymbol{v}\rangle}{\mathrm{d}t}=\tfrac{1}{2}\langle\boldsymbol{v}\times\boldsymbol{B}-\boldsymbol{B}\times\boldsymbol{v}\rangle-\langle\nabla_R\epsilon(R)\rangle\,, \tag{12}$$

where the brackets signify the expectation value of the Schrödinger operators for a solution of (5). The right-hand side of (12) is the expectation value of an operator that looks very much like the Lorentz-force vector. For a diatomic system with stationary nuclei, the Hellmann–Feynman theorem [7] equates the forces acting on the nuclei with the gradient along the internuclear axis of the Born–Oppenheimer eigenvalue $\epsilon(R)$. This term is the "electric" force component occurring in (12). If we allow for zero-point fluctuations of the nuclear motion then the "magnetic" term in (12) will be nonvanishing and the equations of motion for $\langle\boldsymbol{R}\rangle$ are those of a charged particle acted on by a magnetic monopole in addition to the electric force field.

The existence of magnetic monopole fields occurring in simple dynamical systems was first demonstrated by Berry [8], and the appearance of gauge potentials in simple quantum systems was discussed by Wilczek and Zee [9]. Further investigations [5,10] showed that magnetic monopole fields arise naturally in a gauge theory framework. The studies cited above considered the rather special case of a system with a degenerate set of states in the Hilbert space. In the formalism presented here we have relaxed the condition of requiring degenerate channel states, in fact, the PSS equations are most useful when applied to non-degenerate systems. The gauge potentials arising from the non-degenerate mutichannel formalism, presented here, are more general than the ones obtained from a Hilbert space of degenerate states.

Consider two channels in expansion (1). Eq. (5) is now covariant under a local SU(2)×U(1) transfor-

[#1] If the radial potential contains a singularity, then the vanishing of the curl of (9) is no longer a sufficient condition for the existence of a gauge transformation that eliminates the vector potential at al R. However, in the case of a diatomic system the radial vector potential is expected to be regular everywhere.

[#2] In eq. (10) $A(R)$ has a singularity along the z-axis (the Dirac string) and a B field cannot be defined along this axis.

mation at each $\boldsymbol{R}$. The generators of the SU(2) transformations are the Pauli matrices, τ_1, τ_2, τ_3 and the U(1) transformations are generated by the unit matrix, explicitly

$$\tau_0=\begin{pmatrix}1 & 0\\ 0 & 1\end{pmatrix}, \quad \tau_1=\begin{pmatrix}0 & 1\\ 1 & 0\end{pmatrix}, \quad \tau_2=\begin{pmatrix}0 & -\mathrm{i}\\ \mathrm{i} & 0\end{pmatrix}, \quad \tau_3=\begin{pmatrix}1 & 0\\ 0 & -1\end{pmatrix}. \tag{13}$$

The matrices $\mathbf{A}$, V can expressed as a linear combination of these generators with components, A^w, V^w, w being the index of the generators in (13).

We consider two nondegenerate Σ (Σ_a, Σ_b) states in (1). The gauge potentials become

$$A(R)=\mathrm{i}\hat{\boldsymbol{r}}\begin{pmatrix}f_{aa} & f_{ab}\\ f_{ba} & f_{bb}\end{pmatrix}, \quad f_{ab}\equiv\langle\Sigma_a \mathrm{d}/\mathrm{d}R|\Sigma_b\rangle, \quad \mathrm{V}(R)=\begin{pmatrix}\epsilon_a & 0\\ 0 & \epsilon_b\end{pmatrix}. \tag{14}$$

The nonvanishing component of $\mathbf{A}$ is along the radial direction, and is a function of the internuclear distance only. We can eliminate the vector potential by a gauge transformation since the curl of each component of (14) vanishes. First, consider the gauge transformation

$$\mathrm{U}=\exp\begin{pmatrix}-\int^R \mathrm{d}R'\, f_{aa} & 0\\ 0 & -\int^R \mathrm{d}R'\, f_{bb}\end{pmatrix}. \tag{15}$$

This transformation eliminates the diagonal components of the vector potential and is equivalent to setting the phase of the channel states such that they are real functions. We then get

$$\mathrm{A}\to\mathrm{A}'=-\hat{\boldsymbol{r}}f_{ab}\tau_2\,, \tag{16}$$

where f_{ab} is now understood to be evaluated by BO states that are real. Under this transformation V is unchanged. With these potentials eq. (5) is referred to as the adiabatic picture [11]. In it the potential matrix is diagonal, and the vector potential is off-diagonal.

It is useful to introduce the matrix tensor [6] $\mathrm{F}_{\mu\nu}$

$$\mathrm{F}_{\mu\nu}=\partial_\mu \mathrm{A}_\nu-\partial_\nu \mathrm{A}_\mu-\mathrm{i}[\mathrm{A}_\mu, \mathrm{A}_\nu]=\sum \tau_w F^w_{\mu\nu}\,, \tag{17}$$

where $\partial_\mu\equiv(\partial_R, \partial_\theta, \partial_\phi, \partial_t)$, $\mathrm{A}^w_\mu\equiv(A^w_R, A^w_\theta, A^w_\phi, -V^w)$. $F^w_{\mu\nu}$ is a generalization of the electromagnetic antisymmetric field strength tensor. It contains six independent components for each internal space index. Unlike the field strength for abelian gauge potentials, where the commutator term vanishes, $F^w_{\mu\nu}$ has contributions that are nonlinear in the potentials.

The nonvanishing field strengths are

$$F^0_{tR}=\frac{1}{2}\left(\frac{\partial\epsilon_a}{\partial R}+\frac{\partial\epsilon_b}{\partial R}\right), \quad F^3_{tR}=\frac{1}{2}\left(\frac{\partial\epsilon_a}{\partial R}-\frac{\partial\epsilon_b}{\partial R}\right), \quad F^1_{tR}=-f_{ab}(\epsilon_a-\epsilon_b)\,. \tag{18}$$

The component F^1_{tR} is nonlinear in the potentials due to the commutator term in (17). An additional gauge transformation, with

$$\mathrm{U}=\begin{pmatrix}\cos\Omega & -\sin\Omega\\ \sin\Omega & \cos\Omega\end{pmatrix}, \quad \Omega=\int^R \mathrm{d}R'\, f_{ab}\,, \tag{19}$$

results in

$$\mathrm{A}\to\mathrm{A}'=0\,, \quad \mathrm{V}\to\mathrm{V}'=\begin{pmatrix}\epsilon_a\cos^2\Omega+\epsilon_b\sin^2\Omega & (\epsilon_b-\epsilon_a)\cos\Omega\sin\Omega\\ (\epsilon_b-\epsilon_a)\cos\Omega\sin\Omega & \epsilon_a\sin^2\Omega+\epsilon_b\cos^2\Omega\end{pmatrix}. \tag{20}$$

With these potentials (5) describes the diabatic picture [11]. In it the vector potential is transformed away and transitions are driven by the off-diagonal terms of the scalar potential. The field strengths in the diabatic picture are not the same as (18). In non-abelian theories the field strengths are not invariant, but are covariant, under a gauge transformation [6]. In the diabatic picture the nonvanishing field strengths, $F_{R_l} = -\partial_R V$, are linear in the potentials. Thus although the SU(2) gauge group is non-abelian we have found a gauge (we call it the diabatic gauge) where the field strengths act as if they belong to an abelian theory.

If the curl of the vector potential vanishes then we can find a gauge transformation that eliminates **A**, equivalently, we can define a diabatic basis. If the expansion (1) contains channel states with nonvanishing Λ then the curl of the vector potential does not vanish, and we cannot completely eliminate all vector couplings. Under such conditions a diabatic basis is ambiguously defined [12]. We make an analogy with electromagnetism. If a charged particle is in a magnetic field-free region the vector potential can always be locally transformed away. This is not true if the particle is coupled to a vector potential with nonvanishing curl, i.e. a magnetic field.

Finally, consider two channel states composed of a Σ state and a $|\Lambda| = 1$ state, with BO eigenvalues ϵ_a and ϵ_b respectively. We assume that these states approach the $|2p\sigma\rangle$ and $|2p\pi\rangle$ states in the united atom limit. Angular couplings now appear in the **A** matrix, these couplings become large as $R \to 0$, and we can approximate the potentials by evaluating (8) in the united atom limit. Such a description is useful in the theory of the inner-shell excitation of atoms by collision with ions [13]. In this limit we can gauge away the radial coupling terms [2], and the nonvanishing potentials are [2]

$$A_\theta = \frac{1}{\sqrt{2}} \tau_1 , \quad A_\phi = \frac{-\Lambda \sin\theta}{\sqrt{2}} \tau_2 + \frac{\Lambda \cos\theta}{2} (\tau_0 - \tau_3) , \quad V = \tfrac{1}{2}(\epsilon_a + \epsilon_b)\tau_0 + \tfrac{1}{2}(\epsilon_a - \epsilon_b)\tau_3 . \tag{21}$$

These potentials are now truly non-abelian since there exists no gauge transformation where all potentials point in the same direction in the internal space. In this case we cannot define a diabatic picture because the curl of the vector potential does not vanish. Previous studies of systems driven by angular coupling in the united atom limit have ignored the diagonal vector term appearing in (21). Physically, this term is manifest as a magnetic monopole field. The effects of including this term in a study of the excitation of He by collisions with protons is under investigation [2].

These examples illustrate how gauge fields arise naturally in atomic collision theory. The gauge theory formalism allows one to make a direct association between the dynamical PSS equations ant the dynamics of a particle in an electromagnetic field. This similarity is most apparent in the one channel systems discussed above. The discussion leading up to eq. (12) showed that for a molecular system with nonvanishing Λ, the Schrödinger equation for the nuclear motion is equivalent to the Schrödinger equation of a charged particle in the field of a magnetic monopole in addition to the Born–Oppenheimer potential. This allows one to consider a non-perturbative treatment of the rotational states of a diatomic molecule. For multichannel systems, the similarity with electromagnetism is lost, but, only at a superficial level. The PSS equations can now be expressed in a non-abelian gauge theory framework. Non-abelian gauge theories are a generalization of electromagnetism and many aspects of these theories are well known [6]. For systems with many channels numerical solutions of the PSS equations becomes increasingly prohibitive. In such cases a perturbative treatment might be useful, and perturbation techniques developed for non-abelian potentials [6] might be utilized for atomic collision problems.

In a classical description, the gauge field **A** plays a subsidiary role and is considered a non-physical mathematical device. The physical quantities that drive the dynamics are the field strengths defined in (17). In quantum mechanics gauge transformations play a central role and the emphasis is not on the fields but on gauge invariant quantities [6]. This is contrary to the emphasis placed on the **A** matrix in the PSS formalism, where it is usually considered to be the primary agent in the dynamics of the system. If there exists a nonintegrable function **A**, such as considered above, certain observables might be dependent on the path in configuration space that the system traverses. Berry [8] has shown that path-dependent, or geometrical, phase factors occur during the adiabatic development of a variety of simple systems. These Berry phases also appear in atomic collisions [2]. In conclusion, it is our belief that a gauge theoretical perspective may provide a deeper understanding of the dynamical structure of atomic collisions.

I would like to thank A. Dalgarno and R. Jackiw for very helpful comments and suggestions. This work is supported by the US Department of Energy, Office of Basic Energy Sciences, Division of Chemical Sciences.

References

[1] D.R. Bates and R. McCarroll, Poc. R. Soc. A 245 (1958) 175.
[2] B. Zygelman and A. Dalgarno, in preparation (1987).
[3] G. Herzberg and H.C. Longuet-Higgins, Discuss. Faraday Soc. 35 (1963) 77.
[4] C.A. Mead and D.G. Truhlar, J. Chem. Phys. 70 (1979) 2284.
[5] J.M. Moody, A. Shapere and F. Wilczek, Phys. Rev. Lett. 56 (1986) 893.
[6] C. Itzykson and J.-B. Zuber, Quantum field theory (McGraw-Hill, New York, 1980);
R. Jackiw, Rev. Mod. Phys. 52 (1980) 661.
[7] J.C. Slater, Theory of molecules and solids, Vol. 1 (McGraw-Hill, New York, 1963) p. 39;
R.P. Feynman, Phys. Rev. 56 (1939) 340.
[8] M.V. Berry, Proc. R. Soc. A 292 (1984) 45.
[9] F. Wilczek and A. Zee, Phys. Rev. Lett. 52 (1984) 2111.
[10] R. Jackiw, Phys. Rev. Lett. 56 (1986) 2779.
[11] F.T. Smith, Phys. Rev. 179 (1969) 111;
T.G. Heil and A. Dalgarno, J. Phys. B 12 (1979) L557;
B. Zygelman and A. Dalgarno, Phys. Rev. A 33 (1986) 3853.
[12] C.A. Mead and D.G. Truhlar, J. Chem. Phys. 77 (1982) 6090.
[13] D.R. Bates and D. Williams, Proc. Phys. Soc. 83 (1964) 425.

VOLUME 56, NUMBER 24 PHYSICAL REVIEW LETTERS 16 JUNE 1986

Fractional Quantization of Molecular Pseudorotation in Na_3

Guy Delacrétaz,[(1)] Edward R. Grant,[(2)] Robert L. Whetten,[(3)] Ludger Wöste,[(1)]
and Josef W. Zwanziger[(2)]

[(1)]*Institute for Experimental Physics, Swiss Federal Institute for Technology, Lausanne, Switzerland*
[(2)]*Department of Chemistry, Cornell University, Ithaca, New York 14853*
[(3)]*Department of Chemistry and Biochemistry, and Solid State Science Center, University of California, Los Angeles, Los Angeles, California 90024*

(Received 17 March 1986)

Fractional quantization of the adiabatic pseudorotation in an isolated molecule is reported. This result, concerning the large-amplitude pseudorotation in $2^2E'$ Na_3, constitutes the first direct verification of the adiabatic sign-change theorem, and also presents the most complete picture of the level structure and internuclear dynamics of a metal-atom cluster yet given.

PACS numbers: 36.40.+d

The adiabatic theorem of quantum mechanics states that a system in an eigenstate $\Psi_n(R(t))$ responds to slowly varying changes in its parameters $R(t)$ such that the system remains in the same eigenstate, apart from an acquired phase.[1] In an unexpected recent development, Berry[2] discovered a major omission in this theorem, namely that if the parameters complete a circuit in c in a parameter space, then the acquired phase is not simply the familiar dynamical phase $(i\hbar)^{-1}\int E(R(t))dt$. Instead an additional, geometrical phase $\gamma_n(c)$ may result. The origins of this additional phase are anholonomic, that is, they depend only on the geometry of the parameter space and the circuit traversed.

The geometrical phase $\gamma_n(c)$ arises from rather general considerations and thus is relevant to many areas of quantum physics.[2,3] Condensed matter applications have been found in the statistics pertaining to the fractional quantization of the Hall conductance and to linear-chain conductors.[4–6] Wilczek and Zee have presented a generalization of this new theorem to degenerate subspaces, to show how non-Abelian gauge fields and altered quantization conditions arise in simple adiabatic systems.[7]

A further motivating consideration is the behavior of Born-Oppenheimer wave functions in molecules. A significant result first noticed in 1963 by Herzberg and Longuet-Higgins[8] relates to adiabatic excursions of such wave functions in the neighborhood of an electronic degeneracy. They show that, if the internuclear coordinates traverse a circuit within which the state is degenerate with another, then the electronic wave function Ψ_e acquires an additional phase of π, i.e., it changes sign. This sign change, which can now be recognized as a special case $[\gamma_n(c)=\pi]$ of Berry's geometrical phase, applies to the adiabatic electronic states of a large class of molecular systems exhibiting conical intersections. Its corollary is that the stationary vibrational wave function $\chi(R)$ must also change sign when taken through this closed path, to make the total product wave function, $\Psi_e(R)\chi(R)$, single valued. This result explains the long-standing prediction, concerning model systems representing the dynamical Jahn-Teller effect, that the quantum number for the pseudorotation takes on fractional values even in the absence of spin.[9]

Notably lacking are observations on real molecular systems demonstrating the existence of this effect. Especially attractive from the theoretical viewpoint is the high permutational symmetry of X_n systems (atomic clusters). The alkali clusters in particular are often cited as models.[10,11] In 1979 Herrmann *et al.*[12] reported the first absorption spectrum of gas-phase Na_3. These data are too congested to assign, but the recent use of an ultracold cluster beam source has greatly simplified the spectrum,[13,14] revealing detailed vibrational fine structure and rotational band profiles. Figure 1(a) shows the structured part of the absorption spectrum of supersonically cooled Na_3 ($T_{rot} < 10$ K, $T_{vib} < 100$ K) recorded by means of resonant two-photon ionization spectroscopy. The present work focuses on the region from 600 to 625 nm [Fig. 1(b)], presenting a complete assignment of its complex band system. As a direct consequence of this assignment we find compelling evidence for half-odd quantization

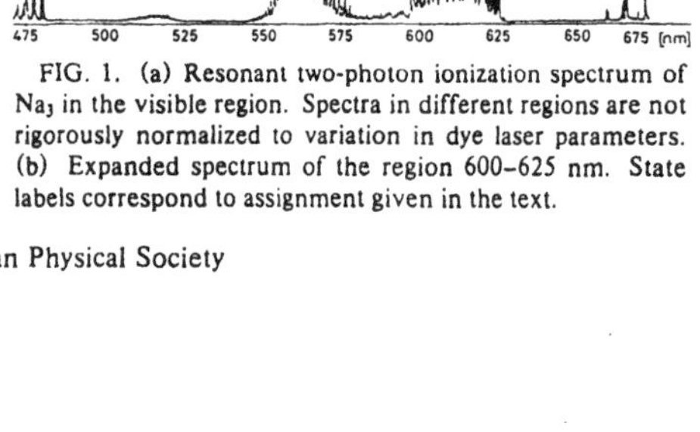

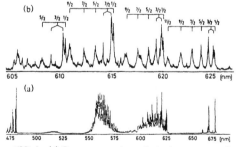

FIG. 1. (a) Resonant two-photon ionization spectrum of Na_3 in the visible region. Spectra in different regions are not rigorously normalized to variation in dye laser parameters. (b) Expanded spectrum of the region 600–625 nm. State labels correspond to assignment given in the text.

VOLUME 56, NUMBER 24 PHYSICAL REVIEW LETTERS 16 JUNE 1986

of the free molecular pseudorotation, thus offering the first experimental confirmation of the sign-change theorem and direct measurement of $\gamma_n(c)$.

Sodium clusters are produced by coexpansion of sodium vapor of moderate partial pressure (10–100 Torr) together with 2–10 atm argon through a 50-μm nozzle. As described elsewhere,[13] this high seeding ratio provides rotational and vibrational dimer temperatures of 7 and 50 K, respectively. The laser excitation and ionization steps are performed with copper-vapor-laser–pumped homebuilt dye lasers[15] (5 kW, 6 kHz, and 30-ns pulse width). This arrangement provides reasonable mass-selected ion current without saturation of the excitation step.

Among the most important characteristics of the richly structured band system in Fig. 1(b) are the following: (1) A long progression composed of nearly equally spaced bands ($\omega \approx 128$ cm^{-1} or 0.016 eV) appears to be split into *doublets*; (2) a series of closely spaced bands fanning out from the doublet and increasing steadily in breadth accompanies each member of the main progression; (3) a much weaker pattern of levels accounting for all remaining bands fits to a harmonic series with $\omega \approx 137$ cm^{-1}.

These features can be qualitatively understood if one ascribes the following properties to the internal coordinates of the three-body system[15]: The system must be strongly distorted from a (D_{3h}) equilateral triangular configuration, to explain the main progression as a set of transitions to a new equilibrium distance in the distortion-symmetry coordinate (e'). The *direction* (phase) of distortion in this two-dimensional mode must not be strongly preferred, yielding a quasifree internal rotation corresponding to the well-known pseudorotation motion (see Fig. 2), which is required to account for the fanning subpattern associated with each distortion state. The weak progression of 137 cm^{-1} remains to be explained by small-amplitude symmetric vibrations (a_1') of the triangle, negligibly coupled to the distortion and pseudorotation coordinates ρ and ϕ.

A symmetry lowering of this type is predicted for an electronic term which is degenerate at the equilateral configuration, because the electronic wave functions can then be superimposed in an unsymmetrical way, so as to favor distorted configurations (Jahn-Teller theorem[16]). For large distortions and deep states, a simple approximate pattern emerges[9]:

$$E(u,j) = (u+\tfrac{1}{2})\omega_0 + Aj^2, \qquad (1)$$

corresponding to oscillations of the distortion amplitude ($u = 0, 1, 2, \ldots$) and internal rotation with characteristic rotor constant $A = \hbar^2/2I$. Here $I = m\rho_0^2$ for an X_3 molecule, where m is the mode's reduced mass and ρ_0 is the equilibrium distortion amplitude. If the electronic wave function is required to change

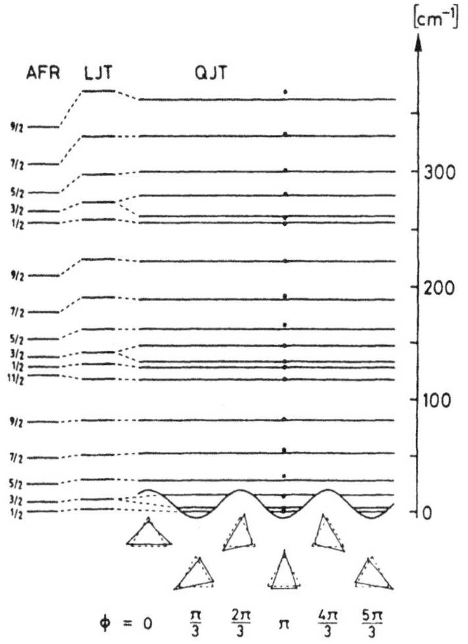

FIG. 2. Comparison of calculated and observed energy levels of $2^2E'$ Na_3. Horizontal lines give calculated values, while points represent observed band maxima. The sinusoidal function gives a cut of the lowest adiabatic surface along the pseudorotation coordinate, and molecular geometries (for arbitrary pseudorotation phase) are given below for surface minima and saddle points. Bond lengths in reduced units are acute 1.14 and 0.76, obtuse 0.91 and 1.24. On the assumption of an equilateral bond distance of 3 Å, apical angles are 39° and 86°, respectively. At left two approximations to the description of this system are given (see text).

sign, then boundary conditions demand $|j| = \frac{1}{2}, \frac{3}{2}, \ldots$. In Fig. 2 the energy-level pattern from this model is given with $\omega_0 = 128$ cm^{-1} and $A = 4$ cm^{-1}. It is readily evident that, although centrifugal terms and localization effects are neglected (see below), only the half-odd spacing accurately reproduces the level pattern. The necessity of half-odd quantum numbers is even more obvious when account of these minor perturbations is made.

A consistent theoretical model of this system, including centrifugal and nonadiabatic effects exactly, can be constructed by a variational solution of the dynamical Jahn-Teller problem. From symmetry alone in a two-state electronic basis $\{|\alpha\rangle, |\beta\rangle\}$, the correct form of the coupling with internuclear motion is (to second order in displacements)[17]

$$\hat{H} = (\hat{T}_N + \tfrac{1}{2}\rho^2) + |\alpha\rangle(k\rho e^{-i\phi} + g\rho^2 e^{2i\phi})\langle\beta| + \text{c.c.} \qquad (2)$$

Here $\hat{T}_N$ is the kinetic energy operator of the e' mode, k is the linear Jahn-Teller (distortion) parameter, and g is the quadratic (localization) parameter, in reduced units where $\omega = \hbar = M = 1$. Setting $\hat{T}_N = 0$ generates adiabatic potential energy surfaces, which for $g = 0$ have free internal rotation in a moat of depth $k^2/2$. For g nonzero, a $\cos(3\phi)$ term modulates this motion.[9] Even with neglect of localization ($g = 0$), a fit of the spectrum by the simple linear Jahn-Teller model is, but for the splitting of the first excited rotor state of each sequence, essentially quantitative (Fig. 2). All states calculated through the first three rotor sequences fit well the adiabatic separation, $\Psi_e(R)\chi(R)$.

For large distortions, some localization or hindering of the free pseudorotation is expected. The minor remaining discrepancies all fit into the class of effects caused by this localization. When (2) is diagonalized in a basis of 400 functions for the parameters $k = 4.04$ and $g = 0.012$, the eigenvalue fit indicated in the right-hand side of Fig. 2 is achieved. Quantitative comparison is made in Table I. In this fit, all observed features find a one-to-one correspondence with calculated levels and the typical deviation is only slightly more than the precision of band maxima measurement.

The spectra also convey additional interesting details about this excited state.

(i) Electronic term symmetry.—The ground state of Na_3 is known to be an E' electronic term (united-atom $1P_1$),[18] and, with strong localization in the distortion coordinate, the ground vibronic level is essentially threefold degenerate (E_1', A_2'). If the excited electronic term is E'' then pseudorotation excitations are of (E'', A_2'', A_1'') type, the last of which is rigorously dipole forbidden, so that only one of the two $|j| = \frac{3}{2}$ levels would be visible in each series. However, if the term is E', then spectroscopically allowed (E', A_2', A_1') levels result in agreement with experiment. Because this band system has intensity appropriate to a favored united-atom transition, the excited state must be of D_2 type ($d_{x^2-y^2}, d_{xy}$).[19] It is unlikely that this is a true Rydberg state, but instead it is probably the first excited term of its type: $1D_2$ or $2^2E'$.

(ii) Adiabatic potential energy surface.—The parameters k and g uniquely determine the energy surface in

TABLE I. Energy levels of the $2^2E'$ system of Na_3.

$\nu - \nu_{00}$ [a] (cm^{-1})	$\nu - \nu_{00}$ [b] ω_0	Calculated[c]	$n(j)$	$\omega_1(a_1')$	Error (cm^{-1})
0	0.00	0.00	0(1/2)		0
2.5	0.02	0.03	0(3/2−)		−1
14.5	0.115	0.13	0(3/2+)		−1.5
32.5	0.255	0.22	0(5/2)		+4
55.0	0.435	0.41	0(7/2)		+2.5
82.5	0.65	0.64	0(9/2 ±)		+1
116.5	0.915	0.92	0(11/2)		−0.5
128	1.01	1.01	1(1/2)		0
133	1.045	1.05	1(3/2−)		−0.5
137	. . .	. . .	. . .	1^1	. . .
146.5	1.155	1.16	1(3/2+)		−0.5
166.5	1.31	1.27	1(5/2)		+0.5
192.5	1.515	1.48	1(7/2)		+4
222.5	1.75	1.75	1(9/2 ±)		0
	. . .	1.96	1(11/2)		
254	2.00	2.02	2(1/2)		−2
259.5	2.045	2.06	2(3/2−)		−1.5
270.5	. . .	. . .	. . .	1^2	. . .
280.5	2.21	2.20	2(3/2+)		+1
301	2.35	2.34	2(5/2)		+1
332	2.615	2.60	2(7/2)		+1.5
369	2.905	2.84	2(9/2 ±)		+6.5
376.5	2.965		3(1/2)		
386.5	3.043		3(3/2−)		
416	3.275				

[a] $\nu_{00} = 15\,996$ cm^{-1}.
[b] $\omega_0 = 127$ cm^{-1}.
[c] With use of Hamiltonian of text [Eq. (2)] with $k = 4.04$, $g = 0.012$.

the e' coordinate space. For deep states only the lowest adiabatic surface is of interest, which has threefold maxima and minima at $\cos(3\phi) = \pm 1$. Minimizing (2) with respect to ρ for these values gives absolute minima (−) and saddle points (+) at $\rho = k/(1 \pm g)$. For the above parameters, the result is a total stabilization energy of 1050 cm^{-1} and a localization energy of 26 cm^{-1}. A cut of the energy surface at $\rho = k$ is given in Fig. 2. The lowest apparent doublet $(\frac{1}{2}, \frac{3}{2})$ of each j sequence is the result of localization, with a tunneling splitting Δ ranging from 3 to 5 cm^{-1}.

(iii) Real-space description of the pseudorotation.—One expresses the equilibrium distortion ρ_0 in atomic units, $\rho_0 = (\hbar/m\omega_0)^{1/2}k$, where for X_3 systems $m = (3M_X)^{1/2} = 8.3$ amu, and $\rho_0 = 0.71$ Å. The displacement amplitude r_i of each atom is therefore $r_i = \rho_0/\sqrt{3} = 0.41$ Å from the equilateral configuration (see Fig. 2).[15]

(iv) Coriolis coupling.—Large-amplitude internal motion implies a nonnegligible coupling with rotation about the center of mass. Accordingly, note that the individual bands have widths increasing linearly with j, the pseudorotation quantum number. Such behavior is the hallmark of the Coriolis effect, which here involves the transformation of the pseudorotation angular momentum into the frame of the molecule rotating with rotation constant B about its center of mass in the plane. Analysis, neglecting centrifugal effects, predicts a K splitting of $4BjK$. This splitting yields precisely the observed pattern if K states are weighted according to $T_{rot} = 10$ K. Broader envelopes at higher temperature account for the loss of pseudorotation structure in less-efficient expansions.[20]

To summarize, the resulting picture of excited Na_3 (Fig. 2) is that of a molecule undergoing slow, weakly hindered internal rotation accompanied by electronic-state evolution such that the electronic wave function, best represented by D_2 orbitals, changes sign upon each complete internal rotation.[8] We have demonstrated the necessity of fractional quantization of this motion. High-resolution experiments will provide better tests of fractional quantization and will also directly verify the proposed Coriolis coupling mechanism. With short laser pulses ($\tau < \Delta^{-1}$) it may be possible to prepare states with a specified quantum phase and observe their temporal evolution directly. We also mention that for large atomic clusters new quantum phases may be found.[10]

This research has benefitted from discussions with G. S. Ezra, D. Lindsay, J. Gole, M. Broyer, M. V. Berry, Ch. Jungen, and K. S. Haber. Two of us (E.R.G.) and (J.W.Z.) acknowledge support through a grant from the U.S. National Science Foundation. One of us (R.L.W.) was the recipient of a Dreyfus Distinguished New Faculty Award.

Note added.—Since this work was completed, Moody, Shapere, and Wilczek[21] have presented specific candidates for experimental detection of $\gamma_n(c)$ in molecular systems, which are unrelated to the present example.

[1] T. Kato, J. Phys. Soc. Jpn. **5**, 435 (1950); M. Born and V. Fock, Z. Phys. **51**, 165 (1928).

[2] M. V. Berry, Proc. Roy. Soc. London, Ser. A **392**, 45 (1984); see also J. H. Hannay, J. Phys. A **18**, 221 (1985); M. V. Berry, J. Phys. A **18**, 15 (1985); C. A. Mead and D. G. Truhlar, J. Chem. Phys. **70**, 2284 (1979).

[3] H. Kuratsuji and S. Iida, Prog. Theor. Phys. **74**, 439 (1985); H. Kuratsuji and S. Iida, Phys. Rev. Lett. **56**, 1003 (1986).

[4] B. Simon, Phys. Rev. Lett. **51**, 2164 (1983).

[5] D. Arovas, J. R. Schrieffer, and F. Wilczek, Phys. Rev. Lett. **53**, 722 (1984).

[6] J. R. Schrieffer, Mol. Cryst. Liq. Cryst. **118**, 57–64 (1985).

[7] F. Wilczek and A. Zee, Phys. Rev. Lett. **52**, 2111 (1984).

[8] G. Herzberg and H. C. Longuet-Higgins, Discuss. Faraday Soc. **35**, 77 (1963); H. C. Longuet-Higgins, Proc. Roy. Soc. London, Ser. A **344**, 147 (1975).

[9] H. C. Longuet-Higgins, Adv. Spectrosc. **2**, 429 (1961).

[10] S. P. Keating and C. A. Mead, J. Chem. Phys. **82**, 5102 (1985), and references within.

[11] C. A. Mead, Chem. Phys. **49**, 23 (1980).

[12] A. Herrmann, M. Hoffmann, S. Leutwyler, E. Schumacher, and L. Wöste, Chem. Phys. Lett. **62**, 216 (1979).

[13] G. Delacrétaz, G. D. Stein, and L. Wöste, to be published.

[14] G. Delacrétaz and L. Wöste, Surf. Sci. **156**, 770 (1985).

[15] M. Broyer, J. Chevaleyre, G. Delacrétaz, and L. Wöste, Appl. Phys. B **35**, 31 (1984).

[16] T. C. Thompson, D. G. Truhlar, and C. A. Mead, J. Chem. Phys. **82**, 2392 (1985).

[17] H. A. Jahn and E. Teller, Proc. Roy. Soc. London, Ser. A **161**, 220 (1937).

[18] D. M. Lindsay and G. A. Thompson, J. Chem. Phys. **77**, 1114 (1982); R. L. Martin and E. R. Davidson, Mol. Phys. **35**, 1713 (1978); J. L. Martins, R. Car, and J. Buttet, J. Chem. Phys. **78**, 5646 (1983).

[19] We adopt for Na_3 the ordering of excited states of W. D. Knight *et al.*, Phys. Rev. Lett. **52**, 2141 (1984). The Na_3 ground term is therefore $^2P_1$, where the splitting of the $P_{\|}$ and P_0 orbitals is noted. This ordering is also given by W. Gerber, Ph.D. thesis, Universität Berne, 1981 (unpublished), without the united-atom labels used here.

[20] H. P. Härri, Ph.D. thesis, Universität Berne, 1983 (unpublished).

[21] J. Moody, A. Shapere, and F. Wilczek, Phys. Rev. Lett. **56**, 893 (1986).

Chapter 5

FRACTIONAL STATISTICS

5

Fractional Statistics

Consider a particle of mass m and charge q moving on a circular ring of radius R. Suppose that the ring is threaded by a thin solenoid carrying flux ϕ. Then the Lagrangian for the particle is[1]

$$L = \tfrac{1}{2}mR^2\dot{\theta}^2 + \frac{q\Phi}{2\pi}\dot{\theta} \tag{5.1}$$

From this we may read off the canonical momentum

$$p_\theta = mR^2\dot{\theta} + \frac{q\Phi}{2\pi} \tag{5.2}$$

and the Hamiltonian

$$H = \frac{1}{2mR^2}\left(p_\theta - \frac{q\Phi}{2\pi}\right)^2 \tag{5.3}$$

The eigenfunctions are

$$\psi_n = e^{in\theta} \tag{5.4}$$

with energies

$$E_n = \frac{1}{2mR^2}\left(n - \frac{q\Phi}{2\pi}\right)^2 \tag{5.5}$$

The energy is easily interpreted as half the square of the kinetic angular momentum, divided by the moment of inertia. Notice that the allowed kinetic angular momenta are spaced by integers—that is, by whole multiples of Planck's constant—but are uniformly displaced from integers by $q\Phi/2\pi$, which might be any real number.

Now all the Lagrangians (5.1), for different values of Φ, yield the same classical equations of motion. Indeed, the term by which they differ is a total derivative, and cannot affect the classical variational principle. As we have seen, however, they lead to different quantum theories. This is not difficult to understand, from the point of view of canonical quantization. To canonically quantize the theory, we need not only to derive the equations of motion but also to impose commutation relations. The commutation relations are imposed between the coordinates and their conjugate momenta, and their content is changed if the definition of these momenta is changed, whether or not the equations of motion are affected. In the simple quantization problem above, we have seen how just such a modification can affect the physical consequences of a given classical Lagrangian.

The situation is slightly puzzling, however, if we consider it from the point of view of path integral, as opposed to canonical, quantization. In path integral quantization, transition amplitudes are calculated by adding the contributions of all possible paths, each weighted by the the exponential of the classical action along the path. Thus it would seem that the transition amplitudes, and thereby the entire physical content of the theory, should be determined by the classical action. On the other hand, we have just seen that different Lagrangians, equivalent at the classical level, can lead to different quantum theories. How did this occur? And could we have forseen this possibility in advance, knowing only the degrees of freedom and the classical limit? The latter question, as we shall see, is of some practical importance, since in proposing effective Lagrangians for complicated systems we often have a clear idea what the important degrees of freedom are, and expectations for their behavior in the classical limit, but a much more shadowy idea of anything more subtle. In such situations, we would like to know whether quantization is ambiguous, and what further considerations are necessary to resolve the ambiguity.

Any reader who has followed us to this point, or kept the title of this book in mind, will naturally suspect that the geometric or Berry phase plays a key role in this sort of problem. Indeed, its quasi-kinematic character, and the fact that it forms the first (order $\hbar^0$) correction to purely classical (order $\hbar^{-1}$) behavior, strongly hint at this role.

Let us return from these generalities to the case at hand. Since we have the explicit Lagrangian (5.1) in front of us, we can easily get to the root of the problem. It is that the classically ignorable $\dot{\theta}$ term is not ignorable in the quantum theory. What does this term do, path by path? It suffices to compare paths that begin at a common position θ_1 at time t_1 and end at a definite position θ_2 at time t_2, since only such paths can interfere. It is easy to see that the effect of the $\dot{\theta}$ term for such paths is to weight their relative contribution by

$$\exp\left[i\frac{q\Phi}{2\pi}\left(\int_{\text{path }1}\dot{\theta}\,dt - \int_{\text{path }2}\dot{\theta}_2\,dt\right)\right] \equiv \exp i\frac{q\Phi}{2\pi}\delta\theta \tag{5.6}$$

Now $\delta\theta$ is a multiple of 2π for paths with common endpoints (and only such paths can interfere). Loosely speaking, $\delta\theta/2\pi$ measures the difference between the number of times the first and second paths winds around the ring, respectively. More precisely, it is the number of times the composite path we get by following the first path from t_1 to t_2 and the inverse of the second path back from t_2 to t_1 winds around the ring.

Working backwards, we can now see very clearly why there was an ambiguity in formulating the path integral. By the argument given immediately above, we can concentrate on closed paths—that is, paths that begin and end at the same point. Such paths fall into distinct, disconnected classes, labeled by the winding number. Because the winding number is an integer for any path, continuous changes in the path cannot alter its value at all. Now, the crucial observation is that the classical Lagrangian cannot guide us in choosing how to weight the relative contributions of disconnected classes of paths. For the classical equations of motion follow from a variational principle that involves only comparisons among infinitesimally nearby paths, and cannot give guidance in comparing disconnected classes of paths. And so, if the closed paths in configuration space fall into disconnected classes, then there is an ambiguity in quantization. In standard mathematical language, we would say that there is a possibility for a single classical Lagrangian to lead to various quantum theories, when the first homotopy group of the configuration space is non-trivial, or in other words when the configuration space is not simply connected.

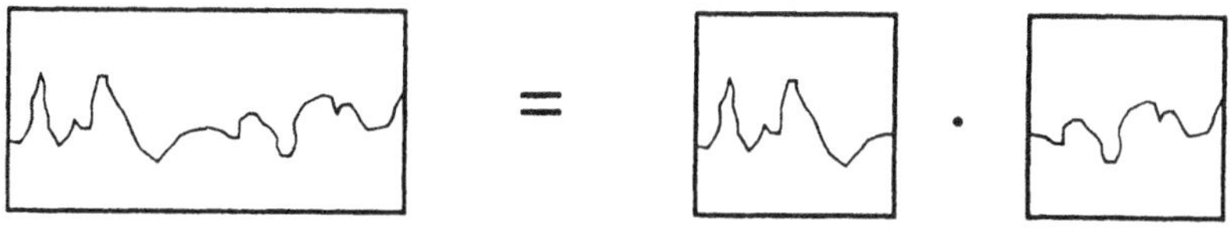

$$e^{iS_{1+2}} = e^{iS_1} \cdot e^{iS_2}$$

Figure 5.1 The amplitude for the composition of two paths is the product of the amplitudes for each path separately.

There is an important principle constraining the choice of relative weightings between different classes of closed paths. It has to do with the "group" aspect of the homotopy group.[2] According to basic principles of quantum mechanics, the amplitude for the composition of two paths must be the product of the amplitude for each path separately (see Figure 5.1). This rule is manifestly obeyed by the usual assignment, that weights with the exponential of the integral of the classical action—since the integrals add, the exponentials multiply. We must make sure that any additional weighting factors obey the rule separately. Now composing paths is precisely the operation that defines the homotopy group. The elements of this group are the disconnected classes of paths, and the product of two such classes is obtained by composing representative paths from each; it is the class of the composed path. Thus if we are assigning extra numerical factors α_π to the paths, we must demand that they obey the rule

$$\alpha_{\pi_1 \circ \pi_2} = \alpha_{\pi_1} \cdot \alpha_{\pi_2} \tag{5.7}$$

or in other words that the factors form a (one-dimensional) representation of the homotopy group. Higher dimensional representations may also have a sensible interpretation; the states have then a non-classical internal degree of freedom, that locates them in the representation space. Some issues that arise in the quantization of theories on non-simply connected spaces are addressed in recent work of Balachandran.[3]

The phase factors we have been discussing certainly deserve to be called geometric—in fact, being topological, they go one step further towards pure kinematics. Are they related to the geometric phases discussed in previous chapters? In fact, the relationship is quite close. Let us illustrate this on our example. The geometric phase between nearby position eigenstates $|\boldsymbol{\theta}\rangle$ and $|\boldsymbol{\theta} + \Delta\boldsymbol{\theta}\rangle$ is, formally

$$\exp i \frac{q\Phi}{2\pi} \Delta\theta \tag{5.8}$$

This phase is, of course, locally integrable, but globally it is not—when we come around adiabatically by 2π, we have accumulated a phase of $e^{iq\Phi}$. Thus the geometric phase around a homotopically non-trivial path in configuration space parametrizes the ambiguity in quantization. The relationship demonstrated here in a simple example is much more general, as we shall see when we come to discuss anomalies and Wess-Zumino terms in Chapter 7.

An interesting implication of the above is that the wave function in general cannot be defined on ordinary configuration space. For as we have seen, the position eigenstates for $\boldsymbol{\theta}$ and $\boldsymbol{\theta} + 2\pi$ are related by a phase factor. Properly then, we should define the wave function with $\boldsymbol{\theta}$ running from $-\infty$ to $+\infty$, with the boundary condition

$$\psi(\theta + 2\pi) = e^{iq\Phi}\psi(\theta) \tag{5.9}$$

In general, the wave function will live on the universal covering space of configuration space (or, if there are internal degrees of freedom, a vector bundle over the covering space) and will obey boundary conditions relating points that project to the same point in configuration space. Actually, we have seen examples of this arise before, abstractly in the case of the monopole bundle (introduction to Chapter 3) and concretely in the context of molecular physics (Chapter 4), where wave functions for half-integral orbital angular momentum appeared. Such angular wave functions cannot be realized as a vector bundle of the familiar, "trivial" kind (concretely, as an array of $2j + 1$

functions on the sphere, the vector spherical harmonics) but live comfortably on a twisted bundle.

A notable application of these ideas arises in the quantum mechanics of identical particles. The configuration space for N identical particles in d-dimensional space is

$$\overbrace{\mathbf{R}^d \times \cdots \mathbf{R}^d}^{N \text{ copies}} \Big/ S_N \tag{5.10}$$

the notation indicates that points differing by a permutation of positions are to be identified. Let us assume also that the amplitude for particles to collide vanishes. (This is found to be true *a posteriori* for anything but bosons, because there is a effective centrifugal barrier. For bosons, the topology of configuration space collapses, but that doesn't matter since the paths are equally weighted anyway.) Then we arrive at the configuration space

$$(\mathbf{R}^d \times \cdots \times \mathbf{R}^d - \Delta) \Big/ S_N \tag{5.11}$$

where Δ is the space of configurations where two or more particles occupy the same point in $\mathbf{R}^d$.

To illustrate, let us consider the case $N = 2$ in more detail. In general, for $d > 2$ the homotopy group of the configuration space for N identical particles is simply the group of permutations. For $d = 2$, however, it is a much more interesting group, the so-called braid group. As we have seen, new possibilities for quantization are associated with representations of the homotopy groups. The one-dimensional representations of the permutation group are of course well known; they are the trivial representation and the "sign" representation, that assigns a factor $+1$ or -1 to even and odd permutations respectively. These two possibilities correspond to quantizing the identical particles as bosons or fermions, respectively. The one-dimensional representations of the braid group are analyzed in the enclosed paper by Wu [5.3]. It turns out that they are parametrized by a single parameter, a complex number of unit modulus, commonly written $e^{i\theta}$. When this parameter is $+1$ we have bosons, when it is -1 we have fermions; the peculiar feature of two spatial dimensions is the possibility of continuous interpolation between these familiar cases.

Higher dimensional unitary representations of the homotopy groups also exist. For $d > 2$, particles quantized in this way are said to obey parastatistics. The higher-dimensional unitary representations of the braid group are, it seems, not completely classified. They arise in conformal field theories, and are a very active topic of recent research.[4]

There is a very simple and attractive way of realizing fractional statistics, closely related to the example discussed above. Indeed the characteristic feature of the problem we encountered in quantizing the charged particle on a flux-enclosing ring was the possibility of interactions which generated phase factors for non-contractible paths. To implement fractional statistics, such phase factors are precisely what we need. All this suggests that fractional statistics—or, in the jargon, non-integrable holonomy factors on configuration space—can be realized dynamically, by a suitable arrangement of fictitious charge and flux. Such a construction has been performed in great generality, both for quantum mechanics and quantum field theory, in the enclosed paper of Arovas, Schrieffer, Wilczek and Zee [5.4], and is discussed further in the article by Arovas in the following chapter [6.2].

Finally, let us mention several occurences of fractional statistics in the natural world.

Effectively two-dimensional tubes of magnetic flux play an important role in the theory of type II superconductors. Indeed, magnetic fields penetrate type II superconductors only in such tubes. The amount of flux carried by a tube is quantized in units of $\Phi_0 = 2\pi/2e$. This is easily understood in terms of our particle on a ring example: the lowest value of the energy can only be achieved for integral values of $q\Phi/2\pi$ in equation (5.5). For a superconductor we are concerned with the wave function for Cooper pairs, so that $2e$ is the relevant charge. Also, the quantization is sharp (for a macroscopic sample) because a very large number of particles are involved in the condensate. Thus the only acceptable way of accomodating magnetic flux, that does not involve energy proportional to the volume of the sample, is to confine it in small flux tubes carrying integral multiples of Φ_0. Now if we consider a single electron—not screened by the condensate—orbiting around a flux tube carrying flux ϕ_0, we see that this composite provides one realization of our "statistical" interaction. Both the flux tube, and the flux tube plus an electron, are bosons. Closely related (but more elaborate and speculative) phenomena occur for particles orbiting string solutions in unified models of particle physics.[5]

The non-linear sigma model, which assigns to each point in space a direction, is used in the description of the low-energy excitations of magnetically ordered systems. In two spatial dimensions, its configuration space is not simply connected, a fact closely connected with the existence of the topological construct known as the Hopf invariant. As a result, the quantization of this classical model of magnetism is ambiguous. This situation is analyzed in the enclosed paper of Wilczek and Zee [5.2]. The correct quantization must be determined by appealing to the microscopic model underlying the effective sigma-model. This model has been a subject of recent research, given some urgency by the importance of two-dimensional magnetism in materials supporting high temperature superconductivity.

Finally, perhaps the most intriguing application of fractional statistics to date concerns the quantized Hall effect. This application is so interesting in itself, and such a nice example of the use of the geometrical phase, that we have devoted the following Chapter exclusively to it.

[1] F. Wilczek, *Phys. Rev. Lett.* **48** (1982) 1144.

[2] N.D. Mermin, "The topological theory of defects in ordered media," *Rev. Mod. Phys.* **51** (1979) 591–648.
Charles Nash and Siddhartha Sen, *Topology and Geometry for Physicists*, London: Academic Press, 1983.

[3] A.P. Balachandran, "Topological Aspects of Quantum Gravity," Syracuse Univ. preprint no. SU–4228–373.

[4] Cumrun Vafa, "Toward Classification of Conformal Theories," *Phys. Lett.* **206B** (1988) 421.
Edward Witten, "Quantum Field Theory and the Jones Polynomial," Inst. for Advanced Study preprint no. IASSNS–HEP–88/33.
Gregory Moore and Nathan Seiberg, "Classical and Quantum Conformal Field Theory," IAS preprint no. IASSNS–HEP–88/35.

[5] Mark Alford and Frank Wilczek, "Aharonov–Bohm Interaction of Cosmic Strings with Matter," Harvard Univ. preprint no. HUTP–88/A047, to appear in Physical Review Letters.

VOLUME 51, NUMBER 25 PHYSICAL REVIEW LETTERS 19 DECEMBER 1983

Linking Numbers, Spin, and Statistics of Solitons

Frank Wilczek
Institute for Theoretical Physics, University of California, Santa Barbara, California 93106

and

A. Zee
Institute for Theoretical Physics, University of California, Santa Barbara, California 93106, and Department of Physics,[a] University of Washington, Seattle, Washington 98195

(Received 24 October 1983)

The spin and statistics of solitons in the (2+1)- and (3+1)-dimensional nonlinear σ models is considered. For the (2+1)-dimensional case, there is the possibility of fractional spin and exotic statistics; for 3+1 dimensions the usual spin-statistics relation is demonstrated. The linking-number interpretation of the Hopf invariant and the use of suspension considerably simplify the analysis.

PACS numbers: 02.40.+m, 11.10.Lm, 11.30.-j, 75.70.-i

The existence of solitons in the (2+1)-dimensional O(3) nonlinear σ model, as first discussed by Belavin and Polyakov,[1] is implied by the homotopy $\pi_2(s_2)=z$. The model is described by the functional

$$E=\int d^2x(1/2f)(\partial_i n^a)^2, \quad i=1,2; \quad a=1,2,3, \tag{1}$$

giving the energy of a static configuration specified by $n^a(\vec{x})$. The "order parameter" n^a is a three-dimensional unit vector: $n^a n^a=1$. If we describe the ground state by $\vec{n}(\vec{x})=(0,0,1)$, then the basic soliton is described by

$$\vec{n}(\vec{x})=(\hat{x}\sin f, \cos f). \tag{2}$$

Here $\hat{x}=\vec{x}/|\vec{x}|$ denotes the two-dimensional unit radial vector and $f(r=|\vec{x}|)$ is a function varying smoothly and monotonically from $f(0)=\pi$ to $f(\infty)=0$ as r increases. We refer to such a topological configuration as a skyrmion.[2] The topological current in this model is

$$J^\mu=(1/8\pi)\epsilon^{\mu\nu\lambda}\epsilon^{abc}n^a\,\partial_\nu n^b\,\partial_\lambda n^c. \tag{3}$$

The space-time indices $\mu, \nu, \ldots$ run over 0, 1, 2. One easily verifies the conservation of this. The topological charge of this current,

$$Q=\int d^2x\,J^0$$
$$=(1/8\pi)\int d^2x\,\epsilon^{ij}\epsilon^{abc}n^a\,\partial_i n^b\,\partial_j n^c, \tag{4}$$

clearly describes the homotopy of the mapping $s_2\to s_2$ for $\vec{n}$ satisfying the boundary condition $\vec{n}(\vec{x}=\infty)=$const. The skyrmion has $Q=1$. By using Bogomolny's inequality, one can solve exactly the problem of minimizing the energy functional for a given Q. Finally, we mention that this model[3] provides a phenomenological description of Heisenberg ferromagnets in a two-dimensional system and thus the phenomena exhibited in this model may conceivably be accessible experimentally.

In this paper we point out that the skyrmion may possess fractional angular momentum and obey peculiar quantum statistics. One of us had previously proposed[4-6] the possibility of fractional angular momentum and of statistics which are neither Bose-Einstein nor Fermi-Dirac. As we will see, the (2+1)-dimensional O(3) nonlinear σ model provides an amusing and explicit field-theoretic realization of these ideas. Our discussion is also related to several other field-theoretic phenomena discovered in recent years.

The relevant mathematics which allows skyrmions to have these peculiar properties is the homotopy $\pi_3(s_2)=z$ [which is perhaps somewhat less obvious than the homotopy $\pi_2(s_2)=z$ responsible for the skyrmion's existence]. It is easy to exhibit the basic Hopf map of $s_3\to s_2$. In fact, physicists should be familiar with this fact from elementary discussions of the Pauli matrices σ^a. Define $n^a=z^\dagger\sigma^a z$ where

$$z=\begin{pmatrix} z_1 \\ z_2 \end{pmatrix}$$

is a complex two-component spinor with the constraint $|z_1|^2+|z_2|^2=1$. Notice that the U(1) transformation $z\to e^{i\theta}z$ leaves n invariant and so the inverse image of any point on s_2 is a circle on s_3. What is not so obvious is the construction of a Hopt invariant to describe $\pi_3(s_2)$ [just as Q describes $\pi_2(s_2)$]. This will be explained below.

Let us first address the physical question of the spin of the skyrmion. To determine the spin, we rotate the skyrmion adiabatically through 2π over a long time period T. According to Feynman, at the end of this rotation the wave function acquires a phase factor e^{iS} where S is the action corresponding to the adiabatic rotation. The angular momentum J of the skyrmion is given by $e^{iS} = e^{i2\pi J}$.

Now, if S has simply the standard form [cf. Eq. (1)]

$$S_0 = \int d^3x (1/2f)(\partial_\mu n^a)^2, \tag{5}$$

then it is easy to see that S_0 is of order $1/T \to 0$ as $T \to \infty$. The skyrmion has $J = 0$. However, we have not taken into account the possibility of including in S a topological term. This possibility is by now familiar from the discussion of the θ vacua[7] in quantum chromodynamics, from studies of three-dimensional Yang-Mills theory and gravity,[8] and from recent work of Witten[9] on strongly interacting skyrmions[10] (based on earlier work of Wess and Zumino[11]). In general, we can have $S = S_0 + \theta H$ where θ is a real parameter and H is the Hopf invariant which we now define.

The conservation of J^μ licenses us to manufacture a "gauge potential" A^μ by the curl equation:

$$J^\mu = \epsilon^{\mu\nu\lambda}\partial_\nu A_\lambda \equiv \tfrac{1}{2}\epsilon^{\mu\nu\lambda}F_{\nu\lambda}. \tag{6}$$

A_μ is defined up to the gauge freedom $A_\mu \to A_\mu - \partial_\mu \Lambda$. Note that A_μ depends nonlocally on $n^a(x)$. In the gauge $\partial A = 0$, we have $A_\mu = -\partial^{-2}\epsilon_{\mu\nu\lambda}\partial_\nu J_\lambda$. [An alternative construction is to write $A_\mu = iz^\dagger \partial_\mu z$. The U(1) phase rotation on z induces the gauge transformation on A_μ.] The Hopt invariant is defined by

$$H = -(1/4\pi)\int d^3x\, \epsilon^{\mu\nu\lambda}A_\mu F_{\nu\lambda}$$
$$= -(1/2\pi)\int d^3x\, A_\mu J^\mu. \tag{7}$$

H is obviously invariant under gauge transformation on A_μ. [We note that this is just the Abelian version of the topological term studied by Deser, Jackiw, and Templeton,[8] but since H is gauge invariant, θ is not quantized. Furthermore, here H is to be regarded as a functional of $\vec{n}(\vec{x})$. In the language of Zumino, Wu, and Zee,[12] H is proportional to $\int \omega_3{}^0$. If $\hat{\mu}$ denotes a four-dimensional index then $\partial_{\hat\rho}\epsilon^{\hat\rho\hat\mu\hat\nu\hat\lambda}A_{\hat\mu}F_{\hat\nu\hat\lambda} = \tfrac{1}{2}\epsilon^{\hat\rho\hat\mu\hat\nu\hat\lambda}F_{\hat\rho\hat\mu}F_{\hat\nu\hat\lambda}$, connecting the Hopf invariant to the chiral anomaly.]

Spatial rotation of a single skrymion is equivalent to an isospin rotation and thus we evaluate H for the time-varying configuration $n_1 \pm in_3 = e^{\pm i\alpha(t)} \times (\hat{n}_1 \pm i\hat{n}_2)(\vec{x})$, $n_3 = \hat{n}_3(\vec{x})$. {Strictly speaking, this defines a map of $s_2 \times [0, 1] \to s_2$.} It is not necessary to know the explicit form of n_a. From Eq. (3) we find

$$J_i = -(1/8\pi)(d\alpha/dt)\epsilon_{ij}\,\partial_j n_3. \tag{8}$$

It suffices to know that $J_0(r) = \epsilon_{ij}\,\partial_i A_j$ is a function of r to determine $A_j = -\epsilon_{jk}x_k g(r)/r^2$ where $g(r) = \int_0^r dr'\, r' J_0(r')$. This and Eq. (8) allow us to determine $A_0 = -(d\alpha/dt)n_3$. Inserting into Eq. (7) we find

$$H = g(\infty)n_3(\infty)[\alpha(T) - \alpha(0)]/2\pi = 1. \tag{9}$$

The skyrmion has angular momentum $\theta/2\pi$.

For a ferromagnet, θ should be determined by the microscopic theory underlying the phenomenological σ model.

It is easy to show that H is a homotopic invariant for $s_3 \to s_2$. Consider a map with $\vec{n}(\vec{x}, t = \infty)$ constant and a small deformation of $\vec{n}$ leaving invariant $\vec{n}(\infty)$. Then

$$\delta J_\mu = \epsilon_{\mu\nu\lambda}\,\partial_\nu \epsilon_{abc} 2n_a\, \delta n^b\, \partial_\lambda n^c$$
$$= \epsilon_{\mu\nu\lambda}\,\partial_\nu \delta A_\lambda \tag{10}$$

and we find

$$\delta H = (-1/2\pi)2\int d^3x\, \delta A_\mu J^\mu = 0.$$

There is a deep theorem[13] which equates the Hopf invariant to the linking number between two curves in R^3. To have a heuristic understanding of this, consider the maps $s_3 \to s_2$. The reverse image of a point in s_2 is a curve in s_3 which by a stereographic projection we can think of as a curve in R^3 (with ∞ identified as one point). Thus, for the basic map given explicitly above, $\vec{n} = (0, 0, 1)$ corresponds to the great circle $|z_1| = 1$, $z_2 = 0$ on s_3, while $\vec{n} = (0, 0, -1)$ corresponds to $z_1 = 0$, $|z_2| = 1$. Write the real components of (z_1, z_2) as $(\cos\psi, \sin\psi\cos\theta, \sin\psi\sin\theta\cos\varphi, \sin\psi\sin\theta\sin\varphi)$ and stereographically project this point to $r(\psi) \times (\cos\theta, \sin\theta\cos\varphi, \sin\theta\sin\varphi)$ in R^3 where $r(\psi)$ ranges monotonically from ∞ to 0 as ψ ranges from 0 to π. We see that the curves corresponding to $\vec{n} = (0, 0, 1)$ and to $\vec{n} = (0, 0, -1)$ link once. The reader may find it amusing to work out the curves corresponding to other points.

Using this linking theorem, we can easily determine the spin and statistics of a skyrmion. Consider the following process in 2+1 dimensions. At some time create a pair of skyrmion and antiskyrmion and pull them apart. Rotate the skyrmion through 2π. Allow the pair to come together. Since at ∞ we have the physical vacuum this defines a map $s_3 \to s_2$. Were the skyrmion not rotated, the map would be homotopically trivial. Here, the corresponding map has Hopt invariant 1. The two curves traced out by two specific values of $\vec{n}$ will be linked once as indicated in Fig. 1.

To determine the statistics obeyed by a skyrmion we consider a process in which we create two skyrmion-antiskyrmion pairs and subsequently

VOLUME 51, NUMBER 25 PHYSICAL REVIEW LETTERS 19 DECEMBER 1983

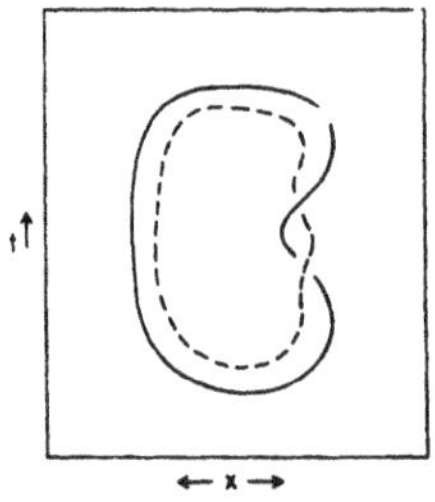

FIG. 1. The creation and annihilation of a skyrmion-antiskyrmion pair, with a 2π rotation of the skyrmion. The two curves correspond to $\vec{n}=(0,0,1)$ and $(1,0,0)$, say.

bring them to annihilation but after interchanging the two skyrmions. We see, by the maneuvering indicated in Fig. 2, that the linking number is 1 for this process. The map of $s_3 \to s_2$ corresponding to this process therefore has Hopf invariant 1. Thus, the skyrmion obeys exotic statistics which interpolates continuously between Bose and Fermi statistics as described in Ref. 6. (By the way, the alternative of directly computing the Hopf integral corresponding to rotating a pair of skyrmions through π appears to be quite difficult.) Note that the discussion there is for a gauge theory. Here, we do not have a gauge theory but, curiously, one can manufacture a gauge potential A_μ.

Given a map $f: s_k \to s_n$ one can always construct[13] a map $\bar{f}: s_{k+1} \to s_{n+1}$ (called the Freudenthal suspension of f) by $\bar{f}(t, (1-t^2)^{1/2}x) = (t, (1-t^2)^{1/2}f(x))$ where $x \in s_k$ and $t \in [0,1]$. This induces a homomorphism[13] $F: \pi_k(s_n) \to \pi_{k+1}(s_{n+1})$ of the homotopy classes of f and $\bar{f}$. Our discussion can thus be "suspended" into (3+1)-dimensional space-time: $\pi_2(s_2) \to \pi_3(s_3) = z$ and $\pi_3(s_2) \to \pi_4(s_3) = z_2$. The first of these is an isomorphism, the second is onto: The suspension of a map $s^3 \to s^2$ to a map $s^4 \to s^3$ is nontrivial if and only if the map has odd Hopf invariant. Since $s_3 = \mathrm{SU}(2)$ manifold, the homotopy $\pi_3(s_3)$ implies the existence of skyrmions in the $\mathrm{SU}(2) \otimes \mathrm{SU}(2)$ nonlinear σ model. The fact that $\pi_4(s_3) = z_2$ allows one to quantize the skyrmion as a spin-$\frac{1}{2}$ fermion as discussed by Witten.[9] It is consistent with the standard three-space angular momentum analysis and with the well-known facts $\pi_1(\mathrm{SO}(2)) = z$ and $\pi_1(\mathrm{SO}(3)) = z_2$ that $\pi_4(s_3)$ is z_2 rather than z.

This material is based upon research supported in part by the National Science Foundation under Grant No. PHY77-27084, supplemented by funds from the National Aeronautics and Space Administration.

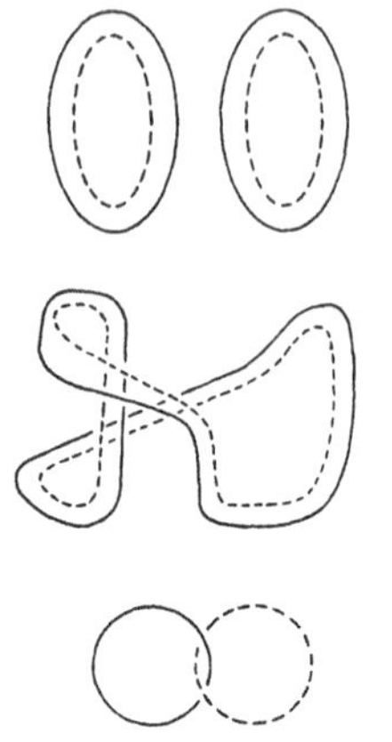

FIG. 2. (a) The creation and annihilation of two skyrmion-antiskyrmion pairs. (b) The process in (a) but with an interchange of the two skyrmions. (c) The two curves in (b) after a homotopic deformation.

[(a)]On leave of absence 1983–1984.

[1]A. A. Belavin and A. M. Polyakov, Pis'ma Zh. Eksp. Teor. Fiz. 22, 503 (1975) [JETP Lett. 22, 245 (1975)].
[2]T. H. R. Skyrme, Proc. Roy. Soc. London, Ser. A 247, 260 (1958).
[3]For a review of the material in this introductory paragraph, see R. Rajaraman, *Solitons and Instantons* (North-Holland, Amsterdam, 1982).
[4]Some aspects of this subject appear to have been anticipated in the remarkable paper of D. Finkelstein and J. Rubinstein, J. Math. Phys. 9, 1762 (1968).
[5]P. Hasenfratz, Phys. Lett. 85B, 338 (1979); J. Schonfeld, Nucl. Phys. B185, 157 (1981).
[6]F. Wilczek, Phys. Rev. Lett. 48, 1144 (1982), and 49, 957 (1982).
[7]G. 't Hooft, Phys. Rev. Lett. 37, 8 (1976); C. Callan, R. Dashen, and D. Gross, Phys. Lett. 63B, 334 (1976); R. Jackiw and C. Rebbi, Phys. Rev. Lett. 37, 172 (1976).
[8]J. Schonfeld, Ref. 5; S. Deser, R. Jackiw, and S. Templeton, Phys. Rev. Lett. 48, 975 (1982), and Ann. Phys. (N.Y.) 140, 372 (1982); see also Y.-S. Wu and A. Zee, to be published.
[9]E. Witten, Nucl. Phys. B223, 422, 433 (1983).
[10]A. P. Balashandran, V. P. Nair, and C. G. Trahern, Phys. Rev. Lett. 49, 1124 (1982).
[11]J. Wess and B. Zumino, Phys. Lett. 37B, 95 (1971).
[12]B. Zumino, Y.-S. Wu, and A. Zee, to be published.
[13]For example, P. J. Hilton, *An Introduction to Homotopy Theory* (Cambridge Univ. Press, Cambridge, England, 1953), Chap. VI.

VOLUME 53, NUMBER 2 PHYSICAL REVIEW LETTERS 9 JULY 1984

Multiparticle Quantum Mechanics Obeying Fractional Statistics

Yong-Shi Wu
Department of Physics, University of Washington, Seattle, Washington 98195
(Received 2 April 1984)

We obtain the rule governing many-body wave functions for particles obeying fractional statistics in two (space) dimensions. It generalizes and continuously interpolates the usual symmetrization and antisymmetrization. Quantum mechanics of more than two particles is discussed and some new features are found.

PACS numbers: 03.65.Ca, 03.65.Ge, 05.30.-d

In two (space) dimensions, there are allowed to be particles of fractional angular momentum or spin.[1,2] If there is a generalized spin-statistics connection, such particles are expected to have unusual (fractional) statistics which continuously interpolates between the normal bosons and fermions. (An example for such interpolation is known in one dimension.[3]) The intriguing problem of how it works is interesting both from the viewpoint of theoretical principles and from the prospect of physical applications. A possible relevance of fractional statistics to the quantized Hall effect has been recently suggested.[4]

Two simple models have been proposed for particles obeying fractional statistics by Wilczek[1,5] and Wilczek and Zee.[2] (See also Ref. 6.) Two-particle quantum mechanics was analyzed in detail. A low-density expansion of the partition function interpolating the standard statistics was obtained. As pointed out in these papers, Feynman's path-integral formulation is a good starting point. However, the formalism in terms of wave functions may be practically more convenient. An immediate problem is the general rule governing the many-body wave functions, namely how to generalize the usual rule to obtain a continuous interpolation between symmetrization and antisymmetrization. In this note I answer this question by deriving the desired rule in the two models mentioned above. As an application, I discuss the quantum mechanics of three particles, not yet touched in the literature. Some new features are found which are not present in the two-particle case.

Anyons revisited.—Following Wilczek,[5] I denote composites formed from charged particles and magnetic flux tubes as anyons, since their spin

$$\Delta = q\Phi/2\pi = \theta/2\pi \tag{1}$$

can take any real values. Here $-q$ is the charge and Φ the flux. That[5] interchange of two anyons leads to a phase $e^{i\theta}$ is an indication of the fractional statistics. We here consider quantum mechanics for more than two anyons.

The Hamiltonian for a charged particle orbiting around a flux tube can be written as

$$H_0 = \frac{1}{2m_q}\left[-i\frac{\partial}{\partial \vec{r}_q} + q\vec{A}(\vec{r}_q - \vec{r}_f)\right]^2 + \frac{1}{2m_f}\left[-i\frac{\partial}{\partial \vec{r}_f} - q\vec{A}(\vec{r}_q - \vec{r}_f)\right]^2. \tag{2}$$

Here we consider the limit in which the size of the flux tube can be neglected. $\vec{r}_q$ and $\vec{r}_f$ are two-dimensional vectors. Let us assume that the flux tube has a finite effective mass m_f in two dimensions. The form (2) has the advantage that the effect of the interaction is confined to the wave function in the relative coordinate. In a regular gauge the vector potential is

$$q\vec{A}(\vec{r}_q - \vec{r}_f) = -q\vec{A}(\vec{r}_f - \vec{r}_q) = (\theta/2\pi)[\vec{n}\times(\vec{r}_q - \vec{r}_f)]/|\vec{r}_q - \vec{r}_f|^2 \tag{3}$$

(with $\vec{n}$ being the unit vector normal to the two-dimensional plane), and the wave function is single-valued everywhere.

Now we proceed to consider n identical anyons and neglect the electrostatic forces between them (i.e., consider the limit $q \to 0$ with $\theta = q\Phi$ fixed). The charged particle in each anyon feels the vector potential of the flux tube in the other. Using the Hamiltonian (1) and applying a procedure similar to that in Goldhaber[7] for the charge-monopole composites, one finds that the anyon-anyon potential is equivalent to that of a charge interacting with twice the flux in one flux tube; namely

$$H = \sum_{i=1}^{n}\frac{1}{2m_a}\left[-i\frac{\partial}{\partial \vec{r}_i} + 2q\sum_{j\neq i}\vec{A}(\vec{r}_i - \vec{r}_j)\right]^2. \tag{4}$$

Let us adopt Eq. (3) for $\vec{A}(\vec{r}_i-\vec{r}_j)$ in the regular gauge in which the wave function ψ is single-valued as in the one anyon case. To eliminate the long-range vector potential between anyons, we make the gauge transformation

$$\psi'(\vec{r}_1,\ldots,\vec{r}_n)=\prod_{i<j}\exp\left\{i\frac{\theta}{\pi}\phi_{ij}\right\}\psi(\vec{r}_1,\ldots,\vec{r}_n), \tag{5}$$

where ϕ_{ij} is the azimuthal angle of the relative vector $\vec{r}_i-\vec{r}_j$. Now the new wave function ψ' satisfies the free Schrödinger equation with no vector potential.

At first sight the multivaluedness of the new wave function ψ' seems to be very discomforting. One can manage to avoid it by imposing appropriate boundary conditions for ψ' on certain cuts in the two-dimensional plane[5] or formulating quantum mechanics on sections on fiber bundles.[8] However, these two methods are very hard to put into practice for more than two anyons. Actually, nothing is wrong with the multivaluedness of the wave function (5). The modulus squared, $|\psi'|^2$, is single-valued, and the multivalued phase factors are just right to keep track of the Aharonov-Bohm effect.[9] In my opinion once one understands the need for extending the notion of a wave function (i.e., not requiring it to be necessarily 2π periodic in ϕ_{ij}), there is no difficulty in accepting and directly using the multivalued wave function (5) as everybody does with the double-valued spinors in three dimensions.[10]

By use of the complex coordinates $z_i=x_i+iy_i$ and $z_i^*=x_i-iy_i$ instead of $\vec{r}_i=(x_i,y_i)$, the wave function (5) can be put into a more elegant form[11]:

$$\psi'(z_i,z_i^*)=\prod_{i<j}(z_i-z_j)^{\theta/\pi}f(z_i,z_i^*), \tag{6}$$

with $f(z_i,z_i^*)=(r_{ij})^{-\theta/\pi}\psi(z_i,z_i^*)$ single-valued. f is totally symmetric (antisymmetric) in the pairs (z_i,z_i^*), if all the fields describing the flux tube and charged particle are bosonic (if the charged particle is fermionic). The equation (6) is the desired rule for many-body wave functions obeying θ statistics.

Solitons in point approximation.—The solitons in the (2+1)-dimensinal O(3) nonlinear sigma model, with a topological action, also provide a model for particles with fractional spin and statistics.[2,6] When widely separated solitons are approximately treated as point particles, the topological term (with the parameter θ) leads to an additional term

$$S'=\int dt\,L',\quad L'=(-\theta/\pi)(d/dt)\textstyle\sum_{i<j}\phi_{ij}, \tag{7}$$

to the ordinary action $S_0=\int dt\,\frac{1}{2}m\sum_i\dot{\vec{r}}_i^{\,2}$. While this term does not affect the equation of motion, it determines the statistics of the particles via path integral.

When one goes from path integral to wave functions, the term (7) also leads to the rule (6) for many-body wave functions associated with usual Hamiltonian containing no peculiar interactions. In fact, the change of ϕ_{ij} can be always written as

$$\phi_{ij}(t)\big|_{t'}^{t''}=2\pi n_{ij}+\phi_{ij}''-\phi_{ij}', \tag{8}$$

with $0\leq\phi_{ij}''-\phi_{ij}'<2\pi$. Thus, the propagator in the n-particle configuration space is a sum of "partial amplitudes," each corresponding to a distinct class of paths having the same winding numbers $\{n_{ij}\}$:

$$K(\vec{r}_i'',t'';\vec{r}_i',t')=\exp\Big[-i\frac{\theta}{\pi}\sum_{i<j}(\phi_{ij}''-\phi_{ij}')\Big]\sum_{n_{ij}}\exp\Big(-i2\theta\sum_{i<j}n_{ij}\Big)\int_{\vec{r}_i'}^{\vec{r}_i''}[\mathscr{D}\vec{r}_i(t)]_{n_{ij}}\exp(iS_0). \tag{9}$$

As usual, a single-valued wave function $\psi(\vec{r}_i,t)$ can be introduced such that

$$\psi(\vec{r}_i'',t'')=\int d\vec{r}_i'K(\vec{r}_i'',t'';\vec{r}_i',t')\psi(\vec{r}_i',t'). \tag{10}$$

We can eliminate the sum in Eq. (9) by introducing a new wave function

$$\tilde{\psi}(\vec{r}_i,t)=\exp\left\{i\frac{\theta}{\pi}\sum_{i<j}\phi_{ij}\right\}\psi(\vec{r}_i,t). \tag{11}$$

Then, corresponding to Eq. (10), now we have

$$\tilde{\psi}(\vec{r}_i'',t'') = \int d\vec{r}_i' \tilde{K}(\vec{r}_i'',t'';\vec{r}_i',t')\tilde{\psi}(\vec{r}_i',t'), \quad (12)$$

$$\tilde{K}(\vec{r}_i'',t'';\vec{r}_i',t') = \int_{\vec{r}_i'}^{\vec{r}_i''} [\mathscr{D}\vec{r}_i(t)]\exp(iS_0). \quad (13)$$

Note that the wave function $\tilde{\psi}(\vec{r}_i,t)$ is single-valued on the universal covering space (or Reimann surface) of the n-particle configuration space. The integration over $\vec{r}_i'$ in Eq. (12) is taken on this covering space. By use of the complex coordinates, it is easy to recover Eq. (6) from Eq. (11).

Another way to derive the same result is the following. The Hamiltonian corresponding to L_0+L' is

$$H=\frac{m}{2}\sum_i \dot{\vec{r}}_i^{\,2}, \quad m\dot{\vec{r}}_i=\vec{p}_i-\frac{\theta}{\pi}\sum_{i<j}\frac{(\vec{r}_i-\vec{r}_j)\times\vec{n}}{|\vec{r}_i-\vec{r}_j|^2}. \quad (14)$$

Here $\vec{p}_i$ is the canonical momentum conjugate to $\vec{r}_i$. It is easy to see that H is the same as given by Eq. (4) together with Eq. (3). We can repeat the same procedure in the last section to arrive at Eq. (6). However, the argument given from Eq. (8) to Eq. (14) has the advantage that it elucidates the relationship between our wave functions and the path integral formulation.

Properties of the wave function (6).—Equation (5) or the rule (6) is invariant under $\theta \to \theta+2\pi$; i.e., fractional statistics is 2π periodic in θ, in agreement with the well-known periodicity of the Aharonov-Bohm effect in the flux or that of the θ parameter in the topological action.

When $\theta=0$ and π, the rule (6) coincides with the standard symmetric or antisymmetric rule. For intermediate θ it gives a continuous interpolation between the two extreme cases. However, when $\theta \neq 0,\pi$, the many-body wave functions are not of the form of products of single-particle wave functions. So generally we expect that the physical quantities of a system of many particles are not simply related to those for one particle.

When $n=2$, from Eq. (6) it is easy to recover the condition[1,5]

$$\psi'(\phi_{12}\pm\pi)=e^{\pm i\theta}\psi'(\phi_{12}). \quad (15)$$

For $n \geq 3$, Eq. (6) exhibits complicated behavior under permutation or interchange of the positions of particles. Complication occurs even when we exchange only two particles in the presence of a third particle. We have to specify along what loop particle 1 moves from $\vec{r}_1$ to $\vec{r}_2$ and particle 2 from $\vec{r}_2$ to $\vec{r}_1$. The resulting phase change will depend on whether the "spectator" 3 is enclosed inside this loop or not. This situation is a reflection of the fact that the configuration space of identical particles is multiply connected. It is the origin of the difficulties pointed out in Refs. 5 and 6 in dealing with more than two particles. The acceptance and direct use of the multivalued wave functions (6) make the many-particle problem accessible to approach, since the complications mentioned above have been simply built into the factors $\prod_{i<j}(z_i-z_j)^{\theta/\pi}$.

Physically, the long-range interactions due to θ statistics are coded in the factors $\prod_{i<j}(z_i-z_j)^{\theta/\pi}$. Moreover, these factors imply the existence of angular momentum barriers between any pair of particles when $\theta \neq 0$. Thus the many-body wave functions are expected to vanish when any two of the particles coincide (if $\theta \neq 0$), although the particles are not fermions for $\theta \neq \pi$.

Three particles, harmonic well.—As an application let us use the many-body wave functions (6) to attack the problem of three identical particles in a harmonic potential. The Schrödinger equation (for n particles with $m=1$) is

$$H\psi=E\psi, \quad H=-\frac{1}{2}\sum_{i=1}^{n}\frac{\partial^2}{\partial z_i \partial z_i^*}+\frac{1}{2}\omega^2\sum_{i=1}^{n} z_i z_i^*, \quad (16)$$

where ψ satisfies the rule (6) with f totally symmetric. (We have omitted the prime on ψ).

The $n=2$ case has been analyzed in Refs. 5 and 12. In our approach we recover the complete set of solutions as follows:

$$\psi=W^{|L|}w^{|l+2\Delta|}L_N^{(|L|)}(2\omega ZZ^*)L_n^{(|l+2\Delta|)}(\tfrac{1}{2}\omega zz^*)\exp[\tfrac{1}{2}\omega(z_1z_1^*+z_2z_2^*)], \quad (17)$$

$$E=(2N+|L|+2n+|l+2\Delta|+2)\omega, \quad (18)$$

where $N,n \geq 0$ are principal quantum numbers for the center-of-mass and relative oscillators respectively; L, $l+2\Delta$ are angular momenta in the center-of-mass and relative coordinates. (l must be even.) $L_M^{(m)}(x)$ are the Laguerre polynomials.[13] We have used the following notation for brevity: $Z=\frac{1}{2}(z_1+z_2)$, $z=z_1-z_2$ and

$$W=\begin{cases}Z \text{ if } L>0,\\ Z^* \text{ if } L<0,\end{cases}\qquad w=\begin{cases}z \text{ if } l+2\Delta>0,\\ z^* \text{ if } l+2\Delta<0.\end{cases} \tag{19}$$

Since θ appears only in the form of $|l+2\Delta|$, the 2π periodicity of θ is made clear. It is also easy to verify the continuous interpolation between the spectrum (including degeneracies) of bosons and that of fermions when θ varies from 0 to π.[12,14]

For $n=3$, we have obtained the following solutions for $0\leqslant\theta<\pi$:

$$\psi=[(z_1-z_2)(z_1-z_3)(z_2-z_3)]^{\theta/\pi}\exp\{-\tfrac{1}{2}\omega r^2\}P, \tag{20}$$

$$P=(z_1+z_2+z_3)^L(z_1-z_2)^l(2z_1-z_2-z_3)^m L_{N_1}^{(L)}(\tfrac{1}{3}\omega R^2)L_{N_2}^{(l+3m+6\Delta-5)}(\tfrac{1}{3}\omega\rho^2)+\text{symmetrization}, \tag{20'}$$

$$E=(2N_1+2N_2+L+l+m+6\Delta+3)\omega, \tag{20''}$$

where all N_1, N_2, L, m, l are nonnegative integers, and l, m such that after symmetrization P does not become identically zero. Moreover, $R^2=|z_1+z_2+z_3|^2$, $r^2=\sum_i|z_i|^2$,

$$\rho^2=|2z_1-z_2-z_3|^2+\text{cyclic permutation}.$$

We note that the parity transformation $z_i\to z_i^*$ and $\theta\to-\theta$ is a good symmetry of the equation (16) and the rule (6). So applying it on the solutions (20) will lead to more solutions (with l, m such that ψ has no singularities at $z_i^*=z_j^*$). We know that this set of solutions does not exhaust those of the problem; e.g., the three-fermion ground state is missing when $\theta=\pi$.

Even so, we are able to see some important features not present in the solutions of two particles. First, for sufficiently small θ, the ground-state energy is $E_0=(3+3\theta/\pi)\omega$. For n particles, it is $E_0=[n+n(n-1)\theta/2\pi]\omega$. Thus, the n dependence of E_0 has a quadratic part which looks like two-body interaction energy. Second, when $\theta=\pi$ the above energy level moves to 6ω, which exceeds the energy of the three-fermion ground state $E_0'=5\omega$. So when θ varies continuously from 0 to π, there must be level crossing and, therefore, the emergence of new ground states at certain values of θ. This effect may lead to interesting phenomena in realistic systems obeying θ statistics when θ can vary under certain circumstances.

To conclude, I stress that though the rule (6) is derived in two concrete mdoels, it is generally true for any fractional statistics in two dimensions, whatever its origins. This will be confirmed in a model-independent formulation in a forthcoming paper.[15]

The author thanks M. Baker, D. Boulware, L. Brown, R. Tao, F. Wilczek, and A. Zee for useful discussions and comments. This work was supported in part by the U. S. Department of Energy.

[1]F. Wilczek, Phys. Rev. Lett. **48**, 1144 (1982). See also M. Peshkin, Phys. Rep. **80**, 376 (1982).

[2]F. Wilczek and A. Zee, Phys. Rev. Lett. **51**, 2250 (1983).

[3]C. N. Yang and C. P. Yang, J. Math. Phys. **10**, 1115 (1969). They have shown how a Bose gas with repulsive two-body delta-function interactions becomes a free Fermi gas for infinite coupling constant.

[4]B. Halperin, Phys. Rev. Lett. **52**, 1583, 2390(E) (1984).

[5]F. Wilczek, Phys. Rev. Lett. **49**, 957 (1982).

[6]F. Wilczek and A. Zee, Institute of Theoretical Physics, University of California, Santa Barbara Report No. NSF-ITP-84-25, 1984 (to be published).

[7]A. S. Goldhaber, Phys. Rev. Lett. **36**, 1122 (1976), and **49**, 905 (1982).

[8]T. T. Wu and C. N. Yang, Phys. Rev. D **12**, 3845 (1975).

[9]Y. Aharonov and D. Bohm, Phys. Rev. **115**, 485 (1959).

[10]We note here that infinite multivaluedness of a wave function can happen only in two dimensions, since $\pi_1(SO(2))=Z$. In three-space only double-valuedness is allowed because $\pi_1(SO(3))=Z_2$.

[11]Here by the notation $\psi(z_i,z_i^*)$, the set $\{(z_i,z_i^*),\ i=1,\ldots,n\}$ is understood. Special wave functions of this form have appeared in Ref. 4. Here we proved that Eq. (6) is the general form of many-body functions obeying fractional statistics.

[12]J. Leinaas and J. Myrlheim, Nuovo Cimento Soc. Ital. Fis. **B37**, 1 (1977).

[13]See, e.g., *Encyclopedic Dictionary of Mathematics*, edited by S. Iyanaga and Y. Kawada (MIT Press, Cambridge, Mass., 1977), Appendix A, Table 20 VI.

[14]The continuous interpolation between the two-body scattering amplitudes of bosons and those of fermions is being discussed by F. Wilczek and A. Zee (private communication).

[15]Y. S. Wu, to be published.

STATISTICAL MECHANICS OF ANYONS

Daniel P. AROVAS

Department of Physics, University of California, Santa Barbara, CA 93106, USA

Robert SCHRIEFFER and Frank WILCZEK

Department of Physics, University of California
and
Institute for Theoretical Physics, Santa Barbara, CA 93106, USA

A. ZEE*

Institute for Advanced Study, Princeton, NJ 08540, USA

Received 31 July 1984

We study the statistical mechanics of a two-dimensional gas of free anyons – particles which interpolate between Bose-Einstein and Fermi-Dirac character. Thermodynamic quantities are discussed in the low-density regime. In particular, the second virial coefficient is evaluated by two different methods and is found to exhibit a simple, periodic, but nonanalytic behavior as a function of the statistics determining parameter.

In two space dimensions, a continuous family of quantum statistics interpolating between bosons and fermions is possible [1, 2]. Two examples of particles obeying exotic statistics have been discussed. The soliton of the (2 + 1)-dimensional O(3) nonlinear σ-model has a spin which is neither integral nor half-odd integral, and obeys a statistics which is neither Bose-Einstein nor Fermi-Dirac [3]. In condensed matter, one finds that the Laughlin quasiparticles [4] in the anomalous quantum Hall effect system also possess fractional charge and fractional statistics [5], a result recently derived from the adiabatic theorem [6].

We first discuss a method [1] by which the statistics of a two-dimensional system of charged particles can be changed (continuously) via the introduction of a fictitious "statistical gauge field." It is well known that the wave function of a charged particle interacting with a magnetic flux tube will acquire a phase change due to the motion of the particle. If the charge is $e^* = \nu e$ and the flux is

* On leave from the University of Washington 1983–1984.

$\phi = \alpha\phi_0 = \alpha hc/e$, a complete revolution will induce a phase change of $e^{i\Delta\gamma}$ with $\Delta\gamma = 2\pi\alpha\nu$. The basic idea of this method is then to introduce a flux tube at the position of each particle whose magnitude ϕ will determine the phase associated with relative particle motion. The gauge field associated with this flux is nondynamical, i.e., its evolution is completely determined by the motion of the particles. It is expected that physical quantities are periodic in the statistics determining parameter α.

In the symmetric gauge, the vector potential due to a flux tube of magnitude $\phi = \alpha\phi_0$ takes the form

$$A(\boldsymbol{r}) = \frac{\alpha\phi_0}{2\pi}\hat{z}x\frac{\boldsymbol{r}}{r^2} = \frac{\alpha\phi_0}{2\pi}\frac{\hat{\theta}}{r} = \frac{\alpha\phi_0}{2\pi}\nabla\theta. \tag{1}$$

The many-particle generalization we seek is therefore*

$$\boldsymbol{A}_\phi(\boldsymbol{r}_i) = \frac{\alpha\phi_0}{2\pi}\hat{z}x\sum_j{}'\frac{\boldsymbol{r}_{ij}}{r_{ij}^2} = \frac{\alpha\phi_0}{2\pi}\sum_j{}'\nabla_i\theta_{ij}, \tag{2}$$

where $\theta_{ij} = \tan^{-1}((y_j - y_i)/(x_j - x_i))$ is the relative angle between i and j, and the prime on the sum indicates that the term $j = i$ is to be excluded. This leads to the following many-body hamiltonian:

$$H(\alpha) = \sum_i \frac{1}{2m}\left(\boldsymbol{p}_i - \frac{e}{c}\boldsymbol{A}_0(\boldsymbol{r}_i) - \frac{e}{c}\boldsymbol{A}_\phi(\boldsymbol{r}_i)\right)^2 + V(\boldsymbol{r}_1,\ldots,\boldsymbol{r}_N), \tag{3}$$

where $\boldsymbol{A}_0$ is the physical vector potential, if present. Suppose one knows an eigenfunction ψ_0 of the bare hamiltonian $H_0 \equiv H(0)$ with energy E_0. Then since

$$\left[\boldsymbol{p}_i, \exp\left(i\frac{\alpha\phi_0}{2\pi}\sum_j{}'\theta_{ij}\right)\right] = \frac{\alpha\phi_0}{2\pi}\sum_j{}'\nabla_i\theta_{ij} = \boldsymbol{A}_\phi(\boldsymbol{r}_i), \tag{4}$$

we see that $\psi_\alpha \equiv \exp(i\alpha\Sigma_{\text{pairs}}\theta_{ij})\psi_0$ is an eigenfunction of $H(\alpha)$ also with energy E_0. The problem is that the function ψ_α will not in general be a single-valued function of its arguments.

For the two-particle problem with no external potential and no interparticle interactions (free particles), the situation becomes eminently tractable. Recall that the wave function can be decomposed into a product $\psi_0(\boldsymbol{r}, \boldsymbol{R}) = \chi(\boldsymbol{R})\xi(\boldsymbol{r})$ where $\boldsymbol{R}$ is the center-of-mass position and $\boldsymbol{r}$ is the relative coordinate vector. We find $\chi(\boldsymbol{R}) = e^{i\boldsymbol{K}\cdot\boldsymbol{R}}$, $\xi(\boldsymbol{r}) = e^{im\theta}J_{|m|}(kr)$, $E = \hbar^2K^2/4M + \hbar^2k^2/M$ (M = particle mass). Imposition of Bose (Fermi) statistics then requires that m be even (odd). The introduction of a statistical gauge field then provides us with new eigenfunctions $\psi_\alpha(\boldsymbol{r}, \boldsymbol{R}) = e^{i\alpha\theta}\psi_0(\boldsymbol{r}, \boldsymbol{R}) = e^{i\boldsymbol{K}\cdot\boldsymbol{R}}e^{i(m+\alpha)\theta}J_{|m|}(kr)$. Here, Bose (Fermi) statistics re-

* Since there is flux-charge interaction as well as charge-flux interaction, $\alpha = 2\phi/\phi_0$ is now *twice* the flux per tube in units of the Dirac quantum.

quires that $m + \alpha \equiv l$ be even (odd) and we obtain wave functions

$$\psi_\alpha(\boldsymbol{r}, \boldsymbol{R}) = e^{i\boldsymbol{K}\cdot\boldsymbol{R}} e^{il\theta} J_{|l-\alpha|}(kr) . \tag{5}$$

If we now introduce a circular boundary at some radius $\tilde{R}$, we find that the allowed energies are

$$\varepsilon_{l,n} = \hbar^2 x^2_{|l-\alpha|,n}/M\tilde{R}^2, \qquad J_\nu(x_{\nu,n}) = 0 . \tag{6}$$

Hence, choosing α to be an odd integer merely shifts the energy spectrum from Bose-like to Fermi-like. Note also that the spectrum is periodic in α with period $\Delta\alpha = 2$. This is in fact true for the N-particle system, although explicit (single-valued) wave functions are difficult to obtain due to the fact that there are $\frac{1}{2}N(N-1)$ relative angles and only $(N-1)$ non-CM angular degrees of freedom. For $N = 2$, these numbers are identical.

This result, eq. (6), can be used to evaluate the second virial coefficient. In two dimensions, it is easily shown that [7]*

$$B(T) = \tfrac{1}{2}A - A^{-1}\lambda_T^4 Z_2 , \tag{7}$$

where A is the area of the system $\lambda_T = (2\pi\hbar^2/MkT)^{1/2}$ is the thermal wavelength, and $Z_2 = \mathrm{Tr}\, e^{-\beta H_2}$ is the two-particle partition function. The virial expansion is an expansion of the equation of state in the density n: $P = nkT[1 + Bn + Cn^2 + \cdots]$. In performing the trace to obtain Z_2, the center-of-mass freedom is trivially separated, yielding a factor $Z_2 = 2A\lambda_T^{-2}\tilde{Z}_2$, where $\tilde{Z}_2$ is now the single particle partition function for the relative coordinate problem: $\tilde{Z}_2 = \mathrm{Tr}_{\mathrm{rel}} e^{-\beta H_{\mathrm{rel}}}$. This will again be area divergent, and it is therefore convenient to calculate the virial coefficient $B(\alpha, T)$ in terms of a known quantity, i.e., $B(2j, T) = -\frac{1}{4}\lambda_T^2$ or $B(2j + 1, T) = +\frac{1}{4}\lambda_T^2$, the familiar result for Bose and Fermi systems, respectively ($j \in \mathbb{Z}$). Thus, we obtain

$$B(\alpha', T) - B(\alpha, T) = 2\lambda_T^2 [\tilde{Z}_2(\alpha) - \tilde{Z}_2(\alpha')] . \tag{8}$$

We now appeal to the result (6). Clearly $B(\alpha, T)$ must be periodic in α with period $\Delta\alpha = 2$. We will take our original particles to have Bose statistics and expand about even and odd values of α. For $\alpha = 2j + \delta$, $|\delta| < 1$, corresponding to quasi-bosons, the allowed values of $|l - \alpha|$ are $|\delta|, 2 \pm \delta, 4 \pm \delta$, etc. For $\alpha = 2j + 1 + \delta$, $|\delta| < 1$, corresponding to quasi-fermions, the allowed values of $|l - \alpha|$ are $1 \pm \delta, 3 \pm \delta$, etc. Since B must be independent of the cutoff $\tilde{R}$ in the limit $\tilde{R} \to \infty$, and since $\tilde{R}$

* To be precise, we should write $B(T) = A[\frac{1}{2} - Z_2/Z_1^2]$ with $Z_1 = \mathrm{Tr}\, e^{-\beta H}$, the single particle partition function. With no external fields, we have $H_1 = p^2/2M$ and $Z_1 = A\lambda_T^{-2}$, which then yields eq. (7)

appears only in the combination $MkT\tilde{R}^2/\hbar^2$, it is desirable to rescale $\tilde{R} \to \sqrt{\hbar^2/MkT}\,\tilde{R}$. Expanding about the Fermi point, we find that (8) and (6) give

$$B(2j+1+\delta,T) = \tfrac{1}{4}\lambda_T^2 + 2\lambda_T^2 \times \lim_{\tilde{R}\to\infty} \sum_{\substack{l=1\\ \text{odd}}}^{\infty} \sum_{n=1}^{\infty} \left[2\,e^{-(x_{l,n}/\tilde{R})^2} - e^{-(x_{l+\delta,n}/\tilde{R})^2} - e^{-(x_{l-\delta,n}/\tilde{R})^2}\right]. \quad (9)$$

The factor in brackets resembles a second derivative. By expanding in δ, one can then perform the l-sum by means of the celebrated Euler-MacLaurin formula [8]. This leaves

$$B(2j+1+\delta,T) = \tfrac{1}{4}\lambda_T^2 - 2\lambda_T^2\delta^2 \lim_{\tilde{R}\to\infty} \tilde{R}^{-2} \sum_{n=1}^{\infty} x_{1,n} \left.\frac{\partial x_{1+s,n}}{\partial s}\right|_{s=0} e^{-(x_{1,n}/\tilde{R})^2}. \quad (10)$$

As $n\to\infty$, $x_{1,n}\to\infty$ and

$$\frac{\partial x_{\nu,n}}{\partial \nu} = \frac{1}{J_{\nu+1}(x_{\nu,n})} \frac{\partial J_\nu(x_{\nu,n})}{\partial \nu} \to \tfrac{1}{2}\pi.$$

The value of n at which this approximation $x_{\nu,n} \sim \frac{1}{2}\nu\pi + n\pi - \frac{1}{4}\pi$ becomes valid is n_0, say, which is certainly independent of $\tilde{R}$. The sum will then be completely dominated by the terms $n_0 \leqslant n < \infty$, the beginning terms being suppressed by the $\tilde{R}^{-2}$ factor. Making this replacement, and writing $\Sigma_n \to \int dx_{1,n}/\pi$, we obtain

$$B(2j+1+\delta,T) = \tfrac{1}{4}\lambda_T^2 - \lambda_T^2\delta^2 \lim_{\tilde{R}\to\infty} \tilde{R}^{-2} \int_{n_0\pi}^{\infty} dx\, x\, e^{-(x/\tilde{R})^2} = \tfrac{1}{4}\lambda_T^2 - \tfrac{1}{2}\delta^2\lambda_T^2. \quad (11)$$

One can check that all other terms in the expansion in δ and in other approximations employed are formally of order $\tilde{R}^{-1}$ as $\tilde{R}\to\infty$. Thus, we predict

$$B(2j+1+\delta,T) = \tfrac{1}{4}\lambda_T^2 (1-2\delta^2)_{\text{per}}, \quad (12)$$

where the subscript indicates that we are to extend this function for $|\delta| > 1$ in a periodic fashion. The complete result has a cusp at Bose values $\alpha = 2j$ due to the required periodic extension. This in fact follows from the general formula (8). The only difference between the Fermi and Bose expansions is the existence of the

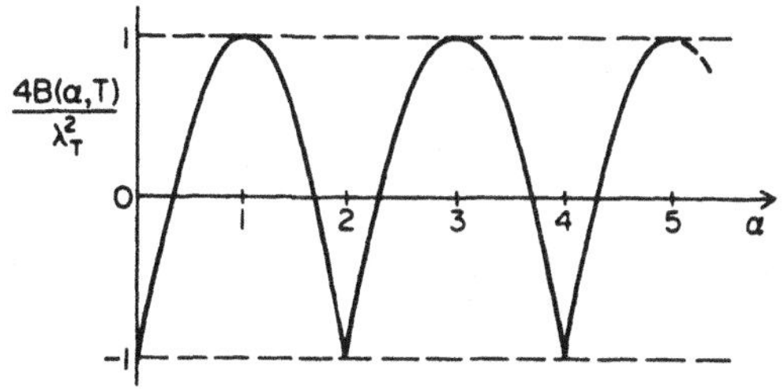

Fig. 1. The second virial coefficient $B(\alpha,T)$ as a function of the statistics determining parameter α (T fixed).

$|l-\alpha| = |\delta|$ term:

$$B(2j+\delta,T) = -\tfrac{1}{4}\lambda_T^2 - \tfrac{1}{2}\delta^2\lambda_T^2 + 2\lambda_T^2 \lim_{\tilde{R}\to\infty} \sum_{n=1}^{\infty} \left[e^{-(x_{0,n}/\tilde{R})^2} - e^{-(x_{|\delta|,n}/\tilde{R})^2} \right]$$

$$= -\tfrac{1}{4}\lambda_T^2 + |\delta|\lambda_T^2 - \tfrac{1}{2}\delta^2\lambda_T^2 . \qquad (13)$$

This is exactly the form predicted above. Thus, we obtain a picture of $B(\alpha,T)$ for fixed T as in fig. 1.

This result is also derivable from a path integral approach. The general lagrangian for the many-body system is

$$L = \sum_i \tfrac{1}{2}M\dot{\boldsymbol{r}}_i^2 + \alpha \sum_{\text{pairs}} \dot{\theta}_{ij} . \qquad (14)$$

For a system of bosons, the partition function takes the form

$$Z_N = \frac{1}{N!}\int d^2r_1 \dots d^2r_N \sum_P \langle \boldsymbol{r}_1 \dots \boldsymbol{r}_N | e^{-\beta H_N} | P\boldsymbol{r}_1 \dots P\boldsymbol{r}_N \rangle , \qquad (15)$$

which may be cast into a path integral form, as done by Feynman [9]. Again, the case $N=2$ is considered, and the CM contribution is directly integrated out. This leaves

$$\tilde{Z}_2 = \tfrac{1}{2}\int d^2r \left[\langle \boldsymbol{r} | e^{-\beta H_{\text{rel}}} | \boldsymbol{r} \rangle + \langle \boldsymbol{r} | e^{-\beta H_{\text{rel}}} | -\boldsymbol{r} \rangle \right] ,$$

$$L_{\text{rel}} = \tfrac{1}{4}M\dot{\boldsymbol{r}}^2 + \alpha\dot{\theta} . \qquad (16)$$

The $\dot{\theta}$ term in L introduces a winding number-dependent phase in the path integral. This problem is in fact equivalent to the Bohm-Aharonov effect [10], the path integral formulation of which was studied extensively by Gerry and Singh [11–14].

The matrix element

$$K(\boldsymbol{r},\boldsymbol{r}';\tau)=\int \mathrm{D}\boldsymbol{r}(t)\mathrm{e}^{i\hbar^{-1}\int_0^\tau \mathrm{d}t' L(t')}=\langle \boldsymbol{r}|\mathrm{e}^{-iH\tau/\hbar}|\boldsymbol{r}'\rangle,$$

$$\boldsymbol{r}(t=0)=\boldsymbol{r},\qquad \boldsymbol{r}(t+\tau)=\boldsymbol{r}',\qquad \beta=i\tau, \tag{17}$$

can be decomposed into a sum over contributions of different homotopy sectors, with $\theta'-\theta\equiv\phi+2\pi n$:

$$K(\boldsymbol{r},\boldsymbol{r}';\tau)=\sum_{n=-\infty}^{\infty}\mathrm{e}^{2\pi i n\alpha}\bar{K}_n(\boldsymbol{r},\boldsymbol{r}';\tau), \tag{18}$$

$$\bar{K}_n(\boldsymbol{r},\boldsymbol{r}';\tau)=\int \mathrm{D}\boldsymbol{r}(t)\,\mathrm{e}^{i\hbar^{-1}\int_0^\tau \mathrm{d}t' L(t')}\delta(\theta'-\theta-\phi-2\pi n), \tag{19}$$

$$\bar{K}_n(\boldsymbol{r},\boldsymbol{r}';\tau)=\frac{M}{4\pi\hbar i\tau}\exp\left[\frac{-M}{4\hbar i\tau}(r^2+r'^2)\right]$$
$$\times\int_{-\infty}^{\infty}\mathrm{d}\lambda\,\mathrm{e}^{i\lambda(\phi+2\pi n)}\mathrm{e}^{i\alpha\phi}I_{|\lambda|}\left(\frac{Mrr'}{2\hbar i\tau}\right), \tag{20}$$

where $I_\nu(Z)$ is the modified Bessel function. In our case, we have $|\boldsymbol{r}'|=|P\boldsymbol{r}|=|\boldsymbol{r}|$, and $\phi=0,\pi$:

$$K(\boldsymbol{r},\boldsymbol{r}';\tau)=\sum_{n=-\infty}^{\infty}\left(\frac{M}{4\pi\hbar i\tau}\right)\exp\left(\frac{-Mr^2}{2\hbar i\tau}\right)\mathrm{e}^{in\phi}I_{|n-\alpha|}\left(\frac{Mr^2}{2\hbar i\tau}\right). \tag{21}$$

Therefore, we arrive at the result

$$\tilde{Z}_2=\tfrac{1}{2}\sum_{\substack{n=-\infty\\ \text{even}}}^{\infty}\int_0^\infty \mathrm{d}x\,\mathrm{e}^{-x}I_{|n-\alpha|}(x). \tag{22}$$

This is, as expected, formally divergent. A convergence factor $\mathrm{e}^{-\varepsilon x}$ is inserted in the integrand, with $\varepsilon\to 0$ at the end of the calculation. We use the result [15]

$$F_\nu(a)\equiv\int_0^\infty \mathrm{d}x\,\mathrm{e}^{-ax}I_\nu(x)=\frac{1}{\sqrt{a^2-1}}\left(a+\sqrt{a^2-1}\right)^{-\nu},$$

$$F_\nu(1+\varepsilon)\to\frac{1}{\sqrt{2\varepsilon}}\left(1+\sqrt{2\varepsilon}\right)^{-\nu}. \tag{23}$$

As before, we appeal to eq. (8). Expanding about Fermi statistics, $\alpha=2j+1+\delta$, and $|n-\alpha| = 1 \pm \delta, 3 \pm \delta$, etc. Thus,

$$B(2j+1+\delta, T) = \tfrac{1}{4}\lambda_T^2 + 2\lambda_T^2 \lim_{\varepsilon\to 0}$$

$$\left[\frac{1}{2}\frac{1}{\sqrt{2\varepsilon}} \sum_{\substack{n=1\\ \text{odd}}}^{\infty} (1+\sqrt{2\varepsilon})^{-n}\left(2-(1+\sqrt{2\varepsilon})^{\delta}-(1+\sqrt{2\varepsilon})^{-\delta}\right)\right]$$

$$= \tfrac{1}{4}\lambda_T^2 - \tfrac{1}{2}\delta^2\lambda_T^2. \tag{24}$$

Expanding about Bose statistics introduces a term $|\delta|\lambda_T^2$ due to the $|n-\alpha| = |\delta|$ piece in the sum. Making the required periodic extension recovers the earlier result of eq. (12).

In some sense, the path integral result is more satisfying, because, although one still is presented with the delicacy involved with extracting the (finite) difference of two divergent expressions, there is no necessity to impose a finite volume constraint, which was originally effected in order to perform the mode counting. One might object to our original calculation on the grounds that the virial coefficient might possibly be sensitive to the manner in which we perform the mode counting, since the dominant terms in the sum of eq. (9) are those at the tail end. As we have seen, this fear is unfounded.

A striking result is the nonanalyticity of eq. (12). It would be interesting to know whether cusps also arise in higher-order virial coefficients.

Due to the proliferation of the number of relative angles, such higher-order virial coefficients are exceedingly difficult to evaluate. In the high density limit, one might consider averaging the statistical flux over the entire system, and then consider the effect of a net statistical uniform magnetic field of magnitude $B = n\alpha\phi_0$, where n is the particle density. It is possible to reproduce the correct form of the free energy to leading order in n in this manner, however, one loses periodicity in α, and only certain values of α actually yield the correct result.

The most significant feature of the statistical interaction is that it is long ranged, hence perturbation expansions in α yield divergences and resummation is necessary, a situation reminiscent of the electron gas. Nevertheless, this statistics transformation process does yield a viable method for interpolating quantum statistics. The representation of a Fermi gas in terms of a Bose gas may be useful in other contexts, such as lattice field theory. Unless the statistical interaction is treated nonperturbatively, however, divergences may be difficult to handle.

Finally, it is interesting to derive the lagrangian of eq. (14) for the solitons of the nonlinear σ-model. Let us briefly recall that the model in question has a unit-vector order parameter $n^a(x)$, $a = 1,2,3$ and a conserved topological current

$$J^{\mu} = \frac{1}{8\pi}\varepsilon^{\mu\nu\lambda}\varepsilon^{abc}n^a\partial_\nu n^b\partial_\lambda n^c. \tag{25}$$

The conservation of J^μ licenses us to manufacture a U(1) "gauge potential" by the curl equation

$$J^\mu = \varepsilon^{\mu\nu\lambda}\partial_\nu A_\lambda . \tag{26}$$

The crucial point is that we could include a topological term

$$H = \frac{\Theta}{2\pi}\int d^3x\, A_\mu J^\mu \tag{27}$$

in the action, with Θ a real number ($\alpha \equiv \Theta/\pi$), which is analogous to the Θ-parameter in quantum chromodynamics. H is, in fact, the Hopf invariant describing maps of S^3 to S^2. In a suitable gauge, such as $\partial A = 0$, we can solve for A_μ and so write H as a nonlocal interaction among the n^a fields. The solitons are bosons for $\Theta = 0$ ($\alpha = 0$) and fermions for $\Theta = \pi$ ($\alpha = 1$). In a more general context, any conserved current J_μ can be coupled to the vector field A_μ. If the only other appearance of A_μ in the lagrangian is the Chern-Simons term [16] $\varepsilon_{\mu\nu\rho}A_\mu\partial_\nu A_\rho$, then A_μ represents a nondynamical field [17] which can be eliminated to give a nonlocal interaction, which will impart anomalous statistics to particles carrying charge associated with the current.

In ref. [3] the statistics of the solitons in this model were determined by invoking the linking number theorem. Here we will determine the statistics directly by interchanging two widely separated solitons, and in the process elucidate the linking number theorem.

For separations large compared to the sizes of the solitons we can approximate the solitons by point particles and the topological current by

$$J^\mu(x) = \sum_a \int d\tau\, \delta^{(3)}(x - q_a(\tau))\frac{dq_a^\mu}{d\tau}, \tag{28}$$

with $a = 1, 2$ and $q_a(\tau)$ describing the trajectories of the two "point solitons." We evaluate H by inserting eq. (28) into eqs. (26) and (27) and keeping only the cross-terms. The divergent self-interaction terms are evidently artifacts of the point approximation. To best understand the situation, we go to euclidean 3-space and think of eq. (26) as one of the time-independent Maxwell's equations $\nabla \times \boldsymbol{B} = \boldsymbol{J}$ with the identification of A_μ as the magnetic field $\boldsymbol{B}$. Then H can clearly be interpreted as the work done on a magnetic monopole moving along the trajectory $q_1(\tau)$ by the magnetic field generated by an electric current flowing along the curve $q_2(\tau)$. With suitable normalization, this is just the number of times curves "1" and "2" wind around each other. We have thus made contact with the explicit form for the linking number between two curves given in mathematical texts [18]. This discussion also defines the linking number between two curves which are not closed.

To evaluate H explicitly, it is easiest to distort one of the curves, say "2," to a straight line $q_2^\mu(\tau) = \tau\delta^{\mu 0}$, as we are allowed to do. We find by eq. (26) that

$A_i = \varepsilon_{ij} x_j / r^2$, $A_0 = 0$, a pure (but topologically nontrivial) gauge. Once again, we could have appealed to (2 + 1)-dimensional electrodynamics, this time interpreting J^0 as B. These remarks make clear that the effect here is essentially the Bohm-Aharonov phenomenon. It is sometimes convenient to transform to a singular gauge wherein $A = 0$ except along string singularities attached to each particle, across which A has a jump discontinuity of constant magnitude.

In summary, the action describing N of these point particles is just

$$S = \int \mathrm{d}t L = \int \mathrm{d}t \left[\tfrac{1}{2} m \sum_{a=1}^{N} \left(\frac{\mathrm{d}\boldsymbol{x}_a}{\mathrm{d}t} \right)^2 + \frac{\Theta}{\pi} \sum_{a<b} \frac{\mathrm{d}}{\mathrm{d}t} \theta_{ab} \right]. \tag{29}$$

Here $\boldsymbol{x}_a$ is a two-dimensional vector locating particle a and θ_{ab} is the angle of particle b relative to particle a, measured from the x-axis, say. The preceding discussion has boiled ΘH down to the second term in this equation. As we have seen, although this term is a total time derivative and appears as an interaction, it determines the statistics of the particles.

In the original model, the solitons have topological charge $Q = \int \mathrm{d}^3 x J_0$ taking on all integer values. The $|Q| > 1$ solitons are unstable against breakup. In writing down eq. (29) we have included only $Q = +1$ particles. It is easy enough, however, to include $Q = -1$ particles as well by noting that the $(+\,-)$ "interaction" has opposite sign from the $(+\,+)$ and $(-\,-)$ "interactions."

DPA would like to thank Stefan Theisen for making the work of Gerry and Singh known to us, and for many useful discussions. This work was supported in part by the National Science Foundation under grants DMR82-16285 and PHY77-27084, supplemented by funds from the National Aeronautics and Space Administration. One of us (DPA) is grateful for the support of an AT&T Bell Laboratories Scholarship.

Note added

This work supersedes the preprint "Interpolating quantum statistics", NSF-ITP-84-25, by two of the authors (F.W. and A.Z.).

References

[1] F. Wilczek, Phys. Rev. Lett. 49 (1982) 957
[2] Y. Wu, US Dept. of Energy preprint 40048-09P4 (1984)
[3] F. Wilczek and A. Zee, Phys. Rev. Lett. 51 (1983) 2250
[4] R.B. Laughlin, Phys. Rev. Lett. 50 (1983) 1395
[5] B.I. Halperin, Phys. Rev. Lett. 52 (1984) 1583
[6] D. Arovas, R. Schrieffer and F. Wilczek, preprint NSF-ITP-84-66, submitted to Phys. Rev. Lett.
[7] J.G. Dash, Films on solid surfaces (Academic Press, 1968)

[8] M. Abramowitz and I. Stegun, Handbook of mathematical functions (Dover, 1972)
[9] R.P. Feynman, Statistical mechanics (Benjamin, 1972)
[10] Y. Aharonov and D. Bohm, Phys. Rev. 115 (1959) 485
[11] C.C. Gerry and V.A. Singh, Phys. Rev. D20 (1979) 2550
[12] A. Inomata and V.A. Singh, J. Math. Phys. 19 (1978) 2318
[13] C.C. Gerry and V.A. Singh, Nuovo Cim. 73B (1983) 161
[14] S.F. Edwards and Y.V. Gulyaev, Proc. Roy. Soc. London A279 (1964) 229
[15] I.S. Gradshteyn and I.M. Ryzhik, Table of integrals, series, and products (Academic Press, 1980)
[16] J. Schonfeld, Nucl. Phys. B185 (1981) 157;
S. Deser, R. Jackiw and S. Templeton, Phys. Rev. Lett. 48 (1982) 975;
Y.Wu, Washington preprint (1983)
[17] C.R. Hagen, Rochester preprint (1983)
[18] H. Flanders, Differential forms (Academic Press, 1963)

Chapter 6

THE QUANTIZED HALL EFFECT

* Original Contribution.

6

The Quantized Hall Effect

The integrally quantized Hall effect[1] (IQHE) and the fractionally quantized Hall effect[2] (FQHE) were discovered in the very special, almost bizarre context of semiconductor heterostructures subjected to huge magnetic fields while held at millikelvin temperatures. However, the theories devised to describe the effects are quite novel and interesting, and it seems increasingly likely that some of the ideas that arise will find a much wider application.

Let us first discuss just what is observed. A layer of electrons may be trapped at the interface between two semiconductors, known as a heterojunction. The electrons in this layer can, to a good approximation, be idealized as a two-dimensional gas with Coulomb repulsion. The quantized Hall effect has to do with the behavior of a two-dimensional electron gas in a strong magnetic field, at low temperatures. It is found that under these conditions the Hall coefficient—that is, the ratio of current flow to transverse potential—behaves in a most peculiar way. To be specific, it is found that as the magnetic field is varied smoothly the Hall coefficient does not vary smoothly, but rather stays constant over finite intervals—"plateaus." The plateaus are separated by intervals of more normal, continuous behavior.

Moreover, the value of the Hall coefficient on the plateaus is found to be

$$R = \frac{h}{\nu e^2} \tag{6.1}$$

where ν is an integer (for the IQHE) or a rational number with odd denominator (FQHE).* Of course, all integers are rational numbers with odd denominators, so the IQHE may be thought of as a special case of the FQHE. However, the most popular theoretical explanations of the two effects are quite different. Here we shall be most interested in the theory of the FQHE. (When we wish to refer to the two effects collectively, we shall speak of the QHE.)

* It is conventional and sometimes illuminating to display Planck's constant h in some of the equations. We shall however often lapse into using $\hbar = 1$ or $h = 2\pi$ without warning.

Part of the interest of the quantized Hall effect is that the Hall coefficient is expressed in terms of fundamental physical quantities. This is quite remarkable, considering that the measurement is made directly on a macroscopic material, with all the complexity and "dirt" that implies. However, there is no question that the relation (6.1) is satisfied to a high degree of accuracy; for the integral Hall effect ($\nu = 1$) the equality has been established to better than a part in ten million.

To appreciate how unusual the QHE is, let us discuss it more quantitatively and contrast it with the more common behavior, which is essentially classical. The Lorentz force law reads

$$m\frac{d\mathbf{v}}{dt} = e\left(\mathbf{E} + \frac{\mathbf{v}}{c} \times \mathbf{B}\right), \tag{6.2}$$

and dissipative processes will lead to the Ohm's law behavior

$$\mathbf{j} = ne\mathbf{v} = \sigma_0\left(\mathbf{E} + \frac{\mathbf{v}}{c} \times \mathbf{B}\right) = \sigma_0\left(\mathbf{E} + \frac{1}{nec}\mathbf{j} \times \mathbf{B}\right) \tag{6.3}$$

Here, n is the electron denstiy and σ_0 is the zero–field conductivity. We are concerned with a situation in which $\mathbf{B}$ and $\mathbf{E}$ are constant fields, with $\mathbf{B}$ pointing perpendicular to the two-dimensional plane within which the electrons are confined while $\mathbf{E}$ lies in that plane. We find then for the current the self-consistent equation

$$\begin{aligned} j_x &= \sigma_0 E_x + \frac{\sigma_0}{nec} j_y B \\ j_y &= \sigma_0 E_y - \frac{\sigma_0}{nec} j_x B \end{aligned} \tag{6.4}$$

or equivalently for the resistivity tensor ρ_{ij}

$$\begin{pmatrix} E_x \\ E_y \end{pmatrix} = \begin{pmatrix} \rho_{xx} & \rho_{xy} \\ \rho_{yx} & \rho_{yy} \end{pmatrix} \begin{pmatrix} j_x \\ j_y \end{pmatrix} = \begin{pmatrix} \rho_0 & B/nec \\ -B/nec & \rho_0 \end{pmatrix} \begin{pmatrix} j_x \\ j_y \end{pmatrix} \tag{6.5}$$

where $\rho_0 = \sigma_0^{-1}$. Notice especially the linear dependence of the Hall resistance ρ_{xy} on B. It is independent of σ_0, reflecting its essentially non-dissipative character.

The basic observation of the quantized Hall effect is that as the density of electrons is varied at fixed magnetic field (or, more practically, if the magnetic field is varied at fixed electron density) the resistivity does not vary continuously, but rather is constant over finite intervals—the above-mentioned plateaus. Now comparing the FQHE and the classical values of the resistivity,

$$\frac{B}{nec} = \frac{h}{\nu e^2} \quad \text{or} \quad \frac{n}{B} = \frac{\nu e}{hc} \tag{6.6}$$

we see that the essence of the FQHE is that on the plateaus n/B is frozen at the rational value ν.

To interpret this further let us recall the basic quantum mechanics of charged particles in magnetic fields. As the textbooks teach us,[3] the continuous spectrum of a free particle breaks up, in the presence of a magnetic field, into discretely spaced, highly degenerate levels known as Landau levels. The density of states in each Landau level is

$$n_0 = \frac{e}{hc} \tag{6.7}$$

per unit area (in units of the magnetic length squared), for each spin. Comparing to Eq.(6.6), we arrive at a simple interpretation of the fraction ν: it is the number of filled Landau levels. The FQHE reveals that certain filling fractions are especially stable, *i.e.*, energetically favorable; it is, in this sense, a commensurability effect.

(Many of the experiments correspond to densities such that only one Laudau level is relevant, and furthermore such that the splitting between electron states having spin aligned or anti-aligned with the magnetic field is so great that only the former need be considered. For simplicity, we shall restrict ourselves to this case. Then the relevant electrons are all identical particles, and ν is less than one.)

With this introduction, we can intelligently begin to discuss the basic theory of the IQHE and FQHE. For those who wish to penetrate more deeply, we would recommend an extremely valuable, authoritative, and accessible account of the field as of 1986.[4]

The IQHE (that is, the case $\nu = 1$) is readily understood at the level of an independent particle model. In the presence of random impurities, the single-particle wave functions fall into two classes: a band of extended states and a set of spatially localized states. When the Fermi energy is in the region of localized states, varying the number of electrons only adds or subtracts localized states, which carry no current. Thus the magnitude of the current is stuck; and furthermore, it is stuck at the full Landau level value, as we can appreciate by turning on the impurities gradually from zero. Indeed, as long as no extended states cross the Fermi energy as the impurity potentials are turned on, the current remains at its value in the absence of impurities, *i.e.*, the full Landau level value. So the IQHE is an inevitable consequence of independent particle behavior, given a modicum of localization theory.

By implication then the FQHE must be an essentially collective, many-body effect. Great insight into its nature followed upon an inspired proposal, by Laughlin, of a variational wave function to describe it.[5] The electron states in the lowest Laudau level, in the symmetric gauge, are of the form

$$\psi(z) = f(z)\, e^{-\frac{1}{4\ell^2}|z|^2} \tag{6.8}$$

where $z = x+iy$, f is an arbitrary analytic function, and $\ell = (eB/c)^{1/2}$ is the magnetic length. Now the complete degeneracy of states in the Landau level means that there is effectively no kinetic energy assocated with motion in the band. There is no inertia, and the particles are effectively infinitely massive. At the same time, they are subject to mutual Coulomb repulsion. To minimize the potential energy, one would like to space the electrons equally, forming a "Wigner crystal".[6] It might seem at first glance that there is no obstruction to doing just this, since it costs nothing to localize the electrons, the usual penalty imposed by the uncertainty principle being inoperative for infinitely massive particles. However, the restriction to the lowest Landau level makes it impossible to localize the electrons strictly. Their position is actually uncertain to within a magnetic length ℓ. At densities high enough so that the electrons at different sites of the putative Wigner crystal have overlapping wave functions, the crystal has every opportunity to melt. Nevertheless, we would expect that the most favorable many-body wave function is one that prevents the electrons from ever getting too close together, and that treats all the electrons symmetrically. Laughlin[7] proposed a specific wave function with these desirable properties, *viz.*

$$\Psi^m(z_1,\ldots,z_N) = \prod_{i<j}(z_i - z_j)^m e^{-\frac{1}{4\ell^2}\sum |z_i|^2} \tag{6.9}$$

up to normalization. Here, m must be an odd number so that Fermi statistics is respected. The Laughlin wave function is analytically tractable, and a number of its properties have been reliably deduced. For large numbers N of electrons, it represents electrons filling a disk of radius $N/2\pi\ell^2$, with density corresponding to filling fraction $1/m$. The zeroes of the wave function as particle coordinates coalesce effectively keeps the particles separated, and cuts down on the unfavorable Coulomb energy. Indeed the Laughlin wave function is known to be an exact ground state for certain model, short-range potentials, and there is good numerical evidence that it is an excellent approximation to the ground state for the actual Coulomb potential.[8]

Qualitatively, the Laughlin wave function represents an incompressible quantum liquid. Incompressibility, it will be seen, is the essence of the fractionally quantized Hall effect. Indeed we have interpreted the remarkable plateaus revealed by experiment as meaning that as n or B is varied, the filling fraction stays fixed. To account for this, we must find that the effect of adding a particle is not to change the filling fraction by a small amount over a large volume, but rather to leave the filling fraction pinned at its favorable value macroscopically, with the deviation from this value carefully localized. Thus, a density perturbation must lead to the creation of localized quasi-particles, with a gap in the spectrum corresponding to the finite energy cost of a quasi-particle. There are strong numerical indications that the quasi-particle (more accurately, quasi-hole) excitations around the Laughlin ground state are well described by the wave function generated by acting on the ground state as follows:

$$\Psi^m_{z_0} \equiv \prod_i (z_i - z_0)\, \Psi^m. \tag{6.10}$$

This creates a quasi-hole at z_0. Effectively, each electron feels a centrifugal barrier at z_0, and is pushed away. Note that the wave function is little disturbed far away from z_0, as required. Note too a certain resemblance between the polynomial factors in equations (6.9) and (6.10); after looking at these, the fact that quasiparticles have the quantum numbers of "fractional electrons" should seem quite plausible.

The Laughlin wave functions describe states at filling fractions $\nu = 1/m$, and indeed these are among the most prominent filling fractions observed experimentally in the FQHE. As the density moves away from a favored value, more and more quasi-particles are created. It is plausible that, when enough of them have been created, these quasi-particles in turn organize themselves into an incompressible quantum liquid, and so on. Implementing this thought, several authors have presented a hierarchical construction of states corresponding to other filling fractions.[9] These constructions, although they certainly contain a core of truth, do not have the compelling simplicity and uniqueness of the Laughlin $1/m$ states.

Several of the enclosed papers make use of the geometric phase to elucidate important aspects of the FQHE. In their brief paper [6.1] Arovas, Schrieffer, and Wilczek derive the fractional charge and fractional statistics of quasiparticles in the FQHE by applying geometric phase techniques to the Laughlin wavefunction. The subsequent contribution [6.2] by Arovas, adapted from his thesis and published here for the first time, gives a much more detailed account of this calculation and other aspects of the quantum mechanics of fractional statistics particles, emphasizing the introduction of a gauge field to capture their dynamics, as we discussed in the previous chapter.

The Laughlin wave function, although it apparently yields an accurate estimate of the ground state energy, and incorporates the correct qualitative physics of the incompressible quantum liquid and its commensurability criterion, is perhaps not the end of all desire. In particular, it does not give a crisp answer to the question "What, precisely, characterizes the quantized Hall state?" comparable to the characterization of magnetic or superconducting states in terms of their order parameters. Also, the wave function is very specially adapted to the problem of electrons in strong magnetic fields, and it is not easy to generalize it to other situations where similar physics may be involved (see below).

In order to remedy these and other limitations, several physicists have attempted to define some kind of order parameter and an effective Landau-Ginzburg type theory for the FQHE.

By far the most substantial and interesting attempts along these lines have been made by Girvin and MacDonald [6.3], and by Read.[10] These authors make essential use of the statistical gauge field that we met in Chapter 5.

There are two issues to address. One is the construction of an effective Lagrangian, that serves to summarize the low-energy excitations and their interactions in a compact fashion. A plausible idea, which is essentially what these authors propose and to some extent derive, is that the effective theory simply consists of fractionally charged particles with fractional statistics. We have seen that these are important qualitative properties of the FQHE quasiparticles, and it is not absurd to suppose that they are the only important long-range properties the effective theory need represent. Simple Lagrangians realizing these ideas are easy to construct, using the techniques discussed in the previous chapter and by Arovas below.

The second issue is the microscopic underpinning of the effective Lagrangian. Girvin and MacDonald have made important progess on this issue. The underlying idea, as we understand it, is to make a singular gauge transformation of the Laughlin wave function—which we know represents an anyon condensate—that removes the statistical phase factors, thus allowing the underlying condensate to reveal itself. Girvin and MacDonald show by explicit calculation that the transformed density matrix has (algebraic) long-range order.

A notable defect of the existing "effective field theories" of the FQHE is their failure to incorporate or to illuminate the commensurability effects. In these theories, the statistical parameter is given from outside, and there is no real understanding of why particular rational values are selected, nor (as far as we know) any connection to the hierarchical constructions. We believe that this failure will only be remedied by a deeper study of topological invariants of the effective gauge fields, that represent the statistical properties of the full wavefunction over the multi-particle configuration space.

As we have mentioned several times above, there are several other contexts in which one expects that insight gained from the quantized Hall effect will lead to progress. A comparatively simple but quite entertaining and possibly important one is the behavior of one-dimensional electron gases in complex topologies—i.e., networks of wires—in strong magnetic fields, as discussed in the enclosed paper of Avron [6.4].

More speculative but very exciting is an idea repeatedly mentioned by Anderson,[11] and given one concrete form by Kalmeyer and Laughlin,[12] that there is a qualitative resemblance between the physics of the FQHE and the physics of several other notoriously difficult problems in condensed matter physics. An outstanding example is the Mott insulator problem. It has been realized since the 1930s that several metallic oxides are insulators even though they contain an odd number of electrons per unit cell. This contradicts the basic principles of band theory, since—because of electron spin—an odd number of electrons per unit cell should lead to a half-filled band, and thus to a metal. If the materials were antiferromagnetically ordered there would be no such difficulty, basically because the size of the unit cell would be doubled (at least). However, although several of the metallic oxides do exhibit antiferromagnetic order, others do not—and in fact in the paradigmatic case of NiO there is a Néel transition from an antiferromagnetic to a disordered state, but the material remains an insulator even above the transition. An attractive hypothesis is that the antiferromagnetic "spin solid" melts to a quantum spin liquid, just as in the FQHE the Wigner crystal melts to an incompressible liquid. The idea of a spin liquid is made more compelling by a variety of observations indicating that localized spin moments do persist through the Néel transition—the spins do not disappear, and in fact their short-range correlations seem to vary little through the transition; it is only the long-range order that disappears.

The Mott insulator problem has long been a skeleton in the closet of condensed matter theory. Due to the lack of a convincing theoretical framework in which to address it, and to the technological unimportance of the materials concerned, the problem has been widely ignored. However the recent discovery of high temperature superconductivity, in which anomalously insulating CuO insulators play a crucial role, has forced this problem back into the light of day. Kalmeyer and Laughlin devised a variational wave function of the FQHE type for a relevant model. Laughlin [6.5] subsequently pointed out that the quasi-particles around the ground state, defined by this variational wave function, carry fractional statistics, as should by now seem quite plausible. In fact, these so-called spinons are half-fermions—the phase i accumulates as one winds around another. Furthermore, it is plausible that holes injected by doping around the half-filled band bind to these spinons, making charged, half-fermionic quasiparticles called holons. Now half-fermions can be thought of as fermions with an additional attractive gauge interaction, and so it is not implausible that they should condense into a BCS type superconducting pair state. In fact, a composite particle composed of two half-fermions is easily seen to possess Bose statistics.

There is a distinct possibility that many-body effects could also occur, and be qualitatively important in electronic systems, well outside the thermodynamic limit, for example giving rise to commensurability effects in macromolecules. This idea is made more plausible by the fact that numerical simulations of the FQHE indicate pronounced energy minima at non-trivial rational fillings even for quite small systems.

The appearance of gauge fields in the effective description of FQHE states suggests that the relatively well-understood physics of the FQHE, including the notion of an incompressible quantum liquid, commensurability, and the associated rich phase structure, may have much to teach us about the even less-understood physics of gauge theories as studied in elementary particle physics. Let us note in particular that the phase structure of lattice quantum electrodynamics with a theta term exhibits some uncanny resemblances to the hierarchical structure of different FQHE states.[13].

The final entry of this chapter, by Wilkinson [6.6], is more closely related to theories of the IQHE. It studies a system known as Harper's equation, which has been used to model the behavior of Bloch electrons in a perpendicular magnetic field. (The precise logical connection to the IQHE is a long story; we refer the reader to the article by Thouless in Ref. 4 for details.) An important feature of Harper's equation is that when the number of magnetic flux quanta per unit cell is rational, an energy gap appears. Wilkinson uses a WKB method to calculate the Bohr–Sommerfeld energy levels. As explained by Kuratsuji and Iida [3.9] in a more general context, to obtain the correct WKB quantization condition it is necessary to include the effect of Berry's phase. This turns out to be a considerably simpler way to derive the beautiful fractal spectrum of Harper's equation than previous methods.

---

[1] K. v. Klitzing, G. Dorda, and M. Pepper, *Phys. Rev. Lett.* **45** (1980) 494.

[2] D.C. Tsui, H.L. Stormer, and A.C. Gossard, *Phys. Rev. Lett.* **48** (1982) 1559.

[3] L.D. Landau and E.M. Lifshitz, *Quantum Mechanics*, Course of Theoretical Physics vol. 3 (Oxford: Pergamon Press, 1977).

[4] R. Prange and S. Girvin, *The Quantum Hall Effect* (Berlin: Springer-Verlag, 1987).

[5] R.B. Laughlin, *Phys. Rev. Lett.* **50** (1983) 1395.

[6] E. Wigner, *Trans. Faraday Soc.* **34** (1938) 678.

[7] R.B. Laughlin, *Phys. Rev. Lett.* **50** (1983) 1395.
Chapter 7 in Ref. 4.

[8] F.D.M. Haldane, Chapter 8 in Ref. 4.

[9] F.D.M. Haldane, *Phys. Rev. Lett.* **51**, (1983) 605;
R. B. Laughlin, *Surface Science* **141**, (1984) 11;
B. Halperin, *Phys. Rev. Lett.* **52**, (1984) 1583.

[10] N. Read, unpublished.

[11] P. Anderson, unpublished.

[12] V. Kalmeyer and R.B. Laughlin, *Phys. Rev. Lett.* **59** (1987) 2095.

[13] J. Cardy *Nucl. Phys.* **B205** (1982) 17;
A. Shapere and F. Wilczek, "Self-dual models with theta terms," to appear in Nuclear Physics B.

VOLUME 53, NUMBER 7 PHYSICAL REVIEW LETTERS 13 AUGUST 1984

Fractional Statistics and the Quantum Hall Effect

Daniel Arovas
Department of Physics, University of California, Santa Barbara, California 93106

and

J. R. Schrieffer and Frank Wilczek
Department of Physics and Institute for Theoretical Physics, University of California, Santa Barbara, California 93106
(Received 18 May 1984)

The statistics of quasiparticles entering the quantum Hall effect are deduced from the adiabatic theorem. These excitations are found to obey fractional statistics, a result closely related to their fractional charge.

PACS numbers: 73.40.Lq, 05.30.-d, 72.20.My

Extensive experimental studies have been carried out[1] on semiconducting heterostructures in the quantum limit $\omega_0\tau >> 1$, where $\omega_0 = eB_0/m$ is the cyclotron frequency and τ is the electronic scattering time. It is found that as the chemical potential μ is varied, the Hall conductance $\sigma_{xy} = I_x/E_y = \nu e^2/h$ shows plateaus at $\nu = n/m$, where n and m are integers with m being odd. The ground state and excitations of a two-dimensional electron gas in a strong magnetic field B_0 have been studied[2-4] in relation to these experiments and it has been found that the free energy shows cusps at filling factors $\nu = n/m$ of the Landau levels. These cusps correspond to the existence of an "incompressible quantum fluid" for given n/m and an energy gap for adding quasiparticles which form an interpenetrating fluid. This quasiparticle fluid in turn condenses to make a new incompressible fluid at the next larger value of n/m, etc.

The charge of the quasiparticles was discussed by Laughlin[2] by using an argument analogous to that used in deducing the fractional charge of solitons in one-dimensional conductors.[5] He concluded for $\nu = 1/m$ that quasiholes and quasiparticles have charges $\pm e^* = \pm e/m$. For example, a quasihole is formed in the incompressible fluid by a two-dimensional bubble of a size such that $1/m$ of an electron is removed. Less clear, however, is the statistics which the quasiparticles satisfy; Fermi, Bose, and fractional statistics having all been proposed. In this Letter, we give a direct method for determining the charge and statistics of the quasiparticles.

In the symmetric gauge $\vec{A}(\vec{r}) = \frac{1}{2}\vec{B}_0 \times \vec{r}$ we consider the Laughlin ground state with filling factor $\nu = 1/m$,

$$\psi_m = \prod_{j<k}(z_j - z_k)^m \exp(-\tfrac{1}{4}\textstyle\sum_l |z_l|^2), \qquad (1)$$

where $z_j = x_j + iy_j$. A state having a quasihole localized at z_0 is given by

$$\psi_m^{+z_0} = N_+ \textstyle\prod_i (z_i - z_0)\psi_m, \qquad (2)$$

while a quasiparticle at z_0 is described by

$$\psi_m^{-z_0} = N_- \textstyle\prod_i (\partial/\partial z_i - z_0/a_0^2)\psi_m, \qquad (3)$$

where $2\pi a_0^2 B_0 = \phi_0 = hc/e$ is the flux quantum and $N_\pm$ are normalizing factors.

To determine the quasiparticle charge e^*, we calculate the change of phase γ of $\psi_m^{+z_0}$ as z_0 adiabatically moves around a circle of radius R enclosing flux ϕ. To determine e^*, γ is set equal to the change of phase,

$$(e^*/\hbar c)\oint \vec{A}\cdot d\vec{l} = 2\pi(e^*/e)\phi/\phi_0, \qquad (4)$$

that a quasiparticle of charge e^* would gain in moving around this loop. As emphasized recently by Berry[6] and by Simon[7] (see also Wilczek and Zee[8] and Schiff[9]), given a Hamiltonian $H(z_0)$ which depends on a parameter z_0, if z_0 slowly transverses a loop, then in addition to the usual phase $\int^t E(t')\,dt'$, where $E(t')$ is the adiabatic energy, an extra phase γ occurs in $\psi(t)$ which is independent of how slowly the path is traversed. $\gamma(t)$ satisfies

$$d\gamma(t)/dt = i\langle\psi(t)|d\psi(t)/dt\rangle \; . \qquad (5)$$

From Eq. (2),

$$\frac{d\psi_m^{+z_0}}{dt} = N_+ \sum_i \frac{d}{dt}\ln[z_i - z_0(t)]\psi_m^{+z_0}, \qquad (6)$$

so that

$$\frac{d\gamma}{dt} = iN_+^2 \left\langle \psi_m^{+z_0}\middle|\frac{d}{dt}\sum_i \ln(z_i - z_0)\middle|\psi_m^{+z_0}\right\rangle. \qquad (7)$$

Since the one-electron density in the presence of

VOLUME 53, NUMBER 7 PHYSICAL REVIEW LETTERS 13 AUGUST 1984

the quasihole is given by

$$\rho^{+z_0}(z) = \langle \psi_m^{+z_0} | \sum_i \delta(z_i - z) | \psi_m^{+z_0} \rangle, \quad (8)$$

we have

$$\frac{d\gamma}{dt} = i \int dx\, dy\, \rho^{+z_0}(z) \frac{d}{dt} \ln[z - z_0(t)], \quad (9)$$

where $z = x + iy$. We write $\rho^{+z_0}(z) = \rho_0 + \delta\rho^{+z_0}(z)$, with $\rho_0 = \nu B/\phi_0$. Concerning the ρ_0 term, if z_0 is integrated in a clockwise sense around a circle of radius R, values of $|z| < R$ contribute $2\pi i$ to the integral while $|z| > R$ contributes zero. Therefore, this contribution to γ is given by

$$\gamma_0 = i \int_{|r| < R} dx\, dy\, \rho_0 2\pi i = -2\pi \langle n \rangle_R = -2\pi\nu\phi/\phi_0, \quad (10)$$

where $\langle n \rangle_R$ is the mean number of electrons in a circle of radius R. Corrections from $\delta\rho$ vanish as $(a_0/R)^2$, where $a_0 = (\hbar c/eB)^{1/2}$ is the magnetic length. This term corresponds to the finite size of the hole.

Comparing with Eq. (4), we find $e^* = \nu e$, in agreement with Laughlin's result. A similar analysis shows that the charge of the quasiparticle $\psi_m^{-z_0}$ is $-e^*$.

To determine the statistics of the quasiparticles, we consider the state with quasiholes at z_a and z_b,

$$\psi_m^{z_a, z_b} = N_{ab} \prod_i (z_i - z_a)(z_i - z_b) \psi_m. \quad (11)$$

As above, we adiabatically carry z_a aroound a closed loop of radius R. If z_b is outside the circle $|z_b| = R$ by a distance $d >> a_0$, the above analysis for γ is unchanged, i.e., $\gamma = -2\pi\nu\phi/\phi_0$. If z_b is inside the loop with $|z_b| - R << -a_0$, the change of $\langle n \rangle_R$ is $-\nu$ and one finds the extra phase $\Delta\gamma = 2\pi\nu$. Therefore, when a quasiparticle adiabatically encircles another quasiparticle an extra "statistical phase"

$$\Delta\gamma = 2\pi\nu \quad (12)$$

is accumulated.[10] For the case $\nu = 1$, $\Delta\gamma = 2\pi$, and the phase for interchanging quasiparticles is $\Delta\gamma/2 = \pi$ corresponding to Fermi statistics. For ν noninteger, $\Delta\gamma$ corresponds to fractional statistics, in agreement with the conclusion of Halperin.[11] Clearly, when ν is noninteger the change of phase $\Delta\gamma$ when a third quasiparticle is in the vicinity will depend on the adiabatic path taken by the quasiparticles as they are interchanged and the pair permutation definition used for Fermi and Bose statistics no longer suffices.

A convenient method for including the statistical phase $\Delta\gamma$ is by adding to the actual vector potential $\vec{A}_0$ a "statistical" vector potential $\vec{A}_\phi$ which has no independent dynamics. $\vec{A}_\phi$ is chosen such that

$$(e^*/\hbar c)\oint \vec{A}_\phi \cdot d\vec{l} = \Delta\gamma = 2\pi\nu, \quad (13)$$

when z_a encirlces z_b. One finds this fictious $\vec{A}_\phi$ to be

$$\vec{A}_\phi(\vec{r} - \vec{r}_b) = \frac{\phi_0 \hat{z} \times (\vec{r} - \vec{r}_b)}{2\pi |\vec{r} - \vec{r}_b|^2} \quad (14)$$

if the quasiparticles are treated as bosons and $\phi_0 \to \phi_0(1 - 1/\nu)$ if they are treated as fermions. Thus, the peculiar statistics can be replaced by a more complicated effective Lagrangian describing particles with conventional statistics.[12]

Finally, we note that if one pierces the plane with a physical flux tube of magnitude ϕ, the above arguments suggest that a charge $\nu e\phi/\phi_0$ is accumulated around the tube, regardless of whether ϕ/ϕ_0 is equal to the ratio of integers.

This work was supported in part by the National Science Foundation through Grant No. DMR82-16285 and No. PHY77-27084, supplemented by funds from the National Aeronautics and Space Administration. One of us (D.A.) is grateful for the support of an AT&T Bell Laboratories Scholarship.

[1]K. von Klitzing, G. Dorda, and M. Pepper, Phys. Rev. Lett. **45**, 494 (1980).

[2]R. B. Laughlin, Phys. Rev. Lett. **50**, 1395 (1983).

[3]F. D. M. Haldane, Phys. Rev. Lett. **51**, 605 (1983).

[4]B. I. Halperin, Institute of Theoretical Physics, University of California, Santa Barbara, Report No. NSF-ITP-83-34, 1983 (to be published).

[5]W. P. Su and J. R. Schrieffer, Phys. Rev. Lett. **46**, 738 (1981).

[6]M. V. Berry, Proc. Roy. Soc. London, Ser. A **392**, 45–57 (1984).

[7]B. Simon, Phys. Rev. Lett. **51**, 2167 (1983).

[8]F. Wilczek and A. Zee, Phys. Rev. Lett. **52**, 2111 (1984).

[9]L. Schiff, *Quantum Mechanics* (McGraw-Hill, New York, 1955), p. 290.

[10]Although ψ is a variational wave function, rather than the actual adiabatic wave function, the statistical properties of the quasiparticles are not expected to be sensitive to this inconsistency. We could regard ψ to be an exact excited-state wave function for a model Hamiltonian.

[11]B. I. Halperin, Phys. Rev. Lett. **52**, 1583, 2390(E) (1984).

[12]F. Wilczek and A. Zee, Institute of Theoretical Physics, University of California, Santa Barbara, Report No. NSF-ITP-84-25, 1984 (to be published).

Topics in Fractional Statistics

Daniel P. Arovas

Department of Physics
University of California at San Diego
La Jolla, CA 92093

ABSTRACT

I review the general theory of quantum mechanical particles with fractional statistics in two space dimensions ('anyons'). The thermodynamics of a low-density gas of anyons is discussed, as well as is the relevance of fractional statistics to quasiparticle excitations in the fractional quantized Hall effect.

Introduction

The physical implications of quantum statistics are numerous and lead to profound consequences for the excitation spectra and thermodynamic properties of all systems. Fermions, which obey the Pauli exclusion principle, fill a Fermi sphere in the $T \to 0$ limit.* Low temperature thermodynamics are then dominated by particle–hole excitations across the Fermi surface, leading to a specific heat $C_{\rm F}^{d=3} \sim \frac{1}{2}\pi^2 N k_B T/T_{\rm F}$, which in the case of metals is much smaller than the Maxwell–Boltzmann value of $\frac{3}{2}Nk_B$ even at room temperature. Bosons, which do not heed the Pauli restriction of at most one quantum per single particle level i, exhibit a ground state in which the lowest such single particle level ($i = 0$) is macroscopically occupied; thermodynamic properties are then related to fluctuations in the occupancy of low–lying excited states. In dimensions $d > 2$, these fluctuations are weak enough to preserve the macroscopic occupancy of the $i = 0$ state, a phenomenon known as Bose condensation.

Even in the dilute gas regime, where the mean interparticle spacing, $n^{-1/2}$, is much larger than the thermal wavelength, $\lambda_T = (2\pi\hbar^2/mk_BT)^{1/2}$, relics of the quantum limit can be identified by examining corrections to the ideal gas law

$$p = nk_BT(1 + \mathrm{B}_2 n + \mathrm{B}_3 n^2 + \cdots). \tag{1}$$

The terms B_l are the ***virial coefficients*** and characterize deviations from ideal gas behavior. For a free gas in two dimensions, $\mathrm{B}_2(T) = \mp\frac{1}{4}\lambda_T^2$, the plus sign applying to the Fermi case, as the Pauli principle effectively pushes fermions away from each other, thus increasing the pressure.

Such well established properties ultimately derive from the respective symmetry and antisymmetry of Bose and Fermi wave functions, and when confronted with the query, "Why is there this sharp Bose–Fermi dichotomy; can no other quantum statistics be formulated?" most learned professors respond that indistinguishability implies that any N–body Hamiltonian will commute with elements σ of the permutation group S_N, and quantum mechanics, being a unitary theory, obliges us to characterize physical states by a one–dimensional representation of S_N, of which

* For the moment, I shall consider noninteracting particles, though the thermodynamic behavior discussed is qualitatively correct for interacting systems within the context of a quasiparticle model.

there are only two : the symmetric, or Bose representation ($\chi_B(\sigma) = 1$), and the antisymmetric, or Fermi representation ($\chi_F(\sigma) = \mathrm{sgn}(\sigma)$). On the other hand, it is often emphasized that wave functions themselves are not physical entities; physical information is conveyed by propagators and matrix elements. It is then natural to ask whether the restrictions on quantum statistics discussed above also apply if one adopts a Feynman path integral approach to quantum mechanics, in which propagators, rather than wave functions, play the fundamental role.

A careful consideration of this issue leads to the possibility of exotic (or 'fractional') statistics in systems of spatial dimension less than three. In this chapter, I will discuss the case $d = 2$, for which a continuous one-parameter family of quantum statistics may be formulated — $d = 2$ also holds the possibility for physical relevance in the case of the fractional quantized Hall effect (FQHE). I will concentrate on the physical consequences of fractional statistics in the dilute gas regime and demonstrate explicitly how certain thermodynamic quantities interpolate between Bose and Fermi behavior as functions of the statistics determining parameter.

Charged Particle–Flux Tube Composites

The essential difference between systems of indistinguishable particles in two and three dimensions is easy to comprehend. In three or more spatial dimensions, no winding can be ascribed to relative particle motion because any two paths between a chosen pair of points in configuration space may be deformed into one another. Only when one descends below $d = 3$ does this relative winding become a well defined concept. It was realized by Wilczek[1] that this result could be exploited by associating to each particle a 'charge' e and a flux tube of strength $\phi = \alpha hc/e$ (see Fig.[1]). The coupling between the charges and the vector potential arising due to the flux tubes will then keep track of the relative winding of particles with an Aharonov–Bohm phase of $e^{-ie\phi/\hbar c}$ per revolution $\Delta\theta_{ij} = 2\pi$. These composites will obey fractional statistics — a feature which prompted Wilczek to name them 'anyons'. The electromagnetic field due to the charges and fluxes is *non-dynamical*, *i.e.* its evolution is completely determined by the motion of the particles.

Consider the quantum mechanics of an electron confined to the two-dimensional plane with a flux tube of strength $\phi = \alpha\phi_o$ piercing the origin ($\phi_o = hc/e$ is the Dirac flux quantum). The vector potential associated with the flux tube may be

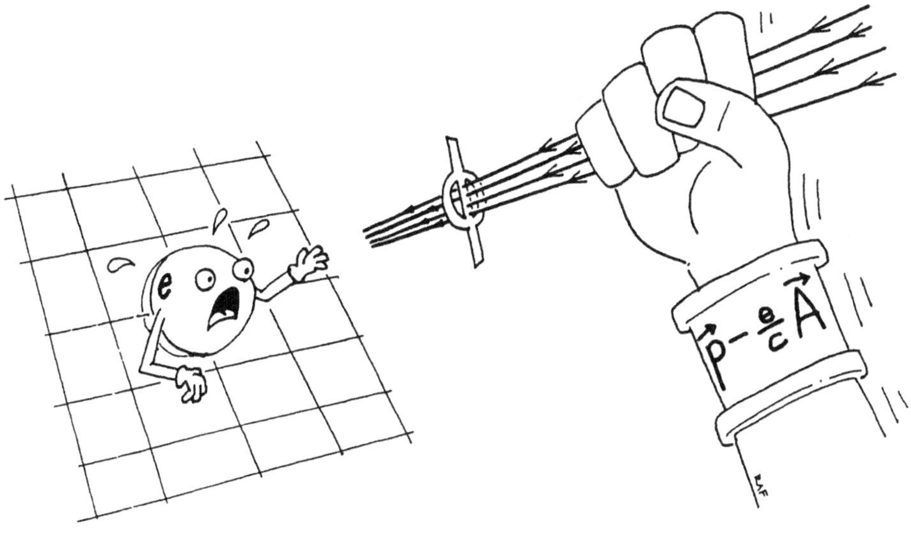

Figure 1. Artist's conception of a charged particle-flux tube composite.

taken as

$$\boldsymbol{A}(\boldsymbol{r}) = \frac{\alpha\phi_o}{2\pi}\frac{\hat{\boldsymbol{z}}\times\boldsymbol{r}}{r^2} = \frac{\alpha\phi_o}{2\pi}\nabla\theta. \tag{2}$$

The field strength, $\boldsymbol{B}(\boldsymbol{r}) = \nabla\times\boldsymbol{A} = \alpha\phi_o\delta(\boldsymbol{r})\hat{\boldsymbol{z}}$, is confined to the interior of the flux tube, which in this toy model is infinitesimally thin.

If the electron is otherwise free, the Hamiltonian is

$$H(\alpha) = \frac{1}{2m}\left(\boldsymbol{p} - \frac{e}{c}\boldsymbol{A}\right)^2 = -\frac{\hbar^2}{2m}\frac{1}{r}\frac{\partial}{\partial r}r\frac{\partial}{\partial r} + \frac{L_z^2(\alpha)}{2mr^2}, \tag{3}$$

where the operator $L(\alpha)$ is the familiar dynamical angular momentum,

$$L_z(\alpha) = e^{i\alpha\theta}\left(-i\hbar\frac{\partial}{\partial\theta}\right)e^{-i\alpha\theta} = \left(-i\hbar\frac{\partial}{\partial\theta} - \alpha\hbar\right). \tag{4}$$

If the system is placed in an eigenstate and the flux ϕ is varied adiabatically, the angular momentum (and hence the energy) will change with ϕ. Put simply, a variation in flux will lead to an azimuthal electric field $\boldsymbol{E}(\boldsymbol{r}) = -(\dot{\alpha}\hbar/er)\hat{\boldsymbol{\theta}}$ by Faraday's law.[1] The rate of change of angular momentum is then $\dot{L}_z = [\boldsymbol{r}\times(e\boldsymbol{E})]_z =$

$-\dot{\alpha}\hbar$, and therefore as the flux is cranked from $\phi = 0$ to $\phi = \alpha\phi_o$, the spectrum of allowed angular momenta changes from $\{L_z = m\}$ to $\{L_z = m - \alpha\}$.

An explicit solution to the eigenvalue equation $H(\alpha)\psi^\alpha = E(\alpha)\psi^\alpha$ is

$$\psi_k^\alpha(r,\theta) = \sum_{m=-\infty}^{\infty} A_m e^{im\theta} J_{|m-\alpha|}(kr) \tag{5}$$

where $J_\nu(x)$ is the Bessel function of the first kind of order ν. By imposing a hard wall constraint at $r = \Lambda$, $\psi_k^\alpha(\Lambda,\theta) = 0$, one can index the various modes by integers l and n:

$$\begin{aligned} \psi_{l,n}^\alpha(r,\theta) &= e^{in\theta} J_{|n-\alpha|}(k_{l,n}r) \\ k_{l,n} &= x_{|n-\alpha|,l}/\Lambda \\ E_{l,n} &= \hbar^2 k_{l,n}^2 \Big/ 2m \\ L_{z\ l,n} &= n - \alpha. \end{aligned} \tag{6}$$

(I have adopted the notation $x_{\nu,l}$ for the lth node of J_ν.)

As expected, the energy spectrum is periodic in α, and an adiabatic increase of α by 1 results in the maps the spectrum back into itself by $\{l,n\} \to \{l-1,n\}$. I would like to emphasize that Eq.(14) constitutes a gauge transformation only when α is an integer. Otherwise, the factors $e^{\pm ie\phi/\hbar c}$ are not single valued, and it is *incorrect*, however tempting, to write $\psi^\alpha = e^{i\alpha\theta}\psi^0$.

Two Anyons

I shall now discuss the quantum mechanics of two anyons. In the absence of an external potential, the Hamiltonian resembles $H = H_1 + H_2$, with

$$H_i = \frac{1}{2m}\left(\boldsymbol{p}_i - \frac{e}{c}\boldsymbol{A}_s(\boldsymbol{r}_i)\right)^2. \tag{7}$$

The 'statistical vector potential' $\boldsymbol{A}_s(\boldsymbol{r}_i)$ felt by particle 1 arises due to the presence of its companion and contains two identical contributions, one due to the interaction of the *charge* of 1 interacting with the *flux* of 2 and the other due to the interaction of the *flux* of 1 with the *charge* of 2. Thus,

$$\boldsymbol{A}_s(\boldsymbol{r}_1) = \frac{\alpha\phi_o}{2\pi}\frac{\hat{\boldsymbol{z}} \times (\boldsymbol{r}_1 - \boldsymbol{r}_2)}{|\boldsymbol{r}_1 - \boldsymbol{r}_2|^2}, \tag{8}$$

where $\alpha = 2\phi/\phi_o$ is *twice* the number of Dirac flux quanta piercing any given particle; a corresponding expression applies for particle 2. By converting to relative and center of mass coordinates $(\boldsymbol{r}, \boldsymbol{R})$, one can easily solve this problem. In fact, the relative coordinate problem is just that of a particle and a flux tube, whose solution is displayed above. The two–body wave functions are then

$$\psi^{\alpha}(\boldsymbol{r}, \boldsymbol{R}) = e^{i\boldsymbol{K}\cdot\boldsymbol{R}} e^{in\theta} J_{|n-\alpha|}(kr). \tag{9}$$

The mode counting may be performed by introducing a potential $V(\boldsymbol{r})$ which forces the wave function to vanish when the interparticle separation exceeds some arbitrary value, Λ, which may later be to be taken to infinity. In this case, the allowed wavevectors are again quantized,

$$\begin{aligned} k_{l,n} &= x_{|n-\alpha|,l}/\Lambda \\ E_{\boldsymbol{K},l,n} &= \frac{\hbar^2 K^2}{4m} + \frac{\hbar^2 k_{l,n}^2}{m}, \end{aligned} \tag{10}$$

and all physical quantities are periodic in α.

If the particles themselves are taken to obey Bose statistics, the allowed values of n are restricted to the even integers. Similarly, Fermi statistics would require that n be odd. Therefore, according to Eq.(10) an adiabatic increase of α by 1 $(\phi \to \phi + \phi_o/2)$ shifts the spectrum from Bose–like to Fermi–like and vice versa. More precisely, the eigenfunctions for a pair of fermions at $\alpha = 1$ are *unitarily equivalent* to those for a pair of bosons at $\alpha = 0$. Hence, no physical measurement could distinguish between the two cases and by the introduction of charge and flux, bosons can be magically transformed into fermions, a process which I shall refer to as 'quantum alchemy'.

A General Recipe for Quantum Alchemy

Consider now an assembly of N particles interacting via arbitrary potentials. Perhaps there is also an external magnetic field $\boldsymbol{B}_{\text{ext}} = \nabla \times \boldsymbol{A}_{\text{ext}}$ present. To change the effective quantum statistics, one introduces the 'statistical' vector potential $\boldsymbol{A}_s$ arising from the flux tubes

$$\boldsymbol{A}_s(\boldsymbol{r}_i) = \frac{\alpha\phi_o}{2\pi} \sum_{\substack{j \\ (j\neq i)}} \frac{\hat{\boldsymbol{z}} \times (\boldsymbol{r}_i - \boldsymbol{r}_j)}{|\boldsymbol{r}_i - \boldsymbol{r}_j|^2} \tag{11}$$

in which case the dynamical momenta become

$$\boldsymbol{\pi}_i^0 = \boldsymbol{p}_i - \frac{e}{c}\boldsymbol{A}_{\text{ext}}(\boldsymbol{r}_i) \to \boldsymbol{\pi}_i = \boldsymbol{p}_i - \frac{e}{c}\boldsymbol{A}_{\text{ext}}(\boldsymbol{r}_i) - \frac{e}{c}\boldsymbol{A}_s(\boldsymbol{r}_i). \tag{12}$$

The many-body Lagrangian one obtains is

$$L = \sum_i \left(\tfrac{1}{2}m\dot{\boldsymbol{r}}_i^2 - \frac{e}{c}\boldsymbol{A}_{\rm ext}(\boldsymbol{r}_i)\cdot\dot{\boldsymbol{r}}_i\right) - V(\boldsymbol{r}_1,\ldots,\boldsymbol{r}_N) - \alpha\hbar\frac{d}{dt}\sum_{i<j}\theta_{ij} \qquad (13).$$

The Hamiltonian is then

$$H = \sum_{i=1}\frac{1}{2m}\left(\boldsymbol{p}_i - \frac{e}{c}\boldsymbol{A}_{\rm ext}(\boldsymbol{r}_i) - \frac{e}{c}\boldsymbol{A}_s(\boldsymbol{r}_i)\right)^2 + V(\boldsymbol{r}_1,\ldots,\boldsymbol{r}_N). \qquad (14)$$

It is again worth emphasizing that although $H(\alpha) = e^{i\alpha\Theta}H(0)e^{-i\alpha\Theta}$, with $\Theta \equiv \sum_{\rm pairs}\theta_{ij}$, the function $\psi^\alpha = e^{i\alpha\Theta}\psi^0$ for general α is not properly single valued and hence does not constitute a solution to the Schrödinger equation with appropriate boundary conditions. The problem of obtaining single valued, N-body wave functions at arbitrary α from the $\alpha = 0$ solutions is in general an extremely difficult one. For $N = 2$, the problem is easily solved due to the separation of center of mass and relative coordinates. For $N = 3$, only a handful of solutions exist.[2] I know of no explicit wave functions describing fractional statistics for more than three particles. The situation gets worse with increasing particle number due to the fact that there are $\frac{1}{2}N(N-1)$ relative angles and only $(N-1)$ non-CM degrees of freedom. For $N = 2$, these numbers are the same.

General Theory

The laws for treating systems of indistinguishable particles by path integration were laid down by Laidlaw and DeWitt,[3] who specialized to the case of $d = 3$. Following Wilczek's[1,4] formulation of fractional statistics in terms of charged particle–flux tube composites, the general $d = 2$ theory was developed by Wu[5] in 1984. Central to the formalism is the application of path integration in multiply connected spaces.[6] The basic idea here is best described by Schulman's example of a free particle on a circle (S^1) parameterized by the angle ϕ. The propagator $K(\phi', t'|\phi, t)$ is written as a sum over paths

$$K(\phi', t'|\phi, t) = \sum_{n=-\infty}^{\infty} A_n \sum_{\phi(t)\in\mathcal{W}_n} e^{iS[\phi(t)]}$$

$$\mathcal{W}_n = \{\text{paths of winding number } n\} \qquad (15)$$

where $S[\phi(t)]$ is the action corresponding to the path $\phi(t)$. The winding number of a path γ is given by the number of times γ passes some arbitrary point ϕ_0, subtracting counterclockwise from clockwise passages. As Schulman points out, there is no *a priori* reason why the amplitudes A_n should be equal, as even with arbitrary A_n, the above expression correctly generates the Schrödinger equation when propagating a wave function over an infinitesimal time interval. Constraints on the A_n are dictated by requirements that the total probability $P = |K|^2$ be independent of ϕ_0, and that K satisfy the composition rule

$$K(\phi'', t''|\phi, t) = \int_0^{2\pi} d\phi' \; K(\phi'', t''|\phi', t')K(\phi', t'|\phi, t). \tag{16}$$

This leads to the form $A_n = e^{i(\delta_0 + n\delta)}$.

The case of path integration on an arbitrary multiply connected manifold M follows by a straightforward generalization of the above discussion. The notion of winding number generalizes to that of homotopy, which is an equivalence relation used by topologists to classify paths† — two paths on M are homotopic if one can be continuously deformed into the other. The section below makes use of only the rudiments of homotopy theory, which are nicely summarized in Schulman's[6] section 23.2, entitled "Rudiments of Homotopy Theory."

The propagator is written as a sum over homotopy classes

$$\begin{aligned} K(q', t'|q, t) &= \sum_{|\mu| \in \pi_1(M)} K_{|\mu|}(q', t'|q, t) \\ &= \sum_{|\mu| \in \pi_1(M)} \chi(|\mu|) \sum_{q(t) \in |\mu|} e^{iS[q(t)]}, \end{aligned} \tag{17}$$

where $q \in M$ is a point, $|\mu| \in \pi_1(M)$ is a homotopy class, and $\pi_1(M)$ is the fundamental group of M. While homotopy among paths from q to q' is a perfectly well defined equivalence relation, it is convenient to label such paths by elements of the loop space of M. This is accomplished by defining a 'standard path mesh' consisting of a set of paths $C(q, q_0)$ from some arbitrary $q_0 \in M$ to every $q \in M$. This construction induces a mapping from $\pi_1(M, q, q')$ to $\pi_1(M, q_0)$:

$$\gamma\colon [0,1] \mapsto M, \qquad \gamma(0) = q, \qquad \gamma(1) = q'$$

and

$$f_{qq'}\colon \pi_1(M, q, q') \mapsto \pi_1(M, q_0)$$

by

† More generally, a homotopy is an equivalence of maps between topological spaces.

$$f_{qq'}(\gamma) = C(q', q_0)^{-1} \circ \gamma \circ C(q, q_0). \tag{18}$$

Mesh invariance of the probability and the composition law require that χ be a unitary representation of $\pi_1(M)$. In the case of the circle,

$$M = \mathrm{S}^1, \qquad \pi_1(\mathrm{S}^1) \cong \mathbf{Z}$$

$$\chi: \mathbf{Z} \mapsto \mathrm{U}(1) \qquad \text{by} \qquad \chi(n) = e^{i(\delta_0 + n\delta)}. \tag{19}$$

The winding number on S^1 used above implicitly defined a path mesh consisting of paths which proceed directly from ϕ_0 to ϕ is a clockwise direction.

Because χ is a unitary representation, any nonabelian structure of $\pi_1(M)$ is left unused. Thus, any element of the commutator subgroup of π_1 will lie in the kernel of the homomorphism χ, and consequently it is only the *abelianized* π_1 (otherwise known as the first homology group H_1) which is needed. While the preceeding statement sounds absolutely marvelous, its content is quite simple. Since $\chi([\mu])$ is a phase, both $[\nu]^{-1}[\mu][\nu]$ and $[\mu]$ map to the same image under χ even though as elements of $\pi_1(M)$ they may be distinct. Roughly speaking, H_k measures the number of 'k-dimensional holes' in the manifold M, and physically, the association of a phase to each element of H_1 corresponds to threading each such hole with a magnetic flux ϕ.‡ The phase accrued in winding about each hole is then $e^{-ie\phi/\hbar c}$, leading to a ϕ-dependent interference between paths of different winding number. Viewed as an element of S^1, $e^{-ie\phi/\hbar c}$ distinguishes one among a continuum of quantum theories indexed by points on the circle.

Intuitively, one can think of this interference by taking a sort of cross product of M with its fundamental group. In this way, paths which start and end at the same point but exhibit different winding are not considered to be homotopic. This notion is but a barbarous interpretation of what algebraic topologists call a *universal covering space*. Every manifold M possesses a unique universal covering $\tilde{M}$ and covering projection p with $\tilde{M}$ simply connected and $\mathrm{p}: \tilde{M} \mapsto M$ a local homeomorphism. As discussed by Schulman, once one agrees on a specific choice of preimage p^{-1} for every point on M, any path on M may be lifted to $\tilde{M}$. Consider two paths in $\tilde{M}$ each emanating from $\tilde{x}$ and concluding at distinct points $\tilde{y}_a$ and $\tilde{y}_b$, respectively. If $\mathrm{p}(\tilde{x}) = x$ and $\mathrm{p}(\tilde{y}_a) = \mathrm{p}(\tilde{y}_b) = y$, these paths may be viewed as the lifts of

‡ Technically, one may associate a different flux ϕ_a with each generator a of H_1. Think about puncturing the plane at n distinct points.

two nonhomotopic paths in M. In this way, the restricted propagator $K_{[\mu]}(q',t'|q,t)$ may be calculated by summing over *all* paths on the universal covering space from $\tilde{q}$ to a particular $\tilde{q}_{[\mu]}$.

These concepts are nicely illustrated by the simple example of the 2–torus, $\mathbf{T}^2 = \mathrm{S}^1 \times \mathrm{S}^1$, shown in Fig.[2]. As one would expect, $\pi_1(\mathbf{T}^2) \cong H_1(\mathbf{T}^2) \cong \mathbf{Z} \times \mathbf{Z}$, and the universal covering space is the plane, $\mathbf{R}^2$. The canonical projection p is defined as

$$\mathrm{p}: \mathbf{R}^2 \mapsto \mathbf{T}^2 \qquad \text{by} \qquad \mathrm{p}(x,y) = (e^{ix}, e^{iy}). \tag{20}$$

The path $\{\tilde{\gamma}: [0,1] \mapsto \mathbf{R}^2 \mid \tilde{\gamma}(t) = (8\pi t, -2\pi t)\}$ is the lift of a path $\gamma: [0,1] \mapsto \mathbf{T}^2$ with winding numbers $n_1 = 4$ and $n_2 = -1$. In general, the phase $\chi(\gamma)$ may be defined in terms of the lifted path $\tilde{\gamma} = \mathrm{p}^{-1}(\gamma)$:

$$\begin{aligned} \chi(\gamma) &= \exp\left(-i\frac{e}{\hbar c}\int_{\tilde{\gamma}} \boldsymbol{A}\cdot d\boldsymbol{l}\right) \\ \boldsymbol{A} &= (\phi_1, \phi_2). \end{aligned} \tag{21}$$

Notice that in the above equation, $\chi(\gamma)$ depends continuously on the endpoints, which seems to violate the requirement that χ be a homomorphism of the *discrete* group $\pi_1(M)$ onto U(1). Indeed, by judicious use of the path mesh, this apparent inconsistency could be eliminated. But there really is no problem at all, for the respective propagators $K(q',t'|q,t)$ will differ only by a $U(1)$–valued function of q_0 (the arbitrary base point), q, and q'. Since this will not affect the probability $P = |K|^2$, it is permissible to redefine the propagator with this phase included.

Configuration Space

Consider now a set of N particles which are confined to a manifold M. M may or may not be simply connected, and $d = \dim(M)$ is as yet unspecified, though the easiest case to imagine is of course $M = \mathbf{R}^d$. For distinguishable particles, the N–particle configuration space is simply the N–fold product $\mathcal{D}_N(M) \equiv M \times M \times \cdots \times M$. For *in*distinguishable particles, one might expect the appropriate space to be $\mathcal{D}_N(M)/S_N$. Unfortunately, such a space is not a manifold. The problem lies in the coincidence points, *i.e.* the set $\mathcal{C}_N(M) \equiv \{(p_1,\ldots,p_N) \in \mathcal{D}_N(M) \mid p_j = p_k \text{ for some } j,k\}$.* To see this, consider the simpler case of two indistinguishable

* In general, when one takes the quotient of any manifold M with a discrete group G, the resulting space will not be a manifold if M contains points which are fixed under the action of elements of G (aside from the identity).

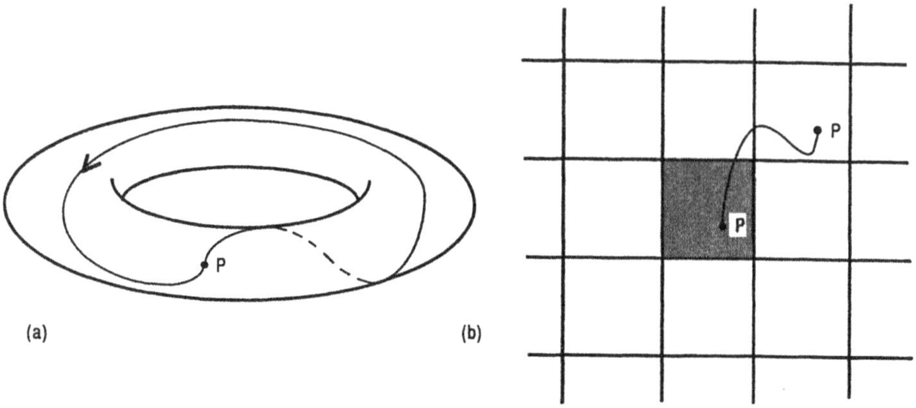

Figure 2. (a) The path shown on the torus is noncontractible, *i.e.* it cannot be continuously deformed to a single point. (b) When the curve is lifted to the universal covering space of the torus (the plane), the end points lie in different principal regions. Any such curve may be characterized by an element (n_1, n_2) of the fundamental group of the torus, $\pi_1(\mathbf{T}^2) \cong \mathbf{Z}^2$, where n_1 and n_2 measure the total number of boundary crossings between squares in the horizontal and vertical directions, respectively.

particles on a line. The candidate configuration space is then $\mathbf{R} \times \mathbf{R}/S_2$, which is simply the set of points in $\mathbf{R}^2$ on or below the diagonal. The boundary imposed by the diagonal ruins the manifold structure. To cope with this mathematical nuisance, Laidlaw and DeWitt proposed that all coincidence points be excluded from $\mathcal{D}_N(M)$ before 'modding out' by S_N. The resulting configuration space is then $\mathcal{I}_N(M) \equiv (\mathcal{D}_N(M) - \mathcal{C}_N(M))\,/S_N$. Physically, this situation would be realized only if the particles had infinitely hard cores, and although this doesn't seem to affect the generality of the discussion in any essential way, I know of no proof of this.

It is the space $\mathcal{I}_N(M)$ which enters into the many particle path integral. Paths in this space may be classified by elements of the loop space $\mathcal{B}_N(M) \equiv \pi_1(\mathcal{I}_N(M))$, as previously discussed. The structure of the loop space will determine what are the allowed phase factors $\chi([\mu])$ which multiply the restricted propagators $K_{[\mu]}$ in the sum of Eq.(17).

For $d > 2$, $\mathcal{D}_N(M)$ is simply connected, and $\mathcal{B}_N(M)$ is isomorphic to S_N, the only unitary representations of which are χ_B and χ_F. Thus, only Bose and Fermi statistics are allowed for spatial dimension three or higher. This was the main conclusion of Laidlaw and DeWitt. In two dimensions, however, $\mathcal{D}_N(M)$ is multiply connected, and $\mathcal{B}_N(M)$ has the structure of an infinite nonabelian group, known in mathematical parlance as the full N-string braid group on M.

The structure of the braid groups for several two-dimensional manifolds has been investigated by Fadell and Van Buskirk,[7] Birman,[8] and others.[9] For the relevant case of $M = \mathbf{R}^2$, there are $N-1$ generators σ_i of $\mathcal{B}_N(\mathbf{R}^2)$ which obey the relations

$$\begin{aligned} \sigma_j\sigma_k &= \sigma_k\sigma_j, \qquad |k-j| \geq 2 \\ \sigma_i\sigma_{i+1}\sigma_i &= \sigma_{i+1}\sigma_i\sigma_{i+1}, \qquad i = 1,\ldots,N-2 \end{aligned} \tag{22}$$

from which it is clear that any representation χ must satisfy $\chi(\sigma_i) = \chi(\sigma_{i+1})$ for all i, *i.e.* all generators σ_i map to *the same* unitary phase $e^{-i\pi\alpha}$ under the homomorphism χ. Therefore, there exists a continuous one-parameter family of unitary representations of $\mathcal{B}_N(\mathbf{R}^2)$ indexed by α.

There is a simple physical interpretation, due to Wu, of the algebra of Eq.(22). Consider a path $\gamma: [0,1] \mapsto \mathcal{I}_N(\mathbf{R}^2)$. This path corresponds to a set of N 'world-lines' in the three-dimensional slab $[0,1] \times \mathbf{R}^2$. These world-lines will in general interweave and be tangled like strings; elements of the braid group will describe the nature of this web. The association of an element of $\mathcal{B}_N(\mathbf{R}^2)$ with the path γ proceeds in several steps. First, obtain a labeling $1,\ldots,N$ for the particles by projecting them onto the x-axis. Thus, the label 1 applies to that particle whose abscissa is the smallest. This labeling scheme is possible at all times t during the trajectory, except for those moments when two particles have abscissae which coincide. Such events are termed *crossings*; it is obvious that only particles with consecutive labels can cross. To each crossing one associates a factor $\sigma_n^{\pm 1}(t_x)$, where n and $n+1$ participate in the crossing at time t_x, and the plus sign is used if the ordinate for n is greater than the ordinate for $n+1$ at the crossing. The resulting product is then time ordered, with earlier time arguments appearing to the right. Finally, the time labels are discarded, and one is left with a product of σ_i's and σ_j^{-1}'s, that is to say, an element of $\mathcal{B}_N(\mathbf{R}^2)$.

As discussed previously, not all of the detailed structure of $\pi_1\left(\mathcal{B}_N(\mathbf{R}^2)\right)$ is needed in defining the propagator. The phase $\chi(\gamma)$ will register a factor $e^{-i\pi\alpha}$ for each crossing, and hence it is only the 'net winding' of all the particles, measured by the sum of the angular differences $\sum_{\text{pairs}} \Delta\theta_{ij} \equiv \sum_{\text{pairs}} \int dt\, \dot{\theta}_{ij}$, to which $\chi(\gamma)$

is sensitive.†

I will now describe the general form for the propagator on $\mathcal{I}_N(\mathbf{R}^2)$. Appealing to the earlier example of the torus, the propagator $K(q',t'|q,t)$ between $q = \{\boldsymbol{r}_1,\ldots,\boldsymbol{r}_N\}$ and $q' = \{\boldsymbol{r}'_1,\ldots,\boldsymbol{r}_{N'}\}$ is expressed as a sum over all paths $\tilde{q}(t)$ in the universal covering space $\tilde{\mathcal{I}}_N(\mathbf{R}^2)$ subject to $p(\tilde{\gamma}(t_k)) = q_k$ for $k = 1,2$. The space $\tilde{\mathcal{I}}_N(\mathbf{R}^2)$ is simply $\mathcal{D}_N(\mathbf{R}^2)$ with the domains of each of the $N(N-1)/2$ relative angles θ_{ij} extended from the interval $[0,2\pi)$ to the entire real line. The canonical projection p then maps each of the θ_{ij} back onto $[0,2\pi)$ by taking the remainder upon dividing by 2π; it also respects the permutation symmetry. Therefore,

$$\begin{aligned} K(q',t'|q,t) &= \sum_{\sigma\in S_N}\sum_{|\mu|}\chi(|\mu|)K_{|\mu|}(\sigma q',t'|q,t) \\ \sigma q' &\equiv \{\boldsymbol{r}'_{\sigma(1)},\ldots,\boldsymbol{r}'_{\sigma(N)}\}. \end{aligned} \tag{23}$$

If the dynamics are determined by a Lagrangian $L(q,\dot{q})$, the resulting path integral is given by

$$K^{\alpha}(q',t'|q,t) = \sum_{\sigma\in S_N}\int \mathcal{D}q(\tau)\,\exp\left[i\int_q^{\sigma q'} d\tau\,\left(L - \alpha\frac{d}{d\tau}\sum_{i<j}\theta_{ij}(\tau)\right)\right]. \tag{24}$$

It is clear that the parameter α determines the statistics of the particles. Consider the diagonal elements of K^{α} for which $q = q'$. With $\alpha = 0$, all permutations carry the same signature and the Bose propagator is recovered. When $\alpha = 1$, the phase χ is given by the Fermi factor

$$(-1)^{\{\#\text{ of pairwise exchanges}\}}.$$

(Note also that the probability $P = |K|^2$ is periodic under $\alpha \to \alpha + 2$.) For noninteger α, K^{α} interpolates between these two familiar cases, and the statistics are said to be *fractional.* Note also that the Lagrangian defined by Eq.(24) is identical to that derived earlier in Eq.(13).

It should be stressed that α couples to a boundary term, namely the net winding number, and consequently it does not appear in the equations of motion.

† Since θ_{ij} is not single valued, $\dot{\theta}_{ij}dt$ cannot be regarded as an exact differential. The idea here is that there is a distinction between $\Delta\theta_{ij} = 0$ and $\Delta\theta_{ij} = 6\pi$, for example.

Anyon Thermodynamics

All thermodynamic properties I will discuss derive from the behavior of the grand partition function,

$$\begin{aligned}\Xi(T,A,\mu) = e^{-\beta\Omega} &= \mathrm{Tr}\, e^{-\beta(H-\mu N)} \\ &= \sum_{N=0}^{\infty} z^N \mathrm{Z}_N,\end{aligned} \tag{25}$$

where μ is the chemical potential, $\beta = 1/k_B T$ the inverse temperature, and A the area of the system. Adhering to convention, I shall abbreviate $z = e^{\beta\mu}$ as the fugacity and Z_N as the N-particle (ordinary) partition function. Partial derivatives of the grand potential Ω with respect to its arguments yield entropy, pressure, and particle number:

$$\begin{aligned}&d\Omega = -SdT - pdA - Nd\mu \\ &S = -\left.\frac{\partial\Omega}{\partial T}\right|_{A,\mu} \qquad p = -\left.\frac{\partial\Omega}{\partial A}\right|_{\mu,T} \qquad N = -\left.\frac{\partial\Omega}{\partial\mu}\right|_{T,A}.\end{aligned} \tag{26}$$

In order to represent the pressure as a function $p(T,A,N)$, as in Eq.(1), one must turn a thermodynamic crank which inverts the defining relations of Eq.(26):

$$\begin{aligned}p/k_BT = -\Omega/Ak_BT &= \sum_{l=1}^{\infty} \mathrm{b}_l(T,A)z^l \\ n = N/A &= \sum_{l=1}^{\infty} l\mathrm{b}_l(T,A)z^l \\ \Longrightarrow \quad p/k_BT &= \sum_{l=1}^{\infty} \mathrm{B}_l(T,A)n^l\end{aligned} \tag{27}$$

The virial coefficients $\mathrm{B}_l(T,A)$ are so obtained. For classical interacting systems, the coefficients $\mathrm{b}_l(T,A)$ are the well known Mayer cluster integrals. In general, the b_l may be expressed in terms of the first l many-body partition functions. For example, the second virial coefficient, which determines the lowest order corrections to ideal gas behavior, is given by

$$\mathrm{B}_2(T,A) = A\left(\tfrac{1}{2} - \mathrm{Z}_2/\mathrm{Z}_1{}^2\right). \tag{28}$$

The thermodynamics of the ideal quantum gases is discussed in many textbooks. It is essentially an academic exercise to work out the low density quantum corrections to Maxwell–Boltzmann statistics — an exercise rendered more academic by the fact that no such systems have ever been observed in nature. Perhaps the closest that workers have come to isolating such quantum corrections in two-dimensional systems is in studies of ^{4}He and ^{3}He films adsorbed on Grafoil.[10] Even in these systems, however, it is the interparticle potential *in conjunction* with the statistics that determines the thermodynamic behavior, rather than pure quantum corrections alone. For example, the measured heat capacity exhibits small deviations from the Dulong–Petit law $C = Nk_B$. A truncated virial expansion would read

$$C/Nk_B = 1 - n\beta^2 \frac{d^2 \mathrm{B}_2}{d\beta^2} \tag{29}$$

and, since the ideal quantum gas would give $\mathrm{B}_2 = \mp\frac{1}{4}\lambda_T^2 \propto \beta$, there would be *no* correction to C at this order. The helium atoms are, of course, interacting, and experience a hard core repulsion together with a weakly attractive van der Waals tail. As discussed by Dash,[11] a ^{4}He pair, whose ground state relative coordinate wave function is an s-wave, is more able to sample the repulsive core than a ^{3}He pair, which obeys the exclusion principle, leading to qualitatively different physics. With the interparticle potential properly treated, the second virial coefficient will indeed give a contribution to Eq.(29).

I shall be concerned with *purely* quantum effects in the thermodynamics of particles obeying fractional statistics. Since the statistical interaction is a many-body one (see Eq.(24) above), the problem is highly nontrivial. In fact, I shall only discuss the second virial coefficient, as higher order terms in the virial expansion require detailed knowledge of three-anyon dynamics, which is not yet available.

The thermodynamic functions for two-dimensional ideal quantum gases may be calculated from the following formulae:

$$\begin{aligned} \Omega(T,A,z) &= \mp A k_B T \lambda_T^{-2} \varsigma_2(\pm z) \\ n(T,z) &= \mp \lambda_T^{-2} \ln(1 \mp z) \\ F(n,T) = \Omega + \mu N &= \mp A k_B T \lambda_T^{-2} \varsigma_2(1 - e^{\mp n\lambda_T^2}) + nAk_B T \ln\left(\mp(e^{\mp n\lambda_T^2} - 1)\right) \end{aligned} \tag{30}$$

where I write the generalized Riemann zeta function

$$\varsigma_p(z) \equiv \sum_{n=1}^{\infty} \frac{z^n}{n^p}. \tag{31}$$

(In all formulae such as Eq.(30), the top sign shall refer to bosons, and the bottom sign to fermions.) The dimensionless parameter which characterizes these functions is the scaled density $\varrho \equiv n\lambda_T^2$, which is the mean number of particles inside a square thermal wavelength. In the quantum limit $\varrho \to \infty$, F tends to the ground state energy, which is finite for fermions and zero for bosons. The Maxwell–Boltzmann limit ($\varrho \to 0$) is one of low density and/or high temperature, and is thus entropy dominated, the free energy tending to negative infinity as

$$F/Nk_BT = \ln(\varrho) - 1 + \sum_{k=1}^{\infty} \frac{1}{k}(\mathrm{B}_{k+1}\lambda_T^{-2k})\varrho^k. \tag{32}$$

The virial expansion converges well in the Maxwell Boltzmann limit. ‡

It would be nice to have explicit formulae for general α that could interpolate between the Fermi and Bose behavior described by Eq.(30). One would then witness a spectacular change in low temperature properties as the ground state changes from a filled Fermi disk to a Bose condensate. Even with no interparticle potential, I unfortunately see little prospect for such a complete solution, due to the extreme nastiness of the general many–anyon problem. Here I will describe a calculation of $\mathrm{B}_2(T)$ which itself evidences some peculiar nonanalytic structure.*

Anyons in the Low Density Limit

My aim here is to calculate $\mathrm{B}_2(\alpha, T)$ for the anyon gas. According to Eq.(28), this requires an evaluation of the two–particle partition function, Z_2 — this is possible due to the separability of H_2 into H_{rel} and H_{CM}. For the purpose of increased generality, I shall solve this problem in the presence of a constant magnetic field of arbitrary strength. The two–body Hamiltonian is written as a sum

$$\begin{aligned} H_2 &= H_{\mathrm{CM}} + H_{\mathrm{rel}} \\ H_{\mathrm{CM}} &= \frac{1}{2M}\left(\boldsymbol{P} + \tfrac{1}{2}M\omega_c\hat{\boldsymbol{z}} \times \boldsymbol{R}\right)^2 \\ H_{\mathrm{rel}} &= \frac{1}{2\mu}\left(\boldsymbol{p} + \tfrac{1}{2}\mu\omega_c\hat{\boldsymbol{z}} \times \boldsymbol{r} + \alpha\hbar\frac{\hat{\boldsymbol{z}} \times \boldsymbol{r}}{r^2}\right)^2 \end{aligned} \tag{33}$$

‡ It is a straightforward task to derive $\tilde{\mathrm{B}}_3 = -1/36$, $\tilde{\mathrm{B}}_4 = 0$, $\tilde{\mathrm{B}}_5 = -1/3600$, $\tilde{\mathrm{B}}_6 = 0$, $\tilde{\mathrm{B}}_7 = 1/211680$, $\tilde{\mathrm{B}}_8 = 0$, $\tilde{\mathrm{B}}_9 = 1/10886400$, *etc*, for $\tilde{\mathrm{B}}_n \equiv \mathrm{B}_n\lambda_T^{2-2n}$. These results apply to both the Fermi and Bose cases. I thank MACSYMA for these numbers.

* In the thermodynamic limit, one defines $\mathrm{B}_2(T) = \lim_{A\to\infty} \mathrm{B}_2(T, A)$.

where $M = 2m$ is the total mass, $\mu = \frac{1}{2}m$ is the reduced mass (not to be confused with the chemical potential), and $\omega_c = eB/mc$ is the cyclotron energy. The general expression for the propagator, Eq.(24), is now continued to imaginary time:

$$Z_2 = \mathrm{Tr}\, e^{-\beta H_2} = \tfrac{1}{2}\int d^2R\, d^2r\, \left[\langle \boldsymbol{R},\boldsymbol{r}\,|\,e^{-\beta H_2}\,|\,\boldsymbol{R},\boldsymbol{r}\rangle + \langle \boldsymbol{R},-\boldsymbol{r}\,|\,e^{-\beta H_2}\,|\,\boldsymbol{R},\boldsymbol{r}\rangle\right] \tag{34}$$

The center of mass contribution thereby factors out, yielding

$$Z_{\mathrm{CM}} = 2Z_1 = 2A\lambda_T^{-2}(\tfrac{1}{2}\beta\hbar\omega_c)\,\mathrm{csch}(\tfrac{1}{2}\beta\hbar\omega_c). \tag{35}$$

To obtain Z_{rel}, one needs the propagator $K^{\alpha}(\boldsymbol{r}',\boldsymbol{r}'';\tau)$ for a single particle in the presence of a flux tube. All the formalism necessary to tackle this problem was derived by Edwards and Gulyaev,[12] Peak and Inomata,[13] and others[14] in addressing the issue of path integration in polar coordinates, a somewhat subtle affair. A brief, but complete review is given in appendix A.

According to Eqs.(A.21–A.24), the imaginary time relative propagator is

$$\begin{aligned} K(\boldsymbol{r}',\boldsymbol{r}'';-i\beta\hbar) &= \frac{\mu\omega_c}{4\pi\hbar}\,\mathrm{csch}(\tfrac{1}{2}\beta\hbar\omega_c) \\ &\times \exp\left[\frac{-\mu\omega_c}{4\hbar}\,\mathrm{ctnh}(\tfrac{1}{2}\beta\hbar\omega_c)(r'^2+r''^2)\right] e^{-\frac{1}{2}\beta\hbar\omega_c\alpha} \\ &\times \sum_{n=-\infty}^{\infty} e^{-\frac{1}{2}\beta\hbar\omega_c n} e^{in(\theta''-\theta')} I_{|n+\alpha|}\left[\frac{\mu\omega_c}{2\hbar}r'r''\,\mathrm{csch}(\tfrac{1}{2}\beta\hbar\omega_c)\right], \end{aligned} \tag{36}$$

where $I_\nu(x)$ is the modified Bessel function of the first kind of order ν. I now define

$$\begin{aligned} F_{\pm}(-i\beta\hbar) &\equiv \int d^2r\, K(\boldsymbol{r},\pm\boldsymbol{r};-i\beta\hbar) \\ &= \tfrac{1}{2}e^{-\alpha\Delta}\sum_{n=-\infty}^{\infty}(\pm 1)^n e^{-n\Delta}\int_0^{\infty} dx\, e^{-x\cosh\Delta} I_{|n+\alpha|}(x) \end{aligned} \tag{37}$$

and

$$\Delta \equiv \tfrac{1}{2}\beta\hbar\omega_c. \tag{38}$$

Notice that $F_{\pm}$ is periodic in α with period 2. The partition function Z_{rel} may be written

$$Z_{\mathrm{rel}} = \tfrac{1}{2}\left[F_+(-i\beta\hbar) \pm F_-(-i\beta\hbar)\right] \tag{39}$$

with the top sign applying when the fiducial ($\alpha = 0$) statistics refer to bosons. Integrals such as those arising in Eq.(37) may be evaluated exactly:[15]

$$\int_0^{\infty} dx\, e^{-ax} I_\nu(x) = \frac{1}{\sqrt{a^2-1}}\left(a+\sqrt{a^2-1}\right)^{-\nu}. \tag{40}$$

Since the partition function scales with the size of the system, one expects the sum in Eq.(37) to diverge. Indeed this is so. Consequently, in evaluating the second virial coefficient via Eq.(38), one is presented with the delicate task of extracting the finite difference of two divergent expressions (namely $\frac{1}{2}A$ and AZ_2/Z_1^2). In order to accomplish this, I shall employ a simple regularization procedure which renders finite all integrals such as those encountered in Eq.(37). This amounts to introducing a factor

$$\exp\left(-\epsilon\frac{\mu\omega_c r^2}{2\hbar}\operatorname{csch}\Delta\right) \tag{41}$$

under every integral $\int d^2r$. The subtraction shall then be performed on the regularized quantities, with $\epsilon \to 0$ at the end of the calculation.

To demonstrate how the regularization procedure works, I shall first evaluate $B_2(\alpha = 0, T)$ within this scheme. In this case, the identity

$$\sum_{n=-\infty}^{\infty}(\pm 1)^n e^{-n\Delta} I_{|n|}(x) = e^{\pm x\cosh\Delta} \tag{42}$$

gives rise to the simple expressions

$$\begin{aligned} F_+(-i\beta\hbar) &= \tfrac{1}{2}\int_0^\infty dx \\ F_-(-i\beta\hbar) &= \tfrac{1}{2}\int_0^\infty dx\, e^{-2x\cosh\Delta}, \end{aligned} \tag{43}$$

which, upon regularization, become

$$\begin{aligned} F_+(-i\beta\hbar) &\longrightarrow \tfrac{1}{2}\int_0^\infty dx\, e^{-\epsilon x} = \frac{1}{2\epsilon} \\ F_-(-i\beta\hbar) &\longrightarrow \tfrac{1}{2}\int_0^\infty dx\, e^{-x(\epsilon+2\cosh\Delta)} = \frac{1}{2(\epsilon+2\cosh\Delta)}. \end{aligned} \tag{44}$$

This means that

$$Z_{\rm rel} = \frac{1}{4\epsilon} \pm \frac{1}{4(\epsilon+2\cosh\Delta)}. \tag{45}$$

Now, the single particle partition function in this representation is

$$\begin{aligned} Z_1 &= \int d^2r\, K(\mathbf{r},\mathbf{r};-i\beta\hbar)\Big|_{\mu=m} \\ &\longrightarrow \int_0^\infty dx\, e^{-\epsilon x} = \frac{1}{\epsilon}, \end{aligned} \tag{46}$$

and the area of the system is regularized to

$$A \longrightarrow \int d^2r\, \exp\left(-\epsilon\frac{\mu\omega_c r^2}{2\hbar}\operatorname{csch}\Delta\right) = \frac{1}{\epsilon}\lambda_T^2\left(\frac{\sinh\Delta}{\Delta}\right). \tag{47}$$

Using Eq.(28), one derives the second virial coefficient:

$$\begin{aligned} B_2(\alpha=0,\Delta,T) &= \frac{A}{2Z_1}\left[Z_1 - 4Z_{\rm rel}\right] \\ &= \mp\tfrac{1}{4}\lambda_T^2\left(\frac{\tanh\Delta}{\Delta}\right). \end{aligned} \tag{48}$$

An identical conclusion is reached by a direct expansion of the thermodynamic functions of an ideal quantum gas in a magnetic field. Note that the $\Delta \to 0$ limit leads to the correct result.

For nonzero α, the situation is more complicated. Defining

$$\begin{aligned} w &\equiv (\epsilon + \cosh\Delta) + \sqrt{(\epsilon+\cosh\Delta)^2 - 1} \\ &= e^{\Delta}\left(1 + \epsilon\,{\rm csch}\Delta - \tfrac{1}{2}\epsilon^2 e^{-\Delta}\,{\rm csch}^3\Delta + \cdots\right), \end{aligned} \tag{49}$$

the regularized value of $F_\pm$ takes the form

$$F_\pm(-i\beta\hbar) = \tfrac{1}{2}e^{-\alpha\Delta}\left[(\epsilon+\cosh\Delta)^2 - 1\right]^{-1/2}\sum_{n=-\infty}^{\infty}(\pm 1)^n e^{-n\Delta}w^{-|n+\alpha|}. \tag{50}$$

The calculation is carried out by restricting α to one of two intervals: $\alpha \in [-1,0]$ or $\alpha \in [0,1]$. The case of arbitrary $\alpha \in \mathbf{R}$ may then be recovered by periodic extension. I shall refer to these two regions as '<' and '>', respectively. The relative coordinate contribution, $Z_{\rm rel}$, is then

$$Z_{\rm rel}^< = \frac{w}{w^2-1}\left\{(we^{-\Delta})^{\alpha}\left(\frac{w^2e^{-2\Delta}}{w^2e^{-2\Delta}-1}\right) + (we^{\Delta})^{-\alpha}\left(\frac{1}{w^2e^{2\Delta}-1}\right)\right\} \tag{51}$$

and

$$Z_{\rm rel}^> = \frac{w}{w^2-1}\left\{(we^{-\Delta})^{\alpha}\left(\frac{1}{w^2e^{-2\Delta}-1}\right) + (we^{\Delta})^{-\alpha}\left(\frac{w^2e^{2\Delta}}{w^2e^{2\Delta}-1}\right)\right\}. \tag{52}$$

In obtaining B_2, one must take care to isolate terms of order ϵ as well as terms of order unity before taking the $\epsilon \to 0$ limit. The second virial coefficient is given by

$$B_2^{>/<}(\alpha,\Delta,T) = \tfrac{1}{4}\lambda_T^2\Delta^{-1}\left[{\rm ctnh}(\Delta) - 2(\alpha\mp 1) - 2e^{-2(\alpha\mp1)\Delta}\,{\rm csch}2\Delta\right], \tag{53}$$

with the top sign applying to region '>', the bottom sign to region '<', and where the $\alpha = 0$ statistics correspond to bosons. One may check explicitly that the special cases $\alpha = 0$ and $\alpha = \pm 1$ lead to the proper ideal quantum gas results of Eq.(48).

Thus, I have derived explicit thermodynamic formulae describing the interpolation of quantum statistics between Bose and Fermi character. The α-dependence

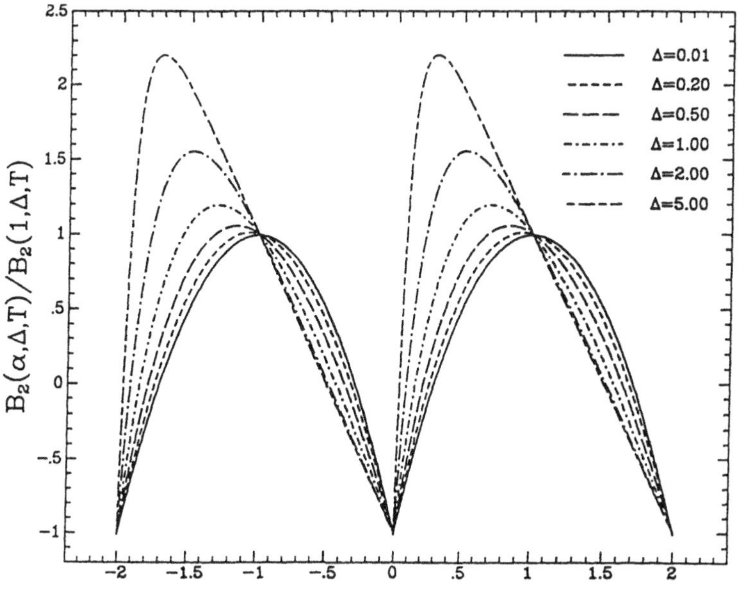

Figure 3. The second virial coefficient $B_2(\alpha, \Delta, T)$ as a function of the statistics determining parameter, α, and scaled by the Fermi value $B_2(1, \Delta, T)$. The asymmetry in the plots is associated with the choice of alignment ($\pm\hat{\mathbf{z}}$) for the magnetic field. All curves are periodic under $\alpha \to \alpha + 2$, with cusps occurring at even values of α.

of $B_2(\alpha, T)$ at various values of Δ is shown in Fig.[3]. When there is no magnetic field present, Eq.(53) reduces to

$$B_2^{>/<}(\alpha, \Delta, T)\Big|_{\Delta=0} = \tfrac{1}{4}\lambda_T^2 \left[1 - 2(\alpha \mp 1)^2\right], \tag{54}$$

which is simply a section of a parabola. The author and coworkers have also obtained the formula of Eq.(54) by a different method[16] which entails performing a Boltzmann-weighted sum of the relative wave function phase shifts $\delta_m(k)$:[11]

$$B_2(\alpha, T) = \mp\tfrac{1}{4}\lambda_T^2 - \frac{2}{\pi}\lambda_T^2 \int_0^\infty dk \sum_{m \text{ even/odd}} e^{-\beta\hbar^2 k^2/2\mu} \frac{\partial}{\partial k}\delta_m(k), \tag{55}$$

where $\delta_m(k)$ is the phase shift for the mth partial wave.

The virial coefficient $B_2(\alpha, T)\Big|_{\Delta=0}$ is symmetric under the inversion $\alpha \to -\alpha$. This is clear upon inspection of equations [21] and [24], for the general many-body Hamiltonian is invariant under the combined operations of inversion in α and complex conjugation. The introduction of a finite magnetic field then breaks this symmetry, which is why the $\Delta \neq 0$ curves of Fig.[3] are skewed toward the left.† The $B \neq 0$ Hamiltonian does satisfy

$$H^*(-B, -\alpha) = H(B, \alpha) \tag{56}$$

so all thermodynamic properties will remain invariant under simultaneous inversion of both the field and the statistics determining parameter.

A particularly striking feature is the cusp present at all even integers, those values of α corresponding to bosonic behavior. I will describe the nature of this effect in the case of zero external field by appealing to the relative coordinate energy spectrum of Eq.(20). Recall that if the fiducial statistics are bosonic, only even m states are allowed. Now for small $|\alpha|$, every $m = 0$ partial wave is nondegenerate, as the allowed wavevectors are $k_{l,0} = \Lambda^{-1} x_{|\alpha|,l}$. All states of nonzero angular momentum are, however, *quasidegenerate*: $k_{l,n} \approx k_{l,-n}$. The situation is completely different for α close to one, due to the absence of the odd m states ($m = 1$ in particular), consequently *all* states evidence this quasidegeneracy. This means that whenever α is an odd integer, every state in the sum of Eq.(55) will possess a mate such that their added phase shifts will *cancel* to order (α), producing the smooth quadratic behavior shown in Fig.[3]. That there are no higher order corrections is an artifact of the 'free anyon' model I have considered.

Fractional Statistics and the Quantized Hall Effect

It may seem rather pointless to consider the issue of fractional statistics when our world is clearly not two-dimensional. Even when large anisotropies in couplings effectively reduce the dimensionality of a system, for instance, the condensed matter physicist is ultimately interested in assemblies of electrons and nucleons, for which exotic statistics is a moot issue. The hope, of course, is that certain *elementary excitations* of some quasi-two-dimensional system might behave as if they obeyed fractional statistics. In order that the winding number be a well defined quantity, it seems necessary that any putative anyonic excitations be spatially *localized* and

† If the field direction were to be reversed, the curves would then be skewed toward the right.

possess a form factor which is attenuated on some microscopic scale. Thus, extended wave-like objects (*e.g.* phonons), which can be localized only at a high cost in kinetic energy, are poor candidate anyons, and intuition drives one to consider topologically stable local excitations (*e.g.* vortices) and similar entities. Here, I shall discuss the relevance of fractional statistics to the quasiparticle excitations predicted in the fractional quantized Hall effect.‡

Recall that Laughlin's[17] picture of the fractional quantized Hall effect is based on a discrete set of Jastrow-type *Ansatz* wave functions of the form

$$\Psi_m(z_1,\ldots,z_N) = \mathcal{N}\prod_{j<k}^{N}(z_j - z_k)^m \prod_{l=1}^{N} e^{-|z_l|^2/4}, \tag{57}$$

where m is an odd integer, $z_j = x_j + iy_j$ is the complex coordinate of particle j, $\mathcal{N}$ is a normalization constant, and where the magnetic length $\ell = \sqrt{\hbar c/eB}$ has been set to unity. Ψ_m describes an incompressible fluid state at filling fraction $\nu = 1/m$ (provided $m \lesssim 70$). The associated quasielectron and quasihole wave functions, respectively, are given by

$$\begin{aligned}
\tilde{\Psi}_m[\eta] &= \mathcal{N}(\eta)\prod_{l=1}^{N} e^{-\bar{z}_l z_l/4}\prod_{i=1}^{N}\left(2\frac{\partial}{\partial z_i} - \bar{\eta}\right)\prod_{j<k}^{N}(z_j - z_k)^m \\
\tilde{\Psi}_m[\xi] &= \mathcal{N}(\xi)\prod_{l=1}^{N} e^{-\bar{z}_l z_l/4}\prod_{i=1}^{N}(z_i - \xi)\prod_{j<k}^{N}(z_j - z_k)^m .
\end{aligned} \tag{58}$$

The charge of these excitations was also discussed by Laughlin, who employed an argument analogous to that used in deducing the fractional charge of solitons in one-dimensional conductors. He concluded that for $\nu = 1/m$, the quasielectron and quasihole have charges $\mp e^* = \mp e/m$. These excitations are localized within a microscopic region whose size is dictated by the magnetic length and the filling fraction. The excitations described by Eq.(58) are centered at $\boldsymbol{r} = \boldsymbol{\eta}$ (quasielectron) and $\boldsymbol{r} = \boldsymbol{\xi}$ (quasihole), respectively. Roughly speaking, a quasihole in the incompressible fluid resembles a 'bubble' of such a size that $1/m$ of an electron is absent. Less clear, however, is the statistics which the excitations satisfy — Fermi,[17] Bose,[18] and

‡ I shall refer to excitations in the generic sense as 'quasiparticles'. Specifically, an excitation corresponding to a localized density depletion will be called a 'quasihole', while a local density accumulation will be referred to as a 'quasielectron'.

fractional[19] statistics have all been proposed. Below, I shall describe a method due to myself and coworkers[20] which allows a direct determination of both the charge e^* and statistics α of the excitations.

We determine the charge e^* by evaluating the phase change γ_C accrued by $\tilde{\Psi}_m[\xi]$ as ξ adiabatically moves around a loop of radius R enclosing a flux ϕ. The charge is obtained when γ_C is set equal to the change in phase

$$\gamma_C = \frac{e^*}{\hbar c} \oint \boldsymbol{A} \cdot dl = 2\pi \frac{e^*}{e} \frac{\phi}{\phi_o}, \tag{59}$$

that an excitation of charge e^* would gain in moving around this loop. The phase γ_C may be calculated via the adiabatic theorem, *treating the quasielectron or quasihole position as an adiabatic parameter.**

There is a simple yet profound result due to Berry[21] which relates the adiabatic phase accumulated over such a closed path to a geometric quantity which is *independent* of how slowly the path is traversed. Specifically, let $\{H_\lambda\}$ denote a family of Hamiltonians indexed by the parameter λ. If one executes a closed loop in parameter space sufficiently slowly, the full time–dependent solution to the Schrödinger equation, $\Phi(t)$, will be related to the adiabatic (time–independent) solutions ψ_λ according to†

$$\Phi(t) = \exp\left(-i \int^t dt' \, E(t')\right) e^{i\gamma(t)} \psi(t), \tag{60}$$

where $E(t)$ is the adiabatic energy, $\lambda(t)$ describes the path, and where I have loosely defined the implicit time–dependent functions

$$\begin{aligned} \psi(t) &\equiv \psi_{\lambda(t)} \\ E(t) &\equiv E_{\lambda(t)} \\ \gamma(t) &\equiv \gamma_{\lambda(t)}. \end{aligned} \tag{61}$$

The 'extra' phase $\gamma(t)$ satisfies

$$\frac{d}{dt}\gamma(t) = i \langle \psi(t) | \frac{d}{dt} | \psi(t) \rangle . \tag{62}$$

* Although $\tilde{\Psi}_m[\xi]$ is a 'variational' wave function, rather than an actual adiabatic wave function, the conclusions obtained are not expected to be sensitive to this inconsistency. We could regard $\tilde{\Psi}_m[\xi]$ to be an exact excited state wave function for a model Hamiltonian.

† I assume that the spectrum of adiabatic wave functions ψ_λ is nondegenerate.

If $\lambda(t)$ describes a *closed* path, then the phase difference

$$\gamma_C \equiv \gamma(T) - \gamma(0) = i \oint dt \, \langle \psi(t) \,|\, \frac{d}{dt} \,|\, \psi(t) \rangle \tag{63}$$

has a beautiful geometric interpretation, as noticed by Simon[22] (see also Wilczek and Zee[23] and Schiff[24]).‡

According to Eq.(58),

$$\frac{d}{dt}\tilde{\Psi}_m[\boldsymbol{\xi}] = \left[\frac{d}{dt} \ln \mathcal{N}[\boldsymbol{\xi}(t)] + \sum_{i=1}^{N} \frac{d}{dt} \ln[z_i - \xi(t)] \right] \tilde{\Psi}_m[\boldsymbol{\xi}], \tag{64}$$

so that

$$\frac{d}{dt}\gamma(t) = i\frac{d}{dt} \ln \mathcal{N}(\boldsymbol{\xi}) + i \langle \tilde{\Psi}_m[\boldsymbol{\xi}] \,|\, \frac{d}{dt} \sum_{i=1}^{N} \ln(z_i - \xi) \,|\, \tilde{\Psi}_m[\boldsymbol{\xi}] \rangle . \tag{65}$$

Using the single particle density in the presence of the quasihole,

$$n_{\boldsymbol{\xi}}(\mathbf{r}) = \langle \tilde{\Psi}_m[\boldsymbol{\xi}] \,|\, \sum_{i=1}^{N} \delta(\mathbf{r} - \mathbf{r}_i) \,|\, \tilde{\Psi}_m[\boldsymbol{\xi}] \rangle , \tag{66}$$

we obtain

$$\frac{d\gamma}{dt} = i\frac{d}{dt} \ln \mathcal{N}[\boldsymbol{\xi}(t)] + i \int d^2r \, n_{\boldsymbol{\xi}}(\mathbf{r}) \, \frac{d}{dt} \ln[z - \xi(t)]. \tag{67}$$

Since the normalization constant $\mathcal{N}(\boldsymbol{\xi})$ is a single–valued function of its argument, it will not contribute to the integral expression Eq.(63) for γ_C. We now write $n_{\boldsymbol{\xi}}(\mathbf{r}) = n_o + \delta n_{\boldsymbol{\xi}}(\mathbf{r})$, where $n_o = \nu/2\pi$ is the density in the Laughlin ground state Ψ_m, and $\delta n_{\boldsymbol{\xi}}(\mathbf{r})$ is concentrated about the point $\mathbf{r} = \boldsymbol{\xi}$.* Concerning the n_o term, if ξ is integrated in a clockwise sense around a circle of radius R, one finds

$$\begin{aligned} \oint_{|\xi|=R} dt \, \frac{d}{dt} \ln[z - \xi(t)] &= \oint_{|\xi|=R} d\xi \, \frac{1}{\xi - z} \\ &= 2\pi i \, \theta(R - |z|), \end{aligned} \tag{68}$$

‡ The quantity γ_C is the line integral of a connection form of a $U(1)$–bundle whose fibers are the adiabatic wave functions. This connection form is given by $\omega = \langle \psi_\lambda \,|\, \frac{\partial}{\partial \lambda} \,|\, \psi_\lambda \rangle \, d\lambda$; the fiber metric is defined by the standard inner product. By Stokes' theorem, γ_C can be expressed as a surface integral of the curvature form $\Omega = d\omega + \omega \wedge \omega$. If the parameter space is two–dimensional, then Ω is related to the first Chern form $c_1 = \frac{i}{2\pi}\Omega$, whose integral over the entire parameter space Λ is an integer topological invariant of the bundle.

* The uniform density $n_o = \nu/2\pi$ of the trial ground state and the localized nature of the excitations are easily deduced from the plasma analogy of Laughlin.

where $\theta(x)$ is a step function. Substituting this result into Eq.(67), we find

$$\gamma_C = i\int^R d^2r\, 2\pi i\, n_o = -2\pi\langle N\rangle_R = -2\pi\nu\phi/\phi_o, \tag{69}$$

where $\langle N\rangle_R$ denotes the mean number of electrons inside a circle of radius R. We originally expected that corrections to Eq.(69) arising from the $\delta n_{\xi}(\mathbf{r})$ term would be of order a_0^2/R^2, where a_0 is the size of the quasihole, and that such effects would be irrelevant in the low density limit, where the average separation between quasiholes is much larger than the quasihole size. In fact, a much stronger result has been obtained by Haldane[25], who has shown that this correction vanishes identically as a consequence of the rotational symmetry of the quasiparticle. Comparing with Eq.(59), we find $e^* = \nu e$, in agreement with Laughlin's result.

A similar analysis shows that the charge of the quasielectron is $e^* = -\nu e$, although one must exercise some caution in dealing with the partial derivative operators in Eq.(58). The adiabatic phase accumulated by a quasielectron is

$$\frac{d}{dt}\gamma(t) = i\frac{d}{dt}\ln \mathcal{N}(\boldsymbol{\eta}) + i\frac{d}{dt}\sum_{i=1}^{N}\langle\, \tilde{\Psi}_m[\boldsymbol{\eta}]\,|$$

$$\times \exp\left(-\tfrac{1}{4}\sum_{l=1}^{N}|z_l|^2\right)\ln\left(2\frac{\partial}{\partial z_i}-\bar{\eta}\right)\exp\left(\tfrac{1}{4}\sum_{l'=1}^{N}|z_{l'}|^2\right)\,|\,\tilde{\Psi}_m[\boldsymbol{\eta}]\,\rangle\,. \tag{70}$$

The above matrix element is a many–particle generalization of the Barg-mann–Fock space inner product,

$$\langle\!\langle\, g\,|\,\hat{O}\left(2\frac{\partial}{\partial z}, z\right)|\,f\,\rangle\!\rangle \equiv \int\frac{dx\,dy}{2\pi}\,e^{-\bar{z}z/2}\,\overline{g(z)}\,\hat{O}\left(2\frac{\partial}{\partial z}, z\right) f(z). \tag{71}$$

It is easy to show[26] that if the operator $\hat{O}$ is *normal ordered* such that all partial derivatives $\partial/\partial z$ appear to the *left* of all complex coordinates z, then the formal replacement $2\partial/\partial z \rightarrow \bar{z}$ is allowed, *i.e.*

$$\langle\!\langle\, g\,|:\hat{O}\left(2\frac{\partial}{\partial z}, z\right):|\,f\,\rangle\!\rangle = \langle\!\langle\, g\,|\,\hat{O}(\bar{z}, z)\,|\,f\,\rangle\!\rangle\,. \tag{72}$$

Making this substitution in Eq.(70), one recovers the result of Eq.(67), with $z\rightarrow\bar{z}$ and $\xi\rightarrow\bar{\eta}$. The charge of the quasielectron follows immediately.

To determine the statistics of the excitations, we consider the state with quasiholes at $\boldsymbol{\xi}$ and $\boldsymbol{\xi}'$,

$$\tilde{\Psi}_m[\boldsymbol{\xi},\boldsymbol{\xi}'] = \mathcal{N}(\boldsymbol{\xi},\boldsymbol{\xi}')\prod_{i=1}^{N}[(z_i-\xi)(z_i-\xi')]\,\Psi_m. \tag{73}$$

As above, we adiabatically carry ξ around a closed loop of radius R. If ξ' lies outside the circle $|\xi| = R$ by a distance $d \gg a_0$, the above analysis is unchanged, *i.e.* $\gamma_C = -2\pi\nu\phi/\phi_o$. If, however, ξ' lies inside this loop and $R - |\xi'| \ll a_0$, there is a deficit in $\langle N\rangle_R$ of $-\nu$, and the phase accrued is $\gamma'_C = \gamma_C + 2\pi\nu$. Therefore, when a quasihole adiabatically encircles another quasihole, an extra 'statistical phase'

$$\Delta\gamma_C = 2\pi\nu \tag{74}$$

is accumulated. Again, an analogous result holds for quasielectron. For the case of the filled Landau level, $\nu = 1$, $\Delta\gamma_C = 2\pi$, and the phase obtained upon interchanging quasiholes is $\Delta\gamma_C/2 = \pi$, corresponding to Fermi statistics. For ν noninteger, we identify the quantity $\Delta\gamma_C/2\pi$ with the statistics determining parameter, α, and the excitations obey fractional statistics, in agreement with the conclusion of Halperin.[19] Clearly, when ν is nonintegral, the change of phase $\Delta\gamma_C$ accumulated when a *third* particle is in the vicinity will depend on the adiabatic path taken by the excitations as they are interchanged, and the permutation definition used for Bose and Fermi statistics no longer suffices.

It has been noted by Su[27] and others that the above derivation is sensitive only to certain very general properties of Ψ_m and $\tilde{\Psi}_m[\boldsymbol{\xi}]$, and that the same conclusions apply whenever the ground state Ψ is of uniform density and the quasiparticles are localized and of the form given in Eq.(58).

There is an unfortunate confusion involving the nature of the elementary excitations which has found its way into the literature.[28] *Although they may obey fractional statistics, the quasiparticles* **do not** *carry a magnetic flux $\phi = \nu\phi_o$ as do Wilczek's charged particle-flux tube composites.* The exotic statistics derive from the sharp fractional charge of the excitations and the incompressibility of the ground state, and there are *no* trapped flux lines in the Hall plasma.

While the derivation is certainly suggestive, its interpretation is not completely clear. For instance, what is the meaning of treating the quasiparticle location as an adiabatic parameter? I believe that this *is* the correct procedure, and I will now present a simple analogous result which seems to justify this assumption. Consider the single particle coherent state $|\boldsymbol{R}\rangle$,

$$\varphi_{\boldsymbol{R}}(\boldsymbol{r}) = \langle \boldsymbol{r} \,|\, \boldsymbol{R} \rangle = \frac{1}{\sqrt{2\pi}}\, e^{-(\boldsymbol{r}-\boldsymbol{R})^2/4}\, e^{-i\boldsymbol{r}\times\boldsymbol{R}\cdot\hat{\boldsymbol{z}}/2}. \tag{75}$$

If the guiding center is adiabatically transported around a closed loop, it is easy to show that the Berry phase satisfies

$$\frac{d}{dt}\gamma(t) = i\,\langle\, \boldsymbol{R}(t) \,|\, \frac{d}{dt} \,|\, \boldsymbol{R}(t) \,\rangle \;=\; -\frac{1}{\phi_o}\frac{d\phi}{dt}, \tag{76}$$

and adiabatically dragging the electron about a closed path leads to a cumulative Berry phase of $-\phi/\phi_o$, from which the correct electron charge is recovered. Thus, in this trivial example, treating the position as an adiabatic parameter leads to the appropriate results.

Halperin's 'Pseudo Wave Function'

Halperin[19] originally used fractional statistics to derive the hierarchy scheme of rational filling fractions based on successive levels of Laughlin condensates. In this manner, states of density $\nu \neq 1/m$ could be described as an underlying primitive $(1/m)$ Laughlin condensate plus a condensate of quasielectrons or quasiholes. If such a composite state does not fully saturate the density, quasiparticles of the 'second level' can be added and themselves condense, and so on, until the desired rational fraction $\nu = p/q$ is reached. As expected, the gap decreases sharply as one climbs up this tree of hierarchical composite states, and experimental conditions will determine how far up the tree the system may progress and still remain condensed.

Each excitation condensate at a level s of the hierarchy is defined by a 'pseudo wave function', $\boldsymbol{\Phi}_s$, whose arguments are the positions $Z_j = X_j \pm iY_j$ of the excitations (the plus sign being taken for quasiholes, and the minus sign for quasielectrons, at every level except $s = 0$). I shall henceforth refer to the condensed entities at stage s in the hierarchy as 's-particles' and thereby distinguish them from 's-excitations', which will be used to refer to defects in the s-condensate. The explicit form of $\boldsymbol{\Phi}_s$ is

$$\begin{aligned}
\boldsymbol{\Phi}_s(\boldsymbol{R}_1,\ldots,\boldsymbol{R}_N) &= P_s(Z_1,\ldots,Z_N)\,Q_s(Z_1,\ldots,Z_N)\prod_{l=1}^{N} e^{-|q_s|\bar{Z}_l Z_l/4} \\
P_s(Z_1,\ldots,Z_N) &= \prod_{j<k}^{N}(Z_j - Z_k)^{2p_s+1} \\
Q_s(Z_1,\ldots,Z_N) &= \prod_{j<k}^{N}(Z_j - Z_k)^{-\sigma_{s+1}/m_s},
\end{aligned} \tag{77}$$

where the excitation charge is $e_s = \pm q_s e$, σ_{s+1} denotes the type of s-particle ($+1$ for electron-like objects, and -1 for hole-like objects), and $m_s > 1$ is a rational number which will be determined by iteration in s. The polynomial Q_s determines the statistics of the entities described by the pseudo wave function — interchange of two s-particles in Eq.(77) leads to a phase of $e^{\pm i\pi m_s}$. With p_{s+1} restricted to the positive integers, multiplication by the (symmetric) polynomial P_s will introduce additional correlations in Φ_s without altering the statistics. For $m_s \neq 1$, the function Φ_s is multiple-valued and thus is formally well defined only on the universal covering of the configuration space for the M s-particles. This last feature of Φ_s is not a particularly important one, however, as far as Halperin's analysis goes.

Halperin interprets Φ_s as the probability amplitude for finding an $(s-1)$-excitation at each of the N positions $(\boldsymbol{R}_1, \ldots, \boldsymbol{R}_N)$, provided that these $(s-1)$-excitations are sufficiently well separated so that their cores do not overlap.† Aside from this latter restriction, Φ_s may be considered to be a conventional, quantum mechanical wave function.

Invoking Laughlin's plasma analogy, the number density of s-particles in the state Φ_s is found to be

$$\begin{aligned} n_s &= |q_s|/2\pi m_{s+1} \\ m_{s+1} &\equiv 2p_{s+1} - \sigma_{s+1}/m_s. \end{aligned} \tag{78}$$

The actual charge of the s-particles at level s is $\tilde{e}_s = \sigma_{s+1} q_s e$. Clearly the true electronic filling fraction, ν_{s+1}, in this state is

$$\nu_{s+1} = \nu_s + \sigma_{s+1} q_s |q_s|/m_{s+1}. \tag{79}$$

By applying the quasielectron (quasihole) operators of Eq.(58),

$$\begin{aligned} \hat{O}_+[\bar{\eta}] &= \prod_{i=1}^{N}\left(2\frac{\partial}{\partial Z_i} - \bar{\eta}\right) \\ \hat{O}_-[\xi] &= \prod_{i=1}^{N}(Z_i - \xi), \end{aligned} \tag{80}$$

one accumulates a surplus (deficit) of $1/m_{s+1}$ s-particles about the point $Z = \eta$ ($Z = \xi$). These s-*excitations* then comprise the fundamental $(s+1)$-particles of

† Since the s-particles are in general not point particles, as are the electrons of the primitive Laughlin condensates, the pseudo wave function has no meaning when any two s-particles lie too close to each other.

the next level — this procedure defines the basic link in the hierarchy chain. This reasoning leads to an iterative equation for the q_s:

$$q_{s+1} = \sigma_{s+1} q_s / m_{s+1}. \tag{81}$$

At the base of the hierarchy, the initial conditions are $\nu_0 = 0$, $q_0 = m_0 = \sigma_1 = 1$. By specifying a sequence $\{p_s, \sigma_s\}$, the formulae [78–81] then determine a corresponding sequence of rational filling fractions. By expressing the filling fractions in terms of continued fractions, one recovers the hierarchy scheme of Haldane.[18]

Halperin has derived the starting expression of Eq.(77) in a somewhat systematic fashion by introducing additional correlations between *pairs* of electrons in Laughlin's primitive trial state. However, it may be more than merely a curious fact that the normalization integral, $\mathcal{N}(\xi_1, \ldots, \xi_M)$, of the excited state pseudo wave function

$$\tilde{\Phi}_s(\xi_1, \ldots, \xi_M; \boldsymbol{R}_1, \ldots, \boldsymbol{R}_N) = \mathcal{N}_s(\xi_1, \ldots, \xi_M) \prod_{\alpha=1}^{M} \hat{O}_-[\xi_\alpha]\, \Phi_s(\boldsymbol{R}_1, \ldots, \boldsymbol{R}_N) \tag{82}$$

is actually *equal*, apart from an overall constant, to the modulus of the next pseudo wave function in the hierarchy, $\Phi_{s+1}|_{p_{s+2}=0}$.

The normalization integral for quasiholes is given by

$$|\mathcal{N}_s(\xi_1, \ldots, \xi_M)|^{-2} = \int \prod_{i=1}^{N} d^2R_i \prod_{i=1}^{N} \prod_{\alpha=1}^{M} |Z_i - \xi_\alpha|^2 \, |\Phi_s(\boldsymbol{R}_1, \ldots, \boldsymbol{R}_N)|^2. \tag{83}$$

Differentiation with respect to ξ_β yields

$$\begin{aligned} -2\frac{\partial}{\partial \xi_\beta} \ln|\mathcal{N}_s| &= \int \prod_{i=1}^{N} d^2R_i \left(\sum_{j=1}^{N} \frac{1}{\xi_\beta - Z_j} \right) |\tilde{\Phi}_s|^2 \\ &= -\int d^2r \left(\frac{1}{Z - \xi_\beta} \right) n(\boldsymbol{R}), \end{aligned} \tag{84}$$

where $n(\boldsymbol{R})$ is the number density of the fundamental level s particles. If the excitations are few in number, this equation may be analyzed using the same reasoning employed in calculating Berry's phase for the Laughlin quasihole. The result is

$$\begin{aligned} -2\frac{\partial}{\partial \xi_\beta} \ln|\mathcal{N}_s| &= \pi n_s \bar{\xi}_\beta + \frac{1}{2m_{s+1}} \sum_{\substack{\alpha=1 \\ \alpha\neq\beta}}^{M} \left(\frac{1}{\xi_\alpha - \xi_\beta} \right) \\ |\mathcal{N}_s(\xi_1, \ldots, \xi_M)| &= \mathcal{A} \prod_{\alpha<\beta}^{M} |\xi_\alpha - \xi_\beta|^{1/m_{s+1}} \prod_{\gamma=1}^{M} \exp\left(-|q_s|\bar{\xi}_\gamma \xi_\gamma / 4m_{s+1}\right), \end{aligned} \tag{85}$$

where $\mathcal{A}$ is a constant. Had I considered quasielectron excitations, the statistics determining exponent, $1/m_{s+1}$, would have appeared with the opposite sign — the type of excitation thus defines the quantity σ_{s+2}. Thus, the normalization integral at level s in the hierarchy is simply the modulus of the pseudo wave function at level $s+1$, with $p_{s+2} = 0$. Again, one may introduce additional correlations without altering the statistics by choosing a nonzero integer for p_{s+2}, thereby recovering the general expression for the level $s+1$ pseudo wave function.

Laughlin was kind enough to tell me that this result is trivial.[29] Consider, for example, the expression

$$\Upsilon(\boldsymbol{\xi}_1, \ldots, \boldsymbol{\xi}_M; \mathbf{r}_1, \ldots, \mathbf{r}_N) \equiv \prod_{\alpha<\beta}^{M} |\xi_\alpha - \xi_\beta|^{2/m} \prod_{j<k}^{N} |z_j - z_k|^{2m} \times \prod_{i=1}^{N} \prod_{\delta=1}^{M} |z_i - \xi_\delta|^2 \prod_{\gamma=1}^{M} e^{-\bar{\xi}_\gamma \xi_\gamma / 2m} \prod_{l=1}^{N} e^{-\bar{z}z/2}, \quad (86)$$

which may be interpreted à la Laughlin as the classical probability distribution of a plasma consisting of N particles of charge m and M particles of charge 1 at a temperature $T = m$ interacting via logarithmic potentials, including the usual neutralizing background. If one integrates out the 'electrons' (charge m particles) in the dilute ξ-particle limit, one expects to obtain the partition function for a constrained system (constrained in that the ξ coordinates are fixed), and this should resemble $e^{-M\mu_\xi/T}$, where μ_ξ is the chemical potential for the ξ-particles. But then

$$\int \Upsilon \, d^2r_1 \cdots d^2r_N = |\mathcal{N}(\boldsymbol{\xi}_1, \ldots, \boldsymbol{\xi}_M)|^{-2} \prod_{\alpha<\beta}^{M} |\xi_\alpha - \xi_\beta|^{2/m} \prod_{\gamma=1}^{M} e^{-\bar{\xi}_\gamma \xi_\gamma / 2m} = e^{-M\mu_\xi/T}, \quad (87)$$

from which the pseudo wave function modulus is obtained immediately. This result holds in the dilute gas limit where the ξ-particles are equally likely to be anywhere in the plasma except within a Debye length of each other. At any rate, the result is the same as that of Eq.(85).

This manner in which the pseudo wave function is derived is very suggestive of something, although what that something is I do not know.

Selection Rules for Anyon Production

Tao[30] has used the fractional statistics of the excitations to argue for the odd denominator rule. This may be putting the cart a bit before the proverbial horse, however Tao's analysis does lead to some selection rules for anyon production. Suppose, for example, that an electron decays into a number of anyons, or, more generally, that p electrons decay into q identical anyons. Without loss of generality, p and q can be taken to be relatively prime. Consider now two groups of q such anyons apiece. Interchange of these groups must give a phase

$$\Delta\gamma = (-1)^{p^2} = e^{i\pi p^2}, \tag{88}$$

because each group derives from p electrons. On the other hand, if the two assemblies are viewed as consisting of anyons, the statistical phase will be

$$\Delta\gamma = e^{i\alpha\pi q^2} = e^{i\pi pq}, \tag{89}$$

if the statistics determining parameter is tied to the charge by $\alpha = e^*/e$. The two formulae of Eqs.(88,89) are incompatible if q is even, leading to the selection rule that fermion–to–anyon decays must proceed via an odd–p, odd–q channel or an even–p, odd–q channel. If the parent particle is a boson, one finds that the boson-to-anyon decays may proceed by either an odd–p, even–q channel or an even–p, odd–q channel.

While the virial coefficient calculations described earlier are somewhat relevant to the fractional quantum Hall effect, they do not take into account the Coulomb interaction between the excitations — I have also neglected effects due to the finite size of the quasielectron and quasihole. In fact, in real systems, the one–body potential along the semiconductor interface is presumed strong enough to *pin* the excitations (this is the qualitative picture of the Hall plateaux). It is conceivable that a charged impurity may bind m fractionally charged excitations ($e^* = \pm e/m$) and thus form an 'atom'. The energy levels, or shell structure, of this atom would presumably be sensitive to the statistics of the excitations.

One might also look for fractional statistics numerically. For example, if one examines FQHE clusters away from $\nu = \frac{1}{3}$, the ground state will resemble a quiescent Laughlin-type liquid with a certain number of quasiparticle excitations present. The energy levels could then be compared to those of a system of noninteracting quasiparticles, for which the statistics of the excitations will have definite consequences.

Summary

In two dimensions, a peculiar formulation of quantum statistics can be implemented if one associates to each particle a fictitious charge $\tilde{e}$ and a flux tube of strength $\phi = \alpha hc/2\tilde{e}$. Evaluating the many-body propagator by a path integral, a fictitious Aharonov–Bohm phase is accrued when particles wind around one another; physical quantities are found to be periodic under $\alpha \to \alpha + 2$. This procedure defines one among a continuum of possible (2+1)-dimensional quantum theories indexed by α, which, due to the aforementioned periodicity, may be restricted to the range $[0, 2)$. Conventional statistics arise when α is an integer: $\alpha = 0$ for bosons, and $\alpha = 1$ for fermions. For general α, the particles are referred to as *anyons*.

Due to the long-ranged nature of this 'statistical' interaction, it is in general difficult to calculate anything interesting within this formalism. In this chapter, I have discussed the solution to the two-anyon problem, which may be used to understand the lowest order quantum corrections to anyon thermodynamics. Specifically, the second virial coefficient, $B_2(\alpha, T)$ was evaluated, and was found to evidence nonanalytic behavior as a function of the statistics determining parameter, nonetheless properly interpolating between its corresponding Bose and Fermi values.

I have also elucidated some of the arguments as to why the Laughlin quasiparticles in the fractional quantized Hall effect should obey fractional statistics. In this case, the statistics determining parameter is itself determined by the ground state filling fraction: $\alpha = \nu \equiv nhc/eB$. While such a conclusion is in conformity with the hierarchical condensate scheme,[18,19] no simple experimental (or numerical) test has yet been performed which could test this prediction.

Acknowledgements

I have greatly benefited from discussions with R. MacKenzie, J. R. Schrieffer, and F. Wilczek. This work is excerpted from my Ph.D. thesis (University of California at Santa Barbara, 1986) I am grateful to AT&T Bell Laboratories for their support while this work was in progress.

APPENDIX A: PROPAGATOR IN THE PRESENCE OF A FLUX TUBE

In this appendix, I shall discuss the method of path integration in polar coordinates and its application to the quantum mechanical propagator of a free particle in the presence of a flux tube.

What is desired is an expression for the propagator

$$K^{\alpha}(\boldsymbol{r}'',\boldsymbol{r}';T) = \int \mathcal{D}\boldsymbol{r}(t)\, \exp\left[iS(\boldsymbol{r}'',\boldsymbol{r}')/\hbar\right], \tag{A.1}$$

with $T \equiv t'' - t'$; the label α will hereafter be suppressed. The technique involves isolating the contribution of a particular homotopy sector to the path integral (here, the homotopy index is simply the winding number of a path about the origin). The constraint

$$\int_{t'}^{t''} dt\, \dot{\theta} = \phi \tag{A.2}$$

is imposed on the propagator by weighing each path with a δ-function:

$$K_{\phi}(\boldsymbol{r}'',\boldsymbol{r}';T) = \int \mathcal{D}\boldsymbol{r}(t)\, e^{iS(\boldsymbol{r}'',\boldsymbol{r}')/\hbar}. \tag{A.3}$$

The above expression will be nonzero *only* for those ϕ which satisfy $\phi = \theta'' - \theta' + 2\pi n$, and the complete propagator is clearly

$$K(\boldsymbol{r}'',\boldsymbol{r}';T) = \int d\phi\, K_{\phi}(\boldsymbol{r}'',\boldsymbol{r}';T). \tag{A.3}$$

The quantity K_{ϕ} is known as the *constrained propagator*. Using an integral representation of the δ-function, one finds

$$\begin{aligned} K_{\phi}(\boldsymbol{r}'',\boldsymbol{r}';T) &= \int_{-\infty}^{\infty} \frac{d\lambda}{2\pi}\, e^{i\lambda\phi} \int \mathcal{D}\boldsymbol{r}(t)\, e^{iS_{\lambda}(\boldsymbol{r}'',\boldsymbol{r}')/\hbar} \\ S_{\lambda}(\boldsymbol{r}'',\boldsymbol{r}') &= \int_{t'}^{t''} dt\, \left[L(\boldsymbol{r},\dot{\boldsymbol{r}};t) - \lambda\hbar\dot{\theta}\right] \equiv \int_{t'}^{t''} dt\, L_{\lambda}(\boldsymbol{r},\dot{\boldsymbol{r}},t). \end{aligned} \tag{A.4}$$

The relative coordinate problem in which I am interested is defined by the Lagrangian

$$L(\boldsymbol{r},\dot{\boldsymbol{r}}) = \tfrac{1}{2}\mu\dot{\boldsymbol{r}}^2 - \tfrac{1}{2}\omega_c r^2\dot{\theta} - \alpha\dot{\theta}. \tag{A.5}$$

This is cast into a more manageable form by redefining θ:

$$\begin{aligned} \vartheta &\equiv \theta - \tfrac{1}{2}\omega_c t \\ L_{\lambda} &= \tfrac{1}{2}\mu\dot{\boldsymbol{r}}^2 - \frac{1}{8}\mu r^2\omega_c^2 - (\alpha+\lambda)\dot{\vartheta} - \tfrac{1}{2}(\alpha+\lambda)\hbar\omega_c. \end{aligned} \tag{A.6}$$

The path integral is calculated by the usual discretization procedure:

$$\begin{aligned} K_\phi(\mathbf{r}'',\mathbf{r}';T) &= \lim_{N\to\infty}\left(\frac{\mu}{2\pi i\hbar\epsilon}\right)^N \int\frac{d\lambda}{2\pi}\,e^{i\lambda\phi} \\ &\quad\times\int d^2r_1\cdots\int d^2r_N\,\exp\left[i\sum_{j=1}^{N+1}S_\lambda(\mathbf{r}_j,\mathbf{r}_{j-1})/\hbar\right], \\ S_\lambda(\mathbf{r}_j,\mathbf{r}_{j-1}) &\approx \epsilon L_\lambda\left[\frac{1}{\epsilon}(\mathbf{r}_j-\mathbf{r}_{j-1}),\mathbf{r}_j\right] \\ &= \frac{\mu}{2\epsilon}\left(r_j^2+r_{j-1}^2-\tfrac{1}{4}r_j^2\omega_c^2\epsilon^2\right) \\ &\quad -\frac{\mu}{\epsilon}r_jr_{j-1}\cos(\vartheta_j-\vartheta_{j-1})-\bar\lambda\hbar(\vartheta_j-\vartheta_{j-1})-\tfrac{1}{2}\bar\lambda\hbar\omega_c\epsilon, \end{aligned} \tag{A.7}$$

with $\bar\lambda=\lambda+\alpha$, $\epsilon=T/(N+1)$, $\mathbf{r}_0=\mathbf{r}'$, and $\mathbf{r}_{N+1}=\mathbf{r}''$. What makes this problem difficult is the periodicity of the cosine and the finite limits on the ϑ integrals. To properly account for all terms of order ϵ^2, the approximation

$$\cos(\Delta\vartheta)-a\epsilon\Delta\vartheta\approx\cos(\Delta\vartheta-a\epsilon)+\tfrac{1}{2}a^2\epsilon^2 \tag{A.8}$$

is employed. This yields

$$\begin{aligned} iS(\mathbf{r}_j,\mathbf{r}_{j-1})/\hbar &= \frac{i\mu}{2\hbar\epsilon}(r_j^2+r_{j-1}^2-\tfrac{1}{4}r_j^2\omega_c^2\epsilon^2)-\tfrac{1}{2}\bar\lambda\hbar\omega_c\epsilon-\tfrac{1}{2}i\bar\lambda^2\frac{\hbar\epsilon}{\mu r_jr_{j-1}} \\ &\quad -\frac{i\mu r_jr_{j-1}}{\hbar\epsilon}\cos\left(\vartheta_j-\vartheta_{j-1}-\bar\lambda\frac{\hbar\epsilon}{\mu r_jr_{j-1}}\right). \end{aligned} \tag{A.9}$$

Further progress is made by appealing to the generating function

$$\exp\left[\tfrac{1}{2}z\left(s+s^{-1}\right)\right]=\sum_{m=-\infty}^{\infty}s^mI_m(z) \tag{A.10}$$

for the modified Bessel function of the first kind. $I_\nu(x)$ behaves asymptotically as

$$I_\nu(x)\sim\frac{e^x}{\sqrt{2\pi x}}\left\{1-\tfrac{1}{2}(\nu^2-\tfrac{1}{4})\frac{1}{x}+\cdots\right\}, \tag{A.11}$$

which leads to the following limit:

$$\lim_{\epsilon\to0}\exp\left(-\tfrac{1}{2}i\bar\lambda^2\frac{\epsilon}{x}\right)\exp\left[-i\frac{x}{\epsilon}\cos\left(\theta-\bar\lambda\frac{\epsilon}{x}\right)\right]=\sum_{m=-\infty}^{\infty}e^{im\theta}I_{|m+\bar\lambda|}\left(\frac{x}{i\epsilon}\right). \tag{A.12}$$

The constrained propagator, then, resembles

$$\begin{aligned} K_\phi(\mathbf{r}'',\mathbf{r}';T) &= \lim_{N\to\infty}\left(\frac{\mu}{2\pi i\hbar\epsilon}\right)^N e^{-i\alpha\phi}\int\frac{d\bar\lambda}{2\pi}\,e^{i\bar\lambda(\phi-\frac{1}{2}\omega_cT)} \\ &\quad\times\int d^2r_1\cdots\int d^2r_N\prod_{j=1}^{N+1}\sum_{m_j=-\infty}^{\infty}\left\{e^{im_j(\vartheta_j-\vartheta_{j-1})}\right. \\ &\quad\left.\times\exp\left[\frac{i\mu}{2\hbar\epsilon}\left(r_j^2+r_{j-1}^2-\tfrac{1}{4}r_j^2\omega_c^2\epsilon^2\right)\right]I_{|m_j+\bar\lambda|}\left(\frac{\mu r_jr_{j-1}}{i\hbar\epsilon}\right)\right\}. \end{aligned} \tag{A.13}$$

At this point, the ϑ_j integrations may be carried out. The expression partly collapses due to the orthonormality relation

$$\int_{-\pi}^{\pi} \frac{d\vartheta}{2\pi} e^{i(m'-m)\vartheta} = \delta_{m,m'}, \tag{A.14}$$

and one obtains

$$\begin{aligned} K_\phi(\boldsymbol{r}'',\boldsymbol{r}';T) = {} & \lim_{N\to\infty} \left(\frac{\mu}{i\hbar\epsilon}\right)^N e^{-ia\phi} \int \frac{d\bar{\lambda}}{2\pi} e^{i\bar{\lambda}(\phi-\frac{1}{2}\omega_c T)} \\ & \times \int d^2r_1 \cdots \int d^2r_N \prod_{j=1}^{N+1} \left\{ \sum_{m=-\infty}^{\infty} e^{im(\vartheta_{N+1}-\vartheta_0)} \right. \\ & \left. \times \exp\left[\frac{i\mu}{2\hbar\epsilon}\left(r_j^2 + r_{j-1}^2 - \tfrac{1}{4} r_j^2 \omega_c^2 \epsilon^2\right)\right] I_{|m+\bar{\lambda}|}\left(\frac{\mu r_j r_{j-1}}{i\hbar\epsilon}\right)\right\}. \end{aligned} \tag{A.15}$$

This still looks sufficiently ugly so as to be rather intimidating. However, a truly remarkable result due to Peak and Inomata[13] saves the day. By iterating the formula

$$\int_0^\infty dx\, e^{i\sigma x} I_\nu(-ia\sqrt{x}) I_\nu(-ib\sqrt{x}) = \frac{i}{\sigma} \exp\left[\frac{-i(a^2+b^2)}{4\sigma}\right] I_\nu\left(\frac{iab}{2\sigma}\right) \tag{A.16}$$

N times ($\mathrm{Re}(\nu) > -1$), one finds

$$\begin{aligned} & \int_0^\infty \prod_{k=1}^N dr_k\, r_k \exp\left(ia\sum_{j=1}^N r_j^2\right) \prod_{j=1}^{N+1} I_\nu(-ibr_j r_{j-1}) = \\ & \prod_{k=1}^N \left(\frac{i}{2a_k}\right) \exp\left\{-i\left[r'^2 \sum_{j=1}^N \frac{b_j^2}{4a_j} + r''^2 \frac{b^2}{a_{N+1}}\right]\right\} I_\nu(-ib_{N+1} r' r''), \end{aligned} \tag{A.17}$$

with

$$\begin{aligned} a_1 &= a, \qquad a_{j+1} = a - \frac{b^2}{4a_j}, \qquad j \geq 1, \\ b_1 &= b, \qquad b_{j+1} = \prod_{k=1}^{j} \frac{b}{2a_k}, \qquad j \geq 1. \end{aligned} \tag{A.18}$$

Peak and Inomata showed that the $N \to \infty$ limit leads to

$$\begin{aligned} \lim_{N\to\infty} \left(\prod_{j=1}^N \frac{b}{2a_j}\right) &= \tfrac{1}{2}\mu\omega_c \csc(\tfrac{1}{2}\omega_c T) \\ \lim_{N\to\infty} \left(\tfrac{1}{2}b - \sum_{j=1}^N \frac{b_j^2}{4a_j}\right) &= \tfrac{1}{4}\mu\omega_c \operatorname{ctn}(\tfrac{1}{2}\omega_c T) \\ \lim_{N\to\infty} \left(\tfrac{1}{2}b - \frac{b^2}{4a_N}\right) &= \tfrac{1}{4}\mu\omega_c \operatorname{ctn}(\tfrac{1}{2}\omega_c T) \end{aligned} \tag{A.19}$$

where

$$
\begin{aligned}
a &\equiv \frac{\mu}{\epsilon}(1 - \frac{1}{8}\omega_c^2\epsilon^2) \\
b &\equiv \frac{\mu}{\epsilon}.
\end{aligned}
\tag{A.20}
$$

Using this result, the constrained propagator may finally be written as

$$
\begin{aligned}
K_\phi(\mathbf{r}'', \mathbf{r}'; T) &= e^{-ia\phi} \int \frac{d\bar{\lambda}}{2\pi}\, e^{i\bar{\lambda}(\phi - \frac{1}{2}\omega_c T)} \sum_{m=-\infty}^{\infty} e^{im(\theta'' - \theta' - \frac{1}{2}\omega_c T)} Q_{|m+\bar{\lambda}|}(r'', r'; T) \\
&= e^{-ia\phi} \sum_{n=-\infty}^{\infty} \delta(\theta'' - \theta' - \phi + 2\pi n) \int d\bar{\lambda}\, e^{i\bar{\lambda}(\phi - \frac{1}{2}\omega_c T)} Q_{|\bar{\lambda}|}(r'', r'; T)
\end{aligned}
\tag{A.21}
$$

where the radial function $Q_\beta(r'', r'; T)$ is defined to be

$$
\begin{aligned}
Q_\beta(r'', r'; T) \equiv \frac{\mu\omega_c}{4\pi i\hbar} &\csc(\tfrac{1}{2}\omega_c T) \\
&\times \exp\left[\frac{i\mu\omega_c}{4\hbar} \operatorname{ctn}(\tfrac{1}{2}\omega_c T)(r'^2 + r''^2)\right] \\
&\times I_\beta\left(\frac{\mu\omega_c}{2i\hbar} r' r'' \csc(\tfrac{1}{2}\omega_c T)\right).
\end{aligned}
\tag{A.22}
$$

In the limit of zero field, the radial function becomes

$$
\lim_{\omega_c \to 0} Q_\beta(r'', r'; T) = \frac{\mu}{2\pi i\hbar T} \exp\left[\frac{i\mu}{2\hbar T}(r'^2 + r''^2)\right] I_\beta\left(\frac{\mu r' r''}{i\hbar T}\right). \tag{A.23}
$$

As advertised, the winding constraint $\int dt\, \dot{\theta} = \phi$ renders the expression (A.21) for K_ϕ zero unless $\phi = \theta'' - \theta' + 2\pi n$. The complete propagator, K, is obtained by integrating over this constraint:

$$
K(\mathbf{r}'', \mathbf{r}'; T) = e^{-i\alpha\omega_c T/2} \sum_{n=-\infty}^{\infty} e^{in(\theta'' - \theta' - \frac{1}{2}\omega_c T)} Q_{|n+\alpha|}(r'', r'; T). \tag{A.24}
$$

Of course, taking both $\omega_c \to 0$ *and* $\alpha = 0$ in Eq.(A.24) recovers the familiar free particle propagator,

$$
\lim_{\omega_c \to 0} K(\mathbf{r}'', \mathbf{r}'; T)\Big|_{\alpha=0} = \frac{\mu}{2\pi i\hbar T} \exp\left[i(\mathbf{r}'' - \mathbf{r}')^2/2\hbar T\right]. \tag{A.25}
$$

I wish to stress that this appendix is meant as a technical review and that all the formulae herein have been derived elsewhere.

REFERENCES

[1] Frank Wilczek, *Phys. Rev. Lett.* **48**, 1144 (1982).

[2] Yong-Shi Wu, *Phys. Rev. Lett.* **53**, 111 (1984).

[3] Michael G. G. Laidlaw and Cécile Morette DeWitt, *Phys. Rev. D* **3**, 6 (1971).

[4] Frank Wilczek, *Phys. Rev. Lett.* **49**, 957 (1982).

[5] Yong-Shi Wu, *Phys. Rev. Lett.* **52**, 2103 (1984).

[6] L. Schulman, *Techniques and Applications of Path Integration*, (Wiley, New York, 1981). Chapter 27 contains material on coherent state path integration, while chapter 23 gives an excellent account of path integration on multiply connected spaces.

[7] Edward Fadell and James Van Buskirk, *Duke Math J.* **29**, 243 (1962).

[8] Joan S. Birman, *Comm. Pure and App. Math.* **22**, 41 (1969).

[9] See, for example, R. Fox and L. Neuwirth, *Math. Scand.* **10**, 119 (1962).

[10] D. C. Hickernell, E. O. McLean, and O. E. Vilches, *Phys. Rev. Lett.* **28**, 789 (1972); M. Bretz, J. G. Dash, D. C. Hickernell, E. O. McLean, and O. E. Vilches, *Phys. Rev. A* **8**, 1589 (1973) and *Phys. Rev. A* **9**, 2814 (1974).

[11] J. G. Dash, *Films on Solid Surfaces*, (Academic Press, New York, 1975).

[12] S. F. Edwards and Y. V. Gulyaev, *Proc. Roy. Soc. London* **A279**, 229 (1964).

[13] D. Peak and A. Inomata, *J. Math. Phys.* **10**, 1422 (1969).

[14] Akira Inomata and Vijay A. Singh, *J. Math. Phys.* **19**, 2318 (1978); Christopher C. Gerry and Vijay A. Singh, *Phys. Rev. D* **20**, 2550 (1979); C. C. Gerry and V. A. Singh, *Il Nuovo Cimento* **73B**, 161 (1983).

[15] I. S. Gradshteyn and I. M. Ryzhik, *Table of Integrals, Series, and Products*, (Academic Press, New York, 1980). The result (6.611.4) is particularly useful for those interested in the material contained in appendix 5A of this thesis.

[16] Daniel P. Arovas, Robert Schrieffer, Frank Wilczek, and A. Zee, *Nucl. Phys.* **B251** [**FS13**], 117 (1985).

[17] R. B. Laughlin, *Phys. Rev. Lett.* **50**, 1395 (1983).

[18] F. D. M. Haldane, *Phys. Rev. Lett.* **51**, 605 (1983).

[19] B. I. Halperin, *Phys. Rev. Lett.* **52**, 1583 (1984).

[20] Daniel Arovas, J. R. Schrieffer, and Frank Wilczek, Phys. Rev. Lett. 53, 722, 1984.

[21] M. V. Berry, *Proc. Roy. Soc. London* **A392**, 45 (1984).

[22] B. Simon, *Phys. Rev. Lett.* **51**, 2167 (1983).

[23] Frank Wilczek and A. Zee, *Phys. Rev. Lett.* **52**, 2111 (1984).

[24] L. Schiff, *Quantum Mechanics*, (McGraw-Hill, New York, 1955), p. 290.

[25] F. D. M. Haldane, private communications.

[26] S. M. Girvin and Terrence Jach, *Phys. Rev. B* **29**, 5617 (1984).

[27] W. P. Su, Univ. of Illinois (Urbana) Preprint, 1985.

[28] M. H. Friedman, J. B. Sokoloff, A. Widom, and Y. N. Srivastava, *Phys. Rev. Lett.* **52**, 1587 (1984).

[29] R. B. Laughlin, private communications.

[30] R. Tao, USC Preprint No. 085/003 (unpublished), 1985; R. Tao, *J. Phys. C* **18**, L1003 (1985).

VOLUME 58, NUMBER 12 PHYSICAL REVIEW LETTERS 23 MARCH 1987

Off-Diagonal Long-Range Order, Oblique Confinement, and the Fractional Quantum Hall Effect

S. M. Girvin
Surface Science Division, National Bureau of Standards, Gaithersburg, Maryland 20899

and

A. H. MacDonald
National Research Council, Ottawa, Ontario, Canada K1A OR6

(Received 24 November 1986)

We demonstrate the existence of a novel type of off-diagonal long-range order in the fractional-quantum-Hall-effect ground state. This is revealed for the case of fractional filling factor $\nu = 1/m$ by application of Wilczek's "anyon" gauge transformation to attach m quantized flux tubes to each particle. The binding of the zeros of the wave function to the particles in the fractional quantum Hall effect is a (2+1)-dimensional analog of *oblique confinement* in which a condensation occurs, not of ordinary particles, but rather of composite objects consisting of particles and gauge flux tubes.

PACS numbers: 72.20.My, 71.45.Gm, 73.40.Lq

A remarkable amount of progress has recently been made in our understanding of the fractional quantum Hall effect (FQHE)[1] following upon the seminal paper by Laughlin.[2] There remains, however, a major unsolved problem which centers on whether or not there exists an order parameter associated with some type of symmetry breaking.[3-6] The apparent symmetry breaking associated with the discrete degeneracy of the ground state in the Landau gauge[5] is an artifact of the toroidal geometry[6,7] and is not an issue here. Rather, the questions that we are addressing have been motivated by the analogies which have been observed to exist[4,8] between the FQHE and superfluidity and by recent progress towards a phenomenological Ginsburg-Landau picture of the FQHE.[4] Further motivation has come from the development of the correlated-ring-exchange theory of Kivelson *et al.*[9] (see also Chui, Hakim, and Ma,[10] and Chui,[10] and Baskaran[11]). The existence of infinitely large ring exchanges is a signal of broken gauge symmetry in superfluid helium[12] and is suggestive of something similar in the FQHE. However, the concept of ring exchanges on large length scales has not as yet been fully reconciled with Laughlin's (essentially exact[7]) variational wave functions which focus on the short-distance behavior of the two-particle correlation function. Furthermore it is clear that we cannot have an ordinary broken gauge symmetry since the particle density (which is conjugate to the phase) is ever more sharply defined as the length scale increases. The purpose of this Letter is to unify all these points of view by demonstrating the existence of a novel type of off-diagonal long-range order (ODLRO) in the FQHE ground state.

In second quantization the one-body density matrix is given by

$$\rho(z,z') = \sum_{m,n} \varphi_m^*(z)\varphi_n(z')\langle 0|c_n^\dagger c_m|0\rangle, \tag{1}$$

where $\varphi_n(z)$ is the nth lowest-Landau-level single-particle orbital[1] in the symmetric gauge, and z is a complex representation of the particle position vector in units of the magnetic length.[1] It is an unusual feature of this problem that there is a unique single-particle state for each angular momentum. Hence by making only the assumption that the ground state is isotropic and homogeneous we may deduce $\langle 0|c_n^\dagger c_m|0\rangle = \nu\delta_{nm}$, and thereby obtain the powerful identity:

$$\rho(z,z') = \nu g(z,z') = (\nu/2\pi)\exp(-\tfrac{1}{4}|z-z'|^2)\exp[\tfrac{1}{4}(z^*z'-zz'^*)], \tag{2}$$

where $g(z,z')$ is the ordinary single-particle Green's function.[13]

We see from (2) that the density matrix is short ranged with a characteristic scale given by the magnetic length, just as occurs in superconducting films in a magnetic field.[14] The same result can be obtained within first quantization via the expression

$$\rho(z,z') = \frac{N}{Z}\int d^2z_2 \cdots d^2z_N \Psi^*(z,z_2,\ldots,z_N)\Psi(z',z_2,\ldots,z_N), \tag{3}$$

where Z is the norm of Ψ.

If the lowest Landau level has filling factor $\nu = 1/m$ and the interaction is a short-ranged repulsion, then in the low-electron mass limit,[7] the *exact* ground-state wave function is given by Laughlin's expression:

$$\Psi(z_1,\ldots,z_N) = \prod_{i<j}(z_i-z_j)^m \exp\left(-\tfrac{1}{4}\sum_k |z_k|^2\right). \tag{4}$$

Laughlin's plasma analogy[2,15] proves that the ground state is a liquid of uniform density so that Eq. (2) is valid. The rapid phase oscillations of the integrand in (3) cause ρ to be short ranged. There is, nevertheless, a peculiar type of long-range order hidden in the ground state. For reasons which will become clear below, this order is revealed by considering the singular gauge field used in the study of "anyons"[16,17]:

$$\mathcal{A}_j(z_j) = \frac{\lambda\Phi_0}{2\pi}\sum_{i\neq j}\nabla_j \operatorname{Im}\ln(z_j - z_i), \quad (5)$$

where $\Phi_0 = hc/e$ is the quantum of flux and λ is a constant. The addition of this vector potential to the Hamiltonian is not a true gauge transformation since a flux tube is attached to each particle. If, however, $\lambda = m$ where m is an integer, the net effect is just a change in the phase of the wave function:

$$\Psi_{\text{new}} = \exp\left(-im\sum_{i<j}\operatorname{Im}\ln(z_i - z_j)\right)\Psi_{\text{old}}. \quad (6)$$

Application of (6) to the Laughlin wave function (4) yields

$$\Psi(z_1, \ldots, z_N) = \prod_{i<j}|z_i - z_j|^m \exp\left(-\tfrac{1}{4}\sum_k |z_k|^2\right), \quad (7)$$

which is purely real and is symmetric under particle exchange for both even and odd m. Hence we have the remarkable result that both fermion and boson systems map into bosons in this singular gauge.

Substituting (7) into (3) and using Laughlins's plasma analogy,[2,15] a little algebra shows that the singular-gauge density matrix $\tilde\rho$ can be expressed as

$$\tilde\rho(z,z') = (\upsilon/2\pi)\exp[-\beta\Delta f(z,z')]\,|z - z'|^{-m/2}, \quad (8)$$

where $\beta \equiv 2/m$, and $\Delta f(z,z')$ is the difference in free energy between two impurities of charge $m/2$ (located at z and z') and a single impurity of charge m (with arbitrary location). Because of complete screening of the impurities by the plasma, the free-energy difference $\Delta f(z,z')$ rapidly approaches a constant as $|z - z'| \to \infty$. This proves the existence of ODLRO[18] characterized by an exponent $\beta^{-1} = m/2$ equal to the plasma "temperature." For $m = 1$ the asymptotic value of Δf can be found exactly: $\beta\Delta f_\infty = -0.03942$. For other values of m, $\beta\Delta f(z,z')$ can be estimated either by use of the ion-disk approximation[2,15] or the static (linear response) susceptibility of the (classical) plasma calculated from the known static structure factor[8] (see Fig. 1).

The rigorous and quantitative results we have obtained above are valid for the case of short-range repulsive interactions for which Laughlin's wave function is exact. We now wish to use these results for a qualitative examination of more general cases and to deepen our understanding of the ODLRO. We begin by noting that $\tilde\rho$ can be rewritten in the ordinary gauge as

$$\tilde\rho(z,z') = \frac{N}{Z}\int d^2z_2 \cdots d^2z_N \exp\left[-i\frac{e}{\hbar c}\int_{\mathbf{z}}^{\mathbf{z}'} d\mathbf{r}\cdot\mathcal{A}_1\right]\Psi^*(z,z_2,\ldots,z_N)\Psi(z',z_2,\ldots,z_N), \quad (9)$$

where $\mathbf{z}$ and $\mathbf{z}'$ are vector representations of z and z'. The line integral in (9) is multiple valued but its exponential is single valued because the flux tubes are quantized. The additional phases introduced by the singular gauge transformation will cancel the phases in Ψ nearly everywhere, and produce ODLRO in $\tilde\rho$ if and only if the zeros of Ψ (which must necessarily be present because of the magnetic field[19]) are bound to the particles. Thus ODLRO in $\tilde\rho$ *always* signals a condensation of the zeros onto the particles (independent of whether or not the composite-particle occupation of the lowest momentum state diverges[18]). Because the gauge field $\mathcal{A}_1$ depends on the positions of *all* the particles, $\tilde\rho$ differs not just in the phase but in *magnitude* from ρ. This multiparticle object, which explicitly exhibits ODLRO, is very reminiscent of the topological order parameter in the XY model[20] and related gauge models[21,22] and is intimately connected with the frustrated XY model which arises in the correlated-ring-exchange theory.[9]

For short-range interactions, the zeros of Ψ are directly on the particles and the associated phase factors are exactly canceled by the gauge term in (9) [see Eq. (7)]. As the range of the interaction increases, $m - 1$ of the zeros move away from the particles but remain nearby

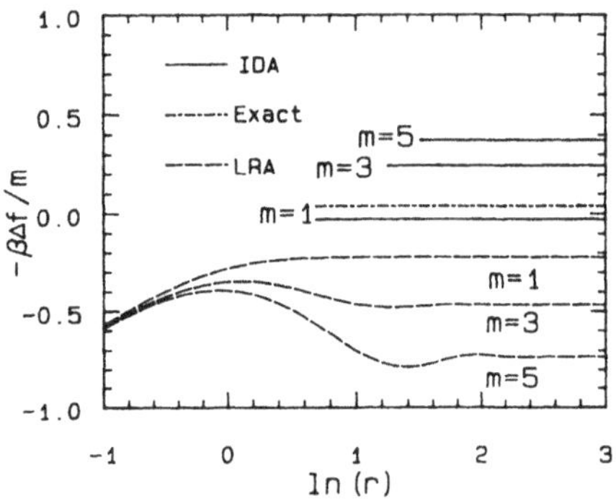

FIG. 1. Plot of $-\beta\Delta f(z,z')/m$ vs $r \equiv |z - z'|$ for filling factor $\upsilon = 1/m$. LRA is linear-response approximation, IDA is ion-disk approximation (shown only for separations exceeding the sum of the ion-disk radii). Because the plasma is strongly coupled, the IDA is quite accurate at $m = 1$ and improves further with increasing m. The LRA is less accurate at $m = 1$ and worsens with increasing m.

and bound to them.[7,23] The gauge and wave function phase factors in (9) now appear in the form of the bound vortex-antivortex pairs. We expect such bound pairs *not* to destroy the ODLRO and speculate (based on our understanding of the Kosterlitz-Thouless transition[20]) that the effect is at most to renormalize the exponent of the power law in (8). As the range of the potential is increased still further, numerical computations[7] indicate that a critical point is reached at which the gap rather suddenly collapses and the overlap between Laughlin's state and the true ground state drops quickly to zero. We propose that this gap collapse corresponds to the unbinding of the vortices and hence to the loss of ODLRO and the onset of short-range behavior of $\tilde{\rho}(z,z')$. Recall that the distinguishing feature of the FQHE state is its long wavelength excitation gap. At least within the single-mode approximation,[8] this gap can only exist when the ground state is homogeneous and the two-point correlation function exhibits perfect screening:

$$M_1 \equiv (\nu/2\pi)\int d^2r(r^2/2)[g(r)-1] = -1.$$

In the analog plasma problem, the zeros of Ψ act like point charges seen by each particle and the M_1 sum rule implies that electrons see each other as charge-$(m = 1/\nu)$ objects; i.e., that m zeros are bound to each electron. Thus (within the single-mode approximation) there is a one-to-one correspondence between the existence of ODLRO and the occurrence of the FQHE.

The exact nature of the gap-collapse transition, which occurs when the range of the potential is increased,[7] is not understood at present. However, it has been proven[8] that the M_1 sum rule is satisfied by every homogeneous and isotropic state in the lowest Landau level. Hence the vortex unbinding should be a first-order transition to a state which breaks rotation symmetry (like the Tao-Thouless state[24]) and/or translation symmetry (like the Wigner crystal[4,8]). We know that as a function of *temperature* (for fixed interaction potential) there can be no Kosterlitz-Thouless transition[20] since isolated vortices (quasiparticles) cost only a finite energy in this system[4,25] (see, however, Ref. 10).

Further insight into the gap collapse can be obtained by considering the exceptional case of Laughlin's wave function with $m > 70$. In this case the zeros are still rigorously bound to the particles so that the analog plasma contains long-range forces (and $\tilde{\rho}$ exhibits ODLRO), but the plasma "temperature" has dropped below the freezing point.[2,15] If such a state exhibits (sufficiently[10]) long-range positional correlations, the FQHE would be destroyed by a gapless Goldstone mode associated with the broken translation symmetry. Hence in this exceptional case the normal connection between ODLRO and the FQHE would be broken by a gap collapse due to positional order at a finite wave vector.

The existence of ODLRO in $\tilde{\rho}$ is the type of infrared property which suggests that a field-theoretic approach to the FQHE would be viable. It is clear from the results presented here that the binding of the zeros of Ψ to the particles can be viewed as a condensation,[18] not of ordinary particles, but rather of composite objects consisting of a particle and m flux tubes. (We emphasize that these are *not* real flux tubes, but merely consequences of the singular gauge. The assumption that electrons can bind real flux tubes[26] is easily shown to be unphysical.[27]) The analog of this result for hierarchical daughter states of the Laughlin states[7,15] would be a condensation of composite objects consisting of n particles and m flux tubes (cf. Halperin's "pair" wave functions[19]). This seems closely analogous to the phenomenon of *oblique confinement*[22] and it ought to be possible to derive the appropriate field theory from first principles by use of this idea.

Since the singular gauge maps the problem onto interacting bosons, coherent-state path integration[28] may prove useful. A step in this direction has been taken recently in the form of a Landau-Ginsburg theory which was developed on phenomenological grounds.[4] In the static limit, the action has the "θ vacuum" form

$$S = \int d^2r\,|(-i\nabla+\mathbf{a})\psi(\mathbf{r})|^2 + i\phi(\mathbf{r})(\psi^*\psi-1) - i(\theta/8\pi^2)(\phi\nabla\times\mathbf{a}+\mathbf{a}\times\nabla\phi), \tag{10}$$

where $\mathbf{a}$ is not the physical vector potential but an effective gauge field[4] representing frustration due to density deviations away from the quantized Laughlin density and ϕ is a scalar potential which couples both to the charge density and the "flux" density. From (10) the equation of motion for $\mathbf{a}$ is (in the static case):

$$\theta\nabla\times\mathbf{a} = (\psi^*\psi - 1). \tag{11}$$

This equation and the parameter θ, which determines the charge carried by an isolated vortex, originally had to be chosen phenomenologically.[4] Now, however, it can be justified by examination of Eq. (5) which shows that the curl of $\mathcal{A}_j$ is proportional to the density of particles. If we identify $\mathbf{a}$ in (10) and (11) as

$$\mathbf{a} = \mathcal{A}_1 + \mathbf{A}, \tag{12}$$

where $\mathbf{A}$ is the physical vector potential and we take $\psi^*\psi$ as the particle density relative to the density in the Laughlin state, then Eq. (11) follows from (5) with the θ angle being given by $\theta = 2\pi/m$. This yields[4] the correct charge of an isolated vortex (Laughlin quasiparticle) of $q^* = 1/m$. The connection between this result and the Berry phase argument of Arovas *et al.*[29] should be noted (see also Semenoff and Sodano[30]). To summarize, it is the strong phase fluctuations induced by the frustration

VOLUME 58, NUMBER 12 PHYSICAL REVIEW LETTERS 23 MARCH 1987

associated with density variations [Eq. (11)] which pin the density at rational fractional values and account for the differences between the FQHE and ordinary superfluidity.[4]

We believe that these results shed considerable light on the FQHE, unify the different pictures of the effect, and emphasize the topological nature of the order in the zero-temperature state of the FQHE. The present picture leads to several predictions which can be tested by numerical computations by use of methods very similar to those now in use.[31] ODLRO will be found only in states exhibiting an excitation gap. The decay of the singular-gauge density matrix will be controlled by the distribution of distances of the zeros of the wave function from the particles. This distribution, which can be artificially varied by changing the model interaction, directly determines the short-range behavior of the density-density correlation function and hence the ground-state energy.[7,23]

The authors would like to express their thanks to C. Kallin, S. Kivelson, and R. Morf for useful conversations and suggestions.

[1]*The Quantum Hall Effect*, edited by R. E. Prange and S. M. Girvin (Springer-Verlag, New York, 1986).

[2]R. B. Laughlin, Phys. Rev. Lett. **50**, 1395 (1983).

[3]P. W. Anderson, Phys. Rev. B **28**, 2264 (1983).

[4]S. M. Girvin, in Chap. 10 of Ref. 1.

[5]R. Tao and Yong-Shi Wu, Phys. Rev. B **30**, 1097 (1984).

[6]D. J. Thouless, Phys. Rev. B **31**, 8305 (1985).

[7]F. D. M. Haldane, in Chap. 8 of Ref. 1.

[8]S. M. Girvin, A. H. MacDonald, and P. M. Platzman, Phys. Rev. Lett. **54**, 581 (1985), and Phys. Rev. B **33**, 2481 (1986); S. M. Girvin in Chap. 9 of Ref. 1.

[9]S. Kivelson, C. Kallin, D. P. Arovas, and J. R. Schrieffer, Phys. Rev. Lett. **56**, 873 (1986).

[10]S. T. Chui, T. M. Hakim, and K. B. Ma, Phys. Rev. B **33**, 7110 (1986); S. T. Chui, unpublished.

[11]G. Baskaran, Phys. Rev. Lett. **56**, 2716 (1986), and unpublished.

[12]R. P. Feynman, Phys. Rev. **91**, 1291 (1953).

[13]S. M. Girvin and T. Jach, Phys. Rev. B **29**, 5617 (1984).

[14]E. Brézin, D. R. Nelson, and A. Thiaville, Phys. Rev. B **31**, 7124 (1985).

[15]R. B. Laughlin, in Chap. 7 of Ref. 1.

[16]F. Wilczek, Phys. Rev. Lett. **49**, 957 (1982).

[17]D. P. Arovas, J. R. Schrieffer, F. Wilczek, and A. Zee, Nucl. Phys. B **251**, 117 (1985).

[18]We refer to this as ODLRO or condensation because of the slow power-law decay even though the largest eigenvalue $\lambda \equiv \int d^2z\,\tilde{\rho}(z,z')$ of the density matrix diverges only for $m \lesssim 4$ [see C. N. Yang, Rev. Mod. Phys. **34**, 694 (1962)].

[19]B. I. Halperin, Helv. Phys. Acta **56**, 75 (1983).

[20]J. V. José, L. P. Kadanoff, S. Kirkpatrick, and D. R. Nelson, Phys. Rev. B **16**, 1217 (1977).

[21]J. B. Kogut, Rev. Mod. Phys. **51**, 659 (1979).

[22]J. L. Cardy and E. Rabinovici, Nucl. Phys. B **205**, 1 (1982); J. L. Cardy, Nucl. Phys. B **205**, 17 (1982).

[23]D. J. Yoshioka, Phys. Rev. B **29**, 6833 (1984).

[24]R. Tao and D. J. Thouless, Phys. Rev. B **28**, 1142 (1983). The symmetric gauge version of this state exhibits threefold rotational symmetry.

[25]A. M. Chang, in Chap. 6 in Ref. 1.

[26]M. H. Friedman, J. B. Sokoloff, A. Widom, and Y. N. Srivastava, Phys. Rev. Lett. **52**, 1587 (1984), and **53**, 2592 (1984).

[27]F. D. M. Haldane and L. Chen, Phys. Rev. Lett. **53**, 2591 (1984).

[28]L. S. Schulman, *Techniques and Applications of Path Integration* (Wiley, New York, 1981).

[29]D. Arovas, J. R. Schrieffer, and F. Wilczek, Phys. Rev. Lett. **53**, 722 (1984).

[30]G. Semenoff and P. Sodano, Phys. Rev. Lett. **57**, 1195 (1986).

[31]F. D. M. Haldane and E. H. Rezayi, Phys. Rev. Lett. **54**, 237 (1985), and Phys. Rev. B **31**, 2529 (1985); F. C. Zhang, V. Z. Vulovic, Y. Guo, and S. Das Sarma, Phys. Rev. B **32**, 6920 (1985); G. Fano, F. Ortolani, and E. Colombo, Phys. Rev. B **34**, 2670 (1986).

Note added: E. H. Rezayi and F. D. M. Haldane have recently succeeded in computing the singular-gauge density matrix for a small number of particles on a sphere. The results are in complete accord with the discussion in the penultimate paragraph of this paper. See: E. H. Rezayi and F. D. M. Haldane, *Phys. Rev. Lett.* Vol. **61** (1988) 1985.

VOLUME 58, NUMBER 20 PHYSICAL REVIEW LETTERS 18 MAY 1987

Quantum Conductance in Networks

J. E. Avron, A. Raveh, and B. Zur
Department of Physics, Technion, Haifa 32000, Israel
(Received 21 November 1986)

We consider the quantum transport in networks. Arguments similar to those for the quantum Hall effect show that the averaged transport coefficients are quantized. Numerical calculations for a three-loop network yield the values 0, 1, and −1, depending on the fluxes threading the loops and the quantum state of the net. We characterize the conductance properties of such networks. We also discuss general properties of the transport coefficients in general multiloop networks.

PACS numbers: 73.60.Aq, 02.40.+m, 72.20.My

It is known that there is a range of circumstances where the Hall conductance, at low temperatures, is a nonzero integer.[1] It is natural to inquire whether there are other systems with integer nonzero conductances. As we shall explain, networks are such systems: A network with L loops has $L(L-1)/2$ integer conductances which characterize the quantum state of the system and reflect its multiconnectivity. Like the Hall conductance, they are nondissipative and can be either holelike or electronlike, but unlike the Hall effect this does not reflect any band-structure properties.

The transport coefficients of the network, $2\pi g_{lm}$, are defined as the charge transported around loop l when the flux threading the mth loop, ϕ_m, increases adiabatically by 2π, the unit of quantum flux. Within linear-response theory, it turns out[2] that this is equivalent to the (time-averaged) ratio of the current in loop l to an infinitesimal emf acting on loop m.[3,4] We shall concentrate on the cases where the network has three loops and l and m are distinct. We shall also assume throughout that the fluxes are changed sufficiently slowly for the adiabatic limit to hold. In particular the energy levels of the network are assumed to have nonvanishing gaps and we exclude situations where levels cross. Under these conditions, which guarantee no dissipation (dissipation arises when $l=m$ and the adiabatic limit does not hold), the nondiagonal conductances have nonlocal features. Also, the quantum (coherence) effects discussed below require temperatures which are low compared with a typical gap energy. Since energy gaps scale like $(\text{length})^{-2}$ this favors small networks. This dictates temperatures in the millikelvin range and emf in microvolts for mesoscopic networks. Quantum coherence effects associated with the dissipative conductance in single mesoscopic loops, including nonlocal effects, are discussed by Sharvin and Sharvin.[5] As yet, there are no experiments nor theory on the transport coefficients in two- or three-loop networks.

Consider, for example, a three-loop network made of mesoscopic, thin (metallic) wires (Fig. 1). Each loop is threaded by an independent flux tube ϕ_j, $j=1,2,3$. In comparison with the Hall effect, ϕ_3 plays the role of the magnetic field on the sample, ϕ_1 can be thought of as a time-dependent flux replacing a battery, and ϕ_2 is the analog of Laughlin's flux tube. g_{12} is then the analog of the Hall conductance. Because of the analogy one may expect that g_{12} will be quantized and will be a nontrivial (antisymmetric) function of ϕ_3. This, as we shall see, is essentially correct provided that suitable averaging is introduced: Let

$$\langle g_{lm}\rangle(\phi)\equiv\frac{1}{2\pi}\int_0^{2\pi}d\phi_m\, g_{lm}(\phi)$$

be the conductance averaged over the flux in the current loop m (ϕ denotes collectively the three fluxes). $\langle g_{12}\rangle(\phi_3)$ (or any other permutation of 1, 2, and 3) is an antisymmetric steplike function of ϕ_3 with steps at integer heights. This holds in great generality (i.e., even with electron-electron interaction, and also for thick wires and for more complicated networks) provided the system is in a pure quantum state which does not become degenerate as ϕ_1 and ϕ_2 are varied. It is a consequence of the fact that Kubo's formula for the (averaged) conductance has a topological interpretation being a first Chern number.[2,6-10]

Here we shall describe parts of our numerical results and sketch the general theoretical structure. Details shall be presented elsewhere.[11]

From the theory of superconducting networks[12] it is known that the analysis of the Schrödinger equation for the network of Fig. 1 (with one-dimensional wires) reduces to the study of 5×5 matrices (5 is the number of

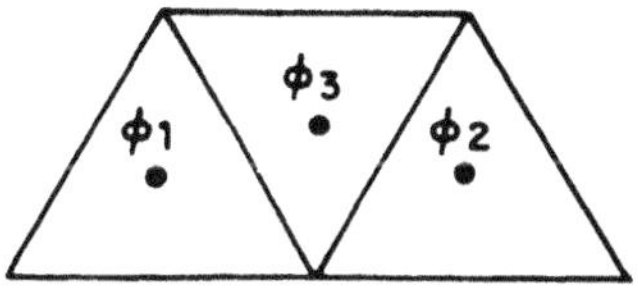

FIG. 1. Three-loop network with seven edges and five vertices. Each loop is threaded by a flux tube. The Hamiltonian for the network with point junctions is a 5×5 matrix. This network has nonzero quantized conductances.

vertices in the network) of the tight-binding type. It is therefore not surprising that the computation of the transport coefficients of the network also reduces to a (5×5)-matrix problem. The details of the reduction shall be given elsewhere.[11] In the matrix description the wave function sits on the vertices of the network. Consider the tight-binding Hamiltonian

$$H(v,v';\phi)=n_v\delta_{vv'}+(1-\delta_{vv'})\textstyle\sum_b[v,b][v',b]\exp(-i[v,b]\gamma_b). \tag{1}$$

v and v' are vertex indices. n_v is the "coordination number" of the vertex v. b is a (directed) edge index, and $[v,b]$ is the incidence matrix, i.e., $[v,b]$ is 1 if the edge b points into v, -1 if it points out of v, and 0 otherwise. $\gamma_b\equiv\int_b A$, where A is the vector potential associated to the fluxes ϕ. The lengths of all the edges b is set equal to one. $H(\phi)$ is identical to the de Gennes–Alexander[12] network Hamiltonian except on the diagonal. This slight modification makes it somewhat easier to handle. (The de Gennes–Alexander Hamiltonian gives an implicit eigenvalue problem.) Because of our interest in topological invariants the difference is presumably immaterial.

Diagonalizing the Hamiltonian one finds that the Chern number in the ground state, defined by Eq. (3) below, is

$$\langle g_{12}\rangle(\phi_3)=\begin{cases}0 & \text{for } -\pi/3<\phi_3\,\mathrm{mod}\,2\pi<\pi/3\,,\\ 1 & \text{for } \pi/3<\phi_3\,\mathrm{mod}\,2\pi<\pi,\\ -1 & \text{for } -\pi<\phi_3\,\mathrm{mod}\,2\pi<-\pi/3.\end{cases} \tag{2}$$

For the excited states one finds qualiatively similar, i.e., nontrivial, antisymmetric, periodic steplike functions that take the values 0, 1, and -1. One also finds that $\langle g_{13}\rangle(\phi_2)=\langle g_{23}\rangle(\phi_1)=0$ identically for all the states. Because of the topological nature of the results the fact that the network is made of three equilateral triangles is immaterial and one finds the same qualitative features in any network which is a deformation of Fig. 1.

To get a complete description and insight into the results we have to introduce some formalism. This is also necessary in order to describe the actual computation.

$H(\phi)$, the exact Schrödinger operator of the network, depends parametrically on the fluxes ϕ. For fixed ϕ, it has discrete spectrum. Because of the periodicity in the fluxes the parameter space can be identified with T^3, the three-torus, i.e., we can identify ϕ_j with $\phi_j+2\pi$.[13]

Let $P(\phi)$ denote a projection on a spectral subspace of $H(\phi)$ and C be a closed, two-dimensional surface in T^3 (equal to a closed two-chain). Suppose that $P(\phi)$ is smooth on C. It is a standard fact that the Chern number,[2,6] $\mathrm{Ch}(P,C)\equiv(i/2\pi)\int_C \mathrm{Tr}[dP\,P\,dP]$, is an integer.

If the initial state of the system is given by $P(\phi)$ and there is no level crossing on $T_{lm}(\phi)$ (the two-dimensional slice of the three-torus going through ϕ and indexed by l and m), then Kubo's formula reads[2,7-10] (see Ref. 10 for a rigorous derivation)

$$\langle g_{lm}\rangle(\phi)=\mathrm{Ch}(P,T_{lm}(\phi)). \tag{3}$$

It follows[11] that $\langle g\rangle_{lm}(\phi)$ is gauge invariant, periodic in ϕ, antisymmetric in l and m, and independent of ϕ_l and ϕ_m, and is quantized to be an integer. Also, in the absence of magnetic fields besides ϕ, which we shall assume, time reversal leads to the Onsager relation $\langle g_{lm}\rangle(\phi)=-\langle g_{lm}\rangle(-\phi)$.

It is known that the Chern numbers are closely related to degeneracies. Let D_q be the set of points where the qth gap in the energy spectrum closes. According to the von Neumann–Wigner theorem[14] D_q is a discrete set. The second homology group of T^3/D_q is spanned by three two-tori, T_{12}, T_{23}, and T_{31}, and $|D_q|$ oriented two-spheres $S(\delta)$ that surround $\delta\in D_q$. An arbitrary closed two-chain in T^3/D_q can be written as a sum of the basic spheres and tori with integer coefficients. This relation lifts to a relation for the Chern numbers. It follows that the set of $3+\sum|D_q|$ Chern numbers contains all the information about the net.

Relations among the basic Chern numbers follow from the following facts:

(1) $\sum_{\delta\in D_q}S(\delta)$ is homologous to zero, so that $\sum_{\delta\in D_q}\mathrm{Ch}(P_q,S(\delta))=0$, for all q, where P_q is the projection on the spectral subspace with energies up to the qth level.

(2) The set D_q is invariant under inversion $\phi\to-\phi$, and for any δ in D_q, $\mathrm{Ch}(P_q,S(\delta))=\mathrm{Ch}(P_q,S(-\delta))$.

For any closed two-chain c which is invariant under inversion, $\mathrm{Ch}(P_q,c)=0$. (4) If $H(\phi)$ is a periodic $n\times n$ matrix, P_n is the identity and so all its Chern numbers vanish.

For the Hamiltonian of Eq. (1) the set of points of degeneracy and their Chern numbers are given in Table I. Because of (2) above only points in the half-cube with $0\le\phi_3\le\pi$ are listed. For the three basic tori we find

$$\mathrm{Ch}(P_q,T_{23}(\phi_1))=\mathrm{Ch}(P_q,T_{13}(\phi_2))=0,\qquad \mathrm{Ch}(P_q,T_{12}(\pi/2))=\delta(q,1). \tag{4}$$

This gives a complete characterization of the nondissipa-

TABLE I. Chern numbers for spheres surrounding points of degeneracy in the network of Fig. 1. $a\equiv\arccos(\frac{1}{2}-\sqrt{2})$.

Gap	Coordinates	Chern number
1	$(\frac{2}{3},\frac{2}{3},\frac{1}{3})\pi$	1
1	$(\frac{4}{3},\frac{2}{3},1)\pi$	-1
2	$(-a,a,0)$	1
2	$(a,a,a/2)$	-1
3	$(\frac{4}{3},\frac{4}{3},\frac{2}{3})\pi$	1
3	$(\frac{2}{3},\frac{4}{3},1)\pi$	-1

tive averaged conductance of the three-loop network of Eq. (1).

Equation (3), which relates the Chern numbers with the transport coefficients, is known to hold for the full Schrödinger equation of the network. We shall now describe how to extend this to matrix Hamiltonians. We shall consider here the case of matrix Hamiltonians where the wave function sits on the bonds of the network. The case where it sits on vertices is more complicated and shall be dealt with elsewhere.[11]

In a one-dimensional Schrödinger equation of a network the wave function is $\Psi_b(x_b)$, $0 \le x_b \le 1$. The coordinate x_b is measured in the b direction. On b, Ψ solves the one-dimensional Schrödinger equation in a gauge potential and therefore has the form

$$\Psi_b(x) = [T(\psi_+,\psi_-)](x) \equiv \exp[i\gamma_b(x)][\psi_+(b)\exp(ikx) + \psi_-(b)\exp(-ikx)], \tag{5}$$

where $\gamma_b(x) \equiv \int^x A$. $\psi_+(b)$ and $\psi_-(b)$ denote the amplitudes of the forward- and backward-moving waves on the bond b. The linear operator T is introduced for later purposes. A vertex with n_v edges connected to it is described by a unitary $n_v \times n_v$ scattering matrix, S_v,[15] which maps the incoming waves on the outgoing waves, $S_v \Psi_{v,\mathrm{in}} = \psi_{v,\mathrm{out}}$. $\Psi_{v,\mathrm{in/out}}$ are n_v-dimensional vectors of complex numbers:

$$\begin{aligned} \Psi_{v,\mathrm{in}}(b) &\equiv \delta(1,[v,b])\psi_+(b)\exp[ik + i\gamma_b] + \delta(-1,[v,b])\psi_-(b), \\ \Psi_{v,\mathrm{out}}(b) &\equiv \delta(1,[v,b])\psi_-(b)\exp[-ik + i\gamma_b] + \delta(-1,[v,b])\psi_+(b). \end{aligned} \tag{6}$$

b runs over the n_v edges associated to v. The unitarity of S_v guarantees that current is conserved at each vertex.

A basic tool is this: Consider $T: H_1 \to H_2$, where $H_{1,2}$ are two Hilbert spaces (not necessarily of the same dimension). Let Q be an orthogonal projection on H_2, and suppose that $TT^*Q = Q$ (T^* denotes the adjoint of T). Suppose that $QdTT^*$ is smooth and globally defined. Then $P \equiv T^*QT$ is a projection on H_1 and the Chern numbers for the two projections coincide.

We apply this to H_1, the finite-dimensional complex vector space, and H_2, the Hilbert space of functions. The map T from C^2 to $L^2[0,1]$ is given by Eq. (5). The scalar product in C^2, induced by the scalar product in L^2, has the "Riemann metric" A where

$$A_{ij} \equiv \delta_{ij} + [(1-\delta_{ij})/k]\exp(-ik\epsilon_{ij})\sin(k), \quad i,j = 1,2.$$

ϵ_{ij} is the completely antisymmetric tensor. The metric is independent of the gauge field and is nonsingular provided $k \neq 0$. One finds[11] for dT^*T

$$dT^*T = d(A^{-1})^*A - i(A^{-1})^* \begin{pmatrix} \int [d\gamma(x) + x\,dk]dx & \int [d\gamma(x) - x\,dk]\exp(-2ikx)dx \\ \int [d\gamma(x) + x\,dk]\exp(2ikx)dx & \int [d\gamma(x) - x\,dk]dx \end{pmatrix}. \tag{7}$$

γ is linear in ϕ, so that $d\gamma$ is independent of ϕ. Q, k, A^{-1}, and $QdTT^*$ are all smooth in ϕ provided no levels cross and k does not vanish. This establishes the equality of the Chern numbers.

In summary, networks have quantized averaged conductances which are nontrivial in networks with three or more loops. The computation of Chern numbers for network Hamiltonians with one-dimensional connecting links reduces to the study of finite matrices. Finally, homology provides a convenient and compact way of presenting the conductance functions of networks.

This work was supported by the Israel Academy of Sciences, Minerva, and the U.S.-Israel Binational Science Foundation Grant No. 84-00376. We are indebted to A. Libchaber for a conversation that started this work, to Y. Imry, S. Lipson, R. Seiler, S. Shtrikman, and J. Zak for discussions, and to D. Vollhardt for help with the references.

[1]K. von Klitzing, G. Dorda, and M. Pepper, Phys. Rev. Lett. **45**, 494 (1980); H. L. Störmer *et al.* Phys. Rev. Lett. **56**, 85 (1986).

[2]J. E. Avron and R. Seiler, Phys. Rev. Lett. **54**, 259 (1985).

[3]Loop currents I_l, $l \in L$, are related to the usual edge currents I_e, $e \in E$, by $I_e = \sum_{f \in F}[e,f]I_{bf}$. $[e,f]$ is the incidence matrix of the graph

[4]R. B. Laughlin, Phys Rev. B **23**, 5632 (1981).

[5]D. Yu. Sharvin and Yu. V. Sharvin, Pis'ma Zh. Eksp. Teor. Fiz. **34**, 285 (1981) [JETP Lett. **34**, 272 (1981)]; R. A. Webb *et al.*, Phys. Rev Lett. **51**, 690 (1983); M. Büttiker *et al.*, Phys. Lett. **96A**, 365 (1983); C. P. Umbach *et al.*, Phys. Rev. B **30**, 4048 (1984); Y. Gefen, Y. Imry, and M. Azbel, Phys. Rev. Lett. **52**, 129 (1984); V. Chandrasekhar *et al.*, Phys. Rev. Lett. **55**, 1610 (1985); R. A. Webb *et al.*, Phys. Rev. Lett. **54**, 2696 (1985); M. Büttiker *et al.*, Phys. Rev. B **31**, 6207 (1985); M. Büttiker, Phys. Rev. Lett. **57**, 1761 (1986); A. D. Benoit *et al.*, Phys. Rev. Lett **56**, 1765 (1986).

[6]D. J. Thouless, M. Kohmoto, M. P. Nightingale, and M. den Nijs, Phys. Rev. Lett. **49**, 405 (1982).

[7]Q. Niu and D. Thouless, J. Phys. A **17**, 2453 (1984).

[8]Q. Niu, D. J. Thouless, and Y. S. Wu, Phys. Rev. B **31**,

3372 (1985).

[9]R. Tao and F. D. M. Haldane, Phys. Rev. B **33**, 3844 (1986).

[10]J. E. Avron, R. Seiler, and L. Yaffe, Commun. Math. Phys. (to be published).

[11]J. E. Avron, A. Raveh, and B. Zur, to be published.

[12]P. G. de Gennes, C. R. Acad. Sci., Ser. B **292**, 9, 279 (1981); R. Rammal and G. Toulouse, Phys. Rev. Lett. **49**, 1194 (1982); S. Alexander, Phys. Rev. B **27**, 1541 (1983); E. Domany *et al.*, Phys. Rev. B **26**, 3110 (1983).

[13]The canonical choice of $H(\phi)$ is not periodic in ϕ. There is, however, a choice that makes $H(\phi)$ periodic. See Ref. 7 for more details.

[14]J. von Neumann and E. Wigner, Phys. Z. **30**, 467 (1929).

[15]P. W. Anderson *et al.*, Phys. Rev. B **22**, 3519 (1980); B. Shapiro, Phys. Rev. Lett. **50**, 747 (1983).

VOLUME 60, NUMBER 25 PHYSICAL REVIEW LETTERS 20 JUNE 1988

Superconducting Ground State of Noninteracting Particles Obeying Fractional Statistics

R. B. Laughlin

Department of Physics, Stanford University, Stanford, California 94305, and
University of California, Lawrence Livermore National Laboratory, Livermore, California 94550

(Received 28 January 1988)

In a previous paper, Kalmeyer and Laughlin argued that the elementary excitations of the original Anderson resonating-valence-bond model might obey fractional statistics. In this paper, it is shown that an ideal gas of such particles is a new kind of superconductor.

PACS numbers: 74.65.+n, 05.30.−d, 67.40.−w, 75.10.Jm

In a recent Letter,[1] Kalmeyer and I proposed that the ground state of the frustrated Heisenberg antiferromagnet in two dimensions and the fractional quantum Hall state for bosons might be the "same," in the sense that the two systems could be adiabatically evolved into one another without crossing a phase boundary. Whether or not this is the case is not presently clear. Indeed, the existence of a spin-liquid state of *any* spin-$\frac{1}{2}$ antiferromagnet in two dimensions has not been demonstrated. However, the case for a phase boundary's not being crossed is sufficiently strong that it is appropriate to ask what the consequences would be if this occurred. Adiabatic evolution is a particularly useful concept in the study of fractional quantum Hall "matter." So long as the energy gap remains intact, the "charge" of its fractionally charged excitations remains *exact* and the concomitant long-range forces between them, their fractional statistics, remain operative. This is why the fractional quantum Hall effect is so stable and reproducible. The persistence of the gap under evolution of the fractional quantum Hall problem into the magnet problem would allow us to make exact statements about the magnet without knowing *anything* about its Hamiltonian. In particular, the excitation spectrum of the magnet would be almost identical to that proposed by Kivelson, Rokhsar, and Sethna,[2] and completely within the spirit of the Anderson resonating-valence-bond idea,[3,4] except for one crucial detail: Both the chargeless spin-$\frac{1}{2}$ excitations, the "spinons," and the charged spinless excitations, the "holons," would obey $\frac{1}{2}$ fractional statistics.[5,6] The purpose of this Letter is to point out that this overlooked property may well account for high-temperature superconductivity.

Kalmeyer and I found the magnetic analog of the charge-$\frac{1}{3}$ quasiparticle of the fractional quantum Hall effect to be a spin-$\frac{1}{2}$ excitation, well described qualitatively as a spin-down electron on site j surrounded by an otherwise featureless spin liquid. This particle is our version of the "spinon." Like the quasiparticle of the fractional quantum Hall state, it carries a "charge," that is, its spin, that is in a deep and fundamental sense fractional. In the limit that the antiferromagnetic interactions are turned off, the excitation spectrum of the magnet is purely bosonic. Spin-$\frac{1}{2}$ particles occur because these "elementary" excitations are fractionalized: Half the boson is deposited in the sample interior and half at the boundary. It was first pointed out by Halperin[6] that, in the fractional quantum Hall effect, the fractionalization of the electron charge e into the quasiparticle charge $\frac{1}{3}e$ causes the quasiparticle to obey $\frac{1}{3}$ fractional statics. That is, each quasiparticle acts as though it were a boson carrying a magnetic solenoid containing magnetic flux $\frac{1}{3} \times hc/e$. This fact, deduced by Halperin from the experimentally observed fractional quantum Hall hierarchical states, was later shown by me[7] to follow from the analytic properties of the quasiparticle wave functions. It arises physically because the states available to the multiquasiparticle system must be enumerated differently from those available to fermions or bosons. In other words, it comes from counting. Now, it is clear by inspection that the preferred nature of this representation does not care about the existence of a lattice. Thus the validity of our identification clearly predicts that spinons obey $\frac{1}{2}$ statistics.

Let us now imagine doping this lattice with holes. The most natural way to do this, in my opinion, is first to make a spinon, thus fixing the spin on site j, and then remove the electron possessing that spin. It is necessary to make the spinon first because an electron cannot be removed before its spin state is known. If one simply rips an "up" electron from site j, one tacitly projects the ground state onto the set of states with the jth spin up, thus creating an excitation with spin 1. This may be thought of as a pair of spinons in close proximity. Unless the interaction between spinons is attractive and sufficiently large (Kalmeyer and I found it to be repulsive[1]), to make this "spin wave" will be more expensive energetically than to make an isolated spinon. Given that this occurs, the resulting spinless particle, the "holon," should also exhibit $\frac{1}{2}$ fractional statistics because it is a composite of a spinon and a fermion.

Assume now that we have a gas of such holons obeying fractional statistics. What are its properties expected to be? This question was addressed to some extent by Arovas *et al.*,[8] who computed the second virial coefficient of an ideal gas of particles obeying fractional statistics as a function of the fraction ν. Not surprisingly, they found a smooth interpolation between the case of fermions, which acts like a classical gas with *repulsive* interactions, and that of bosons, which acts like a classical gas with *attractive* interactions. Thus, if we insist on thinking of these particles as fermions, we must conclude that there is an enormous attractive force between them. This is also evident when one considers the low-temperature properties. Fermions at density ρ have a large degeneracy pressure, and thus a large internal energy, while bosons have neither. Since fractional-statistics particles are in between, they have, *vis-à-vis* fermions, attractive forces comparable in scale to the Fermi energy. It is also important that spinless particles obeying fractional statistics cannot undergo Bose condensation. They are not bosons. However, if the fraction is $\frac{1}{2}$, then *pairs* of particles are bosons.

There is therefore good reason to suspect that a gas of particles obeying $\frac{1}{2}$ statistics might actually be a superconductor with a charge-2 order parameter. Let us investigate this possibility by considering a gas of fractional-statistics particles described by the free-particle Hamiltonian

$$\mathcal{H}=\sum_j^N \frac{p_j^2}{2m}. \tag{1}$$

Any eigenstate of this Hamiltonian may be written in the manner

$$\Psi(z_1,\dots,z_N) = \left[\prod_{j<k}^N \frac{(z_j-z_k)^\nu}{|z_j-z_k|^\nu}\right]\Phi(z_1,\dots,z_N), \tag{2}$$

where z_j denotes the position of the jth particle in the x-y plane expressed as a complex number, $\nu=\frac{1}{2}$, and Φ is a Fermi wave function. This is the singular gauge transformation first discussed by Wilczek.[5] If we have an eigenstate Ψ satisfying $\mathcal{H}\Psi=E\Psi$, then Φ satisfies $\mathcal{H}'\Phi=E\Phi$ where

$$\mathcal{H}'=\sum_j^N \frac{1}{2m}|\mathbf{p}_j+\mathbf{A}_j|^2, \tag{3}$$

and

$$\mathbf{A}_j(\mathbf{r}_j)=\nu\sum_{k\neq j}\hat{\mathbf{z}}\times\mathbf{r}_{jk}/|\mathbf{r}_{jk}|^2. \tag{4}$$

Thus, in the Fermi representation, each particle appears to carry a magnetic solenoid with it as it moves around in the sample. The vector potential felt by a particle is then the sum of the vector potentials generated by all the other particles. Because particles obeying $\frac{1}{2}$ statistics behave like fermions, in the sense that they possess degeneracy pressure, let us attempt to solve this problem in the Hartree-Fock approximation: We make a variational wave function that is a single Slater determinant constructed of orbitals $\phi_j(z)$ and minimize the expected energy. The orbitals then obey equations of the form

$$\mathcal{H}_{\rm HF}\phi_j(z)=\lambda_j\phi_j(z), \tag{5}$$

where $\mathcal{H}_{\rm HF}$ is the first variation of $\langle\mathcal{H}'\rangle$ and λ_j is a Lagrange multiplier. The latter has the physical sense of a partial derivative of the total energy with respect to occupancy of the jth orbital. Since, in the mean-field sense, each particle must see a uniform density of magnetic solenoids carrying flux $\nu hc/e$, it is reasonable to guess the solution to be Landau levels, with the magnetic length a_0 related to the particle density ρ by $a_0^2=(2\pi\nu\rho)^{-1}$. Self-consistency is achieved when the lowest $1/\nu$ Landau levels are filled. Thus, the fractions $\nu=1,\frac{1}{2},\frac{1}{3},\dots$ are special cases in which a gap opens up in the fermionic spectrum.

Let us now test these equations in a case for which we know the answer, namely $\nu=1$, the noninteracting Bose gas. If the variational procedure describes this limit correctly, there is good reason to trust its predictions for $\nu=\frac{1}{2}$. Evaluating the self-consistent field with one Landau level filled, I obtain

$$\mathcal{H}_{\rm HF}=\tfrac{1}{2}E_0+(-E_0-\tfrac{1}{4})\Pi_0+\sum_{n\neq0}\left[n+\sum_{k=1}^{n}(-1)^k\binom{n}{k}\left(\frac{1}{4}\sum_{l=1}^{k}\frac{1}{l}-\frac{1}{2k}\right)+\frac{1}{4(n+1)}\right]\Pi_n, \tag{6}$$

with

$$E_0=\int_0^\infty r^{-1}[1-e^{-r^2/2}]e^{-\alpha r}dr, \tag{7}$$

in units of the equivalent cyclotron frequency $\hbar\omega_c=2\pi\nu(\hbar^2/m)\rho$, where Π_n denotes the projector onto the nth landau level, and α is a regulation parameter, effectively the inverse of the sample radius. Since $\mathcal{H}_{\rm HF}$ preserves Landau-level index, the state we guessed is a true variational minimum. Note, however, the logarithmic divergence in the Lagrange-multiplier spectrum, implying that the cost to inject either a "particle" or an "antiparticle" is arbitrarily large. This is absolutely the correct result. The noninteracting Bose gas has no low-lying fermionic excitations. The fact that these divergences are logarithmic suggests that the relevant excitations are actually quantum vortices. That this is, in fact, the case may be seen by our imagining an extra particle to be placed at the origin and calculating the expected current density $\langle\mathbf{J}(r)\rangle$. The current-density operator may be written $\mathbf{J}(r)=m^{-1}(\mathbf{p}+\mathbf{A}_{\rm old}+\Delta\mathbf{A})$, where $\mathbf{A}_{\rm old}$ is the vector potential in the absence of the extra particle and $\Delta\mathbf{A}$ is the vector potential generated by a solenoid at the origin. Since $\langle \mathbf{p}+\mathbf{A}_{\text{old}}\rangle = 0$, the current density must just be the particle density at r times $\Delta\mathbf{A}$, or a vortex of magnetic strength hc/e.

The expected energy of the ground state state is $N/4$ in these units. This is considerably higher than the correct answer of zero. This discrepancy is due to the fact that the wave function is forced by its construction to go to zero when the particles come together. It is thus more appropriate for the description of real helium than noninteracting bosons. It should also be noted that this behavior is actually required of the $\nu = \frac{1}{2}$ wave function. Let us observe finally that the broken symmetry characteristic of a superfluid is not expressly exhibited by the Hartree-Fock ground state. This is as expected. Is was shown by Bogoliubov[9] that the broken symmetry of a Bose gas is absent unless the bosons interact. All that is required for the symmetry to break is a weak interparticle repulsion and the presence in the "unperturbed" Bose gas of a collective mode dispersing quadratically with the mass of the bare particles. In the present case, it is easy to see that the variational solution possesses a collective mode that disperses quadratically. Since $\langle \mathcal{H}'\rangle/N$ is proportional to the particle density, the pressure is constant, and thus the bulk modulus is zero. It is a straightforward matter to calculate the mass of this mode by the magnetoexciton procedure of Kallin and Halperin.[10] My preliminary results give a value of approximately $\frac{1}{2}$ the bare mass. The precise value of this mass is not so important as the fact that it is of order unity. The collective mode may be thought of both as a density wave and as a magnetoexciton consisting of a hole in the lowest Landau level and a particle in the first excited Landau level, bound together by a logarithmic potential

Let us now turn to the case of interest, $\nu = \frac{1}{2}$. It is so similar to the $\nu = 1$ case that there is little to say. Assuming two Landau levels filled, I obtain

$$\mathcal{H}_{\text{HF}} = \tfrac{11}{16} + \tfrac{1}{4}E_0 + (-\tfrac{1}{2}E_0 - \tfrac{1}{8})\Pi_0 + (-\tfrac{1}{2}E_0 + \tfrac{29}{24})\Pi_1$$
$$+ \sum_{N\geq 2}\left[n + \sum_{k=1}^{n}\binom{n}{k}(-1)^k\left(\frac{1-k}{4}\sum_{l=1}^{k}\frac{1}{l} - \frac{1}{4k}\right) + \frac{3}{8(n+1)} - \frac{1}{8(n+2)}\right]\Pi_n. \quad (8)$$

Thus, we again have a true variational solution with vortexlike fermionic excitations. Repeating the arguments for $\nu = 1$, I find that the flux quantum to which the vortices correspond is $hc/2e$, exactly as expected of a charge-2 superfluid. Once again, a soft collective mode will mix into the ground state to break the symmetry when repulsive interactions are introduced. Thus, the ground state is a superfluid very similar to liquid helium except that the charge of its order parameter is 2.

While considerable work needs to be done to quantify this picture, some of its implications may be seen at a glance. By far the most important is that a normal-metal state, in the sense of Fermi-liquid theory, does not exist, just as Anderson[4] suggested. A corollary is that the occurrence of superconductivity does not have anything to do with self-consistent opening of an energy gap in the tunneling spectrum, as occurs in the BCS theory. Indeed, I find that tunneling cannot even be understood outside the context of the creation of spinons by the tunneling event. It should be noted that this is also consistent with Anderson's views.[11] A critical prediction is that an energy gap *must* occur in the spin-wave spectrum, the spin analog of the collective mode[12] of the fractional quantum Hall state. This is because the presence or absence of this gap is precisely the difference between the disordered and ordered states.

In summary, it is possible that high-T_c superconductivity can be accounted for by the following simple idea: The force mediated by the spins of the Mott insulator is not an attractive potential, but rather an attractive *vector* potential.

I gratefully acknowledge numerous helpful conversations with S. Kivelson, J. Sethna, V. Kalmeyer, L. Susskind, A. L. Fetter, P. W. Anderson, F. Wilczek, B. I. Halperin, J. R. Schrieffer, T. H. Geballe, M. R. Beasley, and A. Kapitulnik. This work was supported primarily by the National Science Foundation under Grant No. DMR-85-10062 and by the National Science Foundation–Materials Research Laboratory Program through the Center for Materials Research at Stanford University. Additional support was provided by the U.S. Department of Energy through the Lawrence Livermore National Laboratory under Contract No. W-7405-Eng-48.

[1]V. Kalmeyer and R. B. Laughlin, Phys. Rev. Lett. **59**, 2095 (1987).

[2]S. Kivelson, D. Rokhsar, and J. Sethna, Phys. Rev. B **35**, 8865 (1987).

[3]P. W. Anderson, Mater. Res. Bull. **8**, 153 (1973).

[4]P. W. Anderson, Science **236**, 1196 (1987); P. W. Anderson, G. Baskeran, Z. Zou, and T. Hsu, Phys. Rev. Lett. **58**, 2790 (1987).

[5]F. Wilczek, Phys. Rev. Lett. **49**, 957 (1982); F. Wilczek and A. Zee, Phys. Rev. Lett. **51**, 2250 (1983).

[6]B. I. Halperin, Phys. Rev. Lett. **52**, 1583 (1984).

[7]R. B. Laughlin, in *The Quantum Hall Effect*, edited by

R. E. Prange and S. M. Girvin (Springer-Verlag, New York, 1987), p. 233.

[8]D. P. Arovas, R. Schrieffer, F. Wilczek, and A. Zee, Nucl. Phys. **B251 [FS13]**, 117 (1985).

[9]N. N. Bogoliubov, J. Phys. (Moscow) **11**, 23 (1947); a good discussion of this may be found in A. L. Fetter and J. D. Walecka, *Quantum Theory of Many-Particle Systems* (McGraw-Hill, New York, 1971), p. 313.

[10]C. Kallin and B. I. Halperin, Phys. Rev. B **30**, 5655 (1984).

[11]P. W. Anderson, private communication.

[12]S. M. Girvin, A. H. MacDonald, and P. M. Platzman, Phys. Rev. Lett. **54**, 581 (1985).

J. Phys. A: Math. Gen. 17 (1984) 3459-3476. Printed in Great Britain

An example of phase holonomy in WKB theory

Michael Wilkinson
H H Wills Physics Laboratory, Royal Fort, Tyndall Avenue, Bristol, BS8 1TL, UK†

Received 4 June 1984

Abstract. This paper discusses the application of WKB theory to Harper's equation

$$\psi_{n+1}+\psi_{n-1}+2\alpha\cos(2\pi\beta n+\delta)\psi_n=E\psi_n,$$

in the case in which β is very close to a rational number, p/q.

The WKB wavefunction for this system is a vector valued quantity, proportional to an eigenvector $\boldsymbol{u}$ of a matrix $\hat{H}(x,p)$, which is parametrised by the phase space coordinates x and p. The complex phase of $\boldsymbol{u}$ is determined by a non-holonomic connection rule; when transported around a cycle and in phase space, $\boldsymbol{u}$ is multiplied by a phase factor $e^{i\gamma_c}$. This phase change manifests itself as a modification of the Bohr–Sommerfeld quantisation condition.

1. Introduction

This paper describes an unusual form of Bohr–Sommerfeld quantisation, involving a holonomy argument. As well as being interesting in its own right, the method discussed here can be applied to the difficult problem of finding the Bohr–Sommerfeld quantisation condition for Bloch electrons in a magnetic field. The system treated in this paper is a simplified model for this problem, which is often called Harper's equation. This model will be introduced in § 2; the remainder of this introduction will describe the principle of the method.

For the system considered, the WKB wavefunction can be thought of as a vector-valued quantity, given by

$$\psi(x)=A(x)\boldsymbol{u}(x)\exp\left(\frac{i}{\hbar}\int^{x}p(x')\,dx'\right) \tag{1.1}$$

where $p(x)$ and $A(x)$ are slowly varying functions, and the vector $\boldsymbol{u}$ is a solution of the eigenvalue equation

$$\hat{H}(x,p)\boldsymbol{u}=\varepsilon\boldsymbol{u}. \tag{1.2}$$

In equation (1.2), ε is the energy of the solution $\psi(x)$, and $\hat{H}$ a complex Hermitian matrix which is a function of two parameters x and p. Since the energy $E=\varepsilon(x,p)$ is a constant for a given solution, equation (1.2) gives both $\boldsymbol{u}$ and p as functions of x, as in (1.1). The curves in the x-p plane defined by $E=\varepsilon(x,p)=$ constant are called phase trajectories. When the phase trajectories given by (1.2) are closed orbits, then a solution $\psi(x)$ must remain single-valued when it is traced around the phase trajectory.

† Address after September 1st 1984: Department of Physics, California Institute of Technology, Pasadena, California 91125, USA

This condition is only satisfied for certain values of E, which are determined by a Bohr–Sommerfeld quantisation condition.

In equation (1.2), the eigenvector $\boldsymbol{u}(x, p)$ is determined only up to a complex-valued multiplying constant, or, if $\boldsymbol{u}$ is assumed to be normalised, up to a complex phase factor $e^{i\theta}$. This phase factor can be determined by requiring that the amplitude $A(x)$ in (1.1) be real. Given this condition on $A(x)$, it will be shown how the WKB theory for the system leads to a connection formula, by means of which the vector $\boldsymbol{u}$ can be transported through the phase space with its phase fully determined. It turns out that this phase connection is non-holonomic, so that when $\boldsymbol{u}$ is transported clockwise around a closed circuit in phase space, it is multiplied by a phase factor $e^{i\gamma}$.

This phase factor affects the Bohr–Sommerfeld quantisation condition. Consider the phase change of the solution (1.1) after making one circuit of a closed phase trajectory. This has contributions from the oscillatory term, from a pair of turning points where $p(x)=0$ and $A(x)$ diverges, plus a contribution $\gamma(E)$ from the phase factor evaluated for a phase trajectory of energy E. The condition for the wavefunction ψ to be single valued is therefore

$$2\pi n = \frac{1}{\hbar}\oint_{\varepsilon=E_n} p(x)\,dx + \frac{\pi}{2} + \frac{\pi}{2} + \gamma(E_n), \tag{1.3}$$

or

$$\oint_{\varepsilon=E_n} p(x)\,dx = [2\pi(n+\tfrac{1}{2}) - \gamma(E_n)]\hbar. \tag{1.4}$$

This equation (1.4) is the Bohr–Sommerfeld quantisation condition determining the eigenvalues E_n of the system.

The plan of this paper is as follows. Section 2 introduces the system under consideration and discusses how WKB theory can be applied to this system. Section 3 derives an asymptotic formula for the product of a string of slowly varying transfer matrices. Section 4 applies this formula to the WKB problem for Harper's equation, and § 5 obtains the Bohr–Sommerfeld quantisation condition. Section 6 summarises the theoretical results and compares them with numerical values, and § 7 discusses the connections between this work and recent work on adiabatic theory and the quantised Hall effect.

2. WKB analysis of Harper's equation

The system analysed in this paper is Harper's equation

$$\psi_{n+1} + \psi_{n-1} + 2\alpha\cos(2\pi\beta n + \delta)\psi_n = E\psi_n, \tag{2.1}$$

which is frequently used in models for Bloch electrons in a magnetic field, and as a model for electrons in an incommensurate potential (Harper 1955, Simon 1982). As pointed out by Sokoloff (1981), solutions of (2.1) can be obtained by a WKB method whenever β is sufficiently close to a rational number, p/q (where p and q are coprime integers). The condition for WKB theory to be applicable is

$$|q^2\Delta\beta| \ll 1, \qquad \Delta\beta = \beta - p/q, \tag{2.2}$$

and for almost all β, there exist values of p/q for which $|q^2\Delta\beta|$ is arbitrarily small. This follows from a property of continued fractions (Khinchin 1964).

Before describing how WKB theory can be applied to (2.2), it will be useful to consider the case $\beta = p/q$, so that the coefficients of the difference equation (2.1) are periodic with period q. In this case, therefore, Bloch's theorem applies and exact solutions can be obtained; the Bloch solution has the form

$$\psi_n = e^{ikn} u_n(\delta, k), \tag{2.3}$$

where u_n is periodic with period q

$$u_{n+q} = u_n. \tag{2.4}$$

This result can also be written in terms of a set of Fourier amplitudes for u_n

$$\psi_n = e^{ikn} \sum_{m=0}^{q-1} a_m \exp(2\pi i pmn/q), \tag{2.5}$$

which will prove more useful for some purposes.

From equation (2.1), it can be seen that the q-component vectors u_n or a_m can be determined as eigenvectors of a $q \times q$ complex Hermitian matrix. To distinguish these matrices and vectors from some two-component vectors and 2×2 matrices which will be introduced later, a quantum mechanical notation will be used;

$$\hat{H}(\delta, k) | u(\delta, k) \rangle = \varepsilon | u(\delta, k) \rangle, \tag{2.6}$$

where the matrix elements of $\hat{H}$ are given by

$$\begin{pmatrix} 2\alpha\cos(2\pi\beta+\delta) & e^{ik} & & & & e^{-ik} \\ \cdot & \cdot & \cdot & & 0 & \\ & e^{-ik} & 2\alpha\cos(2\pi\beta n+\delta) & \cdot & e^{ik} & \\ & 0 & & \cdot & \cdot & \cdot \\ e^{ik} & & & & e^{-ik} & 2\alpha\cos(2\pi\beta q+\delta) \end{pmatrix} \begin{pmatrix} u_1 \\ \vdots \\ u_n \\ \vdots \\ u_q \end{pmatrix} = \varepsilon \begin{pmatrix} u_1 \\ \vdots \\ u_n \\ \vdots \\ u_q \end{pmatrix}. \tag{2.7}$$

It is fairly easy to show that the eigenvalues E are periodic in both δ and k with period $2\pi/q$; in fact the q eigenvalues are given by the equation

$$f(E) = \cos qk + \alpha^q \cos q\delta, \tag{2.8}$$

where f is a qth degree polynomial (Wilkinson 1984). The q different sheets of $\varepsilon(\delta, k)$ normally do not touch each other. When q is even, however, one pair of sheets of $\varepsilon(\delta, k)$ does touch at isolated points in the $\delta - k$ plane (Bellissard and Simon 1982).

There is another, complementary, method for analysing equation (2.1) when β is rational (i.e. $\Delta\beta = 0$); this is the transfer matrix method. It is easy to see that equation (2.1) can be written in the form

$$\begin{pmatrix} \psi_{n+1} \\ \psi_n \end{pmatrix} = \tilde{T}(x_n, E) \begin{pmatrix} \psi_n \\ \psi_{n-1} \end{pmatrix}, \tag{2.9}$$

where

$$\tilde{T}(x, E) = \begin{pmatrix} E - 2\alpha\cos x & -1 \\ 1 & 0 \end{pmatrix},$$

$$x_n = 2\pi\beta n + \delta. \tag{2.10}$$

Consider a transfer matrix $\tilde{M}(x, E)$ describing a 'jump' of q steps

$$\begin{pmatrix} \psi_{(n+1)q+1} \\ \psi_{(n+1)q} \end{pmatrix} = \tilde{M}(x_n, E)\begin{pmatrix} \psi_{nq+1} \\ \psi_{nq} \end{pmatrix} \tag{2.11}$$

where now

$$\tilde{M}(x, E) = \tilde{T}[x+(q-1)\beta, E] \ldots \tilde{T}(x+\beta, E)\tilde{T}(x, E),$$
$$x_n = 2\pi\beta qn + \delta = 2\pi pn + \delta. \tag{2.12}$$

Now the transfer matrix $\tilde{M}$ is independent of n. The eigenvalue condition on E is then just that the eigenvalues of the transfer matrix $\tilde{M}$ lie on the unit circle. Since $\tilde{T}$, and therefore $\tilde{M}$, both satisfy

$$\det \tilde{T} = \det \tilde{M} = 1, \tag{2.13}$$

this condition becomes

$$2\cos k = \mathrm{Tr}\, \tilde{M}(\delta, E). \tag{2.14}$$

Having found the eigenvalues $E(\delta, k)$ using (2.14), the wavefunctions can be generated by means of the formula (2.9).

To summarise: there are two approaches to solving (2.1) when β is rational; one, which will be termed the Bloch picture, involves solving a $q \times q$ Hermitean eigenvalue equation, the other, which will be termed the Floquet picture, involves considering products of q 2×2 transfer matrices. The rest of this section will show how WKB methods can be applied when β is close to a rational number. First the application of the WKB method within the Bloch picture will be described. This has previously been attempted by Sokoloff (1981); it cannot be carried through to yield a full solution, but is worth describing since it is easier to understand because it is closer to ordinary WKB methods. Finally, the application of the WKB method in the Floquet picture will be described. This is harder to visualise, but does lead to a full solution of the problem.

When $\Delta\beta$ is small, the solution must 'locally' look like a solution of the form (2.3). On a 'global' scale, however, there is a slow change in the phase parameter δ; in the region of the amplitude ψ_n the effective phase δ' is

$$\delta' = \delta + n\hbar/q, \tag{2.15}$$

where

$$\hbar = 2\pi\Delta\beta q. \tag{2.16}$$

The symbol $\hbar$ is used in (2.15) because this quantity will be the small parameter of the WKB theory. In the neighbourhood of the amplitude ψ_n, the solution resembles a solution of the form (2.3) or (2.4) with δ replaced by δ'.

The Bloch wavevector, k, now varies slowly with n: the energy E is still given by equation (2.7), and is a constant for a given solution, so that (2.7) defines an implicit relationship between k and δ. The energy E should now be considered to depend on $\hbar$ as well as δ and k,

$$E = \varepsilon(\delta', k; \hbar) = \varepsilon_0(\delta', k) + \hbar\varepsilon_1(\delta', k). \tag{2.17}$$

since β in (2.7) depends on $\hbar$. The term of order $\hbar$ in (2.17) will be important in what follows.

Following Sokoloff (1981), equation (2.1) is written in the form

$$\psi(x+\hbar/q)+\psi(x-\hbar/q)+2\alpha\cos(2\pi px/\hbar+x-x_0)\psi(x)=E\psi(x), \tag{2.18}$$

where

$$x_0=(2\pi p\delta/q) \bmod 2\pi, \qquad \psi_n=\psi(x_n), \qquad x_n=n\hbar/q+\delta. \tag{2.19}$$

By comparison with equation (2.5), this suggests a trial solution of the form

$$\psi(x)=A(x)\exp(\mathrm{i}S(x)/\hbar)\sum_{m=0}^{q-1}a_m(x)\exp(2\pi \mathrm{i}pmx/\hbar). \tag{2.20}$$

This trial solution corresponds to the abstract solution introduced in equation (1.1). The role of the vector $\boldsymbol{u}$ in (1.1) is played by the set of Fourier coefficients a_m in (2.20). These coefficients are easily shown, by substituting (2.20) into (2.18), to satisfy the equation

$$\alpha\,\mathrm{e}^{-\mathrm{i}x}a_{m+1}+\alpha\,\mathrm{e}^{\mathrm{i}x}a_{m-1}+2\cos[(2\pi pm+S')/q]a_m=Ea_m, \tag{2.21}$$

which is an eigenvalue equation for E corresponding to (1.2).

The next step in Sokoloff's approach to the WKB theory of Harper's equation is to expand $\psi(x\pm\hbar/q)$ in (2.20) in powers of $\hbar$, and insert the result into (2.18). Unfortunately, this does not lead to a consistent result; if the calculation is carried out correctly it is found that q independent equations are obtained which $A(x)$ should satisfy. (The solution which Sokoloff obtains for $A(x)$ is easily found to be incorrect.)

It turns out that a full solution of the WKB problem can be obtained in the Floquet picture, however. The transfer matrices $\tilde{M}(x, E)$ introduced in (2.11) are now no longer independent of n, but provided (2.2) is satisfied, these transfer matrices are at least slowly varying. It is possible to calculate the product of a string of slowly varying matrices;

$$\tilde{G}(x,x';\hbar)=\tilde{M}_E(x,\hbar)\tilde{M}_E(x-\hbar,\hbar)\ldots\tilde{M}_E(x'+\hbar,\hbar)\tilde{M}_E(x',\hbar),$$

$$\tilde{M}_E(x,\hbar)=\tilde{T}\{x+2\pi p(q-1)/q+[(q-1)/q]\hbar, E\}\ldots\tilde{T}(x+2\pi p/q+\hbar/q, E)\tilde{T}(x,E),$$

$$\tilde{T}(x,E)=\begin{pmatrix}E-2\alpha\cos x & -1\\ 1 & 0\end{pmatrix}. \tag{2.22}$$

A simple formula for the product $\tilde{G}(x,x';\hbar)$ will be derived in § 3.

Before going on to discuss the WKB theory in detail, it will be helpful to describe briefly the final results of the calculation.

Suppose that β is a low denominator rational number, $\beta_0=p/q$. The spectrum then consists of q bands (the centre two bands touch if q is even), and E is a periodic function of the Bloch wavevector k and position parameter δ; $E=\varepsilon(\delta,k)$, with q branches, one for each band.

When β is close to β_0, $\beta=\beta_0+\hbar/2\pi q$, then WKB theory can be applied and δ and k become the position and momentum coordinates of the phase-space ($\delta\to x$, $qk\to p=S'$). The dispersion relation $E=\varepsilon(\delta,k)$ for a given band becomes the classical Hamiltonian; $H(x,p)=\varepsilon(x,kq)$.

When the phase trajectories (contours of $E=H(x,p)$) are closed orbits, the energies of the eigenstates are restricted by a Bohr–Sommerfeld quantisation condition. Some contours of a typical $H(x,p)$ are shown in figure 1 for the case $\alpha=1$, when (by symmetry) all the phase trajectories are closed orbits. Each of the q bands of the

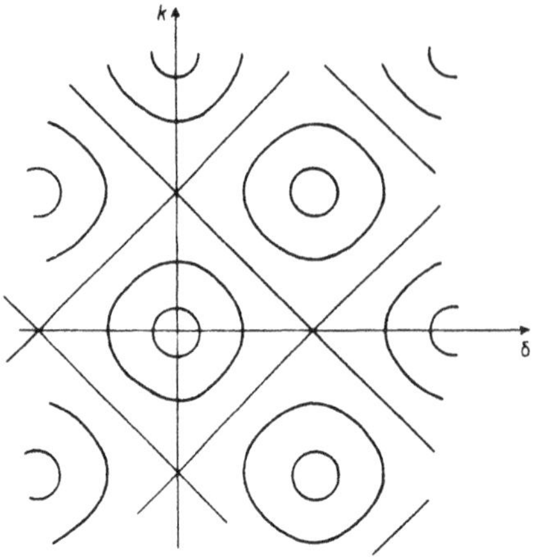

Figure 1. Phase trajectories of the classical Hamiltonian when $\alpha = 1$. All the phase trajectories are closed orbits.

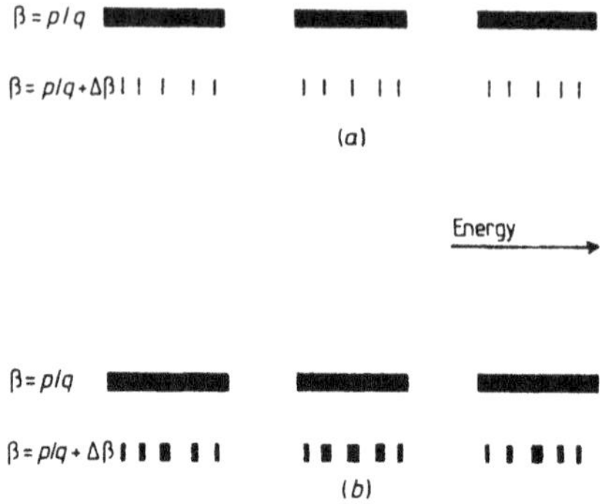

Figure 2. (*a*) Bloch bands of the commensurate system, $\beta = p/q$, become Bohr–Sommerfeld quantised levels when β is changed by a small amount. (*b*) The Bohr–Sommerfeld quantised levels are in practice broadened slightly by tunnelling effects.

spectrum is then split into a number of Bohr–Sommerfeld quantised levels. This situation is shown schematically in figure 2(*a*).

These Bohr–Sommerfeld quantised levels are not truly discrete; since the classical Hamiltonian $H(x, p)$ is a periodic function of x and p, the energy levels are broadened slightly by tunnelling between degenerate states, as illustrated schematically in figure 2(*b*). These tunnelling effects are discussed in detail in Wilkinson (1984) for the case $\beta_0 = 0$. Since the broadening of the levels due to tunnelling vanishes very rapidly as $\hbar \to 0$, as

$$E_{\text{tunnelling}} \simeq \exp(-\text{constant}/\hbar), \tag{2.23}$$

it does still make sense to obtain an asymptotic formula for the Bohr–Sommerfeld quantisation condition.

3. Adiabatic matrix strings

This section derives a formula for the product of a string of slowly varying matrices, $\tilde{M}$. An asymptotic formula is obtained for the product $\tilde{G}$, defined by

$$\tilde{G}(x, x'; \hbar) = \prod_{\substack{n=0 \\ x''=x'+n\hbar}}^{\mathcal{N}} \tilde{M}(x'', \hbar), \qquad \mathcal{N} = \frac{|x - x'|}{\hbar}, \tag{3.1}$$

i.e.

$$\tilde{G}(x, x'; \hbar) = \tilde{M}(x, \hbar)\tilde{M}(x - \hbar, \hbar) \ldots \tilde{M}(x' + \hbar, \hbar) M(x', \hbar),$$

in the limit $\hbar \to 0$. Note that the matrices $\tilde{M}$ depend on the slowness parameter of the adiabatic change, $\hbar$, as well as the variable x.

It is assumed that the matrices $\tilde{M}(x,\hbar)$ can be diagonalised

$$\tilde{M}(x,\hbar)=\tilde{X}^{-1}(x,\hbar)\tilde{D}(x,\hbar)\tilde{X}(x,\hbar), \tag{3.2}$$

where $\tilde{D}(x,\hbar)$ is diagonal, and that all the eigenvalues are distinct and lie on the unit circle.

Now (3.1) can be written

$$\tilde{G}(x,x';\hbar)=\tilde{X}^{-1}(x,\hbar)\tilde{g}(x,x';\hbar)\tilde{X}(x',\hbar) \tag{3.3}$$

where

$$\begin{aligned}\tilde{g}(x,x';\hbar)=\tilde{D}(x,\hbar)[\tilde{1}+\hbar\tilde{V}(x-\hbar,\hbar)]\tilde{D}(x-\hbar,\hbar)\\ \times[\tilde{1}+\hbar\tilde{V}(x-2\hbar,\hbar)]\ldots[\tilde{1}+\hbar\tilde{V}(x',\hbar)]\tilde{D}(x',\hbar),\end{aligned} \tag{3.4}$$

and

$$\tilde{1}+\hbar\tilde{V}(x,\hbar)=\tilde{X}(x+\hbar,\hbar)\tilde{X}^{-1}(x,\hbar). \tag{3.5}$$

Throughout the calculation presented here, it will be sufficient to use the approximation

$$\begin{aligned}\tilde{V}(x)&=[\mathrm{d}\tilde{X}(x,\hbar)/\mathrm{d}x]\tilde{X}^{-1}(x,\hbar)+\mathrm{O}(\hbar)\\ &=[\mathrm{d}\tilde{X}(x,0)/\mathrm{d}x]\tilde{X}^{-1}(x,0)+\mathrm{O}(\hbar).\end{aligned} \tag{3.6}$$

(The second equation of (3.6) shows that, when calculating $\tilde{V}$, the dependence of $\tilde{X}$ on $\hbar$ can be neglected and will not be shown in subsequent equations.) Now, using the notation

$$\tilde{g}_0(x,x';\hbar)=\tilde{D}(x,\hbar)\tilde{D}(x-\hbar,\hbar)\ldots\tilde{D}(x'+\hbar,\hbar)\tilde{D}(x',\hbar), \tag{3.7}$$

and ordering the expansion of equation (3.4) in powers of $\hbar$

$$\tilde{g}(x,x';\hbar)=\tilde{g}_0(x,x';\hbar)+\hbar\sum_{\substack{n=0\\x''=x'+n\hbar}}^{\mathcal{N}}\tilde{g}_0(x,x''+\hbar)\tilde{V}(x'')\tilde{g}_0(x'',x';\hbar)+\mathrm{O}(\hbar^2V^2), \tag{3.8}$$

leads to an exact but implicit equation for $\tilde{g}$

$$\tilde{g}(x,x';\hbar)=\tilde{g}_0(x,x';\hbar)+\hbar\sum_{\substack{n=0\\x''=x'+n\hbar}}^{\mathcal{N}}\tilde{g}_0(x,x''+\hbar,\hbar)\tilde{V}(x'')\tilde{g}(x,x';\hbar). \tag{3.9}$$

An asymptotic solution of (3.9) will now be sought in the form

$$\tilde{g}(x,x';\hbar)=\tilde{f}(x,x')\tilde{g}_0(x,x';\hbar), \tag{3.10}$$

where $\tilde{f}(x,x')$ is diagonal. This trial solution is an adiabatic approximation; it expresses the expectation that when $\hbar$ is small, so that $\tilde{M}$ varies slowly, an eigenvector $\boldsymbol{u}_i(x')$ of $M(x',\hbar)$ becomes, upon multiplying by $G(x,x';\hbar)$, the corresponding eigenvector $\boldsymbol{u}_i(x)$ of $M(x,\hbar)$. Before going any further, it is useful to define diagonal matrices $\tilde{S}$ and $\tilde{v}$ as follows

$$\tilde{D}(x,\hbar)=\exp[\mathrm{i}\tilde{S}'(x,\hbar)], \tag{3.11}$$

$$\tilde{S}(x,x';\hbar)=\int_{x'}^{x}\mathrm{d}x''\,\tilde{S}'(x'',\hbar),$$

$$\tilde{v}_{ij}(x)=\begin{cases}\tilde{V}_{ij}(x) & i=j\\ 0 & i\neq j.\end{cases} \tag{3.12}$$

Now, substituting (3.10) into (3.9), and making use of the definitions (3.11), (3.12)

$$
\begin{aligned}
g_{ij}(x, x'; \hbar) - g_{0ij}(x, x'; \hbar)
&= \hbar \sum_{x''} g_{0ij}(x, x''+\hbar; \hbar)\, V_{ij}(x'')\, g_{0ij}(x'', x'; \hbar) f_j(x'', x') \\
&= \hbar \sum_{x''} \exp\left(i \sum_{u=x''+n\hbar}^{x} S_i'(u, \hbar) \right) V_{ij}(x'') f_j(x'', x') \exp\left(i \sum_{u=x'+n\hbar}^{x''} S_j'(u, \hbar) \right) \\
&= \hbar \exp[\tfrac{1}{2}i(S_i'(x) + S_j'(x'))] \sum_{x''} \exp[(i/\hbar) S_i(x, x''+\hbar/2; \hbar)]\, V_{ij}(x'') f_j(x'', x') \\
&\quad \times \exp[(i/\hbar) S_j(x''+\hbar/2, x'; \hbar)] + O(\hbar) \\
&= \exp[\tfrac{1}{2}i(S_i'(x) + S_j'(x'))] \times \int_{x'}^{x} dx''\; V_{ij}(x'') f_j(x'', x') \\
&\quad \times \exp[(i/\hbar)(S_i(x, x''; \hbar) + S_j(x'', x'; \hbar))] + O(\hbar).
\end{aligned} \tag{3.13}
$$

For terms with $i \neq j$, the integrand in (3.13) contains a rapidly oscillating term and gives a contribution of $O(\hbar)$, whereas for $i = j$ it gives a finite contribution. Therefore

$$
\tilde{g}(x, x'; \hbar) = \tilde{g}_0(x, x'; \hbar) \left[\tilde{1} + \int_{x'}^{x} dx''\, \tilde{f}(x'', x') \tilde{v}(x'') \right] + O(\hbar), \tag{3.14}
$$

since only the diagonal elements of (3.13) remain. This justifies the use of the adiabatic approximation (3.10). From (3.10) and (3.14), $\tilde{f}$ satisfies

$$
\tilde{f}(x, x') = \tilde{1} + \int_{x'}^{x} dx''\, \tilde{f}(x'', x') \tilde{v}(x''), \tag{3.15}
$$

which gives

$$
\tilde{f}(x, x') = \exp\left(\int_{x'}^{x} dx''\; \tilde{v}(x'') \right), \tag{3.16}
$$

where both $\tilde{f}$ and $\tilde{v}$ are diagonal. Therefore the central result of this section, the formula for $\tilde{G}(x, x'; \hbar)$, is found to be

$$
\begin{aligned}
\tilde{G}(x, x'; \hbar) &= \tilde{X}^{-1}(x) \exp\left(\int_{x'}^{x} dx''\; \tilde{v}(x'') \right) \tilde{g}_0(x, x'; \hbar) \tilde{X}(x') + O(\hbar), \\
\tilde{g}_0(x, x'; \hbar) &= \exp[\tfrac{1}{2}i(\tilde{S}'(x) + \tilde{S}'(x'))] \exp[(i/\hbar) S(x, x'; \hbar)] + O(\hbar).
\end{aligned} \tag{3.17}
$$

In this result the dependences of some quantities on $\hbar$ have not been shown, since they are not important at this order of accuracy.

The remainder of this section will discuss a slight simplification of (3.17) which is possible when the transfer matrices preserve some quantity j, which will be called the current. This is usually the case in one-dimensional quantum mechanical problems. For any two vectors $\boldsymbol{\phi}$, $\boldsymbol{\psi}$ the current j is given by

$$
j_{\phi,\psi} = \boldsymbol{\phi}^{*\mathrm{T}} \tilde{J} \boldsymbol{\psi} \tag{3.18}
$$

(where $\tilde{J}$ is a constant matrix). If j is preserved under the action of a transfer matrix $\tilde{M}$, then

$$
j_{M\phi, M\psi} = (\tilde{M}\boldsymbol{\phi})^{\mathrm{T}*} \tilde{J} (\tilde{M}\boldsymbol{\psi}) = j_{\phi,\psi}, \tag{3.19}
$$

so that M satisfies

$$\tilde{M}^{T*}\tilde{J}\tilde{M}=\tilde{J}. \tag{3.20}$$

The transfer matrices introduced in § 2 will be shown later to have this property. It can be shown that properties of $\tilde{M}$ that are preserved on multiplying matrices together are indeed preserved by the formula (3.17) for the product $\tilde{G}$; i.e. if it is real, unimodular, or satisfies the current conservation (3.20) then the approximate formula (3.17) for the product also has these properties.

It will be useful to get an expression for $\tilde{f}(x, x')$ in terms of the eigenvectors of $\tilde{M}(x, \hbar)$. Let $\boldsymbol{u}_i(x)$ and $\boldsymbol{v}_i(x)$ be right and left eigenvectors of $\tilde{M}(x, \hbar)$:

$$\tilde{M}\boldsymbol{u}_i=\lambda_i\boldsymbol{u}_i, \qquad \boldsymbol{v}_i\tilde{M}=\lambda_i\boldsymbol{v}_i. \tag{3.21}$$

The $\boldsymbol{u}_i$ are proportional to the columns of $\tilde{X}^{-1}$ and the $\boldsymbol{v}_i$ to the rows of $\tilde{X}$, so that

$$\boldsymbol{u}_i\cdot\boldsymbol{v}_j=N_i\delta_{ij}. \tag{3.22}$$

For matrices $\tilde{M}$ that satisfy (3.20), a useful relationship can be found connecting the left and right eigenvectors: from (3.20) it is easy to show that

$$(\tilde{J}\boldsymbol{u}_i)^{*T}\tilde{M}=(\tilde{J}\boldsymbol{u}_i)^{*T}\lambda_i^{*-1}, \tag{3.23}$$

so that the eigenvalues and eigenvectors come in pairs, related by

$$\lambda_{i'}=\lambda_i^{*-1}, \qquad \boldsymbol{v}_{i'}=(\tilde{J}\boldsymbol{u}_i)^{*T}. \tag{3.24}$$

Since the eigenvalues λ_i are all on the unit circle for the transfer matrices considered here,

$$\boldsymbol{v}_i=(\tilde{J}\boldsymbol{u}_i)^{*T}. \tag{3.25}$$

Now collecting equations (3.16), (3.12), (3.6), (3.22), (3.25), a simple and useful formula can be given for $f_i\ (x, x')$, in terms of the eigenvector $\boldsymbol{u}_i(x)$

$$\begin{aligned} f_i(x,x') &= \exp\left[\int_{x'}^{x} dx''(d\tilde{X}/dx|_{x''}\tilde{X}^{-1}(x''))_{ii}\right] \\ &= \exp\left[-\int_{x'}^{x} dx''(\boldsymbol{u}^{*T}\tilde{J}\,d\boldsymbol{u}/dx)/(\boldsymbol{u}^{T*}\tilde{J}\boldsymbol{u})|_{x''}\right] \\ &= \exp\left[-\int_{x'}^{x} dx''\, j_{u(du/dx)}/j_{uu}|_{x''}\right], \end{aligned} \tag{3.26}$$

where $\boldsymbol{u}=\boldsymbol{u}_i(x'')$ is the ith eigenvector of $M\ (x, \hbar)$.

4. Solution of the WKB problem for Harper's equation

This section uses the central results of § 3, equations (3.17) and (3.26), to solve the problem of finding a satisfactory WKB theory for Harper's equation (2.1).

The WKB solution required is of the form of equation (1.1)

$$\psi(x)=A(x)\boldsymbol{u}(x)\exp\left(\frac{i}{\hbar}\int^{x}p(x')\,dx'\right). \tag{4.1}$$

Locally, the solution can be described by a set of q amplitudes, either amplitudes of a Bloch function u_n, or, equivalently, the Fourier components a_m of the Bloch function, as in equation (2.5). Alternatively, in the Floquet picture, $\boldsymbol{\psi}$ and $\boldsymbol{u}$ would be specified by just two amplitudes on a pair of adjacent lattice sites.

In the Floquet picture, by equation (3.17), if the wavefunction $\boldsymbol{\psi}(x')$ is an eigenvector $\boldsymbol{u}_i(x')$ at position x', then at x it is given by

$$\begin{aligned}\boldsymbol{\psi}(x) &= g_i(x, x'; \hbar)\boldsymbol{u}_i(x)\\ &= f_i(x, x')g_{0i}(x, x'; \hbar)\boldsymbol{u}_i(x).\end{aligned} \tag{4.2}$$

The $g_{0i}(x, x'; \hbar)$ term in (4.2), which can be written

$$g_{0i}(x, x'; \hbar) = \exp[\tfrac{1}{2}\mathrm{i}(S_i'(x) + S_i'(x'))] \exp\left((\mathrm{i}/\hbar)\int_x^{x'} \mathrm{d}x'' S_i'(x'', \hbar)\right), \tag{4.3}$$

can be associated with the oscillatory term in (4.1), and the $f_i(x, x')$ term with the amplitude $A(x)$.

The derivative $S'(x, \hbar)$ of the 'action' $S(x, x', \hbar)$ therefore plays the role of the momentum p of the phase space. The energy E is given as a function of $x(=\delta)$ and $S'(=kq)$ by (2.6) and (2.7) in the Bloch picture, or alternatively by (2.14) in the Floquet picture. The first order term in $\hbar$ in the relations between E, x and S' must be retained since $\hbar$ appears in the denominator of the argument of the exponential in (4.3).

It can be seen that the transfer matrix (2.10) satisfies the current conservation property (3.20), with

$$\tilde{J} = \begin{pmatrix} 0 & i \\ -i & 0 \end{pmatrix}. \tag{4.4}$$

The current $j_{\phi,\psi}$ can equivalently be calculated for the corresponding q dimensional vectors $|\phi\rangle$, $|\psi\rangle$

$$j_{\phi,\psi} = \langle\phi|\hat{J}|\psi\rangle, \tag{4.5}$$

where in the direct representation (by means of the Bloch function, u_n) the matrix elements of $\tilde{J}$ are given by

$$J_{nn'} = \frac{\mathrm{i}}{q}\begin{pmatrix} 0 & \mathrm{e}^{-\mathrm{i}S'/q} & & & & \mathrm{e}^{\mathrm{i}S'/q} \\ \cdot & \ddots & \ddots & & 0 & \\ & \cdot & \mathrm{e}^{\mathrm{i}S'/q} & 0 & \mathrm{e}^{-\mathrm{i}S'/q} & \\ & & & \ddots & \ddots & \\ & 0 & & & \ddots & \ddots \\ \mathrm{e}^{-\mathrm{i}S'/q} & & & & \mathrm{e}^{\mathrm{i}S'/q} & 0 \end{pmatrix}. \tag{4.6}$$

It is also useful to note that the eigenvalues and eigenvectors of the transfer matrices $\tilde{M}(x, E, \hbar)$ come in complex conjugate pairs, corresponding to points related by $\pm S'$ in the x–S' plane. The eigenvalues and eigenvectors are therefore real when $S' = 0$:

$$\boldsymbol{u}(x, S') = \boldsymbol{u}^*(x, -S'). \tag{4.7}$$

If the eigenvectors $\boldsymbol{u}_i$ are given as functions of x, then $A(x)$ for the solution (4.2) is given by (cf (3.26))

$$A(x) = f_i(x, x') = \exp\left(-\int_{x'}^{x} \mathrm{d}x''\, j_{u(\mathrm{d}u/\mathrm{d}x)}/j_{uu}\right), \tag{4.8}$$

and is in general complex. Alternatively $A(x)$ could be chosen to be real; this condition then defines a connection rule for the phases of the eigenvector $\boldsymbol{u}(x)$. If the $\boldsymbol{u}(x)$ are given with some arbitrary phase, then the correct phase is established by means of a transformation

$$\boldsymbol{u}(x) \to \boldsymbol{u}'(x) = \exp(\mathrm{i}\phi(x))\boldsymbol{u}(x), \tag{4.9}$$

so that the current matrix element $j_{u(du/dx)}$ is transformed according to the equation

$$j_{u\,(du/dx)} \to j_{u'\,(du'/dx)} = j_{u\,(du/dx)} + \mathrm{i}(\mathrm{d}\phi/\mathrm{d}x)j_{uu}. \tag{4.10}$$

Then by a suitable choice of $\phi(x)$, $j_{u(du/dx)}$ can be made to satisfy

$$\mathrm{Im}(j_{u'\,(du'/dx)}) = 0, \tag{4.11}$$

so that by (4.8), $A(x)$ is now real.

5. The Bohr–Sommerfeld quantisation rule

This section derives the Bohr–Sommerfeld quantisation rule (1.3), which is the condition for single-valuedness of the WKB solution under continuation around a closed phase trajectory in the x, S' plane.

In this section it will be useful to consider the eigenvector $\boldsymbol{u}'(x, S')$ to be a given single-valued function defined on the phase plane. If $A(x)$ is to be a real function, the eigenvector $\boldsymbol{u}'$ must be multiplied by a phase factor so that the modified eigenvector $\boldsymbol{u}(x, S')$ satisfies the connection formula (4.11), i.e.

$$\mathrm{Im}[(j_{u(\partial u/\partial x)})\Delta x + (j_{u(\partial u/\partial S')})\Delta S'] = \mathrm{Im}(j_{u\nabla u}) \cdot \Delta \boldsymbol{X} = 0, \tag{5.1}$$

for transport of u by a vector $\Delta \boldsymbol{X} = (\Delta x, \Delta S')$ in the phase plane. This connection (5.1) is non-holonomic, and on transporting $\boldsymbol{u}$ around a closed circuit C in phase space, it is multiplied by a phase factor $e^{i\gamma_c}$, given by

$$\begin{aligned}\gamma_c &= \mathrm{Im} \oint_C (j_{u'\nabla u'}/j_{u'u'}) \cdot \mathrm{d}\boldsymbol{X} \\ &= \mathrm{Im} \oint_C j_{u'(du'/dx)}/j_{u'u'}\,\mathrm{d}x.\end{aligned} \tag{5.2}$$

On transporting $\boldsymbol{u}$ around a phase trajectory of energy E, there is thus a phase change $e^{i\gamma(E)}$, where

$$\begin{aligned}\gamma(E) &= \mathrm{Im} \oint_{\varepsilon = E} (j_{u'\nabla u'}/j_{u'u'}) \cdot \mathrm{d}\boldsymbol{X} \\ &= \mathrm{Im} \oint_{\varepsilon = E} j_{u'(du'/dx)}/j_{u'u'}\,\mathrm{d}x.\end{aligned} \tag{5.3}$$

This phase change makes a contribution to the Bohr–Sommerfeld quantisation formula.

In order to describe correctly the continuation of the solution around the phase trajectory of energy E, it is necessary to consider carefully what happens at the classical turning points, where $S' = 0 \bmod 2\pi$. On the lines $S' = 0$, the current j_{uu} is zero, since by (4.7), $\boldsymbol{u}(x, 0)$ is real. In the neighbourhood of the line $S' = 0$, $j_{uu}(x, S')$ takes the form

$$j_{uu} = a(x)S' + \mathrm{O}(S'^2), \tag{5.4}$$

for some function $a(x)$. Now consider the form of $j_{u\nabla u}$ near the line $S'=0$. Consider the result

$$j_{u+\Delta u, u+\Delta u} = j_{uu} + 2\,\mathrm{Re}\, j_{u,\Delta u} + \mathrm{O}(\Delta u^2) \tag{5.5}$$

(which uses the fact that the current operator is Hermitian), and take $\boldsymbol{u}=\boldsymbol{u}(x,0)$ and $\boldsymbol{u}+\Delta\boldsymbol{u}=\boldsymbol{u}(x+\Delta x, \Delta S')$. Then using (5.5) and ignoring $\mathrm{O}(\Delta u^2)$

$$j_{u+\Delta u, u+\Delta u} = 2\,\mathrm{Re}\, j_{u\Delta u} = 2\,\mathrm{Re}\, j_{u\nabla u}\cdot \boldsymbol{\Delta X} = a(x+\Delta x)\Delta S' \tag{5.6}$$

so that, near $S'=0$

$$\mathrm{Re}(j_{u\nabla u}) = (\mathrm{O}, \tfrac{1}{2}a(x)) + \mathrm{O}(x) + \mathrm{O}(S'). \tag{5.7}$$

Now, from (4.8), the amplitude $A(x)$ (constrained to be real) is given by

$$A(x) = \exp\left(-\mathrm{Re}\int^x [j_{u(\mathrm{d}u/\mathrm{d}x)}/j_{uu}]|_{x'}\,\mathrm{d}x'\right). \tag{5.8}$$

Near the line $S'=0$, therefore,

$$A(x) \simeq \exp\left[\int^x -\tfrac{1}{2}a(x')/(a(x')S'(x'))\,.\,(\mathrm{d}S'/\mathrm{d}x')\,\mathrm{d}x'\right] \tag{5.9}$$

$$A(x) \simeq \mathrm{constant}[S'(x)]^{-1/2}.$$

Thus $A(x)$ diverges at a classical turning point, x_0, where $S'=0$. Near this point the form of the phase trajectory is given by

$$S'^2 = \mathrm{constant}(x-x_0), \tag{5.10}$$

so that as x_0 is approached from within the classically allowed region, $A(x)$ diverges as

$$A(x) \sim \mathrm{constant}(x-x_0)^{-1/4}. \tag{5.11}$$

(Of course there is not a real divergence of the exact solution, only in the WKB approximation; the assumption used in § 3 that the eigenvalues of the transfer matrix are distinct breaks down when $S'=0$.) The divergence of $A(x)$ given by (5.11) is of exactly the same type as is encountered in ordinary WKB problems at first-order turning points, and any of the usual arguments (e.g. continuation in the complex plane, see Landau and Lifshitz (1958)) show that an extra phase change of $\pi/2$ must be included for each of the two classical turning points of the phase trajectory.

The final contribution to the phase change of $\psi(x)$ is from the phase integral term: this is

$$\frac{1}{\hbar}\oint_{\varepsilon(x,S')=E} S'(x,\hbar)\,\mathrm{d}x. \tag{5.12}$$

As noted earlier, the correction to $\varepsilon(x,s')$ of first order in $\hbar$ must be retained when evaluating (5.12), since $\hbar$ appears in the denominator. Collecting together all these contributions to the phase gives the Bohr–Sommerfeld quantisation rule for the system.

$$2\pi n = \frac{1}{\hbar}\oint_{\varepsilon(x,S')=E} S'(x,\hbar)\,\mathrm{d}x + \pi + \gamma(E). \tag{5.13}$$

Finally there are two important points which must be mentioned. Firstly, because j_{uu} is zero on the line $S'=0$, the integrand in the formula for $\gamma(E)$ diverges at the classical turning point as $(x-x_0)^{-1/2}$. The integral $\gamma(E)$ remains finite, but does not tend to zero as E approaches a maximum or minimum of $\varepsilon(x, S')$, and the phase trajectory shrinks to a point. Instead, $\gamma(E)$ tends to a finite limit γ_0 at the top or bottom of a band. This limiting value of $\gamma(E)$ at the band edges is calculated in the appendix.

Secondly, there are some special cases which should be mentioned. When the rational number p/q to which β approximates is zero, then both the phase $\gamma(E)$ and the $\hbar$ dependent corrections to $\varepsilon(x, S')$ vanish, and the Bohr–Sommerfeld quantisation condition takes the usual form. This case is discussed in detail in Wilkinson (1984). There are also some simplifications which occur when $p/q=\frac{1}{2}$, and it is only for p/q with denominators greater than two that all the effects described in this paper are seen.

6. Summary and comparison with numerical results

In this section some comparisons will be made of eigenvalues calculated using the Bohr–Sommerfeld quantisation rule with those calculated exactly. Firstly, however, the important formulae are collected together and summarised.

The parameter β in equation (2.1) is written

$$\beta=\beta_0+\Delta\beta=p/q+\hbar/2\pi q. \tag{6.1}$$

The energy E is considered to be a function $\varepsilon(x, S')$ of the phase plane coordinates x and S'. The relationship between E, x and S' is, in the Bloch picture, given by the eigenvalue equation

$$\hat{H}(x, S')|u\rangle=E|u\rangle, \tag{6.2}$$

and the matrix elements of $\hat{H}$ are given by (2.7), with $\delta\to x$ and $k\to S'/q$:

$$H_{nn'}(x, S')$$

$$=\begin{pmatrix} 2\alpha\cos(x+2\pi\beta) & e^{iS'/q} & & & & & e^{-iS'/q} \\ & \ddots & \ddots & & & 0 & \\ & & e^{-iS'/q} & 2\alpha\cos(x+2\pi\beta n) & \ddots & e^{iS'/q} & \\ & 0 & & & \ddots & \ddots & \ddots \\ e^{iS'/q} & & & & & e^{-iS'/q} & 2\alpha\cos(x+2\pi\beta q) \end{pmatrix}. \tag{6.3}$$

Equivalently, in the Floquet picture, this relationship is given by

$$2\cos S'=\operatorname{Tr}\tilde{M}_E(x,\hbar) \tag{6.4}$$

where

$$\tilde{M}_E(x,\hbar)=\tilde{T}(x+2\pi\beta(q-1), E)\ldots\tilde{T}(x+2\pi\beta, E)\tilde{T}(x, E),$$

$$\tilde{T}(x, E)=\begin{pmatrix} E-2\alpha\cos x & -1 \\ 1 & 0 \end{pmatrix}. \tag{6.5}$$

The phase change $\gamma(E)$ is given by a line integral in phase-space around a phase trajectory

$$\gamma(E)=\operatorname{Im}\oint_{\varepsilon(x,S')=E}(j_{u\nabla u}/j_{uu})\cdot d\boldsymbol{X}. \tag{6.6}$$

In the Bloch picture, the vector $\boldsymbol{u}(x, S')$ is an eigenvector of $\hat{H}(x, S')$, given by (6.3), and in the Floquet picture $\boldsymbol{u}(x, S')$ is an eigenvector of the transfer matrix $\tilde{M}(x, \varepsilon(x, S'), \hbar)$. The matrix elements of the current operator are, in the Bloch picture

$$J_{nn'} = \frac{\mathrm{i}}{q}\begin{pmatrix} 0 & e^{-\mathrm{i}S'/q} & & & & & e^{\mathrm{i}S'/q} \\ & \ddots & \ddots & & & 0 & \\ & & e^{\mathrm{i}S'/q} & 0 & e^{-\mathrm{i}S'/q} & & \\ & 0 & & \ddots & \ddots & \ddots & \\ e^{-\mathrm{i}S'/q} & & & & & e^{\mathrm{i}S'/q} & 0 \end{pmatrix}, \tag{6.7}$$

and in the Floquet picture

$$\tilde{J} = \begin{pmatrix} 0 & \mathrm{i} \\ -\mathrm{i} & 0 \end{pmatrix}. \tag{6.8}$$

The final result, the Bohr–Sommerfeld quantisation condition, is (cf 5.13)

$$\oint_{\varepsilon = E} S'(x, E, \hbar)\,\mathrm{d}x = 2\pi[n + \tfrac{1}{2} - (1/2\pi)\,\mathrm{sign}(\hbar)\gamma(E)] \cdot |\hbar| \tag{6.9}$$

(remember that $\hbar$ can be negative for this system).

Now the theoretical predictions of this paper will be compared with some numerical results.

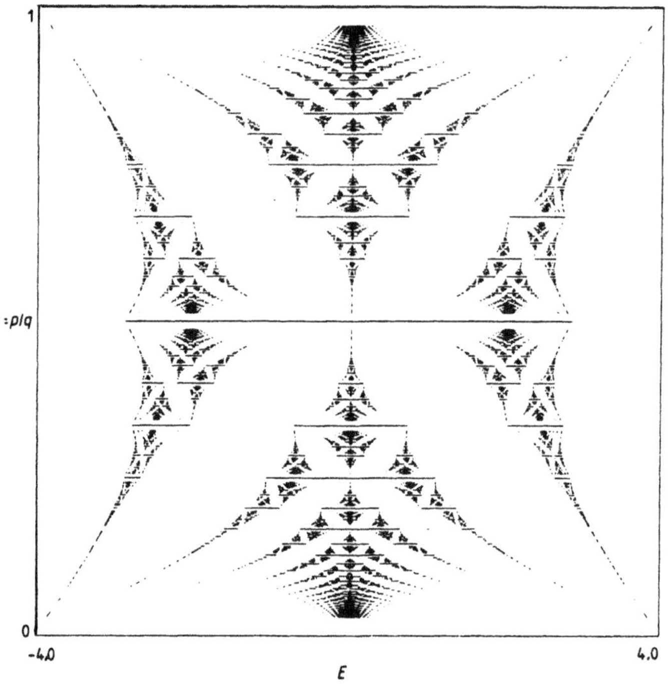

Figure 3. A plot of the spectrum of Harper's equation, plotted for every rational $\beta = p/q$ with $q \leq 40$. This picture illustrates the situation shown schematically in figure 2.

First, figure 3 gives a general illustration of the ideas discussed in this paper. This picture, originally published by Hofstadter (1976), is a plot of the spectrum (i.e. the support of $\varepsilon(\delta, k)$) of Harper's equation for every rational value of β with denominator q less than 40. There are q energy bands (the central two touching when q is even). When β is close to a low-denominator rational, p_0/q_0, then the q bands are very narrow and cluster into q_0 groups, which correspond to the q_0 bands when $\beta = \beta_0 = p_0/q_0$. These q narrow bands are the Bohr–Sommerfeld levels discussed in this paper, slightly broadened by tunnelling effects.

It is also easy to see by inspection of figure 3 that, except for $\beta_0 = 0$, 1 and $\beta_0 = \frac{1}{2}$, the Bohr–Sommerfeld quantisation is not of the usual form, i.e. action $= 2\pi(n + 1/2)\hbar$, since the pattern is not symmetrical about and below the lines $\beta = \beta_0$ (cf equation (6.9), which is not symmetric under $\hbar \to -\hbar$).

Next, table 1 gives some comparisons of energy levels predicted using formulae (6.1)–(6.9), E_{pr}, with exact levels, E_{ex} obtained by the same method as Hofstadter, using high precision arithmetic. In all cases except one the broadening of these levels by tunnelling is smaller than the six-digit precision of the eigenvalues E_{ex}. The meaning of the columns p_0, q_0, is given by equation (6.1), n is the quantum number of the Bohr quantised level (as in equation (6.9)) and n_b labels the energy band within which this level lies. The subscripts t and b of n_b mean that n is counted from the top and bottom of the energy band respectively. Finally, ΔE is the separation of the closest neighbouring energy level, and gives a scale against which the error $|E_{ex} - E_{pr}|$ of the predictions should be compared.

Finally, table 2 shows how the error in the predicted levels decreases very rapidly as $\hbar$ decreases. These results suggest that the error of the prediction is $O(\hbar^3)$ compared to the separation of levels which decreases as $O(\hbar)$.

Table 1. Comparison of predicted energy levels E_{pr} with exact levels E_{ex}. For full description see § 6 of text. For all values in this table, $\alpha = 1$.

p_0	q_0	$\Delta\beta$	n_b	n	E_{pr}	E_{ex}	$\lvert E_{pr} - E_{ex}\rvert$	ΔE	$\gamma(E)$
1	3	1/200	1t	1	−1.99435	−1.99277	0.00158	0.08395	1.6791
1	3	1/200	1t	6	−2.29301	−2.29209	0.00092	0.04154	0.6868
1	3	1/200	1t	10	−2.40938	−2.40860	0.00078	0.02295	−0.4313
1	3	1/200	2b	1	−0.61889	−0.62003	0.00114	0.10129	−1.6406
1	3	1/200	3t	1	2.70954	2.71090	0.00136	0.03379	−2.2567
1	3	1/200	3t	5	2.58201	2.58442	0.00241	0.02839	−3.5757
1	4	1/300	2t	1	−0.28742	−0.28171	0.00621	0.10782	1.4364
1	4	1/300	2t	5	−0.58727	−0.58578	0.00148	0.04299	0.0375
1	3	2/387	1t	1	−1.99414	−1.99252	0.00162	0.08631	1.6786
1	3	2/387	1t	7	−2.33540	−2.33427 to −2.33432	0.001	0.035	0.4193
3	7	1/1960	3b	1	−1.58070	−1.58068	0.00002	0.00623	−0.5570
3	7	1/1960	3b	5	−1.60464	−1.60460	0.00004	0.00557	−0.6380

7. Concluding remarks

This paper has demonstrated a novel type of Bohr–Sommerfeld quantisation, involving a non-holonomic connection rule for transporting the eigenvectors $\boldsymbol{u}$ of the matrix-valued Hamiltonian function $\hat{H}(x, p)$ around a circuit in phase space.

Table 2. Illustrating the rapid improvement of the predictions as $\hbar$ decreases. All the results in this table refer to the case $\alpha = 1$, $p_0 = 1$, $q_0 = 3$, $n_b = 3t$. Using equation (A9), the limiting value of $\gamma(E)$ at the edge of the band concerned is predicted to be $\gamma_0 = -1.8200$. The values of $\gamma(E)$ for the $n = 1$ states approach this limiting value.

n	$\Delta\beta$	E_{pr}	E_{ex}	$\lvert E_{pr} - E_{ex}\rvert$	ΔE	$\gamma(E)$
1	1/100	2.68295	2.69043	0.00748	0.06637	−2.9737
	1/200	2.70954	2.71090	0.00136	0.03379	−2.2567
	1/400	2.72120	2.72140	0.00020	0.01710	−1.9760
	1/800	2.72671	2.72671	0.00000	0.00859	−1.8554
5	1/200	2.58201	2.58442	0.00241	0.02839	−3.5757
	1/400	2.65491	2.65514	0.00023	0.01575	−2.3862
	1/800	2.69286	2.69288	0.00002	0.00826	−2.0428

The phase change γ of the eigenvector has been determined in terms of the line integral of the connection (5.2). In principle, γ could also be expressed as the integral of the curvature of the connection over the area enclosed by C. In practice, however, this is not useful, since this curvature is singular on the line $S' = 0$, and in any case for a computer calculation of $\gamma(E)$ the line integral is much easier to evaluate.

The method given in this paper is easily adapted to the problem of determining the Bohr–Sommerfeld quantisation condition for Bloch electrons in a weak magnetic field. It is well known that this condition takes the form

$$\mathscr{A}_k = 2\pi(eB/\hbar)(n+\Gamma), \tag{7.1}$$

where $\mathscr{A}_k$ is the area enclosed by a section through the Fermi surface perpendicular to the magnetic field (Onsager 1952). The constant Γ is not determined by Onsager's argument, and has previously only been determined exactly by very elaborate methods based on the effective Hamiltonian approach (see e.g. Roth 1966). The main result is that, for crystals with centres of inversion, Γ is always equal to $\frac{1}{2}$ (plus terms of higher order in the magnetic field), but when there is not a centre of inversion there is an additional component of Γ given by an integral analogous to (5.3).

It is worthwhile to note that two other results have appeared recently which involve a non-holonomic connection rule for the phase of the eigenvector of a matrix.

Firstly Mead and Truhlar (1979) and Berry (1984) calculate the phase change of the wavefunction of a system after being varied slowly around a cyclic path in the space of some parameters of the system. In terms of the matrix product calculation of § 3 of this paper, this corresponds to considering a string of slowly varying unitary evolution operators, i.e. to the case $\tilde{J} = \tilde{1}$ in equation (3.20).

Secondly Thouless *et al* (1982) have considered the quantised Hall effect in samples with a weak periodic potential with a rational number of flux quanta per unit cell. They show that the Hall conductance of a full sub-band is $e^2/2\pi h$ times the phase change when the wavefunction is transported around the edge of the magnetic Brillouin zone using the adiabatic connection rule of Mead and Berry. Because the magnetic Brillouin zone is topologically a torus, they are able to show that this phase change is 2π times an integer. Simon (1983) has exhibited a connection between the work of Berry and Mead, and that of Thouless *et al*, and has emphasised the importance of the idea of a non-holonomic connection.

The formula given in § 3 for the product of a string of slowly varying matrices is a very general result, and may have many uses other than those considered here.

Acknowledgment

I wish to thank the UK Science and Engineering Research Council for a postgraduate studentship.

Appendix

This appendix demonstrates the result that $\gamma(E)$ approaches a finite limit γ_0 at the top or bottom of an energy band, by calculating this limit in terms of the transfer matrices.

The transfer matrix is assumed to be known in the form of a series expansion:

$$\tilde{M}(x, E, \hbar) = \begin{pmatrix} A & B \\ C & D \end{pmatrix} = \begin{pmatrix} A_0 & B_0 \\ C_0 & D_0 \end{pmatrix}$$
$$+(x-x_0)\begin{pmatrix} a_x & b_x \\ c_x & d_x \end{pmatrix} + (E-E_0)\begin{pmatrix} a_E & b_E \\ c_E & d_E \end{pmatrix} + \hbar\begin{pmatrix} a_\hbar & b_\hbar \\ c_\hbar & d_\hbar \end{pmatrix} + \ldots, \tag{A1}$$

where x_0 and E_0 are the values of x and E corresponding to the top or bottom of the energy band $E = \varepsilon(x, s')$ when $\hbar = 0$. Also, it is assumed that the first few coefficients in the expansion of Tr $\tilde{M}$ are known

$$2\cos S' = \mathrm{Tr}\, \tilde{M}(x, E, \hbar) = 2 + r(x-x_0)^2 - s(E-E_0) - t\hbar. \tag{A2}$$

From equation (2.8), it can be seen that $r = q^2\alpha^q$. From (A2), the momentum S' is given by

$$S'^2 = s(E-E_0) - r(x-x_0)^2 - t\hbar, \tag{A3}$$

in the neighbourhood of the band edge. The action $S(E)$ for the phase trajectory of energy E is given by

$$S(E) = \oint_{\varepsilon = E} S'\,\mathrm{d}x = (\pi/\alpha^q q)[s(E-E_0) - t\hbar]. \tag{A4}$$

Now the phase change $\gamma(E)$ of the eigenvector $\boldsymbol{u}$ of $\tilde{M}$ will be found, in the neighbourhood of the energy E_0. The transfer matrix $\tilde{M}$ of (A1) can be diagonalised as follows

$$\tilde{M}(x, E, \hbar) = \tilde{X}^{-1}\tilde{D}\tilde{X}, \qquad \tilde{D} = \begin{pmatrix} \mathrm{e}^{\mathrm{i}S'} & 0 \\ 0 & \mathrm{e}^{-\mathrm{i}S'} \end{pmatrix} \tag{A5}$$
$$\tilde{X} = \begin{pmatrix} C & D-\mathrm{e}^{-\mathrm{i}S'} \\ -C & \mathrm{e}^{\mathrm{i}S'}-D \end{pmatrix}, \qquad \tilde{X}^{-1} = \frac{1}{2\mathrm{i}C\sin S'}\begin{pmatrix} \mathrm{e}^{\mathrm{i}S'}-D & \mathrm{e}^{-\mathrm{i}S'}-D \\ C & C \end{pmatrix}.$$

Now, for the eigenvector corresponding to the eigenvalue $\mathrm{e}^{\mathrm{i}S'}$, the differential element of the phase change γ is (retaining only lowest order terms)

$$\mathrm{d}\gamma_1 = \mathrm{Im}\,\mathrm{d}\tilde{V}_{11} = \mathrm{Im}(\mathrm{d}\tilde{X}\tilde{X}^{-1})_{11} = \frac{D-1}{2CS'}\mathrm{d}C - \frac{\mathrm{d}D}{2S'}. \tag{A6}$$

Using the relationship (A3) between x, S' and E, for any quantity $A = A(x, E)$

$$\left(\frac{\partial A}{\partial S'}\right)_x = \frac{2S'}{s}\left(\frac{\partial A}{\partial E}\right)_x, \qquad \left(\frac{\partial A}{\partial x}\right)_{S'} = \left(\frac{\partial A}{\partial x}\right)_E + \frac{2\alpha^q q^2(x-x_0)}{s}\left(\frac{\partial A}{\partial E}\right)_x. \tag{A7}$$

Using (A7) in (A6), to lowest order

$$\begin{aligned} d\gamma_1 &= \gamma_x \, dx + \gamma_{S'} \, dS', \\ \gamma_x &= \{[(D_0-1)/2C_0]c_x - \tfrac{1}{2} d_x\}/S' + O(1) = \kappa/S' + O(1), \\ \gamma_{S'} &= O(1). \end{aligned} \tag{A8}$$

Also, it is easy to show that $d\gamma_2(x, -S') = d\gamma_1(x, S')$. The phase change is then given by, in the limits x, S', $\hbar \to 0$:

$$\gamma(E) \simeq \gamma_0 = \oint_{\varepsilon=E} \gamma_x \, dx + \gamma_{S'} \, dS' \simeq \oint \kappa/S' \, dx \tag{A9}$$

i.e.

$$\gamma_0 = 2\pi\kappa/\alpha^{q/2}q, \qquad \kappa = [(D_0-1)/2C_0]c_x - \tfrac{1}{2} d_x. \tag{A10}$$

The Bohr–Sommerfeld quantisation condition in the neighbourhood of the band edge then becomes (using (5.13), (A4), (A9))

$$E = E_0 + (1/s)\cdot 2\alpha^q q\hbar[n + \tfrac{1}{2} - (\kappa/\alpha^{q/2}q)\,\mathrm{sign}(\hbar) + t/2\pi], \tag{A11}$$

where the level number satisfies the inequality

$$n > (\kappa/\alpha^{q/2}q)\cdot \mathrm{sign}(\hbar) - \tfrac{1}{2}. \tag{A12}$$

References

Bellissard J and Simon B 1982 *J. Funct. Anal.* **48** 408–19
Berry M V 1984 *Proc. R. Soc.* A**392** 45–57
Harper P G 1955 *Proc. Phys. Soc.* A**68** 874–8
Hofstadter D R 1976 *Phys. Rev.* B**14** 2239–49
Khinchin A Ya 1964 *Continued Fractions* (Chicago: University Press)
Landau L D and Lifshitz E M 1958 *Quantum Mechanics* ch 7 (Oxford: Pergamon)
Mead C A 1979 *J. Chem. Phys.* **70** 2276–83
Mead C A and Truhlar D G 1979 *J. Chem. Phys.* **70** 2284–96
Onsager L 1952 *Phil. Mag.* **43** 1006–8
Roth L M 1966 *Phys. Rev.* **145** 434–48
Simon B 1982 *Adv. Appl. Maths.* **3** 463–90
—— 1983 *Phys. Rev. Lett.* **51** 2167–70
Sokoloff J B 1981 *Phys. Rev.* B**23** 2039–41
Thouless D J, Kohmoto M, Nightingale M P and den Nijs M 1982 *Phys. Rev. Lett.* **49** 405–9
Wilkinson M 1984 *Proc. R. Soc.* A**391** 305–50

Current address: Department of Physics and Applied Physics, John Anderson Building, University of Strathclyde, Glasgow G4 ONG, Scotland, United Kingdom.

Chapter 7

WESS-ZUMINO TERMS AND ANOMALIES

7

Wess–Zumino Terms and Anomalies

In this section we look at two topics with origins in quantum field theory, Wess-Zumino terms and anomalies, from the special point of view afforded by geometric phases. We trust that readers who have followed us this far will find such an approach especially accessible. The fundamental ideas have close analogies with phenomena we have already met in previous chapters, and have found applications well outside their original scope, so that we hope that readers with other backgrounds will not be scared away. Most of the papers we have selected for this section have explicitly pedagogical intentions. Also, we have tried to choose a graded cross-section, so that a determined reader can reach the heights by manageable steps. Since the papers do speak so well for themselves, in this introduction we shall merely define a few of the special terms and make some general observations. Those who wish to delve further may wish to consult Refs. [1] and [2].

The original Wess–Zumino term was invented, by Wess and Zumino,[3] in constructing an effective action for mesons. Its purpose was to describe certain interactions, such as $\pi^0 \rightarrow \gamma\gamma$, that the naive effective action failed to include. Since then, it has been generalized to a number of other situations. Without writing a Wess–Zumino term down explicitly (see [7.2] for details) we can summarize some of its most important properties:

(i) The Wess–Zumino term is first order in time derivatives, and persists in the adiabatic limit.

(ii) Its coefficient is quantized, for topological reasons. This is a generalization of the famous Dirac quantization condition, which shows that the existence of magnetic monopoles is only consistent with the principles of quantum mechanics, if the magnetic charges of all particles are integer multiples of a certain basic charge.

(iii) It plays a crucial role in determining the kinematics of quantization. It directly modifies the canonical momentum, and, furthermore, the value of its coefficient modulo 2 dictates whether certain solitons of the model, which represent baryons, are to be quantized as bosons or fermions.

(iv) It results from "integrating out" very high-energy or high-mass states (quarks, in this case). These states may leave behind unusual quantum numbers (the elementary solitons possess baryon number 1) for collective states of the effective low-energy theory.

(v) It has a singular, gauge non-invariant form when written down explicitly as a term in the effective Lagrangian. However, it may also be expressed in a manifestly invariant and non-singular form by adding an extra spatial dimension. That is, one considers a five-dimensional manifold $\mathcal{M}$ whose boundary is four-dimensional (compact Euclidean) spacetime, and writes the corresponding piece of the action as the integral over $\mathcal{M}$ of a well-behaved total divergence.

Reader, you will recognize that any one of these properties could also refer to the term we added to the effective Born-Oppenheimer Lagrangian in the introduction to Chapter 4 and [4.3], in order to describe the effect of Berry's phase on nuclei in diatomic molecules. We saw there that, because the typical splitting between different electronic configurations is very large in comparison with nuclear splittings, we can validly integrate out the electrons in the Born-Oppenheimer approximation, but there remains a highly non-trivial residue of these lost degrees of freedom. The relative phases of the electron ground-state configurations for different nuclear positions must be taken into account when sewing them together to form wavefunctions for the nuclear degrees of freedom. The net effect of the phases is to generate an extra term in the effective nuclear action

$$S_{\mathbf{phase}} = \int A(R) \cdot \dot{R}\, dt = \int A \cdot dR \tag{7.1}$$

involving a magnetic monopole gauge potential. This term satisfies properties (i) and (iv). It satisfies (ii) by calculation (ultimately because of the quantization condition of electronic angular momentum), and (iii) because, as we saw in Chapter 4, it alters the quantization of the nuclear angular momentum. Finally, because of the Dirac string singularity, the term (7.1) in the Lagrangian is not gauge invariant, although the integrated action along a closed loop is invariant up to the addition of a multiple of 2π when the Dirac quantization condition is satisfied. One way to see this is by using Stokes' theorem to write

$$S_{\mathbf{phase}} = e \iint_S dR_i dR_j\, F_{ij} \tag{7.2}$$

where S is a surface whose boundary is the original path in parameter space and $F_{ij} = g\epsilon_{ijk} r^k/|r|^3$ is the gauge-invariant magnetic field of a monopole. The phase is now manifestly gauge-invariant, but seems to depend on the bounding surface. However, the Dirac condition that the product of electric and magnetic charge eg be a multiple of $1/2$, implies that the ambiguity is a multiple of 2π, and does not affect the phase $e^{iS_{\text{Phase}}}$.

Given that the term (7.1) satisfies all of the properties enumerated above, perhaps we should refer to it as a particular example of a Wess–Zumino term. Viewed in this light, the WZ term is nothing more than a field-theoretic generalization of a very basic feature of the Born–Oppenheimer approximation. This point of view is discussed further, in light of explicit examples, in the enclosed paper by Stone [7.1]. Aitchison's paper [7.2] gives a more leisurely and detailed account of the connection between Berry's phase and Wess–Zumino terms, and includes several explicit examples.

One often employs a type of Born–Oppenheimer approximation in the quantization of collective states—solitons—in quantum field theory. Consider, for instance, a theory in which light scalar fields interact with much heavier fermions. One expects that as long as the scalar fields vary slowly in space-time, there will be a large gap between the ground state of the fermion system and any other state of this system, and furthermore that this ground state may be constructed by appropriately patching together locally determined ground-states. The situation is altogether analogous to the one we just met, in connection with the quantization of diatoms. In the paper included here, Goldstone and Wilczek [7.3] analyze the field theory problem. They show that in this situation the scalar field configurations do generally inherit quantum numbers from the fermions. Although they did not use this language, the charges and currents they compute can be summarized by the effective Lagrangian one would obtain for the scalar fields by integrating out the fermions. This effective Lagrangian would contain a term of Wess–Zumino type, associating fermion quantum numbers to certain configurations, and instructing us to quantize these as fermions.

It was in the context of showing how a Lagrangian containing only (bosonic) meson degrees of freedom—which on general grounds is expected to be the effective theory for QCD at low energies—could also contain (fermionic) baryon degrees of freedom as collective excitations, that Witten wrote the remarkable and very influential paper reprinted here [7.4].

Next let us say a few words concerning anomalies. A symmetry is said to be anomalous if it exists at the level of the classical theory, but is inevitably spoiled upon quantization. A particularly well-studied class of anomalies occurs in quantizing theories containing fermions, and is closely related to Wess–Zumino terms as discussed above. One fruitful way of looking at these anomalies is very reminiscent of the Born–Oppenheimer approximation, if we think of the fermions as "fast" and the boson fields as "slow" degrees of freedom. This separation need have nothing to do with the relative mass scales of the two types of fields; here, it is really just a convenience that allows us to construct the theory in stages, by first defining the fermion states for fixed configurations of the other fields, performing the integration over fermionic variables, and then quantizing the resulting effective Lagrangian for the bosonic fields. In particular, we need to define the Dirac sea associated with each Bose field configuration, and patch these together. Now it is often found that holonomy, of the kind that by now should seem quite familiar, can arise on the configuration space. To maintain classical symmetries of the system, we would like to relate (parallel transport) the Dirac seas for different configurations by making symmetry transformations. It is often found, however, that a non-integrable phase holonomy prevents us from doing this consistently. It can also occur that the transport law demanded by one symmetry conflicts with that demanded by another, so that at most one of them can be maintained on quantization.

Some elementary but illuminating remarks introducing anomalies are contained in the paper by Jackiw [1.2]. The paper by Alvarez-Gaumé and Nelson [7.5] is a lovely but very sophisticated treatment from the point of view of Berry's phase. We recommend that you read all the other papers in this section before attempting this one; after that preparation we think you will find it accessible and rewarding.

Anomalies can be good: as will be discussed below, some have even been observed experimentally. On the other hand, anomalies in local gauge symmetries are widely believed to be unacceptable—a belief we share. For local gauge invariance is necessary to insure that longitudinal photons, and their relative gluons of other kinds, are not present in the physical spectum. And they'd better not be, because they are ghosts—the probability of creating one, if not zero, is negative! So cancellation of anomalies in all local gauge symmetries is usually imposed as a constraint on possible quantum field theories. However, anomalous *global* symmetries do not make a field theory inconsistent; they simply get broken by the quantization process. When fermions associated with a global anomaly are integrated out, the effective action for the remaining degrees of freedom will generally include a Wess-Zumino term. Its role is to account for the above-mentioned holonomy.

In conclusion, it may be appropriate to say a few words about the physical applications of anomalies. They find several important applications in QCD, the modern theory of the strong interaction. First of all there is the original application, to the description of the decay $\pi^0 \rightarrow \gamma\gamma$. A naive application of chiral symmetry, which had been spectacularly successful in describing much of the low-energy phenomenology of pions, led to the very poor prediction of a vanishing rate for this decay.[4] It was therefore tremendously important when Adler, and Bell and Jackiw[5] demonstrated that an anomaly spoiled chiral symmetry in this decay. Moreover, Adler[6] traced this anomaly down to a single graph, and showed that the decay rate was profoundly and quantitatively related to the fundamental field content of whatever theory described the strong interaction. The successful quantitative prediction of the $\pi^0 \to 2\gamma$ rate, which is sensitive to the color and fractional charge assignments of quarks, still stands as one of the most remarkable triumphs of quantum field theory. Recently, the related process $\gamma\gamma \to \pi\gamma$ has also been measured, and found to agree with the anomaly-based prediction.

QCD with massless quarks, which is a good approximation for most hadronic physics, is formally scale invariant. However this classical symmetry is spoiled upon quantization. The "scale anomaly" is parametrized by an effective coupling constant that varies depending upon the characteristic energy and momentum scales of the process under consideration.[7] Of course, it was the observed fact that scale invariance is approximately but not exactly realized among strongly interacting particles that led to the discovery of asymptotic freedom and QCD in the first place,[8] and to many successful quantitative predictions for high-energy processes.

QCD with massless quarks also seems to contain another symmetry, axial baryon number, that is not observed in nature. Furthermore, there is no sign of its being spontaneously broken (that is, there is no light Nambu-Goldstone boson with the appropriate quantum numbers).[9] It was therefore most welcome when t'Hooft[10] demonstrated that this apparent symmetry is in fact anomalous, that is not a symmetry the quantized theory at all. With this discovery, the match between the symmetries predicted by QCD and the observed symmetries of the strong interaction in nature became perfect, and the last significant barrier to acceptance of QCD as the theory of the strong interaction fell.

Because anomalies are often sensitive to arbitrarily massive degrees of freedom, and because they depend only on very general features of the theory involved, they have played an important role in guiding speculations concerning physics beyond the standard model. The cancellation of anomalies in electroweak gauge symmetries occurs in a non-trivial manner between quarks and leptons. This was one of the first suggestions of deep connections among these particles, and inspired efforts toward unification. It also goes at least part of the way toward explaining why quarks and leptons occur in repeating families, each with the same basic structure.

Finally, much of the recent interest in string theory was stimulated by the discovery of Green and Schwarz[11] that certain anomalies which spoiled the consistency of the theory cancelled for special choices of space-time dimension and internal gauge group (10 and SO(32), respectively). Their mechanism for cancellation of anomalies also worked at a formal level for the exceptional group $E_8 \times E_8$; however at the time no way of incorporating this gauge group into a string theory was known. Within a few weeks, inspired by this cancellation, Gross, Harvey, Martinec and Rohm[12] discovered the heterotic string construction, which does incorporate $E_8 \times E_8$. Today, most attempts to connect string theory with observed reality start from the heterotic string.

---

[1] S. Treiman, R. Jackiw, B. Zumino, and E. Witten, *Current Algebra and Anomalies* (Princeton: University Press, 1985)

[2] A.P. Balachandran, "Wess–Zumino Terms and Quantum Symmetries," and A. Niemi, "Quantum Holonomy," in M. Jezabek and M. Praszałowicz, eds., *Skyrmions and Anomalies* (Singapore: World Scientific, 1987); A. Dhar and S. Wadia, *Phys. Rev. Lett. bf 52* (1984) 959.

[3] J. Wess and B. Zumino, *Phys. Lett.* **36B** (1971) 95.

[4] D. G Sutherland, *Nucl. Phys.* **B2** (1967) 433.

[5] S. Adler, *Phys. Rev.* **177** (1969) 2426.
J. S. Bell and R. Jackiw, *Nuovo Cimento* **60A** (1969) 47.

[6] S. Adler, in *Lectures on Elementary Particles and Quantum Field Theory*, Proc. 1970 Brandeis Summer Institute, ed. S. Deser *et al.* (Cambridge: MIT Press,1970).

[7] D. Gross, in *Methods in Field Theory*, Proc. of 1975 Les Houches Summer School, ed. R. Balian and J. Zinn–Justin (Amsterdam: North–Holland, 1976).

[8] D. Gross and F. Wilczek, *Phys. Rev. Lett.* **30** (1973) 1343;
H. Politzer, *Phys. Rev. Lett.* **30** (1973) 1346.

[9] S. Weinberg, *Phys. Rev.* **D11** (1975) 3583.

[10] G. 't Hooft, *Phys. Rev. Lett.* **59B** (1976) 172.

[11] M. Green and J. Schwarz, *Phys. Lett.* **149B** (1984) 285.

[12] D.J. Gross, J. Harvey, E. Martinec, and R. Rohm, *Nucl. Phys.* **B256** (1985) 253–284.

PHYSICAL REVIEW D VOLUME 33, NUMBER 4 15 FEBRUARY 1986

Born-Oppenheimer approximation and the origin of Wess-Zumino terms: Some quantum-mechanical examples

Michael Stone
Department of Physics, University of Illinois at Urbana-Champaign, 1110 West Green Street, Urbana, Illinois 61801
(Received 30 September 1985)

I provide some simple quantum-mechanical examples in which the Berry phase gives rise to Wess-Zumino terms. The connection with the Born-Oppenheimer approximation is also discussed.

I. INTRODUCTION

When, in a path-integral description of a quantum field theory, we have a system of Fermi fields interacting with bosons our first reaction is to eliminate the Grassmann variables needed for the fermions, integrating them out to produce the Mathews-Salam determinant.[1] This determinant is a functional of the Bose fields and acts back on them in the form of an effective action. Often, especially when the fermions are heavy, the effective action just renormalizes the parameters of the Bose action,[2] but sometimes the results are more startling: Symmetries may be lost and the theory may even be inconsistent. These unexpected effects are due to terms in the effective action of the type discussed by Wess and Zumino.[3] They encapsulate all the anomalies of the system. A great deal of progress has recently been made in understanding the topological character and origins of these terms.[4,5] In particular, Nelson and Alvarez-Gaume[6] have explained their origin in a Hamiltonian context.

It seems worthwhile to show just how simple the explanation in Ref. 6 is by giving a few examples from quantum mechanics. In this simple context one can calculate everything yet there is still enough structure to retain the essentials of their ideas.

In Sec. II, I shall exhibit some nontrivial fiber bundles that originate naturally in extremely simple models. In Sec. III, I will quantize them by a path-integral route, which mimics what we do in a field theory. Then we repeat the quantization in the Born-Oppenheimer approximation where we will note some deficiencies in the accounts of this method given in textbooks. A conventional solution will be presented in the Appendix so that the reader can verify that nothing inconsistent with the usual rules of quantum mechanics is being perpetrated.

II. MONOPOLES FROM FERMIONS

Consider the family of Hamiltonians $H(\hat{n})$, labeled by points on S^2 (i.e., $\hat{n}$ is a three-dimensional unit vector),

$$H(n)=\mu\hat{n}\cdot\sigma \; . \tag{2.1}$$

Here σ are the usual Pauli matrices. For each $\hat{n}$ there is a two-dimensional Hilbert space. Let us select a basis vector in each one corresponding to the eigenvalue $+\mu$. We can obtain these eigenstates by the use of spin-projection operators

$$|\psi_+^{(1)}\rangle=N\tfrac{1}{2}(1+\hat{n}\cdot\sigma)\begin{bmatrix}1\\0\end{bmatrix}=\begin{bmatrix}\cos\theta/2\\ \sin(\theta/2)e^{i\phi}\end{bmatrix},$$
$$|\psi_+^{(2)}\rangle=N\tfrac{1}{2}(1+\hat{n}\cdot\sigma)\begin{bmatrix}0\\1\end{bmatrix}=\begin{bmatrix}\cos(\theta/2)e^{-i\phi}\\ \sin\theta/2\end{bmatrix}, \tag{2.2}$$

where N is a real normalization factor and $\hat{n}=(\sin\theta\cos\phi,\sin\theta\sin\phi,\cos\theta)$. Clearly,

$$|\psi_+^{(1)}(\hat{n})\rangle=e^{i\phi}|\psi_+^{(2)}(\hat{n})\rangle \; . \tag{2.3}$$

We need (at least) two choices of $|\psi_+\rangle$ because $|\psi_+^{(1)}\rangle$ is a reasonable eigenvector everywhere except at the south pole (where the phase is ambiguous) and $|\psi_+^{(2)}\rangle$ is reasonable everywhere except at the north pole. No one choice of phase is smooth everywhere on S^2; the one-dimensional rays at each $\hat{n}$ fit together to form a nontrivial bundle[7] with (2.3) as the transition function between two patches (if we had not insisted on choosing subspaces the total Hilbert space bundle *is* trivial). We can characterize the eigenspace bundle by putting a connection on it which is compatible with (2.3). One such connection which will arise naturally in Sec. III is

$$iA_+^{(1)}=\langle\psi_+^{(1)}(\hat{n})|d\psi_+^{(1)}(\hat{n})\rangle=\frac{i}{2}(-\cos\theta+1)d\phi \; , \tag{2.4a}$$

$$iA_+^{(2)}=\langle\psi_+^{(2)}(\hat{n})|d\psi_+^{(2)}(\hat{n})\rangle=\frac{i}{2}(-\cos\theta-1)d\phi \; . \tag{2.4b}$$

$A_+^{(1)}$ and $A_+^{(2)}$ differ by a gauge transformation,

$$A_+^{(1)}=A_+^{(2)}+d\phi \; , \tag{2.5}$$

so both have the same curvature

$$F_+=dA_+=\tfrac{1}{2}\sin\theta\, d\theta\wedge d\phi=\tfrac{1}{2}d(\text{area}) \tag{2.6}$$

(the A_-,F_- corresponding to eigenvalue $-\mu$ just differ in sign from these). We recognize the gauge potentials and fields of the monopole bundle[7] with $\int F=2\pi$.

Another example comes from restricting $\hat{n}$ to lie in the (x,z) plane so $H(\hat{n})$ is real. The eigenvectors can be chosen real,

$$\psi_+=\begin{bmatrix}\cos\theta/2\\ \sin\theta/2\end{bmatrix}, \quad \psi_-=\begin{bmatrix}\sin\theta/2\\ -\cos\theta/2\end{bmatrix}, \tag{2.7}$$

so A is identically zero—but we have to use the transition function

$$|\psi_\pm(2\pi)\rangle = -|\psi_\pm(0)\rangle$$

to make the choice single valued. The bundle is still twisted, being the Möbius strip.

The twisted nature of these bundles is only a curiosity so far; it will become important when we raise $\hat{n}$ to the status of a dynamical variable in Sec. III.

III. A SPINNING SOLENOID

We will study the Hamiltonian

$$H = \frac{1}{2I}L^2 - \mu\sigma\cdot\hat{n} , \tag{3.1}$$

where L is the angular momentum operator that generates rotations of $\hat{n}$. One could imagine constructing a Heath-Robinson-type device which would be described by this Hamiltonian: A long thin solenoid rotating about its center of mass would have $\frac{1}{2}L^2/I$ as its Hamiltonian and placing a spin-$\frac{1}{2}$ particle with magnetic moment μ at this center of mass would produce the second term. When μ is small the two systems would spin independently. As μ becomes large the spin will become slaved to the direction of the solenoid and its spin-$\frac{1}{2}$ nature will be transferred to the orbital dynamics of the solenoid. It is the large-μ case which we will consider here because I wish to describe the induction of Wess-Zumino terms by the failure of the naive decoupling theorem[8] and large μ corresponds to a large fermion mass.

A path integral for (3.1) is

$$\int d[\hat{n}]d[\bar{\psi}]d[\psi]\delta(n^2-1) \times\exp\left[-\int\left(\frac{I\dot{n}^2}{2}+\bar{\psi}(\partial_\tau-\mu\hat{n}\cdot\sigma)\psi\right)d\tau\right] . \tag{3.2}$$

The Fermi operators allow four states: no spin, one spin up, one spin down, and two spins—one of each, up and down. The ground state for large μ will be one spin aligned against the field. We will evaluate the fermion determinant by noting that large μ enables us to use the adiabatic theorem since the ground state is well separated from any other state to which transitions may be stimulated by the motion of $\hat{n}$.

The adiabatic theorem[9] says that for slowly varying $H(t)$ we can approximate the solution of the time-dependent Schrödinger equation

$$i\partial_t|\psi\rangle = H(t)|\psi\rangle \tag{3.3}$$

in terms of eigenstates $|\psi^0\rangle$ of the "snapshot" Hamiltonian

$$H(t)|\psi^0(t)\rangle = E(t)|\psi^0(t)\rangle \tag{3.4}$$

as

$$|\psi(t)\rangle = \exp\left[-i\int_0^t E(t')dt'+i\gamma(t)\right]|\psi^0(t)\rangle , \tag{3.5}$$

where γ is the "Berry phase"[9,10] obeying

$$i\frac{d\gamma}{dt}+\left\langle\psi^0\left|\frac{d}{dt}\psi^0\right.\right\rangle = 0 . \tag{3.6}$$

The determinant for a closed path Γ of duration β is just the vacuum-vacuum amplitude so

$$\det[\partial_t+H(n(t))] = \exp\left[-\int_0^\beta E(t)dt+i\gamma(\Gamma)\right] . \tag{3.7}$$

We have changed to Euclidean time in conformity with (3.2). (Note γ, as a phase, does not change.) The effect of the spin is to change the $\hat{n}$ path integral to

$$\int d[\hat{n}]\delta(n^2-1)\exp\left[-\int\left(\frac{I\dot{n}^2}{2}+E(t)+\left\langle\psi(\hat{n})\left|\frac{\partial\psi}{\partial t}(\hat{n})\right.\right\rangle\right)dt\right] \tag{3.8}$$

$$=\text{const}\times\int d[n]\delta(n^2-1)\exp\left[-\int\left(\frac{I\dot{n}^2}{2}+iA_+(\hat{n})\cdot\dot{n}\right)dt\right] . \tag{3.9}$$

The dynamics of $\hat{n}$ has become the motion of a charged particle on a sphere, moving under the influence of a magnetic monopole at the center. The $A\cdot\dot{n}$ in (3.9) is the simple example of a "Wess-Zumino" term introduced by Witten in Ref. 4. It is not rotationally invariant and cannot be written globally without introducing string singularities although its dynamical effects are both rotationally invariant and nonsingular. In this example, unlike Ref. 4, the Wess-Zumino term has been induced dynamically instead of being inserted by hand and is a consequence of the phase ambiguities and bundle structure of Sec. II.

This result is sufficiently remarkable as to require a separate derivation (or two). One can obtain the same physics without the path integral by using the Born-Oppenheimer approximation. This says that if we have a system of slowly moving "nuclear" coordinates R, and some fast moving "electronic" coordinates r whose Hamiltonian $H_e(R)$ is parametrized by R, then the wave function $\Phi(r,R)$ can again be expressed in terms of solutions to the "snapshot" Hamiltonian

$$H_e(R)\psi(r,R) = E(R)\psi(r,R) \tag{3.10}$$

as

$$\Phi(r,R) = \phi(R)\psi(r,R) , \tag{3.11}$$

and ϕ obeys the modified Schrödinger equation

$$-\frac{\hbar^2}{2M}\nabla^2\phi+[E(R)+V(R)]\phi(R) = i\hbar\frac{\partial}{\partial t}\phi(R) \tag{3.12}$$

where ∇ is a covariant derivative:

$$\nabla_\mu = \frac{\partial}{\partial R^\mu}+\left\langle\psi\left|\frac{\partial\psi}{\partial R^\mu}\right.\right\rangle = \partial_\mu+iA_\mu . \tag{3.13}$$

This approximation has exactly the same physics as the adiabatic theorem and is made by ignoring the same sets of matrix elements. It is worth noting that textbooks usually omit the gauge potentials. They have been discussed in the literature,[11] however, and are crucial here.

This factorization of the total wave function shows up the source of the gauge dependence of $\phi(R)$. The total wave function is gauge independent—but a change of phase of $\psi(r,R)$ must be compensated for by a change of phase in $\phi(R)$ which thus inherits the phase problems from the integrated-out fermions. The path integral, with the fermions replaced by the effective action, is the path integral for $\phi(R)$, *not* for the total wave function.

In our case $\phi(R)$ is the wave function for free motion on a sphere around the monopole, i.e., the monopole harmonics, so the eigenstates of (3.1) are

$$\langle \hat{\mathbf{n}}, s|\phi\rangle = [D^j_{m,1/2}(\theta,\phi,\psi)]^*\langle s|\psi_+(\hat{\mathbf{n}})\rangle \ , \quad (3.14)$$

$$E_{jm} = \frac{1}{2I}j(j+1) + \text{const} \ , \qquad \text{degeneracy} = 2j+1 \ . \quad (3.15)$$

Just as the ordinary Y^l_m are equal to $[D^l_{m0}(\theta,\phi,0)]^*$ (the D's being the ordinary rotation matrices in Euler angle parametrization), the monopole harmonics for a particle of charge (q times unit charge) in orbit about a monopole are $[D^j_{m,q/2}(\theta,\phi,\psi)]^*$. The dependence on the angle ψ reflects the gauge dependence. One must choose an angle ψ for each θ,ϕ and one cannot make this choice [of a point in $SU(2)=S^3$ under the inverse of the Hopf map: $S^3\rightarrow S^2$] without introducing gauge patches. That a constrained spin is both classically and quantum-mechanically equivalent to motion about a monopole has been discussed by Leinaas in Ref. 12.

The second example, restricting the motion to the (x,z) plane, can be dealt with similarly. It is essentially the same as the problem of quantizing the collective modes of the SU(2) Skyrmion. There is no Wess-Zumino term since $A=0$ but the homotopy group of the space of paths is $\pi_1(S_1)=Z$ and we have to choose phases for the different classes of configurations. The quantum mechanics clearly requires $(-1)^n$, where n is the element of $\pi(S_1)$ involved. In the Born-Oppenheimer approach the states are

$$\langle\theta,s|j\rangle = e^{ij\theta}\langle s|\psi_+(\theta)\rangle, \quad j=n+\tfrac{1}{2} \ . \quad (3.16)$$

Each factor is double valued but the product is not.

IV. DISCUSSION

These quantum-mechanical models provide a simple realization of how attempting to decouple a degree of freedom, by making the energy gap too large to excite it, can fail and effects are left behind. This is what happens in a quantum field theory when one makes fermion masses large in the way discussed in Ref. 8. These models are also useful for thinking about the non-Abelian [or $\pi_2(G_3)$] and global [or $\pi_1(G_3)$] anomalies in the spirit of Ref. 6 (which was the main influence behind this paper). The analogy can be stretched too far, however: The analysis presented here is only valid in the adiabatic limit. If the coupling between the spin and the solenoid is such that the spin can be excited, the excited state has a Berry phase with the opposite sign and the topological effects are washed out (in other words, our total Hilbert space is a trivial product of the spin and configuration spaces and it is only in the adiabatic limit that we are forced into a subspace with nontrivial twist). This does not happen in a quantum field theory which is much more subtle—at least in the cases of interest. The Berry phase in the field theory arises as a property of the Dirac sea; the ground state and all excitations built on it have the same Berry phase, provided that the vacuum structure is not completely disrupted, and so the Wess-Zumino terms do not depend on the adiabatic limit which is then relegated to being a convenient trick to compute them.[13]

ACKNOWLEDGMENTS

This work was supported by the National Science Foundation Grant No. DMR84-15063. I would also like to thank NORDITA for hospitality and my colleagues at Urbana for encouraging me to give the talks on Wess-Zumino terms which led to my seeking a pedagogical example that even I could understand.

APPENDIX: CONVENTIONAL APPROACH TO QUANTUM MECHANICS

The problem of the solenoid coupled to a spin can, of course, be solved in a conventional manner. I shall do this here.

The Hamiltonian is

$$H=\frac{1}{2I}\mathbf{L}^2-\mu\hat{\mathbf{n}}\cdot\sigma \quad (A1)$$

and is clearly invariant under combined rotations of σ and $\hat{\mathbf{n}}$ so $[H,\mathbf{J}]=0$, $\mathbf{J}=\mathbf{L}+\frac{1}{2}\sigma$. This suggests using the states

$$|j-\tfrac{1}{2},j,m\rangle=\frac{1}{\sqrt{2j}}[(j+m)^{1/2}|j-\tfrac{1}{2},m-\tfrac{1}{2}\rangle|\tfrac{1}{2}\rangle+(j-m)^{1/2}|j-\tfrac{1}{2},m+\tfrac{1}{2}\rangle|-\tfrac{1}{2}\rangle] \ ,$$
$$|j+\tfrac{1}{2},j,m\rangle=\frac{1}{\sqrt{2j+2}}[-(j-m+1)^{1/2}|j+\tfrac{1}{2},m-\tfrac{1}{2}\rangle|\tfrac{1}{2}\rangle+(j+m+1)^{1/2}|j+\tfrac{1}{2},m+\tfrac{1}{2}\rangle|-\tfrac{1}{2}\rangle] \ , \quad (A2)$$

which are eigenstates of L^2, J^2, and J_z expressed in terms of eigenstates of L^2, L_z, and S_z.

The operator $(\sigma\cdot\hat{\mathbf{n}})$ is odd under $\hat{\mathbf{n}}\rightarrow-\hat{\mathbf{n}}$, commutes with J, and obeys $(\sigma\cdot\hat{\mathbf{n}})^2=1$. So

$$\sigma\cdot\hat{\mathbf{n}}|j\pm\tfrac{1}{2},j,m\rangle=-|j\mp\tfrac{1}{2},j,m\rangle \ . \quad (A3)$$

(The minus sign requires some computation since the properties of $\sigma\cdot\hat{\mathbf{n}}$ quoted above could have given ±1.) Restricted to the space with fixed j,m, H becomes

$$H_{jm}=\frac{1}{2I}[j(j+1)+\tfrac{1}{4}]+\begin{pmatrix}(j+\frac{1}{2})/2I & \mu \\ \mu & -(j+\frac{1}{2})/2I\end{pmatrix} \quad (A4)$$

so

$$E_{jm}=\frac{1}{2I}[j(j+1)+\tfrac{1}{4}]\pm\left[\frac{(j+1/2)^2}{4I^2}+\mu^2\right]^{1/2} \ . \quad (A5)$$

For large μ the ground state is

$$|E^-_{jm},j,m\rangle=\frac{1}{\sqrt{2}}(|j+\tfrac{1}{2},j,m\rangle-|j-\tfrac{1}{2},j,m\rangle) \ . \quad (A6)$$

We can express this in the factorized form of (3.14) by applying some rotation operators: Write

$$|E^-_{jm},j,m\rangle=\begin{pmatrix}\phi_{+1/2}(\hat{\mathbf{n}})\\ \phi_{-1/2}(\hat{\mathbf{n}})\end{pmatrix} \ .$$

where $\phi_s(\hat{\mathbf{n}}) = \langle \hat{\mathbf{n}}, s | E_{jm}^{-}, j, m \rangle$. The action of a rotation operator $U(R)$ is

$$U(R)|\hat{\mathbf{n}},s\rangle = |R\hat{\mathbf{n}},s'\rangle D^{1/2}_{s's}(R) \;,$$
$$U(R)|E_{jm}^{-},j,m\rangle = |E_{jm}^{-},jm'\rangle D^{J}_{m'm}(R) \;, \tag{A7}$$

so the matrix element $\langle \hat{\mathbf{n}},s|U(R)|E_{jm}^{-},j,m\rangle$ can be evaluated in two ways to give

$$\langle \hat{\mathbf{n}},s|E_{jm}^{-},j,m'\rangle D^{J}_{m'm}(R) = D^{1/2}_{ss'}\langle R^{-1}\hat{\mathbf{n}},s'|E_{jm}^{-},jm\rangle$$

so

$$\langle \hat{\mathbf{n}},s|E_{jm}^{-},j,m\rangle = D^{1/2}_{ss'}(R)\langle R^{-1}\hat{\mathbf{n}},s'|E_{jm}^{-}j,m'\rangle D^{J}_{m'm}(R^{-1}) \;. \tag{A8}$$

Arrange $R^{-1}\hat{\mathbf{n}}$ to be the north pole and use the properties of the angular momentum states,

$$\langle \theta,\phi|J \pm \tfrac{1}{2}, m \pm \tfrac{1}{2}\rangle = Y^{J\pm 1/2}_{m\pm 1/2}(\theta,\phi)$$

at the north pole

$$Y^{l}_{m}(\theta=0,\phi) = \delta_{m,0}\left(\frac{2l+1}{4\pi}\right)^{1/2}$$

to see that

$$\langle R^{-1}\hat{\mathbf{n}},s'|E_{jm}^{-},j,m\rangle \propto \delta_{s'm}\delta_{m\,1/2} \;.$$

Using this we can write

$$\langle \hat{\mathbf{n}},s|E_{jm}^{-},j,m\rangle = \langle s|\psi_{+}(R)\rangle D^{J}_{1/2,m}(R^{-1})$$
$$= [D^{J}_{m,1/2}(R)]^{*}\langle s|\psi_{+}(R)\rangle \tag{A9}$$

as promised by Eq. (3.14). Different choices of R, the rotation which takes the north pole to $\hat{\mathbf{n}}$, give rise to a different distribution of the phases between the two factors.

---

[1]P. T. Mathews and A. Salam, Nuovo Cimento 12, 563 (1954); 2, 120 (1955).

[2]T. Appelquist and J. Carazzone, Phys. Rev. D 11, 2856 (1975).

[3]J. Wess and B. Zumino, Phys. Lett. 37B, 95 (1971).

[4]E. Witten, Nucl. Phys. B223, 422 (1983).

[5]There have been many papers on this topic. See B. Zumino, Nucl. Phys. B253, 477 (1985), and references therein.

[6]P. Nelson and L. Alvarez-Gaume, Commun. Math. Phys. 99, 103 (1985).

[7]One does not need to know the language of fiber bundles to understand this paper but it helps to put things in context. A good review is T. Eguchi, P. Gilkey, and A. J. Hanson, Phys. Rep. 66, 213 (1980).

[8]E. d'Hoker and E. Farhi, Nucl. Phys. B248, 59 (1984); B248, 77 (1984).

[9]See L. I. Schiff, *Quantum Mechanics*, 3rd ed. (McGraw-Hill, New York, 1968), p. 289.

[10]M. Berry, Proc. R. Soc. London A392, 45 (1984); B. Simon, Phys. Rev. Lett. 51, 2167 (1983).

[11]C. A. Mead and D. G. Truhlar, J. Chem. Phys. 70, 2284 (1979).

[12]J. M. Leinaas, Phys. Scr. 17, 483 (1978).

[13]Similar points are made in C. Gomez, Phys. Rev. D 32, 2235 (1985).

Fractional Quantum Numbers on Solitons

Jeffrey Goldstone[(a)]
Stanford Linear Accelerator Center, Stanford University, Stanford, California 94305

and

Frank Wilczek
Institute for Theoretical Physics, University of California, Santa Barbara, California 93106
(Received 9 July 1981)

A method is proposed to calculate quantum numbers on solitons in quantum field theory. The method is checked on previously known examples and, in a special model, by other methods. It is found, for example, that the fermion number on kinks in one dimension or on magnetic monopoles in three dimensions is, in general, a transcendental function of the coupling constant of the theories.

PACS numbers: 11.10.Lm, 11.10.Np

Peculiar quantum numbers have been found to be associated with solitons in several contexts: (i) The soliton provides, of course, a different background than the usual vacuum around which to quantize other fields. The difference between these "vacuum polarizations" may induce unusual quantum numbers localized on the soliton.[1-3] (ii) Solitons may require unusual boundary conditions on the fields interacting with them, in particular leading to conversion of internal quantum numbers into rotational quantum numbers.[4-6] (iii) In the case of dyons, there is classically a family of solitons with arbitrary electric charge. The determination of which of these are in the physical spectrum requires quantum-mechanical considerations and brings in the θ parameter of non-Abelian gauge theories.[7,8]

At present all these phenomena seem distinct although there are suggestive relationships. In this note, we shall concentrate on (i), proposing a general method of analysis and working out a few examples.

VOLUME 47, NUMBER 14 PHYSICAL REVIEW LETTERS 5 OCTOBER 1981

An intuitively appealing, and perhaps physically realizable, example of the phenomena we are addressing are the fractionally charged solitons on polyacetylene.[2,3,9] A caricature model of a polyacetylene molecule is shown in Fig. 1(a)—in the ground state we have alternating single and double bonds, which may be arranged in two inequivalent but degenerate forms A and B. If there is an imperfection, as shown in Fig. 1(b), we go from A on the left-hand side to B on the right-hand side. This configuration cannot be brought to either pure A or pure B by any finite rearrangement of electrons, and so it will relax to a stable configuration—a soliton. If we put two imperfections together, as in Fig. 1(c), we find a configuration which begins and ends as A. Compared to the corresponding segment of pure A, it is missing one bond. If we add an electron to the two-imperfection strand, we can deform this configuration by a finite rearrangement into a pure A strand. (We are pretending, for simplicity, that each bond represents a single electron instead of a pair.) Interpreting this, we see that a two-soliton state is equivalent to the ground state if we add an electron. Thus, by symmetry, each separated soliton must carry electron number $-\frac{1}{2}$ (and electric charge $+\frac{1}{2}e$).

We can relate these stick-figure pictures of polyacetylene to field theory as follows: Let $d_1 > d_2$ be the internuclear distances characterizing single and double bonds, respectively. Define a scalar field which is a function of the link i by $\varphi_i = (-1)^i(d - \frac{1}{2}d_1 - \frac{1}{2}d_2)$, where d is the internuclear distance for link i. Thus in the A configuration $\varphi_i = \frac{1}{2}(d_1 - d_2)$ (independent of i), in the B configuration $\varphi_i = -\frac{1}{2}(d_1 - d_2)$, and in the soliton configuration φ_i interpolates between these values. Now we can show that it makes sense to approximate φ_i by a continuum field and the interactions of the electrons with φ (a charge-density wave) by $\mathcal{L}_I = g\bar{\psi}\gamma^5\varphi\psi$; furthermore the electrons can be treated for present purposes (near the Fermi energy) as relativistic particles.

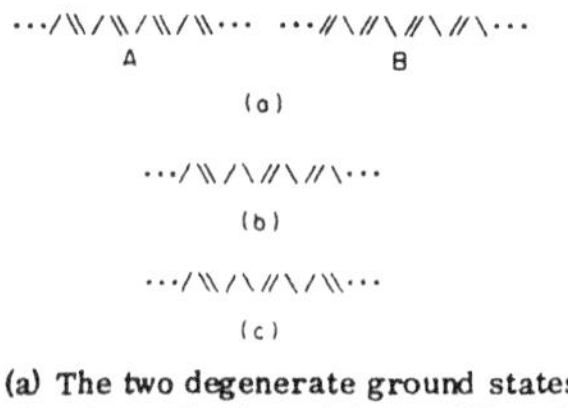

FIG. 1. (a) The two degenerate ground states for electronic structure of polyacetylene. (b) An imperfection interpolating between the two ground states. (c) A chain with two imperfections.

In this formulation, we make contact with the work of Jackiw and Rebbi.[1] They found that the spectrum of the Dirac equation in the presence of a soliton contains a zero-energy solution. By symmetry, this solution is composed of (projects onto) half a positive-energy and half a negative-energy solution with respect to the normal ground state. Thus if we fill the zero-energy level, we have a soliton state with electron number $+\frac{1}{2}$; if we leave it empty, the electron number is $-\frac{1}{2}$.

Su and Schrieffer have described a generalization,[10] which occurs in a chain with a repeating unit of single-single-double bonds, as in Fig. 2. A slight modification of the discussion of Fig. 1 shows that we now have solitons which can be added in triples to give the normal ground state, deficient by one electron. We expect the electron number of a single soliton to be $-\frac{1}{3}$.

A field theoretic model must now have essentially new features. Jackiw and Rebbi emphasized that in their model the Dirac equation in the presence of a soliton has a charge-conjugation symmetry, and then their interpretation of the zero modes cannot account for any charges other than half-integral. Thus we will consider models where the background destroys all symmetries which interchange positive- and negative-energy solutions of the Dirac equation.

Our method of calculating the soliton quantum number will be to imagine building up the soliton by slow changes in fields, starting from the ground state. In order to reach the solitons by slow changes, we may have to enlarge the field space during intermediate stages, as we shall see. In any case, for slow variations of fields in space and time, we can readily compute the flow of the appropriate charge in the no-particle state. We then simply integrate to find the accumulated charge on the soliton.

Let us illustrate these remarks on a concrete example. We consider, in 1 + 1 dimensions, massless fermions interacting with two scalar fields φ_1 and φ_2 as follows:

$$\mathcal{L}_I = g\bar{\psi}(\varphi_1 + i\gamma_5\varphi_2)\psi . \qquad (1)$$

Now if φ_1 and φ_2 are slowly varying in space and

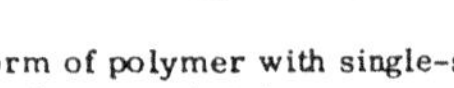

FIG. 2. A form of polymer with single-single-double bond pattern in the ground state.

VOLUME 47, NUMBER 14 PHYSICAL REVIEW LETTERS 5 OCTOBER 1981

time, i.e., their gradients are $\ll g(\varphi_1^2+\varphi_2^2)^{1/2}$, we may conveniently calculate the change in the expected value of $j^\mu=\bar\psi\gamma^\mu\psi$ in the no-particle state by considering the Feynman graph of Fig. 3. Since the interaction (1) is chirally invariant, we may first suppose that only $\varphi_1\neq 0$ at a given point, and then express the result in a chirally symmetric form. We then need only do a very simple calculation for an effectively massive fermion to find

$$\langle j^\mu\rangle=\frac{1}{2\pi}\epsilon^{\mu\nu}\epsilon_{ab}\frac{\varphi_a\partial_\nu\varphi_b}{|\varphi|^2}$$
$$=\frac{1}{2\pi}\epsilon^{\mu\nu}\partial_\nu\tan^{-1}\frac{\varphi_2}{\varphi_1}. \qquad (2)$$

If the scalar fields do not propagate (they represent very massive particles) more complicated graphs need not be considered.

If in the end we reach the soliton state by slow changes, we need only to evaluate (2) to find the fermion number charge on the soliton. It is important to remark that the resulting state will be a true eigenstate of the charge, not a superposition of states of different charge (even though we only derived an expectation value). For this it is only necessary to note that there are no degenerate states of different charge. In this the localized charge on a soliton differs from, for instance, the "localized charges" of $\frac{1}{2}$ on the top and bottom of an ammonia ion.

Two general features of the result deserve comment. First, the divergence $\partial_\mu j^\mu$ vanishes identically, reflecting the conservation of fermion number. Second, the charge $Q=\int j^0 dx^1=(2\pi)^{-1}\times\Delta(\tan^{-1}\varphi_2/\varphi_1)$ is independent of the coupling constant g and depends only on the values of φ_1 and φ_2 at spatial infinity.

We can represent a massive fermion by fixing $\varphi_1=m/g$. If the theory supports a soliton for which $\varphi_2(x)\to\pm v$ as $x\to\pm\infty$, we find

$$Q=\pi^{-1}\tan^{-1}(gv/m). \qquad (3)$$

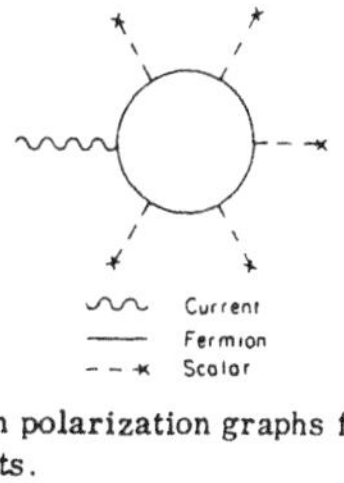

FIG. 3. Vacuum polarization graphs for evaluation of induced currents.

Notice that this is a transcendental function of the couplings! As $m\to 0$, we find $Q\to\frac{1}{2}$; this is the Jackiw-Rebbi case of a single (linear) scalar coupling. The limit $m\to 0$ is delicate just because there are two degenerate states of charge $\pm\frac{1}{2}$ in the limit. If we take $m=0$ from the beginning, adiabatic changes will fill these equally on an average. The current would vanish. A slight perturbation lifts the degeneracy. Of course the charge $-\frac{1}{2}$ state is reached by letting $m\to 0$ through negative values.

A field theory version of the chains of Figs. 1 and 2 is the interaction

$$\mathcal{L}_I=g\bar\psi e^{i\theta\gamma_5}\psi \qquad (4)$$

for which we find

$$\langle j^\mu\rangle=(2\pi)^{-1}\epsilon^{\mu\nu}\partial_\nu\theta,\quad Q=(2\pi)^{-1}\Delta\theta. \qquad (5)$$

The solitons with θ varying from 0 to π (so two together give $0\to 2\pi\sim 0$, equivalent to vacuum) have charge $\frac{1}{2}$; with θ varying from 0 to $2\pi/3$, charge $\frac{1}{3}$; etc.

Some 1+1 dimensional models become especially transparent if the method of bosonization is employed. In 1+1 dimensions, one can rewrite fermion fields as nonlocal expressions in boson fields.[11] Some bilinears transform in a simple local way, however:

$$i\bar\psi\gamma^\mu\partial_\mu\psi\to\tfrac{1}{2}\partial_\mu\varphi\partial^\mu\varphi,$$
$$\bar\psi\gamma^\mu\psi\to\epsilon^{\mu\nu}\partial_\nu\varphi/\sqrt{\pi},$$
$$\bar\psi\psi\to\mu\cos 2\sqrt{\pi}\,\varphi,$$
$$i\bar\psi\gamma_5\psi\to\mu\sin 2\sqrt{\pi}\,\varphi$$

(μ is an arbitrary scale parameter). Thus the interaction (4) becomes in this representation $\mathcal{L}_I=g\mu\cos(2\sqrt{\pi}\,\varphi-\theta)$. Now if θ in a soliton varies by $\Delta\theta$ from $-\infty$ to $+\infty$, the potential $-\mathcal{L}$ is minimized when $\varphi=\theta/2\sqrt{\pi}$; in particular, $\Delta\varphi=\Delta\theta/2\sqrt{\pi}$. Integrating $\bar\psi\gamma^0\psi=\partial_1\varphi/\sqrt{\pi}$, we find the charge $\Delta\theta/2\pi$, as from our earlier derivation.

Although the σ model proper does not support finite-energy solitons, we can consider a fermion interacting with external fields of this type. This proves useful as a warmup for the gauge theory monopoles to be discussed shortly.

The interaction Lagrangian is of standard form

$$\mathcal{L}_I=g\bar\psi(\varphi_0+i\vec\varphi\cdot\vec\tau\gamma_5)\psi$$

with ψ an isodoublet fermion field. We compute the induced current as in the 1+1 dimensional examples, from graphs as in Fig. 3. A straight-

VOLUME 47, NUMBER 14 PHYSICAL REVIEW LETTERS 5 OCTOBER 1981

forward calculation leads to

$$\langle j^\mu\rangle = \frac{1}{12\pi^2|\varphi|^4}\epsilon^{\mu\alpha\beta\gamma}\epsilon_{dabc}\,\varphi_d\,\partial_\alpha\varphi_a\,\partial_\beta\varphi_b\,\partial_\gamma\varphi_c\,. \qquad (6)$$

With this form, $\partial_\mu\langle j^\mu\rangle \equiv 0$. This, of course, indicates that only the behavior at spatial infinity determines the charge, since changes in the fields in a finite volume lead only to current flows in a finite volume and therefore do not change the total charge. In fact, if we take $\varphi_0 = m/g$, $\varphi_a = \hat{\varphi}_a(\vec{x})f(t)$, $a = 1, 2, 3$, where $\hat{\varphi}_a(\vec{x}) \to vx_a/|x|$ as $|x| \to \infty$; and evaluate the current flow at infinity, we find a fermion number

$$\pi^{-1}(\theta - \sin\theta\cos\theta),\quad \tan\theta = gv/m, \qquad (7)$$

which $\to \frac{1}{2}$ as $m \to 0$.

We may extend this analysis in a simple way to the monopole solutions of non-Abelian gauge theories by simply gauging the SU(2) ⊗ SU(2) chiral symmetry of our σ model. In the end, we can specialize by setting the axial gauge fields to zero, and fixing a fermion mass ($\varphi_0 = \text{const}$).

The expression (6) for the current is changed in the first instance by the conversion of ordinary to covariant derivatives, $\partial \to \nabla \equiv \partial + eA$. This is not sufficient, however, since this minimally modified current is not conserved. The current

$$\langle j^\mu\rangle = \frac{1}{12\pi^2}\epsilon^{\mu\alpha\beta\gamma}\epsilon_{dabc}\left[\frac{\varphi_d}{|\varphi|^4}(\nabla_\alpha\varphi)_a(\nabla_\beta\varphi)_b(\nabla_\gamma\varphi)_c + \tfrac{3}{4}eF_{\alpha\beta,ab}\frac{\varphi_d}{|\varphi|^2}(\nabla_\gamma\varphi)_c\right] \qquad (8)$$

obeys $\partial_\mu\langle j^\mu\rangle \equiv (-e^2/128\pi^2)\epsilon^{\alpha\beta\gamma\delta}\epsilon_{abcd}F_{\alpha\beta,ab}F_{\gamma\delta,cd}$. This is the expected anomaly and vanishes when we have only vector gauge fields as in the monopole. The coefficient of the second term in (8) can be checked by the evaluation of the diagram in Fig. 3 with one gauge field vertex inserted.

We now take φ as before and $A_{ab} = \hat{A}_{ab}(\vec{x})$, $A_{a0} = 0$, $a, b = 1, 2, 3$, where $\hat{\varphi}$ and $\hat{A}$ are the monopole fields, and find the current flow at infinity. Since $(\nabla_i\varphi)_a = 0$ at infinity, the only contribution comes from taking $\gamma = d = 0$ in the second term of (8) and gives for the fermion number

$$(e\Phi/4\pi^2)\tan^{-1}(gv/m)\,, \qquad (9)$$

where Φ is the magnetic flux out of the sphere at infinity. Since $e\Phi = 4\pi$, this gives fermion number $\frac{1}{2}$ when $m \to 0$!

The direct utility of our results for particle physics is highly problematical. Even if magnetic monopoles were found, their fermion number is not a reasonable quantity in standard theories. [In principle, we could imagine coupling a U(1) gauge field to the fermion number, and so the calculation is not entirely content free!] We do think that the results are an interesting curiosity in quantum field theory and as such may eventually be useful. It is likely that kindred, but experimentally accessible, effects do arise in condensed matter systems.

We are especially grateful to J. R. Schrieffer for interesting us in this problem and to L. Susskind for reminding us of the bosonization method.

This work was started while the first author was at the Santa Barbara Institute. This work was supported in part by the U. S. Department of Energy under Contract No. DE-AC03-SF00515 and by the National Science Foundation under Grant No. PHY 77-27084.

[a]Permanent address: Center for Theoretical Physics, Massachusetts Institute of Technology, Cambridge, Mass. 02139

[1]R. Jackiw and C. Rebbi, Phys. Rev. D 13, 3398 (1976).

[2]W. P. Su, J. R. Schrieffer, and A. J. Heeger, Phys. Rev. Lett. 42, 1698 (1979), and Phys. Rev. B 22, 2099 (1980).

[3]R. Jackiw and J. R. Schrieffer, Santa Barbara Report No. NSF-ITP-81-01 (to be published).

[4]P. Hasenfratz and G. 't Hooft, Phys. Rev. Lett. 36, 1119 (1976).

[5]R. Jackiw and C. Rebbi, Phys. Rev. Lett. 36, 1116 (1976).

[6]F. Wilczek, to be published.

[7]E. Witten, Phys. Lett. 86B, 283 (1979).

[8]F. Wilczek, to be published.

[9]For a discussion of both theoretical and experimental aspects of polyacetylene, see A. Heeger, Comments Solid State Phys. 10, 53 (1981).

[10]W. P. Su and J. R. Schrieffer, Phys. Rev. Lett. 46, 738 (1981).

[11]See, e.g., S. Mandelstam, Phys. Rev. D 11, 3026 (1975); S. Coleman, R. Jackiw, and L. Susskind, Ann. Phys. (N.Y.) 93, 267 (1975).

Nuclear Physics B223 (1983) 422–432

GLOBAL ASPECTS OF CURRENT ALGEBRA

Edward WITTEN*

Joseph Henry Laboratories, Princeton University, Princeton, New Jersey 08544, USA

Received 4 March 1983

A new mathematical framework for the Wess-Zumino chiral effective action is described. It is shown that this action obeys an a priori quantization law, analogous to Dirac's quantization of magnetic change. It incorporates in current algebra both perturbative and non-perturbative anomalies.

The purpose of this paper is to clarify an old but relatively obscure aspect of current algebra: the Wess-Zumino effective lagrangian [1] which summarizes the effects of anomalies in current algebra. As we will see, this effective lagrangian has unexpected analogies to some 2 + 1 dimensional models discussed recently by Deser et al. [2] and to a recently noted SU(2) anomaly [3]. There also are connections with work of Balachandran et al. [4].

For definiteness we will consider a theory with $SU(3)_L \times SU(3)_R$ symmetry spontaneously broken down to the diagonal SU(3). We will ignore explicit symmetry-breaking perturbations, such as quark bare masses. With $SU(3)_L \times SU(3)_R$ broken to diagonal SU(3), the vacuum states of the theory are in one to one correspondence with points in the SU(3) manifold. Correspondingly, the low-energy dynamics can be conveniently described by introducing a field $U(x^\alpha)$ that transforms in a so-called non-linear realization of $SU(3)_L \times SU(3)_R$. For each space-time point x^α, $U(x^\alpha)$ is an element of SU(3): a 3×3 unitary matrix of determinant one. Under an $SU(3)_L \times SU(3)_R$ transformation by unitary matrices (A, B), U transforms as $U \to AUB^{-1}$.

The effective lagrangian for U must have $SU(3)_L \times SU(3)_R$ symmetry, and, to describe correctly the low-energy limit, it must have the smallest possible number of derivatives. The unique choice with only two derivatives is

$$\mathcal{L} = \tfrac{1}{16} F_\pi^2 \int d^4x \, \mathrm{Tr}\, \partial_\mu U \partial_\mu U^{-1}, \tag{1}$$

* Supported in part by NSF Grant PHY80-19754.

where experiment indicates $F_\pi \simeq 190$ MeV. The perturbative expansion of U is

$$U = 1 + \frac{2i}{F_\pi} \sum_{a=1}^{8} \lambda^a \pi^a + \cdots, \tag{2}$$

where λ^a (normalized so $\mathrm{Tr}\,\lambda^a\lambda^b = 2\delta^{ab}$) are the SU(3) generators and π^a are the Goldstone boson fields.

This effective lagrangian is known to incorporate all relevant symmetries of QCD. All current algebra theorems governing the extreme low-energy limit of Goldstone boson S-matrix elements can be recovered from the tree approximation to it. What is less well known, perhaps, is that (1) possesses an extra discrete symmetry that is *not* a symmetry of QCD.

The lagrangian (1) is invariant under $U \leftrightarrow U^{\mathrm{T}}$. In terms of pions this is $\pi^0 \leftrightarrow \pi^0$, $\pi^+ \leftrightarrow \pi^-$; it is ordinary charge conjugation. (1) is also invariant under the naive parity operation $x \leftrightarrow -x$, $t \leftrightarrow t$, $U \leftrightarrow U$. We will call this P_0. And finally, (1) is invariant under $U \leftrightarrow U^{-1}$. Comparing with eq. (2), we see that this latter operation is equivalent to $\pi^a \leftrightarrow -\pi^a$, $a = 1, \ldots, 8$. This is the operation that counts modulo two the number of bosons, N_{B}, so we will call it $(-1)^{N_{\mathrm{B}}}$.

Certainly, $(-1)^{N_{\mathrm{B}}}$ is not a symmetry of QCD. The problem is the following. QCD is parity invariant only if the Goldstone bosons are treated as pseudoscalars. The parity operation in QCD corresponds to $x \leftrightarrow -x$, $t \leftrightarrow t$, $U \leftrightarrow U^{-1}$. This is $P = P_0(-1)^{N_{\mathrm{B}}}$. QCD is invariant under P but not under P_0 or $(-1)^{N_{\mathrm{B}}}$ separately. The simplest process that respects all bona fide symmetries of QCD but violates P_0 and $(-1)^{N_{\mathrm{B}}}$ is $\mathrm{K}^+\mathrm{K}^- \to \pi^+\pi^0\pi^-$ (note that the ϕ meson decays to both $\mathrm{K}^+\mathrm{K}^-$ and $\pi^+\pi^0\pi^-$). It is natural to ask whether there is a simple way to add a higher-order term to (1) to obtain a lagrangian that obeys *only* the appropriate symmetries.

The Euler-Lagrangian equation derived from (1) can be written

$$\partial_\mu\left(\tfrac{1}{8}F_\pi^2 U^{-1}\partial_\mu U\right) = 0. \tag{3}$$

Let us try to add a suitable extra term to this equation. A Lorentz-invariant term that violates P_0 must contain the Levi-Civita symbol $\varepsilon_{\mu\nu\alpha\beta}$. In the spirit of current algebra, we wish a term with the smallest possible number of derivatives, since, in the low-energy limit, the derivatives of U are small. There is a unique P_0-violating term with only four derivatives. We can generalize (3) to

$$\partial_\mu\left(\tfrac{1}{8}F_\pi^2 U^{-1}\partial_\mu U\right) + \lambda\varepsilon^{\mu\nu\alpha\beta}U^{-1}(\partial_\mu U)U^{-1}(\partial_\nu U)U^{-1}(\partial_\alpha U)U^{-1}(\partial_\beta U) = 0, \tag{4}$$

λ being a constant. Although it violates P_0, (4) can be seen to respect $P = P_0(-1)^{N_{\mathrm{B}}}$.

Can eq. (4) be derived from a lagrangian? Here we find trouble. The only pseudoscalar of dimension four would seem to be $\varepsilon^{\mu\nu\alpha\beta}\mathrm{Tr}\,U^{-1}(\partial_\mu U)\cdot U^{-1}(\partial_\nu U)U^{-1}(\partial_\alpha U)U^{-1}(\partial_\beta U)$, but this vanishes, by antisymmetry of $\varepsilon^{\mu\nu\alpha\beta}$ and cyclic symmetry of the trace. Nevertheless, as we will see, there is a lagrangian.

Let us consider a simple problem of the same sort. Consider a particle of mass m constrained to move on an ordinary two-dimensional sphere of radius one. The lagrangian is $\mathcal{L} = \frac{1}{2}m\int \mathrm{d}t\, \dot{x}_i^2$ and the equation of motion is $m\ddot{x}_i + mx_i(\Sigma_k \dot{x}_k^2) = 0$; the constraint is $\Sigma x_i^2 = 1$. This system respects the symmetries $t \leftrightarrow -t$ and separately $x_i \leftrightarrow -x_i$. If we want an equation that is only invariant under the combined operation $t \leftrightarrow -t$, $x_i \leftrightarrow x_i$, the simplest choice is

$$m\ddot{x}_i + mx_i\left(\sum_k \dot{x}_k^2\right) = \alpha\varepsilon_{ijk}x_j\dot{x}_k, \tag{5}$$

where α is a constant. To derive this equation from a lagrangian is again troublesome. There is no obvious term whose variation equals the right-hand side (since $\varepsilon_{ijk}x_i x_j \dot{x}_k = 0$).

However, this problem has a well-known solution. The right-hand side of (5) can be understood as the Lorentz force for an electric charge interacting with a magnetic monopole located at the center of the sphere. Introducing a vector potential A such that $\nabla \times A = x/|x|^3$, the action for our problem is

$$I = \int \left(\tfrac{1}{2}m\dot{x}_i^2 + \alpha A_i \dot{x}_i\right) \mathrm{d}t. \tag{6}$$

This lagrangian is problematical because A_i contains a Dirac string and certainly does not respect the symmetries of our problem. To explore this quantum mechanically let us consider the simplest form of the Feynman path integral, $\mathrm{Tr}\,\mathrm{e}^{-\beta H} = \int \mathrm{d}x_i(t)\mathrm{e}^{-I}$. In e^{-I} the troublesome term is

$$\exp\left(i\alpha\int_\gamma A_i\,\mathrm{d}x^i\right), \tag{7}$$

where the integration goes over the particle orbit γ: a closed orbit if we discuss the simplest object $\mathrm{Tr}\,\mathrm{e}^{-\beta H}$.

By Gauss's law we can eliminate the vector potential from (7) in favor of the magnetic field. In fact, the closed orbit γ of fig. 1a is the boundary of a disc D, and by Gauss's law we can write (7) in terms of the magnetic flux through D:

$$\exp\left(i\alpha\int_\gamma A_i\,\mathrm{d}x^i\right) = \exp\left(i\alpha\int_{\mathrm{D}} F_{ij}\,\mathrm{d}\Sigma^{ij}\right). \tag{8}$$

The precise mathematical statement here is that since $\pi_1(\mathrm{S}^2) = 0$, the circle γ in S^2 is the boundary of a disc D (or more exactly, a mapping γ of a circle into S^2 can be extended to a mapping of a disc into S^2).

The right-hand side of (8) is manifestly well defined, unlike the left-hand side, which suffers from a Dirac string. We could try to use the right-hand side of (8) in a Feynman path integral. There is only one problem: D isn't unique. The curve γ also bounds the disc D′ (fig. 1c). There is no consistent way to decide whether to choose

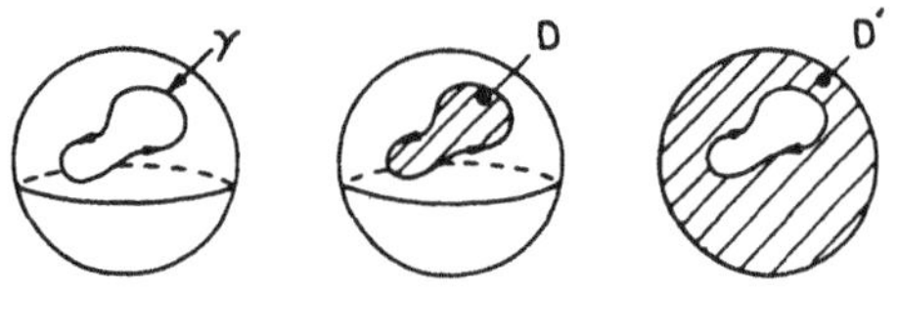

(a) (b) (c)

Fig. 1. A particle orbit γ on the two-sphere (part (a)) bounds the discs D (part (b)) and D′ (part (c)).

D or D′ (the curve γ could continuously be looped around the sphere or turned inside out). Working with D′ we would get

$$\exp\left(i\alpha\int_{\gamma} A_i\,\mathrm{d}x^i\right) = \exp\left(-i\alpha\int_{\mathrm{D}'} F_{ij}\,\mathrm{d}\Sigma^{ij}\right), \tag{9}$$

where a crucial minus sign on the right-hand side of (9) appears because γ bounds D in a right-hand sense, but bounds D′ in a left-hand sense. If we are to introduce the right-hand side of (8) or (9) in a Feynman path integral, we must require that they be equal. This is equivalent to

$$1 = \exp\left(i\alpha\int_{\mathrm{D}+\mathrm{D}'} F_{ij}\,\mathrm{d}\Sigma^{ij}\right). \tag{10}$$

Since D + D′ is the whole two sphere S^2, and $\int_{S^2} F_{ij}\,\mathrm{d}\Sigma^{ij} = 4\pi$, (10) is obeyed if and only if α is an integer or half-integer. This is Dirac's quantization condition for the product of electric and magnetic charges.

Now let us return to our original problem. We imagine space-time to be a very large four-dimensional sphere M. A given non-linear sigma model field U is a mapping of M into the SU(3) manifold (fig. 2a). Since $\pi_4(\mathrm{SU}(3)) = 0$, the four-sphere in SU(3) defined by U(x) is the boundary of a five-dimensional disc Q.

By analogy with the previous problem, let us try to find some object that can be integrated over Q to define an action functional. On the SU(3) manifold there is a unique fifth rank antisymmetric tensor ω_{ijklm} that is invariant under $\mathrm{SU(3)_L} \times \mathrm{SU(3)_R}$*. Analogous to the right-hand side of eq. (8), we define

$$\Gamma = \int_{\mathrm{Q}} \omega_{ijklm}\,\mathrm{d}\Sigma^{ijklm}. \tag{11}$$

* Let us first try to define ω at $U = 1$; it can then be extended to the whole SU(3) manifold by an $\mathrm{SU(3)_L} \times \mathrm{SU(3)_R}$ transformation. At $U = 1$, ω must be invariant under the diagonal subgroup of $\mathrm{SU(3)_L} \times \mathrm{SU(3)_R}$ that leaves fixed $U = 1$. The tangent space to the SU(3) manifold at $U = 1$ can be identified with the Lie algebra of SU(3). So ω, at $U = 1$, defines a fifth-order antisymmetric invariant in the SU(3) Lie algebra. There is only one such invariant. Given five SU(3) generators A, B, C, D and E, the one such invariant is $\mathrm{Tr}\,ABCDE - \mathrm{Tr}\,BACDE \pm$ permutations. The $\mathrm{SU(3)_L} \times \mathrm{SU(3)_R}$ invariant ω so defined has zero curl ($\partial_i \omega_{jklmn} \pm$ permutations $= 0$) and for this reason (11) is invariant under infinitesimal variations of Q; there arises only the topological problem discussed in the text.

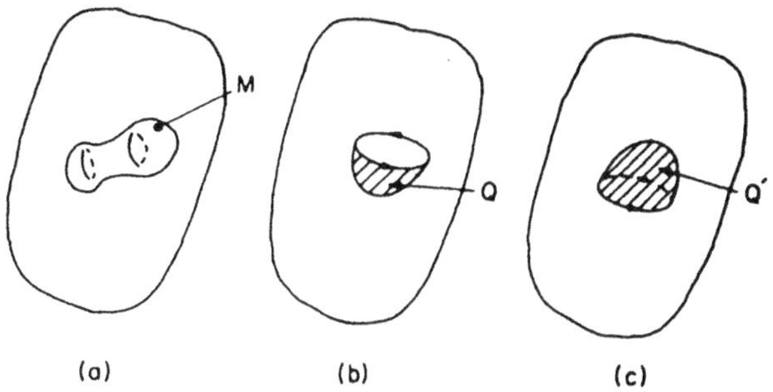

Fig. 2. Space-time, a four-sphere, is mapped into the SU(3) manifold. In part (a), space-time is symbolically denoted as a two sphere. In parts (b) and (c), space-time is reduced to a circle that bounds the discs Q and Q'. The SU(3) manifold is symbolized in these sketches by the interior of the oblong.

As before, we hope to include $\exp(i\Gamma)$ in a Feynman path integral. Again, the problem is that Q is not unique. Our four-sphere M is also the boundary of another five-disc Q' (fig. 2c). If we let

$$\Gamma' = -\int_{Q'} \omega_{ijklm} \, d\Sigma^{ijklm}, \tag{12}$$

(with, again, a minus sign because M bounds Q' with opposite orientation) then we must require $\exp(i\Gamma) = \exp(i\Gamma')$ or equivalently $\int_{Q+Q'} \omega_{ijklm} \, d\Sigma^{ijklm} = 2\pi \cdot \text{integer}$. Since Q + Q' is a closed five-dimensional sphere, our requirement is

$$\int_{S} \omega_{ijklm} \, d\Sigma^{ijklm} = 2\pi \cdot \text{integer},$$

for any five-sphere S in the SU(3) manifold.

We thus need the topological classification of mappings of the five-sphere into SU(3). Since $\pi_5(\mathrm{SU}(3)) = Z$, every five sphere in SU(3) is topologically a multiple of a basic five sphere S_0. We normalize ω so that

$$\int_{S_0} \omega_{ijklm} \, d\Sigma^{ijklm} = 2\pi, \tag{13}$$

and then (with Γ in eq. (11)) we may work with the action

$$I = \tfrac{1}{16} F_\pi^2 \int d^4x \, \mathrm{Tr} \, \partial_\mu U \, \partial_\mu U^{-1} + n\Gamma, \tag{14}$$

where n is an arbitrary integer. Γ is, in fact, the Wess-Zumino lagrangian. Only the a priori quantization of n is a new result.

The identification of S_0 and the proper normalization of ω is a subtle mathematical problem. The solution involves a factor of two from the Bott periodicity theorem. Without abstract notation, the result [5] can be stated as follows. Let y^i, $i = 1 \dots 5$ be coordinates for the disc Q. Then on Q (where we need it)

$$\mathrm{d}\Sigma^{ijklm}\,\omega_{ijklm} = -\frac{i}{240\pi^2}\mathrm{d}\Sigma^{ijklm}\left[\mathrm{Tr}\, U^{-1}\frac{\partial U}{\partial y^i}U^{-1}\frac{\partial U}{\partial y^j}U^{-1}\frac{\partial U}{\partial y^k}U^{-1}\frac{\partial U}{\partial y^l}U^{-1}\frac{\partial U}{\partial y^m}\right]. \tag{15}$$

The physical consequences of this can be made more transparent as follows. From eq. (2),

$$U^{-1}\partial_i U = \frac{2i}{F_\pi}\partial_i A + \mathrm{O}(A^2), \qquad \text{where } A = \Sigma\lambda^a\pi^a. \tag{16}$$

So

$$\omega_{ijklm}\mathrm{d}\Sigma^{ijklm} = \frac{2}{15\pi^2 F_\pi^5}\mathrm{d}\Sigma^{ijklm}\mathrm{Tr}\,\partial_i A\,\partial_j A\,\partial_k A\,\partial_l A\,\partial_m A + \mathrm{O}(A^6)$$

$$= \frac{2}{15\pi^2 F_\pi^5}\mathrm{d}\Sigma^{ijklm}\,\partial_i(\mathrm{Tr}\,A\,\partial_j A\,\partial_k A\,\partial_l A\,\partial_m A) + \mathrm{O}(A^6).$$

So $\int_Q \omega_{ijklm}\mathrm{d}\Sigma^{ijklm}$ is (to order A^5 and in fact also in higher orders) the integral of a total divergence which can be expressed by Stokes' theorem as an integral over the boundary of Q. By construction, this boundary is precisely space-time. We have, then,

$$n\Gamma = n\frac{2}{15\pi^2 F_\pi^5}\int \mathrm{d}^4x\,\varepsilon^{\mu\nu\alpha\beta}\mathrm{Tr}\,A\,\partial_\mu A\,\partial_\nu A\,\partial_\alpha A\,\partial_\beta A + \text{higher order terms}. \tag{17}$$

In a hypothetical world of massless kaons and pions, this effective lagrangian rigorously describes the low-energy limit of $K^+K^- \to \pi^+\pi^0\pi^-$*. We reach the remarkable conclusion that in any theory with SU(3)×SU(3) broken to diagonal SU(3), the low-energy limit of the amplitude for this reaction must be (in units given in (17)) an integer.

What is the value of this integer in QCD? Were n to vanish, the practical interest of our discussion would be greatly reduced. It turns out that if N_c is the number of colors (three in the real world) then $n = N_c$. The simplest way to deduce this is a

* Our formula should agree for $n = 1$ with formulas of ref. [1], as later equations make clear. There appears to be a numerical error on p. 97 of ref. [1] ($\frac{1}{6}$ instead of $\frac{2}{15}$).

procedure that is of interest anyway, viz. coupling to electromagnetism, so as to describe the low-energy dynamics of Goldstone bosons and photons.

Let

$$Q=\begin{pmatrix} \frac{2}{3} & & \\ & -\frac{1}{3} & \\ & & -\frac{1}{3} \end{pmatrix}$$

be the usual electric charge matrix of quarks. The functional Γ is invariant under global charge rotations, $U \to U + i\varepsilon[Q,U]$, where ε is a constant. We wish to promote this to a local symmetry, $U \to U + i\varepsilon(x)[Q,U]$, where $\varepsilon(x)$ is an arbitrary function of x. It is necessary, of course, to introduce the photon field A_μ which transforms as $A_\mu \to A_\mu - (1/e)\partial_\mu \varepsilon$; e is the charge of the proton.

Usually a global symmetry can straightforwardly be gauged by replacing derivatives by covariant derivatives, $\partial_\mu \to D_\mu = \partial_\mu + ieA_\mu$. In the case at hand, Γ is not given as the integral of a manifestly $SU(3)_L \times SU(3)_R$ invariant expression, so the standard road to gauging global symmetries of Γ is not available. One can still resort to the trial and error Noether method, widely used in supergravity. Under a local charge rotation, one finds $\Gamma \to \Gamma - \int d^4x\, \partial_\mu \varepsilon J^\mu$ where

$$J^\mu = \frac{1}{48\pi^2}\varepsilon^{\mu\nu\alpha\beta}\mathrm{Tr}\big[Q(\partial_\nu U\, U^{-1})(\partial_\alpha U\, U^{-1})(\partial_\beta U\, U^{-1}) + Q(U^{-1}\partial_\nu U)(U^{-1}\partial_\alpha U)(U^{-1}\partial_\beta U)\big], \tag{18}$$

is the extra term in the electromagnetic current required (from Noether's theorem) due to the addition of Γ to the lagrangian. The first step in the construction of an invariant lagrangian is to add the Noether coupling, $\Gamma \to \Gamma' = \Gamma - e\int d^4x\, A_\mu J^\mu(x)$. This expression is still not gauge invariant, because J^μ is not, but by trial and error one finds that by adding an extra term one can form a gauge invariant functional

$$\tilde{\Gamma}(U, A_\mu) = \Gamma(U) - e\int d^4x\, A_\mu J^\mu + \frac{ie^2}{24\pi^2}\int d^4x\, \varepsilon^{\mu\nu\alpha\beta}(\partial_\mu A_\nu)A_\alpha \times \mathrm{Tr}\big[Q^2(\partial_\beta U)U^{-1} + Q^2U^{-1}(\partial_\beta U) + QUQU^{-1}(\partial_\beta U)U^{-1}\big]. \tag{19}$$

Our gauge invariant lagrangian will then be

$$\mathcal{L} = \tfrac{1}{16}F_\pi^2\int d^4x\, \mathrm{Tr}\, D_\mu U D_\mu U^{-1} + n\tilde{\Gamma}. \tag{20}$$

What value of the integer n will reproduce QCD results?

Here we find a surprise. The last term in (18) has a piece that describes $\pi^0 \to \gamma\gamma$. Expanding U and integrating by parts, (18) has a piece

$$A = \frac{ne^2}{48\pi^2 F_\pi} \pi^0 \varepsilon^{\mu\nu\alpha\beta} F_{\mu\nu} F_{\alpha\beta} . \tag{21}$$

This agrees with the result from QCD triangle diagrams [6] if $n = N_c$, the number of colors. The Noether coupling $-eA_\mu J^\mu$ describes, among other things, a $\gamma\pi^+\pi^0\pi^-$ vertex

$$B = -\tfrac{2}{3} ie \frac{n}{\pi^2 F_\pi^3} \varepsilon^{\mu\nu\alpha\beta} A_\mu \, \partial_\nu \pi^+ \, \partial_\alpha \pi^- \, \partial_\beta \pi^0 . \tag{22}$$

Again this agrees with calculations [7] based on the QCD VAAA anomaly if $n = N_c$. The effective action $N_c \tilde{\Gamma}$ (first constructed in another way by Wess and Zumino) precisely describes all effects of QCD anomalies in low-energy processes with photons and Goldstone bosons.

It is interesting to try to gauge subgroups of $SU(3)_L \times SU(3)_R$ other than electromagnetism. One may have in mind, for instance, applications to the standard weak interaction model. In general, one may try to gauge an arbitrary subgroup H of $SU(3)_L \times SU(3)_R$, with generators K^σ, $\sigma = 1 \dots r$. Each K^σ is a linear combination of generators T_L^σ and T_R^σ of $SU(3)_L$ and $SU(3)_R$, $K^\sigma = T_L^\sigma + T_R^\sigma$. (Either T_L^σ or T_R^σ may vanish for some values of σ.) For any space-time dependent functions $\varepsilon^\sigma(x)$, let $\varepsilon_L = \Sigma_\sigma T_L^\sigma \varepsilon^\sigma(x)$, $\varepsilon_R = \Sigma_\sigma T_R^\sigma \varepsilon^\sigma(x)$. We want an action with local invariance under $U \to U + i(\varepsilon_L(x) U - U \varepsilon_R(x))$.

Naturally, it is necessary to introduce gauge fields $A_\mu^\sigma(x)$, transforming as $A_\mu^\sigma(x) \to A_\mu^\sigma(x) - (1/e_\sigma)\, \partial_\mu \varepsilon^\sigma + f^{\sigma\tau\rho} \varepsilon^\tau A_\mu^\rho$ where e_σ is the coupling constant corresponding to the generator K^σ, and $f^{\sigma\tau\rho}$ are the structure constants of H. It is useful to define $A_{\mu L} = \Sigma_\sigma e_\sigma A_\mu^\sigma T_L^\sigma$, $A_\mu^R = \Sigma_\sigma e_\sigma A_\mu^\sigma T_R^\sigma$.

We have already seen that Γ incorporates the effects of anomalies, so it is not very surprising that a generalization of Γ that is gauge invariant under H exists only if H is a so-called anomaly-free subgroup of $SU(3)_L \times SU(3)_R$. Specifically, one finds that H can be gauged only if for each σ,

$$\mathrm{Tr}(T_L^\sigma)^3 = \mathrm{Tr}(T_R^\sigma)^3 , \tag{23}$$

which is the usual condition for cancellation of anomalies at the quark level.

If (23) is obeyed, a gauge invariant generalization of Γ can be constructed somewhat tediously by trial and error. It is useful to define $U_{\nu L} = (\partial_\nu U) U^{-1}$ and $U_{\nu R} = U^{-1} \partial_\nu U$. The gauge invariant functional then turns out to be

$$\tilde{\Gamma}(A_\mu, U) = \Gamma(U) + \frac{1}{48\pi^2} \int d^4x \, \varepsilon^{\mu\nu\alpha\beta} Z_{\mu\nu\alpha\beta} ,$$

where

$$
\begin{aligned}
Z_{\mu\nu\alpha\beta} = & -\mathrm{Tr}\left[A_{\mu L}U_{\nu L}U_{\alpha L}U_{\beta L} + (\mathrm{L}\to\mathrm{R})\right] \\
& + i\,\mathrm{Tr}\left[\left[(\partial_\mu A_{\nu L})A_{\alpha L} + A_{\mu L}(\partial_\nu A_{\alpha L})\right]U_{\beta L} + (\mathrm{L}\to\mathrm{R})\right] \\
& + i\,\mathrm{Tr}\left[(\partial_\mu A_{\nu R})U^{-1}A_{\alpha L}\,\partial_\beta U + A_{\mu L}U^{-1}(\partial_\nu A_{\alpha R})\,\partial_\beta U\right] \\
& - \tfrac{1}{2}i\,\mathrm{Tr}\left(A_{\mu L}U_{\nu L}A_{\alpha L}U_{\beta L} - (\mathrm{L}\to\mathrm{R})\right) \\
& + i\,\mathrm{Tr}\left[A_{\mu L}UA_{\nu R}U^{-1}U_{\alpha L}U_{\beta L} - A_{\mu R}U^{-1}A_{\nu L}UU_{\alpha R}U_{\beta R}\right] \\
& - \mathrm{Tr}\Big[\left[(\partial_\mu A_{\nu R})A_{\alpha R} + A_{\mu R}(\partial_\nu A_{\alpha R})\right]U^{-1}A_{\beta L}U \\
& \quad - \left[(\partial_\mu A_{\nu L})A_{\alpha L} + A_{\mu L}(\partial_\nu A_{\alpha L})\right]UA_{\beta R}U^{-1}\Big] \\
& \quad - \mathrm{Tr}\left[A_{\mu R}U^{-1}A_{\nu L}UA_{\alpha R}U_{\beta R} + A_{\mu L}UA_{\nu R}U^{-1}A_{\alpha L}U_{\beta L}\right] \\
& \quad - \mathrm{Tr}\left[A_{\mu L}A_{\nu L}U(\partial_\alpha A_{\beta R})U^{-1} + A_{\mu R}A_{\nu R}U^{-1}(\partial_\alpha A_{\beta L})U\right] \\
& \quad - i\,\mathrm{Tr}\Big[A_{\mu R}A_{\nu R}A_{\alpha R}U^{-1}A_{\beta L}U - A_{\mu L}A_{\nu L}A_{\alpha L}UA_{\beta R}U^{-1} \\
& \qquad + \tfrac{1}{2}A_{\mu L}A_{\nu L}UA_{\alpha R}A_{\beta R}U^{-1} + \tfrac{1}{2}A_{\mu R}U^{-1}A_{\nu L}UA_{\alpha R}U^{-1}A_{\beta L}U\Big]. \qquad (24)
\end{aligned}
$$

If eq. (22) for cancellation of anomalies is not obeyed, then the variation of $\tilde{\Gamma}$ under a gauge transformation does not vanish but is

$$
\begin{aligned}
\delta\tilde{\Gamma} = & -\frac{1}{24\pi^2}\int \mathrm{d}^4x\, \varepsilon^{\mu\nu\alpha\beta}\,\mathrm{Tr}\,\varepsilon_{\mathrm{L}}\left[(\partial_\mu A_{\nu L})(\partial_\alpha A_{\beta L}) - \tfrac{1}{2}i\partial_\mu(A_{\nu L}A_{\alpha L}A_{\beta L})\right] \\
& -(\mathrm{L}\to\mathrm{R}), \qquad (25)
\end{aligned}
$$

in agreement with computations at the quark level [8] of the anomalous variation of the effective action under a gauge transformation.

Thus, Γ incorporates all information usually associated with triangle anomalies, including the restriction on what subgroups H of $\mathrm{SU(3)_L \times SU(3)_R}$ can be gauged. However, there is another potential obstruction to the ability to gauge a subgroup of $\mathrm{SU(3)_L \times SU(3)_R}$. This is the non-perturbative anomaly [3] associated with $\pi_4(\mathrm{H})$. Is this anomaly, as well, implicit in Γ? In fact, it is.

Let H be an SU(2) subgroup of $\mathrm{SU(3)_L}$, chosen so that an SU(2) matrix W is embedded in $\mathrm{SU(3)_L}$ as

$$
\hat{W} = \left(\begin{array}{cc|c} \multicolumn{2}{c|}{\multirow{2}{*}{W}} & 0 \\ & & 0 \\ \hline 0 & 0 & 1 \end{array}\right).
$$

This subgroup is free of triangle anomalies, so the functional $\tilde{\Gamma}$ of eq. (23) is invariant under infinitesimal local H transformations.

However, is $\tilde{\Gamma}$ invariant under H transformations that cannot be reached continuously? Since $\pi_4(\mathrm{SU}(2)) = Z_2$, there is one non-trivial homotopy class of SU(2) gauge transformations. Let W be an SU(2) gauge transformation in this non-trivial class. Under $\hat{W}$, $\tilde{\Gamma}$ may at most be shifted by a constant, independent of U and A_μ, because $\delta\tilde{\Gamma}/\delta U$ and $\delta\tilde{\Gamma}/\delta A_\mu$ are gauge-covariant local functionals of U and A_μ. Also $\tilde{\Gamma}$ is invariant under $\hat{W}^2$, since $\hat{W}^2$ is equivalent to the identity in $\pi_4(\mathrm{SU}(2))$, and we know $\tilde{\Gamma}$ is invariant under topologically trivial gauge transformations. This does not quite mean that $\tilde{\Gamma}$ is invariant under W. Since $\tilde{\Gamma}$ is only defined modulo 2π, the fact that $\tilde{\Gamma}$ is invariant under W^2 leaves two possibilities for how $\tilde{\Gamma}$ behaves under W. It may be invariant, or it may be shifted by π.

To choose between these alternatives, it is enough to consider a special case. For instance, it suffices to evaluate $\Delta = \tilde{\Gamma}(U=1,\ A_\mu = 0) - \tilde{\Gamma}(U = \hat{W},\ A_\mu = ie^{-1}(\partial_\mu \hat{W})\hat{W}^{-1})$. It is not difficult to see that in this case the complicated terms involving $\varepsilon^{\mu\nu\alpha\beta}Z_{\mu\nu\alpha\beta}$ vanish, so in fact $\Delta = \Gamma(U=1) - \Gamma(U=\hat{W})$. A detailed calculation shows that

$$\Gamma(U=1) - \Gamma(U=\hat{W}) = \pi . \tag{26}$$

This calculation has some other interesting applications and will be described elsewhere [9].

The Feynman path integral, which contains a factor $\exp(iN_c\tilde{\Gamma})$, hence picks up under W a factor $\exp(iN_c\pi) = (-1)^{N_c}$. It is gauge invariant if N_c is even, but not if N_c is odd. This agrees with the determination of the SU(2) anomaly at the quark level [3]. For under H, the right-handed quarks are singlets. The left-handed quarks consist of one singlet and one doublet per color, so the number of doublets equals N_c. The argument of ref. [3] shows at the quark level that the effective action transforms under W as $(-1)^{N_c}$.

Finally, let us make the following remark, which apart from its intrinsic interest will be useful elsewhere [9]. Consider $\mathrm{SU}(3)_\mathrm{L} \times \mathrm{SU}(3)_\mathrm{R}$ currents defined at the quark level as

$$J^a_{\mu\mathrm{L}} = \bar{q}\lambda^a\gamma_\mu \tfrac{1}{2}(1-\gamma_5)q, \qquad J^a_{\mu\mathrm{R}} = \bar{q}\lambda^a\gamma_\mu \tfrac{1}{2}(1+\gamma_5)q. \tag{27}$$

By analogy with eq. (17), the proper sigma model description of these currents contains pieces

$$J^{\mu a}_\mathrm{L} = \frac{N_c}{48\pi^2}\varepsilon^{\mu\nu\alpha\beta}\mathrm{Tr}\ \lambda^a U_{\nu\mathrm{L}} U_{\alpha\mathrm{L}} U_{\beta\mathrm{L}},$$

$$J^{\mu a}_\mathrm{R} = \frac{N_c}{48\pi^2}\varepsilon^{\mu\nu\alpha\beta}\mathrm{Tr}\ \lambda^a U_{\nu\mathrm{R}} U_{\alpha\mathrm{R}} U_{\beta\mathrm{R}}, \tag{28}$$

corresponding (via Noether's theorem) to the addition to the lagrangian of $N_c\Gamma$. In this discussion, the λ^a should be traceless SU(3) generators. However, let us try to construct an anomalous baryon number current in the same way. We define the baryon number of a quark (whether left-handed or right-handed) to be $1/N_c$, so that an ordinary baryon made from N_c quarks has baryon number one. Replacing λ^a by $1/N_c$, but including contributions of both left-handed and right-handed quarks, the anomalous baryon-number current would be

$$J^\mu = \frac{1}{24\pi^2}\varepsilon^{\mu\nu\alpha\beta}\mathrm{Tr}\, U^{-1}\partial_\nu U\, U^{-1}\partial_\alpha U\, U^{-1}\partial_\beta U. \tag{29}$$

One way to see that this is the proper, and properly normalized, formula is to consider gauging an arbitrary subgroup not of $\mathrm{SU(3)_L \times SU(3)_R}$ but of $\mathrm{SU(3)_L \times SU(3)_R \times U(1)}$, U(1) being baryon number. The gauging of U(1) is accomplished by adding a Noether coupling $-eJ^\mu B_\mu$ plus whatever higher-order terms may be required by gauge invariance. (B_μ is a U(1) gauge field which may be coupled as well to some $\mathrm{SU(3)_L \times SU(3)_R}$ generator.) With J^μ defined in (29), this leads to a generalization of $\tilde{\Gamma}$ that properly reflects anomalous diagrams involving the baryon-number current (for instance, it properly incorporates the anomaly in the baryon number $\mathrm{SU(2)_L - SU(2)_L}$ triangle that leads to baryon non-conservation by instantons in the standard weak interaction model). Eq. (29) may also be extracted from QCD by methods of Goldstone and Wilczek [10].

References

[1] J. Wess and B. Zumino, Phys. Lett. 37B (1971) 95
[2] S. Deser, R. Jackiw and S. Templeton, Phys. Rev. Lett. 48 (1982) 975; Ann. of Phys. 140 (1982) 372
[3] E. Witten, Phys. Lett. 117B (1982) 324
[4] A.P. Balachandran, V.P. Nair and C.G. Trahern, Syracuse University preprint SU-4217-205 (1981)
[5] R. Bott and R. Seeley, Comm. Math. Phys. 62 (1978) 235
[6] S.L. Adler, Phys. Rev. 177 (1969) 2426;
J.S. Bell and R. Jackiw, Nuovo Cim. 60 (1969) 147;
W.A. Bardeen, Phys. Rev. 184 (1969) 1848
[7] S.L. Adler and W.A. Bardeen, Phys. Rev. 182 (1969) 1517;
R. Aviv and A. Zee, Phys. Rev. D5 (1972) 2372
S.L. Adler, B.W. Lee, S.B. Treiman and A. Zee, Phys. Rev. D4 (1971) 3497
[8] D.J. Gross and R. Jackiw, Phys. Rev. D6 (1972) 477
[9] E. Witten, Nucl. Phys. B223 (1983) 433
[10] J. Goldstone and F. Wilczek, Phys. Rev. Lett. 47 (1981) 986

Vol. B18 (1987) ACTA PHYSICA POLONICA No 3

BERRY PHASES, MAGNETIC MONOPOLES, AND WESS-ZUMINO TERMS OR HOW THE SKYRMION GOT ITS SPIN*

By I. J. R. Aitchison

CERN, Geneva, Switzerland

(*Received July 31, 1986*)

An elementary discussion is given of the mechanism whereby the Wess-Zumino term determines the quantization of the Skyrme soliton. The work of Balachandran et al. is drawn upon to make explicit the remark of Wu and Zee that the Wess-Zumino term acts like a monopole in the space of scalar fields of the non-linear σ-model. The origin of the monopole structure, and its influence on quantization, is discussed in terms of the Berry (adiabatic) phase.

PACS numbers: 11.17.+y

1. *Introduction and outline*

The thing about Skyrmions [1] that is surely hardest to understand is how a lump-like solution (soliton) of a classical scalar field theory can, and in some cases even *must*, be quantized as a fermion. How can you add integers together and get a half left over? I want to draw together here some recent work on this subject, which has certainly helped me to understand how this marvellous trick is pulled.

It has of course been known for quite some time that a classical *extended* object (for example, a top [2, 3]) may be quantized as a fermion. A system which provides an explicit model of how this can come about — and one which is directly relevant to Skyrmions — is that of a particle of charge e in motion about a fixed magnetic monopole of strength g. Almost immediately after Dirac's 'monopole' paper [4], Tamm [5] studied the Schrödinger equation for this system, and found that the solutions of the angular equation are the rotation functions $\mathscr{D}^j_{m'm}(\theta, \phi)$, with $m = eg$ (in units $\hbar = c = 1$); when the product eg has the minimum non-zero value

$$eg = \tfrac{1}{2} \tag{1.1}$$

allowed by the Dirac quantization argument [4], or more generally the value $(n+1/2)$, the system has half-odd angular momentum and is a fermion. (Sometimes this circumstance

* This is an expanded version of the second of two lectures given at the XXVI Cracow School of Theoretical Physics, Zakopane, Poland, 1–13 June, 1986.

is used to run the argument the other way — i.e. that quantization of angular momentum yields the Dirac condition — but, in the present context at least, the Dirac condition will be fundamental.) More recently, field-theory examples of the charge-monopole system have been studied, with analogous results [6, 7].

In the two papers which initiated the recent burst of activity on Skyrmions [1] (and much else besides), Witten [8, 9] showed that the Wess-Zumino (W – Z) term [8–10] in the action for the scalar fields ϕ_a (whose solitons are Skyrmions) actually *determines* how these solitons are to be quantized. He obtained the remarkable result that the Skyrmion is a fermion if N_c is odd, and a boson if N_c is even: furthermore, the W – Z term also determines the pattern of spin-SU(3) multiplets ([1/2+, **8**], [3/2+, **10**]...) in the baryon spectrum [9, 11, 12]. Though obviously correct mathematically, these results were nevertheless still hard to explain in physical terms, especially to anyone who did not know what a W – Z term was — and even to those who did[1].

A good deal of light has been shed on this by the work of Balachandran and collaborators [13–17], Berry [18], Stone [19], and Wu and Zee [20, 21]. I shall try to state the major ideas in single sentences, which we will then examine in greater detail in the following sections.

(*i*) The W – Z term is a generalization, to the configuration space of scalar fields ϕ_a, of the charge-monopole interaction term in ordinary configuration space for particles. It acts like a monopole in ϕ-space.

(*ii*) Because Skyrmion field configurations are maps between field space and real space, the monopole structure of the W – Z term in field space induces, for such configurations, monopole structure in real space.

(*iii*) Upon quantization, fermionic behaviour will emerge via the well-known monopole mechanism referred to above.

These sentences state where we are trying to go, but they do not explain (*a*) where the W – Z term itself comes from, or (*b*) why it is like a monopole in ϕ-space. The short answer to (*a*) is: from the very fermion determinant which we studied in the previous lecture, but generalized to $SU(3)_f$, i.e. it is a term in the effective action for the ϕ fields which arises after integrating over the fermions [22, 23]. This is all very well in its way, but it too is mysterious: why does such an exotic term get induced in the boson sector when we integrate out the fermions? The technical answer to *this* is that the underlying fermion theory has anomalies, which can be calculated from single fermion loop diagrams. These diagrams generate effective vertices in the external fields (ϕ_a, gauge fields, etc.) coupled to the fermions. Hence any bosonic action obtained by integrating out the fermions — which is equivalent to summing all single fermion loop diagrams—*must* faithfully represent these anomaly-induced vertices. The W – Z action precisely encodes these anomalous vertices: if we only consider the 'ungauged' W – Z action, which is a function of the $SU(3)_f$ chiral field ϕ alone, we are representing correctly just the $SU(3)_f$ flavour anomalies of the underlying Fermion theory.

[1] For those who know that there is no W-Z term if the flavour group is SU(2), and wonder what happens then, see Section 6.

But anomalies are pretty mysterious too—are we not getting into an infinite regress of 'explanations'? It would be nice to have some kind of quantum mechanical *analogue*, at least, for what is going on. We can get a clue what to look for when we remember that the characteristic thing about anomaly-induced vertices is that they are independent of the fermion mass M; it is precisely this circumstance that allows the anomaly-cancellation mechanism discussed in the previous lecture, to work. Thus these 'anomalous' vertices will still survive in the fermion determinant with the correct coefficients, even as M becomes very large. This means that these particular vertices—or, equivalently, these particular contributions to the induced bosonic action — can be reliably calculated by the derivative expansion technique: $\partial\phi/M$ can be made as small as we like. (Some explicit examples of this way of calculating anomalous vertices are given in Ref. [22].) Now, a very large fermion mass M implies a large *gap* between the negative energy (sea) levels and the positive energy levels. Small values of $\partial\phi/M$ mean that the momenta and energies associated with shese 'slowly' varying ϕ fields are much less than the mass gap, and will therefore not induce tignificant fermionic excitations across the gap — indeed, in the limit of M *very* large, there will be no excitations at all, and we need only deal with the fermion vacuum (ground state).

This state of affairs is something we can find a quantum mechanical (rather than quantum field theoretic) analogue for. It arises quite frequently in many-body physics. Suppose we have a system described in terms of two sets of degrees of freedom: one (which we call $\boldsymbol{r}$) is 'fast moving' with 'large' differences between excitation levels, and the other ($\boldsymbol{R}$) is 'slow moving' with 'small' associated energy differences. We may think of the electronic (fast) and nuclear (slow) degrees of freedom (d.f.s) in a molecule for instance. It should make sense, when considering the $\boldsymbol{r}$ coordinates, to regard the $\boldsymbol{R}$'s as approximately constant; indeed this is called the adiabatic, or Born-Oppenheimer approximation in quantum mechanics. More precisely, if the $\boldsymbol{R}$ were constant, we would simply solve the stationary state Schrödinger equation for the $\boldsymbol{r}$'s, with the $\boldsymbol{R}$'s appearing parametrically:

$$H_{\boldsymbol{r}}(\boldsymbol{R})\psi_n(\boldsymbol{r}, \boldsymbol{R}) = E_n(\boldsymbol{R})\psi_n(\boldsymbol{r}, \boldsymbol{R}). \tag{1.2}$$

In reality, the $\boldsymbol{R}$'s are varying slowly with time, but not quickly enough to induce transitions from one E_n level to another. Thus the system, if started in a particular E_n level, 'stays with it' as the $\boldsymbol{R}$'s change. This is essentially the content of the adiabatic theorem: the 'fast' coordinates stay in the original eigenstate, which however itself changes slowly in response to the slow changes in the $\boldsymbol{R}$ coordinates which appear parametrically. This sounds very much like the situation of our fermion vacuum evolving slowly in response to the slowly varying ϕ's. But where is the quantum-mechanical analogue of the W—Z term? It must correspond to some non-trivial structure left behind in the space of the 'slow' parameters when we adiabatically decouple the 'fast' ones.

Here is where the work of Berry [18], and Kuratsuji and Iida [24] comes in. The adiabatic assumption tells us that, at any time t, the state of the system $|\psi(t)\rangle$ (adopting now a slightly more abstract notation) will essentially be the 'instantaneous' eigenstate $|n(\boldsymbol{R}(t))\rangle$, where

$$H(\boldsymbol{R}(t))\,|n(\boldsymbol{R}(t))\rangle = E_n(\boldsymbol{R}(t))\,|n(\boldsymbol{R}(t))\rangle, \tag{1.3}$$

if it was prepared to be in one of these states $|n(\boldsymbol{R}_0)\rangle$ at $t = 0$, where $\boldsymbol{R}_0 = \boldsymbol{R}(t = 0)$. In fact, $|\psi(t)\rangle$ will be related to $|n(\boldsymbol{R}(t))\rangle$ by a phase factor. What phase factor? The naïve answer would surely be

$$|\psi(t)\rangle = [\exp -i\int_0^t E_n(\boldsymbol{R}(t'))dt'] \cdot |n(\boldsymbol{R}(t))\rangle, \tag{1.4}$$

the expected integrated 'quasi-stationary state' phase. But this in *not* the whole story. An *additional* phase is generated during such an adiabatic change. That this is so in principle has been known for a long time (see, for example, Ref. [25]), but it had tended to be dismissed as unimportant physically ('just a phase'). Berry [18] pointed out a number of cases where the phase could be of considerable physical interest (see also Ref. [26]). In particular, a non-trivial phase can result from a closed path in $\boldsymbol{R}$ space, as we move along $\boldsymbol{R}_0 \to \boldsymbol{R}(t) \to \boldsymbol{R}_0$. Such 'Berry phases' depend on the actual path followed in $\boldsymbol{R}$-space — which may remind us of something...

The Berry phase is easily calculated [18]. We are looking for a solution of

$$H(\boldsymbol{R}(t))\,|\psi(t)\rangle = i\frac{d}{dt}|\psi(t)\rangle, \tag{1.5}$$

and we try the adiabatically-inspired ansatz

$$|\psi(t)\rangle = [\exp -i\int_0^t E_n(\boldsymbol{R}(t'))dt'] \cdot \exp i\gamma_n(t) \cdot |n(\boldsymbol{R}(t))\rangle. \tag{1.6}$$

Inserting (1.6) directly into (1.5) and using (1.3) yields

$$\begin{aligned} \gamma_n(t) &= i\int_0^t \langle n(\boldsymbol{R}(t'))|\frac{d}{dt'}|n(\boldsymbol{R}(t'))\rangle dt' \\ &= i\int_0^t \langle n(\boldsymbol{R})|\nabla_{\boldsymbol{R}} n(\boldsymbol{R})\rangle \cdot d\boldsymbol{R}, \end{aligned} \tag{1.7}$$

the fundamental formula [18] for the Berry phase $\gamma_n(t)$.

Now — having dealt adiabatically with the $\boldsymbol{r}$ d.f.'s this way — let us turn our attention to the slowly varying $\boldsymbol{R}$'s, and consider *them* as quantum d.f.'s, not just classical parameters. In the same adiabatic approximation, we sit in one 'electronic' state n and look for solutions in which the total state function has the product form $\phi_n(\boldsymbol{R})|n(\boldsymbol{R})\rangle$, and ask: what Schrödinger equation does $\phi_n(\boldsymbol{R})$ obey? The answer is very interesting [25]: if $V(\boldsymbol{R})$ is the potential energy relevant to the $\boldsymbol{R}$ d.f.'s alone, then $\phi_n(\boldsymbol{R})$ obeys

$$[\text{'covariant kinetic energy'} + E_n(\boldsymbol{R}) + V(\boldsymbol{R})]\phi_n(\boldsymbol{R}) = i\frac{d}{dt}\phi_n(\boldsymbol{R}), \tag{1.8}$$

where by 'covariant kinetic energy' is meant that the gradient operator $\nabla_{\boldsymbol{R}}$ in the normal $\boldsymbol{R}$-kinetic energy terms is replaced by

$$\nabla_{\boldsymbol{R}} \to \nabla_{\boldsymbol{R}} + \langle n(\boldsymbol{R})|\nabla_{\boldsymbol{R}} n(\boldsymbol{R})\rangle \tag{1.9}$$

$$\equiv \nabla_{\boldsymbol{R}} - iA_n(\boldsymbol{R}), \tag{1.10}$$

where (1.10) follows since the matrix element in (1.8) is easily seen to be pure imaginary. Thus a sort of gauge potential has been induced in $\boldsymbol{R}$-space!

It is clear that this gauge potential is intimately related to the Berry phase; they are two facets of the same subtlety in the adiabatic approximation. Indeed we can see from (1.9) and (1.10) exactly what the local phase invariance corresponding to this 'gauge' structure is: an $\boldsymbol{R}$-dependent phase change on $|n(\boldsymbol{R})\rangle$ induces a change in $\boldsymbol{A}_n$ of (1.10), which in turn causes a precisely compensating phase change in $\phi_n(\boldsymbol{R})$, so that the total state function $\phi_n(\boldsymbol{R})|n(\boldsymbol{R})\rangle$ is locally phase invariant. Thus a distinctly non-trivial structure has appeared in the 'slow' d.f.'s after adiabatic decoupling of the 'fast' d.f.'s. Note, incidentally, that the Berry phase is nothing but

$$\exp i\gamma_n(t) = \exp\left[i\int_0^t \boldsymbol{A}_n(\boldsymbol{R})\cdot d\boldsymbol{R}\right], \tag{1.11}$$

so that we were right to be reminded of the *path-dependence* of the wave function for a particle in an electromagnetic potential $\boldsymbol{A}$.

Thus we are getting nearer to understanding how funny phase factors — which might influence apparent rotational properties [26] — can arise via adiabatic decoupling. We can make closer contact with the field theory if we reformulate the adiabatic approximation in the path integral formalism. This was done by Kuratsuji and Iida [24]. In view of (1.6) and (1.11) we can almost guess what the result must be. We want the dynamics in $\boldsymbol{R}$-space to correspond to a 'particle' moving in an (additional) 'vector potential' $\boldsymbol{A}_n(\boldsymbol{R})$. Thus we expect to find a piece in the effective action $S_{\mathrm{eff},n}$ in $\boldsymbol{R}$-space which corresponds to the effective Lagrangian

$$\mathscr{L}_{\mathrm{eff},n} = \boldsymbol{A}_n(\boldsymbol{R})\cdot\frac{d\boldsymbol{R}}{dt}. \tag{1.12}$$

Indeed, in that case

$$S_{\mathrm{eff},n}(T) = \int_0^T \mathscr{L}_{\mathrm{eff},n}dt = \int_0^T \boldsymbol{A}_n(\boldsymbol{R})\cdot\frac{d\boldsymbol{R}}{dt}\,dt, \tag{1.13}$$

and

$$\exp iS_{\mathrm{eff},n}(T) = \exp i\gamma_n(T). \tag{1.14}$$

This is just what Kuratsuji and Iida obtain. By considering the trace of the evolution operator $\operatorname{tr}\exp(-iHT)$ in the adiabatic approximation, they show that it is given by

$$K_{\mathrm{eff}}(T) = \sum_n \int \mathscr{D}\boldsymbol{R}\,\exp\left\{iS_0 - i\int_0^T E_n(\boldsymbol{R})dt' + i\gamma_n(T)\right\}, \tag{1.15}$$

where $\boldsymbol{R}(T) = \boldsymbol{R}_0$ (since for the trace we want to return to the same state at $t = T$), and where γ_n is now evaluated over closed loops $\boldsymbol{R}_0 \to \boldsymbol{R}(t) \to \boldsymbol{R}(T) = \boldsymbol{R}_0$ in $\boldsymbol{R}$-space:

$$\gamma_n(T) = i \oint \langle n(\boldsymbol{R}) | \nabla_{\boldsymbol{R}} n(\boldsymbol{R}) \rangle \cdot d\boldsymbol{R} = \oint \boldsymbol{A}_n(\boldsymbol{R}) \cdot d\boldsymbol{R} = \oint \mathscr{L}_{\mathrm{eff},n} dt. \tag{1.16}$$

S_0 is the ordinary action for the $\boldsymbol{R}$ coordinates.

Now, finally, how can we understand the *specific* 'monopole-like' structure which corresponds (we have asserted) to the W–Z term? The secret, as Stone [19] pointed out, lies in a beautiful discovery by Berry [18]. We have assumed throughout that the eigenvalues $E_n(\boldsymbol{R})$ were well separated, and certainly not degenerate. But what happens if, for some particular value of the $\boldsymbol{R}$ d.f.'s, say $\boldsymbol{R}^*$, two of the E_n's coalesce? We expect some sort of catastrophe to show up in our adiabatic result. In fact, this point in $\boldsymbol{R}$-space is very likely to be a point at which the vector potential $\boldsymbol{A}_n(\boldsymbol{R})$ is singular! Such a vector potential would imply sources (δ-function singularities in the associated field strengths) — for example, magnetic monopoles. This is exactly what Berry found, explicitly, for the case in which the degeneracy is a spin-type degeneracy, the 'fast' coordinates are spin d.f.'s, and the 'slow' ones are angles describing the orientation of the (real!) magnetic field $\boldsymbol{B}$. The equation corresponding to (1.5) is then

$$\mu B \boldsymbol{S} \cdot \hat{\boldsymbol{B}} |\psi(t)\rangle = i \frac{d}{dt} |\psi(t)\rangle \tag{1.17}$$

and

$$E_n(\boldsymbol{B}) = \tfrac{1}{2} \mu B n, \tag{1.18}$$

where $n/2$ is the spin eigenvalue, which takes $2s+1$ values. Clearly these $2s+1$ states are degenerate when $\boldsymbol{B} = \boldsymbol{0}$ (the point $\boldsymbol{R}^*$). Berry found that the associated $\boldsymbol{A}_n(\boldsymbol{B})$ was precisely that of a monopole in $\boldsymbol{B}$-space located at $\boldsymbol{B} = \boldsymbol{0}$, having strength $-n/2$ (i.e. $eg = -n/2$). Thus spin-type degeneracies cause monopoles to lurk in the 'slow' space.

We can now see why the integral in (1.16) along a closed loop need not vanish. If we convert the line integral in (1.16) by a (multi-dimensional) Stokes theorem to a surface integral over the 'magnetic field' $\boldsymbol{B}_n = \nabla \times \boldsymbol{A}_n$, and thence to a volume integral via Gauss, we would normally get zero since $\operatorname{div} \boldsymbol{B} = 0$. However, for the singular potential corresponding to a monopole $\operatorname{div} \boldsymbol{B}_n \neq 0$ and a closed loop contributes a non-zero result. Actually we can go even further than this. The line integral over a closed loop C becomes

$$\oint_C \boldsymbol{A}_n(\boldsymbol{R}) \cdot d\boldsymbol{R} = \iint_S \boldsymbol{B}_n \cdot d\boldsymbol{S}, \tag{1.19}$$

where $\boldsymbol{S}$ is a surface spanning C. But what surface? Should we take an $\boldsymbol{S}_1$ (see Fig. 1) which is 'above' C, or an $\boldsymbol{S}_2$ which is 'below'? For consistency we must have

$$\iint_{S_1} \boldsymbol{B}_n \cdot d\boldsymbol{S} = \iint_{S_2} \boldsymbol{B}_n \cdot d\boldsymbol{S} + 2N\pi \tag{1.20}$$

(remember that these quantities are all *phases*). Since the normals for $\boldsymbol{S}_2$ and for $\boldsymbol{S}_1$ are oppositely oriented, we see that (1.20) is equivalent to

$$\oiint \boldsymbol{B}_n \cdot d\boldsymbol{S} = 2N\pi, \tag{1.21}$$

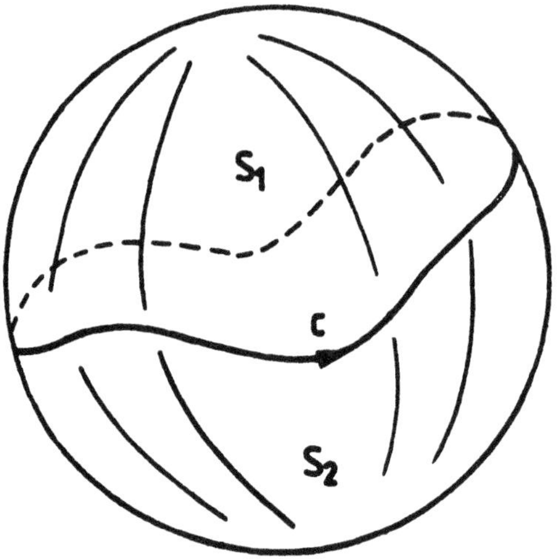

Fig. 1. Two surfaces spanning the curve C on S_2

where the integral is over a *closed* surface surrounding $\boldsymbol{R}$. Thus the total flux out of the 'monopole' is quantized — which is just the Dirac [3] condition ($eg = N \cdot 1/2$, in the real electromagnetic (e.m.) case). The above argument was a poor man's version of a deeper topological treatment, since we relied heavily on tacitly thinking of $\boldsymbol{R}$ as three-dimensional. Nevertheless, the result is correct.

But now notice a remarkable thing: the previous paragraph has shown quite generally that the monopole strength (total flux through a closed surface, divided by 4π) has the quantized value $N/2$, where N is an integer of topological significance. The paragraph before that stated the result that monopole-like structure arose from (1.17), in which the strength of the monopole is $-n/2$, where $n/2$ is the spin eigenvalue. The eigenvalue spectrum of (1.17) near $\boldsymbol{B} = \boldsymbol{0}$ (the point of degeneracy) seems to know something about topology!

We have learned that monopole-like structure can be generated in the 'slow' space $\boldsymbol{R}$ when we adiabatically decouple the 'fast' d.f.'s. Furthermore, the strength of the monopole interaction is an integer (divided by 2), from topological considerations, and this integer corresponds to some label of the energy spectrum of the 'fast' coordinates. This is as near as we are likely to get to a quantum mechanical analogue of the W – Z mystery. The W – Z term results [22, 23] from adiabatically decoupling the ψ's from the π's, starting from a Dirac equation

$$[-i\boldsymbol{\alpha} \cdot \nabla + \beta\mu \exp(i\boldsymbol{\lambda} \cdot \boldsymbol{\pi}\gamma_5 f^{-1})]\psi = i\frac{\partial\psi}{\partial t}, \tag{1.22}$$

which is the analogue of (1.17); μ is a mass parameter, $f \approx 93$ MeV, λ_a ($a = 1, 2, \ldots, 8$) are the Gell-Mann matrices, and the eight π fields are the analogues of the angle variables in $\hat{\boldsymbol{B}}$. The W – Z term (in the fields π) looks like a monopole in π space. Its coefficient is found to be an integer (which is, of course, N_c) by topological considerations [8] exactly analogous to those given above for the $\boldsymbol{R}$-space monopole. There is one gap left to be closed: what is it in the spectrum of the Dirac equation (1.22) that 'knows' about topology (and hence about monopoles)? That is a *deep* question, the answer to which is provided by the mathematical subject called index theory. This way of looking at anomalies (remember?) is called the 'Hamiltonian approach' [27, 28], and is precisely the quantum field theoretical analogue of the quantum mechanical Berry-phase discussion outlined above.

Let us now see how all the foregoing works out in some simple cases.

2. *A simple example*

Consider, following Stone [19], a spin-1/2 particle in a magnetic field $\boldsymbol{B} = B\boldsymbol{n}$, where $\boldsymbol{n}^2 = 1$. The state function for the ('fast') spin d.f.'s satisfies

$$\mu\boldsymbol{\sigma}\cdot\boldsymbol{n}(t)|\psi(t)\rangle = i\frac{d}{dt}|\psi(t)\rangle, \tag{2.1}$$

where the magnitude B of the magnetic field has been absorbed into μ. The 'slow' d.f.'s are $\boldsymbol{n}$, since we shall only consider slow variations of $\boldsymbol{B}$ with fixed B. We consider the large μ limit (cf. large μ in (1.22)), so that the slow changes in $\boldsymbol{n}$ do not cause transitions between the two spin eigenstates $|\uparrow\boldsymbol{n}(t)\rangle$ and $|\downarrow\boldsymbol{n}(t)\rangle$, in the adiabatic approximation. Suppose at $t = 0$ we start in the state $|\uparrow\boldsymbol{n}(0)\rangle$, where $\boldsymbol{n}(0) = (\sin\theta_0\cos\phi_0, \sin\theta_0\sin\phi_0, \cos\theta_0)$. At a general time t, $\boldsymbol{n}(t) = (\sin\theta\cos\phi, \sin\theta\sin\phi, \cos\theta)$, where the time--dependent d.f.'s θ and ϕ will vary over the surface of an S_2. The Berry phase $\gamma_\uparrow(t)$ is

$$\gamma_\uparrow(t) = i\int_0^t \langle\boldsymbol{n}(t')\uparrow\left|\frac{d}{dt'}\right|\uparrow\boldsymbol{n}(t')\rangle dt'. \tag{2.2}$$

We shall calculate this directly using an explicit wave function for $|\uparrow\boldsymbol{n}\rangle$; already here a crucial feature will emerge.

The wave function

$$\langle\theta\phi|\uparrow\boldsymbol{n}+\rangle = \begin{pmatrix}\cos\theta/2\\ \sin\theta/2e^{i\phi}\end{pmatrix} \tag{2.3}$$

is certainly an eigenfunction of

$$\mu\boldsymbol{\sigma}\cdot\boldsymbol{n} = \mu\begin{pmatrix}\cos\theta & \sin\theta e^{-i\phi}\\ \sin\theta e^{i\phi} & -\cos\theta\end{pmatrix} \tag{2.4}$$

with eigenvalue μ. Inserting (2.3) into (2.2) we find

$$\gamma_\uparrow^+(t) = -\int_0^t \tfrac{1}{2}(1-\cos\theta)\frac{d\phi}{dt'}dt' \tag{2.5}$$

for the Berry phase $\gamma_\uparrow^+(t)$, as θ and ϕ vary slowly over S_2. The reason for the + symbols will become clear in a moment.

According to what was advertised in Section 1, the integrand in (2.5) should be closely related to the vector potential of a magnetic monopole of strength $-1/2$, in θ–ϕ space, positioned at the origin. A standard expression for the vector potential of a Dirac monopole of this strength is

$$A_+(r) = \frac{-1}{2r}\frac{1}{z+r}\cdot(-y, x, 0), \tag{2.6}$$

where $x = r\sin\theta\cos\phi$, $y = r\sin\theta\sin\phi$, $z = r\cos\theta$. Thus with $\boldsymbol{r} = (x, y, z)$

$$A_+(\boldsymbol{r})\cdot d\boldsymbol{r} = \frac{-1}{2r(z+r)}(xdy - ydx) = -\tfrac{1}{2}(1-\cos\theta)d\phi. \tag{2.7}$$

Hence indeed (cf. (1.11))

$$\gamma_\uparrow^+(t) = \int_0^t A_+(\boldsymbol{n})\cdot\frac{d\boldsymbol{n}}{dt'}dt' \equiv \int_0^t \mathscr{L}^+_{\text{eff}}(\boldsymbol{n})dt', \tag{2.8}$$

and we see explicitly the 'monopole' character of the phase factor associated with adiabatic motion round the degeneracy point $\boldsymbol{n} = \mathbf{0}$. (We hope the reader will not be confused by the use of $\boldsymbol{n}$ for the slow d.f.'s in this Section, and of n as a label of a 'fast' eigenstate in the previous one.)

However, the wave function (2.3) is ill-defined at $\theta = \pi$ (what is the value of ϕ when $\theta = \pi$?). An alternative choice of $\uparrow$ wave function which is well defined at $\theta = \pi$ is

$$\langle\theta\phi|\uparrow\boldsymbol{n}-\rangle = \begin{pmatrix}\cos\theta/2e^{-i\phi}\\ \sin\theta/2\end{pmatrix}. \tag{2.9}$$

Repeating the above calculations we find that this leads to a Berry phase

$$\gamma_\uparrow^-(t) = -\int_0^t \tfrac{1}{2}(-1-\cos\theta)\frac{d\phi}{dt'}dt', \tag{2.10}$$

which is equivalent to a vector potential

$$A_-(\boldsymbol{r}) = \frac{1}{2r}\cdot\frac{1}{z-r}(-y, x, 0). \tag{2.11}$$

Though good at $\theta = \pi$, (2.9) is ill-defined at $\theta = 0$ — and in fact we are hitting here the famous problem that, for a monopole field, no *single* vector potential exists which is singularity-free over the entire manifold S_2. The A_+ which followed from the choice (2.3) has a singularity along $z = -r$, i.e. the negative z-axis, or $\theta = \pi$. This line of singularities is called a 'Dirac string' [4]. Likewise, the A_- choice has a string along $\theta = 0$. But, comparing (2.5) and (2.10) we see that A_+ and A_- differ by a gradient

$$A_+ - A_- = \nabla\phi, \tag{2.12}$$

that is, by a gauge transformation. Correspondingly,

$$\gamma_t^+ - \gamma_t^- = -[\phi(t) - \phi(0)], \tag{2.13}$$

so that for a closed path on S_2, γ_t is unique.

What we see explicitly here for $A_\pm$ is generally true. Any particular A will have a string singularity somewhere, and by doing a gauge transformation we merely shift the singularity somewhere else. The use of *two* A's (e.g. A_+ and A_-) was advocated by Wu and Yang [29, 30] as a way round the singularity problem, since we can use each in a region (or 'patch') where it is singularity-free, and then connect the two, in a convenient overlap region, by a gauge transformation.

The problem of singularities in the vector potential corresponding to a magnetic monopole would seem to be unavoidable since, if $B = \nabla \times A$ and A is singularity-free, div $B = 0$ and the magnetic charge must be zero. In our ultimate application of the Berry phase concept, the 'slow' d.f.'s will be the meson field variables, which we shall want to quantize. This is analogous to quantizing the n d.f.'s (i.e. θ, ϕ) in the present quantum-mechanical analogue. The presence of the (monopole) singularity at $n = 0$ makes this quantization very awkward, and the Wu–Yang procedure is also not well-adapted to our later purpose.

Remarkably enough, however, it is possible to find a singularity-free Lagrangian for the monopole problem. Indeed, it is given to us automatically by the Berry phase formula, as we shall now describe. We then show how to obtain, from the Berry formula, the elegant Balachandran formalism [13–17], which is ideally suited to the Skyrmion application.

3. *The Hopf fibration of S_2, and the Balachandran Lagrangian*

Let us introduce the notation

$$z = \begin{pmatrix} z_1 \\ z_2 \end{pmatrix} \tag{3.1}$$

for the two-component spinor which is the eigenfunction of (2.1) (e.g. z could be (2.3) or (2.9)). The effective Lagrangian associated with the Berry phase is then

$$\mathscr{L}_{\text{eff}} = iz^\dagger \frac{dz}{dt} \tag{3.2}$$

(cf. (2.8) and (2.10), and check it by trying (2.3) or (2.9) for z). In the two z's considered explicitly so far ((2.3) and (2.9)) only two d.f.'s entered, namely θ and ϕ, the coordinates of a point on the surface of a two-dimensional sphere S_2. We set $\mathscr{L}_{\text{eff}} = -A \cdot dn/dt$ to obtain the potentials A_+ and A_-, also on S_2. However, in principle the spinor z has *three* d.f.'s, since the normalization condition

$$z^\dagger z = 1 = |z_1|^2+|z_2|^2 \tag{3.3}$$

is only one constraint on the two complex numbers z_1, z_2. Indeed, we may in general consider either (2.3) or (2.9) to be multiplied by an arbitrary *phase*, for example

$$z = \begin{pmatrix} \cos\theta/2 & e^{i\chi} \\ \sin\theta/2 & e^{i(\phi+\chi)} \end{pmatrix}. \tag{3.4}$$

The corresponding 'A' must now depend on three d.f.'s, and consequently is not restricted to the surface of an S_2: it turns out, as we shall now see, that it is actually defined on the surface of an S_3, and is non-singular!

Suppose we write

$$\left.\begin{aligned} z_1 &= x_1+ix_2 \\ z_2 &= x_3+ix_4 \end{aligned}\right\}. \tag{3.5}$$

Then $z^+z = 1$ becomes

$$x_1^2+x_2^2+x_3^2+x_4^2 = 1, \tag{3.6}$$

and the x_i are the coordinates of a point on an S_3. Comparing (3.4) and (3.5) we find

$$\left.\begin{aligned} x_1 &= \cos\theta/2\cos\chi, & x_2 &= \cos\theta/2\sin\chi \\ x_3 &= \sin\theta/2\cos(\phi+\chi), & x_4 &= \sin\theta/2\sin(\phi+\chi) \end{aligned}\right\}. \tag{3.7}$$

The metric is

$$ds^2 = \tfrac{1}{4}\,d\theta^2+d\chi^2+\sin^2\theta/2d\phi^2+2\sin^2\theta/2d\phi d\chi. \tag{3.8}$$

It is more convenient to use orthogonal coordinates by introducing

$$\psi = \phi+\chi \tag{3.9}$$

in terms of which (3.8) becomes

$$ds^2 = \tfrac{1}{4}\,d\theta^2+\cos^2\theta/2d\chi^2+\sin^2\theta/2d\psi^2. \tag{3.10}$$

Thus '$A\cdot dn$' now has the form

$$A_\theta\tfrac{1}{2}\,d\theta+A_\chi\cos\theta/2d\chi+A_\psi\sin\theta/2d\psi. \tag{3.11}$$

Inserting (3.4) into (3.2) we find easily,

$$\text{'}\boldsymbol{A}\cdot d\boldsymbol{n}\text{'} = -iz^\dagger dz = d\chi+\tfrac{1}{2}(1-\cos\theta)d\phi \tag{3.12}$$

$$= \cos^2\theta/2d\chi+\sin^2\theta/2d\psi \tag{3.13}$$

whence, via (3.11),

$$A_\theta = 0, \quad A_\chi = \cos\theta/2, \quad A_\psi = \sin\theta/2. \tag{3.14}$$

These potentials are manifestly non-singular. By contrast, the 'S_2' forms (2.7), and the corresponding $\boldsymbol{A}_- \cdot d\boldsymbol{n}$ from (2.11), are singular. Consider, for example (2.7). On S_2 the metric is $ds^2 = d\theta^2 + \sin^2\theta d\phi^2$ and so

$$A_{+,\theta} = 0,\ A_{+,\phi} = -\tfrac{1}{2}\frac{(1-\cos\theta)}{\sin\theta} = -\tfrac{1}{2}\tan\theta/2, \tag{3.15}$$

which is singular (as expected) at $\theta = \pi$. Likewise $\boldsymbol{A}_{-,\phi}$ is singular at $\theta = 0$. In terms of the S_3 coordinates,

$$\boldsymbol{A}_+ \cdot d\boldsymbol{n} = -\tfrac{1}{2}(1-\cos\theta)d\phi = -\sin^2\theta/2d\psi + \sin^2\theta/2d\chi, \tag{3.16}$$

giving

$$A_{+,\theta} = 0, \quad A_{+,\chi} = \frac{\sin^2\theta/2}{\cos\theta/2}, \quad A_{+,\psi} = -\sin\theta/2 \tag{3.17}$$

and $A_{+,\chi}$ is singular at $\theta = \pi$. The S_3 components of $\boldsymbol{A}_-$ can be found similarly, and this time $A_{-,\psi}$ is singular at $\theta = 0$.

Thus a non-singular potential for the monopole can be found provided we enlarge the configuration space from S_2 to S_3, and use the full three d.f.'s available in z. Are we sure that the physics is really the same? The Lagrangian $\mathscr{L}_{\rm eff}$ corresponding to (3.4) is, of course,

$$\mathscr{L}_{\rm eff} = \mathscr{L}^+_{\rm eff}(\boldsymbol{n}) - \dot{\chi}, \tag{3.18}$$

which differs from $\mathscr{L}^+_{\rm eff}(\boldsymbol{n})$ by a total time derivative, and therefore leads to the same equations of motion. From (3.18) we learn that χ is acting like a U(1) gauge d.f. Thus the two d.f.'s of S_2 have been enlarged to three by the addition of a U(1) gauge d.f., χ. This is a well--known construction in mathematics, called the Hopf fibration of S_3. S_3 can be regarded as a principal fibre bundle with base space S_2 and a U(1) structure group. The (Hopf) projection map which takes us from S_3 to S_2 is given explicitly by

$$\boldsymbol{n} = z^{\dagger}\boldsymbol{\sigma} z, \tag{3.19}$$

as can easily be checked (Appendix B). Ryder [31] and Minami [32] were the first to introduce the Hopf map into monopole theory.

From (3.12) it is clear that the $\boldsymbol{A}_+$ potential is obtained (cf. (2.7)) by setting $\chi = 0$, and the $\boldsymbol{A}_-$ one by setting $\chi = -\phi$. Restricting χ in this way is called taking a 'section' of the fibre bundle. These two choices are each called 'local' sections, because they (and the potentials) are not smoothly defined globally over the entire S_2 : $\boldsymbol{A}_+$ is smooth for an upper patch of S_2 excluding $\theta = \pi$, and $\boldsymbol{A}_-$ is smooth for a lower patch excluding $\theta = 0$. It is, in fact, not possible to find any such section which is smooth globally, in this case: a minimum of two is required, as in the explicit examples of $\boldsymbol{A}_\pm$. Mathematically this corresponds to the fact that our (monopole) bundle is non-trivial, or — equivalently — to the fact that S_3 is only locally, but not globally, equivalent to $S_2 \times S_1$. Thus the monopole Lagrangian can be described in a singularity-free way by using a non-trivial bundle over S_2.

The above formulation is not yet quite suitable for our later application to Skyrmion physics. In that case, the d.f.'s in which we shall be interested are actually entries in an SU(3) matrix, and it is hard to see how to generalize z to such a matrix. On the other hand, S_3 is the group manifold of SU(2), and it is quite simple to reformulate the above results in terms of a basic dynamical variable $s(\theta, \phi, \chi) \in SU(2)$; rather than $z(\theta, \phi, \chi)$. This will lead to Balachandran's form for $\mathscr{L}_{\text{eff}}$, which will be directly analogous to the SU(3) case.

We can associate a general SU(2) matrix s with the components z_1, z_2 of z via

$$s = \begin{pmatrix} z_1 & -z_2^* \\ z_2 & z_1^* \end{pmatrix} \tag{3.20}$$

since the condition $|z_1|^2 + |z_2|^2 = 1$ guarantees $s^\dagger s = ss^\dagger = 1$. In terms of s, the $\mathscr{L}_{\text{eff}}$ of (3.2) becomes

$$\mathscr{L}_{\text{eff}} = iz^\dagger \frac{dz}{dt} = \frac{i}{2} \operatorname{tr} (\sigma_3 s^{-1} \dot{s}) \tag{3.21}$$

as may be verified explicitly. Equation (3.21) provides our desired (Balachandran) monopole Lagrangian in terms of $s \in SU(2)$. It is pleasing to see this direct link between the Berry phase and the Balachandran Lagrangian.

The Hopf map can equivalently be described in terms of s. The counterpart of (3.19) is

$$\boldsymbol{\sigma} \cdot \boldsymbol{n} = s\sigma_3 s^{-1} \tag{3.22}$$

(note that $\boldsymbol{n}^2 = 1$ follows automatically upon squaring both sides). Under right multiplication of s by an element of U(1)

$$s \to s \exp i\sigma_3\alpha, \tag{3.23}$$

$$\boldsymbol{\sigma} \cdot \boldsymbol{n} \to s(\exp i\sigma_3\alpha)\sigma_3(\exp - i\sigma_3\alpha)s^{-1} = s\sigma_3 s^{-1} = \boldsymbol{\sigma} \cdot \boldsymbol{n} \tag{3.24}$$

and $\boldsymbol{n}$ is unchanged. Thus the space SU(2)/U(1) of right cosets (3.23) gets mapped by (3.22) into S_2 (see Fig. 2).

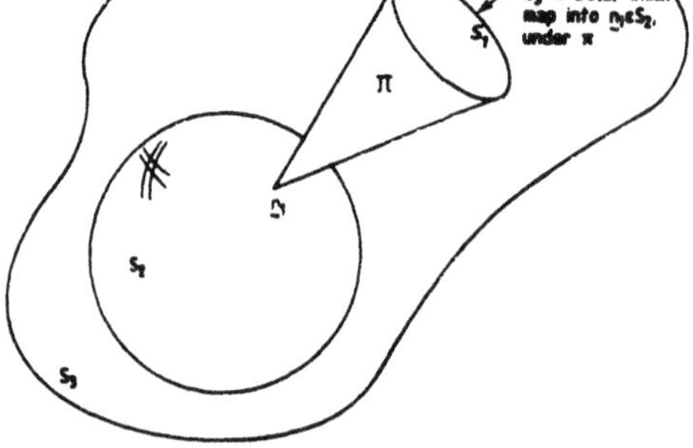

Fig. 2. The one-cycle S_1 of all points in S_3 related to a given point s_1 of S_3 by $s_1 \exp(i\sigma_3\alpha)$, as α varies, is mapped by the Hopf map Π into the single point n_1 of S_2

To gain some confidence with this s-formalism, we can simply insert (2.3) into (3.20), obtaining

$$s_+(\boldsymbol{n}) = \begin{pmatrix} \cos\theta/2 & -\sin\theta/2e^{-i\phi} \\ \sin\theta/2e^{i\phi} & \cos\theta/2 \end{pmatrix} \tag{3.25}$$

whence

$$\mathscr{L}^+_{\text{eff}}(\boldsymbol{n}) = \frac{i}{2}\operatorname{tr}(\sigma_3 s_+^{-1}\dot{s}_+) = \frac{-(1-\cos\theta)}{2}\frac{d\phi}{dt} \tag{3.26}$$

in agreement with (2.8). Alternatively, (2.9) gives

$$s_-(\boldsymbol{n}) = \begin{pmatrix} \cos\theta/2e^{-i\phi} & -\sin\theta/2 \\ \sin\theta/2 & \cos\theta/2e^{i\phi} \end{pmatrix} \tag{3.27}$$

so that

$$s_+(\boldsymbol{n}) = s_-(\boldsymbol{n})e^{i\sigma_3\phi} \tag{3.28a}$$

$$s_+\sigma_3 s_+^{-1} = s_-\sigma_3 s_-^{-1} = \boldsymbol{\sigma}\cdot\boldsymbol{n} \tag{3.28b}$$

$$\mathscr{L}^-_{\text{eff}}(\boldsymbol{n}) = \frac{i}{2}\operatorname{tr}(\sigma_3 s_-^{-1}\dot{s}_-) = \frac{-(-1-\cos\theta)}{2}\frac{d\phi}{dt} \tag{3.28c}$$

$$\mathscr{L}^+_{\text{eff}}(\boldsymbol{n}) - \mathscr{L}^-_{\text{eff}}(\boldsymbol{n}) = -\dot{\phi}. \tag{3.28d}$$

Equation (3.28a) shows that, for given $\boldsymbol{n}$, s_+ and s_- are in the same coset, and get mapped into the same $\boldsymbol{n}$ (3.28b). Equation (3.28d) shows that the corresponding effective Lagrangians differ by a total derivative, and hence lead to the same equations of motion for the $\boldsymbol{n}$ d.f.'s. Indeed, the difference between the choice s_+ and s_- corresponds exactly to the gauge transformation on the associated vector potentials $\boldsymbol{A}_+$ and $\boldsymbol{A}_-$ considered earlier in (2.12).

In general, we may consider now the SU(2) matrix

$$\begin{aligned} s(\theta,\phi,\chi) &= s_+(\theta,\phi)e^{i\sigma_3\chi} \\ &= \begin{pmatrix} \cos\theta/2e^{i\chi} & -\sin\theta/2e^{-i(\phi+\chi)} \\ \sin\theta/2e^{i(\phi+\chi)} & \cos\theta/2e^{-i\chi} \end{pmatrix}, \end{aligned} \tag{3.29}$$

corresponding to (3.4). Then s_+ is the $\chi = 0$ section of this, while s_- is the $\chi = -\phi$ section. The Lagrangian following from (3.29) is, of course,

$$\frac{i}{2}\operatorname{tr}(\sigma_3 s_+^{-1}\dot{s}_+) - \dot{\chi} \tag{3.30}$$

as in (3.18).

In concluding this Section we note (see [31] and [32]) that the foregoing can all be rephrased using the compact formalism of differential forms, and some elementary ideas of homology and cohomology. The potential 1-form is

$$A = iz^{\dagger}dz = -x_1dx_2+x_2dx_1-x_3dx_4+x_4dx_3, \tag{3.31}$$

and the field 2-form B is

$$B = dA = idz^{\dagger} \wedge dz = -2(dx_1 \wedge dx_2+dx_3 \wedge dx_4) = -\tfrac{1}{2}\sin\theta d\theta \wedge d\phi, \tag{3.32}$$

where (3.7) has been used. B is just proportional to the area 2-form of S_2, and

$$\int_{S_2} B = -\tfrac{1}{2}4\pi, \tag{3.33}$$

showing that these potentials and fields indeed correspond to a monopole of strength 1/2. B is certainly closed,

$$dB = 0$$

but it cannot be exact ($B = dA$) on S_2, since if it were we could use Stokes' theorem on S_2 to obtain

$$\int_{S_2} B = \int_{S_2} dA = \int_{\partial S_2} A = 0, \tag{3.34}$$

since S_2 has no boundary; (3.34) would then contradict (3.33). However, if B is regarded as a 2-form on S_3 it is exact, since $H^2(S_3) = 0$, and consequently an A such that $B = dA$ does exist on S_3.

4. *Quantization of the* $\boldsymbol{n}$ *d.f.'s: the Dirac condition again*

We now want to consider, following Balachandran et al. [13, 17], the problem of *quantizing* the d.f.'s $\boldsymbol{n}(\theta,\phi)$ — i.e. we want to promote the 'slow' d.f.'s, which hitherto in Sections 2 and 3 have been parameters, to dynamical variables. In path integral terms, this means — cf. (1.14) — that we want to consider

$$\int \mathscr{D}\boldsymbol{n} \exp\{i\int [\tfrac{1}{2}\dot{\boldsymbol{n}}^2+\mathscr{L}_{\text{eff}}(\boldsymbol{n})]dt\}. \tag{4.1}$$

We know that the quantum theory of the charge-monopole system should only be consistent provided the Dirac condition holds. We are going to see where this arises in the *s*-formalism.

In the previous Section we have seen that the introduction of the new (SU(2)) d.f. χ allowed us to describe the monopole system by a non-singular Lagrangian — and so in quantizing this system we do not have the problem of singularities to contend with. On the other hand, we want the physics to be independent of χ. In the classical theory, as we have seen χ acts like a U(1) gauge d.f., and changing χ is like doing a gauge transformation, under which the equations of motion are invariant. In the quantum theory, we must ensure that a corresponding gauge invariance is correctly implemented. This requirement leads to the Dirac condition.

It is clear that the first term, $1/2\dot{n}^2$, in the Lagrangian of (4.1) is *invariant* under a U(1) gauge transformation

$$s \to s e^{i\sigma_3 \alpha(t)} \tag{4.2}$$

since $\boldsymbol{n}$ remains invariant under (4.2) (see also Appendix C). Thus non-trivial constraints on the theory, associated with the implementation of gauge invariance under (4.2), must arise from the second ('monopole') term. Let us consider a general such term

$$\mathscr{L}_{\text{eff}}(\boldsymbol{n}) = -gi \operatorname{tr}(\sigma_3 s^{-1}\dot{s}), \tag{4.3}$$

where the monopole strength g is not yet determined. Then, under (4.2),

$$\mathscr{L}_{\text{eff}} \to \mathscr{L}_{\text{eff}} + 2g\dot{\alpha}. \tag{4.4}$$

In the quantum theory, s will be promoted to a quantum variable $\hat{s}$, and wave functions will be written as $\Psi(s)$. Consider the infinitesimal (quantum) version of (4.2):

$$\hat{s} \to \hat{s} + i\hat{s}\sigma_3\delta\hat{\alpha}, \tag{4.5}$$

and let $\hat{G}$ be the generator of this transformation so that

$$[\hat{G}, \hat{s}] = \hat{s}\sigma_3. \tag{4.6}$$

Then, from Noether's theorem and (4.4), we deduce

$$\hat{G}\Psi = 2g\Psi \tag{4.7}$$

as a consistency condition on the state functions (it is a kind of 'Gauss Law' associated with gauge invariance under (4.2); see also Appendix C). For finite transformations we then have

$$\Psi'(s) \equiv (e^{i\hat{G}\hat{\alpha}}\Psi)(s) = \Psi(s e^{i\sigma_3\alpha}) = e^{2ig\alpha}\Psi. \tag{4.8}$$

The last two equalities of (4.8) give

$$\Psi(\theta, \phi, \chi+\alpha) = e^{2ig\alpha}\Psi(\theta, \phi, \chi), \tag{4.9}$$

which enforces a kind of 'Bloch' condition on the χ d.f. If we consider the particular case $\alpha = 2\pi$, then since

$$e^{2\pi i\sigma_3} = 1 \tag{4.10}$$

we deduce

$$e^{4\pi i g} = 1 \tag{4.11}$$

and hence

$$g = 0, \pm\tfrac{1}{2}, \pm 1, \ldots, \tag{4.12}$$

which is precisely the Dirac condition. Equation (4.9) is called an 'equivariance' condition on the wave function Ψ: in going from the S_2 of (θ, ϕ) to the S_3 of (θ, ϕ, χ) we have enlarged the configuration space over which our wave functions are to be defined, but an *arbitrary* dependence on the additional variable χ is not consistent with the required gauge invariance (dynamical independence) with respect to χ. Only Ψ's satisfying (4.9) are allowed, with g satisfying (4.12). And, of course, our basic spinor

$$\begin{pmatrix} \cos\theta/2 & e^{i\chi} \\ \sin\theta/2 & e^{i(\phi+\chi)} \end{pmatrix} \tag{4.13}$$

does satisfy (4.12) with $g = -1/2$, the minimum non-trivial magnitude.

A general wave function $\Psi(s)$ can be expressed as a linear combination of the 'top' functions $\mathscr{D}^j_{m'm}(\theta, \phi, \chi)$, which carry irreducible representations of SU(2). It seems obvious from the fact that θ, ϕ, and χ are angles that j should indeed be the angular momentum quantum number; for those who doubt, some further discussion is given in Appendix D. Then

$$\Psi(s) = \sum_{j,m',m} c^j_{m'm}\mathscr{D}^j_{m'm}(\theta, \phi, \chi). \tag{4.14}$$

The constraint (4.9) must now be imposed. If we multiply s from the right by $\exp(i\sigma_3\alpha)$, the $\mathscr{D}$'s get changed by

$$\mathscr{D}^j_{m'm}(s \exp i\sigma_3\alpha) = e^{2\alpha mi}\mathscr{D}^j_{m'm}(s), \tag{4.15}$$

since m is the eigenvalue of $\sigma_3/2$. Thus from (4.15) and (4.9),

$$m = g = 0, \pm\tfrac{1}{2}, \pm 1, \ldots \tag{4.16}$$

and the possibility of 1/2-odd integral spin has emerged (since j is 1/2-odd integral if $2m$ is odd and integral if $2m$ is even). In fact, as stated in Section 1, the system has 1/2-odd angular momentum if the monopole strength g has the value $(n+1/2)$, for integer n.

5. *The Skyrmion case*

Our basic analogy is as follows:

$$\left.\begin{array}{lll} \text{fast d.f.'s : fermion Fock states} & \sim & \text{spin states } |\uparrow\rangle, |\downarrow\rangle \\ \text{fermion vacuum } |0\rangle & \sim & \text{spin state } |\uparrow\rangle \\ \text{slow d.f.'s: Goldstone boson fields } \phi_a & \sim & \text{angular variables } \boldsymbol{n} \\ |0, \phi_a\rangle & \sim & |\uparrow \boldsymbol{n}\rangle \end{array}\right\} \tag{5.1}$$

Just as a monopole structure appeared in the Berry phase associated with $|\uparrow, \boldsymbol{n}\rangle$ for slowly varying $\boldsymbol{n}$, so the W–Z term in the bosonic action is interpreted as a kind of Berry phase for $|0, \phi_a\rangle$.

We begin by introducing the commonly-used notation for the ϕ fields. In the case of $SU(2)_f$, we would have four ϕ's, written as $\phi = (\sigma, \boldsymbol{\pi})$, where $\sigma^2 + \boldsymbol{\pi}^2 = f^2$, quite analogously to $\boldsymbol{n}^2 = 1$. However, this does not generalize to the required $SU(3)_f$ case. Instead, we first rewrite ϕ as

$$\phi = fU, \tag{5.2}$$

where

$$U = \exp(i\boldsymbol{\tau}\cdot\boldsymbol{\pi}/f) \tag{5.3}$$

is a unitary 2×2 matrix. This amounts to a reparametrization of the original σ, $\boldsymbol{\pi}$ in the expression $\phi = (\sigma,\boldsymbol{\pi})$. In $SU(3)_f$, (5.3) is generalized to (cf. (1.22))

$$U = \exp(i\boldsymbol{\lambda}\cdot\boldsymbol{\pi}/f), \tag{5.4}$$

where $\boldsymbol{\pi}$ is understood now to be an 8-component 'angle-type' field. The analogue of (4.1) is then

$$\int \mathscr{D}U \exp\{i\int \mathscr{L}_0(U)dt\} \exp iS_{\text{W-Z}}(U), \tag{5.5}$$

where $\mathscr{L}_0$ is all the rest of the Lagrangian for the U fields, apart from the W–Z term; for example,

$$\mathscr{L}_0 = \tfrac{1}{4}\operatorname{tr}(\partial_\mu U^\dagger \partial^\mu U) + \dots, \tag{5.6}$$

where the dots represent other terms which are necessary to stabilize the soliton, for instance. Finally, the expression for the W–Z action is [8]

$$\exp iS_{\text{W-Z}} = \exp\frac{-iN_c}{240\pi^2}\int \varepsilon^{ijklm}\operatorname{tr}(U^\dagger\partial_i U\, U^\dagger\partial_j U\, U^\dagger\partial_k U\, U^\dagger\partial_l U\, U^\dagger\partial_m U)d^5x. \tag{5.7}$$

The integral in (5.7) is over a 5-dimensional 'disc' whose boundary is 4-dimensional Minkowskian space-time. This disc is the 5-dimensional analogue of the 2-dimensional surfaces considered in (1.19)–(1.21), and N_c is the analogue [8] of the monopole N in (1.21).

Now, we seem a long way from anything like the Balachandran monopole Lagrangian (3.21). However, we can actually make the connection quite explicit, as follows [16, 17]. Instead of treating the *full* quantum-mechanical problem (5.5), in which the whole of the U matrix is treated as a quantum field variable, we perform only a 'semi-classical' quantization. In such an approach, one starts from a solution $U_c(\boldsymbol{r})$ of the static classical field equations in the $SU(2)_f$ case, which is of standard Skyrmion type (cf. (5.3) with $\boldsymbol{\pi} = f\hat{\boldsymbol{r}}\theta(r)$):

$$U_c(\boldsymbol{r}) = \cos\theta(r) + i\boldsymbol{\tau}\cdot\hat{\boldsymbol{r}}\sin\theta(r). \tag{5.8}$$

It is clear that this solution is not rotationally invariant, nor is it invariant under isospin rotations. In fact, there are infinitely many such solutions, related to one another by spatial or isospin rotations, all of which are degenerate in energy since the original Lagrangian *is* invariant under space or isospin rotations. Actually these two kinds of rotation are effectively equivalent for (5.8), since

$$s\tau_i s^{-1} = \tau_j R_{ji}(s) \tag{5.9}$$

for $s \in SU(2)$. The coordinates which distinguish these degenerate classical configurations are the parameters of the matrix s. The semi-classical quantization procedure consists in promoting these d.f.'s into quantum variables $s(t)$. Thus we write

$$U(\boldsymbol{r}, t) = s(t)U_c(\boldsymbol{r})s^{-1}(t). \tag{5.10}$$

Classical quantities will now have a subscript c, and quantum d.f.'s will be distinguished by having no subscript c, instead of by having a '^'. The $s(t)$ will behave just like the s of Sections 3 and 4.

We must now extend (5.8) and hence (5.10) to the $SU(3)_f$ case, or else we get no W−Z term at all [8] (see further Section 6). This means that we have to 'embed' (5.8) inside an SU(3) matrix. The obvious way to do this would seem to be

$$U_c \to \begin{pmatrix} \cos\theta(r) + i\boldsymbol{\tau}\cdot\hat{\boldsymbol{r}}\sin\theta(r) & 0 \\ 0 & 1 \end{pmatrix} \equiv \tilde{U}_c(\boldsymbol{r}) \tag{5.11}$$

(alternative embeddings, which have different physical consequences, are discussed in Refs [12], [16] and [17]). So now,

$$U(\boldsymbol{r}, t) = s(t)\tilde{U}_c(\boldsymbol{r})s^{-1}(t), \quad s \in SU(3)_f. \tag{5.12}$$

We observe at once that $U(\boldsymbol{r}, t)$ is invariant under

$$s \to s e^{iY\alpha(t)}, \tag{5.13}$$

where

$$Y = \tfrac{1}{3}\begin{pmatrix} 1 & 0 & 0 \\ 0 & 1 & 0 \\ 0 & 0 & -2 \end{pmatrix} \tag{5.14}$$

(the normalization is, of course, chosen for convenience). Thus, the configuration space for the s d.f.'s is not SU(3) but rather $SU(3)/U(1)_Y$; this is exactly analogous to our monopole example, where the required configuration space was SU(2)/U(1). In fact, the analogy is very close indeed, for when (5.12) is inserted into (5.7) one finds — after some calculation — that the term involving s (i.e. the piece involving the quantum d.f.'s, in this approximation) is just [16, 17]

$$\mathscr{L}_{\text{W-Z}} = -\tfrac{1}{2} N_c B(U_c)\,\mathrm{tr}\,(Ys^{-1}\dot{s}), \tag{5.15}$$

where $B(U_c)$ is the winding number (= baryon number) of the classical configuration U_c. Equation (5.15) should be compared with (3.21).

We see, from this comparison, that indeed the W−Z term is acting so as to produce, in this semi-classical quantization, exactly a 'monopole in SU(3) space'. The procedure of Section 4 can be transcribed easily to SU(3). The gauge invariance analogous to (4.2) is the invariance of (5.13), under which, however, $\mathscr{L}_{\text{W-Z}}$ changes according to

$$\mathscr{L}_{\text{W-Z}} \to \mathscr{L}_{\text{W-Z}} + \tfrac{1}{3} N_c B\dot{\alpha}. \tag{5.16}$$

In the quantized theory, s and Y are operators, and from Noether's theorem (corresponding to (4.7)) we have

$$\hat{Y}\Psi = \tfrac{1}{3} N_c B \Psi. \tag{5.17}$$

What is the analogue of the quantization constraint (4.12)? For this we note [17] that if we replace s by sh in (5.12), with $h \in SU(2)$, this is equivalent to rotating U_c by some spatial rotation parametrized by h (cf. 5.9)). In particular, consider a rotation by 2π about the 3rd axis. This corresponds to

$$h = \begin{pmatrix} -1 & 0 \\ 0 & -1 \end{pmatrix} \tag{5.18}$$

and thus to the replacement

$$s \to s \begin{pmatrix} -1 & 0 & 0 \\ 0 & -1 & 0 \\ 0 & 0 & 1 \end{pmatrix} = s e^{3\pi i Y}. \tag{5.19}$$

According to (5.17), the allowed Ψ's must then pick up a phase factor

$$e^{i\pi N_c B}, \tag{5.20}$$

and hence *the allowed states for $B = 1$ are fermions if N_c is odd, and bosons if N_c is even*! [9].

The wave functionals Ψ are the SU(3) generalization of the SU(2) rotation functions $\mathscr{D}^j_{m'm}$ [$s \in SU(2)$] — namely

$$\mathscr{D}^{p,q}_{I,I_3,Y;I',I_3'Y'}\,(s \in SU(3)), \tag{5.21}$$

where p and q label the irreducible representation of SU(3). In (5.21) the left-hand group of 'magnetic quantum numbers' refers to transformation properties under *left* multiplication of s by a matrix in SU(3), and hence (cf. (5.12)) to a flavour rotation of U; the right-hand indices refer to right multiplication. But we have already seen that the SU(2) part — in the sense of (5.11) — of any 'right multiplication' matrix corresponds to a spatial rotation. Hence I' and I_3' are actually the real spin and its third component. Now for $B = 1$ and $N_c = 3$ we need the eigenvalue $Y' = 1$ from (5.17): The lowest dimensionality SU(3) representations with $Y' = 1$ are the **8** and **10** (Fig. 3). In the former, the states with $Y' = 1$ have $I' = 1/2$, and hence spin 1/2, while in the latter they have spin 3/2. The left-hand indices give just the flavour quantum numbers corresponding to these SU(3) representations: thus we have an **8** of spin 1/2 and a **10** of spin 3/2.

Further details of Skyrmion quantization are given in Guadagnini [11] and Rabinovici et al. [12]: our concern here has been to place the 'monopole' form (5.15) of $\mathscr{L}_{W\text{-}Z}$ in the context of an adiabatic decoupling problem. From this point of view, the peculiar phase behaviour leading to 'fermion-ness' in the ϕ sector has arisen as a result of non-trivial structure left behind when the fermion vacuum is decoupled adiabatically from

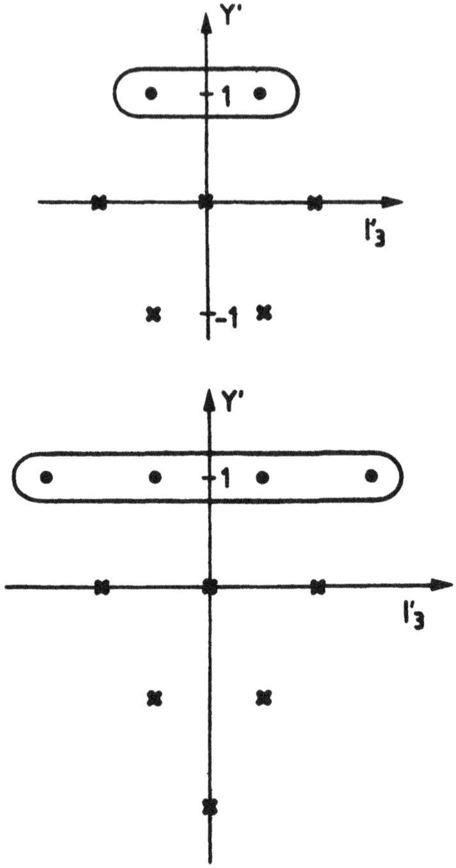

Fig. 3. The **8** and **10** representations of SU(3), showing the two allowed multiplets with $Y' = 1$

the ϕ's. If we use only the ϕ d.f.'s, and integrate the fermions away, we must include a W − Z term which embodies this structure. The ultimate reason that this structure has a 'monopole' form is to be found in the topological approach to anomalies [27, 28].

We may also remark that a similar mechanism holds for Skyrmions in 2+1 dimensions [33]. Here the Wess − Zumino term is replaced by the Hopf term [34–37], which can also be interpreted as arising from integrating out fermions [35]. When the Skyrmion is quantized semi-classically, the angular momentum has the value (*integer* + $\theta/2\pi$), where θ is the coefficient of the Hopf term [38]. The effective Lagrangian in this approximation is exactly analogous to that describing a charged particle moving in two dimensions in the field of a magnetic vortex lying perpendicular to the plane of motion. In that case, the angular momentum can have a value neither integral nor half-odd integral, with consequential 'fractional statistics' [39–41]. Thus just as, in 3+1, the W − Z term acts as a monopole in field space, so in 2+1 the Hopf term acts as a vortex in field space.

6. *Postscript: the case of only two flavours*

The above discussion has been predicated upon the existence of the W–Z term — whose presence determines the quantization of the Skyrmion (fermion if N_c odd, boson if N_c even). But if there are only two flavours, there is no W – Z term: when (5.3) is substituted into (5.7) the SU(2) trace vanishes[1]. Yet there are topological solitons since $\pi_3(SU(2)) = Z$. What determines their quantization?

The answer is that the $B = 1$ soliton can be quantized *either* as a boson *or* as a fermion: there is no restriction involving N_c, and one has to choose the fermionic option by hand [42]. The way in which fermionic quantization is *possible* (but not required) was discussed by Finkelstein [43], Finkelstein and Rubinstein [44], and Williams [45]. One way of putting it is as follows [46]. Since $\pi_4(SU(2)) = Z_2$, time-dependent soliton fields U fall into two distinct homotopy classes of maps from (compactified) space-time to SU(2). Functional integrals over the U's can therefore be separated into two topologically disjoint sectors (analogous to θ-vacua in QCD), corresponding to those U's which can be continuously deformed to the identity, and those which cannot. The contribution from these two sectors to the functional propagator can have a relative + sign or a relative – sign: in the former case the propagator contains all integral spins (bosonic), in the latter half-integral ones (fermionic).

This situation is mathematically the same as that of the spherical top [3], since $\pi_4(O(3)) = Z_2$ also, and the same boson/fermion option therefore exists. In this case one can say, alternatively, that since O(3) is doubly-connected, wave functions on O(3) need not be single-valued. One can define single-valued wave functions by passing to the universal covering space SU(2), but then one has to project back to O(3) via $SU(2) \to O(3) \simeq SU(2)/Z_2$, on which a double-valuedness can appear.

For $N_f > 2$, $\pi_4(SU(N_f)) = 0$ and so this boson/fermion *option* is removed. But then, since $\pi_5(SU(N_f)) = Z$ we have the W – Z addition to the Lagrangian, and the N_c-related quantization is determined.

I am grateful to Jo Zuk for many very helpful discussions; and to Stephen Wilkinson for patient instruction in some of the relevant mathematics, and for carefully reading the manuscript. It is a pleasure to take this opportunity of thanking Drs M. Praszałowicz and W. Słomiński for organising such a stimulating and enjoyable School, and for their warm hospitality.

APPENDIX A

Monopole strength and (U)1 winding number

We have seen that the monopole strength g in

$$\mathscr{L}_{\mathrm{eff}}(\boldsymbol{n}) = gi\,\mathrm{tr}\,(\sigma_3 s^{-1}\dot{s}) \tag{A.1}$$

[1] Alternatively [8], in SU(2) G-parity invariance forbids amplitudes with an odd number of pions, while (5.7) would, if it were non-vanishing, allow them; in SU(3), (5.7) allows $K\bar{K} \to 3\pi$, which is not forbidden by G-parity.

is restricted to the values $g = p/2$ where $p = 0, \pm 1, \pm 2, \ldots$ In this Appendix we will show how p can be interpreted as a winding number associated with the U(1) gauge transformation (4.2).

Consider a *sequence* of gauge transformations

$$s \to s \exp i\sigma_3 \alpha(t) \tag{A.2}$$

parametrized by t, where

$$\alpha(t=0) = 0, \qquad \alpha(t=T) = 2\pi, \tag{A.3}$$

so that we have a closed loop in s-space,

$$s(t=0) = s(t=T). \tag{A.4}$$

Then as we move through this sequence of t-values, the parameter α of the U(1) gauge group goes once round its circle (Fig. A1a).

Corresponding to the gauge transformation (A.2), we have the transformation

$$z = \begin{pmatrix} \cos\theta/2 e^{i\chi} \\ \sin\theta/2 e^{i(\phi+\chi)} \end{pmatrix} \to e^{i\alpha(t)} z \tag{A.5}$$

of the basic ↑ spinor (cf. (3.6) and (3.12)). Thus $\alpha(t)$ is just a variable phase for the associated spinor, and as we go round the sequence of gauge transformations in Fig. A1a, this phase swings round precisely once (Fig. A1b). Meanwhile, what is happening to $\mathscr{L}_{\text{eff}}(\boldsymbol{n})$? This becomes

$${}_{\text{eff}}(\boldsymbol{n}) \to \mathscr{L}_{\text{eff}}(\boldsymbol{n}) - p\dot{\alpha}, \tag{A.6}$$

where $p = 2g$. The associated effective *action* therefore changes by

$$\exp -i \int_0^T p\dot{\alpha}\, dt = \exp(-2\pi i p), \tag{A.7}$$

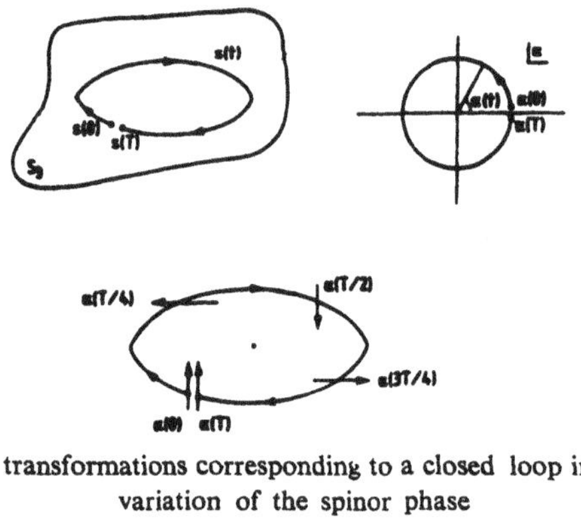

Fig. A1. Sequence of gauge transformations corresponding to a closed loop in s-space, and the associated variation of the spinor phase i.e. its phase swings round p times as we follow the circuit of Fig. A1a. We can therefore interpret p as a winding number which counts the number of rotations of the action phase as we circulate once in α-space (i.e. one circuit in U(1) space).

APPENDIX B

More on the Hopf map

In Section 3 we gave two forms of the Hopf map, one in terms of z

$$\boldsymbol{n} = z^{\dagger}\boldsymbol{\sigma} z, \tag{B.1}$$

and the other in terms of s

$$\boldsymbol{\sigma}\cdot\boldsymbol{n} = s\sigma_3 s^{-1}, \tag{B.2}$$

where

$$z = \begin{pmatrix} z_1 \\ z_2 \end{pmatrix}, \quad |z_1|^2+|z_2|^2 = 1 \tag{B.3}$$

and

$$s = \begin{pmatrix} z_1 & -z_2^* \\ z_2 & z_1^* \end{pmatrix}. \tag{B.4}$$

We make the connection between (B.1) and (B.2) as follows. Let us write $z_1 = x_1 + ix_2$, $z_2 = x_3+ix_4$. Then from (B.1)

$$n_1 = (x_1-ix_2 \quad x_3-ix_4)\begin{pmatrix} 0 & 1 \\ 1 & 0 \end{pmatrix}\begin{pmatrix} x_1+ix_2 \\ x_3+ix_4 \end{pmatrix} = 2(x_1x_3+x_2x_4) \tag{B.5}$$

and

$$n_2 = 2(x_1x_4-x_2x_3) \tag{B.6}$$

$$n_3 = x_1^2+x_2^2-x_3^2-x_4^2, \tag{B.7}$$

while

$$x_1^2+x_2^2+x_3^2+x_4^2 = 1. \tag{B.8}$$

On the other hand, for our s of (3.12), connected to the parameters θ, ϕ of $\boldsymbol{n}$, we had

$$z_1 = \cos\theta/2e^{i\chi}: x_1 = \cos\theta/2\cos\chi, \quad x_2 = \cos\theta/2\sin\chi \tag{B.9}$$

$$z_2 = \sin\theta/2e^{i(\phi+\chi)}: x_3 = \sin\theta/2\cos(\phi+\chi), \quad x_4 = \sin\theta/2\sin(\phi+\chi), \tag{B.10}$$

whence from (B.5)–(B.7)

$$\left.\begin{aligned} n_1 &= \sin\theta\cos\phi \\ n_2 &= \sin\theta\sin\phi \\ n_3 &= \cos\theta \end{aligned}\right\} \tag{B.11}$$

as required.

The matrix s has a simple geometrical interpretation. Consider first the case of

$$s_+(\theta, \phi) = \begin{pmatrix} \cos\theta/2 & -\sin\theta/2e^{-i\phi} \\ \sin\theta/2e^{i\phi} & \cos\theta/2 \end{pmatrix}. \tag{B.12}$$

Let $\hat{\boldsymbol{u}}$ be the unit vector

$$\hat{\boldsymbol{u}} = (-\sin\phi, \cos\phi, 0) \tag{B.13}$$

and consider

$$\exp(-i\boldsymbol{\sigma}\cdot\hat{\boldsymbol{u}}\theta/2) = \cos\theta/2 - i\sin\theta/2\boldsymbol{\sigma}\cdot\hat{\boldsymbol{u}} \tag{B.14}$$

$$= s_+(\theta, \phi). \tag{B.15}$$

This is a rotation of θ about $\hat{\boldsymbol{u}}$, which rotates $\hat{\boldsymbol{z}}$ into $\boldsymbol{n}$ (Fig. B1). So Eq. (B.2), which is equivalent to

$$s_+^{-1}\boldsymbol{\sigma}\cdot\boldsymbol{n}s_+ = \sigma_3 \tag{B.16}$$

in this case, means simply that $\boldsymbol{n}$ has been rotated to be along the 3rd axis. The remaining factor $\exp(i\sigma_3\chi)$ in (3.29) is then just a rotation about the 3 axis (the 'body' axis).

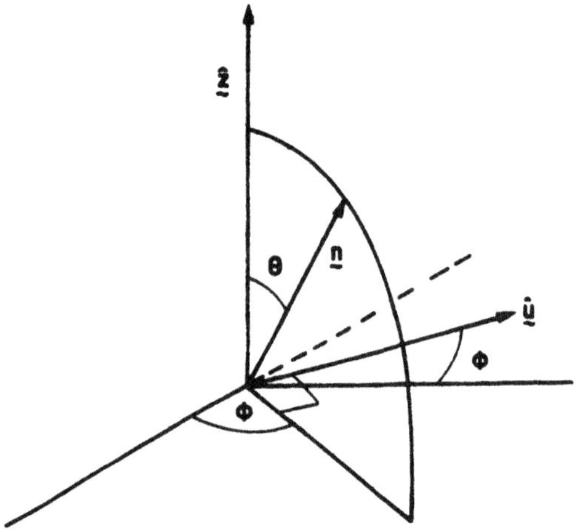

Fig. B1. A rotation of θ about $\hat{\boldsymbol{u}}$ rotates $\hat{\boldsymbol{z}}$ into $\hat{\boldsymbol{n}}$

There is yet one more way of writing the connection between z (or s) and $\boldsymbol{n}$ which the Hopf map enforces. It is

$$\boldsymbol{\sigma}\cdot\boldsymbol{n} = 2zz^\dagger - 1. \tag{B.17}$$

This is easy to verify: using (B.1) for the components of $\boldsymbol{n}$ in terms of z_1, z_2, we find

$$\boldsymbol{\sigma}\cdot\boldsymbol{n} = \begin{pmatrix} |z_1|^2 - |z_2|^2 & 2z_2^* z_1 \\ 2z_1^* z_2 & |z_2|^2 - |z_1|^2 \end{pmatrix} = 2zz^\dagger - 1, \tag{B.18}$$

with the help of $|z_1|^2 + |z_2|^2 = 1$.

APPENDIX C

An alternative quantization for monopoles [12]

Let us consider the monopole action in (4.1),

$$S = \int [\tfrac{1}{2}\dot{\boldsymbol{n}}^2 + \tfrac{1}{2}ip \operatorname{tr}(\sigma_3 s^{-1}\dot{s})]dt \tag{C.1}$$

$$= \int [\tfrac{1}{2}\dot{\boldsymbol{n}}^2 + ipz^\dagger\dot{z}]dt, \tag{C.2}$$

where the second step follows from (3.21), and g has been replaced by $p/2$ (Appendix A). We have not so far considered explicitly the *first* ('kinetic energy') term of (C.2) — let us attend to it now.

We have

$$\tfrac{1}{2}\dot{\boldsymbol{n}}^2 = \tfrac{1}{4}\operatorname{tr}(\boldsymbol{\sigma}\cdot\dot{\boldsymbol{n}})^2 \tag{C.3}$$

$$= 2[(\dot{z}^\dagger\dot{z}) + (z^\dagger\dot{z})^2], \tag{C.4}$$

using (B.17). Let us write

$$a = iz^\dagger\dot{z}; \tag{C.5}$$

then

$$\tfrac{1}{2}\dot{\boldsymbol{n}}^2 = 2\dot{z}^\dagger\dot{z} - 2a^2. \tag{C.6}$$

But also

$$\operatorname{tr}(\dot{s}^\dagger\dot{s}) = 2\dot{z}^\dagger\dot{z}, \tag{C.7}$$

by direct verification. Hence finally we can write this part of the action as

$$\int [\operatorname{tr}(\dot{s}^\dagger\dot{s}) - 2a^2]dt. \tag{C.8}$$

Consider now the behaviour of (C.8) under the U(1) gauge transformation:

$$z \to e^{i\alpha(t)}z, \tag{C.9}$$

which also corresponds to

$$s \to se^{i\alpha(t)\sigma_3}. \tag{C.10}$$

Under (C.9),

$$a \to a - \dot{\alpha}, \tag{C.11}$$

so that

$$-2a^2 \to -2a^2 + 4a\dot{\alpha} - 2\dot{\alpha}^2. \tag{C.12}$$

On the other hand, under (C.10) one finds easily

$$\operatorname{tr}(\dot{s}^\dagger\dot{s}) \to \operatorname{tr}(\dot{s}^\dagger\dot{s}) - 4a\dot{\alpha} + 2\dot{\alpha}^2. \tag{C.13}$$

Thus (C.9) and (C.10) are together an invariance of (C.8), as we stated in Section 4.

We can bring this invariance out by rewriting (C.8) as

$$\int \operatorname{tr}\{[(\vec{\partial}_t - ia\sigma_3)s^\dagger]\,[s(\overleftarrow{\partial}_t + ia\sigma_3)]\} \tag{C.14}$$

as can be simply checked, recalling that $a = \frac{1}{2}\,\mathrm{tr}\,(\sigma_3 s^{-1}\dot{s})$ also. Expression (C.14) is manifestly invariant under the combined transformations (C.9) and (C.10). Thus we are interested in the generating functional

$$Z = \int \mathscr{D}s \exp i[\oint (\mathrm{tr}\,\{[(\vec{\partial}_t - ia\sigma_3)s^\dagger]\,[s(\overleftarrow{\partial}_t + ia\sigma_3)]\} + pa)dt], \tag{C.15}$$

where the action is evaluated over closed loops in s-space. This can be rewritten with the aid of an auxiliary field $A(t)$ [12] as

$$Z \sim \int \mathscr{D}A\mathscr{D}s \exp i[\oint (\mathrm{tr}\,\{[(\vec{\partial}_t - iA\sigma_3)s^\dagger]\,[s(\overleftarrow{\partial}_t + iA\sigma_3)]\} + pA + \tfrac{1}{2}(p/2)^2)dt], \tag{C.16}$$

where a constant

$$\int \mathscr{D}A \exp \{2[A + (\tfrac{1}{4}\,p - a)]^2\}$$

has been ignored.

In (C.16) A acts as an independent gauge field, which changes by $A \to A - \dot{\alpha}$ under the gauge transformation (C.10), so that (C.16) is gauge invariant. We can work in the specific gauge $A = 0$, and require that the equation of motion obtained from the variation with respect to A (i.e. Gauss's law for this case) be realized as a constraint on the physical states. In this gauge the Lagrangian of (C.16) is just

$$\mathscr{L}(A = 0) = \mathrm{tr}\,(\dot{s}^\dagger \dot{s}) + \tfrac{1}{2}(p/2)^2 \tag{C.17}$$

and Gauss's law is

$$\mathrm{tr}\,(-i\sigma_3 s^\dagger \dot{s} + i\dot{s}^\dagger s\sigma_3) = -p. \tag{C.18}$$

Equation (C.18) is the equivalent of (4.7), since the l.h.s. can be identified with the generator of right transformations (C.10), as we discuss further in Appendix D. Indeed, as we also show there, the term $\mathrm{tr}\,(\dot{s}^\dagger \dot{s})$ is precisely $\frac{1}{2}\,\boldsymbol{J}^2$, the square of the angular momentum operator (the motion in $\boldsymbol{r}$ being ignored, only the angles varying). The Hamiltonian in this gauge is therefore

$$H(A = 0) = \tfrac{1}{2}\,\boldsymbol{J}^2 - \tfrac{1}{2}(p/2)^2 \tag{C.19}$$

and the eigenfunctions are again $\mathscr{D}^j_{m'm}(s)$ with m (which carries the right multiplications) restricted to the value $-p/2$, $p = 0, \pm 1, \pm 2, \ldots$. The eigenvalues are $\frac{1}{2}(j(j+1) - (p/2)^2)$, the allowed j being $j = |p/2|, |p/2| + 1, \ldots$.

APPENDIX D

Angular momentum

The connection between the s-formalism and the conventional 'spherical top' formalism can be made explicit by parametrizing s by the *Euler angles* α, β, γ according to

$$s = \begin{pmatrix} e^{i(\alpha+\gamma)/2}\cos\beta/2 & e^{i(\gamma-\alpha)/2}\sin\beta/2 \\ -e^{-i(\gamma-\alpha)/2}\sin\beta/2 & e^{-i(\alpha+\gamma)/2}\cos\beta/2 \end{pmatrix}. \tag{D.1}$$

Straightforward calculation then yields

$$T \equiv \mathrm{tr}\,(\dot{s}^\dagger \dot{s}) = \tfrac{1}{2}\dot{\beta}^2 + \tfrac{1}{2}(\dot{\gamma}+\dot{\alpha}\cos\beta)^2 + \tfrac{1}{2}\dot{\alpha}^2\sin^2\beta, \tag{D.2}$$

which may be compared with the expression for the spherical top kinetic energy given by Edmonds [47], p. 66. The momenta canonically conjugate to α, β, γ are then

$$p_\alpha = \frac{\partial T}{\partial \dot{\alpha}} = \dot{\alpha} + \dot{\gamma}\cos\beta, \quad \text{etc.,} \tag{D.3}$$

and the passage to quantum theory is made by

$$p_\alpha \to -i\frac{\partial}{\partial\alpha}, \quad \text{etc.} \tag{D.4}$$

We find

$$T = -\tfrac{1}{2}\left\{\frac{\partial^2}{\partial\beta^2} + \cot\beta\,\frac{\partial}{\partial\beta} + \mathrm{cosec}^2\beta\,\frac{\partial^2}{\partial\gamma^2} + \frac{1}{\sin^2\beta}\frac{\partial^2}{\partial\alpha^2} - \frac{2\cos\beta}{\sin^2\beta}\frac{\partial^2}{\partial\alpha\partial\gamma}\right\} \tag{D.5}$$

for the operator representing the (rotational) kinetic energy. This is, in fact, precisely the angular kinetic energy

$$T = \tfrac{1}{2}\boldsymbol{J}^2 \tag{D.6}$$

following Edmonds [47]. The eigenfunctions of T are then $\mathscr{D}^j_{m'm}(\alpha, \beta, \gamma)$. Alternative parametrizations of s, such as (3.29), are, of course, also possible.

Finally, we note that

$$i\,\mathrm{tr}\,(\dot{s}^\dagger s\sigma_3 - \sigma_3 s^\dagger \dot{s}) = 2(\dot{\alpha}+\dot{\gamma}\cos\beta) = 2p_\alpha. \tag{D.7}$$

In the quantum theory, p_α is the generator of rotations about the 3-axis, which are represented in terms of s by right transformations

$$s(\alpha, \beta, \gamma) = s(\alpha = 0, \beta, \gamma)e^{i\sigma_3\alpha/2}. \tag{D.8}$$

Thus $2p_\alpha$ is the generator associated with tne transformation (C.10), as claimed in Appendix C, and its eigenvalues should be integral — as indeed is required by the constraint (C.18).

REFERENCES

[1] T. H. R. Skyrme, *Proc. R. Soc. London* **A260**, 127 (1961).
[2] F. Bopp, R. Haag, *Z. Naturforsch.* **5a**, 644 (1950).
[3] L. S. Schulman, *Phys. Rev.* **176**, 1558 (1968); see also L. S. Schulman, *J. Math. Phys.* **12**, 304 (1971).
[4] P. A. M. Dirac, *Proc. R. Soc. London* **A133**, 60 (1931).
[5] I. Tamm, *Z. Phys.* **71**, 141 (1931).

[6] R. Jackiw, C. Rebbi, *Phys. Rev. Lett.* **36**, 1116 (1976); P. Hasenfratz, G. 't Hooft, *Phys. Rev. Lett.* **36**, 1119 (1976).
[7] A. Goldhaber, *Phys. Rev. Lett.* **36**, 1122 (1976).
[8] E. Witten, *Nucl. Phys.* **B223**, 422 (1983).
[9] E. Witten, *Nucl. Phys.* **B223**, 433 (1983).
[10] J. Wess, B. Zumino, *Phys. Lett.* **37B**, 95 (1971).
[11] E. Guadagnini, *Nucl. Phys.* **B236**, 35 (1984).
[12] E. Rabinovici, A. Schwimmer, S. Yankelowicz, *Nucl. Phys.* **B248**, 523 (1984).
[13] A. P. Balachandran, G. Marmo, B.-S. Skagerstam, A. Stern, *Gauge symmetries and fibre bundles*, Springer-Verlag, Berlin etc. 1983.
[14] A. P. Balachandran, G. Marmo, B.-S. Skagerstam, A. Stern, *Nucl. Phys.* **B162**, 385 (1980).
[15] F. Zaccaria et al., *Phys. Rev.* **D27**, 2327 (1983).
[16] A. P. Balachandran, F. Lizzi, V. G. J. Rodgers, A. Stern, *Nucl. Phys.* **B256**, 525 (1985).
[17] A. P. Balachandran, TASI lectures at Yale University, 9 June-15 July 1985.
[18] M. V. Berry, *Proc. R. Soc. London* **A392**, 45 (1984).
[19] M. Stone, Illinois preprint ILL-(TH)-85-#55 (1985).
[20] Y. S. Wu, A. Zee, *Nucl. Phys.* **B258**, 157 (1985).
[21] Y. S. Wu, A. Zee, Santa Barbara preprint NSF-ITP-85-128 (1985).
[22] I. J. R. Aitchison, C. M. Fraser, *Phys. Rev.* **D31**, 2605 (1985).
[23] P. Simic, *Phys. Rev. Lett.* **55**, 40 (1985); and Rockefeller University preprint RU84/B/106 (1984).
[24] H. Kuratsuji, S. Iida, *Prog. Theor. Phys.* **74**, 439 (1985).
[25] A. Messiah, *Quantum mechanics*, North Holland, Amsterdam 1962, Vol. II, pp. 781–800.
[26] F. Wilczek, A. Zee, *Phys. Rev. Lett.* **52**, 2111 (1984).
[27] L. Alvarez-Gaumé, P. Ginsparg, *Nucl. Phys.* **B243**, 449 (1984).
[28] P. Nelson, L. Alvarez-Gaumé, *Commun. Math. Phys.* **99**, 103 (1985).
[29] T. T. Wu, C. N. Yang, *Phys. Rev.* **D12**, 3845 (1975).
[30] T. T. Wu, C. N. Yang, *Nucl. Phys.* **B107**, 365 (1976).
[31] L. H. Ryder, *J. Phys. A: Math. Gen.* **13**, 437 (1980).
[32] M. Minami, *Prog. Theor. Phys.* **62**, 1128 (1979).
[33] A. A. Belavin, A. M. Polyakov, *JETP Lett.* **22**, 245 (1975).
[34] F. Wilczek, A. Zee, *Phys. Rev. Lett.* **51**, 2250 (1983).
[35] A. M. Din, W. J. Zakrzewski, *Phys. Lett.* **146B**, 341 (1984).
[36] Y. S. Wu, A. Zee, *Phys. Lett.* **147B**, 325 (1984).
[37] T. Jaroszewicz, *Phys. Lett.* **146B**, 337 (1984); Harvard preprint HUTP/BO1 (1985).
[38] M. J. Bowick, D. Karabali, L. C. R. Wijewardhana, Yale University preprint YTP 85-20 (1985).
[39] F. Wilczek, *Phys. Rev. Lett.* **48**, 1144 (1982).
[40] F. Wilczek, *Phys. Rev. Lett.* **49**, 957 (1982).
[41] R. Jackiw, A. N. Redlich, *Phys. Rev. Lett.* **50**, 555 (1983).
[42] G. S. Adkins, C. R. Nappi, E. Witten, *Nucl. Phys.* **B228**, 552 (1983).
[43] D. Finkelstein, *J. Math. Phys.* **7**, 1218 (1966).
[44] D. Finkelstein, J. Rubinstein, *J. Math. Phys.* **9**, 1762 (1968).
[45] J. W. Williams, *J. Math. Phys.* **11**, 2611 (1970).
[46] J. S. Dowker, *J. Phys.* **A**, Ser. 2, No. **5**, 936 (1972); see also J. S. Dowker, *Lett. Nuovo Cimento* **4**, 301 (1972), and L. S. Schulman, *J. Math. Phys.* **12**, 304 (1971).
[47] A. R. Edmonds, *Angular momentum in quantum mechanics*, Princeton University Press, Princeton N. J. 1957.

Commun. Math. Phys. 99, 103–114 (1985)

Communications in Mathematical Physics

Hamiltonian Interpretation of Anomalies

Philip Nelson[1]* and Luis Alvarez-Gaumé[2]

1 Institute for Theoretical Physics, University of California, Santa Barbara, CA93106, USA
2 Lyman Laboratory of Physics, Harvard University, Cambridge, MA02138, USA

Abstract. A family of quantum systems parametrized by the points of a compact space can realize its classical symmetries via a new kind of nontrivial ray representation. We show that this phenomenon in fact occurs for the quantum mechanics of fermions in the presence of background gauge fields, and is responsible for both the nonabelian anomaly and Witten's SU(2) anomaly. This provides a hamiltonian interpretation of anomalies: in the affected theories Gauss' law cannot be implemented. The analysis clearly shows why there are no further obstructions corresponding to higher spheres in configuration space, in agreement with a recent result of Atiyah and Singer.

1. Introduction

We say we have an "anomaly" when a symmetry of a classical field theory is not reflected at all in those of the corresponding quantum theory, or more precisely when the full set of classical symmetries cannot be preserved in any of the many possible quantization schemes. When the symmetry in question is an ordinary one such as scale or chiral invariance, we have a straightforward interpretation for the effects of the anomaly in terms of states in Hilbert space: the symmetry in question is absent from the full theory. Coupling constants run; tunneling events do not conserve axial charge. These results are surprising, but not fatal to the theory.

The case of gauged symmetries is very different. Gauge symmetries are properly to be thought of as not being symmetries at all, but rather redundancies in our description of the system [1]. The true configuration space of a (3+1)-dimensional gauge theory is the quotient $\mathscr{C}^3=\mathscr{A}^3/\mathscr{G}^3$ of gauge potentials in $A_0=0$ gauge modulo three-dimensional gauge transformations[1]. When gauge degrees of freedom become anomalous, we find that they are not redundant after all.

* Harvard Society of Fellows. Permanent address: Lyman Laboratory of Physics, Harvard University, Cambridge, MA02138, USA

1 We will sometimes omit the superscript 3

Recently it has become clear that gauge theories with fermion display three different kinds of anomalies, all related to the global topology of the four-dimensional configuration space $\mathscr{C}^4$ by the family index of the Dirac operator $\not{D}^4$. These are the axial U(1) anomaly [the "$\pi_0(\mathscr{G}^3)$ anomaly"], Witten's SU(2) anomaly [2] [from $\pi_1(\mathscr{G}^3)$], and the nonabelian gauge anomaly [3] [from $\pi_2(\mathscr{G}^3)$]. The diversity of the manifestations of these anomalies seems to belie their common origin, however. In the first case we find particle production in the presence of instanton fields [4], breaking of a global symmetry, and no problem with gauge invariance. In the second we find no problem with chiral charge, but instead a nonperturbative failure of *gauge* symmetry, while in the latter the same thing occurs even perturbatively.

What is going on? In the following sections we will attempt to give a hamiltonian picture of the gauge anomalies as simple as the axial anomaly's particle-production interpretation. Essentially the answer will be that in anomalous theories we cannot formulate any Gauss law to constrain the physical states. Along the way we will try to make the above differences a bit less mystifying than they seem in the lagrangian picture. They will all turn out merely to reflect a simple fact about codimension: removing a point from a manifold can sever it into disconnected pieces only if its dimension equals one.

The aim of this paper is expository. We will not find any previously unknown anomalies, but instead will give an approach to understanding them which we have found illuminating. Our point of departure was a remark in [2] which we have generalized to embrace the anomaly of [3] as well[2]. In Sect. 2 we set up our framework and establish our criterion for a global anomaly to exist. In Sects. 3 and 4 we verify the criterion for the cases of [2, 3] respectively, making use of known results from the lagrangian approach. In Sect. 5 we conclude with remarks.

2. Setting Up

It may seem difficult to arrive at a physical interpretation of a problem which renders a gauge theory nonsensical. We know, however, that anomalies do not themselves originate in the gauge sector. We can therefore attempt to quantize a given theory in two steps, starting with the matter fields; at the intermediate point we will have a *family* of quantum systems parametrized by the space of classical background gauge field configurations $\mathscr{A}^3$. Furthermore, the whole collection should realize the classical gauge symmetry *via* unitary operators. The situation is not quite like the usual case of symmetry in quantum mechanics [6], however, since the transformations in question act both on Hilbert space $\mathscr{H}$ and on background configuration space $\mathscr{A}$. They are indeed *bundle maps* of a *family* of Hilbert spaces, $\mathscr{H} \xrightarrow{\pi} \mathscr{A}$. A simple example of such a situation is an ordinary quantum mechanics problem with a Schrödinger particle interacting with a classical rotor degree of freedom $\bar{\varphi}$: for fixed position of the rotor the system has no rotational symmetry, but the full family of theories does have an invariance expressed as a set of isometries, $U_\alpha: \mathscr{H}_{\bar{\varphi}} \to \mathscr{H}_{\bar{\varphi}+\alpha}$.

2 We were also influenced by the work of Rajeev [5]

The notion of families of quantum systems has recently appeared in several papers [7–9]. The phenomenon of "quantum holonomy" discussed in these papers will be crucial to our analysis.

When an ordinary quantum system realizes its classical symmetries, however, it need not do so in the obvious way, by a unitary action of the symmetry group G on $\mathscr{H}$. Instead, Wigner showed [6] that in general we can demand only that $\mathscr{H}$ furnish a projective, or ray, representation of G. When G is a multiply-connected topological group, $\mathscr{H}$ will thus in general have irreducible sectors transforming under $\tilde{G}$, the universal cover of G. This is, of course, the situation with the rotation group, where $\mathscr{H}$ has a sector of odd fermion number transforming as a "double-valued representation of $O(3)$," i.e., as a true representation of spin(3).

The same sort of thing can occur in parametrized families of quantum systems. As a simple example, let us return to the case of the Schrödinger particle and rotor. Constraining the particle to lie on a circle, we have the hamiltonian

$$H=-(\partial_\varphi)^2+b[\delta(\varphi-\tfrac{1}{2}\bar{\varphi})+\delta(\varphi-\tfrac{1}{2}\bar{\varphi}+\pi)]\,, \tag{1}$$

which is continuous for $\bar{\varphi}\in S^1$. For each $\bar{\varphi}$ half of the energy eigenstates of this system are odd under the translation $\varphi\to\varphi+\pi$. Now let $\bar{\varphi}$ vary, and for each value choose a real energy eigenfunction $\psi_{\bar{\varphi}}$ with fixed eigenvalue ε. For the odd states it will be impossible to choose $\psi_{\bar{\varphi}}$ smoothly; as $\bar{\varphi}$ completes a full circuit ψ goes over to its negative. In other words, the odd energy eigenspaces each form twisted line bundles over the parameter space S^1.

Let us attempt to find a unitary action of the symmetry group U(1) on a given odd energy eigenspace $\mathscr{H}_n$ of $\mathscr{H}$. Clearly U_g must map $\mathscr{H}_n$ to itself, but at each point a decision must be made: there is no canonical choice of sign. This raises the possibility that *no* smooth choice may exist. Indeed, any ordinary unitary action of the symmetry group U(1) must take any given ψ at $\bar{\varphi}=0$ and give a nonzero section of $\mathscr{H}$. Since no such section exists, this quantum system cannot realize its U(1) symmetry via an ordinary unitary action.[3] More formally, if the Hilbert bundle $\mathscr{H}$ admits an action of $G=\mathrm{U}(1)$ which projects to the usual action of U(1) on the parameter space S^1, we say it is a "G-bundle" [10]. In this case $\mathscr{H}$ reduces to a new bundle $\bar{\mathscr{H}}$ defined on the quotient $S^1/\mathrm{U}(1)=$ point, and so is trivial. That is, any nontrivial bundle on the base (in our case an energy eigenspace) is not a G-bundle.

If the parameter space consists of many G-orbits it is sufficient to show that any one is nontrivial in order to rule out an ordinary G-action. In any case the key feature which makes possible the unremovable minus sign in the group action is the fact that the orbits are copies of G, which is not simply-connected.

Suppose now that we wish to quantize the rotor degree of freedom as well. The wavefunctions of the complete system can then be taken as complex functions of both φ, the particle position, and $\bar{\varphi}$, the rotor position. Alternatively, however, they can be taken as functions from S^1 into the *space of functions* of φ, that is, as *sections* of the Hilbert bundle $\mathscr{H}$. We will call the complete Hilbert space

3 The reader may well object that we have simply chosen a foolish normalization for the U(1) generator. Indeed the model has another classical $\widetilde{\mathrm{U(1)}}$ symmetry which *is* realized in the usual way. We will return to this point

$\dot{\mathscr{H}} \equiv \Gamma(\mathscr{H})$, the space of sections. It has a subspace spanned by the even eigenfunctions, and on this subspace we can define the unitary operator $\mathscr{U}_\alpha$ by $\mathscr{U}_\alpha \psi = \psi'$, where $\psi'_\varphi = U_\alpha \psi_{\varphi-\alpha}$. On the full $\dot{\mathscr{H}}$, however, we cannot in general define any $\mathscr{U}$.

This is the problem with gauge theory. When we quantize matter in the presence of background gauge fields, the resulting family of quantum theories in general realizes its classical gauge symmetry *via* a perfectly good ray representation. As far as the fermions are concerned there is *nothing wrong* with gauge symmetry. The phases in the ray representation are topologically unremovable; they prevent us from implementing the symmetry at all in the fully quantized theory, and in particular from imposing the constraint of gauge-invariance on the physical quantum states. Equivalently, in the temporal-gauge quantization of gauge theory [11, 12] we require that physical states obey

$$\left(\mathrm{Tr}\left\{T_a \mathbf{D}\cdot\left(\frac{\delta}{\delta \mathbf{A}}\right)\right\} - i\psi^\dagger t_a \psi\right)\Psi = 0\,, \tag{2}$$

which is the infinitesimal version of

$$\Psi[\mathbf{A}^g] = U_g \Psi[\mathbf{A}]\,. \tag{3}$$

But this just says that physical elements of $\dot{\mathscr{H}}$ must be *equivariant* sections of $\mathscr{H}$, or in other words that they must define sections of the reduced bundle $\bar{\mathscr{H}}$ over the true configuration space $\mathscr{C}$. If U_g is only projectively defined, then $\bar{\mathscr{H}}$ is not defined and this requirement makes no sense. If, moreover, the phases which spoil U_g have global topological content and so cannot be removed, then there is no cure for the problem. The theory is then anomalous.

A few remarks are in order before closing this section. We have established the existence of a nontrivial ray representation in a toy model by solving it exactly and noting the behavior of various eigenspaces of the energy globally over the parameter space. This brute-force approach will of course have to be replaced by something more powerful in field theory. Having established that at least one subbundle of $\mathscr{H}$ twists on at least one orbit, we conclude that in the full theory the symmetry is "anomalous," *i.e.*, it cannot be implemented as a true representation. Since the energy eigenspaces were all one-dimensional the only possible twist was the Möbius twist over a noncontractible circle in the symmetry group G. More generally we have to look for twists of higher-dimensional subbundles of $\mathscr{H}$, which will appear over higher-dimensional subspaces of G. In gauge theory, however, it will turn out to be enough to obtain a G-action on the vacuum subbundle, which is one-dimensional, and so there will be no anomalies due to obstructions beyond the first.

One might object that quantum mechanics involves not the real numbers but the complex, and that there are no interesting complex bundles over S^1. We will answer this objection in two different ways in the sequel. For the $\pi_1(\mathscr{G}^3)$ anomaly, it is important for the G-action to preserve the real structure, while for the $\pi_2(\mathscr{G}^3)$ anomaly we indeed must consider two-spheres in $\mathscr{G}$ (as the name implies). The former case resembles the obstruction to placing a spin structure on a space [13], since nontrivial $\pi_1(\mathscr{G}^3)$ implies nontrivial two-cells in $\mathscr{C}^3$, while the latter resembles the obstruction to defining a spin^c structure, since it involves an *integer* (not Z_2) invariant and three-cells in $\mathscr{C}$.

3. Fermions

We begin for simplicity with the theory of [2], an SU(2) gauge theory with a single isodoublet of Weyl fermions. This theory has a Euclidean Dirac operator which is strictly real [2]. Thus the energy eigenstates of the first-quantized theory can be chosen real, and the full second-quantized Hilbert bundle $\mathscr{H}$ has a real structure.[4] Furthermore, the representation matrix appearing in Gauss' law is real, and so the required $\mathscr{G}$-action must respect this real structure. As in our example, it will now suffice to show that the vacuum subbundle, say, is a Möbius bundle over any gauge orbit in order to establish the anomaly.

At each point of gauge configuration space we must now quantize fermions in the given background. This is not, of course, the usual procedure, in which one quantizes *free* fermions and treats gauge interactions perturbatively. Since the SU(2) anomaly is nonperturbative, we must include the gauge fields from the start.

At the first-quantized level we encounter no difficulties. The Hilbert bundle is trivial, and the group action is $U_g v = v'$, where $v'(x) = g(x)v(x)$. Thus we expect any problems to come from second quantization, that is, from the definition of the Dirac sea. Accordingly let us focus our attention first on the vacuum subbundle $\mathscr{H}_0$; we will see that indeed once its $\mathscr{G}$-action has been defined there will be no further problems. Now the Dirac vacuum is defined as the state in Fock space in which all negative-energy states are filled. Since $\not{D}^3$ is gauge-covariant, all of its eigenvalues ε_i are gauge-invariant and $\mathscr{H}_0$ is mapped to itself by any gauge transformation (as indeed is any $\mathscr{H}_\varepsilon$ filled to another Fermi level ε). Actually, though, $\mathscr{H}_0$ is unambiguously defined only on the subset $\mathscr{A}'$ where none of the ε_i vanish. This turns out to be a small but crucial point, since unlike $\mathscr{A}$, which is contractible, $\mathscr{A}'$ has nontrivial topology and so admits the possibility that the vacuum $\mathscr{H}_0$ can be twisted.

To establish the twist we combine the result of Berry [7], which relates twist to degeneracies, with the result of Witten [2], which establishes those degeneracies. Our argument is summarized in Fig. 1. Following Witten, we begin with the generator g^4 of $\pi_4(\mathrm{SU}(2))$ and any point $A_{(0)}$ of $\mathscr{A}^4$, the space of four-dimensional gauge potentials. Take $A_{(0)\mu}(x,t) \equiv 0$. Since $\mathscr{A}^4$ is connected, we can join $A_{(0)}$ to $[A_{(0)}]^{(g^4)}$ by a smooth path $A_{(\tau)}$, $\tau = 0$ to 1. For each τ we now transform $A_{(\tau)}$ by a time-dependent gauge transformation $g_{(\tau)}$ to put it into temporal gauge; call the result $\mathbf{A}'_{(\tau)}$. In particular, $g_{(1)}$ is just $(g^4)^{-1}$, so instead of an open path of vector potentials each periodic in time we now have a closed loop of temporal-gauge histories, each of which ends at $\mathbf{A}'_{(\tau)}(t=\infty) = [0]^{g_{(\tau)}(t=\infty)}$, a three-dimensional gauge transform of $\mathbf{A}'_{(\tau)}(t=-\infty) \equiv 0$.

4 In particular, the vacuum subbundle $\mathscr{H}_0$ gets a real structure

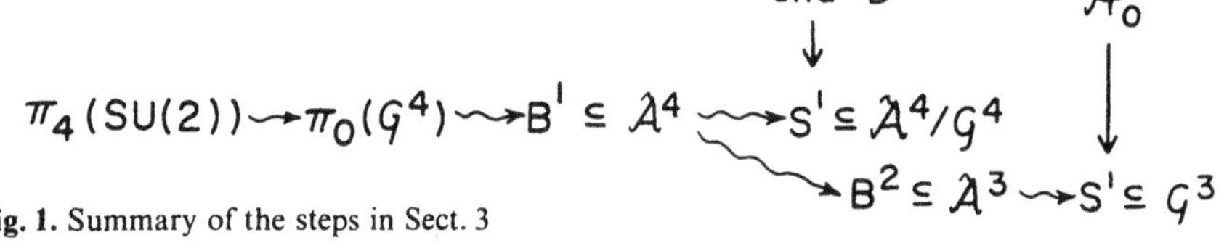

Fig. 1. Summary of the steps in Sect. 3

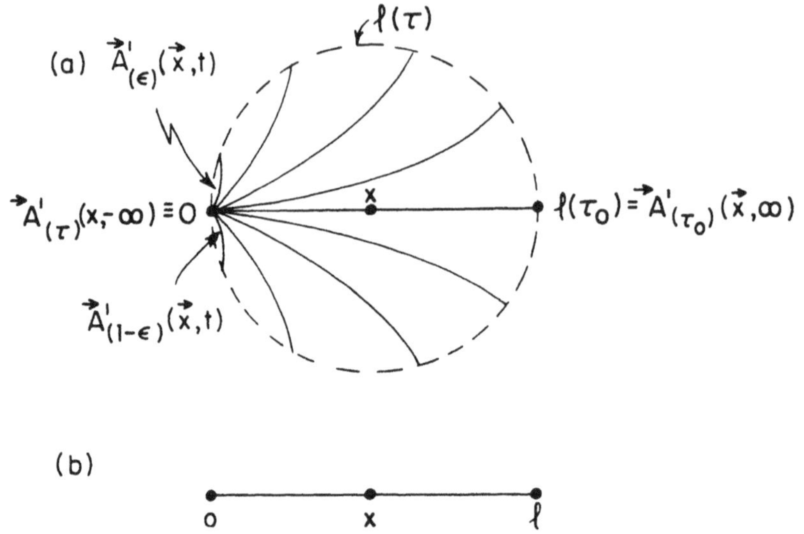

Fig. 2a and b. Disk in $\mathscr{A}^3$ associated to **(a)** SU(2) anomaly, **(b)** axial anomaly

The set of $\mathbf{A}'_{(\tau)}(t)$, $-\infty < t < \infty$, $0 < \tau < 1$ thus forms a disk in $\mathscr{A}^3$ whose rim is a loop $\ell(\tau)$ of gauge transforms of zero (see Fig. 2). Each $\ell(\tau)$ is in $\mathscr{A}'$, since $\not{D}_0$ has no zero modes on compactified space, and so we can restrict $\mathscr{H}_0$ to ℓ. We claim that $\mathscr{H}_0|_\ell$ is in fact twisted. For this to happen, there must be a point x on the disk excluded from $\mathscr{A}'$; that is, there must be a degeneracy at x.

The presence of such a degeneracy follows at once from Witten's argument [2]. From the mod 2 index theorem, $\not{D}^4$ must have a pair of zero modes at some $A_{(\tau_0)}$, and hence for the corresponding $\mathbf{A}'_{(\tau_0)}$ as well. We can take these to be eigenstates $\phi_\pm$ of chirality. Taking Witten's argument one step further, if we choose each $\mathbf{A}'_{(\tau)}$ to vary slowly in t then $\phi_\pm$ must be slowly-varying functions of time times eigenfunctions $\eta^t_\pm$ of the Dirac hamiltonian $H_t \equiv \gamma_0 \not{D}^3_{(\tau_0, t)}$. The energy eigenvalues must pass through zero[5] at some t_0, since $\phi_\pm$ are normalizable zero modes of the Euclidean $\not{D}^4$. Then $x = (\tau_0, t_0)$. Moreover, η^t_- has a CT-conjugated partner of opposite energy and chirality ζ^t_+, leading to the conical arrangement of left-handed energy eigenvalues shown in Fig. 3a. The number of these crossings will be equal, modulo two, to the number of Weyl isodoublets present.

When we second-quantize, the Fermi vacuum ray at each point $\ell(\tau)$ is the ray in Fock space with all negative-energy states filled. Choose a state $|0\rangle_0$ in this ray at 0. We can now attempt to adduce a nonvanishing section of $\mathscr{H}_0|_\ell$ by evolving $|0\rangle_0$ in the slowly-varying backgrounds $\mathbf{A}'_{(\tau)}(t)$ for each τ. By the quantum adiabatic theorem [14], the final state will almost everywhere be almost pure vacuum, and we can project to $\mathscr{H}_0$. This trick fails, however, at y. Here the adiabatic evolution passes through the vertex of the cone in Fig. 3a, producing the particle associated to η_+ and the antiparticle associated to ζ_+. The resulting state has vanishing

5 Here is where the argument fails for the line bundle $\mathscr{H}_\varepsilon$ filled up to some level other than zero, since the index theorem tells us nothing about ε-crossings

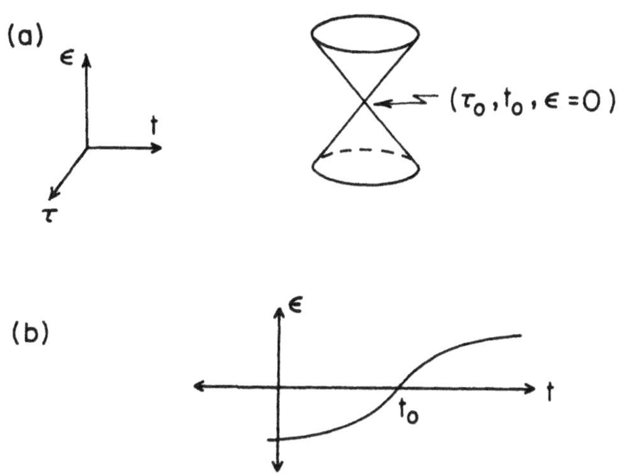

Fig. 3a and b. Eigenvalue behavior near $x=(r_0, t_0)$ for **(a)** SU(2) anomaly, **(b)** axial anomaly

projection to $\mathscr{H}_0$. That is, the putative section "rolls over" near y, out of the plane of $\mathscr{H}_0$ and into an orthogonal direction provided by $\mathscr{H}_{\text{pair}}$. When projected to $\mathscr{H}_0$, it vanishes at y. This reflects the twist of $\mathscr{H}_0$.

This can all be made more precise using the result of [7]: For a loop of real hamiltonians, adiabatic transport around the loop returns to a state which is $(-1)^n$ times the original, if the loop encircles n simple degeneracies. (Note that the adiabatically-continued wavesections of this paragraph and one preceding were chosen only for convenience. Once we know that one section twists, we know they all do.)

While Berry's result is elegant, we have given the pair-production picture as well in order to point up the physical similarities between the present case and the axial anomaly (see Figs. 2b and 3b). Usually the former is thought of in terms of phases, the latter in terms of particle production, but we can see that this really just a matter of emphasis. Particle production is crucial to *both*.[6] In the case of the SU(2) anomaly, however, it occurs only for a special value τ_0; since it is not the generic behavior we do not find an important effect on the vacuum structure. Nevertheless, production is important, as it gives the sign twist which characterizes the anomaly. In the axial anomaly, on the other hand, it is production which is important in suppressing vacuum tunneling [11] while the phases do not matter. After all (Fig. 2b), in this case the rim of the disk is two *points* and so admits no twisted bundles.

Another important qualitative defference between the anomalies also comes from the codimension of x. In the SU(2) case, the level crossing had to be absent for points τ not exactly on τ_0. For this to happen η^t_+ to have a partner ζ^t_+, leading to zero net chirality production for the SU(2) anomaly. No such considerations apply

---

6 J. Goldstone has pointed out to us that our argument for the SU(2) anomaly is similar to one of his, summarized in [12], in which particle production also plays a key role

in the axial anomaly, and indeed (Fig. 3b) only η'_+ or ζ'_+, not both, appears. Thus we get net production of chirality and a global symmetry is broken.

We can summarize the above discussion mathematically [15] by stating that the $\pi_0(\mathscr{G}^3)$ anomaly is given by the simplest invariant of the family index Ind $\not{D}^4$ (a real virtual bundle over $\mathscr{C}^4$), namely its net dimension. Histories $\mathbf{A}'(t)$ for which this is nonzero will have particle production with net change in chirality. The dimension is invariant to perturbations, so production is generic. The $\pi_1(\mathscr{G}^3)$ anomaly comes from the next invariant of Ind $\not{D}^4$, its twist over circles in $\mathscr{C}^4$. We have shown (Fig. 1) that this twist equals that of $\mathscr{H}_0$ over circles in $\mathscr{A}^3$ and so gives the obstruction to finding the $\mathscr{G}^3$-action needed to quantize the theory correctly.[7] For paths in $\mathscr{A}^3$ for which the lowest invariant vanishes, particle production gives no net chirality change and so comes from points where the null space of $\not{D}^4$ jumps; *i.e.*, production is not generic. No further invariants of $\not{D}^4$ are relevant to $\mathscr{H}_0$.

Unlike the $\pi_0(\mathscr{G}^3)$ anomaly, which is an integer, the $\pi_1(\mathscr{G}^3)$ anomaly can be cancelled by adding a second Weyl fermion of either chirality. Now a second pair state becomes degenerate with the vacuum and, by Berry's theorem, there is no sign change as we traverse ℓ.

We can also attempt to evade the anomaly by passing to the cover $\tilde{\mathscr{G}}^3$, as suggested in an earlier footnote. We now get a true $\tilde{\mathscr{G}}$-action on $\mathscr{H}_0$ provided we map the nontrivial element $\hat{g}$ covering the identity to the unitary operator -1, and hence a $\tilde{\mathscr{G}}$ representation on $\mathscr{H}_0$ as in Sect. 2. If we take Gauss' law to mean that wave sections are equivariant under $\tilde{\mathscr{G}}$, however, we must in particular require that they be invariant under $\hat{g}$. Instead, all states have eigenvalue -1 under $\mathscr{U}_{\hat{g}}$! That is, we have succeeded only in defining on $\mathscr{H}$ a ray realization of $\mathscr{G}$ of the type studied in [6]. All states of $\mathscr{H}$ are "fermionic." This construction recovers the formulation of the anomaly given in [2].

We have suggested that the anomaly is a second-quantization phenomenon, preventing us from finding an appropriate family of vacuum states. To go further, let us suppose that we have cancelled the obstruction and so have a well-defined $\mathscr{G}$-action on $\mathscr{H}_0$. To get a $\mathscr{G}$-action on the rest of $\mathscr{H}$, we proceed as usual to define the Fock space creation operators $a_A^{\dagger i}$ on $\mathscr{H}_A$ associated to the eigenfunctions η_A^i with energy $\varepsilon_A^i > 0$. (Similarly, b_A^i creates the mode η_A^j with energy $\varepsilon_A^j < 0$, and we reinterpret b_A^i as a destruction operator.) We can choose η_A^i smoothly in an open set V in $\mathscr{A}^3$, and since there is an unambiguous $\mathscr{G}$-action on first-quantized states we can demand $\eta_{(A^g)}^i(x) = g(x)\eta_A^i(x)$.

Now define

$$U_g(a_A^{\dagger i_1} \dots a_A^{\dagger i_k})|0\rangle_A \equiv a_{(A^g)}^{\dagger i_1} \dots a_{(A^g)}^{\dagger i_k} U_g|0\rangle_A\,, \tag{4}$$

for any vacuum state $|0\rangle_A$, and similarly with the $b^\dagger$. This definition is not arbitrary, but rather is dictated by the requirement that the quantum field built from a and $b^\dagger$ have the same unambiguous transformation law as its first-quantized counterpart. Since there are no phase choices to make, there is no possibility of any obstruction to making them smoothly. Equation (4) defines a $\mathscr{G}$-action on a dense subspace of $\mathscr{H}_A$, $A \in V$. Furthermore, if on some other patch V_1 we choose a different

7 Since these twists are pure torsion, we cannot establish this fact by the use of real characteristic classes

orthonormal expansion η^j_{A1} (still equivariant under $\mathscr{G}$), we end up defining the same $\mathscr{G}$-action for $\mathscr{H}_A$, $A \in V \cap V_1$. Even as we approach a degenerate point, where $\mathscr{H}_0$ is not defined, we can extend this definition. Thus a $\mathscr{G}$-action on $\mathscr{H}_0$ extends without further difficulty to $\mathscr{H}$, and thence as we have seen to the full $\hat{\mathscr{H}}$.

We have therefore found that the higher invariants of Ind $\not{D}^4$, like the lowest one, are irrelevant to implementing Gauss' law. All that matters is the twist of the index over circles. For the case to be discussed in Sect. 4, this agrees with the result of Atiyah and Singer [16], who use the path-integral formulation. It disagrees, however, with [15].

4. The Nonabelian Anomaly

The nonabelian anomaly presents almost no new features. An example of an affected theory is massless QCD with a triplet of left-handed Weyl "quarks." Since $\pi_5(\mathrm{SU}(3)) = \pi_1(\mathscr{G}^4) = Z$, we can consider the loop [17] in $\mathscr{A}^4$ given by transforming zero with each one of the noncontractible loop of $4d$ gauge transformations given by the generator g^5. Again following the procedure outlined in Fig. 1, we then arrive at a three-ball in $\mathscr{A}^3$ whose boundary S^2 consists of three-dimensional gauge transformations of zero. Again by the family index theorem, D^4 generically has a pair of zero modes at one isolated value τ_0, again leading to a conical vanishing of a pair of energy eigenvalues at some x in the interior of the ball. As we follow the trajectory given by τ_0, we again find particle pair production obstructing the definition of a smooth nonvanishing vacuum section on the boundary of the ball. Berry's result for complex hamiltonians now says that indeed $\mathscr{H}_0$ is a twisted (monopole) line bundle over this S^2; its integer invariant is the nonabelian anomaly of the theory. Any action of $\mathscr{G}$ now must have a string singularity somewhere, and so no acceptable version of Gauss' law exists.

Let us now attempt to pass to $\mathscr{U}_g$ as before. Having established that $\mathscr{H}_0$ twists we can now forget about the interior of the orbit $\{\ell(\tau)\}$ and locally define our projective $\mathscr{G}$-action on $\mathscr{H}_0|_\ell$ as follows: Choose an S^2 metric on the orbit ℓ. If g is near the origin of $\mathscr{G}$ and takes P to Q, $P, Q \in \ell$, consider the geodesic from P to Q as a slowly-varying history and evolve any vacuum state $|0\rangle_P$ in this background. Call the result $U_g|0\rangle_P \in \mathscr{H}_0|_Q$. Now suppose that h takes Q to R, also on the orbit; then $(hg)^{-1}$ takes R back to P. By a redefinition of phases we can now arrange for the adiabatic transport on the geodesic triangle so defined to return $|0\rangle_P$ multiplied by $e^{i\Omega/2}$, where Ω is the solid angle subtended by PQR [7]. The $\frac{1}{2}$ is fixed by the requirement that the phase factor be smoothly defined even for large g, h since then Ω is ambiguous by 4π; this "Dirac quantization condition" on the normalization of the anomaly just reflects the fact that the anomaly is quantized due to its origin as a bundle twist.[8]

Thus $U_{hg}^{-1} U_h U_g |0\rangle_P = e^{i\Omega/2}|0\rangle_P$, and so $\mathscr{U}_{hg}^{-1}\mathscr{U}_h\mathscr{U}_g \Psi[P] = e^{i\Omega/2}\Psi[P]$. Choosing P to be any point where Ψ does not vanish we find once again that no state in $\hat{\mathscr{H}}$ is gauge-invariant.

Again we have seen that in the complex case the next-to-lowest invariant of the family index, in this case a two-form on $\mathscr{C}^4$, is the only thing obstructing the

8 See also [17]

definition of a $\mathcal{G}^3$-action on $\mathcal{H}$. Now, however, the obstruction is even more noticeable than in the previous case: since on a sphere we have nontrivial quantum holonomy even on infinitesimal loops,[9] we expect that the $\pi_2(\mathcal{G}^3)$ anomaly should be visible even in perturbation theory. This is of course the case.

5. Remarks

There is an even more direct way to relate the lagrangian derivations of the anomaly to the hamiltonian picture. While it is less physical than the one given above, it does give the quickest way to find the *sign* of the integer invariant in the previous section, something we cannot do by examining the behavior of the energy eigenvalues alone. This sign was irrelevant in the Z_2 case; now we need it in order to recover the anomaly cancellation condition.

The lagrangian derivations show that the fermion partition function $e^{-\Gamma[A]}$ is actually a twisted section on $\mathcal{C}^4$. In particular, it must vanish somewhere. But $e^{-\Gamma[A]}$ is just the vacuum expectation value of the time evolution operator $U(\infty, -\infty)$ in the presence of the time-dependent vector potential A_μ. Gauge transforming to temporal gauge as before, we get

$$\exp - \Gamma[A_{(\tau)}] = {}_{g(\tau)}\langle 0 | U_{\mathbf{A}'_{(\tau)}}(\infty, -\infty) | 0 \rangle_1 \,. \tag{5}$$

Here $\{|0\rangle_g\}$ are a set of vacuum states on the various $\mathcal{H}_{[0]g}$. Now as τ makes a complete circuit in the $\pi_1(\mathcal{G}^3)$ case, $\mathbf{A}'_{(\tau)}$ returns to zero and so does its evolution operator. Since $e^{-\Gamma[A]}$ changes sign, it must be that the $\mathcal{G}$-action is twisted, as we found in Sect. 3. Furthermore, the single vanishing of $e^{-\Gamma[A]}$ which requires that it be twisted is just the signal of pair production again, since at τ_0 the evolved vacuum has no projection onto the transformed vacuum.

Repeating the argument in the case of the nonabelian $\pi_2(\mathcal{G}^3)$ anomaly, we find that not only must the $\mathcal{G}$-action be twisted, the twist in fact agrees in sign with that of the family index bundle. Hence the condition for the cancellation of the anomalous phases is that this bundle have no net twist, in agreement with [17]. In particular, ordinary QCD is safe.

From the hamiltonian point of view, the character of the gauge anomalies is determined by the structure of the possible real or complex line bundles over $\mathcal{G}^3$. Loosely speaking, if over a gauge orbit $\mathcal{H}_0$ contains a unit of "flux" then it cannot be "squeezed" to zero, *i.e.*, the theory does not factor through to one properly defined on $\mathcal{C}^3$. We have shown that the "flux" in a given theory's configuration space can be computed solely in terms of the second invariant of its Ind $\not{D}^4$. The fact that Witten's anomaly appears only for symplectic groups like SU(2), while the nonabelian anomaly appears for unitary groups like SU(3) also comes naturally

9 This is codimension once again: on S^1 there are no interesting paths in a neighborhood of 0 which do not intersect 0. Note however that we do not claim to have obviated the perturbative analysis of gauge anomalies. As is well known, there are anomalies which the global analysis fails to uncover, either in the hamiltonian or lagrangian form. All we are saying is that when the global obstruction *is* present, it is clear why it makes its presence felt in perturbation theory

from our construction, since in order to get interesting real (respectively complex) vacuum bundles over gauge orbits we needed nontrivial $\pi_1(\mathcal{G}^3)$ [respectively $\pi_2(\mathcal{G}^3)$]. This follows for the groups mentioned by the periodicity theorem.

While the higher invariants of the index are not related to gauge anomalies, they may still have interesting physical meaning, just as the lowest one does. The hamiltonian approach may yield further insight into this issue as well.

Note added. Some of the constructions in this paper have already been considered by I. M. Singer; see for example [19]. We thank the referee and Prof. Singer for bringing this work to our attention. After this paper was completed we also received the preprint by Faddeev [18], who discusses similar topics.

Acknowledgements. P. N. would like to thank O. Alvarez, S. Dellapietra, V. Dellapietra, J. Lott, N. Manton, and especially G. Moore for illuminating discussions, and the Institute for Theoretical Physics, Santa Barbara, for its hospitality while this work was being completed. We both thank R. Jackiw for useful discussions, and for bringing to our attention the preprint of Faddeev. This paper is based upon research supported in part by the National Science Foundation under Grant Nos. PHY77-27084 and PHY82-15249, supplemented by funds from the National Aeronautics and Space Administration.

References

1. Babelon, O., Viallet, C.: The Riemannian geometry of the configuration space of gauge theories. Commun. Math. Phys. **81**, 515 (1981)
2. Witten, E.: An SU(2) anomaly. Phys. Lett. **117**B, 324 (1982)
3. Bardeen, W.: Anomalous ward identities in spinor field theories. Phys. Rev. **184**, 1848 (1969)
   Gross, D., Jackiw, R.: Effect of anomalies on quasi-renormalizable theories. Phys. Rev. D**6**, 477 (1972)
4. Coleman, S.: The uses of instantons. In: The whys of subnuclear physics. Zichichi, A. (ed.). New York: Plenum, 1979
   Manton, N.: The Schwinger model and its axial anomaly. Santa Barbara preprint NSF-ITP-84-15 and references therein
5. Rajeev, S.: Fermions from bosons in 3 + 1*d* from anomalous commutators. Phys. Rev. D**29**, 2944 (1984)
6. Wigner, E.: Group theory New York: Academic, 1959; On unitary representations of the inhomogeneous Lorentz group. Ann. Math. **40**, 149 (1939)
7. Berry, M.: Quantal phase factors accompanying adiabatic changes. Proc. R. Soc. Lond. A**392**, 45 (1984)
8. Simon, B.: Holonomy, the quantum adiabatic theorem, and Berry's phase. Phys. Rev. Lett. **51**, 2167 (1983)
9. Wilczek, F., Zee, A.: Appearance of gauge structure in simple dynamical systems. Phys. Rev. Lett. **52**, 2111 (1984)
10. Atiyah, M.: *K-theory* New York: Benjamin 1967
11. Jackiw, R., Rebbi, C.: Vacuum periodicity in Yang-Mills theory. Phys. Rev. Lett. **37**, 172 (1977)
12. Jackiw, R.: Topological investigations of quantized gauge theories. Les Houches lectures, 1983 (MIT preprint CTP-1108)
13. Friedan, D., Windey, P.: Supersymmetric derivation of the Atiyah-Singer index and the chiral anomaly. Nucl. Phys. B**235** 395 (1984)
14. Messiah, A.: Quantum mechanics, Vol. 2. Amsterdam: North-Holland 1962
15. Sumitani, T.: Chiral anomalies and the generalized index theorem. Tokyo preprint UT-KOMABA84-7, 1984

16. Atiyah, M., Singer, I.: Dirac operators coupled to vector potentials. Proc. Nat. Acad. Sci. USA **81**, 2597 (1984)
17. Alvarez-Gaumé, L., Ginsparg, P.: The topological meaning of nonabelian anomalies. Nucl. Phys. B**243**, 449 (1984)
18. Faddeev, L.: Operator anomaly for gauss law. Phys. Lett. **145**B, 81 (1984). For this point of view on family ray representations, see also the recent preprints by R. Jackiw (MIT CTP 1209) and B. Zumino (Santa Barbara NSF-ITP-84-150)
19. Singer, I.: Families of dirac operators with applications to physics, M.I.T. Preprint, to appear in the proceedings of the Conference in Honor of E. Cartan, June 1984

Communicated by A. Jaffe

Received September 27, 1984; in revised form November 30, 1984

Chapter 8

CLASSICAL SYSTEMS

8

Classical Systems

In previous chapters, we have mainly concentrated on geometric phases in the context of quantum physics. The exact dividing line between classical and quantum is not always completely clear, as we mentioned in the cases of optical precession of polarization and nuclear magnetic prescession of spins. In this chapter we discuss geometric phases that are indisputably classical; however we hope you will agree that this in no way detracts from their interest.

Classical geometric phases have been analyzed recently in two contexts. One is immediately suggested by the quantum geometric phase. Completely integrable classical systems are the ones whose quantum limit is most straightforward. Indeed, these systems were the ones treated successfully in the old quantum theory! So we are led to ask, whether the quantum phase has a simple interpretation for these systems. A possible positive answer is suggested, by the fact that integrable systems can be written in terms of action-angle variables. The motion in the angle variable is periodic, and (being canonically conjugate to the Hamiltonian) intimately related to the accumulation of phase in the semiclassical wave function. It is quite natural, therefore, to expect that the shift in the angle variable in response to an adiabatic change of parameters is in a strong sense the classical analog of the quantum phase. This expectation is proved to be correct, and nicely illustrated by examples, in the enclosed papers of Hannay [8.1] and Berry [8.2].

Another context in which geometric phases have been analyzed recently, appears to be less directly related to the quantum phase. It concerns the kinematics of deformable bodies. There is an intrinsic ambiguity in the description of the motion of deformable bodies, that arises simply because different parts of the body move by different amounts. Nevertheless, of course, meaningful questions can be asked about the motion of the body as a whole. For instance, if a swimmer completes a complete cycle of a stroke, coming in the end back to her initial shape, then it is absolutely clear what it means to ask: how far has she gone? how much has she rotated? It is of the utmost convenience, of course, to be able to compare shapes that are not exactly the same. For instance, the laws of motion are most simply formulated infinitesimally, in terms of velocities and accelerations, and therefore necessarily involve different shapes if our swimmer is doing anything at all.

To compare the locations different shapes numerically, we must introduce suitable coordinates—essentially, an *oriented set of axes associated with every possible shape*. Of course the choice of axes is purely a matter of convention, and physical results will not depend on this convention, but intermediate calculations are most easily defined by making some definite choice.

Dear reader, you will recognize that we have heard this song before. In following the phase of an adiabitically evolving quantum state, we had to pick a definite phase for the relevant eigenstate of the Hamiltonian at each intermediate time. This choice was purely conventional, and did not affect the ultimate physical results. In fact we found that different choices produced a gauge transformation, and that the physically meaningful quantities were just the gauge-invariant ones.

In the papers by Shapere and Wilczek [8.3], [8.4], it is demonstrated that a similar kinematic framework applies to the motion of deformable bodies. It is shown, in particular, that swimming at low Reynolds number (the domain of bacteria and *Paramecia*) is a purely geometrical problem, and reduces to the computation of a master gauge field on the space of possible shapes. The same framework also applies to the description of the interesting problem of how falling cats manage to land on their feet, or how a spaceman can reorient himself with nothing to push against.

There are two important ways in which we think these ideas can be fruitfully extended. The mechanical examples treated in the enclosed papers are very special, in that the motion of the self-deforming body is determined entirely by the geometry of the sequence of shape changes in configuration space, and not by how fast this sequence is executed. At low Reynolds number, this is because the motion of the fluid is entirely dominated by friction, so that there is no inertia and pure geometry reigns. For the case of a totally isolated body with zero angular momentum, it occurs because the motion is entirely determined by the conservation equation that relates velocities (including overall rotation) uniquely to changes in position.

In more general mechanical systems, the evolution will of course not be uniquely governed by what happens in configuration space. However, upon passage to phase space the same kinematic framework carries over. Thus we expect that in very general circumstances the motion of deformable bodies through space is governed by an appropriate master gauge field over phase space.

Another generalization, that appears already in a primitive form in the paper [3.3] by Wilczek and Zee, concerns non-conservative systems. For example, we may ask about the result of making slow cyclic changes in external parameters describing a non-conservative oscillator. Since the orbits of such an oscillator, for fixed parameters, are governed by two quantities—

amplitude and phase—the transformation is a two-by-two matrix. Clearly, this matrix has many of the formal properties of the non-abelian geometric phases we have discussed previously. It is associated with a gauge structure on parameter space. A new feature is that the gauge group need no longer be compact, corresponding to the lack of a conserved energy, so that damping or amplification is expected generically.

J. Phys. A: Math. Gen. 18 (1985) 221-230. Printed in Great Britain

Angle variable holonomy in adiabatic excursion of an integrable Hamiltonian

J H Hannay

H H Wills Physics Laboratory, University of Bristol, Tyndall Avenue, Bristol BS8 1TL, UK

Received 15 August 1984

Abstract. If an integrable classical Hamiltonian H describing bound motion depends on parameters which are changed very slowly then the adiabatic theorem states that the action variables I of the motion are conserved. Here the fate of the *angle variables* is analysed. Because of the unavoidable arbitrariness in their definition, angle variables belonging to *distinct* initial and final Hamiltonians cannot generally be compared. However, they *can* be compared if the Hamiltonian is taken on a *closed excursion* in parameter space so that initial and final Hamiltonians are the same. The result shows that the angle variable change arising from such an excursion is not merely the time integral of the instantaneous frequency $\omega \equiv \mathrm{d}H/\mathrm{d}I$, but differs from it by a definite extra angle which depends only on the circuit in parameter space, not on the duration of the process. The 2-form which describes this angle variable holonomy is calculated.

1. Introduction

Holonomy effects, those governed by inexact 'differentials' (1-forms) or 'non-integrable or anholonomic connections' are familiar in physics. Ranging from heat in thermodynamics, to the parallel transport of a vector in a curved space, where they are a fundamental ingredient in a physical theory, they are characterised by the existence of non-zero change in traversing closed circuits. (Strictly, perhaps, this should be called an *an*holonomy but in current usage the reversed emphasis is understood.)

Recently an embarassingly elementary holonomy effect in quantum mechanics that had somehow escaped early discovery was pointed out by Berry (1984). It concerns a quantum Hamiltonian depending on parameters which are changed adiabatically (infinitely slowly) and shows that the *phase* of an eigenstate changes when the Hamiltonian is taken around a closed circuit in parameter space, not only by the dynamical amount $\hbar^{-1}\int E\,\mathrm{d}t$ depending on the duration of the process, but by an extra geometrical amount depending only on the circuit. The fundamental object governing Berry's phase is a phase 2-form in parameter space for which he obtains a formula. The theory was recast in formal mathematical language by Simon (1983).

The present paper arises out of a question posed by Berry, namely what is the semiclassical limit ($\hbar \to 0$) of his phase. In fact, that particular question will not be answered here, though a guess can be made on the basis of the theory below which turns out to be correct when the proper analysis is made (Berry 1985). Instead I shall analyse an *analogous* holonomy effect which arises purely classically. It is analogous in that it arises when a classical Hamiltonian is taken *adiabatically* around a closed circuit in parameter space. Unlike the quantum holonomy though, the classical one

0305-4470/85/020221 + 10\$02.25

has restricted validity. It applies only if the Hamiltonian has just one freedom, or more generally if it has *integrable* classical motion, because it involves the *action angle* variables of the Hamiltonian which only exist in that case. As is recalled in § 2 the action variables are the classical adiabatic invariants of such a system—they are guaranteed *not* to change around the circuit. But the *angle variable*, like the quantum phase, is not constrained to be the same; it is anholonomic, and the object here is to quantify and illustrate this effect. The semiclassical connection between the classical angle holonomy and the quantum phase holonomy was established by Berry (1985) and indeed two of the examples introduced below are there reanalysed in a framework better suited to semiclassical mechanics.

The plan of the paper is this. Restricting attention to one degree of freedom, I shall first recall the classical principle of adiabatic invariance of action (§ 2), present the angle variable holonomy (§ 3), illustrate it (§ 4), and finally generalise the result to an N-freedom integrable system and discuss it (§ 5).

2. Review of classical adiabatic invariance

The standard example which is used to explain adiabatic invariance in classical mechanics is that of the shortening pendulum (figure 1)—a bob swinging on a string which is slowly being drawn up through a hole. Work is done on the pendulum so its energy is not conserved as it would be for a static Hamiltonian, but there is a substitute quantity, the action, which *is* conserved in the adiabatic limit of slow change.

Figure 1. Adiabatic change in the length of a pendulum (from Arnold 1978).

In the phase plane (q, p), the initial, long pendulum has a Hamiltonian $H(q, p, L_{\text{long}})$ with concentric horizontal ellipse shaped contours around one of which the point representing the bob runs depending on its amplitude of oscillation. As the pendulum length $L(t)$ is shortened the contours become squashed horizontally so that finally, the short pendulum has a Hamiltonian $H(q, p, L_{\text{short}})$ with vertical ellipses instead. The bob will again be running around one of them—the question is which one? The answer is not that with the same energy value as initially but that with the same *area*. It is the *action* function $I(q, p, L(t))$, by definition $(2\pi)^{-1}$ times the instantaneous area of the Hamiltonian $H(q, p, L(t))$ contour on which (q, p) lies, which is adiabatically conserved.

The demonstration of this follows immediately from two principles which are most easily expressed not in terms of the individual bob concerned, but of the continuum of 'virtual' bobs chasing each other around the swinging cycle with different phases—a continuous train of dots around the contour of the initial Hamiltonian in the phase plane.

(i) *The adiabatic principle.* If the imposed change is slow enough the bobs will all be affected in the same way and will therefore finally *still be chasing* each other around their new swinging cycle. So the final train of dots once again lies around a *single* contour of the (new) Hamiltonian (rather than being slewed across a range of them as it would in a fast change). The adiabatic principle is otherwise known as the principle of *angle-averaging* (Arnold 1978); each point is supposed to experience the disturbance averaged over all of them.

(ii) *Liouville's theorem.* The area enclosed by any loop of particles in phase space is conserved under all circumstances (whether the Hamiltonian is static, slowly changing or fast changing).

Although the adiabatic principle (i) is well defined and widely realised physically (Landau and Lifshitz 1976), it appears (Arnold 1978) to be surprisingly difficult to eliminate the mathematical loopholes which prevent the simple statement that it holds rigorously in the limit of slow change. It remains an 'assertion of a physical character, i.e. it is untrue without further assumptions'. I shall take it for granted.

3. Angle variable holonomy

The prescription above applies equally well to the slow change of any form of Hamiltonian function $H(q, p)$ in the phase plane (provided, at least, that contours are not forced to change their topology by passage through a saddle point (Robnik and Hannay 1985)). The particle in phase space races around a track (contour of the instantaneous Hamiltonian) of fixed area $2\pi I$ but slowly changing shape. Given this rule of the conservation of action for which contour the particle lies on, it seems natural to ask about the development of the complementary variable, the angle variable, which describes whereabouts on the contour the particle is; to ask, that is, how many circuits it has made.

The instantaneous frequency of traversals' i.e. that which it would obtain if the Hamiltonian was 'frozen' is given by the derivative $(2\pi)^{-1}\,\mathrm{d}H/\mathrm{d}I$, so it is tempting to write the total angle traversed in a time T as simply

$$\int_0^T \frac{\mathrm{d}H(q(t), p(t), t)}{\mathrm{d}I}\,\mathrm{d}t = \int_0^T \frac{\mathrm{d}H(I, t)}{\mathrm{d}I}\,\mathrm{d}t \tag{1}$$

where in the last form H is considered as a function $H(I, t)$ of the area of its contours and adiabatic invariance, $I(t) = I$ constant, is invoked.

The reason (1) is incomplete is of course that the angle variable also changes by virtue of the changing (I, θ) coordinate system in phase space. To reveal the true structure of the situation it is necessary to consider the time dependence of the Hamiltonian function, and (I, θ) coordinate system, as being produced by carrying them along a path $\boldsymbol{R}(t)$ in a parameter space $\boldsymbol{R} \equiv (X, Y, Z\ldots)$ of two or more dimensions in which the functions $H(q, p, \boldsymbol{R})$, $I(q, p, \boldsymbol{R})$, $\theta(q, p, \boldsymbol{R})$ are uniquely defined. The point of making $\boldsymbol{R}$ more than one dimensional is that we shall wish to consider closed excursions $\boldsymbol{R}(T) = \boldsymbol{R}(0)$ in which $\boldsymbol{R}(t)$ executes a loop. With just one parameter, the length of the shortening pendulum for intance, the only way to restore the original length is to reverse the shortening and the holonomy effect is not realised.

In this framework then the exact rates of change of a particle's action and angle are:

$$\dot{I} = -\partial H/\partial\theta(=\text{Zero}) + \dot{\boldsymbol{R}} \cdot \partial I/\partial \boldsymbol{R} \tag{2}$$

$$\dot{\theta} = \partial H/\partial I + \dot{\boldsymbol{R}} \cdot \partial\theta/\partial \boldsymbol{R} \tag{3}$$

where the last terms are the rates of change of action and angle coordinates at a fixed point (q, p) in phase space (figure 2). For non-adiabatic excursion of the Hamiltonian these equations will lead to changes in both I and θ which will depend on the trajectory selected i.e. on the initial values of both I and θ. For adiabatic excursion the equations become

$$\dot{I} = \text{Zero} + \dot{\boldsymbol{R}} \cdot \langle \partial I/\partial \boldsymbol{R}\rangle(=\text{Zero}) \tag{4}$$

$$\dot{\theta} = \partial H/\partial I + \dot{\boldsymbol{R}} \cdot \langle \partial\theta/\partial \boldsymbol{R}\rangle \tag{5}$$

where the average brackets stand for the average around the Hamiltonian contour on which the point lies. Specifically for any function $f(q, p)$ we define a function $\langle f\rangle$ of action I by

$$\langle f\rangle = (2\pi)^{-1} \oint_{\text{contour through } (q,p)} f\,\mathrm{d}\theta \equiv (2\pi)^{-1} \int f(q,p)\delta(I(q,p) - I)\,\mathrm{d}q\,\mathrm{d}p. \tag{6}$$

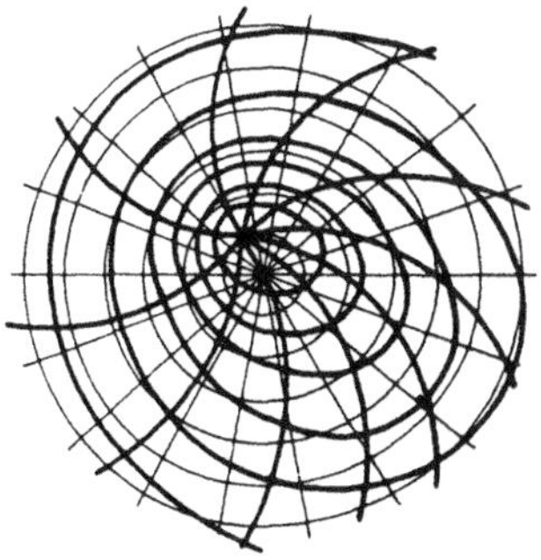

Figure 2. Action-angle coordinates changed with a parameter.

The average in (4) vanishes identically by Liouville's theorem and yields therefore $\dot{I} = 0$ as required. There is no reason however why the average in (5) should vanish and integration of this equation therefore yields the dynamical angle change anticipated in (1) *plus* the extra angle change we are interested in

$$\Delta\theta = \int \dot{\boldsymbol{R}} \cdot \langle \partial\theta(q(t), p(t), \boldsymbol{R}(t))/\partial \boldsymbol{R}\rangle\,\mathrm{d}t = \int \langle \partial\theta/\partial \boldsymbol{R}\rangle \cdot \mathrm{d}\boldsymbol{R}. \tag{7}$$

In the last expression time has been completely eliminated because by definition (6) the average is a function of the conserved, and therefore initial, action I. There is a different field $\langle \partial\theta/\partial \boldsymbol{R}\rangle$ for each I, on which $\Delta\theta$ therefore depends. It does not depend on the initial angle.

Fixing I, then, we now examine the field $\langle \partial\theta/\partial \boldsymbol{R}\rangle$. It depends on the angle variable coordinatisation $\theta(q, p, \boldsymbol{R})$ which is to some extent arbitrary. Unlike the lines of constant action $I(q, p, \boldsymbol{R})$ which for fixed $\boldsymbol{R}$ are fully determined as the contours of the Hamiltonian $H(q, p, \boldsymbol{R})$, the lines of constant angle are only specified once one of them (say $\theta = 0$) is chosen. This one, and thus all the others, may be arbitarily twisted—into a spiral for instance (figure 2). So the angle variable change $\Delta\theta$ inevitably depends upon the angle coordinates chosen for the initial and final parameters $\boldsymbol{R}(0)$ and $\boldsymbol{R}(T)$. Only if they are *identical* coordinate systems, which in turn requires $\boldsymbol{R}(0) = \boldsymbol{R}(T)$ (barring specially favourable circumstances) can one expect to make coordinate independent statements about $\Delta\theta$. The excursions must be closed loops.

That $\Delta\theta$ is generally non-zero for such excursions is the central point of this paper expressing the anholonomy or non-integrability of the angle variable under adiabatic excursions of the Hamiltonian. The equivalent statement in local terms is that while the field $\langle \partial\theta/\partial\boldsymbol{R}\rangle$ is angle coordinate dependent, its 'curl'

$$\partial/\partial\boldsymbol{R} \wedge \langle \partial\theta/\partial\boldsymbol{R}\rangle \tag{8}$$

(or the appropriate generalisation below) is the primitive angle-coordinate-independent object. The assertion is that because the averaging is $\boldsymbol{R}$ dependent, the curl does not vanish identically as it would if the average brackets were absent. Physically speaking then, the field and its curl are rather like the vector potential $\boldsymbol{A}$ and the magnetic field $\boldsymbol{B}$ with their respective dependence on, and independence of, gauge transformations.

The manifestly angle-coordinate-independent expression for the curl is quoted below (9) and derived in the appendix. It is convenient there, and from here on to use the differential form notation appropriate to a structure free parameter space of arbitrary dimensionality. The gradient field $\partial\theta/\partial\boldsymbol{R}$ becomes the 1-form field $\boldsymbol{d}\theta$, and (8) is replaced by the exterior derivative 2-form $\boldsymbol{d}\langle\boldsymbol{d}\theta\rangle$.

The result for this angle 2-form is

$$\boldsymbol{d}\langle\boldsymbol{d}\theta\rangle = \frac{1}{2}\frac{\mathrm{d}^2}{\mathrm{d}I^2}\int_0^{2\pi}\frac{\mathrm{d}\theta'}{2\pi}\int_0^{2\pi}\frac{\mathrm{d}\chi}{2\pi}\chi[\boldsymbol{d}I(\theta'-\chi)\wedge\boldsymbol{d}I(\theta')]. \tag{9}$$

Since the angle variables θ' and χ are integrated over and the bracket in (9) is periodic, this expression is obviously independent of the angle origin chosen, as required. Actually in practice it is usually simplest to choose a specific $\theta(q, p, \boldsymbol{R})$ arbitrarily, and obtain the 2-form by direct evaluation of the left hand side of (9) rather than carry out the integrals on the right.

4. Illustrations

Two illustrations of the geometrical angle holonomy originated in the idea that the simplest realisation of the effect might be to take a particle racing around an elliptical curve either in phase space or in real space, and to rotate the curve slowly and rigidly through one complete turn. The object would then be to demonstrate that the particle had not made the same number of circuits as it would have around the stationary curve, and not merely one less or one more either. It turns out, as we now see, that both the phase space and real space illustrations admit a rather more general analysis.

An important class of Hamiltonians (particularly for quantum mechanical purposes) are quadratic ones—the generalised simple harmonic oscillator,

$$H(q, p) = \tfrac{1}{2}(Xq^2 + 2Yqp + Zp^2) \tag{10}$$

whose contours are concentric oblique ellipses (for $XZ > Y^2$) or hyperbolae ($YZ < Y^2$)

in the phase plane. One could pick an elliptical one and ask that X, Y and Z be changed slowly in such a way as to effect exactly one rotation in the phase plane. It follows from (13) below that the consequent angle change is π times the sum of the ellipse axis ratio and its reciprocal. But the analysis is easily pursued to yield the angle 2-form and hence the result for an arbitrary closed excursion of the Hamiltonian in X, Y, Z space. Let the ellipse be that obtained from a circle by stretching by a factor α along a direction at angle ϕ to the q axis in the phase plane. Then the point on the ellipse at angle ϕ' to the q axis in the phase plane comes from the direction $-\theta+\phi$ on the circle with θ given by

$$\theta = -\tan^{-1}(\alpha \tan(\phi'-\phi)). \tag{11}$$

This is the angle variable on the ellipse measured from the long axis. The negative sign arises from the convention that the angle variable is measured in the opposite sense to ordinary angle. Exterior differentiation (in parameter space α, ϕ) and averaging around the ellipse yields

$$\langle d\theta\rangle = \frac{-1}{2\pi}\int_0^{2\pi} d\phi' \frac{\partial\theta}{\partial\phi'}\left(\frac{\tan(\phi'-\phi)\boldsymbol{d}\alpha - \alpha\sec^2(\phi'-\phi)\boldsymbol{d}\phi}{1+\alpha^2\tan^2(\phi'-\phi)}\right) \tag{12}$$

$$= \text{Zero}\ \boldsymbol{d}\alpha - \tfrac{1}{2}(\alpha+\alpha^{-1})\boldsymbol{d}\phi \tag{13}$$

$$= -\frac{1}{2}\frac{(X+Z)}{(XZ-Y^2)^{1/2}}\boldsymbol{d}\tfrac{1}{2}\tan^{-1}\left(\frac{+2Y}{X-Z}\right) \tag{14}$$

where standard ellipse geometry has been used to relate α (ratio of principal axes) and ϕ to X, Y, Z. Taking the exterior derivative (and using $\boldsymbol{d}\,\boldsymbol{d}\equiv 0$) gives

$$\boldsymbol{d}\langle \boldsymbol{d}\theta\rangle = -\tfrac{1}{4}\boldsymbol{d}\left(\frac{X+Z}{(XZ-Y^2)^{1/2}}\right)\wedge \boldsymbol{d}\tan^{-1}\left(\frac{2Y}{X-Z}\right) \tag{15}$$

$$= \frac{-1}{4(XZ-Y^2)^{3/2}}[(XZ-Y^2)(\boldsymbol{d}X+\boldsymbol{d}Z)-\tfrac{1}{2}(X+Z)(Z\boldsymbol{d}X+X\boldsymbol{d}Z-2Y\boldsymbol{d}Y)]$$

$$\wedge\frac{2(X-Z)\boldsymbol{d}Y-2Y(\boldsymbol{d}X-\boldsymbol{d}Z)}{(X-Z)^2+4Y^2} \tag{16}$$

$$= \frac{-1}{4(XZ-Y^2)^{3/2}}[X(\boldsymbol{d}Y\wedge\boldsymbol{d}Z)+Y(\boldsymbol{d}Z\wedge\boldsymbol{d}X)+Z(\boldsymbol{d}X\wedge\boldsymbol{d}Y)]. \tag{17}$$

The symmetry of the bracket here with respect to the cyclic permutation of X, Y, Z shows that the curl vector field representing the 2-form is a radial one. Moreover since it decays as the inverse square of the 'radius', the phase change around a circuit actually depends only on the circuit of *directions* (i.e. on the radial projection of the circuit onto the unit sphere).

The second illustration involves the rotation of a curve in real space rather than phase space. A wire 'hoop' of area A and perimeter L in a plane with a bead of unit mass sliding frictionlessly around it is slowly turned through one revolution. There is now no need to restrict attention to an elliptical hoop—the shape can be kept general. The analysis of this system is easy in the frame rotating with the hoop. If Ω is the hoop's angular velocity and $\boldsymbol{q}(t)$ the position of the bead in the hoop frame, the canonical momentum $\boldsymbol{p}(t)$ is given by

$$\boldsymbol{p}=\dot{\boldsymbol{q}}+\boldsymbol{\Omega}\wedge\boldsymbol{q} \tag{18}$$

(which is equal to its ordinary momentum in the non-rotating frame). The action is the line integral of $\boldsymbol{p}$ around the hoop

$$I=\frac{1}{2\pi}\oint \boldsymbol{p}\cdot \mathrm{d}\boldsymbol{q} \tag{19}$$

$$=\frac{1}{2\pi}\left(\oint \dot{\boldsymbol{q}}\cdot \mathrm{d}\boldsymbol{q}+2\Omega A\right). \tag{20}$$

This says that the adiabatically conserved quantity is the (spatial) average speed of the bead around the hoop plus $2\Omega A/L$. If Ω is sufficiently small the bead's fluctuations in speed around the length of the hoop are small and the spatial average speed can be taken as the temporal average speed. If Ω rises from zero and falls back again with time integral 2π corresponding to one revolution of the hoop, then the time integral of the average speed is $-4\pi A/L$ greater than it would have been had Ω remained zero. This, then, is the extra distance travelled, or, multiplied by $2\pi/L$, the extra angle change

$$\Delta\theta=-8\pi^2 A/L^2. \tag{21}$$

This result reduces to the expected one $\Delta\theta=-2\pi$ for the circular hoop, where the hoop has simply slipped around, catching up on the bead by one revolution, so that the bead has made one fewer turns in the hoop frame.

The result (21) can be obtained by a variety of alternative means. One of the two used by Berry (1985) is direct integration of the acceleration due to the pseudo-forces in the rotating frame. There the effect emerges entirely as a result of the 'Euler force' $m(\dot{\boldsymbol{\Omega}}\wedge\boldsymbol{q})$. In an analysis of ring gyroscopes (Forder 1984), the result arises from the relativistic lack of synchrony of clocks in the rotating frame.

As a final illustration I consider a rather obvious realisation of the extra angle change. It is a spinning symmetric top with fixed base point whose axis is pulled around a closed circuit enclosing a solid angle Ω. (Gravity plays no role since the axis is driven, not free.) The component of angular velocity ω of the top along its axis is invariant (this time rigorously, not merely adiabatically). So the dynamical part of the top's rotation angle after the circuit is simply ω times the circuit duration. But there is an extra angle which would be present even if the top were not spinning ($\omega=0$), namely that caused by the familiar parallel transport holonomy of a vector on a curved surface, in this case the unit sphere. The angle turned through by a vector, and therefore the extra angle for the top, is the integral of the Gaussian curvature over the area enclosed—that is, the solid angle Ω. In this example the angle 2-form is just the curvature 2-form of the unit sphere.

5. Generalisation and discussion

While the presentation has referred to a Hamiltonian system with one freedom only, it is straightforward, by adding indices in the appendix, to generalise to any N freedom system admitting action angle variables I_j, θ_j $(1\leq j\leq N)$ that is any *integrable* system. All N of the action functions I_j are then adiabatic invariants and each angle variable θ_j suffers extra change on closed adiabatic excursion of the Hamiltonian. The 2-form for this extra change is

$$d\langle d\theta_j\rangle = \frac{1}{2}\frac{\partial^2}{\partial I_j\,\partial I_k}\int_0^{2\pi}\frac{d^N\boldsymbol{\theta}}{(2\pi)^N}\int_0^{2\pi}\frac{d^N\boldsymbol{\chi}}{(2\pi)^N}\chi_l[dI_l(\boldsymbol{\theta}-\boldsymbol{\chi})\wedge dI_k(\boldsymbol{\theta})]. \tag{22}$$

As emphasised in the introduction, while 1-freedom systems are necessarily integrable, N-freedom systems are not. There is no generalisation of (22) to N-freedom systems with the generic mixed structure of chaos and tori, indeed there are *no* known adiabatic invariants in that case.

While it is likely that angle variable changes have long been required and calculated correctly in particular contexts, in celestial mechanics for example, they have probably been found by rather *ad hoc* methods like that used in the rotating hoop illustration earlier. This paper has sought to present the effect in its generality. The other (indeed the original) motivation, the possible connection with the semiclassical limit of Berry's phase $\Delta\phi$ has not been further mentioned, but the standard relationship for the *time* evolution of θ and ϕ in a *static* Hamiltonian

$$\frac{d\theta}{dt} = \frac{dH}{dI} = -\frac{d}{dI}\left(\hbar\frac{d\phi}{dt}\right) \tag{23}$$

suggests the relationship

$$\Delta\theta = -\hbar\, d\phi/dI. \tag{24}$$

This is indeed the conclusion of semiclassical analysis (Berry 1985).

Another connection between the present classical analysis and quantum mechanics, for 1-freedom systems, arises from the trick of treating the wavefunction $\psi(x)$ and its derivative $\psi'(x)$ as 'position' and 'momentum' in a bogus 'phase space'. The time independent Schrödinger equation for given energy E then becomes a Newton's equation for ψ, with x playing the role of 'time', and an x dependent (i.e. 'time' dependent) linear restoring force $(E-V)\psi$. If $V(x)$ is a slowly varying function of x then the quadratic Hamiltonian analysis of the second illustration earlier, is directly applicable. This is one way of viewing a recent analysis of phase holonomy arising not from slow time variation of a quantum Hamiltonian but from slow space variation—that is, phase holonomy in WKB theory (Wilkinson 1984).

Acknowledgments

I am grateful to Dr P J Richens whose elementary explanation of adiabatic invariance I have reproduced in § 2. Also to Professor M V Berry for discussing with me his parallel analyses.

Appendix

The formula (9) for the angle 2-form $d\langle d\theta\rangle$ which is manifestly independent of the arbitrarinesss in the definition of angle variables may be derived as follows. As in the main text, q, p are fixed phase plane coordinates and $I(q, p, \boldsymbol{R})$ and $\theta(q, p, \boldsymbol{R})$ are parameter dependent functions of them. The symbol d stands for exterior differentiation in parameter space $\boldsymbol{R}$ (not phase space q, p). It is convenient for the derivation to distinguish the value, I_0 say, of the conserved action from the action function I.

$$\boldsymbol{d}\langle \boldsymbol{d}\theta\rangle = \boldsymbol{d}\int (\boldsymbol{d}\theta)\delta(I(q,p,\boldsymbol{R})-I_0)\frac{\mathrm{d}q\,\mathrm{d}p}{2\pi} \tag{A1}$$

$$=\int (\boldsymbol{d}I\wedge \boldsymbol{d}\theta)\frac{\mathrm{d}\delta}{\mathrm{d}I}\frac{\mathrm{d}q\,\mathrm{d}p}{2\pi} \tag{A2}$$

$$=\frac{\mathrm{d}}{\mathrm{d}I_0}\int (\boldsymbol{d}\theta\wedge \boldsymbol{d}I)\delta\frac{\mathrm{d}q\,\mathrm{d}p}{2\pi} \tag{A3}$$

$$=\frac{\mathrm{d}}{\mathrm{d}I_0}\langle \boldsymbol{d}\theta\wedge \boldsymbol{d}I\rangle. \tag{A4}$$

The task is now to eliminate $\boldsymbol{d}\theta$ from this average. An expression for $\partial\boldsymbol{d}\theta/\partial\theta$ can be obtained and used. (This derivative is not equal to $\boldsymbol{d}(\partial\theta/\partial\theta)\equiv 0$ because $\boldsymbol{d}$ and $\partial/\partial\theta$ do not commute.) The expression derives from the fact that the variables I; θ remain canonical; $\{I,\theta\}=1$ for all $\boldsymbol{R}$ so that by exterior differentiation

$$0=\{\boldsymbol{d}I,\theta\}+\{I,\boldsymbol{d}\theta\}=\partial\boldsymbol{d}I/\partial I+\partial\boldsymbol{d}\theta/\partial\theta. \tag{A5}$$

Integrating this with respect to θ supplies $\boldsymbol{d}\theta$, which inserted into the average $\langle \boldsymbol{d}\theta\wedge \boldsymbol{d}I\rangle$ gives

$$\langle \boldsymbol{d}\theta\wedge \boldsymbol{d}I\rangle = -\int_0^{2\pi}\left[\left(\int_0^{\theta}\frac{\partial \boldsymbol{d}I}{\partial I}\mathrm{d}\theta'\right)\wedge \boldsymbol{d}I(\theta)\right]\frac{\mathrm{d}\theta}{2\pi} \tag{A6}$$

$$=\int_0^{2\pi}\left(\frac{\partial \boldsymbol{d}I(\theta)}{\partial I}\wedge\int_0^{\theta}\boldsymbol{d}I(\theta')\theta'\right)\frac{\mathrm{d}\theta}{2\pi} \tag{A7}$$

where an integration by parts has been performed, the integrated part vanishing by virtue of the (exact) Liouville relation used in (4)

$$\int_0^{2\pi}\boldsymbol{d}I\frac{\mathrm{d}\theta}{2\pi}=\langle \boldsymbol{d}I\rangle=0. \tag{A8}$$

The expression (A7) is, as can easily be verified; independent of the choice of origin of θ. It can be simplified by inspection. The integrand is the product of two functions evaluated at different angles. Keeping first their *separation* χ fixed and then integrating over all χ (invoking the independence of origin) we obtain

$$\langle \boldsymbol{d}\theta\wedge \boldsymbol{d}I\rangle=\int_0^{2\pi}\left\langle\frac{\partial \boldsymbol{d}I(\theta)}{\partial I}\wedge \boldsymbol{d}I(\theta-\chi)\right\rangle(2\pi-\chi)\frac{\mathrm{d}\chi}{2\pi}. \tag{A9}$$

Some final manipulations now remain to remove the operation $\partial/\partial I$ which is still angle coordinate dependent. On the one hand the 2π term in (A9) can be seen to vanish because the average of the (wedge) product is then the product of the averages, each of which vanish by (A8). On the other hand by the substitution $\chi'=2\pi-\chi$ (A9) can be rewritten

$$\int_0^{2\pi}\left\langle\frac{\partial \boldsymbol{d}I(\theta-\chi')}{\partial I}\wedge \boldsymbol{d}I(\theta)\right\rangle\chi'\frac{\mathrm{d}\chi'}{2\pi} \tag{A10}$$

using the cyclic nature of θ. Taking half the sum of the two expressions, (A9) without the 2π, and (A10), we obtain

$$\langle \boldsymbol{d}\theta \wedge \boldsymbol{d}I\rangle = \frac{1}{2}\int_0^{2\pi}\left\langle -\frac{\boldsymbol{d}I(\theta)}{\partial I}\wedge \boldsymbol{d}I(\theta-\chi) + \frac{\partial \boldsymbol{d}I(\theta-\chi)}{\partial I}\wedge \boldsymbol{d}I(\theta)\right\rangle_\chi \frac{\mathrm{d}\chi}{2\pi} \tag{A11}$$

$$= \frac{1}{2}\frac{\mathrm{d}}{\mathrm{d}I_0}\int_0^{2\pi}\langle \boldsymbol{d}I(\theta-\chi)\wedge \boldsymbol{d}I(\theta)\rangle_\chi \frac{\mathrm{d}\chi}{2\pi} \tag{A12}$$

where the antisymmetry of the wedge product has been used and $\partial/\partial I$ has been extracted from the average brackets and converted to $\mathrm{d}/\mathrm{d}I_0$ as is then legitimate.

Finally then we have the required result

$$\boldsymbol{d}\langle \boldsymbol{d}\theta\rangle = \frac{1}{2}\frac{\mathrm{d}^2}{\mathrm{d}I_0^2}\int_0^{2\pi}\langle \boldsymbol{d}I(\theta-\chi)\wedge \boldsymbol{d}I(\theta)\rangle_\chi \frac{\mathrm{d}\chi}{2\pi}. \tag{A13}$$

The generalised analysis for an N-freedom system is a straightforward matter of adding indices to the above and yields (22). For example (A4) becomes

$$\boldsymbol{d}\langle \boldsymbol{d}\theta_j\rangle = \frac{\mathrm{d}}{\mathrm{d}I_{0l}}\langle \boldsymbol{d}\theta_j \wedge \boldsymbol{d}I_l\rangle. \tag{A14}$$

Since this intermediate expression is often useful in calculations it is worth noting that it can be rewritten, by use of the constancy of the Poisson brackets for all $\boldsymbol{R}$, $\boldsymbol{d}\{I_i, I_j\} = \boldsymbol{d}\{I_i, \theta_j\} = \boldsymbol{d}\{\theta_i, \theta_j\} = 0$, in the more symmetric form

$$\mathrm{d}\langle \boldsymbol{d}\theta_j\rangle = \frac{\mathrm{d}}{\mathrm{d}I_{0j}}\langle \boldsymbol{d}\theta_l \wedge \boldsymbol{d}I_l\rangle. \tag{A15}$$

This form corresponds most closely to that obtained by Berry (1985) who uses q and p as functions of fixed phase space coordinates I, θ instead of *vice versa*. Like (A14), (A15) yields the result (22) directly.

References

Arnold V I 1978 *Mathematical methods of classical mechanics* (New York: Springer)
Berry M V 1984 *Proc. R. Soc.* A **392** 45-57
—— 1985 *J. Phys. A: Math. Gen.* **18** 15-27
Forder P 1984 *J. Phys. A: Math. Gen.* **17** 1343-55
Landau L and Lifshitz, E M 1976 *Mechanics* 3rd edn (Oxford: Pergamon)
Robnik M and Hannay J H 1985 *J. Phys. A: Math. Gen.* To be submitted
Simon B 1983 *Phys. Rev. Lett.* **51** 2167-70
Wilkinson M 1984 *J. Phys. A: Math. Gen.* **17** 3459-76

J. Phys. A: Math. Gen. 18 (1985) 15-27. Printed in Great Britain

Classical adiabatic angles and quantal adiabatic phase

M V Berry
H H Wills Physics Laboratory, Tyndall Avenue, Bristol BS8 1TL, UK

Received 29 May 1984

Abstract. A semiclassical connection is established between quantal and classical properties of a system whose Hamiltonian is slowly cycled by varying its parameters round a circuit. The quantal property is a geometrical phase shift γ_n associated with an eigenstate with quantum numbers $n = \{n_l\}$; the classical property is a shift $\Delta\theta_l(I)$ in the lth angle variable for motion round a phase-space torus with actions $I = \{I_l\}$; the connection is $\Delta\theta_l = -\partial\gamma/\partial n_l$. Two applications are worked out in detail: the generalised harmonic oscillator, with quadratic Hamiltonian whose parameters are the coefficients of q^2, qp and p^2; and the rotated rotator, consisting of a particle sliding freely round a non-circular hoop slowly turned round once in its own plane.

1. Introduction

Consider a quantal or classical system with N freedoms, whose Hamiltonian $H(q, p; X(t))$ depends on a set of slowly changing parameters $X = \{X_\mu\}$ as well as dynamical variables or operators $q = \{q_j\}$, $p = \{p_j\}$ $(i \leq j \leq N)$. The evolution of the system is governed by an adiabatic theorem. In the quantal case (Messiah 1962), this states that a system originally in an eigenstate, labelled by one or more parameters $n = \{n_i\}$, willl remain in the same eigenstate $|n; X(t)\rangle$, with energy $E_n(X(t))$, as the X vary. In the classical case (Dirac 1925), the theorem states that an orbit initially on an N-dimensional phase-space torus with actions $I = \{I_j\}$ (Arnold 1978) will continue to explore the tori with the same values of I (adiabatic invariants), in spite of the changing Hamiltonian corresponding to $X(t)$, provided such tori continue to exist (for example if the system remains integrable for all parameters X).

These well known adiabatic theorems fail to describe an important feature of the evolution, which manifests itself if the Hamiltonian returns to its original form after a (long) time T, i.e. $X(T) = X(0)$. We shall describe such changes as *taking the system round a circuit* C in the space of parameters X.

Quantally, the feature is a *geometrical phase factor* $\exp(i\gamma_n(C))$ accumulated round C by a system in the nth state: if the state is initially $|\Psi(0)\rangle$, then the state at T is

$$|\Psi(T)\rangle = \exp(i\gamma_n(C)) \exp\left(-\frac{i}{\hbar}\int_0^T dt\, E_n(X(t))\right)|\Psi(0)\rangle. \qquad (1)$$

(The second factor contains the familiar dynamical phase, and is present even if the parameters remain constant, and the third factor $|\Psi(0)\rangle$ is an expression of the adiabatic theorem.) A discussion of $\gamma_n(C)$, as well as general formulae and illustrative examples, is given by Berry (1984a) and Simon (1983), commenting on an early version of that paper, explains how the geometrical phase embodies the anholonomy (non-integrable connection) of Hermitian line bundles.

Classically, the feature that the adiabatic theorem does not describe is *shifts* $\Delta\theta(I;C)$ in the *angles* $\theta=\{\theta_j\}$ conjugate to the actions I, in addition to those expected on the basis of the instantaneous frequencies $\omega=\{\omega_j(I;X)\}$: if the initial angles are $\theta(0)$, then after the circuit C the position of the system on its torus I is given by

$$\theta(T)=\theta(0)+\int_0^T \mathrm{d}t\,\omega(I;X(t))+\Delta\theta(I;C). \tag{2}$$

The existence of the $\Delta\theta$ as a general feature of slowly cycled integrable systems was discovered by Hannay (1984), and I will refer to these angle shifts as 'classical adiabatic angles', or as 'Hannay's angles'.

My purpose here is to show by a semiclassical argument that Hannay's angles $\Delta\theta$ are indeed classical analogues of the quantal adiabatic phase $\gamma_n(C)$, and to establish the precise relation between these quantities. This analysis will be presented in § 3. As a preliminary, § 2 will contain an alternative to Hannay's (1985) derivation of his angles. Sections 4 and 5 give a discussion of two families of one-dimensional systems (both suggested by Hannay) for which classical and quantum mechanics can be worked out in detail, thus confirming the correctness of the general theory of § 3. The first system is the 'generalised harmonic oscillator', consisting of a quadratic Hamiltonian for which the coefficients of q^2, qp and p^2 are slowly varied; the second is the 'rotated rotator', consisting of a particle sliding freely round a non-circular hoop which is slowly rotated in its own plane. Two appendices give instructive 'elementary' derivations of Hannay's angles for these systems, based on asymptotic analyses of the corresponding Newtonian equations.

2. Hannay's angles

The evolution of angle variables, which by (2) determine the classical adiabatic angles $\Delta\theta$, can be determined by making a canonical transformation to action-angle variables. This is achieved (Landau and Lifshitz 1976) in terms of a generating function $S^{(\alpha)}(q,I;X(t))$, according to the scheme

$$\{q,p\}\leftarrow S^{(\alpha)}(q,I;X(t))\rightarrow\{\theta,I\}, \tag{3a}$$

$$p^{(\alpha)}=\partial S^{(\alpha)}/\partial q, \qquad \theta^{(\alpha)}=\partial S^{(\alpha)}/\partial I. \tag{3b, c}$$

In these formulae, the superscript α labels the branches of S, a function whose unavoidable multivaluedness reflects the fact that, for a given torus I, q does not uniquely determine p (figure 1).

The new Hamiltonian $\bar{H}$ differs from the old Hamiltonian H in value as well as functional form, because the canonical transformation is time dependent through the slowly changing parameters $X(t)$. In fact

$$\bar{H}(\theta,I,t)=\mathscr{H}(I;X(t))+(\mathrm{d}X/\mathrm{d}t)(\partial/\partial X)S^{(\alpha)}(q,I;X(t)), \tag{4}$$

where

$$\mathscr{H}(I;X(t))\equiv H(q(\theta,I;X(t)),p(\theta,I;X(t));X(t)) \tag{5}$$

is the (angle-independent) 'action' Hamiltonian corresponding to constant X. At

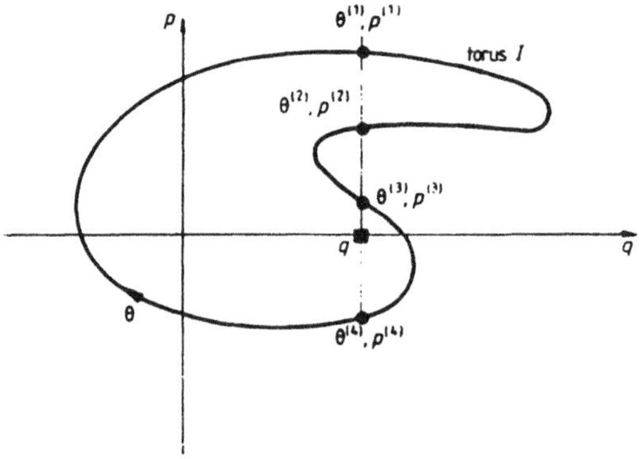

Figure 1. Torus with action $I=(1/2\pi)\oint p\,\mathrm{d}q$ for a system with one freedom, illustrating multivaluedness of mappings from q to p and q to θ.

any time t, the branch α and the value of q occurring in (4) are uniquely defined by θ and I.

To obtain an explicit form for $\bar{H}$, we define the single-valued function

$$\mathscr{S}(\theta, I; X) \equiv S^{(\alpha)}(q(\theta, I; X), I; X), \qquad (0 \leqslant \theta < 2\pi) \tag{6}$$

so that

$$\frac{\partial S^{(\alpha)}}{\partial X} = \frac{\partial \mathscr{S}}{\partial X} - \frac{\partial S^{(\alpha)}}{\partial q}\frac{\partial q}{\partial X} = \frac{\partial \mathscr{S}}{\partial X} - p^{\alpha}\frac{\partial q}{\partial X}. \tag{7}$$

Thus the new Hamiltonian becomes

$$\bar{H}(\theta, I, t) = \mathscr{H}(I; X(t)) + \frac{\mathrm{d}X(t)}{\mathrm{d}t}\left(\frac{\partial \mathscr{S}}{\partial X}(\theta, I; X(t)) - p(\theta, I; X(t))\frac{\partial q}{\partial X}(\theta, I; X(t))\right). \tag{8}$$

This is globally single-valued, because q and p are periodic functions of θ, and the increment of $\mathscr{S}$ round a circuit is

$$\mathscr{S}(\theta + 2\pi, I; X) - \mathscr{S}(\theta, I; X) = \oint p\,\mathrm{d}q = 2\pi I, \tag{9}$$

which does not depend on X.

Hamilton's equation for the angles is

$$\mathrm{d}\theta/\mathrm{d}t = \partial \bar{H}/\partial I. \tag{10}$$

When applied to (8), the first term gives that part of the evolution which would occur even if the parameters remained constant, arising from the frequencies

$$\omega(I; X) = \partial \mathscr{H}(I; X)/\partial I. \tag{11}$$

What we are seeking, however, is the angle shift defined by (2), and this arises from the second term in (8):

$$\Delta\theta = \int_0^T \mathrm{d}t \frac{\mathrm{d}X}{\mathrm{d}t}\frac{\partial}{\partial I}\left(\frac{\partial \mathscr{S}}{\partial X} - p\frac{\partial q}{\partial X}\right). \tag{12}$$

As it stands, this integral is difficult to evaluate because the integrand depends on time implicitly through the changes in θ and I as well as explicitly through the variations of X. It is natural at this point to invoke the adiabatic technique (Arnold 1978) of averaging over the implicit (fast) variations by integrating over the torus at each time t. When applied to the Hamilton equation conjugate to (10), this procedure shows that the actions I remain constant in spite of the (slow) variations in X—which is of course the familiar adiabatic theorem. When applied to (12), it gives

$$\Delta\theta(I;C)=\oint \mathrm{d}X\frac{\partial}{\partial I}\frac{1}{(2\pi)^N}\oint \mathrm{d}\theta\left\{\frac{\partial\mathcal{S}}{\partial X}(\theta,I;X)-p(\theta,I;X)\frac{\partial q}{\partial X}(\theta,I;X)\right\} \tag{13}$$

where

$$\oint \mathrm{d}\theta=\prod_{j=1}^{N}\int_0^{2\pi}\mathrm{d}\theta_j. \tag{14}$$

Equation (13) has the form of a line integral over a single-valued function in parameter space. The first term vanishes because $\partial\mathcal{S}/\partial X$ is a gradient. The second term can be transformed by Stokes' theorem into an integral over any surface A, in parameter space, whose boundary is C. In the language of differential forms (Arnold 1978):

$$\Delta\theta(I;C)=-\frac{\partial}{\partial I}\int_{\partial A=C}W \tag{15}$$

where the *angle 2-form* is expressed in terms of W, given by

$$W=\frac{1}{(2\pi)^N}\oint \mathrm{d}\theta\,\mathrm{d}p\wedge\mathrm{d}q. \tag{16}$$

Of course the forms in this expression are parameter-space forms, not the more familiar phase-space ones. A more explicit expression, for the case where $\{X\mu\}$ can be written as a three-dimensional vector $\boldsymbol{X}$, is

$$\Delta\theta_i(I_j;C)=-\frac{\partial}{\partial I_i}\iint_{\partial A=C}\mathrm{d}\boldsymbol{A}\cdot\boldsymbol{W}(I;\boldsymbol{X}) \tag{17}$$

where $\mathrm{d}\boldsymbol{A}$ denotes the element of area in parameter space and

$$\boldsymbol{W}(I;\boldsymbol{X})=\frac{1}{(2\pi)^N}\oint \mathrm{d}\theta\,\nabla_{\boldsymbol{X}}p_j(\theta,I;\boldsymbol{X})\wedge\nabla_{\boldsymbol{X}}q_j(\theta,I;\boldsymbol{X}). \tag{18}$$

The formulae (15)-(18) for the classical adiabatic angles constitute one version of the expressions obtained by Hannay (1985). It is not difficult to show that Hannay's angles are invariant under parameter-dependent and action-dependent deformations of the (arbitrary) origin from which the angles θ are measured, i.e. under

$$\theta\to\theta+\beta(I;X), \tag{19}$$

provided β is single-valued across the area A in parameter space. This classical invariance corresponds to the invariance of the quantal phase factor under parameter-dependent changes in the phases of the eigenvectors $|n;X\rangle$ (see the appendix of Berry 1984a).

3. Angles and phase: semiclassical theory

The quantal geometrical phase $\gamma_n(C)$ defined by equation (1) will be written in the form

$$\gamma_n(C) = -\iint_{\partial A = C} \mathrm{d}\boldsymbol{A} \cdot \boldsymbol{V}(n; \boldsymbol{X}) \tag{20}$$

where

$$\boldsymbol{V}(n; \boldsymbol{X}) \equiv \operatorname{Im} \nabla_{\boldsymbol{X}} \wedge \langle n; \boldsymbol{X} | \nabla_{\boldsymbol{X}} | n; \boldsymbol{X} \rangle, \tag{21}$$

corresponding to equation (7) of Berry (1984a) and analogous to (17) and (18) of the preceding section. In position representation we define the wavefunction ψ_n by

$$\psi_n(q; \boldsymbol{X}) \equiv \langle q | n; \boldsymbol{X} \rangle, \tag{22}$$

so that the phase 2-form becomes

$$\boldsymbol{V}(n; \boldsymbol{X}) = \operatorname{Im} \nabla_{\boldsymbol{X}} \wedge \int \mathrm{d}q\, \psi_n^*(q; \boldsymbol{X}) \nabla_{\boldsymbol{X}} \psi_n(q; \boldsymbol{X}) \tag{23}$$

where

$$\int \mathrm{d}q \equiv \prod_{j=1}^{N} \int_{-\infty}^{\infty} \mathrm{d}q_j. \tag{24}$$

Semiclassically, ψ_n is associated with a torus whose actions are quantised by the corrected Bohr-Sommerfeld rule (Keller 1958)

$$I_j = (n_j + \sigma_j)\hbar \tag{25}$$

where the σ_j are N constants whose values are unimportant in the present context. The wavefunction is obtained from the torus by projection from phase space to q space, according to the method of Maslov (see Maslov and Fedoriuk (1981) and simplified presentations by Percival (1977) and Berry (1983)):

$$\psi_n(q; \boldsymbol{X}) = \sum a_\alpha(q, I; \boldsymbol{X}) \exp[\mathrm{i}\hbar^{-1} S^{(\alpha)}(q, I; \boldsymbol{X})], \tag{26}$$

where $S^{(\alpha)}$ is the classical generating function (equation (3)), the summation over α corresponds to all branches $p^{(\alpha)}$ contributing at q (figure 1), and the amplitude is given in terms of the projection Jacobian by

$$a_\alpha^2 = \frac{1}{(2\pi)^N} \frac{\mathrm{d}\theta^{(\alpha)}}{\mathrm{d}q} = \frac{1}{(2\pi)^N} \det\left(\frac{\mathrm{d}\theta_i}{\mathrm{d}q_j}\right) \tag{27}$$

(this quantity may be positive or negative, corresponding to $\pi/2$ phase shifts across turning points).

When the wavefunction (26) is substituted into (23), products of contributions from different branches α give rapid oscillations and cancel semiclassically on integrating over q, leaving

$$\boldsymbol{V}(n; \boldsymbol{X}) = \frac{1}{\hbar} \nabla_{\boldsymbol{X}} \wedge \int \mathrm{d}q \frac{1}{(2\pi)^N} \sum_\alpha \frac{\mathrm{d}\theta^{(\alpha)}}{\mathrm{d}q} \nabla_{\boldsymbol{X}} S^{(\alpha)}(q, I; \boldsymbol{X}). \tag{28}$$

Transformation of the variables of integration from q to θ, and use of the formulae

(6) and (7) give

$$V(n;X) = \frac{1}{\hbar}\nabla_X \wedge \frac{1}{(2\pi)^N}\oint d\theta(\nabla_X \mathscr{S} - p\nabla_X q) \tag{29}$$

$$= \frac{-1}{\hbar(2\pi)^N}\oint d\theta\, \nabla_X p_j(\theta, I; X) \wedge \nabla_X q_j(\theta, I; X)$$

$$= -\frac{1}{\hbar}\mathbf{W}(I;X), \tag{30}$$

thus relating the phase 2-form to the angle 2-form (18).

Finally, this relation, together with (17) and (20), immediately gives the connection between Hannay's angles and the geometrical phase:

$$\Delta\theta_l(I;C) = -\hbar\frac{\partial}{\partial I_l}\gamma_n(C) = -\frac{\partial\gamma_n(C)}{\partial n_l}, \tag{31}$$

where the association (25) enables the quantum numbers n to be considered as continuous variables.

4. Example: generalised harmonic oscillator

The classical Hamiltonian for this system with one freedom is

$$H = \tfrac{1}{2}(X(t)q^2 + 2Y(t)qp + Z(t)p^2). \tag{32}$$

The parameters are X, Y, Z; when these are held fixed, H describes oscillatory motion round elliptic contours in the phase plane (figure 2), provided

$$XZ > Y^2, \tag{33}$$

and we assume henceforth that this remains the case as X, Y, Z vary. For given energy E the area of the contour is $2\pi E/(XZ - Y^2)^{1/2}$, and this is $2\pi I$ by definition, so the action Hamiltonian (5) is

$$\mathscr{H}(I;X) = I(XZ - Y^2)^{1/2} \tag{34}$$

giving the action-independent frequency (11) as

$$\omega = (XZ - Y^2)^{1/2}. \tag{35}$$

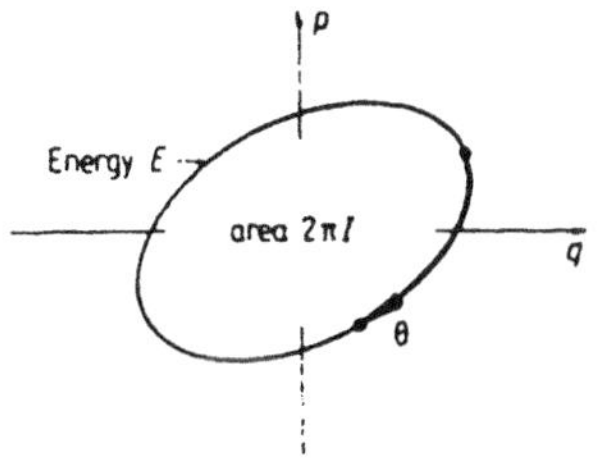

Figure 2. Elliptic phase-plane contour for a generalised harmonic oscillator with Hamiltonian (32).

It will be shown in appendix 1 that when the parameters vary the Hamiltonian (32) describes an oscillator with parametrically forced frequency, whose classical motion, including Hannay's angle, can be determined by the WKB method commonly employed in quantum mechanics. Here the angle 2-form will be calculated from (16), and requires the solution of Hamilton's equations for fixed X, Y, Z in the form $q(\theta, I; X)$ and $p(\theta, I; X)$. Choosing the origin of angle at the positive extreme of the q motion (figure 2) the solutions are

$$q=\left(\frac{2ZI}{\omega}\right)^{1/2}\cos\theta, \qquad p=-\left(\frac{2ZI}{\omega}\right)^{1/2}\left(\frac{Y}{Z}\cos\theta+\frac{\omega}{Z}\sin\theta\right) \tag{36}$$

as can easily be verified. Equation (16) now gives

$$\begin{aligned} W&=-2I\frac{1}{2\pi}\int_0^{2\pi}\mathrm{d}\theta\cos^2\theta\,\mathrm{d}\left(\frac{Y}{Z}\left(\frac{Z}{\omega}\right)^{1/2}\right)\wedge\mathrm{d}\left(\left(\frac{Z}{\omega}\right)^{1/2}\right)\\ &=\frac{I}{2}\mathrm{d}\left(\frac{Z}{(XZ-Y^2)^{1/2}}\right)\wedge\mathrm{d}\left(\frac{Y}{Z}\right). \end{aligned} \tag{37}$$

A little reduction produces the symmetrical form

$$W=-\tfrac{1}{4}I(XZ-Y^2)^{-3/2}(X\,\mathrm{d}Y\wedge\mathrm{d}Z+Y\,\mathrm{d}Z\wedge\mathrm{d}X+Z\,\mathrm{d}X\wedge\mathrm{d}Y). \tag{38}$$

If X, Y, Z are regarded as Cartesian components of a vector $\boldsymbol{X}$, this can be written

$$\boldsymbol{W}(I;\boldsymbol{X})=-\tfrac{1}{4}I\boldsymbol{X}(XZ-Y^2)^{-3/2}. \tag{39}$$

Quantally, (32) corresponds to the Hamiltonian operator

$$\hat{H}=\tfrac{1}{2}[X\hat{q}^2+Y(\hat{q}\hat{p}+\hat{p}\hat{q})+Z\hat{p}^2]. \tag{40}$$

When the parameters are constant, this gives rise to the Schrödinger equation satisfied by the wavefunction (22), namely

$$-\frac{Z\hbar^2}{2}\frac{\mathrm{d}^2\psi_n}{\mathrm{d}q^2}-\mathrm{i}\hbar Yq\frac{\mathrm{d}\psi_n}{\mathrm{d}q}+\left(\frac{Xq^2}{2}-\mathrm{i}\hbar\frac{Y}{2}\right)\psi_n=E_n\psi_n. \tag{41}$$

As is easily verified, the normalised solution can be written as

$$\psi_n(q,\boldsymbol{X})=\frac{(XZ-Y^2)^{1/8}}{Z^{1/4}\hbar^{1/4}}\chi_n\left(\frac{q(XZ-Y^2)^{1/4}}{Z^{1/2}\hbar^{1/2}}\right)\exp\left(\frac{-\mathrm{i}Yq^2}{2Z\hbar}\right), \tag{42}$$

where $\chi_n(\xi)$ are the real, normalised, Hermite functions satisfying

$$\frac{\mathrm{d}^2\chi_n(\xi)}{\mathrm{d}\xi^2}+(2n+1-\xi^2)\chi_n(\xi)=0, \tag{43}$$

and the energies are exactly given by the semiclassical formula

$$E_n=(n+\tfrac{1}{2})\hbar\omega=(n+\tfrac{1}{2})\hbar(XZ-Y^2)^{1/2}. \tag{44}$$

This wavefunction must now be substituted into the exact formula (23) for the phase 2-form, giving

$$\boldsymbol{V}=-\nabla_{\boldsymbol{X}}\wedge\frac{(XZ-Y^2)^{1/4}}{Z^{1/2}\hbar^{1/2}}\int_{-\infty}^{\infty}\mathrm{d}q\,\chi_n^2(\xi)\frac{q^2}{\hbar}\nabla_{\boldsymbol{X}}\left(\frac{Y}{2Z}\right). \tag{45}$$

Transforming the integration variable from q to ξ and using the standard result

$$\int_{-\infty}^{\infty} \mathrm{d}\xi\, \xi^2 \chi_n^2(\xi) = n + \tfrac{1}{2} \tag{46}$$

leads to

$$V = -\frac{(n+\frac{1}{2})}{2} \nabla_X \left(\frac{Z}{(XZ - Y^2)^{1/2}} \right) \wedge \nabla_X \left(\frac{Y}{Z} \right). \tag{47}$$

The semiclassical quantisation rule $(n+\frac{1}{2}) = I/\hbar$ (cf (25)) is exact for this case, and comparison of (47) with the angle 2-form as given by (37) confirms the truth of the central semiclassical relation (30).

5. Example: rotated rotator

A particle of unit mass slides freely round a non-circular hoop, which is slowly turned through one complete rotation, in its own plane, about a centre 0 (figure 3). The rotation can be described by a single parameter X, namely the orientation angle of the hoop, which slowly varies from 0 to 2π. The particle motion in the plane with coordinates $\boldsymbol{q} = (x, y)$ can be described by the Hamiltonian

$$H(\boldsymbol{q}, \boldsymbol{p}; X(t) = \tfrac{1}{2}p_x^2 + \tfrac{1}{2}p_y^2 + V(x \cos X(t) + y \sin X(t), -x \sin X(t) + y \cos X(t)), \tag{48}$$

where V is a confining potential which is zero in a narrow strip centred on the hoop and very large elsewhere.

For fixed X, the particle executes periodic motion, whose constant speed relative to the hoop is the magnitude p of the momentum $\boldsymbol{p}$. The action is

$$I = \tfrac{1}{2}p\mathscr{L} \tag{49}$$

where $\mathscr{L}$ is the length of the hoop, and the angle may be taken as

$$\theta = 2\pi s/\mathscr{L} \tag{50}$$

where s is arc length measured relative to a material point A on the hoop (figure 3). (For this problem with two freedoms there is of course a second pair of action-angle variables, corresponding to transverse vibrations of the particle, but this motion is considered here to have zero amplitude.)

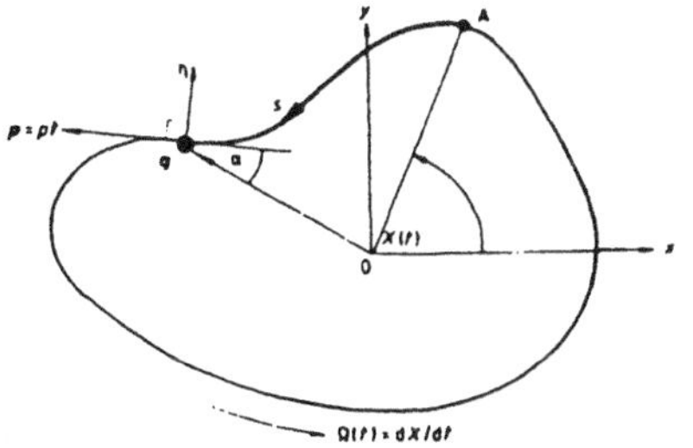

Figure 3. Coordinates and notation for the rotated rotator; the hoop has length $\mathscr{L}$ and area $\mathscr{A}$.

The classical angle shift for a complete turn can be written as a line integral, using (13) with the first term omitted because it gives zero when integrated

$$\Delta\theta = \frac{-1}{2\pi}\frac{\partial}{\partial I}\int_0^{2\pi} \mathrm{d}X \int_0^{2\pi} \mathrm{d}\theta\, p(\theta, I; X)\frac{\partial}{\partial X} q(\theta, I; X). \tag{51}$$

Changing variables from θ to s, and writing the momentum using (49) as

$$p = (2\pi I/\mathscr{L})\boldsymbol{t}, \tag{52}$$

where $\boldsymbol{t}$ is the hoop's unit tangent vector at the point $\boldsymbol{q}$, this becomes

$$\Delta\theta = \frac{-2\pi}{\mathscr{L}^2}\int_0^{2\pi} \mathrm{d}X \int_0^{\mathscr{L}} \mathrm{d}s\, \boldsymbol{t}\cdot\frac{\partial \boldsymbol{q}}{\partial X}. \tag{53}$$

Elementary geometry gives

$$\boldsymbol{t}\cdot \partial\boldsymbol{q}/\partial X = q(s)\sin\alpha(s) \tag{54}$$

where q is the radius of the hoop at s and α the angle between the radius and the tangent; thus $\boldsymbol{t}\cdot\partial\boldsymbol{q}/\partial X$ is independent of the orientation X, and

$$\Delta\theta = \frac{-(2\pi)^2}{\mathscr{L}^2}\int_0^{\mathscr{L}} \mathrm{d}s\, q(s)\sin\alpha(s)$$
$$= -8\pi^2\mathscr{A}/\mathscr{L}^2 \tag{55}$$

where $\mathscr{A}$ is the area of the hoop. A more conventional derivation of this result is given in appendix 2.

Writing Hannay's angle in the form

$$\Delta\theta = -2\pi + 2\pi(1 - 4\pi\mathscr{A}/\mathscr{L}^2), \tag{56}$$

we see that the first term gives the expected phase slippage resulting from the fact that the origin A, from which the angle is measured, has made a complete rotation. The non-trivial aspect of the anholonomy is embodied in the second term. By the isoperimetric inequality this term is never negative, so that for small deviations from circularity the particle appears to have advanced further round the hoop than would be predicted by an argument that neglected anholonomy. Hannay's hoop is thus a detector of complete rotations and, more generally, of *absolute angular displacements*, closely analogous to the ring gyroscopes (Forder 1984) employed to detect angular velocities, and to the Sagnac effect (Post 1967).

Quantally, the rotator eigenstates for fixed X are

$$\psi_n(\boldsymbol{q}; X) = (a(\boldsymbol{q}; X)/\mathscr{L}^{1/2})\exp[2\pi i n s(\boldsymbol{q}; X)/\mathscr{L}] \tag{57}$$

where the amplitude a which confines particles to the hoop can be expressed in terms of a perpendicular coordinate η (figure 3) by

$$a^2 = \delta(\eta(\boldsymbol{q}; X)). \tag{58}$$

The adiabatic phase γ_n, accumulated during one rotation of the hoop, can be calculated as a circuit integral (cf (20) and (21) and Berry 1984a) from

$$\gamma_n = -\mathrm{Im}\int_0^{2\pi} \mathrm{d}X \int_{-\infty}^{\infty} \mathrm{d}x \int_{-\infty}^{\infty} \mathrm{d}y\, \psi_n^*(\boldsymbol{q}; X)\frac{\partial}{\partial X}\psi_n(\boldsymbol{q}; X). \tag{59}$$

Substituting (57) and changing to s, η coordinates gives

$$\gamma_n = \frac{-2\pi n}{\mathscr{L}^2}\int_0^{2\pi} \mathrm{d}X \int_0^{\mathscr{S}} \mathrm{d}s \frac{\partial s}{\partial X}(q; X). \tag{60}$$

Now, inspection of figure 3 shows that

$$\partial s/\partial X = -q \sin\alpha \tag{61}$$

so that

$$\gamma_n = -8\pi^2 n\mathscr{A}/\mathscr{L}^2. \tag{62}$$

But $n = I/\hbar$, so that when compared with the angle shift (55) this result gives another confirmation of the central semiclassical relation (31).

6. Discussion

At the heart of this paper lies the relation (31) between the geometric adiabatic phase shift in quantum mechanics and Hannay's adiabatic angles in classical mechanics. The relation holds at the level of semiclassical approximation, and its applicability is restricted to systems whose classical motion is integrable (for fixed parameters) and whose quantal stationary states are associated with phase-space tori. How realistic is a treatment dependent upon this restriction?

For the case of one freedom, all bound systems are integrable, with orbits on one-dimensional tori (closed energy contours) in the phase plane. It is therefore not unrealistic to consider a family of such systems, and the example of the generalised harmonic oscillator provides an illustration. Nevertheless, caution should be exercised when considering systems possessing more than one torus with the same action, because then barrier penetration effects may cause discordance between the quantal and classical adiabatic theorems, as discussed by Berry (1984b).

For more freedoms, integrability is exceptional, because of the occurrence of chaotic motion (Lichtenberg and Lieberman (1983)). Even in quasi-integrable systems, where most of the phase space is occupied by tori, variation of even one parameter X will, generically, cause the system to pass through many resonant zones where the tori are destroyed. Therefore our arguments (like most theories of semiclassical wavefunctions) apply only to very special systems when $N > 1$. In spite of this, some exceptional families of integrable systems are interesting, as illustrated by the example of the rotated rotator. More generally, the argument for that case could be adapted to apply to slow rigid rotation of any multidimensional integrable system—such as an elliptic or cubical cavity containing particles—or even any non-integrable system with a stable closed orbit.

In the case where classical motion is chaotic, it is not clear what is the classical analogue of the quantal adiabatic phase, or of the phase 2-form. In view of the fact that the latter quantity for state $|n\rangle$ has singularities (in parameter space) where $|n\rangle$ degenerates with the state above or below (Berry 1984a), and also because degeneracies presumably get denser as $\hbar \to 0$, the classical analogue of the phase 2-form at parameters X might give a measure of the average density of degeneracies near X.

Acknowledgments

I thank Dr J H Hannay for telling me about his angles. This research was not supported by any military agency.

Appendix 1. Newtonian asymptotics of generalised harmonic oscillator

The time-dependent Hamiltonian (32) yields equations of motion for q and p. Eliminating p gives the Newtonian equation for the coordinate $q(t)$ as

$$\ddot{q}-(\dot{Z}/Z)\dot{q}+[XZ-Y^2+(\dot{Z}Y-\dot{Y}Z)/Z]q=0 \tag{A1.1}$$

where dots denote time derivatives. Defining a (small) adiabatic parameter ε and a 'slow time' variable τ by

$$X\equiv X(\tau)\text{ etc,}\qquad \tau=\varepsilon t, \tag{A1.2}$$

and eliminating the term in $\dot{q}$ by introducing the new coordinate $Q(\tau)$ defined by

$$q(t)\equiv[Z(\tau)]^{1/2}Q(\tau), \tag{A1.3}$$

gives

$$Q''+\varepsilon^{-2}\{XZ-Y^2+\varepsilon(Z'Y-Y'Z)/Z+\varepsilon^2[\tfrac{1}{2}(Z'/Z)'-\tfrac{1}{4}(Z'/Z)^2]\}Q=0 \tag{A1.4}$$

where primes denote derivatives with respect to τ.

This equation decribes a parametrically driven oscillator whose variable frequency is given by adding to (35) some terms arising from the time-dependence of H. Because of the restriction (33) and the assumed smallness of ε the quantity in curly brackets in (A1.4) never vanishes, so that the motion is always oscillatory and never exponential. Therefore the most elementary form of WKB asymptotics (see e.g. Fröman and Fröman 1965) may be employed, without the complications that would arise from turning points, to determine $Q(\tau)$ for small ε. The leading-order behaviour is

$$Q(\tau)\approx[1+O(\varepsilon)]\cos[\theta(\tau)](XZ-Y^2)^{-1/2} \tag{A1.5}$$

where

$$\theta(\tau)=\theta(0)+\frac{1}{\varepsilon}\int_0^\tau d\tau'(XZ-Y^2)^{1/2}+\tfrac{1}{2}\int_0^\tau d\tau'\frac{(Z'Y-Y'Z)}{Z(XZ-Y^2)^{1/2}}+O(\varepsilon), \tag{A1.6}$$

$\theta(0)$ being the initial phase. Of course these terms correspond precisely to those in equation (2), enabling Hannay's angle to be identified as

$$\Delta\theta=\tfrac{1}{2}\int_0^T dt\frac{(\dot{Z}Y-\dot{Y}Z)}{Z(XZ-Y^2)^{1/2}}. \tag{A1.7}$$

Transforming to a line integral in parameter space and thence, by Stokes' theorem, to a surface integral, gives

$$\Delta\theta=\int_{\partial A=C}\frac{1}{2}\left[d\left(\frac{Y}{Z(XZ-Y^2)^{1/2}}\right)\wedge dZ-d\left(\frac{1}{(XZ-Y^2)^{1/2}}\right)\wedge dY\right] \tag{A1.8}$$

which reduces to

$$\Delta\theta = \tfrac{1}{4}\int_{\partial A = c} \frac{1}{(XZ - Y^2)^{3/2}} (X\,dY \wedge dZ + Y dZ \wedge dX + Z\,dX \wedge dY). \tag{A1.9}$$

This is precisely the angle shift given by the previously obtained (38) and (15).

Appendix 2: Rotated rotator in a rotating frame

In the frame of reference rotating with the angular velocity $\Omega = dX/dt$ of the hoop, the acceleration $\ddot{s}$ is determined by the centrifugal and Euler pseudo-forces. (The Coriolis force acts perpendicular to the motion and thus affects only the normal reaction of the hoop on the particle.) Thus

$$\ddot{s} = \boldsymbol{t}\cdot[-\boldsymbol{\Omega}\wedge(\boldsymbol{\Omega}\wedge \boldsymbol{q}) - \dot{\boldsymbol{\Omega}}\wedge \boldsymbol{q}] \tag{A2.10}$$

where the vector $\boldsymbol{\Omega}$ is normal to the hoop's plane, and $|\boldsymbol{\Omega}| = \Omega$. Referring to figure 3 we see that this can be written

$$\ddot{s}(t) = \Omega^2(t) q(s(t))\, dq(s(t))/ds - \dot{\Omega}(t) q(s(t)) \sin[\alpha(s(t))]. \tag{A2.11}$$

Integrating twice we obtain

$$s(t) = s_0 + p_0 t + \int_0^t dt'(t - t')\{\tfrac{1}{2}\Omega^2(t')\, dq^2(s(t'))/ds)$$
$$-\dot{\Omega}(t') q(s(t')) \sin[\alpha(s(t'))]\} \tag{A2.12}$$

as the equation implicitly determining $s(t)$, where s_0 and p_0 are the initial position and velocity.

Because Ω and $\dot{\Omega}$ are small, the particle makes many circuits while the hoop rotates a little, so that the s-dependent quantities in the square brackets can be replaced by their averages round the hoop, giving the following explicit formula for the position of the particle after the (long) time T in which the hoop turns once:

$$s(T) = s_0 + p_0 T + \int_0^T dt(T - t)\left(\frac{\Omega^2(t)}{2\mathcal{L}}\int_0^{\mathcal{L}} ds \frac{dq(s)}{ds}\right.$$
$$\left.- \frac{\dot{\Omega}(t)}{\mathcal{L}}\int_0^{\mathcal{L}} ds\, q(s) \sin[\alpha(s)]\right). \tag{A2.13}$$

The first hoop integral (from the centrifugal force) vanishes. The entire anholomic effect thus arises, in this formulation, from the Euler force. Use of

$$\int_0^T dt(T - t)\dot{\Omega}(t) = \int_0^T dt\, \Omega(t) = 2\pi \tag{A2.14}$$

gives, finally,

$$S(T) = s_0 + p_0 T - \frac{4\pi\mathcal{A}}{\mathcal{L}}, \tag{A2.15}$$

in exact agreement with the previously calculated angle shift (55).

References

Arnol'd V I 1978 *Mathematical Methods of Classical Mechanics* (New York: Springer)
Berry M V 1983 in *Chaotic Behavior of Deterministic Systems* ed G Iooss, R H G Helleman and R Stora (Amsterdam: North-Holland) pp 171-271
—— 1984a *Proc. R. Soc.* A **392** 45-57
—— 1984b *J. Phys. A: Math. Gen.* **17** 1225-33
Dirac P A M 1925 *Proc. R. Soc.* **107** 725-34
Forder P W 1984 *J. Phys. A: Math. Gen.* **17** 1343-55
Fröman N and Fröman P O 1965 *JWKB Approximation; Contributions to the Theory* (Amsterdam: North-Holland)
Hannay J H 1985 *J. Phys. A: Math. Gen.* **18** in press
Keller J B 1958 *Ann. Phys., NY* **4** 180-8
Landau L D and Lifshitz E M 1976 *Mechanics* (*Course of Theoretical Physics*, vol 1) 3rd edn (Oxford: Pergamon)
Lichtenberg A J and Lieberman M A 1983 *Regular and Stochastic Motion* (New York: Springer)
Maslov V P and Fedoriuk M V 1981 *Semiclassical Approximation in Quantum Mechanics* (Dordrecht: Reidel)
Messiah A 1962 *Quantum Mechanics* (Amsterdam: North-Holland)
Percival I C 1977 *Adv. Chem. Phys.* **36** 1-61
Post E J 1967 *Rev. Mod. Phys.* **39** 475-93
Simon B 1983 *Phys. Rev. Lett.* **51** 2167-70

Gauge Kinematics of Deformable Bodies

Alfred Shapere and Frank Wilczek

Institute for Advanced Study
Princeton, NJ 08540

ABSTRACT

The treatment of the motion of deformable bodies requires a specification of axes for each shape. We present a natural kinematic formulation of this problem in terms of a gauge structure over the space of shapes that the body may assume. As an example, we discuss how deformations of a body with angular momentum zero can result in a change in orientation.

1. Introduction

Gauge potentials figure prominently in the formulation of fundamental physical laws. The abstractness of these laws, however, does not easily lend itself to an intuitive understanding of the concepts involved. Here we argue that gauge potentials arise naturally in a much more mundane, but in return more readily visualized, context—the description of the motion of deformable bodies. We hope that our exposition will provide both an introduction to some of the basic concepts of gauge theories and a useful framework for discussing the kinematics of deformable bodies.

A cat, held upside-down by its feet and released at rest from a suitable height, will almost always manage to land on its feet [1] [2]. A diver leaving the board with no angular momentum may perform several twists and somersaults before hitting the water [3] [4]. In both cases, by executing a sequence of deformations beginning and ending at the same shape, a deformable body with nothing to push against and no angular momentum has undergone a net rotation.

In this note, we will present a convenient and natural context for computing the net rotation of a body, in the absence of external forces and torques, due to a given sequence of deformations. Our starting point is the observation that such rotations have no dependence on the rate at which the deformations are made—the equations governing the motion are invariant under time reparameterizations. Only the geometry of the sequence of deformations matters. We shall show that the rotation of a self-deforming body may be naturally expressed in a purely geometric form, in terms of a gauge potential over configuration space.

A similar kinematic framework was devised recently for the description of another problem involving deformable bodies: swimming at low Reynolds number [5]. In that case, calculation of the gauge potential required the solution of a highly non-trivial hydrodynamic problem, while here we shall be able to write the complete solution in a simple, closed form.

The configuration space of a deformable body is the space of all possible shapes [5]. We should at the outset distinguish between the space of shapes located somewhere in space and the more abstract space of *unlocated* shapes. The latter space may be obtained from the space of shapes *cum* locations by declaring two shapes with different centers-of-mass and orientations to be equivalent. When no external forces act upon a deformable body, then we may always work in its center-of-mass frame, in which case the space of located shapes is just the space of shapes *with orientation* and centered at the origin.

The problem we wish to solve may be stated as follows: what is the net rotation which results when a deformable body goes through a given sequence of unoriented shapes, in the absence of external forces? In other words, given a path in the space of unlocated shapes, what is the corresponding path in the space of located shapes? The problem is intuitively well-defined—if a body changes its shape in some way, a net rotation is induced. This net rotation may be computed by making use of the law of conservation of angular momentum. Thus, if the body begins with some angular momentum L, then it will adjust its orientation in such a way as to preserve L. In general, this constraint is enough to determine fully the net rotation of the body.

These remarks may seem straightforward enough, but if we attempt to formulate them mathematically, we immediately run into a crucial ambiguity. Namely, how can we specify the net rotation of an object which is continuously changing its shape? The situation is illustrated in Fig. 1—in order to talk about the relative orientations of two shapes, we must choose a set of body-fixed axes for each. It would seem that for the problem at hand, there is a natural choice of axes for an arbitrary shape—its three moments of inertia. But even this choice is ambiguous: we must still specify which moments correspond to body-fixed $x-$, $y-$, and $z-$axes. One could then say that the $x-$axis is always the longest and the $z-$axis the shortest, but this choice becomes singular when two or more moments become degenerate, and the choice of axes for the sphere remains completely ambiguous.

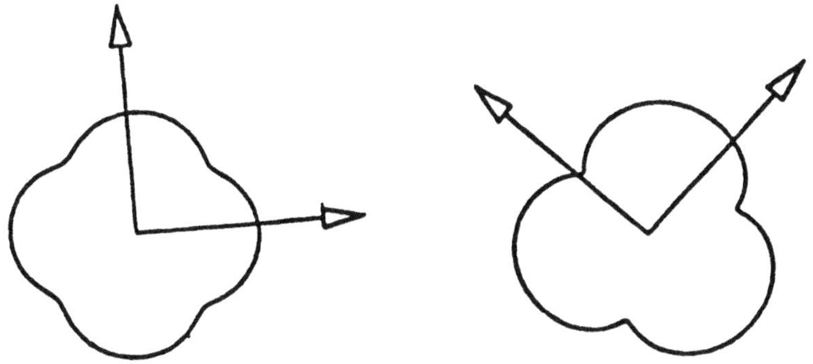

Figure 1. In order to measure the relative orientation of two different shapes, a choice of axes for each shape must be made.

Of course, we should not be too disoriented by this ambiguity, since whatever choice of axes we make, the problem still must have a solution. The distinction between particular choices is really no more than a matter of convenience.

We are faced with an enormous degeneracy of possible kinematic descriptions: at each point in an infinite-dimensional shape space, we must pick a set of reference axes from a space which looks like $SO(3)$. In the following section, we shall develop a formalism which works for any choice of axes, and which makes it easy to translate between different choices. The formalism is based on a *gauge structure* over shape space, and involves a construction known as a non-Abelian gauge potential. Although physicists first used gauge structures in the context of elementary particle theory, they have been shown to arise naturally in many other areas of physics and mathematics [5-9] . The present problem provides a further illustration of this universal concept.

2. Kinematics

Let us suppose that we have chosen a set of standard body-fixed references axes for each possible unoriented shape. Then every unoriented shape is associated to a standard oriented shape, whose $x-$, $y-$, and $z-$axes coincide with our choice of standard reference axes for the shape. For a given sequence of these standard shapes $S_0(t)$, we wish to find the corresponding sequence of physically oriented shapes $S(t)$, which are related by rotations $R(t)$:

$$S(t) = R(t)\, S_0(t) \tag{2.1}$$

$R(t)$ is a 3×3 rotation matrix which depends, in general, on the choice of reference axes for S_0. Indeed, if we make a local change in our standard shapes

$$\tilde{S}_0 = \Omega[S_0]\, S_0 \tag{2.2}$$

then the physical shapes $S(t)$ must be unchanged, so

$$\tilde{R}(t) = R(t)\, \Omega^{-1}[S_0(t)] \tag{2.3}$$

We shall compute $R(t)$ infinitesimally, and integrate to get the net rotation at finite times. We define

$$\frac{dR}{dt} = R\left[R^{-1}\frac{dR}{dt}\right] \equiv RA \tag{2.4}$$

A gives the infinitesimal rotation which results from the infinitesimal deformation of $S_0(t)$. We shall see presently that A is uniquely determined by the shape change. Once A is known, the full rotation at time t may be expressed as a path-ordered exponential

$$\begin{aligned} R(t) &= \overline{\mathrm{P}} \exp \int_0^t A(t')\, dt' \\ &\equiv \mathbf{1} + \int_{0<t'<t} A(t')\, dt' + \int_{0<t'<t''<t}\!\!\int A(t')A(t'')\, dt'dt'' + \cdots \end{aligned} \tag{2.5}$$

where the $\overline{\mathrm{P}}$ indicates that in expanding the exponential integral, all matrices are to be ordered with later times on the right. (For simplicity, we have taken $R(0) = \mathbf{1}$.)

The expression (2.5) is actually invariant under arbitrary time rescalings. Under $t \to \tau(t)$, the measure scales as $dt \to \dot{\tau} dt$, while $A \to A/\dot{\tau}$ since A contains one time derivative. This suggests that we should be able to write Eq. (2.5) in a completely geometric (*i.e.*, time-independent) form. In fact, we can define a ***gauge potential*** (or ***connection***, to mathematicians) over the space of standard shapes S_0, which we shall also denote as A, by

$$A_{\dot{S}_0}[S_0(t)] \equiv A(t) \tag{2.6}$$

This definition requires some explanation. A is defined on the tangent space to S_0—that is, it is a vector field with a component for every direction of shape space—and it takes its values in the Lie algebra of infinitesimal $SO(3)$ rotations. In Eq.(2.6), A is evaluated at a particular shape $S_0(t)$, in the direction $\dot{S}_0$ in which the shape is changing. Thus, for a given infinitesimal deformation of S_0 by δS, the resulting net rotation of the shape is $A_{\delta S}[S_0]$. In terms of a fixed basis of tangent vectors $\{\omega_i\}$ at S_0, we can define components $A_i[S_0] \equiv A_{\omega_i}[S_0]$.

Now for a given path in shape space, the integration in (2.5) may be performed without referring to a time coordinate:

$$R(t) = \overline{\mathrm{P}} \exp \int^{S_0(t)} A[S_0] \cdot dS_0 \tag{2.7}$$

Multiplying this equation on the right by $\Omega^{-1}[S_0(t)]$ and differentiating in the ω_i direction shows that under the "gauge" transformation (2.2), (2.3), A transforms as a non-Abelian gauge potential should:

$$A_i \to \Omega A_i \Omega^{-1} + \Omega \nabla_i \Omega^{-1} \tag{2.8}$$

If one is only interested in rotations resulting from cyclic infinitesimal deformations of a shape S_0, then an approximate evaluation of the path-ordered exponential in Eq. (2.7) is possible. In the expansion of Eq. (2.5), each successive term will be down by a power of ϵ, where ϵ characterizes the size of the deformations. The first order term will vanish for a closed cycle, and the second order term may be written as an expression quadratic in A and linear in the first derivatives of A. It is this term we shall now compute.

Let the standard shapes near S_0 be parametrized by

$$S_0(t) = S_0 + s(t) \tag{2.9}$$

where the $s(t)$ are infinitesimal, of order ϵ. We expand $s(t)$ in terms of a basis of tangent vectors at S_0:

$$s(t) = \sum_i \alpha_i(t)\, \omega_i \tag{2.10}$$

Then the velocity in shape space is:

$$\dot{S}_0(t) = \sum_i \dot{\alpha}_i\, \omega_i \tag{2.11}$$

Now let us expand the gauge potentials to second order:

$$\begin{aligned} A_{\dot{S}_0}[S_0 + s(t)] &\cong A_{\dot{S}_0}[S_0] + \sum_i \frac{\partial A_{\dot{S}_0}}{\partial \omega_i} \alpha_i \\ &\cong \sum_j \left(A_j \dot{\alpha}_j + \sum_i \frac{\partial A_j}{\partial \omega_i} \alpha_i \dot{\alpha}_j \right) \end{aligned} \tag{2.12}$$

In the path ordered exponential integral (2.5) around a closed cycle, the first order term in (2.12) gives no contribution, for it is a total derivative. The second order contributions are terms quadratic in A and linear in its derivatives. Because (2.5) is gauge covariant for a cyclic path, its Taylor expansion in powers of $s(t)$ must also be gauge covariant, order by order. In fact, there is a unique (up to normalization) second order gauge covariant term we can form, which is antisymmetric in its indices i and j:

$$F_{ij} \equiv \frac{\partial A_i}{\partial \omega_j} - \frac{\partial A_j}{\partial \omega_i} + [A_i, A_j] \tag{2.13}$$

(Antisymmetry of F_{ij} means that the reverse cycle leads to the reverse rotation.) We shall call F the *field strength tensor* (or *curvature*) at S_0. It is easily verified that expansion of Eq. (2.5) to second order gives

$$\overline{\mathrm{P}} \exp \oint A\, dt = 1 + \tfrac{1}{2} \oint \sum_{ij} F_{ij}\, \alpha_i \dot{\alpha}_j\, dt \tag{2.14}$$

The field strength tensor, evaluated at a shape S_0, encodes all information on rotations due to arbitrary infinitesimal deformations of S_0.

3. Dynamics: Computing the Gauge Potential

The dynamics of a free self-deforming body is completely determined by the law of angular momentum conservation. The gauge potential A, in turn, completely describes the dynamics. In this section, we derive a general expression for A, for a body with angular momentum zero.

Let us consider a body which is a collection of point masses $m^{(n)}$ at $x^{(n)}$. Then the total angular momentum of the body is

$$L_i = \epsilon_{ijk} \sum_n m^{(n)}\, x_j^{(n)} \dot{x}_k^{(n)} \tag{3.1}$$

where the sums over repeated indices are implicit and all indices run from 1 to 3. Now to each possible configuration of the $x^{(n)}$'s is associated a unique standardly oriented configuration $\tilde{x}^{(n)}$. At time t, the two configurations are related by a rotation:

$$x^{(n)}(t) = R(t)\tilde{x}^{(n)}(t) \tag{3.2}$$

Thus, expressed in terms of $\tilde{x}^{(n)}$ and $R(t)$, the total angular momentum is

$$L_i = \epsilon_{ijk} \sum m^{(n)} \left[R_{jl}\tilde{x}_l^{(n)}\, R_{km}\dot{\tilde{x}}_m^{(n)} + R_{jl}\tilde{x}_l^{(n)}\, \dot{R}_{km}\tilde{x}_m^{(n)} \right] \tag{3.3}$$

To find the gauge potential $A(t) \equiv R^{-1}\dot{R}$, we set $L_i = 0$ and solve. The result, after a few lines of algebra, is

$$A(t)_{ij} = (R^{-1}\dot{R})_{ij} = \epsilon_{ijk}\tilde{I}_{kl}^{-1}\,\tilde{L}_l \tag{3.4}$$

where $\tilde{I}$ is the inertia tensor of the standardly oriented shape $S_0(t)$ and $\tilde{L}$ is the apparent angular momentum of $S_0(t)$ at time t:

$$\tilde{I}_{ij} \equiv \sum_n m^{(n)}\left((\tilde{x}^{(n)})^2\delta_{ij} - \tilde{x}_i^{(n)}\tilde{x}_j^{(n)}\right) \tag{3.5}$$

$$\tilde{L}_i \equiv \epsilon_{ijk}\sum_n m^{(n)}\,\tilde{x}_j^{(n)}\dot{\tilde{x}}_k^{(n)} \tag{3.6}$$

The formulae (2.5) and (3.4) in principle provide a complete, and rather elegant, solution to the problem of computing net rotations of a deformable body. Often, however, it is easier to compute the gauge potential directly, as we shall now do for a simple example.

4. An Example

We now consider the example of two concentric spheres rotating about their common center of mass, as depicted in Fig. 2. The space of possible orientations for each sphere is $SO(3)$ (familiarly parameterized by Euler angles), and the full configuration space of the system is $\mathcal{S} = SO(3)\times SO(3)$. The space $\mathcal{S}_0$ of standard shapes may be thought of as the space of relative orientations of the two spheres; it, too, is isomorphic to $SO(3)$. We shall choose as standard shapes those configurations in which the outer sphere is in a fixed orientation, say, with the north pole pointing in the direction of the positive z-axis and the $\phi = 0$ meridian in the xz-plane.

We can easily write down the gauge potential at an arbitrary point in $\mathcal{S}_0$ up to proportionality. Indeed, any rotation of the inner sphere relative to the outer must be compensated by an opposing rotation of the outer sphere, in order to conserve angular momentum. In the particular basis of standard shapes we have chosen, this rotation of the outer sphere is equal to the net rotation of the system. So, letting J_i be the three generators of relative rotations, the net rotation due to an infinitesimal change of shape $\Omega = \omega_i J_i$ is

$$A = -\alpha\,\Omega \tag{4.1}$$

where α is a proportionality constant between 0 and 1. Whereas the relative orientation of the two spheres at time t is

$$R_0(t) = \overline{\mathrm{P}}\exp\int^t \Omega(t')\,dt' \tag{4.2}$$

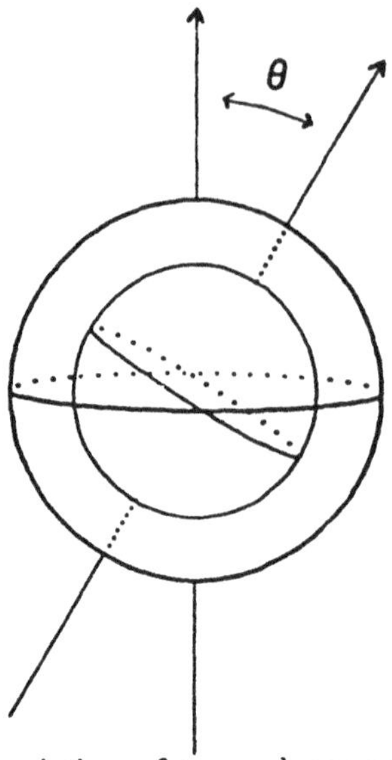

Figure 2. A body consisting of two spheres rotating about a common center of mass. Shapes are labeled by the angle θ between their polar axes.

the net rotation of the system is

$$R(t) = \overline{\mathrm{P}}\ \exp\ - \alpha \int^t \Omega(t')\, dt' \tag{4.3}$$

Now, even if $R_0(T) = \mathbf{1}$, there is no reason for $R(T)$ to be trivial. Thus by purely internal rearrangements, even with nothing to push against, our system can reorient itself.

We may evaluate the path ordered exponential for infinitesimal closed paths (and thus the field strength) by a simple argument. Suppose we rotate the outer sphere about the x-axis and the y-axis successively, by an angle ϵ, and then about the x-axis and the y-axis by $-\epsilon$. Finally, we close the path in shape space with a rotation about the z-axis by $-\epsilon^2$. We come back to the same shape we started at because of the general properties of rotations, as expressed by the equation

$$e^{-i\epsilon^2 J_z}\, e^{-i\epsilon J_y}\, e^{-i\epsilon J_x}\, e^{i\epsilon J_y}\, e^{i\epsilon J_x} = \mathbf{1} \tag{4.4}$$

true to order ϵ^3. According to Eq. (4.1), the net change in orientation of the inner sphere is

$$e^{i\alpha\epsilon^2 J_z}\, e^{i\alpha\epsilon J_y}\, e^{i\alpha\epsilon J_x}\, e^{-i\alpha\epsilon J_y}\, e^{-i\alpha\epsilon J_x} = e^{i(\alpha-\alpha^2)\epsilon^2 J_z} \tag{4.5}$$

The net rotation is $\epsilon^2(\alpha - \alpha^2)$ about the z-axis and the field strength F_{xy} is $(\alpha - \alpha^2)J_z$. More generally, by rotational invariance, the field strength at any point in shape space is

$$F_{ij} = (\alpha - \alpha^2)\epsilon_{ijk} J_k \tag{4.6}$$

Thus, the field strength is a monopole field [6][9] of strength $\alpha - \alpha^2$, located at the origin of the tangent plane at a point of shape space. Note that for $\alpha = 0$ or 1, F_{mn} vanishes, as it should. Furthermore, the net rotation is maximized when $\alpha = \frac{1}{2}$. An easy calculation will reveal that $\alpha = I/(I + I')$, where I and I' are the moments of inertia of the two spheres, so that $\alpha = \frac{1}{2}$ when $I = I'$.

To find α, we begin with the equation for the angular momentum:

$$L = I\dot{\theta} + I'\dot{\theta}' = 0 \tag{4.7}$$

where θ and θ' are the physical orientations of the two spheres in the plane of rotation. If θ' refers to the outer sphere, then the specification of a sequence of standard shapes means that we know $\theta - \theta'$ at all times. From this information, it is easy to obtain $\dot{\theta}'(t)$:

$$\dot{\theta}' = -\frac{I}{I + I'}(\dot{\theta} - \dot{\theta}') \tag{4.8}$$

Comparing with Eq. (4.1) gives

$$\alpha = \frac{I}{I + I'} \tag{4.9}$$

Note that α is always between 0 and 1, as claimed previously.

The foregoing example is readily generalized to more complicated situations. For instance, if we vary the moments of inertia of the spheres with time, the only modification of the above calculation that needs to be made is to bring α under the integral sign in Eq. (4.3):

$$R(t) = \overline{\mathrm{P}} \exp - \int^t \alpha(t')\Omega(t')\, dt' \tag{4.10}$$

Even the most general case of two bodies with arbitrary time-dependent inertia tensors rotating about a common center or mass is hardly any more difficult to write down:

$$R(t) = \overline{\mathrm{P}} \exp - \int^t (I + I')^{-1} I\, \Omega(t')\, dt' \tag{4.11}$$

For two bodies whose centers of mass do not coincide, the solution is more involved. In [3], the rotation of a simple system consisting of two rods joined at a hinge is solved in closed form, with overall orientation expressed as a function of the hinge angle. That result generalizes readily to more complicated systems—we leave it to the interested reader to work out the details.

5. Remarks

As we mentioned in the introduction, there are a variety of contexts where one might wish to find the net rotation of a self-deforming body. The diverse catalog of such bodies includes divers performing multiple twists [3], cats in free fall [1][2], and astronauts and satellites in space [10]. By way of example, we shall briefly consider the application of our ideas to satellites.

There are two primary means of changing the orientation of a satellite—propulsive and mechanical. The former relies on external thrusters to impart an angular momentum to the satellite, which is cancelled by a reverse thrust when the desired orientation is reached. This method has the disadvantage of requiring the initial and final thrusts to be precisely equal and opposite, a problem not shared by the second method, which falls under the general category of rotation by self-deformation. One might implement it, for example, by mounting two perpendicular flywheels near the center of the satellite. The flywheels would be used to generate rotations about the body's x- and y-axes. (A rotation about the z-axis could be generated by a sequence of x and y rotations.) The calculation of Sec. 4 may be taken over almost directly, if the flywheels are constructed so that they share a common center of mass with the rest of the satellite. Otherwise, the calculation may be modified in accord with the remarks at the end of Section 4. Note that no matter how the mechanical approach is implemented, the body's initial and final angular momentum are guaranteed to be the same.

Once the net rotation of a body due to any sequence of deformations is known, it is natural to try to optimize. Thus we ask, what is the most efficient way for a body to change its orientation? The answer will depend on many factors, including the definition of efficiency and constraints on the space of possible shapes. If one considers only infinitesimal deformations of a particular shape S_0, then the methods of [11] may be applied. However, large deformations are trickier, due to the path ordering in Eq. (2.5). The following qualitative observations bring out the subtleties of this problem. First, it is necessary to take account of non-contractible paths in shape space (*e.g.*, rotation of a wheel by 2π). Also in considering large deformations, one might expect that one could increase efficiency, by first deforming to a region of shape space with high curvature (large F_{mn}) and then performing small changes of shape. Where the curvature is large, a little bit of internal motion generates a large motion through space. Conversely, where there is small curvature even large internal motions generate only small motions through space. Hence large curvature configurations are appropriate when one wants to generate gross motions most efficiently, whereas small curvature configurations are appropriate to insure noise immunity for fine motions. This is a design principle that is intuitively evident and quantifiable once the language is understood, but might otherwise be difficult to express. Actual calculations along these lines would be illuminating.

We wish to thank S. Coleman and A. Pines for useful discussions. This research was supported by in part by the National Science Foundation under Grants No. PHY82-17853 and No. PHY-87-14654, supplemented by funds from the National Aeronautics and Space Administration.

References

[1] Cliff Frohlich, "The Physics of Somersaulting and Twisting," *Scientific American* **263**, 155-64 (March 1980).

[2] T.R. Kane and M.P. Scher, " A Dynamical Explanation of the Falling Cat Phenomenon," *J.Solids Struct.* **5**, 663-667 (1969).

[3] Cliff Frohlich, "Do Springboard Divers Violate Angular Momentum Conservation?" *Am. J. Phys.* **47**, 583-92 (1979).

[4] Cliff Frohlich, "Resource Letter PS-1: Physics of Sports," *Am. J. Phys.* **54**, 590-593 (1986).

[5] Alfred Shapere and Frank Wilczek, "Geometry of Self-Propulsion at Low Reynolds Number," *Phys. Rev. Lett.* **58**, 2051 (1987); Santa Barbara Inst. for Theor. Physics preprint No. 86-147, to appear in *Journal of Fluid Mechanics.*

[6] Michael V. Berry, "Quantal Phase Factors Accompanying Adiabatic Changes," *Proc. Roy. Soc. London* **A392**, 2111 (1984).

[7] C.A. Mead and D.G. Truhlar, "On the determination of Born-Oppenheimer nuclear motion wave functions including complications due to conical intersections and identical nuclei," *J.Chem.Phys.* **70**, (1979) 2284-96.

[8] T. Eguchi, P.B. Gilkey and A.J. Hansen, "Gravitation, Gauge Theories, and Differential Geometry," *Phys. Reports* **66**, 213 (1980).

[9] Sidney Coleman, "The Magnetic Monopole Fifty Years Later", in A. Zichichi, ed., *The Unity of the Fundamental Interactions* (New York: Plenum Press, 1983) pp.21-117.

[10] P.G. Smith and T.R. Kane, "The Reorientation of a Human Being in Free Fall." Technical Report No. 171, Stanford University, U.S. Govt. Accession No. N67-31537, 1967 (unpublished).

[11] Alfred Shapere and Frank Wilczek, "Efficiencies of Self-Propulsion at Low Reynolds Number," Santa Barbara ITP preprint No. 87-30, to appear in *J. Fluid Mech.*, January 1989.

J. Fluid Mech. (1989), vol. 198, pp. 557–585
Printed in Great Britain

Geometry of self-propulsion at low Reynolds number

By ALFRED SHAPERE† AND FRANK WILCZEK‡
† Institute for Advanced Study, Princeton, NJ 08540, USA
‡ Institute for Theoretical Physics, University of California, Santa Barbara, CA 93106, USA

(Received 15 April 1987 and in revised form 12 July 1988)

The problem of swimming at low Reynolds number is formulated in terms of a gauge field on the space of shapes. Effective methods for computing this field, by solving a linear boundary-value problem, are described. We employ conformal-mapping techniques to calculate swimming motions for cylinders with a variety of cross-sections. We also determine the net translational motion due to arbitrary infinitesimal deformations of a sphere.

1. Introduction

It has been appreciated for some time that self-propulsion at low Reynolds number is an interesting fluid-dynamical problem of considerable biological importance (Taylor 1951). Dynamics at low Reynolds number has a rather special and unique character. The effects of inertia are negligible in this limit; in the absence of driving forces, bodies are at rest. For this reason, motion at low Reynolds number has been called a realization of Aristotelean mechanics (Purcell 1977).

In the absence of inertia, the motion of a swimmer through a fluid is completely determined by the geometry of the sequence of shapes that the swimmer assumes. It is independent of any variation in the rates at which different parts of the sequence are run through (as long as this rate is slow, of course).

The purely geometrical nature of the problem of self-propulsion at low Reynolds number suggested to us that there should be a natural, attractive mathematical framework for this problem. We believe that we have found such a framework. It is the subject of this paper.

We shall show, in §2, that the problem of self-propulsion at low Reynolds number naturally resolves itself into the computation of a gauge potential field on the space of shapes. The gauge potential A describes the net translation and rotation resulting from an arbitrary infinitesimal deformation of a shape. It takes its values in the Lie algebra of rigid motions in Euclidean space. To find the translation and rotation of a swimmer which changes its shape along a given path in shape space, one computes the (path-ordered) integral of the gauge potential A along this path. We shall describe how to calculate A, in principle, by solving a linear boundary-value problem.

In two dimensions, there are powerful techniques which make explicit calculations of A quite practical for a wide range of shapes. The similarity between the equations of low-Reynolds-number hydrodynamics and of elasticity theory is well known (Rayleigh 1878). In two dimensions, complex-variable methods developed in the context of elasticity theory (Muskhelishvili 1953) can be carried over almost without

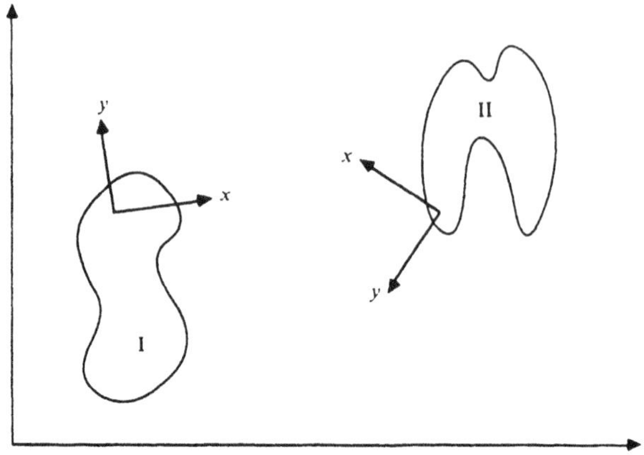

FIGURE 1. In order to measure distances between different shapes, an arbitrary choice of reference frames must be made.

modification (for a review, see Hasimoto & Sano 1980). These techniques are well-adapted to solving our boundary-value problem. The general procedure is outlined, and some examples are presented, in §3. Strokes involving infinitesimal deformations of a circle are analysed completely, as are finite deformations within a restricted space of shapes.

In §4 we discuss the swimming of a nearly spherical organism in three dimensions. Again, the case of infinitesimal deformations can be analysed completely, by using vector spherical harmonics to exploit the symmetry of the problem. In the final section, we discuss possible extensions of the work. Several appendices contain details of our calculations and discussions of mathematical topics.

The calculation of efficiencies, leading to the determination of optimally efficient strokes, is presented in a companion paper (Shapere & Wilczek 1989).

2. Kinematics

2.1. *General framework; gauge structure*

The configuration space of a deformable body is the space of all possible shapes. We should at the outset distinguish between the space of shapes located somewhere in space and the more abstract space of unlocated shapes. The latter space may be obtained from the space of shapes *cum* locations by declaring two shapes with different centres of mass and orientations to be equivalent.

The problem we wish to solve may be stated as follows: what is the net rotation and translation which results when a deformable body goes through a given sequence of unoriented shapes, in the absence of external forces and torques? In other words, given a path in the space of unlocated shapes, what is the corresponding path in the space of located shapes? The problem is intuitively well-defined – if a body changes its shape in some way, a net rotation and translation is induced. The net motion may be found by solving Stokes' equations for the fluid flow, with boundary conditions on the surface of the shape corresponding to the given deformation.

These remarks may seem straightforward enough, but as soon as we try to formulate the problem more specifically, we encounter a crucial ambiguity. The root problem in the kinematics of deformable shapes is displayed in figure 1. We wish to compute the motion: how far has shape I moved, and how has it reoriented itself, in the process of becoming shape II? In this form, the question is clearly ill-posed. Different points on or inside of the boundary may have moved differently.

To quantify the motion, it is necessary to attach a centre and a set of axes to each unlocated shape, as in figure 1. This is equivalent to choosing a 'standard location' for each shape; namely, to each unlocated shape there now corresponds a unique located shape, whose centre and axes are aligned with the origin and coordinate axes of physical space. Once a choice of standard locations for shapes has been made, then we shall say that the rigid motion required to move from I to II is the displacement and rotation necessary to align their centres and axes.

Now let σ parameterize the boundary of a shape $S(\sigma)$, and let $S_0(\sigma)$ be the associated standard shape. (For example, if S is a simply connected shape in three dimensions, we will take $\sigma = (\theta, \phi)$ to be a coordinate on the unit two-sphere, and $S(\sigma)$ to be a map from S^2 into $\mathbf{R}^3$. For two-dimensional (cylindrical) shapes, it will prove to be most convenient to use the complex coordinate on the unit circle, $\sigma = e^{i\theta}$.) Then

$$S(\sigma) = \mathscr{R} S_0(\sigma), \tag{2.1}$$

where $\mathscr{R}$ is a rigid motion. We emphasize that S and S_0 are *parameterized* shapes; different functions $S(\sigma)$ and $S'(\sigma)$ correspond to different shapes, even if their images coincide geometrically.

To make (2.1) more explicit, we introduce a matrix representation for the group of Euclidean motions, of which $\mathscr{R}$ is a member. A three-dimensional rigid motion consisting of a rotation R followed by a translation d may be represented as a 4×4 matrix

$$[R, d] = \begin{pmatrix} R & d \\ 0 & 1 \end{pmatrix}, \tag{2.2}$$

where R is an ordinary 3×3 rotation matrix, d is a 3-component column vector, and 1 is just the 1×1 identity matrix. These matrices obey the correct group algebra

$$[R', d'][R, d] = [R'R, R'd + d'],$$
$$[R, d]^{-1} = [R^{-1}, -R^{-1}d].$$

Vectors v on which $[R, d]$ acts are represented as 4-component column vectors $(v, 1)^{\mathrm{T}}$; then $v \rightarrow Rv + d$. In the notation of (2.1), $\mathscr{R} \equiv [R, d]$ acts on the vector $(S_0(\sigma), 1)^{\mathrm{T}}$, for each σ.

Now in considering the problem of self-propulsion at low Reynolds number, we shall assume that our swimmer can squirm, but not pull itself by its bootstraps. That is, we shall assume it has control over its form (i.e. its standard shape, as defined above), but cannot exert net forces and torques on itself. A swimming stroke is therefore specified by a time-dependent sequence of forms, or equivalently standard shapes $S_0(t)$ (the σ-dependence is implicit). The actual shapes will then be

$$S(t) = \mathscr{R}(t)\, S_0(t), \tag{2.3}$$

where $\mathscr{R}(t)$ is a time-dependent sequence of rigid displacements. Note that we allow shape changes which change both the volume and surface area in our general formulation. These may be constrained at a later stage, although we shall not consider such restrictions in this paper.

The dynamical problem of self-propulsion at low Reynolds number thus resolves itself into the computation of $\mathscr{R}(t)$, given $S_0(t)$. For example, if the stroke is cyclic, i.e. $S_0(t_1) = S_0(t_2)$, then the net motion each cycle induces is $\mathscr{R}(t_2)\,\mathscr{R}(t_1)^{-1}$. In computing this displacement, it is most convenient to begin with infinitesimal motions and to build up finite motions by integrating. So let us define the infinitesimal motion $A(t)$ by

$$\frac{\mathrm{d}\mathscr{R}}{\mathrm{d}t} = \mathscr{R}\left(\mathscr{R}^{-1}\frac{\mathrm{d}\mathscr{R}}{\mathrm{d}t}\right) \equiv \mathscr{R}A. \tag{2.4}$$

As we shall show, A is mathematically a gauge potential†, taking its values in the Lie algebra of the group of rigid motions. For any given infinitesimal change of shape, A describes the net overall translation and rotation which results. As in (2.2), a convenient 4×4 matrix representation for A is the following:

$$A = \begin{pmatrix} A^{\mathrm{rot}} & A^{\mathrm{tr}} \\ 0 & 0 \end{pmatrix}, \tag{2.5}$$

where A^{rot} is a 3×3 generator of rotations and A^{tr} is a 3-component velocity vector.

Given $A(t)$, we can integrate (2.4) to obtain

$$\mathscr{R}(t_2) = \mathscr{R}(t_1)\,\bar{P}\exp\left[\int_{t_1}^{t_2} A(t)\,\mathrm{d}t\right], \tag{2.6}$$

where $\bar{P}$ denotes a 'reverse' path ordering:

$$\bar{P}\exp\left[\int_{t_1}^{t_2} A(t)\,\mathrm{d}t\right] = 1 + \int_{t_1<t<t_2} A(t)\,\mathrm{d}t + \int_{t_1<t<t'<t_2}\!\!\int A(t)\,A(t')\,\mathrm{d}t\,\mathrm{d}t' + \ldots. \tag{2.7}$$

That is, in expanding the exponential in (2.7), products of matrices $A(t)$ are always arranged so that earlier times occur on the left.

(This reverse path ordering should not seem peculiar, since $R(t_1)$ – referring to the earliest time that occurs – appears on the left in (2.6). For those who prefer ordinary ordered integrals, we note that there is a parallel treatment with rotations defined to act on the right:

$$S(t) = S_0(t)\,\mathscr{R}(t), \tag{2.3$'$}$$

$$\frac{\mathrm{d}\mathscr{R}}{\mathrm{d}t} = \left(\frac{\mathrm{d}\mathscr{R}}{\mathrm{d}t}\mathscr{R}^{-1}\right)\mathscr{R} \equiv B\mathscr{R}, \tag{2.4$'$}$$

$$\mathscr{R}(t_2) = \mathscr{R}(t_1)\,P\exp\left[\int_{t_1}^{t_2} B\right]. \tag{2.6$'$}$$

For definiteness, we always employ the first alternative.)

The assignment of centres and axes being arbitrary, we should expect that physical results are independent of this assignment. How does this show up in our formalism? A change in the choice of centres and axes can equally well be thought of as a change (rigid motion) of the standard shapes; let us write

$$\tilde{S}_0 = \Omega(S_0)\,S_0. \tag{2.8}$$

† Gauge potentials have been shown to arise in a variety of contexts outside of particle physics and electromagnetism. For a review see (Shapere & Wilczek 1988).

The physical shapes being unchanged, (2.1) requires us to define

$$\tilde{\mathscr{R}}(t) = \mathscr{R}(t)\,\Omega^{-1}(S_0(t)). \tag{2.9}$$

From this, the transformation law A and for the connecting path integral follow:

$$\left.\begin{aligned} \tilde{A} &= \Omega A \Omega^{-1} + \Omega \frac{\mathrm{d}\Omega^{-1}}{\mathrm{d}t} \\ \tilde{W}_{21} &= \Omega_1\, W_{21}\, \Omega_2^{-1}. \end{aligned}\right\} \tag{2.10}$$

Readers familiar with gauge field theory will recognize these transformation laws. Ω implements a gauge transformation in the space of shapes; A transforms as a gauge potential and W as a Wilson line integral. Our freedom in choosing the assignment of centres and axes shows up as a freedom of gauge choice on the space of standard shapes. The final relationship between physical shapes, i.e. $S(t_2)$ and $S(t_1)$, is manifestly independent of such choices.

The appearance of a gauge structure in the context of low-Reynolds-number fluid mechanics is in fact quite natural. Generally, gauge structures are associated with large redundancies in the description of a physical system. Here, the redundancy is associated with our freedom to choose a standard orientation and location for each possible shape. This gauge structure is a general feature of the mechanics of deformable bodies without inertia.

The gauge potential A has a geometric origin. Namely, A may be viewed as a connection on a fibre bundle over the space of standard shapes. This point of view is discussed in Appendix A.

2.2. *The dynamical problem: determining the gauge potential*

The dynamical problem of self-propulsion at low Reynolds number has been reduced to the calculation of the gauge potential A. Here we outline an effective method for determining A. Later parts of the paper contain many specific examples.

First, let a sequence of forms $S_0(t)$ be given. In general, this sequence of forms does not in itself specify a possible motion according to our hypotheses, for it will involve net forces and torques on the swimmer. The allowed motion, involving the same sequence of forms, will include additional time-dependent rigid displacements. In other words, the actual motion will be the superposition of the given motion sequence $S_0(t)$ and counterflows, corresponding to additional rigid displacements which cancel the forces and torques.

To calculate the counterflow, we solve for the response of the fluid to the trial motion $S_0(t)$. This is given by the solution to the boundary-value problem (Happel & Brenner 1965; Childress 1978)

$$\nabla \cdot v = 0, \tag{2.11}$$

$$\nabla^2(\nabla \times v) = 0, \tag{2.12}$$

$$v\,|_{S_0} = \frac{\delta S_0}{\delta t}. \tag{2.13}$$

Equations (2.11) and (2.12) are the standard equations for incompressible flow at low Reynolds numbers, and (2.13) is the no-slip boundary condition. In interpreting (2.13), it is important to remember that the $S_0(t)$ are really parametrized shapes $S_0(\sigma, t)$, and that the variation is meant to be taken with σ fixed.

The force and torques associated with the trial motion can be inferred from the asymptotic behaviour of v at spatial infinity. The force on the shape is related to the external force on the fluid at spatial infinity, and thence to the asymptotic flow, by the conservation of momentum. Indeed, if σ_{ij} is the stress tensor then the force on the shape is (Batchelor 1970)

$$F_i = \int_{\text{shape}} \sigma_{ij}\, dS_j,$$

but the stress tensor is conserved, $\partial_i \sigma_{ij} = 0$, so this is

$$F_i = -\int_{\infty} \sigma_{ij}\, dS_j.$$

Now the stress tensor is given in terms of the velocity, and only the terms that fall off slowly and have the right symmetry survive. In fact, we shall show in §4 that the force is linearly related to the leading term in the asymptotic flow. A similar argument, leading to a similar conclusion, can be made for the torque.

To cancel these forces and torques, we must correct the motion by subtracting a Stokes' flow corresponding to a rigid displacement of the shape with the same leading behaviour at infinity as our trial solution. The result is the actual fluid motion. By our definition, this rigid displacement is

$$1 \times A(t)\,\delta t. \tag{2.14}$$

This completes our outline of the method for calculating A.

Thus far, we have treated A as a time-dependent quantity. However, the geometric nature of our problem suggests that it should be possible to formulate an answer to it in a completely time-independent way, i.e. to express the integrand in (2.6) in a manner that makes no reference to a time coordinate. Accordingly, we can define an abstract vector field over shape space, which we shall also denote by A, whose projection onto the direction $\delta S_0/\delta t$ at the point $S_0(t)$ is just $A(t)$:

$$A(t) \equiv A_{\dot{S}_0}[S_0(t)]. \tag{2.15}$$

Then the integral of (2.6) is equal to the line integral of A over the path $S_0(t)$ in shape space, and is manifestly independent of how this path is parameterized:

$$\mathscr{R}(t_2) = \mathscr{R}(t_1)\,\bar{P}\exp\left[\int_{S_0(t_1)}^{S_0(t_2)} A[S_0]\,dS_0\right]. \tag{2.16}$$

A may have infinitely many components, one for each direction of shape space, and each of which is a generator of a rigid motion. In terms of a fixed basis of vector fields $\{w_i\}$ over $\mathscr{S}_0$, we may define components $A_i[S_0] \equiv A_{w_i}[S_0]$.

2.3. *Two corollaries*

We now pause to discuss two simple but notable general properties of self-propulsion at low Reynolds number, which are particularly easy to appreciate in our framework.

Generalized scallop theorem Purcell (1977) has emphasized the 'scallop theorem', according to which a simple hinged object such as the one shown in figure 2 cannot swim at low Reynolds number. Any repeatable stroke gives no net motion. In our framework, this is evident because the space of shapes available to this object is simply a bounded line, $0 \leqslant \theta \leqslant 2\pi$. Thus the Wilson line integral encloses no area, and the displacements induced by moving along a segment are cancelled by those accumulated in the return motion. (The theorem assumes that the scallop cannot turn through 2π on its hinge!)

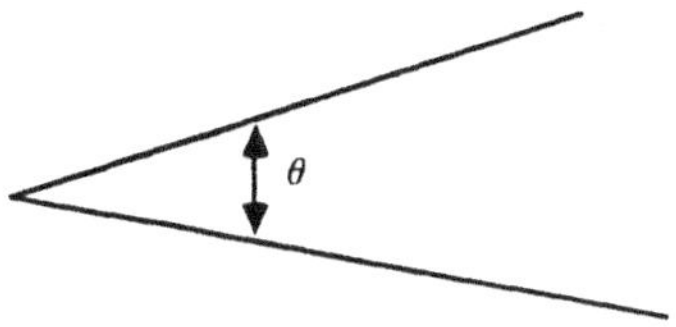

FIGURE 2. A simple hinged animal with one degree of freedom cannot swim.

Helix theorem One cycle of a swimming stroke results in a definite displacement, i.e. translation and rotation. Repeating this cycle will lead to the square of the displacement, and so forth. The result is that the swimmer will trace out a *generalized helix*. To put it more precisely, a true helix is described by

$$x(t) = \exp(t\alpha)\, x(0), \tag{2.17}$$

where α is in the Lie algebra of rigid motions, the infinitesimal displacement which generates the helix. A generalized helix, in our sense, is described by

$$x(t) = \exp(t\alpha)\, R(t)\, x(0), \tag{2.18}$$

where $R(t)$ is some periodic function, with $R(T) = R(0) = \mathbf{1}$. The proof is as follows. Let the period of the cyclic motion be one time unit, and let the rotation and displacement due to one stroke be

$$\exp \alpha = \bar{P} \exp\left[\int_0^1 A\, \mathrm{d}t\right].$$

Then
$$\bar{P} \exp\left[\int_0^t A\, \mathrm{d}t = \exp(t\alpha)\exp([t]-t)\,\alpha \bar{P} \exp\left[\int_{[t]}^t A\, \mathrm{d}t\right]\right.$$
$$\equiv \exp(t\alpha)\, R(t),$$

where $[t]$ denotes the greatest integer $\leqslant t$.

Many swimming micro-organisms have indeed been observed to follow helical trajectories. Some examples of helical paths for flagellar swimmers have been computed and compared with observations by Keller & Rubinow (1976). Helices are ubiquitous in biology; we suspect the mathematical reason is this theorem.

In two dimensions, the helix theorem takes on a peculiar form. It says that cyclic swimming strokes can only lead to net motions which are 'generalized circles'. That is, orbits of the Euclidean group E^2 are circles, and the path of a swimmer will in general be a sort of a squiggly polygon described by (2.18). Motion of this type is depicted in figure 4. In order for the swimmer to avoid going around in circles, the net displacement per cycle must be a pure translation.

2.4. *Infinitesimal deformations*

The case of infinitesimal deformations of a shape is sufficiently important and interesting that it deserves separate comment.

Let the standard shapes be parametrized by

$$S_0(t) = S_0 + s(t), \tag{2.19}$$

where the $s(t)$ are infinitesimal. We expand $s(t)$ in terms of a fixed basis of vector fields on S_0:

$$s(t) = \sum_i \alpha_i(t)\, w_i. \tag{2.20}$$

Then we have for the velocity on $S_0(t)$:

$$v(t) = \frac{\delta S_0(t)}{\delta t} = \sum_i \dot{\alpha}_i w_i. \tag{2.21}$$

Now let us expand the gauge potentials to second order:

$$A_{v(t)}[S_0(t)] \approx A_{v(t)}[S_0] + \sum_i \frac{\partial A_v}{\partial w_i} \dot{\alpha}_i$$

$$\approx \sum_j \left(A_{w_j} \dot{\alpha}_j + \sum_i \frac{\partial A_{w_j}}{\partial w_i} \alpha_i \dot{\alpha}_j \right). \tag{2.22}$$

In the path-ordered exponential integral (2.7) around a cycle, which is the basic object giving the net displacement, the first-order term gives no contribution, for it is a total derivative. The second-order contributions are terms quadratic in A and linear in its derivatives. Because (2.7) is gauge covariant for a cyclic path, its Taylor expansion in powers of $s(t)$ must also be gauge covariant, order by order. In fact, there is a unique (up to normalization) second-order gauge covariant term we can form, which is antisymmetric in the indices i and j:

$$F_{w_i w_j} \equiv \frac{\partial A_{w_i}}{\partial w_j} - \frac{\partial A_{w_j}}{\partial w_i} + [A_{w_i}, A_{w_j}]. \tag{2.23}$$

The physical significance of $F_{w_i w_j}$ is as follows. Suppose we make a sequence of successive deformations of S_0 by ϵw_i, ηw_j, $-\epsilon w_i$, and $-\eta w_j$. Finally, we close the sequence of shapes with the Lie bracket $-\epsilon\eta[w_i, w_j]$. Then the net displacement will be $\epsilon\eta F_{w_i w_j}$. This makes it clear why F must be antisymmetric in its indices, so that the reverse sequence of shapes gives the reverse displacement.

It is easily verified that expansion of (2.7) to second order gives

$$\bar{P} \exp\left[\oint A \, \mathrm{d}t\right] = 1 + \frac{1}{2} \oint \sum_{ij} F_{w_i w_j} \alpha_i \dot{\alpha}_j \, \mathrm{d}t. \tag{2.24}$$

The field strength tensor, evaluated at a shape S_0, thus encodes all information on swimming motions due to arbitrary infinitesimal deformations of S_0.

3. The two-dimensional problem

3.1. *Two-dimensional techniques*

We shall now apply the techniques described in §2 to study the swimming motion of extended bodies at very low Reynolds number. In this section, we restrict our attention to the admittedly unbiological example of an infinitely long cylindrical body of constant cross-section. The boundary-value problem (2.11)–(2.13) then becomes effectively two-dimensional, and may be solved by techniques of complex analysis. The solution is qualitatively similar to the three-dimensional case, yet easier to obtain and to interpret.

After setting up the machinery for handling the general two-dimensional problem, we shall apply it to the computation of the swimming motions of cylinders with some simple cross-sectional shapes. We shall also compute the field strength tensor for a cylinder with circular cross-section, leading to a description of all swimming motions of nearly circular cylinders.

In §2, we found the set of equations that must be satisfied by the velocity field v at low Reynolds number:

$$\nabla \cdot v = 0, \tag{3.1}$$

$$\nabla^2(\nabla \times v) = 0, \tag{3.2}$$

$$v|_s = \frac{\partial S}{\partial t}. \tag{3.3}$$

Let us suppose that the shape S is a cylinder and that the velocity field v contains no z-component, so that the boundary-value problem is two-dimensional. Then the first equation implies that the two-component vector v is the curl of a scalar potential U (possibly multivalued), and

$$\left.\begin{aligned} v = \nabla \times U &\equiv \left(\frac{\partial U}{\partial y}, -\frac{\partial U}{\partial x}\right), \\ \nabla \times v = \nabla \times (\nabla \times U) &= -\nabla^2 U. \end{aligned}\right\} \tag{3.4}$$

Thus U, by (2.2), satisfies the biharmonic equation,

$$\nabla^4 U = 0. \tag{3.5}$$

This equation has been extensively studied in the theory of elasticity in two dimensions. In elastic boundary-value problems, the second partial derivatives of U represent the stresses on an elastic medium (Rayleigh 1878; Hill & Power 1955). Muskhelishvili (1953) (see also England 1971) has applied methods of complex analysis to these problems, with elegant results. His methods have proved equally useful in the context of low-Reynolds-number fluid mechanics (Richardson 1968; Hasimoto & Sano 1980).

One reason complex analysis is so useful in solving the biharmonic equation is that biharmonic functions have a simple representation in terms of analytic functions. Namely, any U satisfying (3.5) may be written in the form

$$\tfrac{1}{2}U(z, \bar{z}) = \bar{z}\phi(z) + z\overline{\phi(z)} + \psi(z) + \overline{\psi(z)}, \tag{3.6}$$

where ϕ and ψ are analytic in $z \equiv x + iy$. As a corollary, we obtain an important representation for the velocity field, written as $v \equiv v_x + iv_y$,

$$\begin{aligned} v(z) = \nabla \times U = -\frac{i}{2}\frac{\partial U}{\partial \bar{z}} &= -i[\phi(z) + z\overline{\phi'(z)} + \overline{\psi'(z)}] \\ &\equiv \phi_1(z) - z\overline{\phi_1'(z)} + \overline{\phi_2(z)}. \end{aligned} \tag{3.7}$$

To discuss the swimming of shapes, we wish to consider an external boundary-value problem for U, with $v = \nabla \times U$ specified on the exterior boundary of a compact region in the plane. Let s represent the complex coordinate z restricted to the boundary. Then given $v(s)$, we wish to find functions ϕ_1, ϕ_2 analytic in the exterior of s such that

$$v(s) = \phi_1(s) - s\overline{\phi_1'(s)} + \overline{\phi_2(s)}. \tag{3.8}$$

The problem is easily solved if S is a circle – we simply equate Fourier coefficients on both sides of (3.8). Although the result has been derived elsewhere (Muskhelishvili

1953), we shall present a derivation in order to establish notation. Suppose we have Fourier expansions

$$\left.\begin{aligned} v(s) &= \sum_{k=-\infty}^{\infty} v_k s^{k+1}, \\ \phi_1(s) &= \sum_{k<0} a_k s^{k+1}, \\ \phi_2(s) &= \sum_{k<-1} b_k s^{k+1}, \end{aligned}\right\} \tag{3.9}$$

where $s = e^{i\theta}$. (Summation over non-positive k ensures that $\phi_1(z)$ and $\phi_2(z)$, and consequently $v(z)$, are finite at infinity. We may take $b_{-1} = 0$ without loss of generality.) Then (3.8) is equivalent to

$$\sum_{k=-\infty}^{\infty} v_k s^{k+1} = \sum_{k<0} a_k s^{k+1} - \sum_{k<0} (k+1)\bar{a}_k s^{-k+1} + \sum_{k<-1} \bar{b}_k s^{-k-1},$$

since $s^{-1} = \bar{s}$. The complete solution is

$$\left.\begin{aligned} a_k &= v_k && (k<0), \\ b_{-2} &= \bar{v}_0, && \\ b_k &= \bar{v}_{-k-2} + (k+3)\, v_{k+2} && (k<-2). \end{aligned}\right\} \tag{3.10}$$

Thus the solutions with $v(s) = \lambda s^{l+1}$ on the circle correspond to

$$\phi_1(s) = 0, \qquad \phi_2(s) = \bar{\lambda} s^{-l-1} \qquad (l > -1), \tag{3.11}$$

$$\phi_1(s) = \lambda, \qquad \phi_2(s) = 0 \qquad (l = -1), \tag{3.12}$$

$$\phi_1(s) = \lambda s^{l+1}, \qquad \phi_2(s) = \lambda(l+1)\, s^{l-1} \qquad (l < -1). \tag{3.13}$$

These may be extended to the entire region of flow, i.e. the exterior of the circle, by substituting $s \to z$ and using the representation (3.7). The results are

$$v = \lambda \bar{z}^{-l-1} \qquad (l > -1), \tag{3.14}$$

$$v = \lambda \qquad (l = -1), \tag{3.15}$$

$$v = \lambda \bar{z}^{l+1} - \bar{\lambda}(l+1)\, \bar{z}^{l-1}(z\bar{z} - 1) \qquad (l < -1). \tag{3.16}$$

This is the complete solution to the boundary-value problem (3.8) when S is a circle.

For those who prefer two-component vector notation, we can successively take λ real or imaginary in (3.14)–(3.16) to obtain a basis of equivalent solutions

$$\left.\begin{aligned} v_l^1(r,\theta) &= r^{-l-1}(\cos(l+1)\theta, \sin(l+1)\theta), \\ v_l^2(r,\theta) &= r^{-l-1}(-\sin(l+1)\theta, \cos(l+1)\theta) \end{aligned}\right\} \text{ for } l \geqslant -1, \tag{3.17}$$

and

$$\left.\begin{aligned} v_l^1 &= r^{l+1}(\cos(l+1)\theta - (l+1)(1-r^{-2})\cos(l-1)\theta, \\ &\quad \sin(l+1)\theta + (l+1)(1-r^{-2})\sin(l-1)\theta), \\ v_l^2 &= r^{l+1}(-\sin(l+1)\theta + (l+1)(1-r^{-2})\sin(l-1)\theta, \\ &\quad \cos(l+1)\theta + (l+1)(1-r^{-2})\cos(l-1)\theta) \end{aligned}\right\} \text{ for } l < -1. \tag{3.18}$$

It will prove useful to form combinations with definite helicity (i.e. simple properties under rotation)

$$w_l^{\pm}(r,\theta) \equiv \frac{1}{\sqrt{2}}(v_l^1 \mp \mathrm{i} v_l^2)$$

$$= r^{-l-1}\,\mathrm{e}^{\pm\mathrm{i}(l+1)\theta}\frac{1}{\sqrt{2}}(1, \mp\mathrm{i}) \qquad (l \geqslant -1)$$

$$= r^{l+1}\left[\mathrm{e}^{\pm\mathrm{i}(l+1)\theta}\frac{1}{\sqrt{2}}(1, \mp\mathrm{i}) - (l+1)(1-r^{-2})\,\mathrm{e}^{\pm\mathrm{i}(l-1)\theta}\frac{1}{\sqrt{2}}(1, \pm\mathrm{i})\right] \qquad (l < -1). \tag{3.19}$$

(It should be kept in mind that the i appearing in (3.19) is not the same as in $z = x+\mathrm{i}y$.) Rotation through α changes these flows by

$$w_l^{\pm} \rightarrow \mathrm{e}^{\pm\mathrm{i}l\alpha} w_l^{\pm}. \tag{3.20}$$

We say that $w_l^{\pm}$ has helicity $\pm l$.

Note that the solutions (3.15) corresponding to translations of the circle involve rigid motion of the fluid as a whole. (Solutions corresponding to rigid rotations ($l = 0$ and $|\lambda| = 1$) fall off slowly, like r^{-1}.) This unphysical behaviour is known as Stokes' paradox, and is a well-known peculiarity of two-dimensional low-Reynolds-number hydrodynamics. Because of our requirement that the external forces and torques vanish, we never encounter these rigid motions of the circle – in fact we determine the gauge potentials precisely by 'subtracting them off'. The fact that, mathematically, rigid motions of the circle give rise to such long-range motions of the fluid is actually a convenience, since it allows us to identify the necessary counterflows, i.e. the gauge potentials, very easily from the asymptotics of a trial flow at infinity. (As we shall see, the story is different for non-circular shapes, and for three-dimensional spheres.)

3.2. *Nearly circular shapes*

Before continuing to build up the general formalism, we pause to work out the important example of nearly circular shapes in detail. This computation has been done previously by Blake (1971*b*), in the case of irrotational strokes symmetric about the axis of propulsion.

To compute the field strength tensor, F, which governs the motion resulting from infinitesimal deformations, we must consider closed paths in two-dimensional subspaces of shape space. Let v_1, v_2 be two velocity fields on the circle and let $\mathcal{R}(\epsilon v_1, \eta v_2)$ be the rotation and translation of the circle induced by the following sequence of motions, as depicted in figure 3:

$$S \rightarrow S+\epsilon v_1 \rightarrow S+\epsilon v_1+\eta v_2 \rightarrow S+\eta v_2 \rightarrow S. \tag{3.21}$$

We work to second order in ϵ, η. Then, by (2.24),

$$\mathcal{R}(\epsilon v_1, \eta v_2) = [1,0]+\epsilon\eta F_{v_1 v_2} \tag{3.22}$$

$F_{v_1 v_2}$ lies in the Lie algebra of rigid motions. F is most easily computed by matching the boundary condition $\eta v_2(\theta)$ on the surface of the circle deformed by $\epsilon v_1(\theta)$. If we

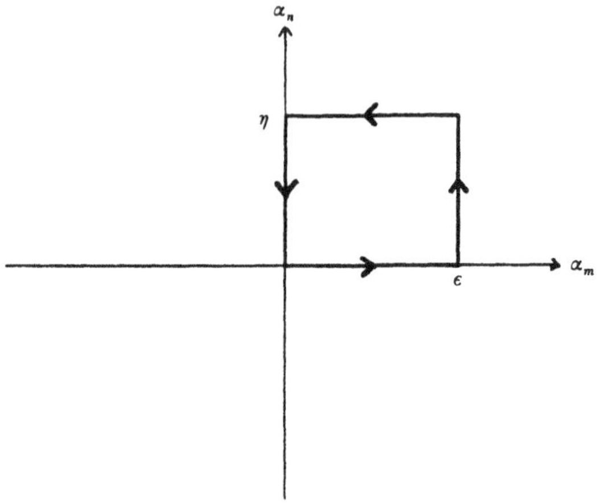

FIGURE 3. An infinitesimal closed path in shape space, coupling modes m and n.

call the resulting velocity field v_{12}, then $F_{v_1 v_2}$ is related to the asymptotics of v_{12} at infinity. In fact it is not hard to see that following our prescriptions we find

$$\left.\begin{aligned} F^{\text{tr}}_{v_1 v_2} &= \lim_{r\to\infty} \int \frac{\mathrm{d}\theta}{2\pi}(v_{12}-v_{21}), \\ F^{\text{rot}}_{v_1 v_2} &= \lim_{r\to\infty} \int \frac{\mathrm{d}\theta}{2\pi} r\times(v_{12}-v_{21}), \end{aligned}\right\} \tag{3.23}$$

where the translational and rotational parts are defined by

$$F_{v_1 v_2} \equiv \begin{pmatrix} 0 & F^{\text{rot}} & F^{\text{tr}}_x \\ -F^{\text{rot}} & 0 & F^{\text{tr}}_y \\ 0 & 0 & 0 \end{pmatrix}. \tag{3.24}$$

and the integral is around a large circle. It remains to compute v_{12}. The boundary condition for v_{12} is

$$\begin{aligned} v_2(\theta) &= v_{12}(x)\,|_{\text{surf}} \\ &\approx v_{12}(r,\theta) + \epsilon(v_1\cdot\nabla)\, v_{12}(r,\theta)\,|_{r=1}. \end{aligned} \tag{3.25}$$

To first order,
$$v_2(\theta) \approx v_{12}(1,\theta) + \epsilon(v_1\cdot\nabla)\, v_2(r,\theta)\,|_{r=1}. \tag{3.26}$$

Thus we can find v_{12}, v_{21}, and F from v_1 and v_2. Putting (3.23) and (3.26) all together, we arrive at the master formulae

$$\left.\begin{aligned} F^{\text{tr}}_{v_1 v_2} &= \int \frac{\mathrm{d}\theta}{2\pi}\{(v_2\cdot\nabla)\, v_1 - (v_1\cdot\nabla)\, v_2\}, \\ F^{\text{rot}}_{v_1 v_2} &= \text{Im} \int \frac{\mathrm{d}\theta}{2\pi} r\times\{(v_2\cdot\nabla)\, v_1 - (v_2\cdot\nabla)\, v_1\}. \end{aligned}\right\} \tag{3.27}$$

Non-trivial hydrodynamics enters these formulas only in that we generally need to determine v_1 and v_2 away from the circle where they are given, in order to evaluate the derivatives.

(It is worth remarking that the formulas (3.27) for the gauge field strength can be generalized to describe tangential deformations of an arbitrary shape. The argument preceding (3.27) implies that

$$\frac{\partial A_{v_1}}{\partial v_2} - \frac{\partial A_{v_2}}{\partial v_1} = A_{[v_1, v_2]},$$

where

$$[v_1, v_2] = (v_1 \cdot \nabla) v_2 - (v_2 \cdot \nabla) v_1$$

is the Lie bracket. Thus the complete field strength is

$$F_{v_1 v_2} = A_{[v_1, v_2]} + [A_{v_1}, A_{v_2}].$$

For the circle and sphere the second term vanishes – but it does not in general. When both v_1 and v_2 are tangential fields, the Lie bracket may be evaluated completely in terms of their values on the shape, with no hydrodynamics. Thus for purely tangential motions – reparametrizations of the boundary, which do not change the bulk shape – the form of A determines F directly.)

After these preliminaries, it is now a matter of straightforward algebra to insert the vector fields (3.19) into the master formula and thus derive F. The results for the translational part are as follows:

$$F^{\text{tr}}_{m^+n^+} = \frac{1}{\sqrt{2}}[-(m+1)\theta_{-m}\delta_{m+n+1}e_- + (n+1)\theta_n\delta_{m+n+1}e_- + (m+1)\theta_{-m}\delta_{m+n-1}e_+ - (n+1)\theta_{-n}\delta_{m+n-1}e_+], \quad (3.28)$$

$$F^{\text{tr}}_{m^-n^-} = \frac{1}{\sqrt{2}}[-(m+1)\theta_{-m}\delta_{m+n+1}e_+ + (n+1)\theta_{-n}\delta_{m+n+1}e_+ + (m+1)\theta_{-m}\delta_{m+n-1}e_- - (n+1)\theta_{-n}\delta_{m+n-1}e_-], \quad (3.29)$$

$$F^{\text{tr}}_{m^+n^-} = \frac{1}{\sqrt{2}}[(m+1)\theta_m\delta_{m-n+1}e_- - (n+1)\theta_n\delta_{-m+n+1}e_+ + (m+1)\theta_{-m}\delta_{m-n-1}e_+ - (n+1)\theta_{-n}\delta_{-m+n-1}e_-], \quad (3.30)$$

$$F^{\text{tr}}_{m^-n^+} = -F^{\text{tr}}_{n^+m^-}. \quad (3.31)$$

where $e_\pm \equiv (1/\sqrt{2})(1, \pm i)$, and θ_n is zero for negative n and 1 for non-negative n. It is understood here that the + and – labels refer to the solutions $w_l^\pm$ in (3.19). The matrix F is antisymmetric; apart from this all components of F which do not appear explicitly in (3.28)–(3.31) vanish. It is of course no accident that the vast majority of the components of F vanish in the helicity basis. Under a rotation through α, $w_l^\pm$ is multiplied by the phase in (3.20), while $e_\pm \to e^{\pm i\alpha}e_\pm$. Since F is linear in its arguments, and everything about our problem is symmetric under rotations, this leads directly to constraints on which components of F may be non-zero.

For the rotational part of F we find

$$F^{\text{rot}}_{m^+n^+} = -[(m+1)\theta_m - (n+1)\theta_n]\delta_{m+n}, \quad (3.32)$$

$$F^{\text{rot}}_{m^-n^-} = [(m+1)\theta_m - (n+1)\theta_n]\delta_{m+n}, \quad (3.33)$$

$$F^{\text{rot}}_{m^+n^-} = -|m+1|\delta_{m-n}, \quad (3.34)$$

$$F^{\text{rot}}_{m^-n^+} = -F^{\text{rot}}_{n^+m^-}. \quad (3.35)$$

An alternative method for computing the components of the field strength, using complex variables throughout, is presented in Appendix B. In Appendix C, we discuss an interpretation of F_{mn} in terms of the Virasoro algebra of conformal deformations.

3.3. *Large deformations*

In the preceding section, we computed the net translation and rotation of a nearly circular cylinder due to an infinitesimal closed sequence of deformations. Here, we shall perform a complementary calculation for deformations of finite size. This will provide a concrete application of the gauge potential formalism we introduced in §2. Because the complexity of the calculation increases with the complexity of the deformations, we shall restrict attention to shapes described by conformal maps of the unit circle with degree $D \leqslant 2$. (The extension to shapes of arbitrary degree will be discussed later.) A sequence of such deformations may be parameterized as

$$S(\sigma, t) = \alpha_0(t)\,\sigma + \alpha_{-2}(t)\,\sigma^{-1} + \alpha_{-3}(t)\,\sigma^{-2} \tag{3.36}$$

where $\sigma = e^{i\theta}$. Here we have taken $\alpha_{-1} = 0$ to 'fix the gauge' with respect to translations. We may also choose orientations for the standard shapes by requiring α_0 to be real and positive. Note that α_2 must vanish if the analytic extension of S to the region of flow is to be conformal at infinity.

To compute the translation and rotation due to the sequence of deformations (3.36), we need to solve the boundary-value problem (3.8) on the exterior of each of the shapes $S(\sigma, t)$ for $0 \leqslant t \leqslant T$. This is most easily accomplished by conformally mapping the exterior of $S(\sigma, t)$ in the z-plane onto the exterior of the unit circle, using $z = S(\zeta)$. Pulled back to the unit circle $\zeta = \sigma$ in the ζ-plane, (3.8) becomes

$$v^*(\sigma) = \dot{S}(\sigma) = \phi_1^*(\sigma) - \frac{S(\sigma)}{S'(\sigma)}\,\overline{\phi_1^{*\prime}(\sigma)} + \overline{\phi_2^*(\sigma)}, \tag{3.37}$$

where $v^*(\sigma) \equiv v(S(\sigma))$ and $\phi^*_{1,2}(\sigma) \equiv \phi_{1,2}(S(\sigma))$. (The asterisk here denotes a pull-back of ϕ, not complex conjugation.)

We may now solve for $\phi_1^*(\zeta, t)$ and $\phi_2^*(\zeta, t)$. Then, if we want to know the actual fluid velocity field, we map back to the physical z-plane to obtain $\phi_{1,2}(z) \equiv \phi^*_{1,2}(S^{-1}(z))$ and use (3.7) to find $v(z)$. More precisely, suppose that we have Laurent expansions in the ζ- and z-planes

$$\phi_1(z) = \sum_{k<0} a_k z^{k+1}, \tag{3.38}$$

$$\phi_2(z) = \sum_{k<-1} b_k z^{k+1}, \tag{3.39}$$

$$\phi_1^*(\zeta) = \sum_{k<0} a_k^* \zeta^{k+1}, \tag{3.40}$$

$$\phi_2^*(\zeta) = \sum_{k<-1} b_k^* \zeta^{k+1}, \tag{3.41}$$

$$z = S(\zeta) = \alpha_0 \zeta + \sum_{k<-1} \alpha_k \zeta^{k+1}, \tag{3.42}$$

$$\zeta = S^{-1}(z) = \frac{z}{\alpha_0} - \frac{\alpha_{-2}}{z} + \dots. \tag{3.43}$$

Then we solve for a_k^* and b_k^* by equating Fourier components in (3.37) and use $S^{-1}(z)$ to express a_k and b_k linearly in terms of a_k^* and b_k^*. For example, the leading coefficients a_{-1} and b_{-2} are

$$a_{-1} = a_{-1}^*, \tag{3.44}$$

$$b_{-2} = \alpha_0 b_{-2}^*. \tag{3.45}$$

These are in fact the only coefficients we need in order to compute the gauge potential

$$A_{\dot{S}}[S(\sigma, t)] \tag{3.46}$$

and consequently the net velocity of the shape at time t.

To proceed, (3.37) gives the following four equations for the leading a_k^* and b_k^* coefficients:

$$\left.\begin{aligned} \dot{\alpha}_{-3}\sigma^{-2} &= a_{-3}^*\sigma^{-2}, \\ \dot{\alpha}_{-2}\sigma^{-1} &= a_{-2}^*\sigma^{-1}, \\ 0 &= a_{-1}^* + \bar{\alpha}_0^{-1}\alpha_{-3}\bar{a}_{-2}^*, \\ \dot{\alpha}_0\sigma &= (\bar{b}_{-2}^* + \bar{\alpha}_0^{-1}\alpha_{-2}\bar{a}_{-2}^* + 2\bar{\alpha}_0^{-1}\alpha_{-3}\bar{a}_{-3}^*)\,\sigma. \end{aligned}\right\} \tag{3.47}$$

These may be solved to yield

$$\left.\begin{aligned} a_{-1} &= a_{-1}^* = -2\alpha_0^{-1}\alpha_{-3}\dot{\bar{\alpha}}_{-2}, \\ b_{-2} &= \alpha_0 b_{-2}^* = \alpha_0\dot{\alpha}_0 - \bar{\alpha}_{-2}\dot{\alpha}_{-2} - 2\bar{\alpha}_{-3}\dot{\alpha}_{-3}, \end{aligned}\right\} \tag{3.48}$$

since α_0 is real.

The constant component of the fluid flow at infinity must be zero in order for the net force on the shape to vanish. So we subtract from our solution $v(z)$ a counterflow a_{-1}, leading to a net translation velocity for the cylinder of

$$A^{\text{tr}} = \alpha_0^{-1}\alpha_{-3}\dot{\bar{\alpha}}_{-2}. \tag{3.49}$$

Similarly, after some algebra involving the equations of motion, the net torque is found to be

$$\begin{aligned} N &= \lim_{r_0\to\infty} \oint \epsilon_{ij}\, x_i\, \sigma_{jk}\, \mathrm{d}S_k \\ &= \lim_{r_0\to\infty} r_0\mu\, \mathrm{Re}\left[\int \bar{z}\frac{\partial v}{\partial \bar{z}}\mathrm{d}\bar{z}\right] \\ &= 8\pi\mu\, \mathrm{Im}\,(b_{-2}). \end{aligned} \tag{3.50}$$

We can cancel this torque with a rotational counterflow of angular velocity ω

$$\left.\begin{aligned} \dot{S}_{\text{rot}}(\sigma, t) &= \mathrm{i}\omega S(\sigma, t) \\ \dot{\alpha}_i &= \mathrm{i}\omega\alpha_i \end{aligned}\right\} \tag{3.51}$$

such that, from (3.48),

$$\mathrm{Im}\,(b_{-2}) = \mathrm{Im}\,\{\mathrm{i}\omega[|\alpha_0|^2 + |\alpha_{-2}|^2 + 2|\alpha_{-3}|^2]\}. \tag{3.52}$$

Solving for ω and using (3.48) we find then net rotational velocity of the shape

$$A^{\text{rot}} = \omega = \mathrm{Im}\left[\frac{\alpha_{-2}\dot{\bar{\alpha}}_{-2} + 2\alpha_{-3}\dot{\bar{\alpha}}_{-3}}{|\alpha_0|^2 + |\alpha_{-2}|^2 + 2|\alpha_{-3}|^2}\right]. \tag{3.53}$$

FIGURE 4. A typical trajectory for a cyclic swimmer in two dimensions. Six complete cycles occur between successive frames.

The complete gauge potential in the notation of §2 is

$$A = \begin{pmatrix} 0 & \omega & \mathrm{Re}\,(A^{\mathrm{tr}}) \\ -\omega & 0 & \mathrm{Im}\,(A^{\mathrm{tr}}) \\ 0 & 0 & 0 \end{pmatrix} \tag{3.54}$$

and the net translation and rotation due to the full sequence of deformations is

$$[R, d] = \bar{P} \exp\left[\int_0^T A(t)\,\mathrm{d}t\right]. \tag{3.55}$$

We have evaluated (3.54) explicitly for a particular sequence of shapes, and plotted the result in figure 4, as a stroboscopic picture. The sequence is a near-circle in the real $(\alpha_{-3}, \alpha_{-2})$-plane, with a small (5%) out-of-phase imaginary component (to produce a net rotation):

$$\left.\begin{aligned} \alpha_0 &= 1, \\ \alpha_{-2} &= 0.3\cos(2\pi t) + \mathrm{i}0.015\sin(2\pi t), \\ \alpha_{-3} &= -0.3\sin(2\pi t) + \mathrm{i}0.015\cos(2\pi t). \end{aligned}\right\} \tag{3.56}$$

6.1 cycles occur between each depicted shape, and the net motion through space is a counterclockwise 'generalized circle'.

For shapes of degree $D > 2$, the procedure for calculating A is the same. One solves $D+2$ linear equations as in (3.47), for $a^*_{-D-1}, \ldots, a^*_{-1}$ and b^*_{-2}, and determines the net translational and rotational velocities analogously.

The solution to our problem, (3.53)–(3.55), demonstrates the usefulness of our kinematic framework, and shows why it is necessary to introduce A in order to compute the net rigid motion. In fact, any solution to the problem we have posed in this section must be given by a path-ordered exponential integral over shape space, of a quantity which transforms under changes of reference axes for shapes as a gauge potential.

4. Squirming spheres

We now wish to study the possible swimming motions of a nearly spherical deformable body. This is a problem of some relevance in the biophysics of animal locomotion (Lighthill 1952 and Blake 1971*a*). In particular, consider a spherical animal which swims by waving a layer of short, densely packed cilia. (For reviews on the subject of self-propulsion of ciliated micro-organisms, see Blake & Sleigh 1975; Brennen & Winet 1977; Childress 1978; Lighthill 1975; Jahn & Votta 1972; Pedley 1975). An exact determination of all the swimming motions of such an animal would be impractical. However, we can usefully approximate the shape of this animal by a quasi-sphere, whose boundary just encloses the cilia. This approximation is known as the envelope model, and it is valid in the dual limit of short cilia relative to the radius of the sphere and dense packing relative to the lengths of the cilia. *Paramecia*, for example, are shaped like elongated spheres of length 200–300 μm (Blake & Sleigh 1975). The cilia are roughly 10 μm long and spaced 2 μm apart over the surface of the organism. Waves produced by the synchronous beating of the cilia are observed to have a frequency of about 30 Hz and a wavelength of 10 μm. The resulting helical trajectory is traversed with a velocity which has been observed to be between 600 and 2500 μm/s. While *Paramecia* are far from perfect spheres, one might hope to obtain at least a qualitative understanding of their swimming patterns. We also expect that our methods can be extended to encompass simple non-spherical shapes such as ellipsoids.

The problem of determining all swimming motions of such an animal lends itself perfectly to solution within the framework we have developed. Namely, if we know the 'field strength tensor' F at the sphere (which we may take to be of unit radius), then we know everything. The computation of $F(S^2)$ parallels that for the cylinder. We first find the general solution of the Navier–Stokes equations as an expansion in terms of vector spherical harmonics. We then solve the boundary-value problem for a slightly deformed sphere and obtain F_{mn} from the asymptotic behaviour of this solution. We shall ignore any constraints on the volume or surface area of the quasi-sphere, other than the limits imposed by the lengths of the cilia.

Since our boundary conditions for the flow $\boldsymbol{v}$ are going to be on the surface of a unit sphere, it is appropriate to expand $\boldsymbol{v}$ in vector spherical harmonics:

$$\boldsymbol{v}(r,\theta,\phi) = \sum_{k,J,L,M} a_{k,JLM}\, r^k\, \boldsymbol{Y}_{JLM}(\theta,\phi). \tag{4.1}$$

The $\boldsymbol{Y}_{JLM}$ are defined in terms of ordinary scalar spherical harmonics by

$$\boldsymbol{Y}_{JLM} = \sum_{m} \sum_{q=-1}^{1} Y_{LM}\langle Lm1q\,|\,L1JM\rangle\, \hat{\boldsymbol{e}}_q, \tag{4.2}$$

where $\hat{\boldsymbol{e}}_{\pm 1} = \mp(\hat{\boldsymbol{e}}_x \pm i\hat{\boldsymbol{e}}_y)/\sqrt{2}, \hat{\boldsymbol{e}}_0 = \hat{\boldsymbol{e}}_z$, and $\langle Lm1q\,|\,L1JM\rangle$ is a spin-1 Clebsch–Gordan coefficient.

We now insert the expansion of (4.1) $\boldsymbol{v}$ into the Navier–Stokes equations (2.11) and (2.12) to find k as a function of JLM. Using standard formulas for the Laplacian, curl, and divergence of $f(r)\, Y_{JLM}$ (see Appendix D) it is straightforward to show that (Lamb 1895)

$$v(r,\theta,\phi) = \sum_{JM} a_{JM}\, r^{-J-2}\, Y_{JJ+1M} + b_{JM}\, r^{-J-1}\, Y_{JJM} + c_{JM}\, r^{-J}\left(Y_{JJ-1M} - \left(\frac{J}{J+1}\right)^{\frac{1}{2}} \frac{2J-1}{2}\, Y_{JJ+1M}\right). \quad (4.3)$$

This expansion should be compared with its two-dimensional analogue, given by (3.14)–(3.16). It is easily checked that the boundary-value problem of matching $\boldsymbol{v}$ with an arbitrary velocity field on a surface is well-determined.

The net velocities due to a given change in shape are found, as in the two-dimensional case, from the condition that the net force and torque on the shape vanish. In fact, the net force and torque are proportional to leading asymptotics of $\boldsymbol{v}$, of order r^{-1} and r^{-2}. To see this, recall that the net force is the surface integral of the fluid stress tensor over the boundary of the shape (Batchelor 1970):

$$F_i = \int_{\text{shape}} \sigma_{ij}\, \mathrm{d}S_j. \quad (4.4)$$

By the divergence theorem and the fact that $\partial_j \sigma_{ij} = 0$ (Stokes' equations), this is equal to the integral of σ_{ij} over a large sphere of radius r

$$F_i = \int_{S^2} \sigma_{ij}\, \hat{n}_j\, r^2\, \mathrm{d}\Omega, \quad (4.5)$$

where $\hat{n}_j$ is a unit outward normal to the sphere. As $r \to \infty$, the only piece of σ_{ij} which survives in the integral is the term proportional to r^{-2}, which comes from the term

$$C_{1M}\, r^{-1}\, Y_{10M} \quad (4.6)$$

in (4.3). Thus, if our solution $\boldsymbol{v}$ contains such a piece, we must subtract it as a 'counterflow' in order to satisfy $\boldsymbol{F} = 0$. The resulting net translational velocity of the shape is accordingly

$$v_{\text{trans}} = -c_{1q}\, \hat{\boldsymbol{e}}_q. \quad (4.7a)$$

(Here and henceforth, we sum implicitly over $q = -1, 0, 1$.) Similarly, the rotation velocity comes from the term in the expansion of $\boldsymbol{v}$ proportional to Y_{01M}:

$$v_{\text{rot}} = a_{0q}\, \hat{\boldsymbol{e}}_q. \quad (4.7b)$$

To compute the translational components of the field strength, we follow the same procedure as in our earlier two-dimensional calculation, leading to the master formula of (3.27). The computation is presented in detail, and compared to earlier results of Blake, in Appendix E. We obtain

$$F^{\text{tr}}_{JLM,\,J'L'M'} = \frac{1}{4\pi} \sum_q \langle JM1q \,|\, J1J'M' \rangle\, \hat{\boldsymbol{e}}_q \Bigg\{ \Big[(J+1)(2J+1) \Big]^{\frac{1}{2}} \delta_{JL-1}\, \delta_{JL'} + \left[\frac{J-1}{J(2J-1)}\right]^{\frac{1}{2}} (2J-3)\, \delta_{J'J+1}\, \delta_{J'L'+1} \Big[(J+1)^{\frac{1}{2}} \delta_{JL-1} - J^{\frac{1}{2}} \delta_{JL+1} \Big] \Bigg\} - \{JLM \leftrightarrow J'L'M'\}, \quad (4.8)$$

It is worth remarking on the similarity of this result to the field strength F^{tr}_{mn} of a circular cylinder, found in §3.2. First, the only non-zero components of F in either case correspond to pairs of modes which are connected by a two- or three-dimensional angular momentum operator. In the three-dimensional case, this is a consequence of the spherical symmetry of the sphere. Thus, if the sphere is rotated by some $\boldsymbol{J}\cdot\hat{\boldsymbol{n}}$, then the translation $\boldsymbol{d}$ due to $\boldsymbol{F}$ must rotate similarly. Rephrasing the argument given in §3.2 for the cylinder now shows that J and J' must differ by at most 1 in order for the sphere to translate.

A second similarity is that $\boldsymbol{F}_{JLM,\,J'L'M'}$ grows linearly with J, for large J. This shows that the swimming motions of spheres and cylinders are quantitatively as well as qualitatively similar, a conclusion which presumably extends to other shapes as well.

The computation of $F^{\mathrm{rot}}_{JLM,\,J'L'M'}$ is similar, although somewhat more involved. F^{rot} is certainly essential in determining the helical motion which results from an arbitrary periodic swimming stroke. However, we expect that any maximally efficient stroke will involve no rotation (see Shapere & Wilczek 1989).

5. Summary and concluding remarks

This paper provides a general kinematic framework for discussing self-propulsion at low Reynolds number. We have formulated this problem in terms of a gauge potential A, which gives the net rigid motion resulting from an arbitrary change of shape. Finite motions due to a sequence of changes of shape are given by a path-ordered exponential integral of A along a path in shape space, and cyclic infinitesimal swimming motions are described by the covariant curl F of A. We have discussed an algorithm for determining A at shapes related to the circle by a conformal map of finite degree, and evaluated A explicitly for all deformations of conformal degree two. Our computations of the field strength of the circular cylinder and of the sphere effectively determine all possible infinitesimal swimming motions of these shapes.

Knowing, as we now do, the motion that results from any infinitesimal cyclic swimming stroke around a circle or a sphere, we may try to find the most efficient strokes. We analyse this problem in an accompanying paper (Shapere & Wilczek 1989). Qualitatively, we find that optimal infinitesimal strokes are wave-like motions symmetric about the axis of propulsion. The waves propagate from front to rear (relative to the direction of motion), achieving a maximum amplitude near the middle.

There are several other directions in which our work should be extended:

It should be possible, following the methods of Muskhelishvili, to calculate F for a variety of two-dimensional shapes, e.g. ellipses. It is quite possible that there is a fairly direct algorithm of calculating F for any conformal image of the circle; we have not examined this closely. In three dimensions, it would be biologically interesting to extend our calculation of F for the sphere to prolate spheroids, by expanding the flow in prolate spheroidal harmonics.

The similarity of the field strengths of the cylinder and the sphere in the high-frequency limit suggests a possible approximation which could apply to arbitrary shapes. Since the flows generated by high-frequency disturbances on the boundary tend to die rapidly with distance, it should be possible to treat them approximately for any shape, by replacing the shape locally with its tangent plane. Such an approximation has been mentioned in the literature (Childress 1978 and references therein), although, to our knowledge, a firm mathematical justification is lacking. A useful application would be to the computation of high-frequency components of F_{mn} for arbitrary shapes.

Finally, a very interesting mathematical generalization is to consider unparametrized shapes. This might be appropriate to describing the motion of moving holes, i.e. oscillating bubbles. Both the kinematics and the boundary conditions have to be rethought to cover such cases; presumably one divides the shape space further by the group of diffeomorphisms, and imposes $v_{\text{tangential}} = 0$. The gauge group then becomes infinite dimensional.

We would like to express our appreciation to Edward Purcell for introducing us to the world of life at low Reynolds number and for his encouragement. We also wish to thank Sidney Coleman, Freeman Dyson, T. J. Pedley, and John C. Taylor for useful discussions, and Larry Romans for his comments on the manuscript. This research was supported in part by the National Science Foundation under Grant No. PHY82-17853, supplemented by funds from the National Aeronautics and Space Administration, at the University of California at Santa Barbara.

Appendix A. Shape space as a fibre bundle

Fibre bundles provide a natural geometric setting for understanding gauge potentials (Choquet-Bruhat, Dewitt-Morrette & Dillard-Bleick 1977; Eguchi, Gilkey & Hansen 1980). In this Appendix, we show how the problem of swimming at low Reynolds number can be formulated in terms of a fibre bundle. This should help to clarify the mathematical origin of our gauge potential, while also providing a nice concrete example of the fibre-bundle concept.

We have been considering the space $\mathscr{S}$ of shapes in $\mathbf{R}^3$ and its quotient modulo the Euclidean group, $\mathscr{S}/E_3$. Given a path of (unlocated) shapes in $\mathscr{S}/E_3$, our problem has been to lift to a path of shapes with locations in $\mathscr{S}$. Stokes' equations (2.11)–(2.13) determine a local rule for lifting the path, in terms of the gauge potential A. A tells us the net velocity of the shape through the fluid corresponding to an infinitesimal change of shape.

In the language of fibre bundles, $\mathscr{S}$ is the bundle, E_3 the fibre, and $\mathscr{S}/E_3$ the base space. A is a connection, a linear map from the tangent space of the base space, $T(\mathscr{S}/E_3)$, to the Lie algebra of E_3. A is defined only locally, relative to a local section – but of course the transport of shapes is defined globally. It is the defining property of G-bundles that under a change of section, $\sigma(x) \to g(x)\,\sigma(x)$, a connection transforms as

$$A \to g A g^{-1} + g\,\mathrm{d}g^{-1}.$$

If it were only a matter of dividing out by translations, the bundle would be topologically trivial: there is a globally smooth way of choosing the centre of a shape; namely to take its centre of mass (or what would be, if it was made from material of constant density). However, choosing orientations does not appear to be so trivial. One might think of aligning the principal axes with the coordinate axes – but in what order? A natural choice is to order them xyz in order of the magnitudes of the moments of inertia; but an ambiguity arises when two moments become equal, and it is not clear that a smooth choice is possible globally.

It would be interesting to study the global topology of the bundle $\mathscr{S}$. The base space $\mathscr{S}/E_3$ seems to have a non-trivial topology, which the bundle inherits. How 'twisted' is the connection A? An answer to this mathematical question might provide us with some qualitative insight into the motion of shapes which undergo large deformations.

Appendix B. The nearly circular cylinder in complex coordinates

In §3.2, we computed the swimming motions of a cylinder with nearly circular cross-section. Our results were given in vector notation in order to make the generalization to three dimensions straightforward. However, the details of the calculation take a somewhat simpler form when presented in the complex variables framework employed elsewhere in §3. In addition, the corresponding calculation for a cylinder of non-circular cross-section is most directly approached using conformal-mapping techniques, which requires the use of complex coordinates. With this motivation, we now calculate, using complex coordinates, the field strength of a circular cylinder.

Our strategy for evaluating F at the unit circle $\zeta = \sigma = e^{i\theta}$ will be as in §3.2. We consider the sequence of motions in (3.21), and directly compute the resulting net translation and rotation, to find $F_{v_1 v_2}$. But now we work in the complex basis for vector fields on the circle given in (3.11)–(3.16). Thus, if

$$v_1(\sigma) = \sigma^{m+1}, \qquad v_2(\sigma) = \sigma^{n+1} \tag{B 1}$$

then we can define the components of F by

$$\mathscr{R}(\epsilon v_1, \eta v_2) = [1,0] + \epsilon\eta F_{mn} + \bar{\epsilon}\eta F_{\bar{m}n} + \epsilon\bar{\eta} F_{m\bar{n}} + \bar{\epsilon}\bar{\eta} F_{\bar{m}\bar{n}}. \tag{B 2}$$

Note that for each m and n, F has four components, because ϵ and η each have two real components. It turns out that the particular holomorphic decomposition given above is computationally the most convenient.

As before, we define $v_{12}(z)$ to be velocity field of the fluid, which results when the boundary condition $\eta v_2(\sigma)$ is applied at the surface of the cylinder $z = s = \sigma + \epsilon v_1(\sigma)$. The complex analogue of (3.27) is

$$\left.\begin{aligned} F^{\mathrm{tr}}_{v_1 v_2} &= \lim_{|z|\to\infty} \oint \frac{dz}{2\pi i z}(v_{12} - v_{21}), \\ F^{\mathrm{rot}}_{v_1 v_2} &= \lim_{|z|\to\infty} \mathrm{Im}\left[\oint \frac{d\bar{z}}{2\pi i}(v_{12} - v_{21})\right]. \end{aligned}\right\} \tag{B 3}$$

Expanding v_{12} as in (3.7),

$$\begin{aligned} v_{12}(z) &= \phi_1(z) - z\overline{\phi_1'(z)} + \overline{\phi_2(z)} \\ &= \sum_{k<0} a_k z^{k+1} - z \sum_{k<0} (k+1)\overline{a_k z^k} + \sum_{k<-1} \overline{b_k z^{k+1}}, \end{aligned} \tag{B 4}$$

we see that the leading asymptotic behaviour of v_{12}, and hence the field strength tensor of the cylinder, is obtained by solving for a_{-1} and b_{-2}, the leading coefficients of ϕ_1 and ϕ_2:

$$\left.\begin{aligned} F^{\mathrm{tr}}_{mn} &= a_{-1}[v_{21}] - a_{-1}[v_{12}], \\ F^{\mathrm{rot}}_{mn} &= \mathrm{Im}\,(b_{-2}[v_{12}] - b_{-2}[v_{21}]). \end{aligned}\right\} \tag{B 5}$$

The first step in solving the boundary-value problem (3.8) for v_{12}, is to pull back to the circle, expressing everything in terms of the circle coordinate σ. To lowest order in ϵ, we get

$$\begin{aligned}
\eta\sigma^{n+1} &= v_{12}(\sigma+\epsilon\sigma^{m+1}) \\
&= v_{12}(\sigma)+\epsilon\sigma^{m+1}\left(\frac{\partial v_{12}}{\partial z}\right)_{z=\sigma}+\bar{\epsilon}\sigma^{-m-1}\left(\frac{\partial v_{12}}{\partial \bar{z}}\right)_{z=\sigma} \\
&= \sum_{k<0} a_k\sigma^{k+1}-\sum_{k<0}(k+1)\overline{a_k}\sigma^{-k+1}+\sum_{k<-1}\overline{b_k}\sigma^{-k-1} \\
&\quad+\epsilon\sigma^{m+1}\left(\sum_{k<0}(k+1)a_k\sigma^k-\sum_{k<0}(k+1)\overline{a_k}\sigma^{-k}\right) \\
&\quad+\bar{\epsilon}\sigma^{-m-1}\left(-\sum_{k<0}k(k+1)\overline{a_k}\sigma^{-k+2}+\sum_{k<-1}(k+1)\overline{b_k}\sigma^{-k}\right).
\end{aligned} \tag{B 6}$$

In the terms on the right-hand side of this equation which are of order ϵ, we may replace a_k $(k \neq -1)$ and b_k $(k \neq -2)$ by their values $a_k^{(0)}$ and $b_k^{(0)}$ when $\epsilon \equiv 0$. Any corrections to a_k and b_k for small ϵ can be ignored in terms which are already small. From (3.10), we have

$$\left.\begin{aligned}
a_k^{(0)} &= \eta\theta_{-n}\delta_{nk}, \\
b_k^{(0)} &= (k+3)\,\eta\theta_{-n}\delta_{n,k+2}+\bar{\eta}\theta_n\delta_{n,-k-2},
\end{aligned}\right\} \tag{B 7}$$

with $\theta_n = 0$ for negative n and $\theta_n = 1$ for non-negative n. We now solve for a_{-1} and F^{tr} by isolating the constant term in the expression (B 6). After some algebra, the result is

$$\left.\begin{aligned}
F^{\text{tr}}_{mn} &= [(n+1)\,\theta_{-n}-(m+1)\,\theta_{-m}]\,\delta_{m+n,-1}, \\
F^{\text{tr}}_{\bar{m}n} &= [-(n+1)\,\theta_n+(m+1)\,\theta_{-m}]\,\delta_{m-n,1}, \\
F^{\text{tr}}_{m\bar{n}} &= [-(n+1)\,\theta_{-n}+(m+1)\,\theta_m]\,\delta_{m-n,-1}, \\
F^{\text{tr}}_{\bar{m}\bar{n}} &= [-(n+1)\,\theta_{-n}+(m+1)\,\theta_{-m}]\,\delta_{m+n,1}.
\end{aligned}\right\} \tag{B 8}$$

Similarly, the term proportional to σ yields an expression for b_{-2} from which we obtain the rotational components of F:

$$\left.\begin{aligned}
F^{\text{rot}}_{mn} &= [-(n+1)\,\theta_{-n}+(m+1)\,\theta_{-m}]\,\delta_{m+n,0}, \\
F^{\text{rot}}_{\bar{m}n} &= |m+1|\,\delta_{m-n,0}, \\
F^{\text{rot}}_{m\bar{n}} &= -|m+1|\,\delta_{n-m,0}, \\
F^{\text{rot}}_{\bar{m}\bar{n}} &= [(n+1)\,\theta_{-n}-(m+1)\,\theta_{-m}]\,\delta_{m+n,0}.
\end{aligned}\right\} \tag{B 9}$$

Note that the net translation and rotation due to a sequence of infinitesimal deformations, determined by F according to (B 2), agree with the corresponding results found in §3.3 for certain large deformations. Consider, for example, the translation $d_{(-2)(-3)}$ associated with the closed path in $(\alpha_{-2}, \alpha_{-3})$ space shown in figure 3:

$$\begin{aligned} d_{(-2)(-3)} &= \bar{P}\exp\left[\oint A^{\mathrm{tr}}\right] \\ &\sim \int_0^T \alpha_{-3}\dot{\bar{\alpha}}_{-2}\,\mathrm{d}t \\ &= -\eta_{-3}\overline{\epsilon_{-2}}. \end{aligned} \tag{B 10}$$

On the other hand, since $m = n+1$, only $F^{\mathrm{tr}}_{\bar{m}n}$ is non-zero for $m = -2$ and $n = -3$, so the net translation according to (B 8) is

$$\begin{aligned} d_{(-2)(-3)} &= F_{\overline{(-2)}(-3)}\overline{\epsilon_{-2}}\,\eta_{-3} \\ &= -\eta_{-3}\overline{\epsilon_{-2}}. \end{aligned} \tag{B 11}$$

In Appendix C, we discuss a fluid-mechanical modification of the algebra of infinitesimal conformal transformations of the circle and its relation to F_{mn} for cylinders.

Appendix C. The Virasoro algebra

In this Appendix we would like to point out a connection between F_{mn} and the two-dimensional Virasoro algebra. This is the Lie algebra of infinitesimal deformations of the unit circle, with infinitely many generators L_n. L_n generates the infinitesimal deformation

$$\exp \epsilon_n L_n: \quad \zeta \to \zeta + \epsilon_n \zeta^{n+1}. \tag{C 1}$$

Note that L_{-1} generates rigid translations and that L_0 generates rigid scale transformations (for ϵ_0 real) and rotations (ϵ_0 imaginary). We shall denote the generator of rotations by $\mathrm{Im}\, L_0$.

A simple computation shows that

$$[L_m, L_n] = (m-n)\, L_{m+n}. \tag{C 2}$$

This algebra is important in string theory and conformal field theory (Shenker 1986). In these contexts, quantization modifies (C 2) by the addition of an 'anomaly' term proportional to $(n^3-n)\,\delta_{m+n,0}$. Our field strength tensor F_{mn} also produces a modification of (C 2). Consider the path in shape space generated by applying L_m, L_n, $-L_m$ and $-L_n$, successively. The failure of this path to close is given by $[L_m, L_n]$. But there is a further failure to close in the actual configuration space of shapes with locations, given by F_{mn}:

$$[L_m, L_n] = (m-n)\, L_{m+n} + F^{\mathrm{tr}}_{mn} L_{-1} + F^{\mathrm{rot}}_{mn}\, \mathrm{Im}\, L_0. \tag{C 3}$$

There are obvious parallels between (C 3) and the anomalous Virasoro algebra, but there is also an important difference. The anomaly which arises in conformal field theories depends cubically on the mode number n, and cannot be 'gauged away' by redefining the L_n. However, our fluid-mechanical modification of the Virasoro algebra, which is linear in n, can be absorbed into the L_n by including compensating rotations and translations which keep the shape centred at the origin.

Appendix D. Vector spherical harmonics

This Appendix contains formulas involving vector spherical harmonics which were used to derive (4.3) (see Edmonds 1957 for details).

$$\nabla^2 f(r)\, \boldsymbol{Y}_{JLM}(\theta,\phi) = \left(\frac{\mathrm{d}^2}{\mathrm{d}r^2} + \frac{2}{r}\frac{\mathrm{d}}{\mathrm{d}r} - \frac{L(L+1)}{r^2}\right) f(r)\, \boldsymbol{Y}_{JLM}(\theta,\phi),$$

$$\nabla \times f(r)\, \boldsymbol{Y}_{JJ+1M} = \mathrm{i}\left(\frac{\mathrm{d}}{\mathrm{d}r} + \frac{J+2}{r}\right) f(r) \left(\frac{J}{2J+1}\right)^{\frac{1}{2}} \boldsymbol{Y}_{JJM},$$

$$\nabla \times f(r)\, \boldsymbol{Y}_{JJM} = \mathrm{i}\left(\frac{\mathrm{d}}{\mathrm{d}r} - \frac{J}{r}\right) f(r) \left(\frac{J}{2J+1}\right)^{\frac{1}{2}} \boldsymbol{Y}_{JJ+1M} + \mathrm{i}\left(\frac{\mathrm{d}}{\mathrm{d}r} + \frac{J+1}{r}\right) f(r) \left(\frac{J+1}{2J+1}\right)^{\frac{1}{2}} \boldsymbol{Y}_{JJ-1M},$$

$$\nabla \times f(r)\, \boldsymbol{Y}_{JJ-1M} = \mathrm{i}\left(\frac{\mathrm{d}}{\mathrm{d}r} - \frac{J-1}{r}\right) f(r) \left(\frac{J+1}{2J+1}\right)^{\frac{1}{2}} \boldsymbol{Y}_{JJM},$$

$$\nabla \cdot f(r)\, \boldsymbol{Y}_{JJ+1M} = -\left(\frac{J+1}{2J+1}\right)^{\frac{1}{2}} \left(\frac{\mathrm{d}}{\mathrm{d}r} + \frac{J+2}{r}\right) f(r)\, Y_{JM},$$

$$\nabla \cdot f(r)\, \boldsymbol{Y}_{JJM} = 0,$$

$$\nabla \cdot f(r)\, \boldsymbol{Y}_{JJ-1M} = \left(\frac{J}{2J+1}\right)^{\frac{1}{2}} \left(\frac{\mathrm{d}}{\mathrm{d}r} - \frac{J-1}{r}\right) f(r) Y_{JM}.$$

Appendix E. Calculation of $F(S^2)$

In this Appendix, we sketch a derivation of (4.8) and compare it to a result of Blake (1971*a*).

First, we compute the constant component $\boldsymbol{v}_{12}$ of the velocity field on the surface of a sphere when the sphere is successively deformed by the (unphysical) fluid velocity fields $\boldsymbol{v}_1 = \epsilon r^k \boldsymbol{Y}_{JLM}$ and $\boldsymbol{v}_2 = \eta r^{k'} \boldsymbol{Y}_{J'L'M'}$:

$$\boldsymbol{v}_{12} \equiv \epsilon\eta \frac{1}{4\pi}\int_{S^2} r^k \boldsymbol{Y}_{JLM} \cdot \nabla(r^{k'} \boldsymbol{Y}_{J'L'M'})\, \mathrm{d}\Omega\,|_{r=1}. \tag{E 1}$$

Next, we consider physical fluid velocity fields (of the form given in (4.3)), whose values on the boundary of the two-sphere are $\boldsymbol{v}_1 = \epsilon \boldsymbol{Y}_{JLM}$ and $\boldsymbol{v}_2 = \eta \boldsymbol{Y}_{J'L'M'}$, i.e.

$$\epsilon \boldsymbol{Y}_{JLM}(\theta,\phi) = \begin{cases} \epsilon r^{-J-2} \boldsymbol{Y}_{JJ+1M} \\ \epsilon r^{-J-1} \boldsymbol{Y}_{JJM} \\ \epsilon r^{-J} \boldsymbol{Y}_{JJ-1M} \\ \quad + \epsilon \left(\dfrac{J}{J+1}\right)^{\frac{1}{2}} \dfrac{2J-1}{2} (r^{-J-2} - r^{-J})\, \boldsymbol{Y}_{JJ+1M} \end{cases} \tag{E 2}$$

for $r = 1$. Using (E 1), it is then straightforward to compute the net translation due to the closed cycle of shapes of figure 3, namely

$$\boldsymbol{d}_{JLM,J'L'M'} = \boldsymbol{F}^{\mathrm{tr}}_{JLM,J'L'M'}\, \epsilon\eta.$$

We evaluate $\boldsymbol{v}_{12}$ using

$$\left.\begin{aligned} \boldsymbol{Y}_{JLM} &= \sum_{mq} Y_{LM}\langle Lm1q\,|\,L1JM\rangle\,\hat{\boldsymbol{e}}_q, \\ \boldsymbol{\nabla} f(r)\, Y_{LM} &= -\left(\frac{L+1}{2L+1}\right)^{\frac{1}{2}}\left(\frac{\partial}{\partial r}-\frac{L}{r}\right) f(r)\, \boldsymbol{Y}_{LL+1M} \\ &\quad + \left(\frac{L}{2L+1}\right)^{\frac{1}{2}}\left(\frac{\partial}{\partial r}+\frac{L+1}{r}\right) f(r)\, \boldsymbol{Y}_{LL-1M}. \end{aligned}\right\} \qquad \text{(E 3)}$$

Expanding the $\boldsymbol{Y}_{JLM}$ and taking the gradient of $Y_{L'M'}$ gives

$$\boldsymbol{v}_{12} = \frac{1}{4\pi}\int \sum_{mq} Y_{Lm}\langle Lm1q\,|\,L1JM\rangle\,\hat{\boldsymbol{e}}_q \cdot \sum_{m'q'}\left[-\left(\frac{L'+1}{2L'+1}\right)^{\frac{1}{2}}(k'-L')\, \boldsymbol{Y}_{L'L'+1m'} + \left(\frac{L'}{2L'+1}\right)^{\frac{1}{2}}(k'+L'+1)\, \boldsymbol{Y}_{L'L'-1m'}\right]\langle L'm'1q\,|\,L'1J'M'\rangle\,\hat{\boldsymbol{e}}_{q'}.$$

We now expand $\boldsymbol{Y}_{L'L'+1m'}$ and $\boldsymbol{Y}_{L'L'-1m'}$ and integrate over the sphere, using orthonormality of spherical harmonics:

$$\begin{aligned} \boldsymbol{v}_{12} = \frac{1}{4\pi}\sum_{mq}\sum_{m'q'} &\langle Lm1q\,|\,L1JM\rangle\langle L'm'1q'\,|\,L'1J'M'\rangle\,\hat{\boldsymbol{e}}_{q'} \\ &\times\left[-\left(\frac{L}{2L-1}\right)^{\frac{1}{2}}(k'-L+1)\,\delta_{LL'+1}\langle Lm1q\,|\,L1L-1m'\rangle \right.\\ &\left. + \left(\frac{L+1}{2L+3}\right)^{\frac{1}{2}}(k'+L+2)\,\delta_{LL'-1}\langle Lm1q\,|\,L1L+1m'\rangle\right]. \end{aligned}$$

Now by completeness and reality of the Clebsch–Gordan coefficients,

$$\sum_{mq}\langle Lm1q\,|\,L1JM\rangle\langle Lm1q\,|\,L1L+1m'\rangle = \delta_{JL+1}\,\delta_{Mm'}.$$

So finally,

$$\boldsymbol{v}_{12} = \frac{1}{4\pi}\left[-\left(\frac{J+1}{2J+1}\right)^{\frac{1}{2}}(k'-L+1)\,\delta_{LL'+1}\,\delta_{JL-1} + \left(\frac{J}{2J+1}\right)^{\frac{1}{2}}(k'+L+2)\,\delta_{LL'-1}\,\delta_{JL+1}\right]\langle JM1q\,|\,J1J'M'\rangle\,\hat{\boldsymbol{e}}_q. \qquad \text{(E 4)}$$

We now consider a physical fluid flow $\boldsymbol{v}_2$. For $\boldsymbol{v}_2 = \eta r^{-J'-2}\boldsymbol{Y}_{J'J'+1M}$ we find a net velocity

$$\begin{aligned} \boldsymbol{v}_{12} \equiv \boldsymbol{v}_{JLM,\,J'L'M'} &= \epsilon\eta\frac{1}{4\pi}\int (\boldsymbol{Y}_{JLM}\cdot\boldsymbol{\nabla})\, r^{-J'-2}\,\boldsymbol{Y}_{J'J'+1M}\,\mathrm{d}\Omega\,|_{r=1} \\ &= \epsilon\eta[(J+1)(2J+1)]^{\frac{1}{2}}\delta_{JL'}\,\delta_{JL-1}\langle JM1q\,|\,J1J'M'\rangle\,\hat{\boldsymbol{e}}_q. \end{aligned} \qquad \text{(E 5)}$$

We get precisely the same answer for $v_2 = \eta r^{-J'-1} Y_{J'J'M'}$. However, the case

$$v_2(\theta, \phi) = \eta Y_{J'J'-1M'}(\theta, \phi)$$
$$= \eta r^{-J'} Y_{J'J'-1M'} + \eta \left(\frac{J}{J+1}\right)^{\frac{1}{2}} \frac{2J-1}{2} (r^{-J-2} - r^{-J})|_{r=1}$$

is slightly more complicated. We find

$$v_{12} \equiv v_{JLM,J'J'-1M'} = \frac{1}{4\pi}\Bigg([(J+1)(2J+1)]^{\frac{1}{2}} \delta_{JL'} \delta_{JL-1}$$
$$+ \left[\frac{J-1}{J(2J+1)}\right]^{\frac{1}{2}} (2J-3)\, \delta_{JL'} [(J+1)^{\frac{1}{2}} \delta_{JL-1} - J^{\frac{1}{2}} \delta_{JL+1}] \Bigg) \langle JM1q | J1J'M' \rangle \hat{e}_q.$$

Putting everything together now yields the field strength tensor

$$F_{JLM,J'L'M'} = v_{JLM,J'L'M'} - v_{J'L'M',JLM}$$

as in (4.8).

In a classic paper, Blake (1971*a*) studied axisymmetric irrotational swimming motions of a sphere. We wish to show that our result reduces to his.

Blake considered deformations of a sphere of radius $a = 1$ of the form

$$R = 1 + \epsilon \sum_{n=2}^{N} \alpha_n(t) P_n(\cos\theta_0), \quad \theta = \theta_0 + \epsilon \sum_{n=1}^{N} \beta_n(t)\, V_n(\cos\theta_0),$$

where θ_0 is the azimuthal coordinate for the undeformed sphere, and (R, θ) are coordinates for the axisymmetrically deformed sphere. P_n is the nth Legendre polynomial and

$$V_n(\cos\theta_0) = \frac{2}{J(J+1)} \frac{\partial}{\partial\theta} P_n(\cos\theta).$$

To lowest order in ϵ, the fluid velocity components in the radial and azimuthal directions at (R, θ) are, respectively,

$$v_R = \dot{R} = \epsilon \sum \dot{\alpha}_n P_n, \qquad v_\theta = R\dot{\theta} = \epsilon \sum \dot{\beta}_n V_n.$$

For such a velocity field, it is clear that the net velocity of the sphere through the fluid will always be in the z-direction. Blake computes it to second order in ϵ:

$$v_{\text{net}} = \epsilon^2 \Bigg\{ \sum_{n=2}^{N-1} \frac{(2n+4)\alpha_n \dot{\beta}_{n+1} - 2n\dot{\alpha}_n \beta_{n+1} - (6n+4)\alpha_{n+1}\dot{\beta}_n - (2n+4)\dot{\alpha}_{n+1}\beta_n}{(2n+1)(2n+3)}$$
$$+ \sum_{n=2}^{N-1} \frac{4(n+2)\beta_n \dot{\beta}_{n+1} - 4n\dot{\beta}_n \beta_{n+1}}{(n+1)(2n+1)(2n+3)}$$
$$- \sum_{n=2}^{N-1} \frac{(n+1)^2 \alpha_n \dot{\alpha}_{n+1} - (n^2 - 4n - 2)\alpha_{n+1}\dot{\alpha}_n}{(2n+1)(2n+3)} \Bigg\}. \qquad \text{(E 6)}$$

(We have neglected terms with $n = 1$, since these depend on Blake's choice of an origin for each shape, i.e. on his choice of gauge.)

We may extract the field strength $F^{\text{tr}}\hat{z}$ from (E 6) by considering closed paths $\epsilon\alpha_n(t)$ and $\epsilon\beta_n(t)$ $(0 \leqslant t \leqslant 1)$ of the type depicted in figure 3 and integrating v_{net} from $t = 0$ to $t = 1$. This yields the non-zero field components

$$\left.\begin{aligned}
F(P_n\hat{\boldsymbol{r}}, P_{n+1}\hat{\boldsymbol{r}}) &= -\int_0^1 \frac{(n+1)^2\alpha_n\dot{\alpha}_{n+1} - (n^2-4n-2)\,\alpha_{n+1}\dot{\alpha}_n}{(2n+1)(2n+3)}\,\mathrm{d}t \\
&= \frac{-2n^2+2n+1}{(2n+1)(2n+3)}, \\
F(P_n\hat{\boldsymbol{r}}, V_{n+1}\hat{\boldsymbol{\theta}}) &= \frac{4n+4}{(2n+1)(2n+3)}, \\
F(V_n\hat{\boldsymbol{\theta}}, P_{n+1}\hat{\boldsymbol{r}}) &= \frac{4n}{(2n+1)(2n+3)}, \\
F(V_n\hat{\boldsymbol{\theta}}, V_{n+1}\hat{\boldsymbol{\theta}}) &= \frac{8}{(2n+1)(2n+3)}.
\end{aligned}\right\} \qquad \text{(E 7)}$$

Note that $F(P_n\hat{\boldsymbol{r}}, P_{n+1}\hat{\boldsymbol{r}})$ is just half the coefficient of the antisymmetric sum $\alpha_n\dot{\alpha}_{n+1} - \dot{\alpha}_n\alpha_{n+1}$. This is because the symmetric sum $\alpha_n\dot{\alpha}_{n+1} + \alpha_{n+1}\dot{\alpha}_n$ is a total time derivative, so that its time integral is zero.

To make contact with our computation of $F(\boldsymbol{Y}_{JLM}, \boldsymbol{Y}_{J'L'M'})$, we must express $P_n\hat{r}$ and $V_n\hat{\theta}$ in terms of $\boldsymbol{Y}_{JLM}$:

$$\left.\begin{aligned}
P_J(\cos\theta)\,\hat{\boldsymbol{r}} &= \left(\frac{4\pi}{2J+1}\right)^{\frac{1}{2}} Y_{J0}(\theta)\,\hat{\boldsymbol{r}} \\
&= (4\pi)^{\frac{1}{2}}\left[-\frac{(J+1)^{\frac{1}{2}}}{2J+1}\,\boldsymbol{Y}_{JJ+10} + \frac{J^{\frac{1}{2}}}{2J+1}\,\boldsymbol{Y}_{JJ-10}\right], \\
V_J(\cos\theta)\,\hat{\boldsymbol{\theta}} &= \frac{2}{J(J+1)}\left(\frac{4\pi}{2J+1}\right)^{\frac{1}{2}} Y'_{J0}(\theta)\,\hat{\boldsymbol{\theta}} \\
&= \frac{2(4\pi)^{\frac{1}{2}}}{2J+1}\left[\frac{1}{(J+1)^{\frac{1}{2}}}\,\boldsymbol{Y}_{JJ+10} + \frac{1}{J^{\frac{1}{2}}}\,\boldsymbol{Y}_{JJ-10}\right].
\end{aligned}\right\} \qquad \text{(E 8)}$$

These follow from the following explicit representations of $\boldsymbol{Y}_{JL0}$ in terms of Y_{J0} (see Arfken 1985):

$$\boldsymbol{Y}_{JJ+10}(r,\theta) = -\left(\frac{J+1}{2J+1}\right)^{\frac{1}{2}} Y_{J0}(\theta)\,\hat{\boldsymbol{r}} - \left(\frac{1}{(J+1)(2J+1)}\right)^{\frac{1}{2}} Y'_{J0}(\theta)\,\hat{\boldsymbol{\theta}},$$

$$\boldsymbol{Y}_{JJ0}(r,\theta) = -\mathrm{i}\left(\frac{1}{J(J+1)}\right)^{\frac{1}{2}} Y'_{J0}(\theta)\,\hat{\boldsymbol{\phi}},$$

$$\boldsymbol{Y}_{JJ-10}(r,\theta) = \left(\frac{J}{2J+1}\right)^{\frac{1}{2}} Y_{J0}(\theta)\,\hat{\boldsymbol{r}} - \left(\frac{1}{J(2J+1)}\right)^{\frac{1}{2}} Y'_{J0}(\theta)\,\hat{\boldsymbol{\theta}}.$$

The last ingredients needed are four components of $F_{JLM,J'L'M'}$ [see (4.8)]:

$$\left.\begin{aligned} F_{JJ+10,J+1J+20} &= \frac{1}{4\pi}[(J+1)(J+2)]^{\frac{1}{2}}, \\ F_{JJ+10,J+1J0} &= \frac{1}{4\pi}(J+1), \\ F_{JJ-10,J+1J+20} &= \frac{1}{4\pi}[J(J+2)]^{\frac{1}{2}}\frac{2J-1}{2J+3}, \\ F_{JJ-10,J+1J0} &= -\frac{1}{4\pi}[J(J+2)]^{\frac{1}{2}}\frac{2J-1}{2J+3}. \end{aligned}\right\} \qquad (\text{E } 9)$$

Here, we have made use of the Clebsch–Gordan coefficients

$$\langle J-1010\,|\,J-11J0\rangle = \left(\frac{J}{2J-1}\right)^{\frac{1}{2}},$$

$$\langle J+1010\,|\,J+11J0\rangle = -\left(\frac{J+1}{2J+3}\right)^{\frac{1}{2}}.$$

We shall now compute $F(P_n\hat{r}, P_{n+1}\hat{r})$ explicitly. Combining (E 8) and (E 9), and using the linearity of F, we obtain

$$\begin{aligned} F(P_n\hat{r}, P_{n+1}\hat{r}) &= 4\pi F\Bigg(-\frac{(J+1)^{\frac{1}{2}}}{2J+1}Y_{JJ+10} + \frac{J^{\frac{1}{2}}}{2J+1}Y_{JJ-10}, \\ &\qquad -\frac{(J+2)^{\frac{1}{2}}}{2J+3}Y_{J+1J+20} + \frac{(J+1)^{\frac{1}{2}}}{2J+3}Y_{J+1J0}\Bigg) \\ &= \frac{-2J^2+2J+1}{(2J+1)(2J+3)}, \end{aligned}$$

in agreement with (E 7). The remaining components of F may be evaluated similarly.

REFERENCES

ARFKEN, G. 1985 *Mathematical Methods for Physicists*. Academic.

BATCHELOR, G. K. 1970 *An Introduction to Fluid Dynamics*. Cambridge University Press.

BLAKE, J. R. 1971*a* A spherical envelope approach to ciliary propulsion. *J. Fluid Mech.* **46**, 119–208.

BLAKE, J. R. 1971*b* Self propulsion due to oscillations on the surface of a cylinder at low Reynolds number. *Bull. Austral. Math. Soc.* **3**, 255–264.

BLAKE, J. R. & SLEIGH, M. A. 1975 Hydromechanical aspects of ciliary propulsion. In *Swimming and Flying in Nature* (ed. T. Y. Wu, C. J. Brokaw and C. Brennen), pp. 185–210. Plenum.

BRENNEN, C. & WINET, H. 1977 Fluid mechanics of propulsion by cilia and flagella. *Ann. Rev. Fluid Mech.* **9**, 339–398.

CHILDRESS, S. 1978 *Mechanics of Swimming and Flying*. Cambridge University Press.

CHOQUET-BRUHAT, Y., DEWITT-MORRETTE, C. & DILLARD-BLEICK, M. 1977 *Analysis, Manifolds, and Topology*. North-Holland.

EDMONDS, A. E. 1957 *Angular Momentum in Quantum Mechanics*. Princeton University Press.

EGUCHI, T., GILKEY, P. B. & HANSEN, A. J. 1980 Gravitation, gauge theories, and differential geometry. *Phys. Rep.* **66**, 213.

ENGLAND, H. 1971 *Complex Variable Methods in Elasticity*. Wiley-Interscience.

HAPPEL, J. & BRENNER, H. 1965 *Low Reynolds Number Hydrodynamics*. Prentice-Hall.

HASIMOTO, H. & SANO, H. 1980 Stokeslets and eddies in creeping flow. *Ann. Rev. Fluid Mech.* **12**, 335–364.

HILL, R. & POWER, G. 1955 Extremum principles for slow viscous flow and the approximate calculation of drag. *Q. J. Mech. Appl. Maths* **9**, 313–319.

JAHN, T. L. & VOTTA, J. J. 1972 Locomotion of protozoa. *Ann. Rev. Fluid Mech.* **4**, 93–116.

KELLER, J. B. & RUBINOW, S. J. 1976 Swimming of flagellated microorganisms. *Biophys. J.* **16**, 151.

LAMB, H. 1895 *Hydrodynamics*. Cambridge University Press.

LIGHTHILL, J. 1952 On the squirming motion of nearly spherical deformable bodies through liquids at very small Reynolds number. *Commun. Pure. Appl. Maths* **5**, 109–118.

LIGHTHILL, J. 1975 *Mathematical Biofluidmechanics*. SIAM.

MUSKHELISHVILI, N. I. 1953 *Some Basic Problems of the Mathematical Theory of Elasticity* (translated by J. R. M. Radok). Noordhoff.

PEDLEY, T. J. (ed.) 1975 *Scale Effects in Animal Locomotion*. Academic.

PURCELL, E. 1977 Life at low Reynolds number. *Ann. J. Phys.* **45**, 3.

RAYLEIGH, LORD 1878 *The Theory of Sound*, vol. 1, chap. 19. Macmillan.

RICHARDSON, S. 1968 Two-dimensional bubbles in slow viscous flows. *J. Fluid Mech.* **33**, 476–493.

SHAPERE, A. & WILCZEK, F. 1988 *Geometric Phases in Physics*. World Scientific (to be published).

SHAPERE, A. & WILCZEK, F. 1989 Efficiencies of self-propulsion at low Reynolds number. *J. Fluid Mech.* **198**, 587–599.

SHENKER, S. 1986 Introduction to conformal and superconformal field theory. In *Unified String Theories* (ed. D. Gross & M. Green). World Scientific.

TAYLOR, G. I. 1951 Analysis of the swimming of microscopic organisms. *Proc. R. Soc. Lond.* A **209**, 447–461.

Chapter 9

ASYMPTOTICS

9

Asymptotics

It seems appropriate that we conclude with another beautiful contribution from Michael Berry. In this paper [9.1], he considers the construction of a systematic iterative approximation to the full phase, for which the geometric phase provides the leading term, in the original context of the spin 1/2 system. An asymptotic series is found, that has some quite remarkable and possibly universal properties. This work illustrates in a very concrete way how focusing on the concept of the geometric phase, has been a fruitful procedure.

One can imagine extending the analysis in several ways, particularly to treat degenerate levels. We believe, that many more attractive discoveries await determined explorers in these directions.

Proc. R. Soc. Lond. A 414, 31–46 (1987)
Printed in Great Britain

Quantum phase corrections from adiabatic iteration

By M. V. Berry, F.R.S.
H. H. Wills Physics Laboratory, Tyndall Avenue, Bristol BS8 1TL, U.K.

(Received 22 April 1987)

The phase change γ acquired by a quantum state $|\psi(t)\rangle$ driven by a hamiltonian $H_0(t)$, which is taken slowly and smoothly round a cycle, is given by a sequence of approximants $\gamma^{(k)}$ obtained by a sequence of unitary transformations. The phase sequence is not a perturbation series in the adiabatic parameter ϵ because each $\gamma^{(k)}$ (except $\gamma^{(0)}$) contains ϵ to infinite order. For spin-$\frac{1}{2}$ systems the iteration can be described in terms of the geometry of parallel transport round loops C_k on the hamiltonian sphere. Non-adiabatic effects (transitions) must cause the sequence of $\gamma^{(k)}$ to diverge. For spin systems with analytic $H_0(t)$ this happens in a universal way: the loops C_k are sinusoidal spirals which shrink as ϵ^k until $k \sim \epsilon^{-1}$ and then grow as $k!$; the smallest loop has a size $\exp\{-1/\epsilon\}$, comparable with the non-adiabaticity.

1. Introduction

It is known (Berry 1984) that the phase of a quantum system whose hamiltonian is taken slowly round a cycle will acquire a geometric contribution, characteristic of the cycle, as well as the familiar dynamical one. The argument assumed that the instantaneous eigenstates were non-degenerate and that the adiabatic theorem could be applied. Two generalizations have removed these restrictions: Wilczek & Zee (1984) allow the instantaneous eigenstates to be degenerate (see also Segert 1987 and Mead 1987); and Aharonov & Anandan (1987) (see also Page 1987) allow the evolution to be non-adiabatic provided the system returns exactly to its initial state (apart from a phase, of course).

My purpose here is to develop a third generalization, going back to the original non-degenerate adiabatic scenario in which the system returns to its original state not exactly but in a close approximation, but now taking into account the finite rate at which the hamiltonian is changed. This leads to a technique for systematically obtaining corrections to the geometric phase. Garrison (1986) has made a start along these lines, by calculating the first phase correction in adiabatic perturbation theory. The method I shall use is not perturbative but iterative, and involves a sequence of unitary transformations chosen so as to make the hamiltonian cling ever more closely to the evolving state; it has the merit of being easy to visualize.

In §2 the phase is defined precisely and the iteration scheme described. It is shown how the phase can be interpreted as geometric or dynamical, depending on the choice of unitary transformation. The simplest non-trivial application (§3) is to a two-state (spin-$\frac{1}{2}$) system, for which the iteration can be formulated explicitly in terms of geometry on the hamiltonian sphere. For finite slowness the evolution of the state will not be perfectly adiabatic, and as will be explained in §4 this implies the eventual divergence of the iteration scheme for the phase; for spin systems the divergence exhibits remarkable universality.

2. Iterated adiabatic anholonomy

Let the state $|\psi_0(t)\rangle$ be driven by a hamiltonian $H_0(t)$ (the suffixes denote the zeroth state of the iteration scheme to be described below). In units with $\hbar = 1$, $|\psi_0\rangle$ satisfies

$$\mathrm{i}|\dot{\psi}_0\rangle = H_0|\psi_0\rangle, \tag{1}$$

where here and hereafter time-dependences are understood where not written explicitly, and dots denote time derivatives. H_0 is taken smoothly round a cycle, i.e. $H_0(+\infty) = H_0(-\infty)$ with all derivatives vanishing as $|t| \to \infty$ (for this it suffices to take H_0 analytic in a strip including the real t axis). Let the (non-degenerate) instantaneous eigenstates of H_0 be $|n_0(t)\rangle$ with energies $E_0(n,t)$, i.e.

$$H_0|n_0\rangle = E_0(n)|n_0\rangle. \tag{2}$$

This defines the $|n_0\rangle$ up to a time-dependent phase which we make unique by demanding that

$$\langle n_0|\dot{n}_0\rangle = 0. \tag{3}$$

With this choice, the eigenstates are parallel-transported, as explained by Simon (1983). Let the system start in the nth eigenstate, i.e.

$$|\psi_0(-\infty)\rangle = |n_0(-\infty)\rangle \equiv |N\rangle. \tag{4}$$

The phase which is the object of study is now defined as

$$\gamma(n) \equiv \operatorname{Im}\ln\langle N|\psi_0(+\infty)\rangle + \int_{-\infty}^{\infty} \mathrm{d}t\, E_0(n,t). \tag{5}$$

This form of writing assumes that the integral over E_0 (which is minus the dynamical phase) converges, or can be made to converge by shifting the energy origin; if not, γ can be defined by a suitable limiting procedure. As defined by (5) the phase is more general than that which arises in the cyclic evolutions of Aharonov & Anandan (1987), because transitions may (and usually do) make $|\langle N|\psi_0(+\infty)\rangle| < 1$, i.e. the final state may be a superposition including states other than the original. Here, however, the emphasis is on cases where such non-adiabatic effects are small. We shall introduce a slowness parameter ϵ, entering H_0 in the combination ϵt, and regard ϵ as small. In the limit $\epsilon = 0$, γ becomes the geometric phase studied previously (Berry 1984).

Now let us follow several other authors (e.g. Avron *et al.* 1987; Anandan & Stodolsky 1987; Mead 1987) and define $U_0(t)$ as the unitary operator generating the eigenstates $|n_0(t)\rangle$ by acting on the original eigenstates $|N\rangle$, i.e.

$$|n_0(t)\rangle = U_0(t)|N\rangle. \tag{6}$$

Because of the parallel-transport law (3), $|n_0(+\infty)\rangle$ differs from $|N\rangle$ by a phase which is precisely the original geometric phase $\gamma_0(n)$ (anholonomy of H_0), so that

$$U_0(+\infty)\,|N\rangle = \exp\{\mathrm{i}\,\gamma_0(n)\}\,|N\rangle. \tag{7}$$

The operator U_0 naturally leads to a new representation of the evolving state $|\psi_0\rangle$ as that state $|\psi_1\rangle$ on which it must act to produce $|\psi_0\rangle$, i.e.

$$|\psi_1\rangle = U_0^\dagger\,|\psi_0\rangle. \tag{8}$$

Thus

$$\langle N\,|\,\psi_0(+\infty)\rangle = \langle N\,|\,U_0(+\infty)\,|\,\psi_1(+\infty)\rangle = \exp\{\mathrm{i}\,\gamma_0\}\,\langle N\,|\,\psi_1(+\infty)\rangle \tag{9}$$

so that the phase (5) now becomes

$$\gamma(n) = \gamma_0(n) + \mathrm{Im}\,\ln\langle N\,|\,\psi_1(+\infty)\rangle + \int_{-\infty}^{\infty} \mathrm{d}t\,E_0(n,t). \tag{10}$$

To proceed further, we need the Schrödinger equation satisfied by $|\psi_1\rangle$. This involves a hamiltonian $H_1(t)$ which differs from H_0, because the transformation U_0 is time-dependent. Thus

$$\mathrm{i}\,|\dot{\psi}_1\rangle = H_1\,|\psi_1\rangle, \tag{11}$$

where

$$H_1 = U_0^\dagger H\,U_0 - \mathrm{i}\,U_0^\dagger \dot{U}_0. \tag{12}$$

It is not difficult to show from (2), (3) and (6) that the matrix elements of H_1 in the $|N\rangle$ representation are

$$\langle M\,|\,H_1\,|\,N\rangle = E_0(n)\,\delta_{MN} - \frac{\mathrm{i}\,\langle m_0\,|\dot{H}_0|\,n_0\rangle}{E_0(n)-E_0(m)}(1-\delta_{MN}). \tag{13}$$

The simplest adiabatic approximation is to neglect the off-diagonal elements on the grounds that $\dot{H}_0$ is of order ϵ. Then (11) gives

$$|\psi_1(t)\rangle \approx \exp\left\{-\mathrm{i}\int_{-\infty}^{t} \mathrm{d}t'\,E_0(n,t')\right\}|N\rangle \tag{14}$$

so that (10) gives

$$\gamma(n) \approx \gamma_0(n). \tag{15}$$

Systematic improvements can, however, be achieved by not neglecting the off-diagonal terms. Instead, $H_1(t)$ is regarded as a new hamiltonian with new eigenstates $|n_1(t)\rangle$ and new eigenvalues $E_1(n,t)$, and the transformation repeated, leading to a new representation $|\psi_2(t)\rangle$ and a further hamiltonian $H_2(t)$. Obviously the procedure can be iterated according to the scheme

$$\left.\begin{aligned} H_{k+1} &= U_k^\dagger H_k\,U_k - \mathrm{i}\,U_k^\dagger\,\dot{U}_k, \\ \text{i.e.}\quad \langle M\,|H_{k+1}|\,N\rangle &= E_k(n)\,\delta_{MN} - \frac{\mathrm{i}\,\langle M_k\,|\dot{H}_k|\,n_k\rangle}{E_k(n)-E_k(m)}(1-\delta_{MN}),\end{aligned}\right\} \tag{16}$$

where

$$U_k\,|N\rangle = |n_k\rangle, \quad H_k\,|n_k\rangle = E_k(n)\,|n_k\rangle, \quad \langle n_k\,|\,\dot{n}_k\rangle = 0. \tag{17}$$

(Iteration does not change the initial states $|N\rangle$, because of the assumed smoothness of H_0.)

At the kth iteration step,

$$\langle N|\psi_0(+\infty)\rangle = \langle N|U_0(+\infty)\,U_1(+\infty)\ldots U_k(+\infty)|\psi_{k+1}(+\infty)\rangle$$
$$= \exp\left\{\mathrm{i}\sum_{j=0}^{k}\gamma_j(n)\right\}\langle N|\psi_{k+1}(+\infty)\rangle, \tag{18}$$

where $\gamma_j(n)$ are the anholonomies of the cycled hamiltonians $H_j(t)$. We can stop at this iteration by neglecting the off-diagonal elements in H_{k+1}. This gives the *kth phase approximant*

$$\gamma(n) \approx \gamma^{(k)}(n) \equiv \sum_{j=0}^{k}\gamma_j(n) + \int_{-\infty}^{\infty}\mathrm{d}t\,[E_0(n,t)-E_k(n,t)]. \tag{19}$$

The sequence of approximants is not a perturbation series in the adiabatic parameter ϵ, because even $\gamma^{(1)}$ involves ϵ to infinitely high order (of course $\gamma^{(0)} = \gamma_0$ is independent of ϵ). Rather, the iterations can be regarded as successive superadiabatic transformations to moving frames (in Hilbert space) attempting to cling ever more closely to the evolving state $|\psi_0\rangle$ (we will see in §4 that the attempts ultimately fail).

It is instructive to digress and consider iteration schemes that are not based on the parallel-transport law (3). An obvious class of alternatives (infinitely many) is to require the $|n_k(t)\rangle$ to return exactly to $|N\rangle$ as $t \to +\infty$. This would eliminate the anholonomies of the U_k, but would change the diagonal elements of the iterated hamiltonians to

$$\langle N|H_{k+1}(t)|N\rangle = E_k(n) - \mathrm{i}\,\langle n_k|\dot{n}_k\rangle, \tag{20}$$

where E_k and $|n_k\rangle$ are of course different from those in the previous scheme. Instead of (19) we would have

$$\gamma(n) \approx \int_{-\infty}^{\infty}\mathrm{d}t\,[E_0(n,t)-E_k(n,t)-\mathrm{Im}\,\langle n_k|\dot{n}_k\rangle], \tag{21}$$

so that in this form of iteration the phase arises from the approximate eigenvalue of H_{k+1}, and so its derivation appears entirely 'dynamical', even for $k = 0$! (γ_0 itself is geometric, regardless of how it is derived, because it does not depend on ϵ). Obviously it is possible to construct intermediate iteration schemes, in which the derivation appears as partly anholonomic and partly dynamical. The precise classical analogue of this interpretational ambiguity can be seen in the contrasting treatments of adiabatic angles by Hannay (1985) (anholonomic) and Berry (1985) (dynamical). The reason for choosing the iteration scheme based on (3) is firstly that it is unique and secondly that the corrections to the hamiltonian at each stage (16) are entirely in the off-diagonal terms and thus higher order in ϵ.

3. Spin-$\frac{1}{2}$ systems

We take

$$H_0(t) = \boldsymbol{R}_0(t)\cdot\boldsymbol{\sigma}, \tag{22}$$

where $\boldsymbol{R}_0 \equiv (X_0\,Y_0, Z_0)$ and $\boldsymbol{\sigma}$ is the vector spin-$\frac{1}{2}$ operator. Thus

$$H_0 = \frac{1}{2}\begin{pmatrix} Z_0 & X_0 - \mathrm{i}\,Y_0 \\ X_0 + \mathrm{i}\,Y_0 & -Z_0 \end{pmatrix}. \tag{23}$$

The transformation (12) will generate a new hamiltonian $H_1(t) = \boldsymbol{R}_1(t)\cdot\boldsymbol{\sigma}$, and the aim of this section is to find the explicit form of this operator.

For simplicity of writing we temporarily omit the suffixes zero. Define unit vectors $\boldsymbol{r}(t)$, $\boldsymbol{v}(t)$, $\boldsymbol{w}(t)$, and positive scalars $R(t)$, $V(t)$, by

$$\boldsymbol{R} \equiv R\boldsymbol{r}, \quad \dot{\boldsymbol{r}} \equiv V\boldsymbol{v}, \quad \boldsymbol{w} \equiv \boldsymbol{r}\times\boldsymbol{v}, \tag{24}$$

and think of $\boldsymbol{r}$ as a radius vector of the unit sphere. Then the cycle of $H_0(t)$ is represented by transport of the triad $\boldsymbol{r}, \boldsymbol{v}, \boldsymbol{w}$ round a circuit C_0 on the sphere (figure 1). The eigenvalues of H are $\pm\frac{1}{2}R(t)$, and the corresponding eigenstates $|n(t)\rangle$ will be denoted by $|\pm(t)\rangle$, or often simply by $|\pm\rangle$; these states depend on $\boldsymbol{r}(t)$ but not $R(t)$.

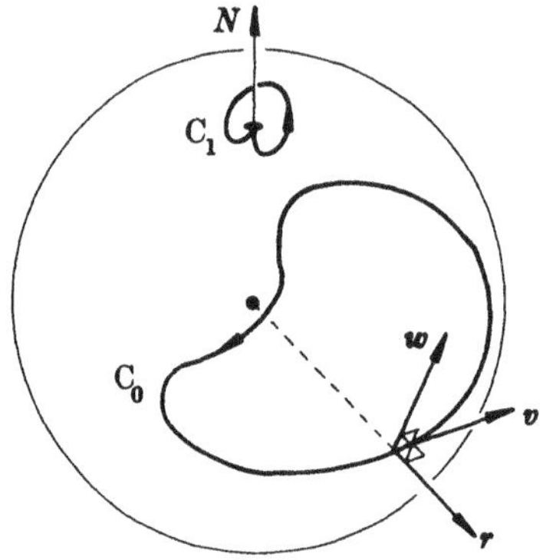

FIGURE 1. Transport of triad $\boldsymbol{r}, \boldsymbol{v}, \boldsymbol{w}$ round initial circuit C_0 on the hamiltonian sphere, and iterated circuit C_1.

The operator $U(t)$ (6) turns the eigenstates $|\pm(-\infty)\rangle$ into the eigenstates $|\pm(t)\rangle$ by a sequence of infinitesimal rotations. It is shown in appendix A that the instantaneous angular velocity of the rotation is uniquely determined by (3), that is $\langle\pm|\dot{\pm}\rangle = 0$, to be the angular velocity $\boldsymbol{\Omega}_{\parallel}(t)$ of a frame *parallel-transported* with $\boldsymbol{r}$ over the spheres. This angular velocity is

$$\boldsymbol{\Omega}_{\parallel} = \boldsymbol{r}\times\dot{\boldsymbol{r}} = V\boldsymbol{w} \tag{25}$$

and must be distinguished from the different angular velocity $\boldsymbol{\Omega}_t$ of the triad $\boldsymbol{r}, \boldsymbol{v}, \boldsymbol{w}$, which is

$$\boldsymbol{\Omega}_t = \boldsymbol{v}\times\dot{\boldsymbol{v}}. \tag{26}$$

Thus $U(t)$ is the time-ordered product

$$U(t) = T\exp\left\{-\mathrm{i}\int_{-\infty}^{t}\mathrm{d}t'\,\boldsymbol{\Omega}_{\parallel}(t')\cdot\boldsymbol{\sigma}\right\}. \tag{27}$$

To find the new operator H_1, we now use (12), which gives

$$H_1 = U^{\dagger}(\boldsymbol{R}-\boldsymbol{\Omega}_{\parallel})\cdot\boldsymbol{\sigma}\,U. \tag{28}$$

This result has a purely classical origin (valid for any spin) in the transformation, to a non-inertial frame moving with $\boldsymbol{R}$ and rotating with angular velocity $\boldsymbol{\Omega}_{\parallel}$, of the equation of motion for the expectation value $\langle\boldsymbol{\sigma}\rangle$. (See, for example, Cina (1986), Suter *et al.* (1987), Anandan & Stodolsky (1987), and, for explicit calculations in terms of Hannay's angle, Berry (1986) and Gozzi & Thacker (1987)). From (13) and (28), the matrix representation in terms of the initial states is

$$H_1 = \begin{pmatrix} \frac{1}{2}R & \mathrm{i}V\langle + |\boldsymbol{\sigma}\cdot\boldsymbol{v}| - \rangle \\ -\mathrm{i}V\langle - |\boldsymbol{\sigma}\cdot\boldsymbol{v}| + \rangle & -\frac{1}{2}R \end{pmatrix}, \tag{29}$$

where use has been made of the fact that the off-diagonal elements of $\boldsymbol{\sigma}\cdot\boldsymbol{r}$ vanish.

To make this explicit we must evaluate $\langle + |\boldsymbol{\sigma}\cdot\boldsymbol{v}| - \rangle$. In Appendix A this is shown to be

$$\mathrm{i}\langle + |\boldsymbol{\sigma}\cdot\boldsymbol{v}| - \rangle = \tfrac{1}{2}\exp\left\{-\mathrm{i}\int_{-\infty}^{t} \mathrm{d}t'\,\Omega(t')\right\}, \tag{30}$$

where

$$\Omega \equiv (\boldsymbol{\Omega}_t - \boldsymbol{\Omega}_{\parallel})\cdot\boldsymbol{r} = \boldsymbol{v}\times\dot{\boldsymbol{v}}\cdot\boldsymbol{r} = \boldsymbol{w}\cdot\dot{\boldsymbol{v}}. \tag{31}$$

Thus Ω measures the rate at which the $\boldsymbol{rvw}$ triad twists about $\boldsymbol{r}$, relative to the parallel-transported frame. The new hamiltonian (29) now becomes $H_1(t) = \boldsymbol{R}_1(t)\cdot\boldsymbol{\sigma}$ where (reinstating the suffixes)

$$Z_1 = R_0, \quad X_1 + \mathrm{i}Y_1 = V_0 \exp\left\{\mathrm{i}\int_{-\infty}^{t} \mathrm{d}t'\,\Omega_0(t')\right\}, \tag{32}$$

As $\boldsymbol{r}_0(t)$ executes the loop C_0 representing H_0, the new unit vector executes a loop C_1 representing H_1 (figure 1). Because V_0 is adiabatically small (it is the speed at which $\boldsymbol{r}_0$ moves on the unit sphere), this new loop is very close to the north pole of the sphere. In fact C_1 resembles a cardioid with a cusp or corner at the pole, because $V_0 \to 0$ as $|t| \to \infty$. Iteration of the map from $\boldsymbol{R}_0$ to $\boldsymbol{R}_1$ gives $\boldsymbol{R}_2(t)$, $\boldsymbol{R}_3(t)$... and hence further loops $\mathrm{C}_2, \mathrm{C}_3, \ldots$. For small ϵ these loops rapidly diminish in size at first, and we might hope that they will continue to do so (especially because $R_{k+1} = (R_k^2 + V_k^2)^{\frac{1}{2}} > R_k$ so iteration takes H further from the degeneracy at $R = 0$). But this hope cannot be realized, as will be explained in §4.

To obtain the phase approximations $\gamma^{(k)}(\pm)$ from (19) we must determine the anholonomies $\gamma_j(\pm)$ from the unitary operators (27). These can be obtained by noting that the effect of all the infinitesimal rotations in the product (27) from $t = -\infty$ to $t = +\infty$ is a spinor rotation about the initial direction $\boldsymbol{r}(-\infty)$ (which of course is the same as the final direction) by the parallel transport angles A_j associated with the loops C_j. These are simply the *solid angles* subtended by the loops at the origin of the sphere, obtained from (31) as the total twist of the parallel frame about the $\boldsymbol{rvw}$ triad, plus the 2π rotation of that triad, namely

$$\begin{aligned} A_j &= 2\pi - \int_{-\infty}^{\infty} \mathrm{d}t\,\Omega_j(t) = 2\pi - \int_{-\infty}^{\infty} \mathrm{d}t\,\boldsymbol{v}\times\dot{\boldsymbol{v}}\cdot\boldsymbol{r} \\ &= 2\pi - \oint_{\mathrm{C}_j} \mathrm{d}\boldsymbol{r}\cdot\boldsymbol{r}''\times\boldsymbol{r}, \end{aligned} \tag{33}$$

where primes denote differentiation with respect to arc length on the sphere.

Thus with the z axis temporarily along $\boldsymbol{r}(\pm\infty)$ we have

$$U_j(+\infty) = \exp\left\{-\tfrac{1}{2}\mathrm{i}A_j\begin{pmatrix}1 & 0\\ 0 & -1\end{pmatrix}\right\} = 1\cos(\tfrac{1}{2}A_j) - \mathrm{i}\begin{pmatrix}1 & 0\\ 0 & -1\end{pmatrix}\sin(\tfrac{1}{2}A_j) \tag{34}$$

so that

$$U_j(+\infty)\begin{pmatrix}a\\ b\end{pmatrix} = \begin{pmatrix}\exp\{-\tfrac{1}{2}\mathrm{i}\,A_j\}\,a\\ \exp\{+\tfrac{1}{2}\mathrm{i}\,A_j\}\,b\end{pmatrix} \tag{35}$$

giving

$$\gamma_j(\pm) = \mp\tfrac{1}{2}A_j \tag{36}$$

as found by Berry (1984). The energies in (19) are simply $E_k(\pm) = \pm\tfrac{1}{2}R_k$, so the phase approximants are

$$\gamma^{(k)}(\pm) = \mp\frac{1}{2}\left\{\sum_{j=0}^{k} A_j + \int_{-\infty}^{\infty} \mathrm{d}t\,[R_k(t) - R_0(t)]\right\}. \tag{37}$$

As illustration, consider the lowest-order approximations to the phase when C_0 is a circle of latitude on the unit sphere, with polar angle θ and azimuth $\phi(t)$ satisfying $\phi(+\infty) - \phi(-\infty) = 2\pi$. Thus

$$\boldsymbol{R}_0 = \boldsymbol{r}_0 = (\sin\theta\cos\phi(t), \sin\theta\sin\phi(t), \cos\theta). \tag{38}$$

The speed (24) on the sphere is

$$V_0 = \sin\theta\dot{\phi} \tag{39}$$

and the twist (31) is

$$\Omega_0 = \cos\theta\dot{\phi} \tag{40}$$

so (32) gives the new loop C_1 as $\boldsymbol{r}_1(t) = \boldsymbol{R}_1(t)/R_1(t)$, where

$$\boldsymbol{R}_1(t) = (\sin\theta\dot{\phi}\cos\{\phi(t)\cos\theta\}, \sin\theta\dot{\phi}\sin\{\phi(t)\cos\theta\}, 1). \tag{41}$$

Now we introduce an adiabatic ϵ by

$$\phi(t) \equiv \Phi(\tau), \quad \tau \equiv \epsilon t. \tag{42}$$

Then it is not difficult to show that the anholonomies A_j in (37) are even in ϵ while the 'dynamical' integrals over R_k are odd. Apart from

$$A_0 = 2\pi(1 - \cos\theta), \tag{43}$$

which is of course independent of ϵ, all the other terms in (37) are of infinite order in ϵ. Thus

$$\int_{-\infty}^{\infty} \mathrm{d}t\,[R_1(t) - R_0(t)] = \frac{1}{\epsilon}\int_{-\infty}^{\infty} \mathrm{d}\tau\,[(1 + \epsilon^2\sin^2\theta\Phi'^2(\tau))^{\frac{1}{2}} - 1], \tag{44}$$

where the prime denotes $\mathrm{d}/\mathrm{d}\tau$. The first three terms of an expansion are contained in the first iteration $\gamma^{(1)}(\pm)$, and are

$$\gamma_{\pm}^{(1)}(\pm) \approx \mp\left\{\pi(1 - \cos\theta) + \tfrac{1}{4}\epsilon\sin^2\theta\int_{-\infty}^{\infty} \mathrm{d}\tau\,\Phi'^2(\tau) + \tfrac{1}{4}\epsilon^2\sin^2\theta\cos\theta\int_{-\infty}^{\infty} \mathrm{d}\tau\,\Phi'^3(\tau)\right\}. \tag{45}$$

The terms originate in A_0, R_1 and A_1 respectively. Garrison (1986) has obtained the first two terms of (45) by adiabatic perturbation (ϵ-expansion).

4. Inevitability of divergence

To get the kth phase approximant (19) we neglected the off-diagonal terms in H_{k+1} and so approximated $\langle N|\psi_{k+1}(+\infty)\rangle$ in (18) by a pure phase factor. This ignores transitions to other states, which will cause the survival probability $|\langle N|\psi_{k+1}(+\infty)\rangle|^2$ to deviate from unity. The transition probability is typically exponentially small (Hwang & Pechukas 1977), i.e.

$$\Delta(\epsilon) \equiv 1-|\langle N|\psi_k(+\infty)\rangle|^2 \sim \exp\{-1/\epsilon\}; \tag{46}$$

$\Delta(\epsilon)$ is independent of the order of iteration k (cf. (18)). The sequence $\gamma^{(k)}(n)$ cannot converge for finite ϵ because this would imply $\Delta = 0$. Therefore the sequence must diverge: the true phase must reflect the non-analyticity of the survival amplitude, unlike the terms in (19), which although of infinite order in ϵ are nevertheless analytic at $\epsilon = 0$.

We expect the terms $\gamma_j(n)$ to get smaller at first and then increase. This is the typical behaviour of an asymptotic expansion (Dingle 1973), and it is reasonable to hope that the best approximant is the one for which $|\gamma^{(k+1)}(n)-\gamma^{(k)}(n)|$ is smallest (excluding perversities such as oscillations in this quantity) and that this value is of the same order as the non-adiabaticity (see Balian *et al.* (1978) for a numerical demonstration of a related phenomenon).

We can illustrate the inevitable divergence with the loops C_j on the unit sphere (figure 1) which represents the successive hamiltonians $H_j(t)$ for a spin-$\frac{1}{2}$ particle. For small ϵ the first iterated loop C_1, and many subsequent ones, will be small and close to the north pole. If we write the radius vector as $\boldsymbol{r} = (x, y, z)$ then $z \approx 1$ and we can approximate the loops as lying in the tangent plane at the pole. Parallel transport is now ordinary euclidean parallel translation, and in the iteration (32) $\int_{-\infty}^{t} \mathrm{d}t'\Omega(t')$ is the direction of the tangent to the loop at t, relative to that at $t = -\infty$. Denoting $x+\mathrm{i}y$ by ζ we find that (32) reduces to the simple iteration

$$\zeta_{j+1}(t) = \dot{\zeta}_j(t)/R_j(t). \tag{47}$$

When investigating the behaviour of the loops thus generated we can set $R_j(t) = 1$, because the successive radii differ little from $R_0(t)$ (cf. 31 which shows that $R_1^2 = R_0^2 + V_0^2$) and can be reduced to unity by the time rescaling $\mathrm{d}t \to R_0(t)\,\mathrm{d}t$. Thus (47) becomes the iterated hodograph transformation of mechanics, namely

$$\zeta_k(t) = \mathrm{d}^k\zeta_0(t)/\mathrm{d}t^{k+1}. \tag{48}$$

The loop C_k is generated in the xy plane by letting t run from $-\infty$ to $+\infty$. To describe adiabatic circuits we consider t to appear in the combination ϵt. Then ζ_k contains a factor ϵ^k and the C_k initially decrease in size. It would be reasonable to expect this decrease to continue, but it does not. The surprising fact is that, for almost all initial loops $\zeta_0(t)$, the C_k ultimately get bigger, and moreover for small ϵ the nature of the increase is *universal*. Furthermore, the winding number of C_k, defined as the number of rotations of the tangent as the loop is traversed, exceeds that of C_{k-1} by $\frac{1}{2}$ (alternate loops have cusps at $\zeta = 0$).

To justify these assertions we begin by recalling the assumption that $H_0(t)$, and hence $\zeta_0(t)$, is smooth, and interpret this as analyticity in a strip about the t axis. Thus the Fourier transform $\bar{\zeta}_0(\omega)$, defined by

$$\zeta_0(t) = \int_{-\infty}^{\infty} \mathrm{d}\omega\, \bar{\zeta}_0(\omega) \exp\{-\mathrm{i}\omega\epsilon t\} \tag{49}$$

decays exponentially as $|\omega| \to \infty$, the exponents being the imaginary parts of the singularities of $\zeta_0(t)$ nearest to the real axis in the upper and lower halves of the ϵt plane. If these singularities are

$$\tau_+ = \tau_{1+} + \mathrm{i}\,\tau_{2+}, \quad \tau_- = \tau_{1-} - \mathrm{i}\,\tau_{2-} \quad (\tau_{2+}, \tau_{2-} > 0) \tag{50}$$

then
$$\bar{\zeta}_0(\omega) \to A_\pm \exp\{\mathrm{i}\omega\tau_{1\pm}\} \exp\{-|\omega|\,\tau_{2\pm}\} \quad \text{as} \quad \omega \to \pm\infty. \tag{51}$$

The iterated loops (49) are given by (48) as

$$\zeta_k(t) = (-\mathrm{i})^k \int_{-\infty}^{\infty} \mathrm{d}\omega\, (\epsilon\omega)^k\, \bar{\zeta}_0(\omega) \exp\{-\mathrm{i}\omega\epsilon t\}. \tag{52}$$

For large k only the asymptotic form (51) of $\bar{\zeta}_0(\omega)$ contributes (because of the $(\epsilon\omega)^k$ factor) so that

$$\zeta_k(t) \to (-\mathrm{i}\epsilon)^k k!\{A_+/[\tau_{2+} - \mathrm{i}(\tau_{1+} - \epsilon t)]^{k+1} + (-1)^k A_-/[\tau_{2-} + \mathrm{i}(\tau_{1-} - \epsilon t)]^{k+1}\}. \tag{53}$$

The term with the smaller of $\tau_{2\pm}$ dominates exponentially, so that after a trivial shift of time origin and redefinition of ϵ as $\epsilon/\min\{\tau_{2\pm}\}$ the kth loop takes the universal form

$$\zeta_k(t) \to \frac{A(\mathrm{i}\epsilon)^k k!}{(1-\mathrm{i}\epsilon t)^{k+1}} = A[\epsilon^k k!/(1+\epsilon^2 t^2)^{\frac{1}{2}(k+1)}] \times \exp\{\mathrm{i}\,[\tfrac{1}{2}k\pi + (k+1)\arctan \epsilon t]\} \quad \text{as} \quad k \to \infty. \tag{54}$$

These universal loops C_k are Maclaurin's sinusoidal spirals (Lawrence 1972), some of which are shown in figure 2. In polar coordinates defined by $r = \mathrm{i}^k \exp\{\mathrm{i}\phi\}$ their equation is

$$r_k(\phi) = A\epsilon^k k! \cos^{k+1}\{\phi/(k+1)\} \tag{55}$$

(C_0 is a circle, C_1 a cardioid and C_2 Cayley's sextic). The maximum radius (at $\phi = 0$) is $r_k = A\,\epsilon^k\,k!$. This decreases at first (because of ϵ^k) but ultimately increases (because of $k!$). The smallest maximum radius occurs when $k \approx \epsilon^{-1}$ and is

$$r_{1/\epsilon}(0) \approx A(2\pi/\epsilon)^{\frac{1}{2}} \exp\{-1/\epsilon\}. \tag{56}$$

The kth loop has winding number $\frac{1}{2}k+1$; the initial and final windings ($t \approx \pm(k+1)/\pi\epsilon$) have radii smaller than the largest radius ($t = 0$) by a factor $[\pi/(k+1)]^{k+1}$.

The universality of the sinusoidal spirals can be described alternatively by saying that these curves are the attractors of the hodograph map in the space of loops. In view of the well-known instability of differentiation, the existence of attractors is remarkable, especially when considered backwards: the almost spirals C_k (k large), when iterated under the inverse map which is an integration and therefore supposedly stabilizing, must diversify into the infinite variety of

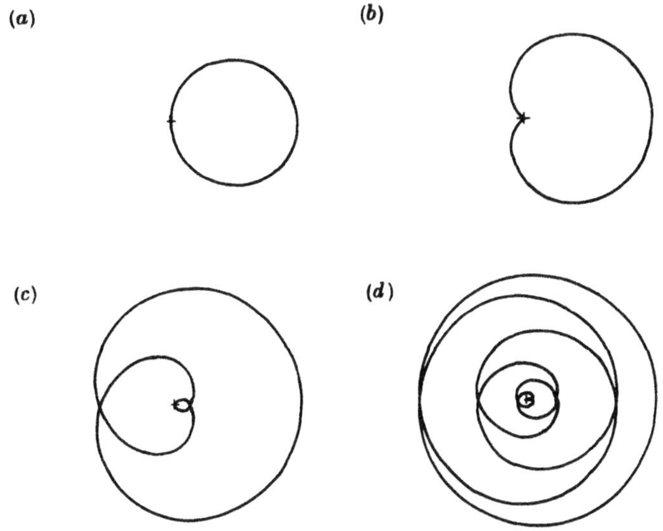

FIGURE 2. Universal loops (55) (sinusoidal spirals) for (a) $k = 0$; (b) $k = 1$; (c) $k = 9$; (d) $k = 50$. The loops are normalized to have the same maximum distance from the origin (which is reached at $t = \pm\infty$). The kth loop has $\frac{1}{2}k+1$ windings (not all visible in (c) and (d) because they are so small).

possible C_0. The resolution of the apparent paradox must lie in the assumed analyticity of C_0.

The asymptotic behaviour of the loops is reflected in the phase approximants $\gamma^{(k)}$. In the north-pole plane approximation (48) the solid angle A_k is just the euclidean area of C_k, namely

$$A_k = \frac{1}{2}\int_{-\infty}^{\infty} \mathrm{d}t \, \mathrm{Im}\, \zeta_k^* \dot{\zeta}_k \tag{57}$$

and the difference between successive radii is (cf. 32)

$$R_{k+1} - R_k = (R_k^2 + V_k^2)^{\frac{1}{2}} - R_k \approx \tfrac{1}{2}|\dot{\zeta}_k|^2 \tag{58}$$

because we have normalized R_k approximately to unity. Thus, roughly, the difference between successive approximants (37) is, with (48),

$$|\gamma^{(k+1)} - \gamma^{(k)}| \approx \frac{1}{4}\int_{-\infty}^{\infty} \mathrm{d}t \{\mathrm{Im}\,(\dot{\zeta}_k^* \ddot{\zeta}_k) + |\dot{\zeta}_k|^2\} \tag{59}$$

With the universal (54) this becomes

$$|\gamma^{(k+1)} - \gamma^{(k)}| \approx \tfrac{1}{4}A^2 [(k+1)!]^2 \epsilon^{2k+1} (\epsilon(k+\tfrac{3}{2}) + 1) I_{k+2}, \tag{60}$$

where $$I_k \equiv \int_{-\infty}^{\infty} \mathrm{d}\tau/(1+\tau^2)^k = 2\int_0^{\frac{1}{2}\pi} \mathrm{d}\theta \cos^{2(k-1)}\theta \rightarrow \left(\frac{\pi}{k}\right)^{\frac{1}{2}} \quad \text{as} \quad k \rightarrow \infty. \tag{61}$$

For small ϵ, (60) falls to a minimum when $k \approx \epsilon^{-1}$, whose value is of order $\exp\{-1/\epsilon\}$. When units are reinstated, the critical k acquires the following physical interpretation: ϵ^{-1} is the ratio of the average transition frequency between the two instantaneous eigenstates and the average rate of adiabatic change of

$\ln H_0(t)$. With this interpretation, $\exp\{-1/\epsilon\}$ is indeed the size of the non-adiabaticity (Hwang & Pechukas 1977).

It is possible to construct loops C_0 whose iterates do not fall into the universality class just discussed. One way is to make $\zeta_0(t)$ have its 'nearest singularity' actually *on* the real t axis whilst possessing all derivatives. For example, if $\zeta_0 \sim 1 - \exp\{-1/\epsilon|t|\}$ the smallest loop occurs for iteration $k \sim \epsilon^{-\frac{1}{2}}$ and has radius of order $\exp\{\epsilon^{-\frac{1}{2}}\}$. The opposite situation is for $\zeta_0(t)$ to be so smooth that its 'nearest singularity' lies infinitely far from the axis. For example, if $\zeta_0 \sim \exp\{-(\epsilon t)^2\}$ the smallest loop occurs for iteration $k \sim \epsilon^{-2}$ and has radius of order $\exp\{\epsilon^{-2}\}$. Another possibility is for the imaginary parts $\tau_{2\pm}$ in (50) to be equal, so that neither of the two contrary loops in (53) dominates; one example is when $\zeta_0(t)$ is real, that is C_0 is simply a back-and-forth swinging enclosing no area (then there is no anholonomy and γ is purely dynamical). It does seem, however, that whatever the form of $\zeta_0(t)$ (provided all derivatives exist, and vanish as $|t| \to \infty$) the loops do eventually grow, but I have not been able to construct a general proof.

The nonadiabaticity, as expressed by the transition probability $\Delta(\epsilon)$ defined by (46), can be estimated by perturbation theory, using $|N\rangle$ as the unperturbed state. For spin-$\frac{1}{2}$ systems it is tempting to employ perturbation theory in the north pole plane approximation, applying it to the hamiltonian for the iteration for which C_k is smallest. However, this gives only a crude approximation, for reasons worth exploring because they illuminate a curious general feature of the adiabatic approximation.

Application of standard time-dependent perturbation theory to the iterated hamiltonian (13) gives the transition probability $\Delta(\epsilon)$ in (46) as the sum of the probabilities of transitions to states $M \neq N$:

$$\Delta(\epsilon) \approx \sum_{M \neq N} \left| \int_{-\infty}^{\infty} \mathrm{d}t \frac{\langle m_0 | \dot{H}_0 | n_0 \rangle}{E_0(n) - E_0(m)} \exp\left\{ \mathrm{i} \int_{-\infty}^{t} \mathrm{d}t' \left[E_0(m) - E_0(n) \right\} \right|^2, \quad (62)$$

For a spin-$\frac{1}{2}$ system, starting (say) in $|+\rangle$, this gives, for the probability that at $t = +\infty$ there has been a transition to $|-\rangle$,

$$\Delta(\epsilon) \approx \frac{1}{4} \left| \int_{-\infty}^{\infty} \frac{V_0(t)}{R_0(t)} \exp\left\{ \mathrm{i} \int_{-\infty}^{t} \mathrm{d}t' \left[\Omega_0(t') - R_0(t') \right] \right\} \right|^2, \quad (63)$$

where use has been made of (29) and (30). Both the preceding formulae are independent of the order of iteration.

In the north-pole plane approximation with $R_0 = 1$, (63) becomes

$$\Delta(\epsilon) \approx \frac{1}{4} \left| \int_{-\infty}^{\infty} \mathrm{d}t\, \zeta_0(t) \exp\{-\mathrm{i}t\} \right|^2. \quad (64)$$

This too is independent of the order of iteration, as can be seen from (48) and integration by parts. Substituting the universal loop (54) gives

$$\begin{aligned} \Delta(\epsilon) &\approx \frac{A^2}{4} \left| \int_{-\infty}^{\infty} \mathrm{d}t\, (1 - \mathrm{i}\epsilon t)^{-1} \exp\{-\mathrm{i}t\} \right|^2 \\ &\left. \begin{aligned} &= \frac{A^2 \pi^2}{\epsilon^2} \exp\{-2/\epsilon\} \quad \text{if} \quad \epsilon > 0 \\ &= 0 \quad \text{if} \quad \epsilon < 0. \end{aligned} \right\} \end{aligned} \quad (65)$$

The two cases $\epsilon > 0$ and $\epsilon < 0$ correspond to opposite senses for the traversals of the loops and hence to an original hamiltonian $H_0(t)$ and its time reverse $H_0(-t)$. It is not surprising that time reversal can lead to very different transition probabilities because almost all the hamiltonians which give rise to anholonomy lack time-reversal symmetry. What *is* surprising – and this is the curious feature mentioned earlier – is that the phase is insensitive to this qualitative distinction to all orders of adiabatic iteration: the approximants (19) merely change sign under time reversal.

However, the extreme difference between $\epsilon > 0$ and $\epsilon < 0$ in (65) is an artefact of the north-pole plane approximation, arising from the fact that for $\epsilon < 0$ the perturbation $(1-\mathrm{i}\epsilon t)^{-1}$ has no negative-frequency components to stimulate the transition from $|+\rangle$ to $|-\rangle$. Without this approximation, but still taking C_0 as a circle, not necessarily small, as given by (38), the perturbation formula (63) gives, on making use of (39) and (40),

$$\Delta(\epsilon) \approx \frac{\sin^2\theta}{4}\left|\int_{-\infty}^{\infty} \mathrm{d}t\,\dot{\phi}(t)\exp\{\mathrm{i}\,[\phi(t)\cos\theta - t]\}\right|^2. \tag{66}$$

Now take

$$\phi(t) = 2\arctan\epsilon t, \tag{67}$$

which gives the same form of cycling as the universal loop (54). Then integration by parts gives

$$\Delta(\epsilon) \approx \frac{\tan^2\theta}{4\epsilon^2}\left|\int_{-\infty}^{\infty}\mathrm{d}\tau\left(\frac{1+\mathrm{i}\tau}{1-\mathrm{i}\tau}\right)^{\cos\theta}\exp\{-\mathrm{i}\tau/\epsilon\}\right|^2. \tag{68}$$

Changing the sign of ϵ has the same effect as changing θ to $\pi-\theta$, as it should because the effects on $|+\rangle$ of time reversal and latitude reversal of C_0 are the same.

In appendix B it is shown that the asymptotic form of (68) for small positive ϵ is

$$\Delta(\epsilon) \approx [\sin\pi\cos\theta)\tan\theta\,\Gamma(1-\cos\theta)2^{\cos\theta}]^2\exp\{-2/\epsilon\}/\epsilon^{2\cos\theta}. \tag{69}$$

Near the poles and the equator this has the limiting forms

$$\left.\begin{aligned} \Delta(\epsilon) &\to 4\pi^2\theta^2\exp\{-2/\epsilon\}/\epsilon^2 && \text{as}\quad \theta\to 0,\\ &\to \pi^2\exp\{-2/\epsilon\} && \text{as}\quad \theta\to\tfrac{1}{2}\pi,\\ &\to \tfrac{1}{16}\pi^2(\pi-\theta)^6\epsilon^2\exp\{-2/\epsilon\} && \text{as}\quad \theta\to\pi. \end{aligned}\right\} \tag{70}$$

The sense in which (65) is a crude approximation is now evident; instead of a discontinuity between $\epsilon > 0$ and $\epsilon < 0$ there is a smooth transition involving θ, which shows that time reversal does not make Δ vanish but reduces its value by ϵ^4.

5. Concluding remarks

The main result of this work is the formula (19) giving the phase approximants obtained from a (unique) succession of unitary transformations clinging ever closer to the evolving state. Successive approximants $\gamma^{(k)}(n)$ are correct to higher orders in the adiabatic parameter ϵ, but (19) is not a power series because each approximant (except the lowest) contains ϵ to infinite order. Nevertheless, at all orders of iteration the scheme neglects non-adiabatic transitions, and because these are of order $\exp\{-1/\epsilon\}$ this quantity, rather than ϵ itself, can be regarded as the adiabatic parameter and then the entire sequence of approximants can be considered to be contained within the lowest-order adiabatic approximation (higher approximations to the exact phase (5) would involve powers of $\exp\{-1/\epsilon\}$).

For spin systems the unitary sequence can be interpreted geometrically as (initially) shrinking loops on the hamiltonian sphere. The loops could be observed, for example, by exploiting the fact that the expectation $\langle\boldsymbol{\sigma}\rangle$ evolves classically according to $\langle\dot{\boldsymbol{\sigma}}\rangle = \boldsymbol{R}_0 \times \langle\boldsymbol{\sigma}\rangle$. Then the unitary sequence corresponds to a sequence of transformations to rotating frames. Stopping at the kth such transformation, making the adiabatic approximation and transforming back to the original reference frame, we find that $\langle\boldsymbol{\sigma}\rangle$ follows not $\boldsymbol{R}_0(t)$ but $\boldsymbol{R}^{(k)}(t)$ which (cf. 28) includes corrections from the angular velocities $\boldsymbol{\Omega}_{\|}(t)$ of the successive frames, i.e.

$$\boldsymbol{R}^{(k)}(t) = \boldsymbol{R}_0(t) - \sum_{j=0}^{k-1} \boldsymbol{\Omega}_{\|j}(t). \tag{71}$$

Now, successive $\boldsymbol{\Omega}_{\|}$ are (initially) smaller (by ϵ), and generate loops C_k (§4) which in their own frames have increasing winding numbers. Therefore the motion of $\langle\boldsymbol{\sigma}\rangle$ is *a sequence of ever-finer nutations* forming a hierarchy reminiscent of Ptolemy's epicycles. Successive orders of iteration correspond to observing the motion with increasing resolution.

Several questions are raised by the divergence of the sequence of phase approximants. One concerns the universality of the shrinking-and-growing of the hamiltonian loops C_k for spin systems, when k is large, ϵ small and $H_0(t)$ analytic. This was derived within the north-pole plane approximation, and slight doubt lingers as to whether the result would survive inclusion of the effects of the curvature of the sphere. Assuming it does, another question is whether the divergence for non-spin systems has a similar adiabatic universality. (Of course for spins the universality we are discussing can last only for iteration numbers not much greater than $k \approx \epsilon^{-1}$; subsequent iterations will take the expanded loops C_k away from the north pole plane. What happens then? Is there an infinite sequence – possibly irregular – of further shrinking and growing when $k \gg \epsilon^{-1}$? Or do the windings continue to increase, leading to ultimate loops C_∞ covering the sphere densely?)

Finally, one wonders whether there are any systems for which the iteration would converge (or stop at some order) because the adiabatic approximation would be exact. This question is prompted by a spatial analogy: the existence of one-dimensional potentials $V(x)$ for which the semiclassical ('adiabatic') approximation is exact for some energies E, so that there is no reflection (i.e. no 'transition'). For example, with arbitrary real 'quantum momentum' $k_{\mathrm{qu}}(x)$ the local plane wave

$$\psi(x) = \exp\left\{\mathrm{i}\int_0^x \mathrm{d}x'\, k_{\mathrm{qu}}(x')\right\}\Big/[k_{\mathrm{qu}}(x)]^{\frac{1}{2}} \tag{72}$$

is an exact solution of

$$\psi''(x)+k_{\mathrm{cl}}^{2}(x)\,\psi(x)=0 \tag{73}$$

provided it is related to the classical momentum $k_{\mathrm{cl}}=[(E-V)]^{\frac{1}{2}}$ by

$$k_{\mathrm{qu}}=[k_{\mathrm{cl}}^{2}+k_{\mathrm{qu}}^{\frac{1}{2}}(k_{\mathrm{qu}}^{-\frac{1}{2}})'']^{\frac{1}{2}}. \tag{74}$$

Then there is no coupling to the reflected wave which is the complex conjugate of (72). We can think of k_{qu} as the outcome of infinite order semiclassical iteration of (74), the lowest (W.K.B.) approximation being $k_{\mathrm{qu}}\approx k_{\mathrm{cl}}$. The analogue of γ is the phase of the transmitted wave (in this case entirely dynamical) referred to the W.K.B. phase, i.e.

$$\begin{aligned}\gamma &= \int_{-\infty}^{\infty}\mathrm{d}x\,(k_{\mathrm{qu}}-k_{\mathrm{cl}})\\ &= \int_{-\infty}^{\infty}\mathrm{d}x\,k_{\mathrm{qu}}^{\frac{1}{2}}(k_{\mathrm{qu}}^{-\frac{1}{2}})''/\{k_{\mathrm{qu}}+[k_{\mathrm{qu}}^{2}-k_{\mathrm{qu}}^{\frac{1}{2}}(k_{\mathrm{qu}}^{-\frac{1}{2}})'']^{\frac{1}{2}}\},\end{aligned} \tag{75}$$

which from (74) is of infinite order in spatial slowness. The analogous adiabatic problem, of constructing a state $|\psi(t)\rangle$ evolving exactly as an eigenstate of some changing $H(t)$, seems much more difficult. Successive iterated hamiltonians would have to commute with each other. This is impossible for non-trivial spin systems, but it is conceivable that in other cases there would occur a conspiracy of the off-diagonal elements $\langle m_k|\dot{H}_k|n_k\rangle$ to allow at least one of the states to evolve adiabatically. (Of course I am not here denying the existence of the cyclic evolutions considered by Aharonov & Anandan 1987, because these involve states that return to themselves without having at every instant to be eigenstates of the driving hamiltonian; the spatial analogy is the Ramsauer–Townsend effect, in which perfect transmission is achieved without requiring $\psi(x)$ to be everywhere locally plane as in (72).)

Appendix A. Spin-½ calculations justifying (27) and (30)

The defining equation for $U(t)$ is (6), which can be written

$$U(t)\,|\pm(-\infty)\rangle=|\pm(t)\rangle. \tag{A 1}$$

Differentiating leads to

$$\dot{U}U^{\dagger}|\pm\rangle=|\dot{\pm}\rangle, \tag{A 2}$$

which on making use of $|+\rangle\langle+|+|-\rangle\langle-|=1$ becomes

$$\dot{U}=(|\dot{+}\rangle\langle+|+|\dot{-}\rangle\langle-|)\,U\equiv BU. \tag{A 3}$$

Then (27) follows if

$$B=-\mathrm{i}V\boldsymbol{w}\cdot\boldsymbol{\sigma}. \tag{A 4}$$

The diagonal elements $\langle\pm|B|\pm\rangle$ vanish because of orthogonality and $\langle\pm|\dot{\pm}\rangle = 0$. The diagonal elements $\langle\pm|\boldsymbol{w}\cdot\boldsymbol{\sigma}|\pm\rangle$ also vanish because $|\pm\rangle$ are eigenstates of $\boldsymbol{r}\cdot\boldsymbol{\sigma}$, and $\boldsymbol{w}$ is perpendicular to $\boldsymbol{r}$. The off-diagonal elements are

$$\langle\pm|B|\mp\rangle = \langle\pm|\dot{\mp}\rangle. \tag{A 5}$$

Differentiating the eigenequations

$$\boldsymbol{r}\cdot\boldsymbol{\sigma}|\pm\rangle = \pm\tfrac{1}{2}|\pm\rangle \tag{A 6}$$

gives

$$\langle\pm|\dot{\mp}\rangle = \mp V\langle\pm|\boldsymbol{v}\cdot\boldsymbol{\sigma}|\mp\rangle. \tag{A 7}$$

Now choose local axes in which $\boldsymbol{r}, \boldsymbol{v}, \boldsymbol{w}$ lie along z, x, y respectively. Then

$$\langle+|\dot{-}\rangle = -\frac{V}{2}(1\quad 0)\begin{pmatrix}0 & 1\\ 1 & 0\end{pmatrix}\begin{pmatrix}0\\1\end{pmatrix} = -\frac{V}{2} \tag{A 8}$$

and

$$\langle+|B|\rangle = -\mathrm{i}V\langle+|\boldsymbol{w}\cdot\boldsymbol{\sigma}|-\rangle = -\mathrm{i}\frac{V}{2}(1\quad 0)\begin{pmatrix}0 & -\mathrm{i}\\ \mathrm{i} & 0\end{pmatrix}\begin{pmatrix}0\\1\end{pmatrix} = \frac{-V}{2}. \tag{A 9}$$

Arguing similarly for $\langle-|\dot{+}\rangle$ we see that the operator equation (A 4) holds for all matrix elements and hence is true, thus implying (27).

To prove (30) we first show that $|\langle+|\boldsymbol{\sigma}\cdot\boldsymbol{v}|-\rangle| = \frac{1}{2}$:

$$\begin{aligned}\langle+|\boldsymbol{\sigma}\cdot\boldsymbol{v}|-\rangle\langle-|\boldsymbol{\sigma}\cdot\boldsymbol{v}|+\rangle &= \langle+|\boldsymbol{\sigma}\cdot\boldsymbol{v}(|-\rangle\langle-|+|+\rangle\langle+|)\boldsymbol{\sigma}\cdot\boldsymbol{v}|+\rangle\\ &= \langle+|(\boldsymbol{\sigma}\cdot\boldsymbol{v})^2|+\rangle = \tfrac{1}{4}.\end{aligned} \tag{A 10}$$

(The first equality is valid because $\langle+|\boldsymbol{\sigma}\cdot\boldsymbol{v}|+\rangle = 0$.) Thus we can write

$$\mathrm{i}\langle+|\boldsymbol{\sigma}\cdot\boldsymbol{v}|-\rangle = \tfrac{1}{2}\exp\{\mathrm{i}\mu(t)\} \tag{A 11}$$

To find an equation for μ, differentiate:

$$\tfrac{1}{2}\dot{\mu}\exp\{\mathrm{i}\mu\} = \langle\dot{+}|\boldsymbol{\sigma}\cdot\boldsymbol{v}|-\rangle + \langle+|\boldsymbol{\sigma}\cdot\dot{\boldsymbol{v}}|-\rangle + \langle+|\boldsymbol{\sigma}\cdot\boldsymbol{v}|\dot{-}\rangle. \tag{A 12}$$

The first and last terms vanish because

$$\langle\dot{\pm}|\boldsymbol{\sigma}\cdot\boldsymbol{v}|\mp\rangle = \langle\dot{\pm}|\pm\rangle\langle\pm|\boldsymbol{\sigma}\cdot\boldsymbol{v}|\mp\rangle + \langle\dot{\pm}|\mp\rangle\langle\mp|\boldsymbol{\sigma}\cdot\boldsymbol{v}|\mp\rangle = 0. \tag{A 13}$$

In the middle term, $\dot{\boldsymbol{v}}$ can be replaced by its component along $\boldsymbol{w}$, which from (31) gives, with (A 11)

$$\dot{\mu} = -\mathrm{i}\Omega\frac{\langle+|\boldsymbol{\sigma}\cdot\boldsymbol{w}|-\rangle}{\langle+|\boldsymbol{\sigma}\cdot\boldsymbol{v}|-\rangle}. \tag{A 14}$$

The matrix elements are just those previously evaluated in (A 7–A 9), so that

$$\dot{\mu} = -\Omega, \tag{A 15}$$

which with (A 10) and (A 11) gives (30).

Appendix B. Asymptotic evaluation of the integral (68)

Because of the exponent in (68), the integration contour can be deformed (for positive ϵ) into the negative half-plane to surround a cut extending from the branch point at $\tau = -\mathrm{i}$ to $\tau = -\mathrm{i}\infty$. On the right side we can take $\tau =$

$-\mathrm{i}+r\exp\{-\frac{1}{2}\mathrm{i}\pi\}$ (r from 0 to ∞) and on the left we can take $\tau = -\mathrm{i}+r\exp\{\frac{3}{2}\mathrm{i}\pi\}$ (r from ∞ to 0). Thus the integral becomes

$$-\mathrm{i}\exp\{-1/\epsilon\}\int_0^\infty \mathrm{d}r\,\frac{\exp\{-r/\epsilon\}(2+r)^{\cos\theta}}{r^{\cos\theta}}[\exp(\mathrm{i}\pi\cos\theta)-\exp(-\mathrm{i}\pi\cos\theta)]$$

$$= \frac{-\mathrm{i}\exp\{-1/\epsilon\}}{\epsilon^{(\cos\theta-1)}}\int_0^\infty \mathrm{d}x\,\frac{\exp\{-x\}(2+\epsilon x)^{\cos\theta}}{x^{\cos\theta}}\,2\mathrm{i}\sin(\pi\cos\theta).$$

For small ϵ the term ϵx can be neglected. The resulting integral is a Γ-function and squaring gives (69).

References

Aharonov, Y. & Anandan, J. 1987 *Phys. Rev. Lett.* **58**, 1593–1596.
Anandan, J. & Stodolsky, L. 1987 *Phys. Rev.* D **35**, 2597–2600.
Avron, J. E., Seiler, R. & Yaffe, L. G. 1987 *Communs math. Phys.* **110**, 33–49.
Balian, R., Parisi, G. & Voros, A. 1978 *Phys. Rev. Lett.* **41** 1141–1144.
Berry, M. V. 1984 *Proc. R. Soc. Lond.* A **392**, 45–57.
Berry, M. V. 1985 *J. Phys.* A **18**, 15–27.
Berry, M. V. 1986 *Adiabatic phase shifts for neutrons and photons*. In *Fundamental aspects of quantum theory* (ed. V. Gorini & A. Frigerio). NATO ASI series vol. 144, pp. 267–278. New York: Plenum.
Cina, J. 1986 *Chem. Phys. Lett.* **132**, 393–395.
Dingle, R. B. 1973 *Asymptotic expansions: their derivation and interpretation*. New York and London: Academic Press.
Garrison, J. C. 1986 Preprint UCRL 94267, Lawrence Livermore Laboratory.
Gozzi, L. E. & Thacker, W. D. 1987 *Phys. Rev.* D **35**, 2388–2398.
Hannay, J. H. 1985 *J. Phys.* A **18**, 221–230.
Hwang, J.-T. & Pechukas, P. 1977 *J. Chem. Phys.* **67**, 4640–4653.
Lawrence, J. D. 1972 *A catalog of special plane curves*. Dover Publications.
Mead, C. A. 1987 *Phys. Rev. Lett.* **59**, 161–164.
Page, D. H. 1987 *Phys. Rev. Lett.* (Submitted.)
Segert, J. 1987 *J. math. Phys.* (In the press.)
Simon, B. 1983 *Phys. Rev. Lett.* **51**, 2167–2170.
Suter, D., Chingas, G., Harris, R. A. & Pines, A. 1987 *Molec. Phys.* (In the press.)
Wilczek, F. & Zee, A. 1984 *Phys. Rev. Lett.* **52**, 2111–2114.